DECISION MAKING IN ENGINEERING DESIGN

Edited by **Kemper E. Lewis, Wei Chen and Linda C. Schmidt**

ASME PRESS

Library of Congress Cataloging-in-Publication Data

Decision making in engineering design / edited by Kemper E. Lewis, Wei Chen, Linda C. Schmidt.
p. cm.
Includes bibliographical references and index.
ISBN 0-7918-0246-9
1. Engineering design--Decision making. I. Lewis, Kemper E. II. Chen, Wei, 1960- III. Schmidt, Linda C.

TA174.D4524 2006
620'.0042--dc22

2006010805

About the Editors

Dr. Kemper Lewis is currently a Professor in the Department of Mechanical and Aerospace Engineering and Executive Director of the New York State Center for Engineering Design and Industrial Innovation (NYSCEDII) at the University at Buffalo—SUNY. His research goal is to develop design decision methods for large-scale systems where designers understand the dynamics of distributed design processes, employ valid and efficient decision-support methods, and use modern simulation, optimization and visualization tools to effectively make complex trade-offs. His research areas include decision-based design, distributed design, reconfigurable systems, multidisciplinary optimization, information technology and scientific visualization.

Dr. Lewis received his B.S. degree in mechanical engineering and a B.A. in mathematics from Duke University in 1992, an M.S. degree in mechanical engineering from Georgia Institute of Technology in 1994 and a Ph.D. in mechanical engineering from the Georgia Institute of Technology in 1996. He was the recipient of the National Science Foundation Faculty Early Career Award, the Society of Automotive Engineer's Teetor Award and the State University of New York Chancellor's Award for Excellence in Teaching.

Dr. Lewis is the author of more than 80 technical papers in various peer-reviewed conferences and journals. He was the guest editor of the *Journal of Engineering Valuation & Cost Analysis* and is currently an Associate Editor of the ASME *Journal of Mechanical Design*. Dr. Lewis is a member of ASME, the American Society of Engineering Education, the International Society for Structural and Multidisciplinary Optimization and the American Society for Quality; he is also an Associate Fellow of the American Institute of Aeronautics & Astronautics (AIAA).

Dr. Wei Chen is currently an Associate Professor in the Department of Mechanical Engineering at Northwestern University. She is the director of the Integrated DEsign Automation Laboratory (IDEAL- http://ideal.mech.northwestern.edu/). Her research goal is to develop rational design methods based on mathematical optimization techniques and statistical methods for use in complex design and manufacturing problems. Her current research involves issues such as robust design, reliability engineering, simulation-based design, multidisciplinary optimization and decision-based design under uncertainty.

Dr. Chen received her B.A. in mechanical engineering from the Shanghai Jiaotong University in China (1988), an M.A. in mechanical engineering from University of Houston (1992) and a Ph.D. in mechanical engineering from Georgia Institute of Technology (1995). Dr. Chen is the recipient of the 1996 U.S. National Science Foundation Faculty Early Career Award, the 1998 American Society of Mechanical Engineers (ASME) Pi Tau Sigma Gold Medal Achievement Award, and the 2006 Society of Automotive Engineer's Teetor Award.

Dr. Chen is the author of more than 90 technical papers in various peer reviewed conferences and journals. She was the guest editor of the *Journal of Engineering Valuation & Cost Analysis*. She is currently an Associate Editor of the ASME *Journal of Mechanical Design* and serves on the editorial board of the *Journal of Engineering Optimization* and the *Journal of Structural & Multidisciplinary Optimization*. Dr. Chen is a member of ASME and the Society of Automotive Engineering (SAE), as well as an Associate Fellow of the American Institute of Aeronautics & Astronautics (AIAA).

Dr. Linda Schmidt is currently an Associate Professor in the Department of Mechanical Engineering at the University of Maryland. Her general research interests and publications are in the areas of mechanical design theory and methodology, mechanism design generation, design generation systems for use during conceptual design, design rationale capture, effective student learning on engineering project design teams, and increasing the retention and success of women in STEM fields.

Dr. Schmidt completed her doctorate in mechanical engineering at Carnegie Mellon University (1995) with research in grammar-based generative design. She holds B.S. (1989) and M.S. (1991) degrees from Iowa State University for work in industrial engineering, specializing in queuing theory and organization research. Dr. Schmidt is a recipient of the 1998 U.S. National Science Foundation Faculty Early Career Award. She co-founded RISE, a summer research experience and first-year college-orientation program for women. RISE won the 2003 Exemplary Program Award from the American College Personnel Association's Commission for Academic Support in Higher Education.

Dr. Schmidt is the author of 50 technical papers published in peer-reviewed conferences and journals. Four of the conference papers were cited for excellence in the field of design theory, and two in research on the impact of roles on student project teams in engineering education. Dr. Schmidt has co-authored two editions of a text on product development and a team training curriculum for faculty using engineering student project teams. She was the guest editor of the *Journal of Engineering Valuation & Cost Analysis* and has served as an Associate Editor of the ASME *Journal of Mechanical Design*. Dr. Schmidt is a member of ASME and the Society of Manufacturing Engineers (SME), as well as the American Association of Engineering Educators (ASEE).

TABLE OF CONTENTS

PREFACE

In 1996 a group of established members of the engineering design research community envisioned a novel Internet-based learning community to explore the design perspective now known as decision-based design (DBD). These senior colleagues recruited us, a trio of assistant professors at the beginning of our careers, to develop the idea of an open, Internet-based workshop and to guide and manage its growth. In addition to their confidence, our colleagues gave us their wholehearted support, participation, advice and guidance. The result is The Open Workshop on Decision-Based Design, a project funded by a series of modest, collaborative grants from the National Science Foundation's Division of Manufacturing and Industrial Innovation, Program of Engineering Design.

This book is a collection of materials fundamental to the study of DBD research. To a large extent, it presents representative research results on DBD developed since the inception of the DBD Workshop. The core topics discussed in the DBD Workshop have helped define the topics of the major sections in this book. The work is presented in a thoughtful order to emphasize the breadth and multidisciplinary nature of DBD research as it is applied to engineering design. At the end of each technical chapter, exercise problems are provided to facilitate learning.

The content and format of this text has been designed to benefit a number of different audiences, including:

- Academic educators who are teaching upper-level or graduate courses in DBD.
- Graduate students who want to learn the state-of-the-art in DBD theory and practice for their research or course work.
- Researchers who are interested in learning the relevant scientific principles that underlie DBD.
- Industrial practitioners who want to understand the foundations and fundamentals of making decisions in product design and want to see clear examples of the effective application of DBD.

It is a major challenge to compile an explanatory text on a topic that is under active research. There are several lively debates ongoing in our research community that are manifested as chapters in this book presenting differing views on aspects of DBD. This is particularly evident in the context of methods to apply DBD principles to engineering design while maintaining academic rigor. We have purposely embraced alternate interpretations of approaches to DBD applications. Sections 2 through 8 begin with a short commentary on their content. Differences in the approaches articulated by authors in each section are highlighted along with the context in which these differences can be understood. We created these section introductions to facilitate scholarly debate on these open issues in engineering design decision-making, even in the classroom. In this way readers and students of DBD will appreciate the books insight into some of the most divergent and modern issues of DBD, such as validation, uncertainty, preferences, distributed design and demand modeling.

We must thank the University at Buffalo—SUNY for hosting the workshop for the past nine years. We acknowledge the support of all our colleagues who were regular speakers, panelists and participants at our face-to-face meetings of the Open Workshop as well as online. Many became collaborators in a variety of research endeavors spun off from the Open Workshop. Many provided welcome intellectual sparring on the issues of design theory research, helping each of us to define and articulate our own positions on decision-making in design and develop separate research programs and identities that will carry us through our careers.

We extend special thanks to Dr. George Hazelrigg, the former program manager of the Engineering Design program in the NSF's DMII Division, who supported the original vision of the Open Workshop. We also must thank Dr. Delcie Durham, the current program manager of the Engineering Design program at NSF, who encouraged us to create this text.

While preparing this book we have had help from many people. Our sincere thanks go to all authors who diligently improved the quality of their individual chapters, who provided constructive review comments on other chapters in the book and who helped us refine the content of section introductions.

The richness and complexity of topics central to understanding the decision-based perspective on design can not be covered in any single volume of work. In the end, our hope is that this book primarily provides learning materials for teaching decision-based design. We also hope that the book archives the research initiated by the Open Workshop on Decision-Based Design. Thank you to all the brilliant researchers and visionaries who have helped make this book and the DBD field a reality.

Kemper Lewis
Wei Chen
Linda Schmidt

SECTION 1

CHAPTER

1

The Need for Design Theory Research

Delcie R. Durham

"Engineers do design" – a factual statement made by many both inside and outside of the engineering community. The statement has been the basis of speeches by William Wulf, President of the National Academy of Engineering; by John Brighton, Assistant Director for Engineering at the National Science Foundation (NSF); and by industry leaders commenting on the current outsourcing of manufacturing and, now increasingly, some engineering design jobs overseas. Design permeates the activities of all engineering disciplines: Civil Engineers working on large-scale infrastructure systems in transportation; bioengineers creating new sensors for human health monitoring; mechanical engineers developing new alternative energy sources and power trains for the hydrogen economy; and electrical engineers linking information and communications networks through new advances in photonics. So if all engineers are already doing design, why do we need a program that supports design theory research? Given that engineering design crosses all the disciplinary domains in engineering, our challenge is to focus on creating the new knowledge, advancing the support tools, and building the necessary principles and foundations into a domain-neutral framework that enables engineers to meet the future needs of society. As a research community, a design research program is needed to continue our work to establish the set of principles that underlie all design, such as:

Design requires a clearly stated objective function.
Design must address the uncertainties within all aspects of the system to better inform the decision-making.

Over the past three decades, design theory research has taken several twists and turns, as computational tools became the standard for how engineers of all disciplines "did design." In an early NSF Workshop report, ***Design Theory '88*** [1], research was categorized into topical areas focused on the design process that included the computational modeling; the cognitive and social aspects; the representations and environments; the analysis tools including optimization and the design "for" such as "for manufacturing." At that time, the NSF program was called Design Theory and Methodology and consisted of three components that essentially captured these five topical areas: The first, Scientifically Sound Theories of Design, established a home for proposals that were directed at creating the scientific basis for the design process. The second, Foundations for Design Environments, was aimed at advancing the understanding of fundamental generic principles that could be used and understood across engineering domains. The third, Design Processes, was focused on the how and why of the design process, including early work on life-cycle concepts and concurrent design.

At this point, you may ask, "So what is new?" The tools certainly have advanced over the years, from early computer-aided design (CAD) through solid modeling capability. The introduction of virtual reality, computer integration engineering, and collaborative and distributed design processes created demands upon the community to focus on how decisions were made, under what conditions and to what purpose. Decision-based design became a major thrust for the research community, with the issues of uncertainty and predictive modeling capability becoming the foci. As with any science, the theories must be put forward, tested for consistency and completeness, and then incorporated (or not) into the framework of the science. This is true, too, for engineering design, if it is to become more than just an ad hoc, intuitive process that is domain-specific. In response, the **Open Workshops on Decision-Based Design** [2], a series of face-to-face and website workshops, addressed the spectrum of issues that were raised.

These activities demonstrated that decision-based design creates a challenging avenue for research that encompasses:

(1) the cognitive "structuring" of a problem
(2) the drive for innovation where the existing "structure" or solution space is ill-defined or insufficient
(3) the need to reduce complexity by mapping to what we know
(4) the consistent use of decision technologies to optimize the decision-making capabilities within the design space we have created.

As socially and technically responsible engineers, we must be able to demonstrate that we have searched and populated the design space with the necessary and appropriate data and information, that we have considered the risks and the odds to an appropriate level, that we have created and/or integrated models that capture the intent of the design (design informatics), that these models can be validated and that we have reduced the potential for unintended outcomes to the best of our capability.

If design were easy, then the following eight sections of this book would be unnecessary. Engineering implies doing something, and this moves us beyond the regime of descriptive, theoretical study into the need for predictive action. This leads to the challenges addressed in sections 2, 3 and 5, where the difficulty often comes down to eliciting the answer to the simple question, "What do you want?" If we could come up with a single equation that represented the design objective, and solve this equation in closed analytical form, then sections 6 and 7 would be redundant, and the differences of perspective would be resolved. If all modeling were predicative rather than descriptive, then computer software tools would take care of all Section 8 validation methods. Finally, if we could just engineer without the consideration of economics, well, that wouldn't be "good" engineering, and so the methods addressed in Section 4 become critical to the realization of viable products and systems.

Finally, in looking toward our future, the vision statement from the recent ***ED 2030: Strategic Planning for Engineering Design*** [3], includes the following: "*In 2030, designers will work synergistically within design environments focused on design not distracted by the underlying computing infrastructure. Designers will interact in task-appropriate, human terms and language with no particular distinction between communicating with another human team member or online computer design tools. Such environments will amplify human creativity leading toward innovation-guided design. Future design tools and methods will not only support analysis and decision–making from a technological point of view, but will also account for psychological, sociological, and anthropological factors based on fundamental understanding of these factors and their interaction. ... Designers will effortlessly and effectively explore vast and complex design spaces. Design will go from incremental changes and improvements to great bold advances. Therefore design will be an exciting activity fully engaging our full human creative abilities.*"

REFERENCES

1. Design Theory, '88, 1989. S. L. Newsome, W.R. Spillers, S. Finger, eds., Springer-Verlag, New York, NY.
2. Open Workshops on Decision-Based Design, http://dbd.eng.buffalo.edu/.
3. ED2030: Strategic Planning for Engineering Design, *2004*. Report on NSF Workshop, March 26–29, AZ.

CHAPTER

2

THE OPEN WORKSHOP ON DECISION-BASED DESIGN

Wei Chen, Kemper E. Lewis and Linda C. Schmidt

2.1 ORIGIN OF THE OPEN WORKSHOP

During the late 1990s, members of the engineering design research community articulated a growing recognition that decisions are a fundamental construct in engineering design. This position and its premise that the study of how engineering designers should make choices during the design represented the foundation of an emerging perspective on design theory called decision-based design (DBD). DBD provides a framework [1] within which the design research community could conceive, articulate, verify and promote theories of design beyond the traditional problem-solving view. As we define here:

> *Decision–based design (DBD) is an approach to engineering design that recognizes the substantial role that decisions play in design and in other engineering activities, largely characterized by ambiguity, uncertainty, risk, and trade-offs. Through the rigorous application of mathematical principles, DBD seeks to improve the degree to which these activities are performed and taught as rational, that is, self-consistent processes.*

The Open Workshop on Decision-Based Design (DBD) was founded in late 1996. The open workshop engaged design theory researchers via electronic and Internet-related technologies as well as face-to-face meetings in scholarly and collegial dialogue to establish a rigorous and common foundation for DBD. Financial support for the Open Workshop on DBD was provided by the National Science Foundation (NSF) from the workshop's inception through November 2005. The goal of the DBD workshop has been to create a learning community focused on defining design from a DBD perspective and investigating the proper role that decisions and decision-making play in engineering design.

Over the years the investment made by our colleagues and the NSF has contributed to the development of a body of scholarly research on DBD. This research and the community built around it are the result of investing, adopting and adapting, where necessary, principles for decision-making from disciplines outside of engineering. This synergy has led to numerous conference papers and journal publications, special editions of journals dedicated to DBD [2] and successful research workshops. Both the Design Automation Conference and the Design Theory Methodology Conference at the ASME Design Engineering Technical Conferences have established technical sessions on DBD for the past few years on various issues within DBD. The role of the Open Workshop on DBD has been that of a catalyst for this growth in scholarly investigation, presentation and debate of ideas.

This book is, to a large extent, a collection of representative research results on DBD developed since the inception of the DBD workshop. This work is a survey of material for the student of DBD, providing insights from some of the researchers who have developed the DBD community. However, we now feel that the field has matured to the point where a textbook has the potential to not only be an effective tool to help teach the principles of DBD, but is a necessity to further the successful progress of the field.

2.2 THE OPEN WORKSHOP ON DECISION-BASED DESIGN

The Open Workshop on Decision-Based Design was launched by a group of 10 researchers in the engineering design community and related fields in November 1996. Specifically, the Open Workshop's stated objectives were to:

- synthesize a sound theory of DBD
- determine the proper role of decision-making in design
- develop consensus on defining the DBD perspective
- establish the role of DBD within design theory and methodology research
- build a repository of foundational materials (e.g., a lexicon, case studies, references, text materials) that illustrate design as decision-making
- establish a useful relationship between the DBD and theories in other science domains such as physics, mathematics, information and management science
- transfer decision support methods and tools into industry

To achieve these goals, the Open Workshop established a website located at http://dbd.eng.buffalo.edu/. This was the discussion forum for the broadest possible audience of design theory researchers, engineering student, scholars and students from related fields, industry representatives and practitioners. At the same time, the website acted as an entry point into the DBD learning community and a repository for DBD materials before they were presented in a collaborative form at conferences and in journals.

The success of the DBD Open Workshop can be attributed to a dual-access strategy. The online Workshop presence provided continuous access to the learning community via the website. Online workshop registration grew to over 540 registered online participants. At the same time, a series of face-to-face meetings with Workshop participants was held to guide research questions, spread discussion on DBD and to engage people interested in the topic who had not yet discovered the website. More about this strategy follows.

2.2.1 Open Workshop Website

The term "open workshop" was coined during this undertaking to describe a web-based platform for interactive discussion on DBD. The DBD Workshop architecture was tailored to perform each workshop function. The Open Workshop home page in Fig. 2.1 shows the links to the site areas disseminating information ("Position Papers," "Reading List," "Meetings," "Related Links" and "Pedagogy") and both disseminating and collecting information in the form of feedback ("Current Questions for Debate," "Open Research Issues" and "Lexicon").

Interest in the DBD Open Workshop has increased steadily since its launch. Traffic on the website increased significantly throughout the years of the workshop. For example, the number of total hits (nondistinct) to the website rose from a few 1,000 per year in the first couple of years to close to 300,000 per year in the final years of the workshop.

The registered participants (over 540) *do not represent* all the people who visited the website without registering. The DBD workshop drew world-wide interest from both design researchers and practitioners. Registrants are from 38 different countries, 87 different American Universities and more than 60 companies, government agencies and laboratories.

2.2.2 Face-to-Face Meetings to Enrich the Online Workshop

Engaging the audience in online participation was a critical challenge in maintaining a viable open workshop. Workshop dialogue had to be sparked and kept relevant. From November 1996 to September 2004, Workshop organizers held 18 face-to-face meetings to supplement, plan and direct the open workshop. Table 2.1 lists the meetings, venues, formats and discussion topics. A review of the timing and location of the face-to-face meetings demonstrates how organizers used scheduled events, like NSF-sponsored workshops, to attract members of a workshop's target audience.

The nature of the face-to-face meetings evolved as the Open Workshop and its companion learning community grew. The first five meetings were set up as a means to determine the content and means of operation of the Open Workshop. The face-to-face meetings were a venue to create working groups and discuss topics areas that were of mutual interest and would be added to the online workshop. The early meetings were often the initial introduction to the DBD community of new members. For that reason, meetings were longer and included introduction sections to update new participants.

After the Open Workshop on DBD had become a known part of the design research community, more emphasis was placed on

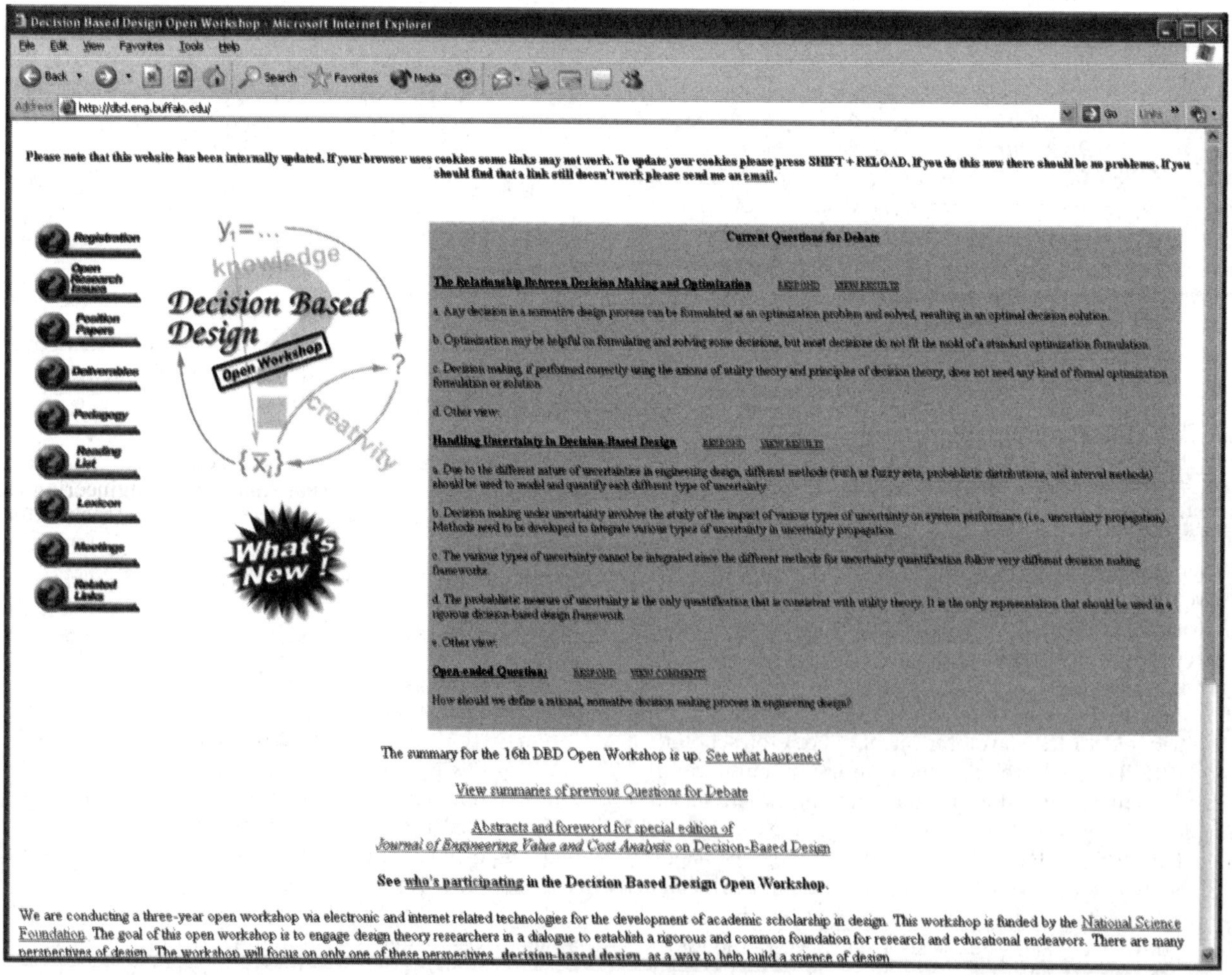

FIG. 2.1 HOME PAGE OF THE OPEN WORKSHOP ON DECISION-BASED DESIGN (DBD)

TABLE 2.1 CONTENTS OF FACE-TO-FACE MEETINGS

Date, Location and Affiliation	Topic and Speakers
1st Meeting: November 22–23, 1996 Georgia Tech, Atlanta, GA The Launch Meeting	Launch Meeting Method of operation Set measures for success of Workshop Identify target audience groups Discuss strategies to secure participation
2nd Meeting: January 10–11, 1997 Seattle, WA NSF Design and Manufacturing Grantees 1997 Conference Site	Steering Meeting with Working Group Sessions Rigor, education and philosophy of DBD Determined deliverables for Workshop Created list of guiding intellectual questions to address. Defining "What is DBD and what isn't?"
3rd Meeting: April 5–6, 1997 Kissimmee, FL SDM Meeting Site	Topics: Tutorial for New Members Presentation of six position papers Report from working group Demonstration of DBD website
4th Meeting: September 13–14, 1997 Sacramento, CA ASME DETC '97 Conference Site	Topics: Tutorial for New Members Progress reports from working groups State-of-the-art in DBD Breakout sessions to create design examples "How can we achieve an effective dialogue on the web?"
5th Meeting: January 5, 1998 Monterrey, Mexico NSF Design and Manufacturing Grantees 1998 Conference Site	Topics: Progress Reports from working groups DBD taxonomy creation Open Workshop website development Educational objectives of the Workshop
6th Meeting: September 12, 1998 Atlanta, GA ASME DETC '98 Conference Site	Topics: Decision Theory 101, D. Thurston, UIUC Managing Risk in DBD, B. Wood, University of Maryland BC Intuitive vs. Analytical Cognition in Decision-Making, A. Kirlik, Georgia Tech Economics of Product Selection, B. Allen University of Minnesota
7th Meeting: January 5, 1999 Long Beach, CA NSF Design and Manufacturing Grantees 1999 Conference Site	Topics: Decision-Based Risk Management, G. Friedman, USC Workshop Participants on Current Work (F. Mistree, Georgia Tech; Y. Jin, USC; L. Schmidt, University of Maryland) Group Discussion on Increasing Website Effectiveness
8th Meeting: Las Vegas, NV September 12, 1999 ASME DETC '99 Conference Site	Topics: Bad Decisions: Experimental Error or Faulty Methodology? D. Saari, Northwestern University Toolkit for DBD Theory, B. Allen, University of Minnesota Decision Model Development in Engineering Design, S. Krishnamurty, University of Massachusetts
9th Meeting: January 3, 2000 Vancouver, BC NSF Design and Manufacturing Grantees 2000 Conference Site	Topic: The Role of DBD Theory in Engineering Design Panelists: Jami Shah, Arizona State Debbie Thurston, UIUC George Hazelrigg, NSF Farrokh Mistree, Georgia Tech
10th Meeting: September 10, 2000 Baltimore, MD ASME DETC '00 Site	Topic: The Role of Decision Analysis in Engineering Design Panelists: David Kazmaer, University of Massachusetts Sundar Krishnamurthy, University of Massachusetts John Renaud, Notre Dame University
11th Meeting: January 7, 2001 Tampa, FL NSF Design and Manufacturing Grantees 2001 Conference Site	Topic: The Practical Perspectives in Decision-Based Design Panelists: Jerry Resigno, Black & Decker Tao Jiang, Ford Motor Company Joe Donndelinger, General Motors Ed Dean, The DFV Group
12th Meeting: September 9, 2001 Pittsburgh, PA ASME DETC '01 Conference Site	Topic: Aggregation of Preference in Engineering Design Panelists: Debbie Thurston, UIUC Achille Messac, RPI Shapour Azarm, University of Maryland Joe Donndelinger, General Motors Kemper Lewis, University of Buffalo
13th Meeting: January 7, 2002 San Juan, PR NSF Design and Manufacturing Grantees 2002 Conference Site	Topic: Research Issues on DBD Theory Development Panelists: Ralph Keeney, USC Beth Allen, University of Minnesota Abhi Deshmukh, University of Massachusetts

(Continued)

TABLE 2.1 Continued

14th Meeting: September 29, 2002 Montreal, Canada ASME DETC '02 Conference Site	Topic: Demand and Preference Modeling in Engineering Design Panelists: Ken Small, UC Irvine Harry Cook, UIUC Jie Cheng, JD Power & Associates Joe Donndelinger, General Motors Panos Papalambros, University of Michigan
15th Meeting: January 6, 2003 Birmingham, AL NSF Design and Manufacturing Grantees 2003 Conference Site	Topic: Model Validation in Engineering Design Panelists: Raphael Haftka, University of Florida George Hazelrigg, NSF Don Saari, UC Irvine Martin Wortman, Texas A&M
16th Meeting: September 2, 2003 Chicago, IL ASME DETC '03 Conference Site	Topic: Perspective on the Role of Engineering Decision-Based Design: Challenges and Opportunities Panelists: Zissimos Mourelatos, Oakland University Debbie Thurston, UIUC
17th Meeting: January 5, 2004 Dallas, TX NSF Design and Manufacturing Grantees 2004 Conference Site	Topic: Decision Management and Evolution of Decision Analysis Speakers: David Ullman, Robust Decisions Inc. Ali Abbs, Stanford University
18th Meeting: September 28, 2004 Salt Lake City, UT ASME DETC '04 Conference Site	Topic: DBD Book Planning Participants: Authors of the chapters included in this book

attracting new online members and facilitating additional online dialogue. To that end, the face-to-face meeting format evolved into more of a town-hall style discussion than a lecture format. A typical meeting included a panel session on pre-assigned topics. Specialists from academia and industry representatives were invited to stimulate discussion on particular subjects. After panelists provide opening remarks, each panel session is followed by an hour-long, moderated, open-floor discussion—this includes questions and answers and ends with closing remarks from each panelist.

Each face-to-face meeting included a discussion section during which participants could interact in real time with each other, invited speakers, panelists and workshop organizers. These sessions generated stimulating conversations regarding numerous DBD-related topics. After each face-to-face meeting the website was updated with an overview of the major questions posed and following discussion. Presentations of panelists were also made available online to all Open Workshop participates. These documents provided an updated view of the common interests in developing the DBD theory as well as the research status in this field. It is noted that the core topics discussed in our Workshop meetings have helped define the topics of the major sections in this book; each section contains multiple research papers in the form of chapters.

2.3 INTERACTION STRATEGIES TO FACILITATE DIALOGUES

The success of the Open Workshop on DBD demonstrated that it is possible to engage a number of researchers in dialogue by holding an Open Workshop on a website. The Open Workshop dialogue was originally organized around a core of foundational research questions. During the last few years, new strategies online were implemented to engage Open Workshop participants' contribution to the website and to attract new participants. These strategies included: (1) a regular electronic *newsletter publishing* schedule; (2) a set of *polling questions* to the website to collect response and stimulate discussion; and (3) evolving our face-to-face meeting format to more of a *town-hall style discussion* than a lecture format. Our modifications to the site and the companion face-to-face meetings enabled more feedback from the site visitors, effectively expanding our user base.

2.3.1 Establishing Fruitful Dialogue

The Open Workshop dialogue was organized around a core of foundational research questions. The questions were refined through the thread of the web discussion and reflection on that discussion. The workshop website included a set of message boards (see example in Fig. 2.2) to promote and record registrant dialogue. The website had a different discussion board for each question. New questions and issues were raised and added to the site as appropriate. Examples of questions that were discussed are:

- What are the key activities designers perform in the design process?
- What are the advantages and limitations of a DBD approach to a design process?
- What is the role of decision-making in design in general, and in the approach you use in design?
- How can an underlying science of design and lexicon of design be developed? What should it look like?
- What are the issues of coordination and collaboration in distributed design?
- How do we teach design effectively?

The manner in which research topics were developed and added to the website has been discussed in the review of face-to-face meetings in Section 2.2.2.

2.3.2 Using Online Polling on Questions for Debate

To effectively collect response and stimulate discussion on debatable research issues, organizers periodically posted polling questions on the Workshop website. The polling questions were prominently displayed on the website's home page. All visitors to the website could answer the question and see the results of

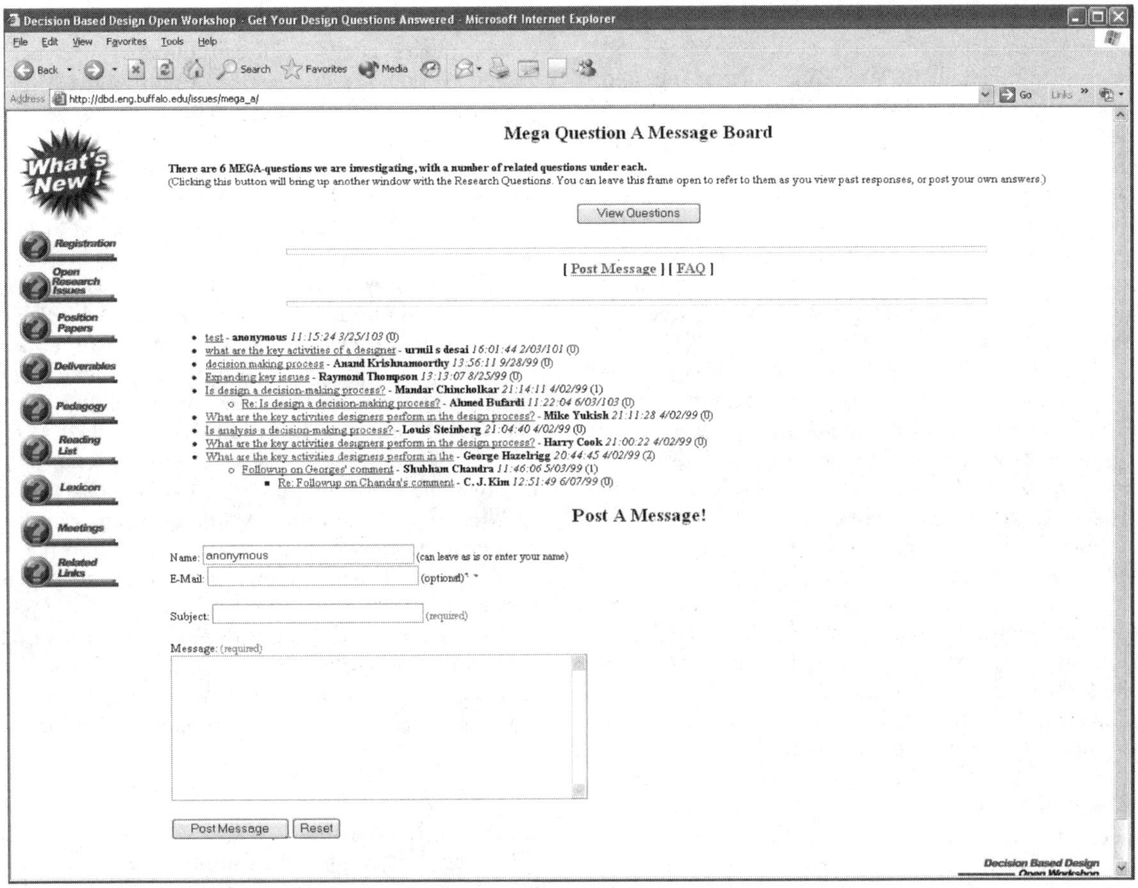

FIG. 2.2 EXAMPLE OF OPEN WORKSHOP'S MESSAGE BOARD

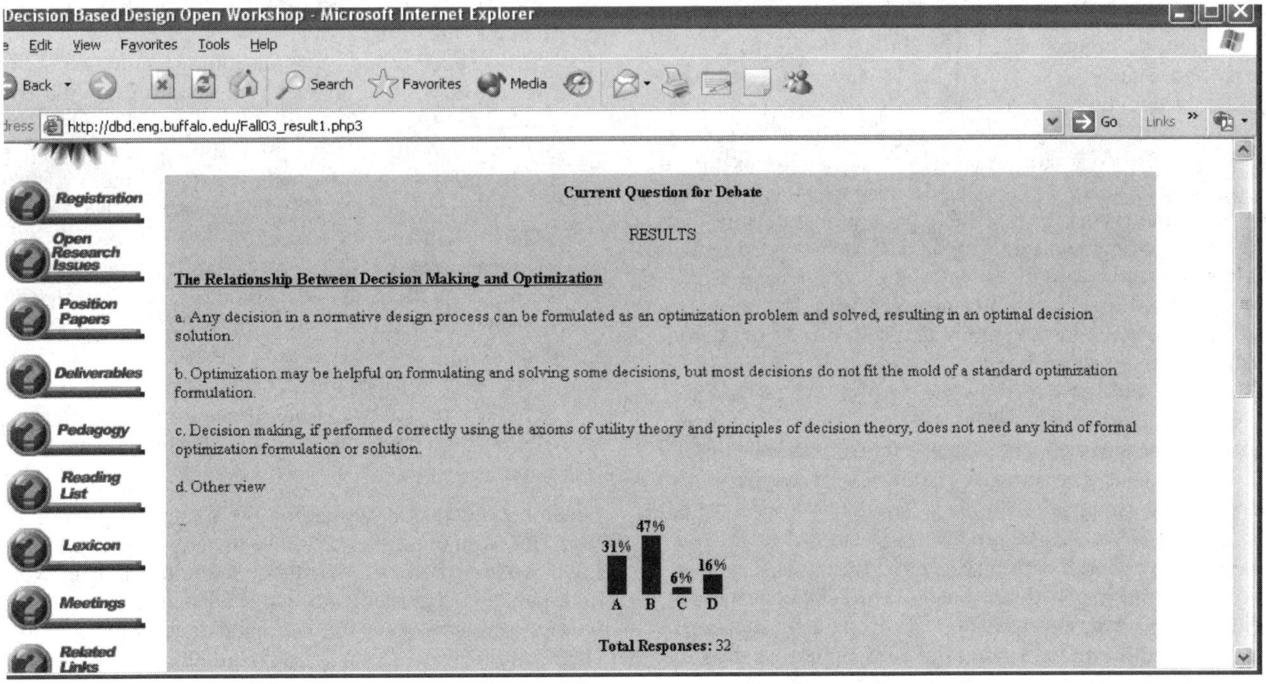

FIG. 2.3 EXAMPLE OF POLLING RESULTS

the poll updated after their own entry (see sample in Fig. 2.3). The polling questions were selected from discussion points raised at previous face-to-face meetings of Workshop participants and summaries of polling results were disseminated in the electronic newsletters as well as included in presentations at subsequent face-to-face meetings.

The Workshop's first set of questions centered on quality function deployment (QFD) and its use as a tool for design decision-making. QFD was chosen as the topic because it figured prominently in the discussion at our 9th face-to-face meeting. The home page asked the main question, **"Is QFD a useful approach in engineering design?"** To explore attitudes of website visitors toward QFD, a series of questions about the method were posted. The polling results showed that there was a core group of researchers who objected to QFD on the grounds of the mathematical flaw in the method. Most respondents were neutral to QFD's use in industry.

The Workshop's second set of questions centered on **The Role of Decision Analysis in Engineering Design**. This tracked with the 10th face-to-face meeting topic. The site received 74 responses to the statement that "Decision analysis is the most important aspect of design." A clear majority of respondents accepted that design is both art and science, implying that it's not all "analysis," and that decision analysis brings both benefits and limitations in its application to the design process. Comments also revealed a sense that the community was not afraid of using quantitative methods in this hybrid (art and science) activity of design.

The third series of Workshop polling questions dealt with **multi-criteria decision-making approaches versus single-criterion approach** in engineering design. The site received 115 responses to the statement that "Existing multicriteria decision-making approaches are flawed. Only single-criterion approaches (such as maximization of the profit) should be used for product design." About 87% of respondents disagreed with that statement, only 7% of respondents agreed with the statement. The overall consensus gained from this question set was that multicriteria decision-making approaches should still play an important role in engineering design even though they have limitations.

The Workshop's fourth set of polling questions centered on **"Is game theory applicable for decision-making in engineering design?"** Again, the topic tracked with the debate on the use of game theory and related discussions in the concurrent face-to-face meetings. The site recorded an average of 78 responses to each of the four views posed. The overall consensus gained from this survey was that the game theory can be applied to decision-making in engineering design. A larger percentage of respondents supported the view that game theory is applicable to design situations that involve different companies compared to those supporting the view that it is applicable to any design situation whenever multiple designers are involved. The respondents expressed diverse views on whether engineering design should be profit-driven or should be performance- and quality-based.

The fifth set of polling questions focused on **"Meeting Customers' Needs in Engineering Design."** The first question dealt with whether the primary goal of design is to meet the customers' needs or to make profit. The second question asked how to capture the preference of a group of customers. The last question asked how to meet both the needs of producer and customers in engineering design. The overall consensus gained from this survey set of questions was that meeting customers' needs is important in engineering design. A large percentage of respondents supported the view that a multi-attribute value function cannot be directly used to capture the preference of a group of customers. However, respondent views were divided on how to simultaneously model the customers' needs and the producer's preference.

The Workshop's final series of polling question were based on the topics covered by the panel speakers in our 16th face-to-face meeting. An on-site survey was conducted among the 32 workshop participants, and the polling question results were posted on the website right after the meeting. On **"Relationship between Decision-Making and Optimization,"** close to half of the respondents believed that "Optimization may be helpful in formulating and solving some decisions, but most decisions do not fit the mold of a standard optimization formulation," while close to a third of the respondents agreed with the statement that "Any decision in a normative design process can be formulated as an optimization problem and solved, resulting in an optimal decision solution." On **"Handling Uncertainty in Decision-Based Design,"** respondents split on the view that "Decision-making under uncertainty involves the study of the impact of various types of uncertainty. Methods need to be developed to integrate various types of uncertainty in uncertainty propagation," and the view that "The probabilistic measure of uncertainty is the only quantification that is consistent with utility theory. It is the only representation that should be used in a rigorous decision-based design framework."

Overall, the workshop organizers found that using the website to accumulate responses to a focused set of questions was informative and proved to be a useful method for engaging online participation.

2.3.3 Education

The workshop organizers piloted the website's educational use by asking for critical feedback on the site from graduate students enrolled in a course on engineering decision-making at workshop organizers' schools. One set of comments was collected from graduate students after they spent one hour at the Open Workshop during the Spring Semester of 2003. Comments included the following:

- "... the site appears to bring together many researchers from industry and academia throughout the globe into a common dialogue regarding DBD."
- "One aspect I find very intriguing is the section entitled, 'Where does DBD fit in?'... Comparing the discussion that took place in last Thursday's class comparing problem-solving and decision-making, it's interesting to see where negotiation fits in and if it's coupled with these two other perspectives."
- "I am amazed that the NSF has an open workshop on DBD."
- "The most interesting part of the site was the discussion board on open research issues. That is, there are people there who are willing to participate in discussions ... I believe for having the best understanding of any subject there's no way better than discussing it with a group of people ..."
- "The reading list of the website can help me narrow my search for useful text books about design decision-making. The position papers inspire new questions in the reader's mind ..."

Several students commented on the issues of agreeing on definitions of common terms and the value that would have in the ongoing discussion. Overall, the students found the Open Workshop to be a rich resource that informed their own opinions on decision-making in engineering. This demonstrated how an Open Workshop can involve students in the ongoing activities of the research community.

2.4 CLOSURE

The Open Workshop on Decision-Based Design has succeeded in focusing research efforts. A DBD presence has been established not only nationwide, but internationally. There is an acute and increasing awareness of the relevance of DBD to studying design and the corresponding research issues. By providing an overview of the workshop activities and strategies in the past nine years, we have also shown that it is possible to engage a number of researchers in dialogue by holding an open workshop on a website. Certainly it is a challenging task that demands active monitoring but the results can be stunning.

Organized based on the core DBD research issues identified through workshop meetings, the following sections contain the research views and results from different authors, a true reflection of the current status of research in DBD.

REFERENCES

1. Hazelrigg, G. A., 1998. "A Framework for Decision-Based Engineering Design," *Journal of Mechanical Design*, Vol. 120, pp. 653–658.
2. Engineering Valuation & Cost Analysis, 2000. Special edition on *Decision-Based Design: Status & Promise*," Vols. 1 & 2.

SECTION

2

DECISION THEORY IN ENGINEERING DESIGN

INTRODUCTION

Decision-making is integral to the engineering design process and is an important element in nearly all phases of design. Viewing engineering design as a decision-making process recognizes the substantial role that decision theory can play in design and other engineering activities. Decision theory articulates the three key elements of decision-making processes as:

- identification of *options* or choices
- development of *expectations* on the outcomes of each choice
- formulation of a system of *values* for rating the outcomes to provide an effective ranking and thereby obtaining the preferred choice.

Correspondingly, engineering decision-making can be viewed as a process of modeling a decision scenario resulting in a mapping from the design option space to the performance attribute space. Subsequently, a utility function is constructed that reflects the designer's (acting on behalf of the decision-maker) preference while considering trade-offs among system attributes and the risk attitude toward uncertainty. This section introduces the fundamental concepts and principles that have long been employed in traditional decision theory and discusses their relevance to engineering decision-making. The fundamentals in decision theory provide the mathematical rigor of decision-based design methods. The chapters included in this section emphasize the areas of preference modeling, design evaluation and trade-offs under uncertainty.

The authors of the first two chapters in this section (Chapters 3 and 4) lay out the normative decision analysis principles that are considered fundamental to decision-based engineering design. In Chapter 3 the axiomatic foundations of utility analysis are presented, followed by the method for calculating expected utility, which reflects the decision-maker's attitude toward risk and uncertainty. In Chapter 4, topics central to the development of decision models are reviewed, with an emphasis on their use and implementation in engineering design. Included at the beginning of Chapter 4 is a set of key lexicons used in decision-based design research.

In the following two chapters (5 and 6), some critical issues of applying decision theory in engineering design are presented, giving readers insight and precautions into the use of conventional decision analysis approaches in design. In Chapter 5, it is vividly demonstrated that the choice of a decision rule can play a surprisingly major role in selecting a design option. Rules for minimizing the likelihood of accepting inferior design choices are described. In Chapter 6, it is argued that the foundations of decision and measurement theory require major corrections. The author questions the validity of von Neumann and Morgenstern's utility theory while proposing a theory of measurement that is demonstrated to provide strong measurement scales.

Collectively, the chapters in this section describe the fundamental methods of applying normative decision analysis principles to engineering design. In its entirety, this section also reveals differing, and sometimes opposing, philosophical viewpoints on the application and appropriateness of using certain decision theory constructs for engineering design. In Chapter 5, the paradox is raised of aggregating preferences using ordinal multi-attribute utility functions, a method introduced in Chapter 4 and supported by a few subsequent chapters in Section 5 ("Views on Aggregating Preferences in Engineering Design"). Views in Chapters 5 and 6 differ regarding the appropriateness of ordinal scales as a foundation for measurement in design theory. An even more basic assumption is challenged in Chapter 6, which questions the use of von Neumann and Morgenstern's utility theory. This theory is considered by many experts to be a useful pillar for decision-based design as described in both Chapters 3 and 4. These differences serve to illustrate the ongoing, scholarly debate within the decision-based design research community. Simultaneously, considering the validity of multiple points of view is one of the greatest challenges encountered in a field of active research and emerging theory.

CHAPTER

3

UTILITY FUNCTION FUNDAMENTALS

Deborah L. Thurston

3.1 INTRODUCTION

The decision-based design community has made impressive progress over the past 20 years. The theory has become better founded. Industrial test-bed applications have grown more sophisticated. The successes and failures of methodologies in the field have informed the research agenda. Our interdisciplinary approach has evolved from that of a dilettante to that of a highly skilled systems analyst—one with a deep understanding of the integration of one or more technical specialties. The most fundamental contribution has been to bring the same mathematical rigor to design decision-making that has long been employed in traditional design analysis, particularly in the areas of preference modeling, design evaluation and trade-offs under uncertainty.

The engineering design community is now keenly aware that decision-making is integral to the design process, rather than an afterthought relegated to others. It is an important element in nearly all phases of design, from defining the problem, synthesizing alternatives, evaluating what is acceptable and what is not, identifying which design elements to work on first, specifying what information is needed and by whom, selecting which alternatives are worth pursuing further and finally configuring the optimal design.

Later chapters in this book will reveal the liveliness of ongoing debates about the pros and cons of alternative DBD approaches. While these debates can be quite interesting, don't let them detract from the central message of this book. In truth, the single greatest contribution of DBD has been to help designers recognize that decision-making has always been integral to the design process; only now we think much more carefully about *how* to make decisions.

3.2 AXIOMATIC BASIS

This chapter presents the fundamentals of decision-based engineering design with utility analysis. The first thing to understand is that in contrast to engineering design analysis, which is a *descriptive* modeling tool, utility analysis is a *normative* modeling tool. Engineering design analysis employs mathematical models of physical systems toward the goal of describing, predicting and thus controlling the behavior of the design artifact. In contrast, while utility analysis also employs mathematical models, its goal is *not* to predict or mimic the choices of human decision-makers, but to help humans make better decisions. Decision theory was originally developed because people are often dissatisfied with the choices they make, and find it difficult to determine which choice best reflects their true preferences. Unaided human decision-making often exhibits inconsistencies, irrationality and suboptimal choices, particularly when complex trade-offs under uncertainty must be made [1]. To remedy these problems, decision theory first postulates a set of "axioms of rational behavior" [2]. From these axioms, it builds mathematical models of a decision maker's *preferences* in such a way as to identify the option that would be chosen if that decision-maker were consistent, rational and unbiased. For the remainder of this chapter, the term "utility" refers to a preference function built on the axiomatic basis originally developed by von Neumann and Morgenstern [2]. The basic axioms and conditions of the most popular approach to multi-attribute utility analysis are well presented elsewhere [3] as well as employed in Chapter 12. Howard and Matheson [4] describe a slightly different approach, but Keeney and Raiffa's [3] approach is the focus here. The following definitions of the axioms is intended only as a most general introduction to an engineering design audience. The reader is referred to von Neumann and Morgenstern [2], Luce and Raiffa [5] and French [6] for a much more thorough treatment. The first three axioms enable one to determine a value function for the purpose of rank ordering of alternatives. They are:

Axiom 1: Completeness of complete order. This means that preferences on the part of the decision-maker exist, and that the decision-maker is capable of stating them. As shown below, either X is preferred to Y, or Y is preferred to X, or the decision-maker is indifferent between X and Y. The symbol $\succ$ means "is preferred to", and ~ means "is equally preferred to".

$$\begin{array}{ll} \text{Either} & X \succ Y \\ \text{or} & X \prec Y \\ \text{or} & X \sim Y \end{array} \qquad \text{Eq. (3.1)}$$

Axiom 2: Transitivity. The decision-maker's rank ordering of preferences should be transitive.

$$\begin{array}{ll} \text{If} & X \succ Y \\ \text{and} & Y \succ Z \\ \text{then} & X \succ Z \end{array} \qquad \text{Eq. (3.2)}$$

Axiom 3: Monotonicity. The decision-maker's preferences over the range of an attribute shall be either monotonically increasing or monotonically decreasing. So, either more of an attribute is always preferred to less, or less of an attribute is always preferred to more. The are many instance where this axiom is violated, for example, deflection in automotive bumper beams might be desirable for shock absorption, but beyond a certain point further deflection is undesirable due to intrusion. In this case, the range of deflection over which the decision analysis should focus should extend only to that point, and not beyond.

The result of the three axioms described above is that any value function can be altered by a monotonic transformation to another strategically equivalent value function. Their purpose is to help the analyst structure the problem in such a way that the resulting rank ordering of alternatives is robust.

While achieving a rank ordering of design alternatives that one can be confident is important, DBD is much more complex than simply choosing among alternatives. Designers need information about the strength of preferences, they need to quantify willingness to make trade-offs, and they need to make decisions and commit resources in the face of uncertainty. Three more axioms help one structure a preference function that reflects these considerations. The axioms for utility functions are:

Axiom 4: Probabilities exist and can be quantified. This is essentially an agreement to employ the concept of probability to model uncertainty. Discrete probabilities can be employed, as well as continuous probability distributions, depending on the situation.

Axiom 5: Monotonicity of Probability. The decision-maker prefers a greater chance at a desirable outcome to a lesser chance.

Axiom 6: Substitution-Independence. This is perhaps the most powerful and most misunderstood axiom. If $X \sim Y$, then X and Y can be substituted for each other in any decision without changing the rank ordering of alternatives. One of the implications is that the decision-maker's degree of preference for outcomes is linear with probability. For example, if a decision-maker is willing to pay \$$X$ for particular 50/50 gamble, then the decision-maker would be willing to pay only \$$X/2$ if the chances of winning are reduced to 25%. Note this does *not* imply that preferences are linear with attribute level, (but only linear with respect to probability.)

Note that the axioms are presented elsewhere as ordering and transitivity, reduction of compound uncertain events, continuity, substitutability, monotonicity and invariance, but their effect for the purposes of DBD are the same.

3.3 UNCERTAINTY AND THE EXPECTED UTILITY CALCULATION

The axioms or "rules for clear thinking" serve two purposes: First, they establish some ground rules for defining "good decision-making," so that we can recognize it when we see it. Second, they help structure the problem in such a way that it becomes relatively straightforward to assess an individual decision-maker's utility function, and to express that utility function in mathematical form. Chapter 12 describes the lottery methods for assessing single-attribute utility functions (that reflect nonlinear preferences over an attribute range and the decision-maker's attitude toward risk) as well as the scaling constants that reflect the decision-maker's willingness to make trade-offs.

A method for employing a beta distribution described in Thurston and Liu [7] demonstrated the use of probabilistic multi-attribute utility analysis for determining the effect of attribute uncertainty on the desirability of alternatives. The assessed single-attribute utility functions reflect the decision-maker's degree of risk aversion for each attribute. Maximizing expected utility captures the full range of uncertainty and the decision-maker's attitude toward risk, unlike much simpler approaches such as minimax, maximin and minimax regret.

Expected utility $E[U(x_j)]$ can be calculated from the single-attribute utility functions $U(x_j)$ and the probability-density functions $f(x_j)$ for each attribute j. Expected utility is calculated using Eq. (3.3):

$$E[U_j] = \int_{x\,\min}^{x\,\max} U_j(x_j) f(x_j) dx_j \qquad \text{Eq. (3.3)}$$

The beta distribution is often appropriate and convenient to characterize the uncertainty associated with each attribute. The input required of the user is fairly straightforward to assess. The beta distribution is part of the theoretical basis for the project evaluation and review technique (PERT) employed to determine the optimal schedule of interdependent tasks with user-estimated uncertainty in completion times.

The required user inputs are the minimum, maximum and most-probable values. A beta random variable distributed on the interval (x_L, x_U) of the lower and upper limits, respectively, on x_j has probability density

$$f(x) = \frac{\Gamma(p+q)}{r\Gamma(p)\Gamma(q)} \left(\frac{x - x_L}{r}\right)^{p-1} \left(\frac{x_U - x}{r}\right)^{q-1} \qquad x_L \le x \le x_U$$

$$= 0 \qquad \text{otherwise} \qquad \text{Eq. (3.4)}$$

where the range is $r = x_L - x_U$. If the shape parameters p and q are chosen to be greater than 1, the distribution is unimodal; if they are equal to 1, the beta distribution degenerates to a uniform distribution. For such a beta variate, the mean μ and the mode m can be calculated by:

$$\mu = x_L + r\frac{p}{p+q} \qquad \text{Eq. (3.5)}$$

and

$$m = x_L + r\frac{p-1}{p+q-2} \qquad \text{Eq. (3.6)}$$

where the mode, m, is sometimes referred to as the most probable or "most likely" value. Here x_L, x_U and m are supplied, so Eq. (3.6) defines the relationship between the shape parameters that will produce the requested mode. As p and q vary, however, the probability mass is distributed in a variety of ways about the mode, as shown in Fig. 3.1.

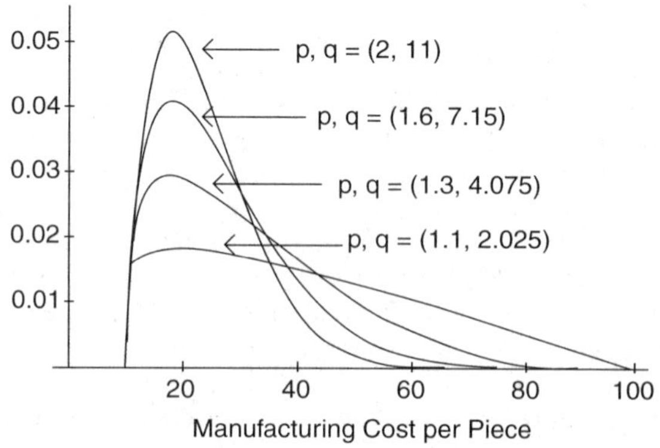

FIG. 3.1 MANUFACTURING COST UNCERTAINTY WITH A RANGE OF UNDERLYING BETA DISTRIBUTION PARAMETERS *p, q*

The expected utility for an attribute whose estimated performance level is characterized by a beta probability-density function, assuming that the interval (x_L, x_U) is contained within (x_{min}, x_{max}), from Eq. (3.3) and (3.4) is:

$$E[U_j(x_j)] = \frac{\Gamma(p+q)}{r\Gamma(p)\Gamma(q)} \int_{x_L}^{x_U} [U_j(x_j)] \left(\frac{x - x_L}{r}\right)^{p-1} \times \left(\frac{x_U - x}{r}\right)^{q-1} dx \qquad \text{Eq. (3.7)}$$

The expected value can be obtained by numerical integration of Eq. (3.7). Thurston and Liu [7] show that if an exponential form of $U(x_j)$ is used and shape parameters p and q are integers greater than or equal to 1, then a closed-form expression can be obtained—it requires little computational effort when the shape parameters are small.

This approach has been employed to investigate the effect of uncertainty on the rank ordering of automotive bumper beam systems [8]. The rank ordering of alternatives was sensitive to the degree of uncertainty associated with manufacturing cost estimates, even when the median (or most likely) estimate and bounds were the same in each case. For a risk-averse decision-maker, utility decreased as the range of uncertainty increased. In addition, the rank ordering of alternatives was dependent on the decision-maker's degree of risk aversion.

3.4 UTILITY ASSESSMENT DIFFICULTIES

This section describes some potential difficulties with the assessment procedure and how they can be overcome. A more comprehensive description of both the real and the misconceived limitations of DBD is provided in Thurston [9].

3.4.1 Difficulties With Utility Assessment Procedure

There are several potential difficulties with the utility assessment procedure, including level of effort required, biases and inconsistencies. Some argue that the level of effort and length of time required to properly formulate and assess a utility function is too great, and that the lottery questions commonly employed are nonintuitive and difficult to understand. The lottery methods have been described for design elsewhere [10], and are also presented in Chapter 12. We argue that "you get what you pay for, and you should pay for only what you need." For most applications in which the subject has collaborated with the decision analyst in defining the design problem and conflicting attribute set, this author's experience has been that the utility assessment procedure takes approximately 1 hour ± 30 minutes, depending on the number of attributes. The payoff is the ability to accurately quantify, for that particular design, the desirability of alternative trade-offs and the effect of uncertainty. At the extreme opposite end of the spectrum of decision tools is coin flipping, which is fast and easy to understand. In between are methods such as the weighted averages often employed in quality function deployment, the analytic hierarchy process and others. In the context of engineering design, the appropriate tool depends on the phase of design and the level of complexity of the design decision.

In addition, some mistakenly argue that the lottery methods force choices to be made under uncertainty, and are therefore not valid for comparing design alternatives where no uncertainty is involved. However, one of the strengths of utility analysis is its ability to model utilities that might be nonlinear over the tolerable attribute range, *whether or not* uncertainty is involved. The inability to reflect nonlinear preferences is a shortcoming of widely used weighted average methods [10]. Probabilities are employed in utility assessment primarily as a mechanism to elicit and measure preferences, and those measurements are accurate as long as the substitution-independence and other axioms are not violated.

A more legitimate (but still resolvable) concern is that the responses to the lottery questions can be subject to systematic distortions. For example, the substitution axiom of rational decision-making indicates that if one is willing to pay $X for a 25% gamble, then one should be willing to pay exactly $2X if the probability of winning the same payoff increases from 25% to 50%. However, the "certainty effect" can sometimes lead the decision-maker to be willing to pay far *more* than $4X if the probability of winning the same payoff is increased from 50% to 100%, or it becomes a "sure thing." It is extremely important to distinguish this unwanted distortion from the very features that one is attempting to model through utility assessment, which are nonlinear preferences over the *attribute* range and risk aversion. Another systematic distortion the lottery assessments can be vulnerable to is an anchoring bias. This is where the measured degree of risk aversion can be inordinately influenced by the probability the analyst happens to employ in the certainty equivalent method, where the designer indicates the point at which he or she is indifferent between a certain outcome and a lottery. One assessment procedure that solves these problems is the lottery equivalent method, where the designer indicates the point at which he or she is indifferent between two lotteries, rather than a lottery and a certainty [11]. Another approach employs pairwise comparison concepts to avoid inconsistencies [12].

The Allais paradox [13] is sometimes cited as an illustration of how a reasonable person might violate one of the axioms (substitutability), rendering the results of the lottery assessment questionable. To make this point, the Allais example employs a combination of extreme differences in outcomes (lotteries whose outcomes range from winning something on the order of $5 million to winning nothing) and very small differences in the probability of outcomes (10% vs. 11%), as shown in Fig. 3.2. Given the choice between *L1* and *L2*, many people prefer *L1*, the "sure thing" of $1,000,000 to *L2*, which would expose them to a 1% chance of ending up with nothing. Assigning $U(\$0) = 0$ and $U(\$5{,}000{,}000) = 1$, then one can write

If $\quad L1 \succ L2$

then $\quad U(\$1M) > 0.10\,U(\$5M) + 0.89\,U(\$1M) + 0.01\,U(\$0) \qquad$ Eq. (3.8)

and $\quad 0.11\,U(\$1M) > 0.10\,U(\$5M)$

Now, given the choice between L3 and L4 shown in Figure 3.1, many of those same people prefer L3, perhaps focusing more on the difference in outcomes ($5,000,000 vs. $1,000,000) than the difference in probabilities of those outcomes (0.10 vs. 0.11). But

if $\quad L3 \succ L4$

and $\quad 0.10\,U(\$5M) + 0.90\,U(0) > 0.11\,U(\$1M) + 0.89\,U(0) \qquad$ Eq. (3.9)

then $\quad 0.10\,U(\$5M) > 0.11\,U(\$1M)$

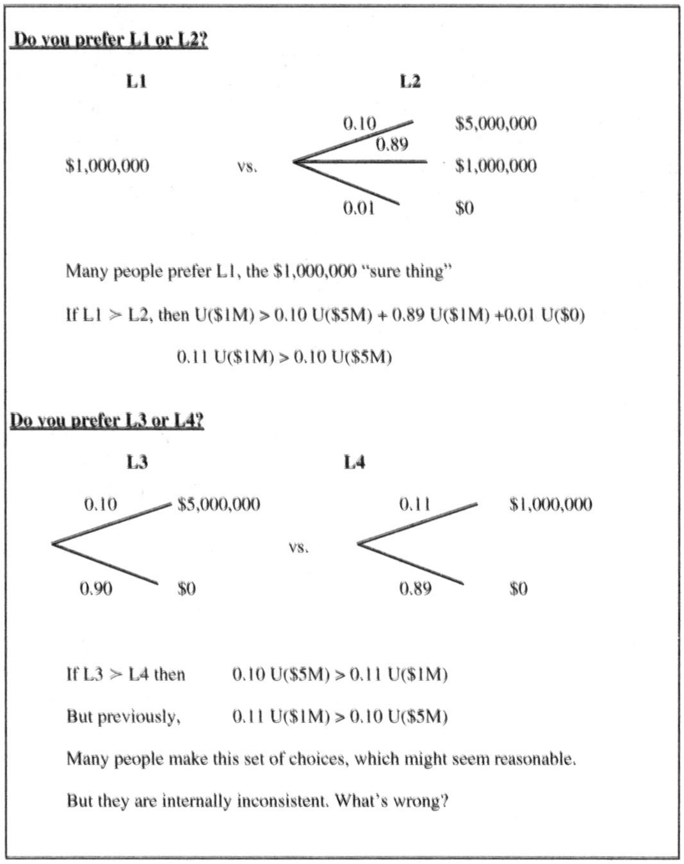

FIG. 3.2 ALLAIS PARADOX

But previously, the opposite result was obtained, which was 0.11 $U(\$1M) > 0.10\ U(\$5M)$!

Since this inconsistency violates the substitutability axiom, one might be led to the conclusion that "the axioms are wrong" and utility analysis cannot be used. The argument would be that since the axioms can be shown to not accurately reflect real preferences in some instances, they should not be employed as decision-making "rules." The more accurate interpretation is that "decision-makers are wrong," and that the paradox merely illustrates one example of internally inconsistent decision-making, which was the whole reason for the development of normative decision theory in the first place! Decision theory is normative rather than descriptive. Its goal is to help people make better decisions by avoiding inconsistency, not to mimic their unaided choices.

Once again, the key lies in a carefully performed utility analysis. Skillful definition of the attributes and their ranges can avoid many problems. To avoid the inconsistency illustrated in the Allais paradox, two actions can be taken. First, note that the example in the paradox initially presents the decision-maker with an option of a certainty equivalent that for most people is quite a significant improvement ($1,000,000) over their current assets position. And both sets of assessments force the decision-maker to consider an extreme range of outcomes, from 0 to $5,000,000. In contrast, most engineering design problems require the designer to consider outcomes over a much smaller absolute range, and the range of the impact from the best outcome to the worst outcome would also be much smaller. So one element of good design utility problem formulation is to define the tolerable attribute ranges as narrowly as possible. For example, when the attribute xj is weight, the lower end of the range should *not* be defined as 0 pounds, but instead as the most optimistic yet realistic estimate of the lowest weight that is technically feasible (given other constraints such as material properties, strength requirements, etc.). Similarly, the upper end of the range should not be defined as the highest weight presented in the alternative set, but rather the highest weight that the decision-maker is willing to tolerate. That limit is defined as that which is the worst tolerable, but is still acceptable.

Note also that the example in the paradox forces the decision-maker to simultaneously compare an extreme range of outcomes ($5,000,000, $1,000,000 and 0) and a very small range of probabilities (0.10 and 0.11). A problem such as this is extremely difficult to think about in an internally consistent manner. More important, it is not typical of most engineering design problems. So in addition to narrowing the range of possible outcomes, another element of good design utility assessment is to employ probabilities in the lottery assessments that the decision-maker can cognitively differentiate in a meaningful way. For example, use p = (0.25 and 0.75) vs. (0.5 and 0.5), rather than (0.89 and 0.11) vs. (0.90 and 0.10).

3.4.2 Biases/Difficulties in Estimating Uncertain Design Performance

Subjective expected utility analysis relies on the decision-maker being able to model uncertainty using either discrete probabilities or a probability-density function $f(xj)$ for an attribute j. A variety of methods are available for estimating such a distribution from a few simple inputs. For example, "most likely," "pessimistic" and "optimistic" values can be used to estimate parameters for the beta distribution as described earlier [7], [8].

However, the process of estimating uncertainties is in itself prone to biases. Kahneman et al. [1] have described heuristics (necessary and useful rules of thumb) that rational people employ during the difficult task of estimating uncertain quantities. They also document the systematic biases to which these heuristics are prone, which can lead to inaccurate or irrational results. For example, a designer might employ the "anchoring and adjustment" heuristic in order to estimate manufacturing cost for a certain design alternative. He or she begins by "anchoring" the estimate on a similar design for which the cost is known, then "adjusts" for other factors. This is a reasonable heuristic. However, it has been shown that estimates can be inordinately influenced by the initial anchor. Amazingly, this bias is even exhibited in experiments where the subjects are shown that the initial "anchor" is obtained from the spin of a roulette wheel, and is thus completely irrelevant to the problem at hand [1].

Another systematic bias is overconfidence; even when probability distributions are employed, the actual distribution is often much broader than estimated. Engineers have grown accustomed to providing definitive expert recommendations in the face of uncertainty. Their initial reaction to the question "How much will it cost?" might be "We have no idea," which really means that he or she cannot say with 100% certainty what the cost will be. However, once they do arrive at an estimate, they tend to overestimate the degree of confidence in their answer. Part of the reason is that "experts" are often asked to provide discrete point estimates for uncertain outcomes, upon which decisions are then made. In contrast, utility theory takes the breadth of the probability distribution into account in determining expected utility.

Methods whose end result is to present the designer with several alternatives and their "probability of success" over a range of outcomes perhaps attempt to avoid these problems, but in effect are throwing the decision back into the designers lap.

3.5 SUMMARY

This chapter has provided a brief introduction to the fundamentals of utility theory. The axioms, or "rules for clear thinking," were presented in generic terms. A method to calculate expected utility employing a relatively easily assessed beta distribution to model uncertainty was presented. Finally, some difficulties with the assessment procedures were presented, along with methods for overcoming them when formulating a design utility problem.

The DBD community has come quite a long way from the days of debating whether or not designers make decisions. Subsequent chapters will reveal the depth and breadth of the contributions made by the community.

REFERENCES

1. Kahneman, D., Slovic, P. and Tversky, A., eds., 1982. *Judgment Under Uncertainty: Heuristics and Biases*, Cambridge University Press.
2. von Neumann, J. and Morgenstern O., 1947. *Theory of Games and Economic Behavior*, Princeton University Press, Princeton, NJ.
3. Keeney, R. L. and Raiffa, H., 1993. *Decision with Multiple Objectives: Preferences and Value Tradeoffs*, Cambridge University Press (first published in 1976 by John Wiley and Sons).
4. Howard, R. A. and J. E. Matheson, eds., 1984. "The Principles and Applications of Decision Analysis," Strategic Decision Group, Menlo Park, CA.
5. Luce, R. D. and Raiffa, H., 1957. *Games and Decisions: Introduction and Critical Survey*, Wiley, New York, NY.
6. French, S., 1986. *Decision Theory: An Introduction to the Mathematics of Rationality*, Wiley, London.
7. Thurston, D. L. and Liu, T., 1991. "Design Evaluation of Multiple Attribute Under Uncertainty," *Systems Automation: Research and Applications*, Vol. 1, No. 2.
8. Tian, Y., Thurston, D., and J., Carnahan, 1994. "Incorporating End-Users' Attitudes Towards Uncertainty into an Expert System," *ASME J. of Mech. Des.*, Vol. 116, pp. 493–500.
9. Thurston, D. L., 2001. "Real and Misconceived Limitations to Decision Based Design with Utility Analysis," *ASME J. of Mech. Des.*, Vol. 123, No. 2.
10. Thurston, D. L., 1991. "A Formal Method for Subjective Design Evaluation with Multiple Attributes," *Res. in Engg. Des.*, Vol. 3, No. 2.
11. McCord, M. and R., de Neufville, 1986. "Lottery Equivalents: Reduction of the Certainty Effect in Utility Assessment, *Mgmt. Sci.*, Vol. 32, pp. 56–60.
12. Wan, J. and Krishnamurthy, S., 1998. "Towards a Consistent Preference Representation in Engineering Design," *Proc., ASME Des. Theory and Meth. Conf.*
13. Allais, M., 1953. "Le Comportemente de l'homme rationnel devant le risque: critique des postulates et axiomes de l'ecole americaine," *Econometrica*, Vol. 21, pp. 503–46.

PROBLEMS

3.1 Jason, your risk-averse roommate, has asked you to help him assess his utility function for money over the range +\$2,000 to −\$500. Show a set of three lottery questions you could use to assess Jason's utility function (using either the certainty equivalent or probability equi-valent method), including Jason's risk-averse responses. Plot the results of his responses where the x-axis is dollars and the y-axis is utility of dollars.

3.2 Ritesh currently has assets of \$1,000 and his utility function for dollars is $U(x) = \ln(x)$. He is considering a business deal with two possible outcomes. He will either gain \$3,000 with a probability of 0.4, or lose \$800 with a probability of 0.6.

a. What is the most he would be willing to pay for the deal?
b. After he has purchased the deal for half the amount he had been willing to pay, what is the smallest amount for which he would sell the deal?

3.3 You are performing a utility assessment in order to help a designer make design decisions. It is not going well. She has exhibited the Allais paradox on lotteries over x = \$cost ranging from \$0 to \$1 million.

10 pts: What are three possible implications for your utility assessment?

10 pts: Show three things you might do to solve the problem. Feel free to make (and describe) any assumptions necessary to illustrate your point.

CHAPTER

4

Normative Decision Analysis in Engineering Design

Sundar Krishnamurty

NOMENCLATURE [4, 5, 6, 8, 17]

Decision analysis = a combination of philosophy, methodology, practice and application useful in a formal introduction of logic and preferences to the decisions of the world [5]. "Decision Analysis" is a structured way of thinking about how the action taken in the current decision would lead to a result.

Decision = an irrevocable allocation of resources; the only thing one can control is the decision and how one can go about that decision.

Objective = indicates the direction in which the designer should strive to do better.

Attribute = characteristic of design performance; also a measure of objective achievement.

Alternative = a particular set of controllable design variables, which will lead to °particular attribute performances. The purpose of engineering design is to find the alternative with highest "value" or "utility."

Weight/scaling constant = a measure of relative influence of each attribute on the entire design excellence. In value trade-offs, "weight" is used, while in utility, "scaling constant" is used.

Value = a numerical quantity to illustrate goodness of attributes under certainty.

Preferential independence = attribute Y is preferentially independent of the complementary attribute(s) Z if the conditional preferences of y' and y'' given z' do not depend on z'.

Mutual preferential independence = indicates that every subset of attributes is preferentially independent of its complementary set.

Value theory = deals with trade-off under certainty in mul-tiattribute problems. Mutual preferential independence is generally tested in value assessment. SMARTS and SMARTER methods use value theory.

Utility = a numerical quantity to illustrate the goodness of attributes under uncertainty.

Lottery assessment = a method used to elicit design preferences in expected utility theory. Based on *vN-M* axioms, a certainty is found to represent a lottery indifferently, then a mathematical expression is developed for single-preference (utility) function, or a scaling constant is computed.

Utility independence = attribute(s) Y is utility independent of Z when conditional preferences for lotteries on Y given z do not depend on the particular level of z.

Note: The utility independence is stricter; it requires preferential independence, however, the reverse is not guaranteed.

Mutual utility independence = indicates that every subset of attributes is utility-independent of its complementary set.

Additive utility independence = attribute Y and Z are additive; independent if the paired preference comparison of any two lotteries, defined by two joint probability on $(Y * Z)$, depends only on their marginal probability distribution. Additive utility independence ensures utility independence, but the reverse is generally not true.

Expected utility theory = deals with trade-offs under uncertainty in multi-attribute problems. Mutual utility independence is generally tested. The preference is assessed through lottery questions.

Utility function = is a mathematical mapping of the decision-maker's attitude toward risk.

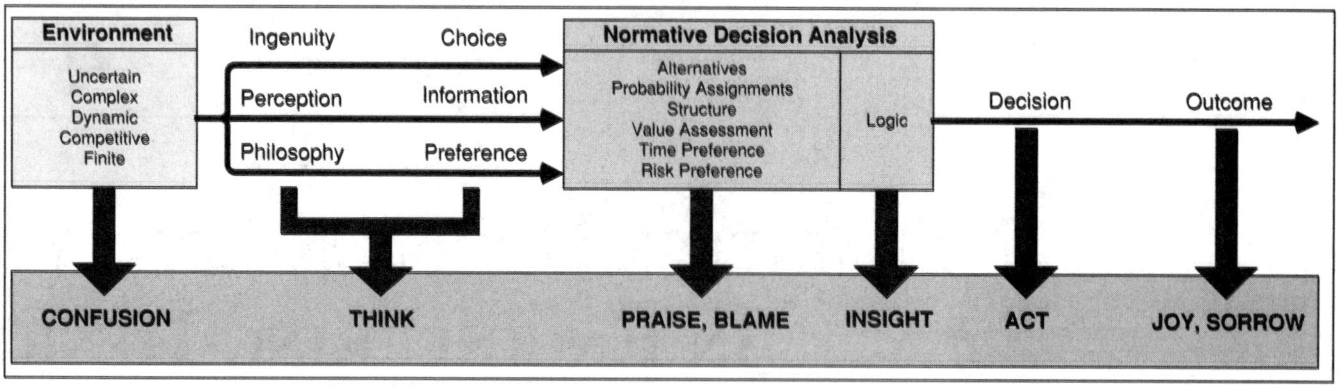

FIG. 4.1 NORMATIVE DECISION ANALYSIS (DR. HOWARD)

4.1 DECISION ANALYSIS OVERVIEW

The basis of making design decisions can be found in decision analysis. Fundamental to decision analysis is the concept of value, which measures what is preferred or desirable about a design [1]. This is the underlying principle behind <u>d</u>ecision-<u>b</u>ased <u>d</u>esign (***DBD***), which states that engineering design is a decision-making based design exploration task mainly involving three important factors: (1) human values; (2) uncertainty; and (3) risk [1–4]. While DBD has seen substantial growth in recent years, decision analysis itself is an already matured field and has been widely applied to many fields, including engineering. Seminal works relating to system engineering were published by Dr. Howard in the 1960s [5–7]. He has documented a clear overview of the normative decision analysis process and how it can be applied to system engineering. Figure 4.1 presents a schematic representation of his normative decision analysis procedure. This self-explanatory figure captures the essence of the decision-making process and how it relates to human thoughts, feelings and decisions. Paraphrasing Dr. Howard, normative decision analysis does not eliminate judgment, intuition, feelings or opinion. Instead, it provides a mathematical framework to quantify them and express them in a form where logic can operate on them, instead of being buried in a human mind, where we cannot get access to them.

The purpose of decision analysis is to achieve a rational course of action by capturing the structure of a problem relationship and by treating the uncertainty through subjective probability and of attitude towards risk using expected utility theory. The underlying concepts found here are universal, and arguably, they are more relevant today as the computational efforts required to execute design decisions are becoming more feasible due to improved computing capabilities.

Expected utility theory is a normative decision analysis approach with three main components: options, expectations and value, where the decision rule is that the preferred decision is that option with an expectation of the highest value (utility). It is based on the premise that the preference structure can be represented by real-valued functions and can provide a normative analytical method for obtaining the utility value ("desirability") of a design with the rule of "the larger the better" [8]. Five major steps associated with this technique are [8]:

(1) Identification of significant design attributes and the generation of design alternatives
(2) Verification of relevant attribute independence conditions
(3) Evaluation of single-attribute utility (SAU) functions and trade-off preferences
(4) Aggregation of SAUs into a system multi-attribute utility (MAU) function
(5) Selection of the alternative with the highest MAU value by rank-ordering alternatives

Here, the mechanism to get preference structure is based on the notion of the lottery, referred to as a von Neumann-Morgenstern (vN-M) lottery, and by employing the certainty equivalent, which is the value at which the decision-maker is indifferent to a lottery between the best and the worst [8]. The lottery questions provide the basis for describing the logic between attribute bounds, where analytical function formulations are typically used to complete the preference structure description. Similarly, lottery questions form the basis for eliciting trade-off information among attributes [8].

A decision process begins with the formulation of design alternatives (Figure 4.2). These alternatives are then used to identify the attribute space and features that are critical to evaluating alternatives and their performances. The preference assessment step enables determination of single-attribute functions for each attribute from its attribute space and feature. The results are then aggregated toward establishing a multi-attribute function to serve as a super criterion through elicitation of trade-off preferences among the individual attributes. Ranking all alternatives according to the super criterion results in the determination of the final optimal from an overall perspective, reflecting a decision-maker's preferences under conditions of uncertainty.

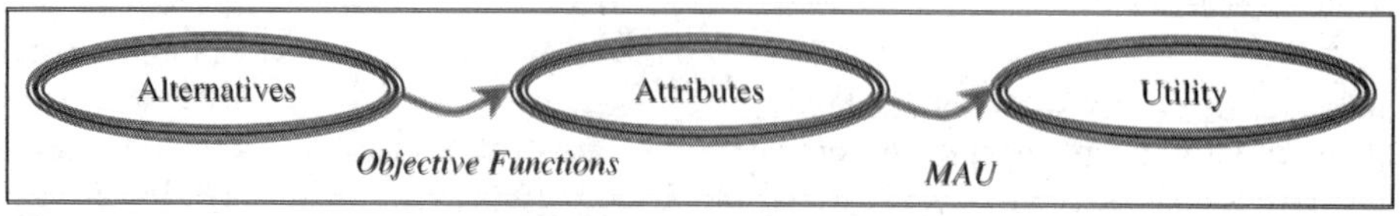

FIG. 4.2 DECISION-BASED DESIGN PROCESS

4.2 DECISION ANALYSIS IN ENGINEERING DESIGN

Engineering design refers to a variety of activities that lead to the definitions of a product, a process and their related issues, such as the manufacturing and production system for products and processes. The many stages involved in those activities will include concept exploration, model selection and configuration design, as well as detailed parametric design. Finding the best or the most optimal solution will often require an examination of several alternatives and their relative performances. Problems in engineering design are often highly nonlinear and complex. They usually involve multiple design attributes based on potentially conflicting technical and economical performance and other related requirements. Furthermore, one has to deal with the issues concerning uncertainty and risk. For these reasons, difficulties often encountered can be: (1) limitless possible design configurations and options; (2) uncertainty in design performance due to process, model, computational uncertainties; (3) no commensurate value measurement of different attributes that incorporate designer's intent and preferences. Consequently, engineering design problems may not always be suitable to the straightforward application of the existing linear or nonlinear optimization techniques. More important, design can never be reduced to a prescriptive procedure exclusive of human input, or say, human decisions. Alternatively, decision analysis principles can provide a foundation for the development of decision models of complex, dynamic engineering problems in the presence of uncertainty, as well as facilitate the identification of optimal design decisions.

In the context of engineering design, the quintessence of a decision analysis approach can be stated as first modeling the decision scenario, resulting in a mapping from the design option space to the performance attribute space, and then constructing an overall utility function that reflects the designer's preference information, including trade-offs among system attributes [8, 9]. When dealing with engineering design problems, one can observe certain fundamental characteristics that are typical to decision situations in engineering design. For example, the alternative sets are generally discrete and limited in traditional decision-making situations. In engineering design, and in parametric design in particular, there may be continuous alternatives and limitless design options. Similarly, unlike in traditional decision-making, which mostly deals with the alternative sets of existing option domain, alternative domains in engineering design may not always be known *a priori*. Furthermore, in addition to the use of experience, expectation and prior information as in traditional decision-making, engineering design may often require use of simulation and computational models. However, the main difference is the fact that the objective in traditional decision-making is often to "***pick up***" the best. However, in engineering design, the goal is often to "***make***" the best, requiring exploration of design alternatives using optimization techniques for the purpose of maximizing the utility (super criterion value) [10].

A successful implementation of any decision analysis approach will require a careful study of critical issues such as attribute formulation, design decision assumptions, and the process of preference elicitation and formulation. The following sections discuss these topics in detail in the context of engineering design.

4.2.1 Design Attribute Formulation

Design attribute formulation refers to selection of attributes, setting up attribute bounds, and selecting the logic to represent the designer's preferences between the attribute bounds that will be reflected in the attribute features. Attributes are measures of objectives. They are indicators of the performance of the objectives in any given design scenario. Decision-making in engineering design will require an essential step of generating and searching design alternatives. It is important that the set of alternatives does not contain an alternative that is dominated by another alternative. Not only is a dominated alternative useless, but from a psychological point of view, it may mislead the decision-maker into choosing the dominating solution that will be "***unfair***" to other non-inferior alternatives. Therefore, though it may not be necessary to include all non-inferior options, care should be taken to ensure that all inferior options are excluded. However, if a mathematical expression is already formulated to construct a super criterion using techniques based on expected value theory, modern optimization techniques can be employed to achieve an optimal option without worrying about creating a new alternative domain or its relative merits.

The book by Raiffa and Keeney [8] provides a set of criteria for attribute set selection to represent a problem [8]. The criteria can be summarized as follows: (1) Attribute set should be complete, so that all the important aspects are covered and reflected in the design problem formulation; (2) it should be operational, so that design decision analysis can be meaningfully implemented; (3) it should be non-redundant, so that double counting of impacts can be avoided; and (4) the set should be kept minimal, as a small problem dimension is preferred for simplicity.

4.2.2 Attribute Bounds

To describe completely the consequences of any of the possible outcomes of action in a complex decision problem requires specified focus region(s). The goal here can be stated as finding the appropriate single-attribute range that is useful and in a manageable form. For example, it should be comprehensive enough to indicate what a certain attribute level means (its utility), and if that certain level attribute is achieved, what can be the expected performance in terms of the resulting design. Note that while any option with an attribute level below the least preferred would be treated as total failure, an attribute level above the best preferred would still be considered. Although, according to normative analysis of attribute range effects, the range will not influence final options if the designer does not change preferences; in practical situations it may pay to handle attribute ranges with great care [11]. According to research in behavioral decision-making [11, 12], in a multi-attribute value/utility model, the attribute range may influence attribute weight/scaling constant assessment (range sensitivity of attribute range). That is, range may have some effect on the decision-maker's preference expression with respect to assessment method. It is the authors' experience that in difficult design environments, single-objective optimization techniques can provide great insight into attribute behaviors that can be of use in selecting their range bounds.

4.2.3 Attribute Features: Monotonicity vs. Non-Monotonicity

Attribute features refer to the mathematical form of the designer's logic for the attributes between their bounds. If an attribute has a monotonic characteristic, this preference assessment task is relatively simple. In engineering design, strict monotonicity is a very reasonable assumption for single-attribute functions. Note that the above monotonicity condition is for the attributes with respect to their preferences. This is not to be confused with the attribute characteristics with respect to design inputs.

For example, in the beam design problem, the attribute deflection may be highly nonlinear with respect to design variables such as the load condition, beam cross-section, length, etc. However, a designer's preference with respect to that deflection attribute will most likely be monotonic, i.e., the designer will prefer smaller deflection to larger deflection in a monotonic manner. Generally, two possible conditions of monotonicity exist. The mathematical expressions are:

(1) $[x_1 > x_2] \Leftrightarrow [U(x_1) > U(x_2)]$ implies monotonic increase. It generally forms a maximizing problem, i.e., the more of the attribute, the better. Note that when the attribute increases, the utility/value of the corresponding attribute increases, too.
(2) $[x_1 > x_2] \Leftrightarrow [U(x_1) < U(x_2)]$ implies monotonic decrease. It can be interpreted as a minimizing problem, i.e., the less the better. By increasing the attribute, the utility/value decreases.

Sometimes it is possible that an attribute needs to be considered in a "the nominal, the better" sense, i.e., the designer prefers a "target value" fixed at a certain level and considers any deviation from this level as a reduction in the goodness of the attribute. For example, tire pressure of cars cannot be too large as it would reduce the comfort of the car; however, if the pressure is too small, it will reduce maneuverability of the car. In these non-monotonic attribute situations, it is possible to rechoose attribute expressions that may have monotonicity. For example, monotonicity can be achieved by including two attributes such as maneuverability and comfort instead of the tire pressure attribute. Clearly, the monotonicity or non-monotonic condition issue is closely related to attribute selection. However, in cases where such a re-selection of attributes is not possible, methods based on mathematical transformation are typically used. Another interesting scenario that may lead to non-monotonicity is related to direction, mathematically positive or negative. In a mechanism design case study, if an attribute is structural error and the objective is to minimize it, the structural error can have two possible directions and they can be mathematically expressed as positive/negative errors. Here, a statement of "zero structural error is preferred" will result in a non-monotonic situation.

For non-monotonic functions, the range of attribute can be divided into intervals so that preferences are monotonic in each interval. An alternative way to deal with such problems is to reformulate them so that we can transform non-monotonic into monotonic by choosing proper attribute expression. For instance, the attribute in the mechanism design example can be expressed as "the absolute structural error." But if we transit the utility function between monotonic and non-monotonic, although it simplifies the assessment, the decision-maker may suffer information loss, such as the smoothness characteristic or asymmetry. A common non-monotonic situation is illustrated in Figure 4.3. Its equivalent transformed one is shown in Figure 4.4, resulting in a loss of asymmetric preference information.

4.2.4 Decision Scenarios: Certainty vs. Uncertainty

Decision-making, in general, can be undertaken in deterministic and non-deterministic scenarios. Value theory based methods, such as SMARTS/SMARTER, deal with deterministic design cases while their complementary expected utility theory methods deal with design under uncertainty. Note that much of the logic and rationale found in deterministic methods can serve as a road map for dealing with uncertainties as well.

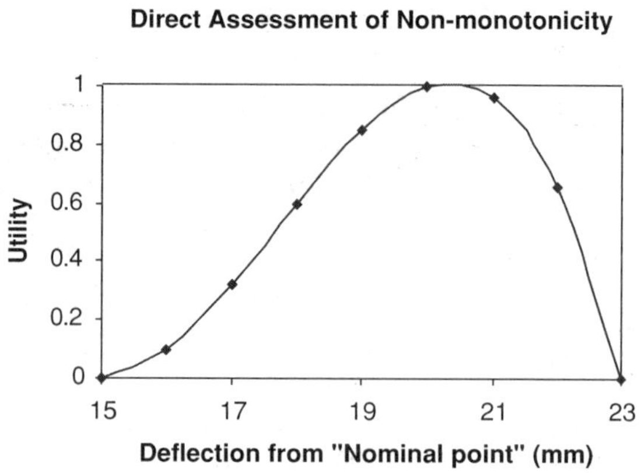

FIG. 4.3 SAU IN NON-MONOTONICITY CONDITION

Uncertainty is an inherent characteristic of decision-making in engineering design, which can be due to inadequate understanding, incomplete information and undifferentiated alternatives. Generally, the attitude to a risk situation can be divided into three categories; risk averse, risk prone and risk neutral (Figure 4.5).

(1) Risk Neutral
The designer is considered risk neutral if the utility function is linear, and the expected consequence is equal to the certainty equivalent ($U[E(x)] = E[U(x)]$). Traditional optimization techniques that by default follow the risk-neutral attitude in all situations can be interpreted as equivalent to no human subjective input involvement, and as such, do not incorporate the risk attitude preferences of the designer.
(2) Risk Averse
The designer is considered risk averse if the utility function is concave ($U[E(x)] > E[U(x)]$), indicating that the expected consequence is greater than the certainty equivalent. A risk-averse designer prefers to behave conservatively. In engineering design, due to safety considerations, utility assessment is expected to be risk averse in most design scenarios. Risk aversion may also be due to a lack of confidence in achieving an improvement in an attribute.
(3) Risk Prone
Contrary to risk aversion, when the utility function is convex ($U[E(x)] < E[U(x)]$), the expected consequence is less than certainty equivalent. In this case, the designer is considered risk prone. When a designer believes the performance of the attributes will meet basic requirements without doubt, he/she may demonstrate a conditionally risk-prone attitude.

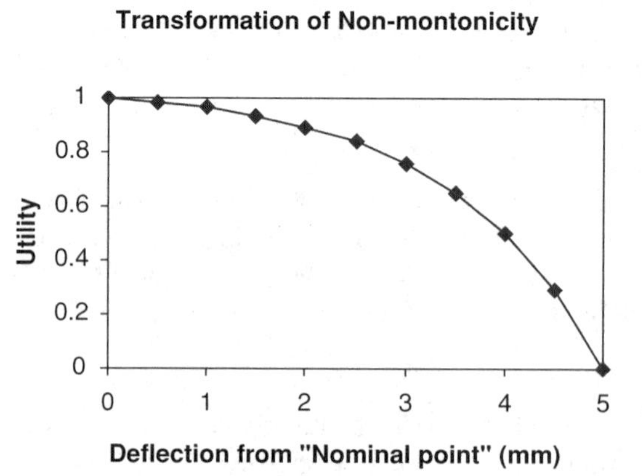

FIG. 4.4 TRANSFORMED MONOTONIC SAU

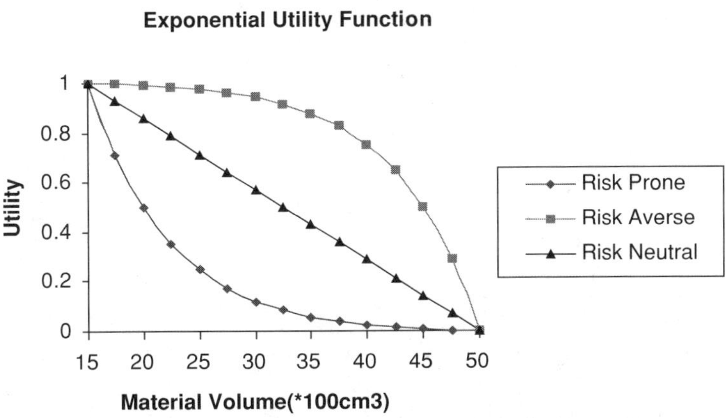

FIG. 4.5 RISK ATTITUDE IN DECREASING MONOTONICITY

It is interesting to note that Taguchi's philosophy-based *S/N* metric is indeed representative of a risk-averse approach for the signal (mean) and a conditionally risk-prone attitude for noise (standard deviation) [13].

4.3 PREFERENCE ASSESSMENT IN ENGINEERING DESIGN

This is arguably the most significant and most complex step in a decision-based engineering design process. Currently prevailing preference assessment techniques for multi-attribute scenarios can be categorized as based on either deterministic value theory or expected utility theory. For clarification, in this paper, value refers to decision with certainty and utility refers to decision with uncertainty. Preferences can be assessed through employment of implicit and/or explicit articulation of a designer's priorities using normative and descriptive points of view. Preference assessment can be defined as determining the single-attribute function, and then developing a mechanism to generate its aggregate multi-attribute function that can serve as a super-criterion metric for evaluating and selecting the desired design. One may choose a different theory and a corresponding assessment procedure for different design scenarios. The selection of a procedure may influence the outcome, and it is conceivable that different procedures may lead to different solutions for the same set of preferences. The best fit for a given problem may depend on not only the design scenario, but on the set of assumptions as well. Furthermore, it may depend on a designer's cognitive work on assessments, as well as his or her subjective expressions of the problem formulation.

4.3.1 Value Theory Based Preference Assessment

Value theory is a formal deterministic theory to quantitatively express the strength of preferences [9, 14]. Central to the decision theory is the concept of value that represents the measurement of what is good or desirable in a design. It is an inherent property of engineering designs, i.e., engineers design a product so that it generates maximum value. In a multi-attribute decision-making scenario, the designers trade off attribute values to achieve the best combination of engineering elements so as to maximize its total potential value. In this setup, the single-attribute value (SAV) functions can be generally obtained using one of the following methods: (1) Direct estimation—the value magnitude is set at some attribute levels directly; (2) variation of numerical estimation—the upper and lower levels are arbitrarily assigned values, the other levels are assigned relative to the former assignment; (3) indifferent bisection, a halfway value between the upper and lower attribute levels is found; and (4) different standard sequence, where the range is divided in such a way that the sequence of stimuli are equally spaced and the increments in preference strength going through any interval are equal [8, 9]. Similarly, the aggregate multi-attribute value measurement can be assessed using: (1) direct estimation, where the designer assigns values to alternatives; (2) swing weights, where the designer ranks the relative importance of moving from the worst to the best level of each attribute; and (3) trade-off weights, where the designer finds an indifference point among options (all attribute vectors have same values). Note: Of these methods, research has shown that except for the direct value assignment, all other methods are range-sensitive [11].

For the purposes of illustration, the SMARTS/SMARTER method developed by Dr. Edwards, a premier researcher in this field, is considered [15]. The concepts behind the original ***S***imple ***M***ulti-***A***ttribute ***R***ating ***T***echnique (***SMART***) and its later revisions include simple linear approximations to SAV functions, and an additive aggregate model using swing weights for multi-attribute value functions. An overview of this approach is shown in Fig. 4.6.

The main difference between SMARTS and SMARTER methods is that SMARTS method uses swing weight assessment procedures, while the SMARTER method does not need weight assessment. In this setup, rank order centroid (ROC) weights are assigned directly from the knowledge on the weight rank order. ROC weights use simple mathematics such as centroidal relationship and a simplex algorithm to turn ranking of weights into weights [10, 15, 16]. The general form of the weight expression for a system with *k* attributes is:

$$w_k = (1/k)\sum_{i=1}^{k}(1/i) \qquad \text{Eq. (4.1)}$$

While most engineering problems may not be adequately addressed using the deterministic value theory approach, it can be argued that the spirit and techniques contained in value theory can play a significant role in the understanding of the design process and the expected design consequences.

4.3.2 Expected Utility-Theory-Based Preference Assessment

The value-based deterministic model is for design scenarios in which no uncertainty and risk are considered, while its complementary expected utility theory is a probabilistic model that provides an analytical approach for decision-making under conditions

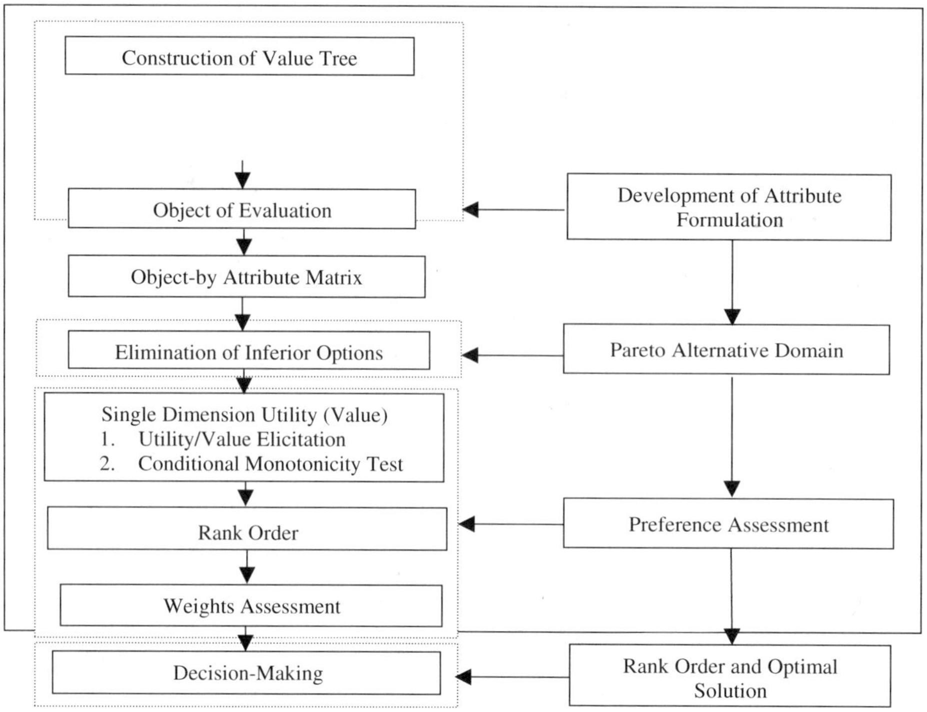

FIG. 4.6 AN OVERVIEW OF SMARTS/SMARTER

of uncertainty and risk. Details on expected utility theory and their fundamental importance to the engineering design process in providing a valid measure of utility under uncertainty and risk can be found in the works by Dr. Hazelrigg [3, 17].

The mechanism to get preference structure in expected utility theory is based on the notion of the lottery, referred to as a von Neumann-Morgenstern lottery that is built upon six axioms. The basic idea is that the designer keeps varying certainty or probability over lottery until an indifference point between lottery and certainty is found. The certainty equivalent of a lottery is the value at which the decision-maker is indifferent to a lottery between the best and the worst [3, 8].

While lottery questions certainly provide the basis for describing the logic between attribute bounds, analytical function formulations are typically used to complete the preference structure description in SAU formulation. While several analytical forms ranging from polynomial to logarithmic functions can be considered, it has been shown that the exponential function often meets most of the conditions in practice. This is due to the fact that the exponential form provides a constant risk premium in all conditions, and it can be shown to work well in the whole attribute range [18].

An exponential formulation can be written as: $U(x) = a + b^* \exp(c^*x)$, where "*c*" is risk coefficient, and "*a*" and "*b*" are parameters that guarantee the resulting utility is normalized between 0 and 1. The risk coefficient reflects the degree of risk averse (prone) and the risk premium is a measure of risk attitude, reflecting the rate at which risk attitude changes with a different attribute level. If we assume that the engineering designer has the same degree of risk premium in the whole attribute range, then the exponential formulation and linear formulation are the only formulations that will fit that behavior [18]. In fact, linear forms are indeed another broadly used formulation of utility and value functions. Using appropriate small piecewise divisions, the linear function formulation can be used to satisfy desired accuracy requirements. Other formulations like quadratic utility function and logarithmic utility function may be feasible. However, if a utility scale of 0-1 interval is preferred, these formulations are not recommended because they may have problems in normalization.

Development of SAU As stated before, SAU functions are typically obtained by analyzing the designer's considerations of a set of lottery questions based on the concept of the certainty equivalent. This concept treats SAU as a monotonic function with a utility of $U_{\text{best}} = 1$ defined to be the most preferred value for an attribute function and $U_{\text{worst}} = 0$ for the least preference. SAU functions are then developed to describe the designer's compromise between the two extremes based on his/her priority-reflected answers on the lottery questions. Specifically, the concept of certainty equivalent is used, where a *certainty* value is regarded as a guaranteed result compared to the lottery between the two extreme values in which there is a probability p_o of obtaining the best value and a probability of $1-p_o$ of obtaining the worst value. A probability of $p_o = 1$ causes the designer to choose the lottery where, similarly, a value of $p_o = 0$ will lead to the selection of the certainty. That value of p_o at which the designer is indifferent to the certainty or the lottery is characterized as the utility of the certainty.

Figure 4.7 illustrates the certainty equivalent regarding the sectional area of an I-beam [10]. Here, the *certainty* of 500cm^2 is a guaranteed result compared to the lottery between two extreme attribute values in which there is a probability p_o of obtaining the best value of 250 cm^2 and a probability of $1-p_o$ of obtaining the worst value of 900 cm^2. In this example, the best value may correspond to the lowest cross-sectional area permissible from a stress-based safety point of view, whereas the highest value can be interpreted as the highest area allowable from a cost perspective. A probability of $p_o=1$ causes the designer to choose the lottery where, similarly, a value of $p_o=0$ will lead to the selection of the certainty. The value of p_o at which the designer is indifferent to the certainty equivalent or the lottery is characterized as the utility

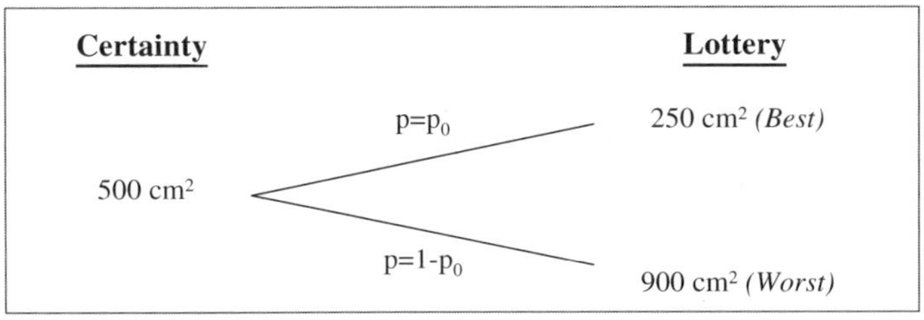

FIG. 4.7 TYPICAL LOTTERY QUESTION IN SAU ASSESSMENT

of the certainty equivalent. In effect, the utility of the certainty equivalent is equal to the mathematical expectation in the lottery, i.e., $u(500\text{ cm}^2) = p_o{*}u(250\text{ cm}^2) + (1-p_o)u(900\text{ cm}^2)$. For instance, an indifferent point at p_o=0.6 will result in the utility function value of $U(500)=0.6$.

Development of MAU The mathematical combination of the assessed SAU functions through the scaling constants yields the MAU function, which provides a utility function for the overall design with all attributes considered simultaneously. The scaling constants reflect the designer's preferences on the attributes, and they can be acquired based on scaling constant lottery questions and preference independence questions [8]. For the formulation of the MAU function, additive and multiplicative formulations are commonly considered. The main advantage of the additive utility function is its relative simplicity, but the necessary assumptions can be restrictive. In addition, it is difficult to determine whether the requisite assumptions are reasonable in a specific engineering design problem. This is due to the fact that the assumptions are stated in terms of the preferences for probability distributions over consequences, with more than one attribute simultaneously varying. No interaction of the designer's preference for various amounts of the attributes is allowed. If additive independence assumption is not applicable, then one may consider multiplicative utility function, or a multilinear function. A typical multiplicative MAU function can be expressed as follows [8]:

$$U(x) = \frac{1}{K}\left[\left[\prod_{i=1}^{n}(Kk_iU_i(x_i)+1)\right]-1\right] \qquad \text{Eq. (4.2)}$$

where:

$$1+k = \prod_{i=1}^{n}(1+Kk_i) \qquad \text{Eq. (4.3)}$$

If scaling factor $K = 0$, indicating no interaction of attribute preference, this formulation is equivalent to its additive form. Note that through proper transformation, multiplicative form can always be shown to be equivalent to its complementary additive model [17]. For a two-attribute (*sectional area and vertical deflection*) I-beam problem, assuming say *Sectional area > vertical deflection*, a lottery setup can be formulated as shown in Figure 4.8. Here, the best case scenario may be the result of individual optimal values from a Pareto set, while the worst case may represent the values of the other criteria corresponding to that optimal set.

4.3.3 Example: A Beam Design Problem

The development of decision models for parametric design optimization is illustrated with the aid of a simple beam design problem involving two conflicting criteria and a single constraint [10, 19, 20]. The goal of this problem is to determine the geometric cross-sectional dimensions of an I-beam (Fig. 4.9) that will simultaneously minimize cross-sectional area and vertical deflection while satisfying a stress constraint under given loads.

The various parameter values for the problem are:

Allowable bending stress of the beam material = 16 kN/cm²
Young's modulus of elasticity (E) = 2^*10^4 kN/cm²
Maximal bending forces P = 600 kN and Q = 50 kN
Length of the beam (L) = 200 cm

The problem objectives/constraints of area, deflection and stress can be expressed mathematically as follows:

Minimize cross-sectional area:

$$f_1(\mathbf{x}) = 2x_2x_4 + x_3(x_1 - 2x_4) \qquad \text{Eq. (4.4)}$$

Minimize vertical deflection: $f2(\mathbf{x}) = PL^3/48EI$

$$f_2(\mathbf{x}) = \frac{5{,}000}{\left[\left(\frac{1}{12}\right)x_3(x_1-2x_4)^3 + \left(\frac{2}{12}\right)x_2x_4{}^3 + 2x_2x_4\left(\frac{x_1-x_4}{2}\right)^2\right]} \qquad \text{Eq. (4.5)}$$

Subject to the stress constraint:

$$f_3(\mathbf{x}) = \frac{180{,}000x_1}{x_3(x_1-2x_4)^3 + 2x_2x_4\left(4x_4^2 + 3x_1(x_1-2x_4)\right)} + \frac{15{,}000x_2}{(x_1-2x_4)x_3^3 + 2x_4x_2^3} \le 16 \qquad \text{Eq. (4.6)}$$

This problem is also subject to the following material geometric constraints of:

$$10 \le x_1 \le 50\text{ cm},\ 10 \le x_2 \le 50\text{ cm},\ 0.9 \le x_3 \le 5\text{ cm},\ 0.9 \le x_4 \le 5\text{ cm} \qquad \text{Eq. (4.7)}$$

Decision Model Formulation The first step in the proper application of the SAU functions is the determination of the attribute

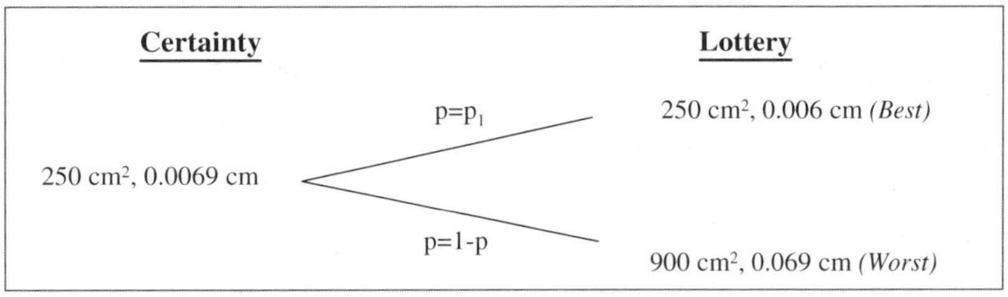

FIG. 4.8 A SCALING CONSTANT LOTTERY QUESTION FOR A TWO-ATTRIBUTE PROBLEM

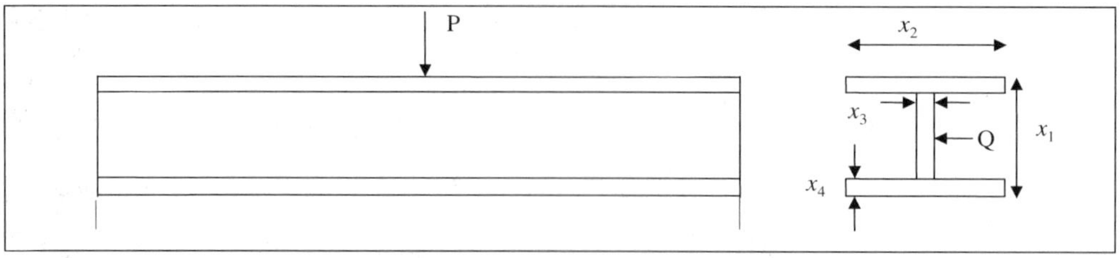

FIG. 4.9 BEAM DESIGN PROBLEM

bounds that correspond to the feasible design space. There are a few strategies in defining design space to aid in obtaining precise value or utility descriptions without the computational burden. The typical procedures include:

(1) Using *global* performance extreme values on each attribute with SAU between 0 and 1
(2) Using *current* best and worst performance values on each attribute with SAU between 0 and 1
(3) Using *acceptable* performance bounds on each attribute with SAU between 0 and 1

The third procedure is typically recommended due to its proven practicability and engineering significance. The best bounds for the attribute set can be found by optimizing each of the two attributes of area and deflection under stress constraint using single-attribute feasible optimization procedures. Identification of a feasible domain using the constraint treatment process of Kunjur and Krishnamurty [20] results in the following SAU and variables bound:

Area: $$250 \le f_I(\mathbf{x}) \le 900 \text{ cm}^2 \qquad \text{Eq. (4.8)}$$

Deflection: $$0.005 \le f_2(\mathbf{x}) \le 0.03 \text{ cm} \qquad \text{Eq. (4.9)}$$

$$56.7 \le x_1 \le 50, \quad 36.7 \le x_2 \le 50, \quad 0.9 \le x_3 \le 5, \quad 3.6 \le x_4 \le 5 \qquad \text{Eq. (4.10)}$$

These upper and lower attribute values provide the bounds for the proper lottery questions from which the SAUs are formulated. The area SAU is constructed using the values from Fig. 4.7, and Fig. 4.10 shows the deflection attribute lottery question setup.

In this example, the indifferent probability p_0 is set at 0.75, indicating a risk-adverse concave utility function with $U(0.015) = 0.75$. This utility evaluation along with the boundary utilities of $U(0.005) = 1$ and $U(0.03) = 0$ provide for three points on the utility function. Using an exponential of the form $u(x) = ae^{bx} + c$ yields the following SAUs for the area and deflection attributes:

Area: $$U_1(f_1) = 2.028 - 0.7917e^{0.001045 f_1} \qquad \text{Eq. (4.11)}$$

Deflection: $$U_2(f_2) = 1.308 - 0.231e^{57.8 f_2} \qquad \text{Eq. (4.12)}$$

Figure 4.11 graphically shows the designer's nonlinear preference for deflection as a mapping of the utility, lost or gained, when altering the beam deflection and its dependence on where that change takes place.

Note that although in this case of two attributes, the acceptable bounds of the attributes can be easily identified based on the single-attribute feasible optimal solutions, such identification for problems involving three or more attributes becomes more complex and will need further consideration. Also, the designer, according to his/her experiences and/or expectations, can further modify these acceptable levels. Usually the smaller attribute interval will contribute to a higher precision of the SAU assessment and an improved effectiveness of the design process.

Assuming mutual utility independence between design attributes with reasonable accuracy and effectiveness, a multiplicative MAU can be constructed using standard lottery questions. For example, Figure 4.8 shows the setup for the evaluation of deflection scaling constant. The scaling constants of area and deflection are set as 0.3 and 0.5, respectively, reflecting the decision-maker's trade-off preference. The above individual scaling constants yield a normalized scaling constant value of 1.33. This results in the following MAU function:

$$U(f_i, f_2) = 0.7519 * [0.3990\,(2.028 - 0.7917e^{0.001045 f_1}) + 1] \\ * [0.6650\,(1.308 - 0.231e^{57.8 f_2}) + 1] - 1 \qquad \text{Eq. (4.13)}$$

Constraint Handling In the previous formulation, only the objectives were constructed using attribute formulation. Alternatively, it is possible to view design constraints as design attributes as well. In these cases, when an attribute is in a certain range, the designer is satisfied (feasible solution), indicating that any attribute level in this range should receive the highest possible value/utility. However, if the attribute is outside the range, then the solution is considered infeasible and correspondingly assigned a value/utility near 0. The mathematical expression that describes this attribute

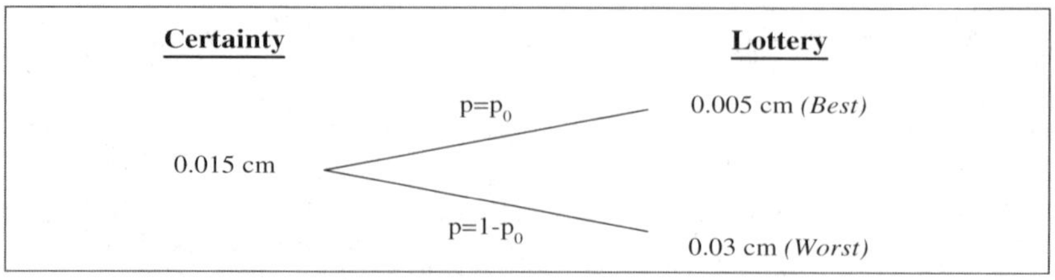

FIG. 4.10 SAU DEFLECTION LOTTERY QUESTION

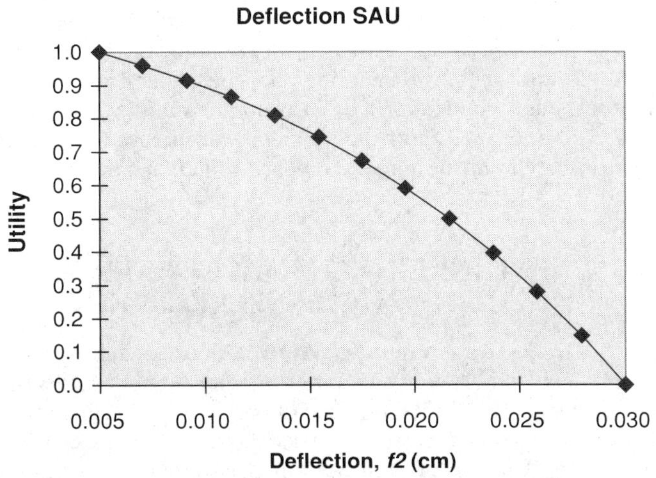

FIG. 4.11 SAU FUNCTION $U_2(F_2)$ FOR DEFLECTION

feature will be a step function. If the designer chooses to allow some flexibility in the design constraint, it can be shown that a Boltzmann sigmoidal function is a valid representation of this scenario [6]. The formulation can be modeled as follows:

$$U(x) = \frac{1}{\left(1 + e^{\frac{(X_{0.5} - x)}{\text{slope}}}\right)} \quad \text{Eq. (4.14)}$$

Where $X_{0.5}$-constraint value at which the utility is 0.5, which can be the threshold for constraint violation. Slope can be adjusted to approximate the shape of a step function. For example, handling the bending stress constraint as an attribute, the following range can be identified:

Bending Stress: $0 \le f_3(\boldsymbol{x}) \le 32$ kN/cm^2 Eq. (4.15)

These bounds are determined based on the constraint condition. Per the definition of a constraint, when $f_3(x)$ is below 16 kN/cm^2, the decision-maker is satisfied and thus any value in this domain should receive a utility near 1. Thus, this procedure recognizes that a stress value of 5 or 10 kN/cm^2 will not alter the designer's decision by assigning him or her both utility values near 1. Similarly, any constraint evaluation above 16 kN/cm^2 results in a utility near 0. Rather than using a step -function to describe this behavior, a SAU based on a Boltzmann sigmoidal function can be implemented to yield the following SAU equation:

Stress: $$U_3(f_3) = 0 + \frac{1}{\left(1 + e^{\frac{(16.9 - f_3)}{-0.4443}}\right)} \quad \text{Eq. (4.16)}$$

Figure 4.12 shows this equation graphically. As the stress reaches the constraint value (dashed line), the utility begins to drop sharply and eventually reaches 0 at about 18 kN/cm^2. Rather than an abrupt change of utility associated with a step function from 1 to 0 at the feasible boundary, the Boltzmann Sigmoidal function allows for a more gradual utility descent, which proves more appropriate in cases where a constraint is not absolute and slight violations may be permitted to achieve better designs. The slope of the decent is based on the response of the decision-maker to the constraint lottery questions.

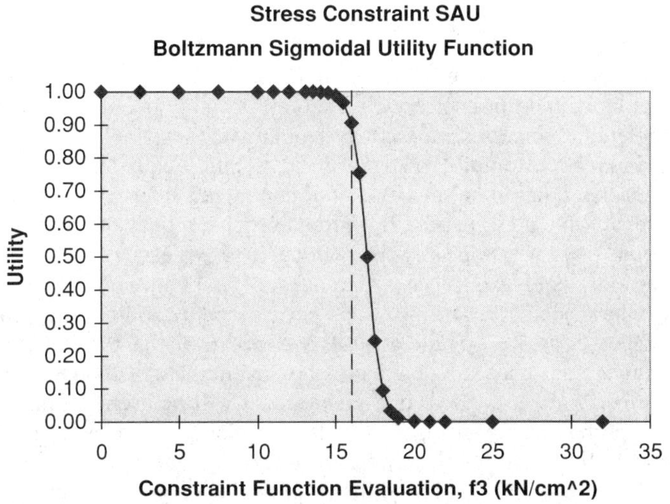

FIG. 4.12 SAU FUNCTION $U_3(F_3)$ FOR STRESS

When the constraints are treated as special attributes and incorporated into the MAU function, we can handle them as one-way parametric dependence from attributes to constraints. And, the final superior function can be expressed as:

$$U(X, C) = \{U(X^0, C^*) + [1 - U(X^0, C^*)]U(X \mid C^*)\}U(C) \quad \text{Eq. (4.17)}$$

where C^*-no constraints violation; $U(C)$-utility function of constraints; and $U(X^0, C^*)$-value when all attributes are at their least preferred levels but the constraints are satisfied. It has been shown that such a representation will ensure that the utility of any alternative associated with a constraint SAU value of 0 will also be 0. More details on this formulation can be found in [21, 22].

4.3.4 Preference Inconsistencies and Preference Errors

Researchers have shown that a lack of precise knowledge while answering trade-off questions can lead to *preference inconsistencies* and *preference errors* [23–26]. *Preference inconsistencies* refer to possible contradictions or intransitivities, and *preference errors* refer to confusion or indefinite responses. They can occur due to: (1) the utility functions not reflecting the designer's risk behavior accurately; and/or (2) the scaling constants violating constraints. At a higher level, the inconsistency can be observed as the inability of the decision model to capture the proposed relationship [24]. Note: In this setup, the trade-off questions are mainly answered based on intuition or experience rather than on any experiment or knowledge supported by deductive reasoning. Such direct quantitative evaluation can potentially lead to a high occurrence of inconsistent and inexact preference information [26, 27].

A major task in constructing such an information-based preference modeling process is in the setting up of the lottery elicitation strategy that is consistent with the axioms of decision analysis. Experience with utility function assessments has shown that the above lottery process often can be subject to considerable preference inconsistencies, such as possible contradictions or intransitivities and rank-reversal problems [26, 27]. This may be due to a lack of precise knowledge of the gradient directions of the value functions, or due to the uncertainty and/or risk conditions, as well as a lack of appropriate inherent logic [27]. It can then be stated that a potential limitation of the straightforward application of the elicitation process is that it precludes the use of any useful information that may be obtained from using deductive reasoning from an available set of real outcomes. Alternatively, one can identify a modified lottery elicitation procedure, one that ensures preference consistency based on an obtainable set of existing information. For example, a range for the certainty equivalent can be considered based on the concept of confidence interval (CI) from probability and statistics [28]. From both theoretical and practical points of view, it appears to be reasonable to express a range for the certainty equivalent, where such bounds can be initially specified by the designer according to his/her experience, observations or expectations, etc. To reiterate, this does not mean that the certainty equivalent is uncertain, but it provides an alternate method to elicit the certainty equivalent in a manner that will be useful in bringing preference consistency. Another possibility is to request the range bounds through reliability index-like side intervals (say, +/−1%, or 3%, 5%) around the single value response. Similarly, descriptions can be derived on the trade-off weights or scaling constants by set inclusions in the form of $c \leq w_i/w_j \leq d$. The statement "$c \leq w_i/w_j \leq d$" can be interpreted as follows: "given a particular population of designs, the difference in performance between the best and worst alternative with respect to attribute *i* is at least *c* times and at most *d* times important as that with respect to attribute *j*." One can argue that such an approach can thus provide considerable modeling flexibility that can be used to ensure consistency with alternative ranking. Irrespective of the modifications, the resulting method must ensure satisfaction of utility monotonicity conditions, boundary conditions, specified confidence interval range conditions, as well as transitivity condition on comparison relations among the outcomes to ensure no rank reversal. Recognizing the fundamental problem of rank reversal and the controversial nature of approaches based on rational pairwise comparisons of alternatives, the employment of such methods must guarantee that the work at all stages is consistent with the axioms of *vN-M* [3].

Focusing on the efforts toward an easy-to-use, yet consistent, preference representation in engineering design, Jie and Krishnamurty have proposed an information-based preference modeling process to efficiently aid in engineering design decision-making under conditions of uncertainty and risk [28]. A strong motivation for this work is to ensure consistency in preference modeling and to minimize cognitive burden in constructing such models. A method is likely to be regarded as user-friendly if only a few easy questions are asked and answered, and the designer can observe the dynamic change/effects interactively whenever he/she makes a meaningful response. The characteristics of human cognitive process appear to indicate that it may be more difficult for the designer to provide all the inclusive information at one time than to consider them one by one in a certain sequence [29]. Alternatively, a preferred decision aid is one that can guide the designer in drawing his/her preference structure gradually to avoid the possible cognitive confusion and to reduce the recovery cost once the information inconsistency occurs. Accordingly, it may be appropriate to develop a method where the designer can initially set a range for the uncertainty and/or risk conditions, and then iteratively augment and validate them using additional knowledge such as the outcome comparison pairs, over which the designer may have more confidence. In such systematic information elicitation patterns, the designer's cognitive burden is naturally reduced and the reliability of the results can be enhanced. In general, the more information, especially derived from experiments or simulation, can lead to more accurate and more robust solutions. Furthermore, such an approach can efficiently support the designer through the entire process of design decision-making by providing strong and clear feedback information concerning the consistency of any new input response. For example, by incorporating a linear regression model, the infeasibility of the problems can be used to indicate inconsistency in the designer's comparison judgments [28].

4.4 ENGINEERING MODELING FROM A DECISION ANALYSIS PERSPECTIVE

In his works, Dr. Howard defines two important functions of a decision analyst: (1) construction of preference models (decision models) by addressing the problem of preference measurement by capturing in quantitative terms what the decision-maker wants; and (2) construction of predictive models by accurately capturing the structure of the problem. The previous sections focused on the construction of decision models. However, engineering design decisions are often complex and require results from modeling and simulation to construct the structure of the problem. As such, predictive models will always involve assumptions and idealizations and cannot be expected to represent nature perfectly. We can then define engineering models as abstractions, or idealizations, of the physical world designed and developed for the purposes of mimicking system behavior [30–34]. In the context of engineering design, they play the role of providing structure to the design problem. Typically, models are simplified representations of reality, primarily because reality is very hard to model precisely due to a lack of sufficient knowledge about the system. Second, it may be practically impossible to build an elaborate model and use it in the design optimization cycle simply because of the high computation and the expense involved. Furthermore, even data collection to build such elaborate models may be prohibitively expensive. It is then apparent that some type models, whether they are iconic, symbolic or surrogate meta-models, become necessary to achieve design decisions [30, 31]. The fundamental question then is how to build such engineering models that are computationally efficient, sufficiently accurate and meaningful, especially from the viewpoint of their utilization in the subsequent engineering design phase. To start, physics-based engineering analysis models can often be difficult to build and validate, and the problem becomes more complicated and computationally intensive in the case of predictive models that can enable reliable and accurate design mapping of the system. This is particularly true in numerically generated analysis models such as the engineering mechanics models using finite-element analysis, the empirically constructed process models in manufacturing and the models of fluid flow used in computational fluid mechanics. An excellent discussion on modeling and the role of information in model building in the design process can be found in [35].

4.4.1 Engineering Model Validation

Often, model validation is the only way of ensuring accuracy and reliability and avoiding possible errors due to simplified, inaccurate or inappropriate models. Literature on verification and validation of simulation models can be found in [36–39]. A model validation process will need to address the basic questions of: (1) how to initiate the information gathering phase and initial model building; (2) how to incorporate preference information that will ensure that resulting design decisions using such models will be robust, reliable and accurate; (3) how to select optimally informative sample points to test the model, recognizing that the model

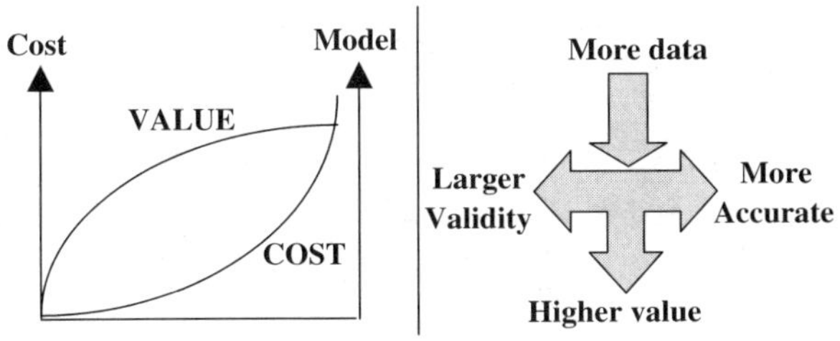

FIG. 4.13 VALUE OF INFORMATION VS. COST OF INFORMATION

cannot be tested at every location in the design set; and (4) how to capture new information and use it to update model fidelity.

Model validation is usually defined to mean "substantiation that a computerized model within its domain of applicability possesses a satisfactory range of accuracy consistent with the intended application of the model" [30, 31]. Published works related to model validation from a risk and reliability perspective can be found in [40–42]. A model is usually developed for a specific purpose or application, and its validity should therefore be determined with respect to that goal. It then becomes apparent that if a simulation-based predictive model is used to obtain results for evaluating the value of a particular design, then that model should be validated with regard to its ability to evaluate the values of all potential design solutions with sufficient reliability and accuracy. The accuracy and reliability here will refer to the ability of the model/surrogate to mimic the expected reality closely. More data used in the construction of the surrogate model implies larger validity and higher accuracy for the model. However, this also implies higher cost for the designer and thus there is a clear "trade-off" between cost and accuracy in a model-building and validation process (Fig. 4.13).

4.4.2 Bayesian Analysis in Model Building

The purpose of models in design is to capture the structure of the problem and to enable design decisions based on normative decision analysis principles. It is therefore important that the use of such models takes into account the many causes for the errors in the prediction of system performance. Note that the models can only be expected to be as accurate as the set of data available to build them. And the only reliable and provable mechanism for model validation is to actually build the design from the predictive model results, based on the designer's preferences and by checking its performance. However, this is a dilemma as the most desired design outcomes cannot be found a priori unless the perfect predictive model exists. On the other hand, a predictive model cannot be perfected until it can be validated against the expected outcomes from the engineering decisions. Therefore, in the absence of perfect information or clairvoyance, it can be reasoned that the best approach is to study predictive model building as a trade-off between the cost and value of information. In the context of assessing model appropriateness, a predictive model can be viewed as the best trade-off between the quest for more accurate results and reduction of analysis costs under conditions of uncertainty, while considering the expected payoffs from the resulting design using such a model [43, 44].

Here, an engineering model assessment framework based on Bayesian analysis offers a unique approach to handle modeling errors from a DBD perspective. Research on the Bayesian analysis in engineering design can be found in the works by Howard [5–8] and Raiffa [45]. Bayesian methods are all based on making inferences and projections using our current state of information. By combining prior information about a system with the likelihood of its behavior, it is possible to predict the system behavior at instances where we have no experience. The heart of Bayesian techniques lies in the celebrated inversion formula,

$$P(H \mid e) = \frac{P(e \mid H)P(H)}{P(e)} \qquad \text{Eq. (4.18)}$$

which states that the belief we accord in a hypothesis H upon obtaining evidence e can be computed by multiplying our previous belief $P(H)$ by the likelihood $P(e \mid H)$ that e will materialize if H is true. $P(H \mid e)$ is sometimes called the posterior probability (or simply posterior); and $P(H)$ is called the prior probability (or prior). One of the attractive features of Bayes' updating rule is its amenability to recursive and incremental computation schemes, which results in:

$$P(H \mid e_n, e) = p(H \mid e_n)\frac{P(e \mid e_n, H)}{P(e \mid e_n)} \qquad \text{Eq. (4.19)}$$

where $P(H|e_n)$ assumes the role of the prior probability in the computation of new impact. This completely summarizes the past experience, and for updating, it only needs to be combined with the likelihood function $P(e|e_n,H)$, which measures the probability of the new datum e, given the hypothesis and the past observations.

Bayesian analysis can thus provide formalism to representation of probability and conditional probability of states and consequences corresponding to actions, which can then be combined and manipulated at all decision points of a decision-flow design problem according to the rules of probability theory. In this scenario, current state of information about the performance of the system can be stated as *prior* information. *A priori* information refers to the information obtained before data collection begins. This information can be obtained from first principles, application of physical laws to simplified systems or empirical correlation obtained from previous experiments on the system. Any new information obtained can be combined with the already existing information to create a better state of information known as *posterior* information. For example, if the system performance (as simulated by computer models) can be modeled as a Gaussian stochastic process, the prior and posterior distributions can be captured using mean, standard deviation and covariance information [46]. Thus, for any stage of data collection, a posterior mean and posterior standard deviation describing the model can be estimated as a measure of uncertainty of that prediction. The simulated performance and the predicted performance can then be compared and the updated model can be evaluated for its accuracy and resolution. There have been many suggested ways of measuring the accuracy of any model. A simple measure of accuracy would be percentage errors of the prediction (model) from the actual computer simulation. Resolution is the ability of a model to distinguish between alternatives even when differences between such alternatives are considerably smaller than a model's accuracy as indicated by its uncertainty band. In the engineering design context, it can be stated that resolution translates to the uncertainty in design decision-making and, hence, resolution of the model in the context of the proposed work can be captured in the uncertainty band for the prediction of the design performance from the model.

Several research questions need to be addressed in developing implementation strategies based on Bayesian analysis. While it is proven that the recursive and incremental feature of the Bayesian updating rule makes it amenable to develop probabilistic models that can act as the normative rule for updating human value-based expectations in response to evidence, much work needs to be done before it can be implemented in a large-scale computationally viable environment. Implementation strategies need to be developed to uniquely enable encoding of information from several related issues, such as estimation of the overall design error given the tools, variables and the design process cost, or optimal design cost given the acceptable error or convergence of the design process. Related research topics include the establishment of guidelines and metrics to quantify the effect that modeling simplifications have on the predicted design behavior, as well as the verification of the basic approach's independence with respect to mathematical tools.

4.5 SUMMARY

Normative decision analysis principles can provide valuable insight into advancing the state of knowledge on rational design decisions in engineering design as well as enable a better understanding of their consequences from an overall design perspective. From a practical point of view, a decision-based engineering design approach offers a formal strategy to reduce the multiple attributes in an engineering design problem to a single overall utility function in a probabilistic sense, which reflects the designer's intent and preferences under conditions of uncertainty. This chapter presents a detailed review of some of the topics central to the understanding and implementation of decision-analysis-based techniques in engineering design. Issues that are typical in a decision-based engineering design situation, such as how to build preference models and how to validate predictive design models, are discussed. Attribute space formulation, design scenario classification and the significance of the selection of a preference assessment method in the resulting decision model formulation are discussed in the context of engineering design. The chapter concludes with a discussion on engineering models from a design decision perspective, followed by an overview of Bayesian analysis that can form the basis for a rational approach for model validation in engineering design.

REFERENCES

1. Siddall, J. N., 1972. *Analytical Decision-Making in Engineering Design*, Prentice-Hall, Inc., Englewood Cliffs, NJ.
2. Starr, M. K., 1963. *Product Design and Decision Theory*, Prentice-Hall, Inc., Englewood Cliffs, NJ.
3. Hazerlrigg, G. A., 1996. *Systems Engineering: A New Framework for Engineering Design*, ASME Dynamic Systems and Control Division, Vol. 60, pp. 39–46.
4. Decision-Based Design Open Workshop, http://www.eng.buffalo.edu/Research/DBD/.
5. Howard, R. A., 1968. "The Foundations of Decision Analysis," *IEEE Trans. on Sys. Sci. and Cybernetics*, Vol. 4, pp. 211–219.
6. Howard, R., 1973. "Decision Analysis in System Engineering," *Sys. Concepts–Lectures on Contemporary Approaches to Sys., Sys. Engg. and Analysis Ser.*, F. Ralph and Miles, Jr., eds., pp. 51–85, Wiley-Interscience.
7. Howard, R., 1965. "Bayesian Decision Models for Systems Engineering," *IEEE Trans. on Sys. Sci. and Cybernetics*, Ssc-1, pp. 36–41.
8. Raiffa, H. and Keeney, R. L., 1976. *Decisions with Multiple Attributes: Preferences and Value Tradeoffs*, Wiley and Sons, New York, NY.
9. von Winterfeldt, D. and Edwards, W., 1986. *Decision Analysis and Behavioral Research*, Cambridge University Press.
10. Gold, S. and Krishnamurty, S., 1997. "Tradeoffs In Robust Engineering Design," *ASME Des. Engrg. Tech. Conf., DETC97/DAC-3757*, Sacramento, CA.
11. Fischer, G. W., 1995. "Range Sensitivity of Attribute Weights in Multiattribute Value Models," *Organizational Behavior and Human Decision Process*, Vol. 62, pp. 252–266.
12. Keen, G., 1996. "Perspective of Behavior Decision Making: Some Critical Notes," *Organizational Behavior and Human Decision Process*, Vol. 65, pp. 169–178.
13. Iyer, H.V. and Krishnamurty. S., 1998. "A Preference-Based Robust Design Metric," *ASME Des. Engrg. Tech. Conf., DETC98/DAC-5625*, Atlanta, GA.
14. Thurston, D. L., 1991. "A Formal Method for Subjective Design Evaluation with Multiple Attributes," *Res. in Engrg. Des.*, Vol. 3, pp. 105–122.
15. Edwards, W., 1994. "SMARTS and SMARTER: Improved Simple Method for Multiattribute Utility Measurement," *Organizational Behavior and Human Decision Process*, Vol. 60, pp. 306–325.
16. Srivastava, J., Beach, L.R. and Connolly, T., 1995. "Do Ranks Suffice? A Comparison of Alternative Weighting Approaches in Value Elicitation," *Organizational Behavior and Human Decision Process*, Vol. 63, pp. 112–116.
17. Hazelrigg, G. A., 1996. *System Engineering: An Approach to Information-Based Design*, Prentice Hall, Upper Saddle River, NJ.
18. Kirkwood, C. W., 1997. *Strategic Decision Making: Multiobjective Decision Analysis with Spreadsheets*, Duxbury Press, Belmont, CA.
19. Osyczka, A., 1985. "Multicriteria Optimization for Engineering Design," *Design Optimization*, J.S. Gero, ed., Academic Press, pp. 193–227.
20. Kunjur A. and Krishnamurty, S., 1997. "A Robust Multi-Criteria Optimization Approach," *Mechanisms & Machine Theory*, 32, (7), pp. 797–810.
21. Tang, X. and Krishnamurty, S., 2000. "On Decision Model Development in Engineering Design," *Special issue on Decision-Based Design: Status and Promise, Engineering Valuation and Cost Analysis*, Vol. 3, pp. 131–149.
22. Iyer, H. V., Tang, X. and Krishnamurty, S., 1999, "S. Constraint Handling and Iterative Attribute Model Building in Decision-Based Engineering Design," *ASME Des. Engrg. Tech. Conf., DETC99/DAC-8582*, Las Vegas, NV.
23. Fischer, G. W., 1979. "Utility Models for Multiple Objective Decisions: Do They Accurately Represent Human Preferences?" *Decision Sci.*, Vol. 10, pp. 451–479.
24. Larichev, Q. I., Moshkovich, H. M., Mechitov, A. I. and Olson, D. L., 1993. "Experiments Comparing Qualitative Approaches to Rank Ordering of Multiattribute Alternatives," *Multi-criteria Decision Analysis*, 2(1), pp. 5–26.
25. Belton, V., 1986. "A Comparison of the Analytic Hierarchy Process and a Simple Multiattribute Value Function," *European J. of Operational Search*, Vol. 26, pp. 7–21.
26. Olson, D. L. and Moshkovich H. M., 1995. "Consistency and Accuracy in Decision Aids: Experiments with Four Multiattribute Systems," *Decision Sci.*, 26 (6) pp. 723–748.
27. Badinelli, R. and Baker, J. R., 1990. "Multiple Attribute Decision Making with Inexact Value-Function Assessment," *Decision Sci.*, Vol. 21, pp. 318–336.
28. Jie, W. and Krishnamurty, S., 2001. "Learning-based Preference Modeling in Engineering Design Decision-Making," *ASME J. of Mech. Des.*, 123(2), pp. 191–198.
29. Ranyard, R., Crozier, W. R. and Svenson, O., 1997. *Decision Making: Cognitive Models and Explanations*, Routledge Press, New York, NY.
30. Hazelrigg, G. A., 1999. "On the Role and Use of Mathematical Models in Engineering Design," *J. of Mech. Des.*, Vol. 121, pp. 336–341.
31. Hazelrigg, G. A., 2003. "Thoughts on Model Validation for Engineering Design," *ASME Des. Engrg. Tech. Conf., DETC2003/DTM-48633*, Chicago, IL.
32. Haugen, E. B., 1980. *Probabilistic Mechanical Design*, John Wiley & Sons, New York, NY.
33. Simpson, T. W., 1998. Ph.D. Dissertation, *George W. Woodruff School of Mechanical Engineering, Georgia Institute of Technology.*
34. Draper, D., 1995. "Assessment and Propagation of Model Uncertainty," *J. of the Royal Statistical Soc.*, Vol. 1, pp. 45–97.
35. McAdams, D. A. and Dym, C. A., 2004. "Modeling and Information in the Design Process," *ASME Des. Engrg. Tech. Conf., DETC2004-57101*, Salt Lake City, Utah.

36. Sargent, R. G., 1998. "Verification and Validation of Simulation Models," *1998 Winter Simulation Conf.*, pp. 121–130.
37. Hills, R. G. and Trucano T.G., 1999. "Statistical Validation of Engineering and Scientific Models: Background," *Computational Phys. Res. and Dev.*, Sandia National Laboratories, pp. 1099–1256.
38. Kelton, W. D., 1999. "Designing Simulation Experiments," *1999 Winter Simulation Conf.*, pp. 33–38.
39. Kleijnen, J. P. C., 1999. "Validation of Models: Statistical Techniques and Data Availability," *1999 Winter Simulation Conf.*, pp. 647–654.
40. French, S., 1995. "Uncertainty and Imprecision: Modeling and Analysis," *J.* of the *Oper. Res. Soc.*, Vol. 2, pp. 70–79.
41. Haimes, Y. Y., 1998. *Risk Modeling, Assessment and Management*, Wiley-Interscience, New York, NY.
42. Wahlstrom, B., 1994. "Models, Modeling and Modelers: An Application to Risk Analysis," *European J. of Operational Res.*, pp. 477–487.
43. Chandrashekar, N. and Krishnamurty, S., 2002. "Bayesian Evaluation of Engineering Models," *ASME Des. Engrg. Tech. Conf., DETC2002/DAC-34141*, Montreal, Canada.
44. Wilmes, G. and Krishnamurty, S., 2004. "Preference-Based Updating of Kriging Surrogate Models," *AIAA MDO Conf., AIAA-2004-4485*, Albany, NY.
45. Raiffa, K., 1970. *Decision Analysis Introductory Lectures on Choices under Uncertainty*, Addison-Wesley, Reading, MA.
46. Pacheco, J. E., Amon, C. H. and Finger, S., 2001. "Developing Bayesian Surrogates for Use in Preliminary Design," *ASME Des. Tech. Conf., DETC2001/DTM-21701*, Pittsburgh, PA.

PROBLEMS

4.1.

a. How does normative decision-making differ from descriptive decision-making?
b. Comment on Dr. Howard's Normative Decision Analysis process shown in Figure 4.1. What changes would you recommend to this process from an engineering design perspective?

4.2.

a. Write a program to automatically calculate the SAU functions of the beam design problem and verify the coefficient values of the area and displacement SAU functions.
b. Write a program to automatically calculate the multiplicative MAU function of the beam design problem and verify the coefficient values of the scaling constants.
c. Do you agree with the statement "multiplicative utility models can be reduced to equivalent additive utility models?" Comment with an example.

You can find additional information on this topic in [3].

d. Can robust optimal design be studied as a problem in decision-making requiring trade-offs between mean and variance attributes? If so, can you view Taguchi's philosophy-based design metrics using signal-to-noise (SN) ratios as empirical applications of decision-making under uncertainty with a priori sets of attribute trade-off values?

4.3.

a. What do you see as the purpose(s) of engineering models?
b. Suggest a simplified third-order polynomial surrogate model for the sixth-order polynomial problem below:

$$f(x) = 5x^6 - 36x^5 + 82.5x^4 - 60x^3 + 36$$

Using optimization concepts, discuss the appropriateness of your simplified model from a design decision perspective.

CHAPTER

5

FUNDAMENTALS AND IMPLICATIONS OF DECISION-MAKING*

Donald G. Saari

5.1 INTRODUCTION

As W. Wulf, president of the National Academy of Engineering, emphasized during a talk at the NSF (12/8/1999), making decisions is a crucial aspect of engineering. At all stages of design, with almost all engineering issues, decisions must be made. They may involve a choice of materials, options, approaches, selection of members for a team and just about all aspects of daily work. Decisions are crucial for those widely discussed approaches such as Quality Function Deployment (QFD), which provide guidelines on how to schedule and coordinate decision-making. Without argument, making accurate, good decisions is a crucial component of engineering.

A way to underscore the importance of "good decisions" is to reflect on the effects of "bad decisions." Unfortunately, bad decisions—particularly if subtle—may be recognized only after the fact as manifested by their consequences. For instance, an inappropriate or overly cautious decision made during the early stages of design can be propagated, resulting in lost opportunities (of exploring other options) and even inferior outcomes. At any stage of design or engineering, even minor decision errors contribute to inefficiencies and a decrease in effectiveness. Faulty decisions can cause products to fail to reach attainable levels, an increase in cost, a decline in customer appeal and many other problems—such as the decision-maker being fired. Good decision-making is central in our quest to achieve competitive advantage and engineering excellence.

The need for accurate decisions is manifested by our search for accurate information: We do this by performing tests, designing prototypes, carrying out extensive computer simulations, among other approaches. But even if we can assemble all of the appropriate accurate and complete information to make a decision, we can still make a bad decision. When this occurs, we tend to blame the data, such as the possibility of experimental error, the ever-present uncertainties and the effects of the many unknown variables. Rarely is blame assigned to the decision rule. What a serious mistake!

The point of this chapter is to demonstrate that *the choice of a decision rule can play a surprisingly major role in causing and, inadvertedly, accepting inferior choices.* I also describe how to identify and create rules to minimize the likelihood of these difficulties.

5.2 SELECTING WEIGHTS

Often a decision problem involves selecting "weights," maybe of the

$$\sum_{j=1}^{n} \lambda_j \nabla U_j \qquad \text{Eq. (5.1)}$$

form, where the objective is to select the correct choice of λ_j's. How do you do this? Does the choice really matter?

To bring home the message that the choice of the weights, and of a decision rule, is a significant concern, I describe the simpler but closely related problem of analyzing voting rules. The advantage is that a theory for selecting voting rules exists, and it allows us to understand all of the problems that occur. Of importance to this chapter, these same difficulties arise with engineering decisions, so the conceptual arguments used to select the appropriate voting rule also help us select engineering decision rules as well as weights such as for Eq. (5.1): I will describe what needs to be done.

5.2.1 Selecting a Beverage

To start with a simple example [1], suppose when 15 people select a common beverage for a party, six prefer milk to wine to beer; denote this as $M \succ W \succ B$ (see Table 5.1). Let the group preferences be

TABLE 5.1 GROUP PREFERENCES

Number	Preferences	Number	Preferences
6	$M \succ W \succ B$	4	$W \succ B \succ M$
5	$B \succ W \succ M$	—	—

By using our standard plurality vote, where each voter votes for his or her favored alternative, the outcome is $M \succ B \succ W$ with the 6:5:4 tally.

What does this selection ranking mean? Does it mean, for instance, that this group prefers milk to wine, and milk to beer?

* This chapter is based on several invited talks, including a tutorial presented at the 1999 ASME meeting in Las Vegas, NV, and lectures at a 2001 NSF workshop on decision analysis in engineering that were held at the University of California at Irvine. This research was supported by NSF grant DMI-0233798.

Both assertions reflect how decisions are used in actual practice, but neither is true here. Instead, for each comparison, a simple count proves that milk is the *loser* by the overwhelming 9:6 (60%) margin. Similarly, this group outcome, which is free from all possible experimental error, data and parameter uncertainty, or other input factors, does not even mean that these people prefer beer to wine. Instead, as a count proves, they really prefer wine to beer by a 10:5 (67%) vote. Indeed, comparing the alternatives in pairs suggests that the "correct outcome" is the *reversed ranking* $W > B > M$.

So far both milk and wine can "win" by using an appropriate decision rule. To make beer the "optimal choice," use a runoff: drop the last-place alternative from the first election and select the majority winner of the remaining two alternatives. In the beverage example, wine is dropped at the first stage and beer beats milk in the runoff to be victorious.

To summarize, even though the fixed data is free of any error or uncertainty,

- Milk wins with the plurality vote
- Beer wins with the runoff rule
- Wine wins when alternatives are compared pairwise

Clearly, *each* of the three alternatives cannot be optimal, but each is designated as "the best" with an appropriate decision rule. Stated in another manner, beware: *an "optimal choice" may more accurately reflect the choice of a decision, assessment or optimization rule rather than the data.* Although this troubling behavior occurs surprisingly often, many decision-makers are totally unaware of this danger. The message for us is that the choice of the decision rule is crucial for engineering.

5.2.2 Engineering Decisions

For an example involving engineering concerns, suppose we want to select one of the alternatives M, B or W based on how they rank over 15 criteria. To be specific, suppose we want to choose an alternator coming from a plant in Madison, Buffalo or Washington. Or, to raise the economic stakes, suppose a new plant is to be located in Madison, Buffalo or Washington. A standard approach is to rank the three alternatives according to various criteria—in the location problem this might involve tax advantages, availability of labor, parts, transportation, etc. (A criterion more important than others may count as, say, "two criteria.") Suppose that the rankings of the locations are as specified by Table 5.1. With money and effort spent to assemble accurate and full data, what should be the decision?

I posed this problem to several people from the industry who were at various decision-making levels. The following describes three types of answers I received.

- *Madison* is the optimal choice: To achieve excellence, use a maximax "best of the best" approach. Rank locations according to the number of (weighted) criteria over which they are top-ranked. This corresponds to the plurality vote.
- *Buffalo* should be selected; to avoid error, use an iterative approach involving stages. Drop the "inferior" choice of Washington at the end of the first stage, and then compare the remaining two alternatives. This decision approach is equivalent to a runoff.
- *Washington* is the best choice because it avoids "disaster." Use the conservative "avoid the worst of the worst" approach: penalize the worst alternative for each criterion by assigning a point for each of the first two ranked alternatives. (This is equivalent to "vote for two candidates.") Washington would also be selected by comparing alternatives pairwise.

Not all three alternators, or three locations, can simultaneously be the "optimal choice." Nevertheless, each alternative is designated as the "best" with an appropriate multi-criteria decision rule. The ambiguity in the choice has nothing to do with faulty data; it results from the choice of a decision rule. Even more, notice that these described rules are in actual use. The stakes and problematic nature of decision-making are further illustrated by the disturbing coincidence that Milwaukee, the winner of the "best of the best" philosophy, is designated as the location to definitely avoid, with the "avoid the worst of the worst" approach's $W > B > M$ ranking. The message is clear: "bad decisions" can arise—even with the best information—by using an inappropriate decision rule. A disturbing reality is that the decision-maker may never realize that the choice of the decision rule sabotaged the decision. The choice of an "appropriate rule" is addressed below.

5.3 THE LEVEL OF DIFFICULTY

While disturbing, the above merely hints at what can go wrong: what follows is a selective review of what else can happen with standard decision rules and their combinations. All of these disturbing outcomes extend to other rules: Indeed, expect nonlinear rules to inherit all of the difficulties of the linear methods while generating still new kinds of problems. (Theoretical support comes from Arrow's Theorem [2] and extensions [3], [1, 4, 5, 6, 7]). Moreover, as illustrated above, each voting method becomes a decision method by identifying how a particular voter and a particular criterion rank the alternatives. As the following discussion reflects fundamental problems that can arise with any aggregation rule [7, 8], expect them to arise in statistical rules [9], nonlinear multicriteria decision methods, the use of QFD and so forth. After describing how serious the situation can be, positive results are given.

Definition 5.1: *A positional voting (decision) rule over n alternatives is defined by the weight vector* $w^n = (w_1, w_2, \ldots, w_n)$ *where* $w_1 \geq w_2 \geq \ldots \geq w_n$ *and* $w_1 > w_n$*. In using these weights, for each voter (for each criterion),* w_j *points are assigned to the jth ranked alternative. The alternatives are ranked according to sums of the assigned weights.*

To find the "best" positional rule we need to find weights where the outcomes best reflect the intent of the data (or voter preferences). Notice the close similarity of this problem with the selection of λ's in Eq. (5.1), or the choice of weights with assessment rules.

To connect this definition with commonly used election rules, the widely used plurality vote, or "best of the best" decision rule, is defined by the positional vector $(1, 0, \ldots, 0)$. The *antiplurality* voting rule, or "avoid the worst of the worst" decision method, is defined by $(1, 1, \ldots, 1, 0)$. The *Borda Count for n alternatives* is defined by $(n-1, n-2, \ldots, 1, 0)$. (So, the Borda Count agrees with the standard "four-point grading system" where an A is worth four points, a B is three, etc.) Although Definition 5.1 requires assigning larger weights to more favored alternatives, all results hold by assigning smaller weights to more favored alternatives; e.g., giving one point to a top-ranked alternative, two to the second-ranked, ... is an equivalent form of the Borda Count (BC). Of course the same "smaller is better" convention must be applied to interpret the outcome.

5.3.1 Differences among Positional Methods

The large literature describing problems with positional methods starts with the work of Borda [10] in 1770 (his paper was published over a decade later), continues with the 1785 work of Condorcet [11] and escalated significantly after Arrow's [2] seminal 1952 book. A sense of the size and diversity of this literature comes from J. Kelly's extensive bibliography [12]. The following results subsume and extend that literature, which is relevant for our discussion. The first statement asserts that the problems introduced with the beverage example get worse by adding alternatives.

Theorem 5.1 [13, 14]: *For $n \geq 3$ alternatives, select any $n-1$ rankings of the alternatives. These rankings need not be related in any manner. Next, select a positional rule for each ranking. The only condition imposed on these methods is that the associated $\mathbf{w}^n$ vectors are linearly independent. There exist examples of data (describing rankings for criteria) where the decision outcome for each rule is the assigned ranking.*

For $n \geq 2$ alternatives, choose any k satisfying $1 \leq k \leq (n-1)(n-1)!$. There exist data examples where the rankings coming from all possible choices of $\mathbf{w}^n$ weights define precisely k strict (i.e., involving no ties) rankings. Indeed, for $n \geq 3$, there are data sets where each alternative is the "winner" with appropriate choices of positional decision rules.

According to this troubling assertion, whatever two choices of $\mathbf{w}^3$weights are selected, it is possible to encounter data where the rankings for the three alternatives are precisely the opposite of one another. [So expect such problems to happen with different λ choices for Eq. (5.1.)] Indeed, this behavior occurs with the beverage example data where the "best of the best" ranking is $M \succ B \succ W$ while the ranking for the antiplurality "avoid the worst of the worst" method, is the reversed $W \succ B \succ M$. With just elementary algebra [1], it can be shown that with different weights this data generates seven different rankings where four $(4 = ((3-1)(3-1)!)$ of them are strict (no ties).

To appreciate the magnitude of the problem, the second part of this theorem ensures that with only 10 alternatives, data sets exist for which millions upon millions of different decision rankings are generated—the data remains fixed, so all of the different outcomes are caused strictly by the choice of the decision rules. This behavior is illustrated with the four-alternative, 10-criteria example [8] given in Table 5.2:

TABLE 5.2 STRUCTURE OF GROUP PREFERENCES

Number	Preference	Number	Preference
2	$A \succ B \succ C \succ D$	2	$C \succ B \succ D \succ A$
1	$A \succ C \succ D \succ B$	3	$D \succ B \succ C \succ A$
2	$B \succ D \succ C \succ A$	—	—

where A wins with the plurality vote (1, 0, 0, 0), B wins by voting for two alternatives [i.e., with (1, 1, 0, 0)], C wins with the antiplurality (1, 1, 1, 0) rule and D wins with the Borda Count $\mathbf{w}^4 = (3, 2, 1, 0)$, so each alternative "wins" with an appropriate rule. Even more $3(3!) = 18$, different decision rankings emerge by using different rules; [15] shows how to find all rankings for the associated rules. There is nothing striking about the data from Tables 5.1 and 5.2, so anticipate unexpected decision behavior to arise even with innocuous-appearing data and a limited number of criteria (10 in this case). Again, a decision outcome may more accurately reflect the choice of a decision rule rather than the data—unwittingly, "optimal decisions" may be far from optimal.

But do these illustrations represent highly concocted, essentially isolated examples, or do they describe a general phenomenon that must be taken seriously? As a preface, it is easy to envision engineering settings where all criteria enjoy the same ranking. With unanimity settings, there is no need to use a decision rule: use the obvious outcome. Decision rules are needed only when we need to eliminate doubt about the interpretation of data: this more general setting is addressed by the following theorem. But Tables 5.1 and 5.2 counsel caution about deciding when "doubt exists": even seemingly minor differences in the data can generate serious conflict in the decision outcomes.

Theorem 5.2 [16]: *Assume there are no restrictions on the number of criteria, or voters. For three alternatives, if the data is distributed with any IID probability distribution with a mean of complete indifference, then, in the limit and with probability 0.69, at least three different rankings occur with different positional rules.*

So, more than two-thirds of the cases in this neutral setting can have dubious decision conclusions. Rather than an isolated phenomenon, it is reasonable to expect situations where engineering decisions more accurately reflect the choice of the decision rule rather than the carefully assembled data. The likelihood of difficulties significantly increases with more alternatives.

A way to challenge this theoretical conclusion is to question why we don't observe these oddities in actual elections. They exist: examples are easy to find during most election seasons [1, 4, 5], but most people are unaware of them. The reason is that to bother looking for worrisome examples, we must first know that they can exist and then how to find them. Similarly in engineering decisions, only after recognizing that the choice of the decision rule can influence the outcome are we motivated to search for examples. Otherwise, we tend to rely on alternative explanations involving data errors and uncertainty. With confidence I predict that once we recognize the role of decision rules, many engineering examples will be discovered.

5.3.2 Dropping or Adding Alternatives

A related problem involves how we use and interpret decision rankings. For instance, a natural way to reduce the number of alternatives is to drop "inferior choices." To explain the rationale, suppose a decision rule defines the $A \succ B \succ C \succ D$ ranking. Suppose, for various reasons, alternative C no longer is feasible. As C is nearly bottom-ranked, it seems reasonable to expect the ranking of the remaining alternatives to be $A \succ B \succ D$. Is this correct?

The beverage example raises doubts about this tacit assumption: each pair's pairwise ranking is the opposite of what it is in the plurality ranking. This phenomenon becomes even more complicated with more alternatives. To illustrate with the data set [17] (Table 5.3):

TABLE 5.3

Number	Preference	Number	Preference
3	$A \succ C \succ B \succ D$	2	$C \succ B \succ D \succ A$
6	$A \succ D \succ C \succ B$	5	$C \succ D \succ B \succ A$
3	$B \succ C \succ D \succ A$	2	$D \succ B \succ C \succ A$
5	$B \succ D \succ C \succ A$	4	$D \succ C \succ B \succ A$

the plurality $A \succ B \succ C \succ D$ outcome (9:8:7:6 tally) identifies A as the top alternative. Is it? If *any* alternative or pair of alternatives is dropped, the new "best of the best" ranking *flips* to agree with the reversed $D \succ C \succ B \succ A$. (For instance, if C is dropped, the $D \succ B \succ A$ outcome has a 11:10:9 tally. If both B and D are dropped, the $C \succ A$ outcome has a 21:9 tally.) It is arguable that D, not A, is the optimal alternative even though D is plurality bottom-ranked.

It turns out (see [18] for the special case of the plurality vote and dropping one candidate, and [13] for the general statement) that this problem plagues all possible positional rules. To be explicit, specify a ranking for the set $\{a_1, a_2, ..., a_n\}$ of n alternatives; for instance, $a_1 \succ a_2 \succ ... \succ a_n$. Drop any one alternative, say a_n and specify *any* ranking for the remaining set, maybe the reversed $a_{n-1} \succ ... \succ a_1$. Continue this process of dropping one alternative, and supplying a ranking for the remaining set—this choice can be selected randomly or even to create a particularly perverse example. Then, for each set of alternatives, specify the positional decision rule to be used to determine the outcome. The result is that a data set can be constructed whereby for each of the nested sets, the specified rule defines the specified outcome. In other words, no matter what positional rules we use, do not expect consistency when alternatives are dropped in a nested-set structure.

Now go beyond the nested-set scenario to consider all possible subsets. While some results are even more frustrating, other results finally promise hope.

Theorem 5.3 [20]: *With $n \geq 3$ alternatives, for each subset of two or more alternatives, select a ranking and a positional rule. For almost all choices of positional methods, a data set can be constructed where the outcome for each set is its specified ranking.*

A special set of positional rules avoids this negative conclusion. In particular, using the Borda Count with all subsets of alternatives minimizes the number and kinds of paradoxical behavior that can occur.

This result asserts that with most choices of decision rules, extremely wild examples can result, which cast significant doubt on the reliability of any outcome. For instance, we can construct data sets where $a_1 \succ a_2 \succ ... \succ a_n$ is the ranking of the "best of the best" (plurality) rule. Then if *any* alternative is dropped, the "best of the best" outcome reverses. But by dropping *any two* alternatives, the outcome reverses again to agree with the original ranking, only to reverse once more if any three alternatives are dropped, and so forth.

Another major point is that these decision oddities occur with almost all choices of weights. [This comment suggests exercising considerable care when selecting the λ's in Eq. (5.1.)] While this is not part of the formal statement, it follows from the arguments outlined later that these negative results are robust and occur with a positive probability.

On the other hand, this theorem introduces a measure of hope with its assertion that the Borda Count is the unique way to determine rankings that provides the maximal consistency. To explain, suppose the Borda Count changes a ranking in unexpected ways if an alternative is dropped: *The precise same paradox arises with all other choices of weights* (but, maybe with a different data set). In the other direction, each of the other choices of weights allows a data set to be constructed that leads to wild ranking changes when alternatives are dropped, but these wild changes in the decision outcomes can never occur with the Borda Count. For an illustration, the Borda Count *never* admits the reversal behavior described in the paragraph following the theorem. (All Borda Count rankings for the Table 5.3 data—all triplets and pairs—agree with the full BC ranking of $D \succ C \succ B \succ A$.) This suggests, as discussed next, that the Borda Count may be the "optimal" choice of a positional decision rule.

5.4 CONSISTENCY

While it has been recognized since the 18th century that the Borda Count enjoys certain desirable properties, only recently [15] has it been established that, the Borda Count is the unique positional rule that can be trusted, and why. In describing these results, I develop intuition as to why only the Borda Count provides consistency in outcomes. In part, I do so by showing how the Borda method relates to pairwise voting. A small listing of the Borda properties is given.

5.4.1 Pairwise Votes

To understand the relationship between the pairwise and Borda votes, consider how a voter with preferences $A \succ B \succ C$ casts his pairwise votes when voting over the three possible pairs. This is indicated in Table 5.4, where a dash in a row represents that the alternative is not in the indicated pair.

Over the three pairs, the voter casts a total of 2 points for A, 1 point for B and 0 for C: this is precisely what the Borda Count assigns each alternative with this ranking. The same behavior holds for any number of alternatives; in other words, with n alternatives, the $(n-1)$ points assigned to a top-ranked alternative reflects the $(n-1)$ times this alternative receives a vote in all $n(n-1)/2$ pairwise comparisons, the $(n-2)$ points assigned to a second-ranked alternative reflects the $(n-2)$ times this alternative is top-ranked over all pairs, and so forth. It can be shown that no other weighted system satisfies this relationship. (Of course, instead of assigning 2, 1, 0 points, respectively, to a top-second-, and bottom-ranked alternative, we could assign 320, 220, 120 points, respectively—where the differences between weights is the same—and obtain the same properties.)

This argument identifies the Borda Count as the natural generalization of the pairwise vote: it aggregates results that arise over all pairs. As a consequence of this structure, the Borda outcome can be directly determined from the pairwise tallies by adding the number of points an alternative receives in each pairwise comparison. To illustrate, if A beat B with a 37 to 23 tally and A beats C with a 41 to 19 tally and C beats B with a 31 to 29 tally. The Borda outcome is $A \succ B \succ C$ (which conflicts with the pairwise rankings) with the $(37 + 41) : (23 + 29) : (19 + 31)$ tallies. Some consequences of this behavior follow.

Theorem 5.4: *For three alternatives, the Borda Count never elects the candidate that loses all pairwise elections. (Borda [10]) For any number of candidates, the Borda Count never has bottom-ranked the candidate who wins all pairwise comparisons. In fact, the Borda Count always ranks a candidate who wins all pairwise comparisons above the candidate who loses all such comparisons. (Nanson [19])*

For any number of alternatives, only the Borda Count satisfies these properties, as well as other favorable comparisons with the pairwise rankings. Indeed, for any other positional method, rank the pairs in any manner; specify a ranking for the n alternatives,

TABLE 5.4 PAIRWISE VOTES

Pairs	A	B	C
$\{A, B\}$	1	0	—
$\{A, C\}$	1	—	0
$\{B, C\}$	—	1	0
Total	2	1	0

it can even be the opposite of the pairwise rankings. There exists data sets where the rankings of the rule and the pairs are as specified. (Saari [20])

While this result provides added support for Borda's method, it also suggests that there can be conflict between the pairwise and the Borda rankings. When this happens, which rule should be used? This question, which has been debated for the last two centuries, has only recently been answered in favor of the Borda Count [15, 21]. The explanation follows.

5.4.2 Cancellations

A way to explain why Borda's method has highly favorable features while so many other rules are inconsistent is to borrow basic principles from physics. When finding the total effect of the three forces acting on the body in Fig. 5.1(a), suppose we follow the lead of a particularly poor beginning student by emphasizing different pairs of forces, such as *A* and *B*, then *B* and *C*. We know this will lead to an incorrect, distorted answer: by emphasizing "parts," the approach fails to recognize a "global" natural cancellation. Indeed, by considering all forces as an ensemble to identify cancellations, the force resolution uses the obvious 120° symmetry cancellation to leave a single force *C*" acting downward on the body.

A surprisingly similar effect occurs in decision problems: certain collections of preferences define natural cancellations. *All* difficulties with decision rules occur when a rule fails to recognize and use these cancellations. As with the force problem, this failure to recognize natural "cancellations" causes the rule to generate distorted outcomes. Examples of "natural cancellations of data" follow.

The first example of a natural cancellation of data uses what I call a "ranking disk." As indicated in Fig. 5.1(b), attach to a fixed background a disk that freely rotates about its center. Equally spaced along its circular boundary place the "ranking numbers" 1, 2, ..., n. To represent a ranking r with n-alternatives, place each alternative's name on the fixed background next to its ranking number. This is the first ranking. Rotate the disk clockwise until number 1 points to the next candidate—the new position defines a second ranking. Continue until n rankings are defined. These n rankings define what I call *the Condorcet n-tuple*.

To illustrate with the three alternatives listed in Fig. 5.1(b), the first ranking is $A \succ B \succ C$. By rotating the disk so the "1" now is next to *B*, we obtain the ranking $B \succ C \succ A$. The third rotation defines the final $C \succ A \succ B$ ranking for the *Condorcet triplet*. By construction, no alternative has a favored status over any other; each alternative is in first, second and third place precisely once. The comparison with the force diagram is striking: The Condorcet triplet is based on a 120° symmetry [more generally, the Condorcet n-tuple has a $2\pi/n$ symmetry] that should cancel to define a complete tie.

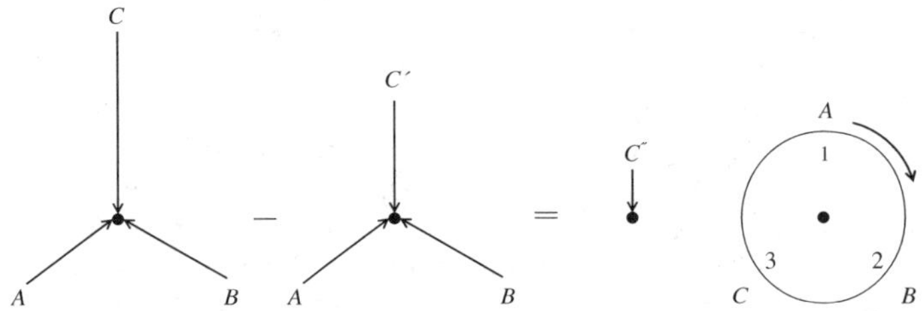

FIG. 5.1 FINDING RANKING CANCELLATIONS: (A) CANCELLING FORCES; (B) RANKING DISK

This completely tied $A \sim B \sim C$ decision ranking holds for all positional rules. Indeed, for the weights $(w_1, w_2, 0)$, each alternative receives $w_1 + w_2$ points because it is in first and second place once. In contrast, the pairwise vote concentrates on only portions of the data—each pair—so, like the beginning student confronting a force diagram, it fails to recognize this broader symmetry. As a consequence, the pairwise outcome is the cycle $A \succ B$, $B \succ C$, $C \succ A$ that frustrates any decision analysis as it does not allow an optimal choice. (For a detailed discussion about pairwise comparisons in engineering, see [22].)

Another natural symmetry classification of data involves a complete reversal. Here, for each ranking of the alternatives, another criterion delivers the exact opposite ranking. An example would be $A \succ B \succ C$ and $C \succ B \succ A$. Think of this as a husband and wife deciding to go to the beach rather than voting because their directly opposite beliefs will only cancel. Again, borrowing from physics what happens with directly opposing forces with equal magnitudes, we must expect a vote cancellation ensuring a complete tie. Indeed, the pairwise vote (and, by the above relationship between the *BC* and pairwise tallies), the Borda Count, always delivers the anticipated complete tie. However, *no positional method other than the BC has this tied outcome*. Instead, *A* and *C* each receive w_1 points and *B* receives $2w_2$ points. So, if $w_1 > 2w_2$, as with the plurality vote and the "best of the best," the outcome is $A \sim C \succ B$. However, should $2w_2 > w_1$, as true for the "avoid the worst of the worst" analysis, then the ranking is the opposite $B \succ A \sim C$.

Surprisingly, for three alternatives, *these are the only possible combination of data causing distortions and differences in positional and pairwise decision outcomes* [21]. In other words (as illustrated in the next section), all differences in three alternative decision outcomes are caused when the data possesses these symmetries and rules are used that cannot recognize and cancel these symmetries. For instance, to create the beverage example Table 5.1, I started with a setting where: 1 prefers $M \succ W \succ B$ while 4 prefer $W \succ B \succ M$.

Here, the reasonable outcome is $W \succ B \succ M$. To create conflict, I then added the symmetric data (which should cancel to create a tie) where: 5 prefer $M \succ W \succ B$ and 5 prefer the reversed $B \succ W \succ M$.

Rules incapable of recognizing and canceling this symmetry will introduce a bias in the outcome-this is the source of the conflict. (Similarly, the reversal symmetry found in the Table 5.3 data partially causes that conflict.)

With more alternatives, there are more data symmetries that can generate other problems. For our purposes, the important fact is that only the Borda Count places all of these symmetries in its kernel. Stated in other terms, Borda's method is the only rule that recognizes the different symmetry arrangements and cancels these forces: this result holds for any number of alternatives Saari [15]. This is the complete explanation of the Borda Count's desirable properties; in contrast, no other rule can handle all of these symmetries, so they introduce a bias that creates conflicting and distorted conclusions.

5.4.3 Constructing Examples

A way to illustrate the power of the BC is to create examples so disturbing that they probably will startle colleagues in economics or political science. Start with data involving three criteria where there is no disagreement about the outcome:

$$2 \text{ have } A \succ B \succ C,\ 1 \text{ has } B \succ A \succ C. \qquad \text{Eq. (5.2)}$$

As C is bottom-ranked over all criteria, C should be bottom-ranked in the decision ranking. Moreover, C's lowly status over all criteria means that this data set really defines a two-alternative decision, where A is superior over two criteria while B is superior only over one; e.g., the outcome should be $A \succ B \succ C$. This is the outcome for the pairwise vote and all positional rules *except* the "avoid the worst of the worst" approach with its $A \sim B \succ C$ outcome.

To generate an example where the pairwise vote differs from the positional outcomes, the above discussion shows that we must add Condorcet triplets. Suppose data comes from six more criteria that defines the following Condorcet triplet (which interchanges B and C in the ranking disk description): each ranking is specified by *two* criteria.

$$A \succ C \succ B,\ C \succ B \succ A,\ B \succ A \succ C \qquad \text{Eq. (5.3)}$$

Adding this data to the original set does not affect the positional rankings, so they retain the $A \succ B \succ C$ ranking coming from the Eq. (5.2) data. The pairwise outcomes, however, now define the conflicting $B \succ A$, $B \succ C$, $A \succ C$ outcomes, which falsely suggest that B is "better" than A. In other words, the only difference between the Borda and pairwise outcomes is that the Borda Count recognizes and cancels the Condorcet effect; the pairwise votes do not. This assertion holds for any number of alternatives.

Theorem 5.5 [15]: *For any number of alternatives, any difference between the pairwise and the Borda rankings is caused by Condorcet components in the data. If all Condorcet components are removed from the data, the pairwise and Borda rankings agree.*

Now let's augment our example to make the plurality ranking conflict with Borda Count ranking. The only way to do so is to add data illustrating complete reversal symmetry. By adding rankings from 10 additional criteria (Table 5.5)

TABLE 5.5 ADDITIONAL RANKINGS

Number	Ranking	Ranking
3	$C \succ A \succ B$	$B \succ A \succ C$
2	$C \succ B \succ A$	$A \succ B \succ C$

we have a setting where the C is the plurality winner, while the Borda outcome (because Borda cancels this reversal symmetry effect) keeps the original $A \succ B \succ C$ ranking. Adding the three data sets creates an example where:

- The Borda Count ranking is $A \succ B \succ C$.
- The pairwise rankings of $B \succ A$, $B \succ C$, $A \succ C$ conflict with the Borda ranking.
- The plurality outcome of $C \succ A \sim B$ adds still further conflict.
- By carrying out the algebraic computations (where $w_1 = 1$, $w_2 = x$, $w_3 = 0$ it follows that using other choices of weights define the decision rankings of $C \succ A \succ B$, $A \sim C \succ B$, $A \succ C \succ B$, $A \succ B \sim C$, and $A \succ B \succ C$. Thus, this data set has six positional rankings where none of them has B top-ranked; B is the alternative that beats all other alternatives in pairwise comparisons.

All examples used in this paper were constructed in this manner. (I show in Saari [1, 4, 21] how to do this in general.)

5.4.4 Selecting Weights in Engineering Problems

The same principle extends to engineering problems. As an illustration, a way to select the λ's in Eq. (5.1) is to first determine the different engineering settings where the outcome should be a tie. By scaling, we can assume such a neutral setting is captured by $\sum_{j=1}^{n}\lambda_j \nabla U_j = 0$. So, by first identifying neutral settings where the $\{\nabla U_j\}$ terms should define a cancellation, we obtain a set of equations for the λ values. The reason for doing so is simple: if the λ's are not chosen in this manner, they will always introduce a bias for any other settings that includes this neutral one.

The first step of identifying all neutral settings is where engineering considerations come into play. The engineering conditions that define "neutrality" lead to algebraic expressions involving the λ's. The λ weights are selected by solving this system.

5.4.5 Basic Conclusions

To justify my earlier comment that almost all voting decision rules have some distortion in their outcomes for almost all data sets, I represent the data space with n alternatives with a $n!$-dimensional space where each dimension represents a particular ranking. A data set is represented by a point in this space: each component value indicates the number of criteria with that particular ranking.

The approach involves finding a coordinate system that reflects the causes of distortions. To do so, the analysis for n-alternatives [15] extends the above discussion by identifying all data symmetries that should define cancellations: this endows the data space with a particular coordinate system. Each direction in this coordinate system defines a different data symmetry. To illustrate with three alternatives [21], a 2-D subspace is defined by the complete reversal symmetries, a 1-D subspace is defined by the Condorcet symmetries, and a 1-D subspace is defined by the data sets that correspond to where the same number of criteria have each possible ranking. As this defines 4 of the $3! = 6$ dimensions, we should expect that there is a 2-D subspace of data sets that is free of these symmetry properties. This is the case.

More generally, for n alternatives, the $n!$-dimensional data space has a $(n-1)$-dimensional subspace of data that is free from all data symmetry that should define a tie. On this subspace, all positional and pairwise rules always have the same decision outcome—there is no disagreement of any kind. But the small dimension of this space means that most of the data space offers cancellation opportunities. So, a rule that is not capable of reacting to these cancellations must provide a distorted decision outcome whenever the data is tainted by these symmetries. Dimensional considerations alone prove that *it is highly unlikely (in the limit, it has probability of 0) to avoid distortions in the decision conclusions of these rules.*

To be precise, state that an outcome is "distorted" if the tallies differ from what they would be after removing all canceling symmetries.

Theorem 5.6: *For all non-BC rules, almost all data leads to a distorted outcome. The BC outcome over all n alternatives is the only rule that never has distorted outcomes.*

There are two approaches to resolve these decision challenges. One is a filter argument that uses a vector analysis argument to strip these symmetry parts from the data and then uses the reduced data. Dimensional considerations, where just six alternatives generate a 720-dimensional vector analysis problem, prove that this approach is not realistic. The alternative is to use the Borda Count because it is the only rule that recognizes the different symmetry configurations of data and cancels them.

5.5 SUMMARY

In the quest for making accurate decisions, the choice of a decision rule is a crucial variable. Among positional rules, or rules that use these methods, only the Borda Count offers reliable conclusions. For all other rules, the outcomes for almost all data sets suffer some distortion. This distortion is caused strictly by the inability of the rules to extract and then cancel certain data symmetries; by not doing so, the rule introduces a bias in the outcome.

What should be done for more general engineering decisions, such as those involving an Eq. (5.1) component where weights must be selected? Interestingly, the answer depends on the particular problem being considered. But the underlying principle is clear. First, identify configurations of settings where no alternative is favored, settings where it is arguable that the outcome should be a complete tie. The choice of the weights must be made so that in these settings, a tie does occur. Any other choice of the weights is guaranteed to distort the outcome for any setting that includes even parts of a neutral setting.

REFERENCES

1. Saari, D. G., 1995. *Basic Geometry of Voting*, Springer-Verlag, New York, NY.
2. Arrow, K. J., 1963. *Social Choice and Individual Values*, 2nd Ed., Wiley, New York, NY.
3. Arrow, K. J. and Raynaud, H., 1986. *Social Choice and Multicriterion Decision-Making*, MIT Press.
4. Saari, D. G., 2001. *Chaotic Elections! A Mathematician Looks at Voting*, American Mathematical Society, Providence, RI.
5. Saari, D. G., 2001. *Decisions and Elections: Explaining the Unexpected*, Cambridge University Press, New York, NY.
6. Saari, D. G., 1998. "Connecting and Resolving Sen's and Arrow's Theorems," *Social Choice & Welfare*, Vol. 15, pp. 239–261.
7. Saari, D. G., 1995. "A Chaotic Exploration of Aggregation Paradoxes," *SIAM Review*, Vol. 37, pp. 37–52.
8. Saari, D. G., 1999. "More Chaos; But, in Voting and Apportionments?" *Perspective, Proc., Nat. Acad. of Sci.*, Vol. 96, pp. 10568–10571.
9. Haunsperger, D., 1992. "Dictionaries of Paradoxes for Statistical Tests on k Samples," *J. Am. Stat. Assoc.*, Vol. 87, pp. 149–155.
10. Borda, J. C., 1782. *Mémoire Sur Les Élections au Scrutin*, Histoire de l'Académie Royale des Sciences, Paris.
11. Condorcet, M., 1785. *Éssai sur L'application de L'analyse à La Probabilité des D´ecisions Rendues à la Pluralité des Voix*, Paris.
12. Kelly, J., 2004. www.maxwell.syr.edu/maxpages/faculty/jskelly/biblioho.htm.
13. Saari, D. G., 1984. "The Ultimate of Chaos Resulting From Weighted Voting Systems," *Advances in App. Math.*, Vol. 5, pp. 286–308.
14. Saari, D. G., 1992. "Millions of Election Outcomes From a Single Profile," *Social Choice & Welfare*, Vol. 9, pp. 277–306.
15. Saari, D. G., 2000. "Mathematical Structure of Voting Paradoxes I: Pairwise Vote; Mathematical Structure of Voting Paradoxes II: positional voting," *Economic Theory* Vol. 15, pp. 1–103.
16. Saari, D. G. and Tataru, M., 1999, "The Likelihood of Dubious Election Outcomes," *Economic Theory*, Vol. 13, pp. 345–363.
17. Saari, D. G., 2005. "The Profile Structure for Luce's Choice Axiom," *J. Math. Psych.* Vol. 49, pp. 226–253.
18. Fishburn, P., 1981. "Inverted Orders for Monotone Scoring Rules," *Discrete App. Math.*, Vol. 3, pp. 27–36.
19. Nanson, E. J., 1882. "Methods of Election," *Trans. Proc. Roy. Soc. Victoria*, Vol. 18, pp. 197–240.
20. Saari, D. G., 1989. "A Dictionary for Voting Paradoxes," *J. Eco. Theory*, Vol. 48, pp. 443–475.
21. Saari, D. G., 1999. "Explaining All Three-Alternative Voting Outcomes," *J. Eco. Theory*, Vol. 87, pp. 313–355.
22. Saari, D. G. and Sieberg, K., 2004. "Are Partwise Comparisons Reliable?" *Res. in Engrg. Des.*, Vol. 15, pp. 62–71.

PROBLEMS

The purpose of the following problems is to help the reader develop intuition about the kinds of data that can give rise to dubious, and conflicting, decision outcomes.

5.1 In the beginning of this chapter, it is asserted that the beverage example gives rise to seven different election rankings with changes in the choice of the positional procedure. To prove this, notice that any positional scores $(w_1, w_2, 0)$ can be scaled to become $[(w_1/w_1), (w_2/w_1), 0]$, or $(1, s, 0)$, where $0 \le s \le 1$. By tallying the beverage example ballots with $(1, s, 0)$, find all seven election rankings and the choices of s that define each of them.

5.2 To illustrate Theorem 5.1, find an example where the plurality ranking is $A \succ B \succ C \succ D \succ E$, but if E is dropped, it becomes $D \succ C \succ B \succ A$, and if A is dropped, it becomes $B \succ C \succ D$, even though in a pairwise vote $C \succ B$. (Hint: Analyze the example illustrating the "dropping or adding alternatives" section. For instance, if D is dropped, two of these six voters now vote for B and four vote for C: this leads to the $C \succ B \succ A$ ranking. So construct a tree to determine which alternatives should be second- and third-ranked to achieve a desired outcome.)

5.3 Use the same idea as for Problem 5.2 to create an example involving four alternatives where the plurality ranking is $A \succ B \succ C \succ D$, but if *any* alternative is dropped, the plurality ranking of the three-alternative set is consistent with $D \succ C \succ B \succ A$, but all pairwise rankings are consistent with the original four-candidate ranking.

5.4 Starting with one criterion satisfying $B \succ A \succ C$, use the material in the "constructing examples" subsection to create an example where the Borda Count outcome is $B \succ A \succ C$, the pairwise outcome is the cycle $A \succ B$, $B \succ C$, $C \succ A$, the plurality outcome is $C \succ B \succ A$, and the "vote for two" outcome is $A \succ B \succ C$. Namely, each alternative can be the "winner" with an appropriate decision rule, but the pairwise approach is useless as it defines a cycle.

5.5 Show that if weights $(w_1, w_2, \ldots, w_n, w_j \ge w_{j+1}$ and $w_1 \succ w_n$, are assigned to a voter's first, second, . . . , last and nth ranked candidates, the same election ranking always occurs if ballots are tallied with: $1/(w_1 - w_n)[(w_1 - w_n, w_2 - w_n, \ldots, w_n - w_n)]$.

Next, prove that over all possible pairs, a voter with $a_1 \succ a_2 \succ \ldots \succ a_n$ ranking will vote $n - j$ times for candidate a_j. In other words, the vote is the same as for the Borda Count.

5.6 Use the ranking disk with the starting ranking of $A \succ B \succ C \succ D \succ E \succ F$ to find the corresponding Condorcet six-tuple. Next, suppose that only the first three rankings from this six-tuple are the rankings for three criteria. Suppose the decision rule is to compare rankings pairwise: the losing alternative is dropped and the winning alternative is moved on to be compared with another alternative. This means there are five steps—the winning alternative in the last step is the overall "winner." Show that for each of the six alternatives, there is an ordering so that this alternative will be the "winner." (Hint: compute the pairwise rankings for adjacent alternatives in this listing.)

5.7 It is possible for data to be cyclic; e.g., we might have six criteria supporting $A \succ B$, four supporting $B \succ C$ and five supporting $C \succ A$. What should be the ranking? A way to handle this problem is to use comment preceding Theorem 5.4 to compute the Borda Count via the number of pairwise points an alternative receives. Use this approach to determine the Borda Count ranking for this data. Explain the answer in terms of the Condorcet triplet information described earlier. (For more information on this, see [22].)

5.8 A discussion and research problem is to take an actual engineering problem that involves weights and determine how to select the weights. Using the above argument, the approach reduces to determining what configurations of the problem are neutral—they define a setting where it is impossible to select one alternative over another because they all should be equal. Select such an engineering problem and carry out the analysis.

CHAPTER

6

Preference Modeling in Engineering Design

Jonathan Barzilai

PART 1 SELECTION AND PREFERENCE–THE FOUNDATIONS

We establish that none of the classical theories of choice can serve as a proper foundation for decision theory (and hence for decision-based design) and we construct a new theory of measurement that provides such a foundation. A proper theory of selection in engineering design cannot be founded on scales and variables to which the mathematical operations of addition and multiplication are not applicable. Yet it has not been proven that addition and multiplication are applicable to von Neumann and Morgenstern's utility scales or to any scales based on classical decision theory (which, in turn, is based on the classical theory of measurement). In fact, addition and multiplication are not applicable to utility scales, value scales, ordinal voting scales or any scales based on the classical theory of measurement whether the underlying variables are physical or subjective (i.e., psychological).

Selection is an important problem in engineering design (see [1], Chapter 3). By definition, selection means making choices and choice is synonymous to preference since we choose those objects that are preferred. Therefore, the scientific foundation of selection in engineering design (and elsewhere) is the measurement of preference. Consequently, our goal is the construction of preference scales that serve similar purposes as scales for measurement of physical variables such as *time, energy* and *position*. In Part 1 of this chapter, we consider the issues of the mathematical foundations for scale construction.

6.1 THE PURPOSE OF MEASUREMENT

Our starting point is the observation that the purpose of representing variables by scales is to enable the application of mathematical operations to these scales. Indeed, the analysis that follows (which is based on this observation) explains why these scales are typically numerical.

6.2 CLASSIFICATION OF SCALES

Since the purpose of the construction of scales is to enable the application of mathematics to them, we classify scales by the type of mathematical operations that they enable. We use the terms *proper scales* to denote scales to which the operations of addition and multiplication apply, and *weak scales* to denote scales to which these operations do not apply. This partition is of fundamental importance and we shall see, after formally defining the operations of addition and multiplication, that the related mathematical structures are fields, one-dimensional vector spaces and affine spaces (straight lines in affine geometry).

Although we will further refine the classification of proper scales, the key element of our analysis is the distinction between proper and weak scales and we note that it follows from the theory presented here (see Section 6.8) that all the models of the classical theory of measurement (e.g., [2], [3] and [4]) are weak because they are based on operations that do not correspond to addition and multiplication as well as for other reasons. To re-emphasize, even in the case of physical measurement, the models of the classical theory produce scales that do not enable the operations of addition and multiplication. Physics as well as other sciences cannot be developed without the mathematical tools of calculus for which the operations of addition and multiplication are required.

6.3 ON THE MEASUREMENT OF SUBJECTIVE PROPERTIES

In the case of physical variables, the set of scales is uniquely determined by the set of objects and the property under measurement. In other words, scale construction requires specifying only the set of objects and the property under measurement. In the social sciences, the systems under measurement include a person or persons so that the property under measurement is associated with a human being and, in this sense, is personal, psychological or subjective.

Except that in the case of subjective properties the specification of the property under measurement includes the specification of the "owner" of the property (for example, we must specify whose preference is being measured), the mathematical modeling of measurement of subjective properties does not differ from that of physical ones. Among other things, this implies that there is no basis for the distinction between value and utility scales (e.g., [5]) or between von Neumann and Morgenstern's utility scales [6] and Luce and Raiffa's utility scales [7].

6.4 UTILITY THEORY CANNOT SERVE AS A FOUNDATION

Since some researchers advocate utility theory as the foundation for DBD (e.g., [8]), it is important to establish that utility theory cannot serve as a foundation for any scientific theory.

6.4.1 The von Neumann and Morgenstern Axioms

By "modern utility theory" (e.g., [9, Section 1.3] and Coombs *et al.* [10, p. 122]), we mean the utility theory of von Neumann and Morgenstern [6, Section 3.5–3.6] and its later variants.

In essence, von Neumann and Morgenstern study a set of objects A equipped with an operation (i.e., a function) and order that satisfy certain assumptions. The operation is of the form $f(\alpha,a,b)$, where a and b are objects in A, α is a real number, and $c = f(\alpha,a,b)$ is an object in A.

The main result of von Neumann and Morgenstern is an *existence* and *uniqueness* theorem for isomorphisms that reflect the structure of the set A into a set B equipped with order and a corresponding operation $g[\alpha, s(a), s(b)]$, where $a \to s(a)$, $b \to s(b)$ and $f(\alpha,a,b) \to g[\alpha, s(a), s(b)]$. This framework does not address explicitly the issues of utility scale *construction* and, in Section 6.4.4, we shall see that there are difficulties with this construction. When the set B is equipped with the operations of addition and multiplication, and in particular in the case of the real numbers, these isomorphisms are of the form

$$f(\alpha, a_1, a_0) \to g(\alpha, s_1, s_0) = \alpha s_1 + (1-\alpha)s_0 \qquad \text{Eq. (6.1)}$$

6.4.2 Barzilai's Paradox—Utility's Intrinsic Self-Contradiction

Utility theory does not impose constraints on the values of preference scales for prizes. However, the interpretation of the utility operation in terms of lotteries is required in the construction of these scales, and this interpretation constrains the values of utility scales for lotteries. The theory permits lotteries that are prizes (e.g., Raiffa's "neat example" [7, pp. 26–27]) and this leads to a contradiction since an object may be both a prize, which is not constrained, and a lottery, which is constrained.

For example, suppose the prizes A and C are assigned by a decision-maker the utility values $u(A) = 0$, and $u(C) = 1$ and let D be the lottery $D = \{(0.6, A);(0.4, C)\}$. According to utility theory (see e.g., [5]) $u(D) = 0.6u(A) + 0.4\, u(C) = 0.4$, so that the value of $u(D)$ is determined by the other given parameters and the decision-maker has no discretion as to its value.

Now suppose that the decision-maker assigns the value $u(B) = 0.5$ to the prize B, and is offered an additional prize E. According to utility theory, there are no constraints on the possible utility values for prizes so that the value of $u(E)$ is at the discretion of the decision-maker and is not dictated by the theory. The decision-maker then assigns the utility value $u(E) = 0.8$.

Since utility theory allows prizes that are lottery tickets, suppose that the prize E is the lottery $E = \{(0.6, A);(0.4, C)\}$. It follows that $D = E$, yet the utility value of this object is either 0.8 or 0.4 depending on whether we label the object $\{(0.6, A);(0.4, C)\}$ a prize or a lottery. That is, we have $u(D) = 0.4 \neq 0.8 = u(E)$ where D and E are the same object! In other words, the utility value of the object $\{(0.6, A);(0.4, C)\}$ depends on its label. Note that $u(D) < u(B)$ and $u(E) > u(B)$ yet $D = E$, so that the object $\{(0.6, A);(0.4, C)\}$ is rejected in favour of B if it is labeled a lottery and accepted as preferred to B if it is labeled a prize.

6.4.3 Utility is Not a Normative Theory

According to utility theorists, utility is a normative theory (see e.g., [8, Section 1] and [11, p. 254]). Specifically, Coombs *et al.* [10, p. 123]) state that "utility theory was developed as a prescriptive theory" and Howard [12] advocates this position in strong religious terms.

However, von Neumann and Morgenstern's utility theory as well as its later variants (e.g., [7, Section 2.5], [9, pp. 7–9], [10, pp. 122–129], [13, Chapter 5], [14, p. 195]) are mathematical theories. These theories are of the form $P \to Q$, that is, *if* the assumptions P hold *then* the conclusions Q follow. In other words, mathematical theories are not of the form *"Thou Shall Assume P,"* but rather "*if* you assume P." As a result, the claim that utility theory is normative has no basis in mathematical logic nor in modern utility theory since mathematical theories do not dictate to decision-makers what sets of assumptions they *should* satisfy.

6.4.4 The von Neumann and Morgenstern Structure is Not Operational

The construction of utility functions requires the interpretation of the operation $f(\alpha, a_1, a_0)$ as a lottery on the prizes a_1, a_0 with probabilities $\alpha, 1-\alpha$, respectively. The utility of a prize a is then assigned the value α where $a = f(\alpha, a_1, a_0)$, $u(a_1) = 1$ and $u(a_0) = 0$.

In order for $f(\alpha, a_1, a_0)$ to be an operation, it must be single-valued. Presumably with this in mind, von Neumann and Morgenstern interpret the relation of equality on elements of the set A as *true identity:* In [6, A.1.1–2, p. 617] they remark in the hope of "dispelling possible misunderstanding" that "[w]e do not axiomatize the relation =, but interpret it as *true identity.*" Under this interpretation, equality of the form $a = f(\alpha, a_1, a_0)$ cannot hold if a is a prize that is not a lottery since these are not identical objects. Consequently, von Neumann and Morgenstern's interpretation of their axioms does not enable the practical construction of utility functions.

Possibly for this reason, later variants of utility theory (e.g., [7]) interpret equality as indifference rather than true identity. This interpretation requires the extension of the set A to contain the lotteries in addition to the prizes. In this model, lotteries are elements of the set A rather than an operation on A, so that this extended set is no longer equipped with any operations but rather with the relations of order and indifference (see e.g., [10, p. 122]). This utility structure is not homomorphic (and therefore is not equivalent) to the von Neumann and Morgenstern structure, and the utility functions it generates are weak (i.e., do not enable the operations of addition and multiplication) and only enable the relation of order despite their "interval" type of uniqueness.

6.4.5 Utility Models are Weak

Modern utility models (e.g., [7, Section 2.5], [9, pp. 7–9], [10, pp. 122–129], [13, Ch. 5]) are not equivalent to the model of von Neumann and Morgenstern and *The Principle of Reflection* (see Section 6.8) implies that all utility models are weak. Despite the fact that they produce "interval" scales, none of these models enables the operations of addition and multiplication, although these models enable order and some of them also enable the operation $g(\alpha, s_1, s_0) = \alpha s_1 + (1-\alpha)s_0$.

The model of von Neumann and Morgenstern produces weak scales because it differs from the one-dimensional affine space model. For example, instead of the *two binary* operations of addition and multiplication, this model is equipped with *one* compound *ternary* operation.

As a result, it is not homomorphic—and is not equivalent—to the homogeneous field model which is required for proper scales.

6.4.6 Implications for Game Theory

It follows from the above that utility theory cannot serve as a foundation for game theory. Although utility scales may be replaced with strong preference scales, some difficulties remain. Since preference is subjective, i.e., personal, the notion of "transferable utility" is self-contradictory as it appears to mean that the players' utility functions are identical. In fact, von Neumann and Morgenstern do not provide a clear path from their utility axioms to the notion of transferable utility and this missing link does not seem to exist elsewhere in the literature. Further, since each player attempts to maximize the utility of his payoff, the objective function of the minimax operation in the case of a zero-sum two-person game is a utility function (see e.g., [7, p. 89]) and, since utility functions are not unique, it is not clear what "zero-sum" means in this case. In addition, since utility functions are unique only up to additive and (positive) multiplicative constants, "the value of the game" depends on these constants and is in fact undefined, since by changing these constants any real number can be made to be "the value of the game." It follows that von Neumann and Morgenstern's concept of the characteristic function in the general n-person case is undefined as well. Note also that for utility functions the concept of the sum of game values for two different coalitions (e.g., [6, p. 241]) is undefined because the operation of addition—as opposed to the expected value operation—is not applicable to utility functions. Since von Neumann and Morgenstern's solution of the game depends on the concept of the characteristic function, it (and any other solution that depends on this concept) is not properly defined as well.

6.5 ORDINAL SCALES AND PAIRWISE COMPARISON CHARTS

The work of Saari [15, 16] on Arrow's impossibility theorem and ordinal voting systems, as well as the position advocated by Dym et al. [17], appear to suggest the replacement of utility theory with ordinal theories as a foundation for the selection problem, although Saari does not provide a reason for abandoning utility theory for this purpose.

Concerning Arrow's impossibility theorem, we note that the construction of preference scales cannot be founded on negative results. (Examples of negative results are the impossibility of trisecting an arbitrary given angle by ruler and compass alone and the insolvability of quintic equations by radicals.) Negative results indicate that a solution cannot be found following a given path and, in this sense, are terminal. Although they may lead us to a successful investigation of alternative paths, no scientific theory can be founded on negative results.

Concerning ordinal scales, we note that they enable the relation of order but not the operations of addition and multiplication. Further, the concept of "cancellation" is applicable in algebraic systems with inverse elements but is inapplicable in ordinal systems [16, p. 1]. Since ordinal scales are weak they cannot serve as foundations for scientific disciplines.

Ordinal scales do not enable the operations of addition and multiplication, and the concepts of cancellation and trade-off do not apply to them. To appreciate the practical implications of ignoring differences and ratios, consider the following example: Two competing designs for a new passenger airplane are compared with respect to their range, fuel consumption and the number of passengers they can carry. Suppose that design A is superior to B with respect to range and fuel consumption but is inferior to B with respect to the number of passengers. Since A is better than B twice while B is better than A once, design A will be chosen over B based on ordinal counting procedures. These procedures ignore the question "by how much is A better than B?" Indeed, these procedures will indicate a preference for A even if B performs *slightly less* well as A on range and fuel consumption but can carry *twice* the number of passengers as A. Note that the concept of "slightly less" is applicable to proper scales but is not applicable to ordinal ones. In our example, because the concepts of *difference, slight difference, large* difference or *twice* are inapplicable in the ordinal methodologies advocated by Saari [15, 16] and Dym et al. [17], these methodologies lead to an unacceptable "cancellation" or "trade-off" of a slight advantage in fuel consumption against a large advantage in the number of passengers.

PART 2 STRONG MEASUREMENT SCALES

In Barzilai [18, 19] we developed a new theory of measurement, which is outlined below. The most important elements of this theory are:

- Recognition of the purpose of measurement
- A new classification of measurement models by the mathematical operations that are enabled on the resulting scales
- The Principle of Reflection
- Homogeneity considerations

6.6 THE NEW CLASSIFICATION

The essence of measurement is the construction of a mathematical system that serves as a model for a given empirical system. The purpose of this construction is to enable the application of mathematical operations to scale values within the mathematical system.

In particular, we are interested in models that (1) enable the application of the operations of addition and multiplication (including subtraction and division) to scale values; (2) enable the modeling of an order relation on the objects; and (3) enable the application of calculus to scale values, i.e., closure under the limit operation. (For example, in statistics, the definition of standard deviation requires the use of the square root function and the computation of this function requires the limit operation of calculus.) We use the term *strong models* to denote such models and *strong scales* to denote scales produced by strong models.

We also use the terms *proper scales* to denote scales to which the operations of addition and multiplication apply, and *weak scales* to denote scales to which the operations of addition and multiplication do not apply. Strong scales are proper but proper scales may or may not be strong, i.e., proper scales enable addition and multiplication but may not enable order and calculus.

6.7 THE MAIN RESULT

The main result of the new theory is that there is only one model of strong measurement for preference. It also follows from *the Principle of Reflection* that all the models of the classical theory of measurement generate weak scales to which the operations of addition and multiplication are not applicable.

Specifically, in order to enable the operations of addition and multiplication, the relation of order and the application of calculus on preference scales, the objects must be mapped into the real one-dimensional homogeneous field, i.e., a one-dimensional affine space. Furthermore, the set of objects must be a subset of the points in an empirical one-dimensional ordered homogeneous field over the real numbers.

6.8 THE PRINCIPLE OF REFLECTION

Consider the measurement of *length* and suppose that we can only carry out ordinal measurement on a set of objects, that is, for any pair of objects we can determine which one is longer or whether they are equal in length (in which case we can order the objects by their length). This may be due to a deficiency in the state of technology (appropriate tools are not available) or in the state of science (the state of knowledge and understanding of the empirical or mathematical system is insufficient). We can still construct scales (functions) that map the empirical objects into the real numbers, but although the real numbers admit many operations and relations, the only relation on ordinal scale values that is relevant to the property under measurement is the relation of order. Specifically, the operations of addition and multiplication can be carried out on the range of such scales since the range is a subset of the real numbers, but such operations are extraneous because they do not reflect corresponding empirical operations. Extraneous operations may not be carried out on scales and scale values—they are irrelevant and inapplicable; their application to scale values is a modeling error.

The principle of reflection is an essential element of modeling that has not been recognized in the classical theory of measurement. It states that operations within the mathematical system are applicable *if and only if* they reflect corresponding operations within the empirical system. In technical terms, in order for the mathematical system to be a valid model of the empirical one, the mathematical system must be homomorphic to the empirical system (a homomorphism is a structure-preserving mapping). A mathematical operation is a valid element of the model only if it is the homomorphic image of an empirical operation. Other operations are not applicable to scales and scale values.

By the principle of reflection, a necessary condition for the applicability of an operation on scales and scale values is the existence of a corresponding empirical operation (the homomorphic pre-image of the mathematical operation). That is, the principle of reflection applies in both directions and a given operation is applicable to the mathematical image only if the empirical system is equipped with a corresponding operation.

6.9 GROUPS AND FIELDS

In this and the next section we summarize the construction of proper scales in homogeneous fields. The applicability of the operations of *addition and multiplication* plays a central role in the theory that underlies the practical construction of *preference* scales. Sets that are equipped with the operations of addition and multiplication are studied in abstract algebra and are called *fields*. We define fields in terms of groups that are single-operation structures.

A *group* is a set G with a single operation that satisfies the following requirements (i.e., axioms or assumptions):

- The operation is *closed:* the result of applying the operation to any two elements a and b in G is another element c in G. We use the notation $c = a \circ b$ and since the operation is applicable to pairs of elements of G, it is said to be a binary operation.
- The operation is *associative:* for any elements in G, $(a \circ b) \circ c = a \circ (b \circ c)$.
- The group has an *identity:* there is an element e of G such that $a \circ e = a$ for any element a in G.
- *Inverse elements:* for any element a in G, the equation $a \circ x = e$ has a unique solution x in G.

If $a \circ b = b \circ a$ for all elements of a group, the group is called *commutative*. We reemphasize that a group is an algebraic structure with *a single operation* and we also note that a group is not a homogeneous structure because it contains an element, namely its identity, which is unlike any other element of the group since the identity of a group G is the only element of the group that satisfies $a \circ e = a$ for all a in G.

A *field* is a set F with two operations that satisfy the following assumptions:

- The set F with one of the operations is a commutative group. This operation is called *addition* and the identity of the additive group is called zero (denoted "0").
- The set of all nonzero elements of F is a commutative group under the other operation on F. That operation is called *multiplication* and the multiplicative identity is called one (denoted "1").
- For any element a of the field, $a \times 0 = 0$.
- For any elements of the field the *distributive* law $a \times (b + c) = (a \times b) + (a \times c)$ holds.

Two operations are called addition and multiplication only if they are related to one another by satisfying all the requirements of a field; a single operation on a set is not termed addition nor multiplication. The additive inverse of the element a is denoted $-a$, and the multiplicative inverse of a nonzero element is denoted a^{-1} or $1/a$. Subtraction and division are defined by $a - b = a + (-b)$ and $a/b = a \times b^{-1}$.

As we saw, modeling a single-operation structure by a structure with two operations is a modeling error. Specifically, a group may be modeled by a homomorphic group and a field may be modeled by a homomorphic field, but modeling an empirical group by a field is an error. "Hölder's theorem" (see e.g., [3, Section 3.2.1]) deals with ordered groups. Models that are based on ordered groups rather than ordered fields are weak. The operations of addition and multiplication are not applicable to scales constructed on the basis of such models.

6.10 HOMOGENEOUS FIELDS

A homogeneous structure is a mathematical structure (a set with operations and relations) that does not have special elements. In other words, a homogeneous structure is a structure whose elements are indistinguishable from one another. A field is not a homogeneous structure since the additive and multiplicative identities of a field are unique and distinguishable. A homogeneous empirical structure (physical or subjective) must be modeled by a corresponding (homomorphic) mathematical structure. This requires us to define the structures of a *homogeneous field* and a *partially homogeneous field*. By a homogeneous field we mean a one-dimensional affine space, while a one-dimensional vector space is a partially homogeneous field (it is homogeneous with respect to the multiplicative identity but not with respect to the additive one). Formally, a vector space is a pair of sets (*V, F*) together with associated operations as follows. The elements of *F* are termed *scalars* and *F* is a field. The elements of *V* are termed *vectors* and *V* is a commutative group under an operation termed vector addition. These sets and operations are connected by the additional requirement that for any scalars $j,k \in F$ and vectors $u,v \in V$ the scalar product $k \cdot v \in V$ is defined and satisfies, in the usual notation:

$$(j+k)v = jv + kv \qquad \text{Eq. (6.2)}$$

$$k(u+v) = ku + kv \qquad \text{Eq. (6.3)}$$

$$(jk)v = j(kv) \qquad \text{Eq. (6.4)}$$

$$1 \cdot v = v. \qquad \text{Eq. (6.5)}$$

An *affine space* (or a *point space*) is a triplet of sets (*P, V, F*) together with associated operations as follows: The pair (*V, F*) is a vector space. The elements of *P* are termed *points* and two functions are defined on points: a one-to-one and onto function $h: P \to V$ and a "difference" function $\Delta: P^2 \to V$ that is defined by $\Delta(a, b) = h(a) - h(b)$. Note that this difference mapping is not a closed operation on *P*: although points and vectors can be identified through the one-to-one correspondence $h: P \to V$, the sets of points and vectors are equipped with different operations. Formally, the operations of addition and multiplication are not defined on points. If $\Delta(a,b) = v$, it is convenient to say that the difference between the points *a* and *b* is the vector *v*. Accordingly, we say that a point space is equipped with the operations of (vector) addition and (scalar) multiplication *on point differences*. Note that in an affine space no point is distinguishable from any other.

The dimension of the affine space (*P, V, F*) is the dimension of the vector space *V*. By a homogeneous field we mean a one-dimensional affine space. A homogeneous field is therefore an affine space (*P, V, F*) such that for any pair of vectors $u,v \in V$ where $v \neq 0$, there exists a unique scalar $\alpha \in F$ so that $u = \alpha v$. In a homogeneous field (*P, V, F*) the set *P* is termed a *straight line* and the vectors and points are said to be collinear. Division in a homogeneous field is defined as follows. For $u,v \in V$, $u / v = \alpha$ means that $u = \alpha v$ provided that $v \neq 0$. Therefore, in an affine space, the expression $\Delta(a,b) / \Delta(c,d)$ for the points $a,b,c,d \in P$ where $\Delta(c,d) \neq 0$, is defined and is a scalar:

$$\frac{\Delta(a,b)}{\Delta(c,d)} \in F \qquad \text{Eq. (6.6)}$$

if and only if the space is one-dimensional, i.e., it is a straight line or a homogeneous field. When the space is a straight line, by definition, $\Delta(a,b)/\Delta(c,d) = \alpha$ [where $\Delta(c,d) \neq 0$] means that $\Delta(a,b) = \alpha\Delta(c,d)$ ·

REFERENCES

1. Dym, C. L. and Little, P., 1999. *Engineering Design: A Project-Based Introduction,* Wiley.
2. Luce, R.D., Krantz, D.H., Suppes, P. and Tversky, A., 1990. *Foundations of Measurement*, Vol. 3, Academic Press.
3. Roberts, F.S., 1979. *Measurement Theory*, Addison-Wesley.
4. Narens, L., 1985. *Abstract Measurement Theory*, MIT Press.
5. Keeney, R.L. and Raiffa, H., 1976. *Decisions With Multiple Objectives*, Wiley.
6. Neumann, J. von and Morgenstern, O., 1953. *Theory of Games and Economic Behavior*, 3rd ed., Princeton University Press.
7. Luce, R. D. and Raiffa, H., 1957. *Games and Decisions*, Wiley.
8. Thurston, D.L., 2001, "Real and Misconceived Limitations to Decision Based Design with Utility Theory," *Trans., ASME*, Vol. 123, pp. 176–182.
9. Fishburn, P.C., 1964. *Decision and Value Theory*, Wiley.
10. Coombs, C.H., Dawes, R.M. and Tversky, A., 1970. *Mathematical Psychology: An Elementary Introduction*, Prentice-Hall.
11. Edwards, W. ed., 1992, *Utility Theories: Measurements and Applications*, Kluwer.
12. Howard, R.A., 1992. "In Praise of the Old Time Religion," in *Utility Theories: Measurements and Applications*, W. Edwards, ed., Kluwer.
13. French, S. 1988. *Decision Theory*, Ellis Horwood.
14. Luce, R.D., 2000. *Utility of Gains and Losses*, Erlbaum.
15. Saari, D.G., 1995. *Basic Geometry of Voting*, Springer.
16. Saari, D.G. and Sieberg, K.K., "Are Part Wise Comparisons Reliable?"
17. Dym, C.L., Wood, W.H. and Scott, M.J., 2002. "Rank Ordering Engineering Designs: Pairwise Comparison Charts and Borda Counts," *Res. in Engg. Des.*, Vol. 13, pp. 236–242.
18. Barzilai, J., 2005, "Measurement and Preference Function Modelling," *Int. Trans. in Operational Res.*, Vol. 12, pp. 173–183.
19. Barzilai, J., 2004. "Notes on Utility Theory," *Proc., IEEE Int. Conf. on Sys., Man, and Cybernetics*, pp. 1000–1005.

SECTION

3

CONCEPT GENERATION

INTRODUCTION

The first step in decision-based design is generating a set of alternative designs from which to make selections. This is the most critical of the design steps; without concept generation there is no design; on the other hand, limited resources for design evaluation require the generation of a set of most-likely-to-succeed design alternatives. The creation of such a set is seldom, if ever, a trivial task. It is beyond the scope (and length limitation) of the book to include a review chapter on popular or representative approaches to alternative generation, a topic widely studied in the field of engineering design. Most research in DBD conveniently presumes that existing concept generation methods can be directly employed, or that a decision-maker has a pool of design alternatives ready for examination, evaluation, comparison and selection. On the other hand, research has been conducted on incorporating decision principles into an alternative generation process and that work is included here. This section is intended to enlighten students about this critically important initial stage of design and encourage them to further explore approaches to alternative generation under a DBD perspective.

This short section presents two complementary topics related to alternative generation under a DBD perspective. The foundational principles of DBD can be incorporated into some aspects of the alternative generation process, improving a designer's initial pool of options. This is the case of the work presented in Chapter 7, "Stimulating Creative Design Alternatives Using Customer Values." A specific description of procedures that help generate innovative and effective product ideas in the initial stages of a design process can be found in Chapter 7.

In Chapter 8, "Generating Design Alternatives Across Abstraction Levels," a methodology is described for using decision-making concepts (e.g., probabilistic design modeling, value functions, expected value, decision-making under uncertainty and information value theory) to control the creation of design alternatives across multiple abstraction levels. In this context DBD results in a set of cascading decisions that enable refinement of design candidates and the initial requirements from which they were derived.

Note that the material in this section is a small sample of more general work that is available on methods for creating design alternatives. Readers are encouraged to compare other methods to those presented here to see how a decision-based approach impacts the process.

CHAPTER

7

STIMULATING CREATIVE DESIGN ALTERNATIVES USING CUSTOMER VALUES*

Ralph L. Keeney

INTRODUCTION

A company designs and sells products to achieve its objectives. These objectives include maximizing profits and market share, which pleases stockholders and allows better financial reward for employees. The company is also interested in pleasing customers, which enhances sales, and in providing a stimulating, enjoyable workplace that pleases employees.

Visualize an iterative process that begins with the question "What could we design?" and ends with "our degree of success." Figure 7.1 illustrates this process on a high level and indicates that it is a process driven by decisions [8, 15, 20, 22]. Many decisions affect how successful a company is. At the very beginning, decisions must be made to provide the conceptual design for the product. The decisions specify the product properties and benefits as well as many aspects of its production and delivery to customers. The process usually begins with the creative generation of a rough conceptual design based on perceived customer needs [7]. This design is then honed through decisions and appraisal cycles to produce a more detailed conceptual design.

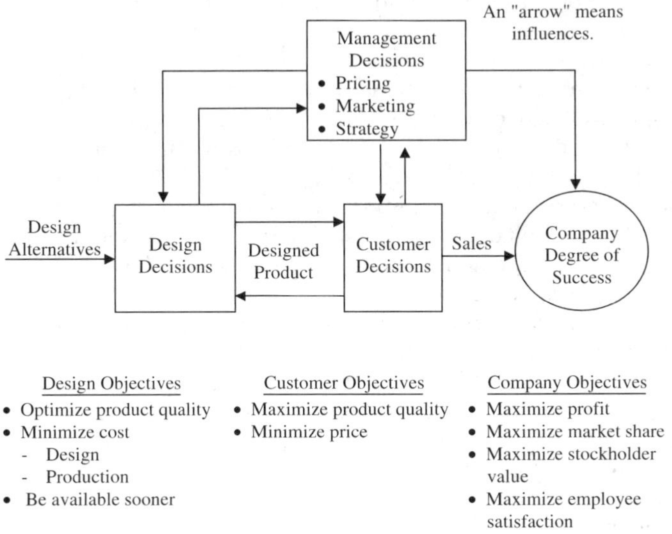

FIG. 7.1 A GENERAL MODEL OF THE DESIGN DECISION PROCESS

Subsequent to selecting a conceptual design, there are many design decisions that eventually lead to a product. For an extensive review of research on product development decisions, see [14]. At the same time, many other company management decisions about pricing, marketing, advertising and strategy influence both the product design and its availability for prospective customers to consider. Each prospective customer then makes the decision on whether or not to purchase the product. Finally, the degree of company success is determined by the profits and market share resulting from the collective customer response and all previous company decisions.

Since a chosen alternative can be no better than the best in the set from which it is picked, we would often be in a better position if we had many alternative potential conceptual designs to choose from. Thus, it is important to generate a set of worthwhile alternatives for conceptual designs. Creating these alternatives is the topic of this paper.

Given several conceptual design alternatives, they should be systematically compared to select the best one. Many approaches have been suggested to evaluate such alternatives. They range from informal to structured mathematical evaluation [4, 5, 9, 18, 21, 23, 24, 28] and include several new Web-based methods [3]. However selected, if the chosen alternative is better than any other existing designs, you have made a significant contribution.

The literature on design, and other decision processes, gives less attention to the creation of alternatives than to the evaluation of those created alternatives. Existing literature suggests general procedures to create alternatives such as brainstorming, neglecting constraints, using analogies, or progressive techniques [1, 2, 12, 19]. However, some literature has been more specific in suggesting and illustrating that identifying customers' concerns or needs can aid the alternative creation process [6, 10, 11, 17, 25]. For custom products needed to meet very specialized needs, much of the design process can be turned over to actual customers [26, 27].

This paper focuses on the very beginning of the design process, going from "no ideas" to "some ideas," which hopefully include "some potentially great design concepts." The approach first identifies customer values as broadly and deeply as we can, using in-depth personal discussions. Then, these values are organized and structured to provide a basis to stimulate the creation of design alternatives. Several procedures are described to facilitate such creative thought processes. The approach is illustrated with two cases.

7.1 DESIGN DECISIONS

What constitutes an excellent design process? Answer: one that results in higher-quality products that are cheaper to design and produce and available sooner. The less a product costs to develop and produce, the better it is from the company's viewpoint. This allows it to be more competitive and to make more money. Also, other things being equal, it is preferred to have the product available sooner. If for some reason it was desirable to hold back introduction, this could be done. On the other hand, you can't go ahead with production or implementation when the product is not ready.

Quality of a design is difficult to define, as quality is naturally in the eyes of the beholder. From the designer's perspective, quality should mean those features and aspects of the product that are more highly valued by potential customers. Hence, it is the customers' concept of quality that is fundamental, and from this we derive the implications for quality in the design process.

So how can one obtain the definition of "quality" for a specific potential product? The answer is simple: you ask prospective customers what aspects of the product are important to them. Product quality is determined using the values of prospective customers. It is their values that count, because their values are the basis for their choice to purchase or not. To have a quality product, you need a great design. Design quality is determined by balancing design objectives, but these objectives must be recognized as means to a great product.

An individual's responses indicate what is important about a conceptual product and represent his or her values regarding the product. In this sense, values refer to any aspects of a potential product that could influence the likelihood that customers would purchase it. These values may be stated as features, characteristics, needs, wants, concerns, criteria or unwanted aspects. What is critical is that the designers can understand the meaning of each of the values.

To adequately define quality in a specific case, you should interview a number of prospective customers, say between 1 and 1,000 depending on the product, to determine values. You want to stimulate customers to think hard about their responses with probing questions. These questions might be as simple as "How?" and "Why?" For instance, if you are designing a cellular phone, one prospective customer may say that safety of phone use is important. You might inquire about how safety is influenced by the design. The prospective customer may respond that the necessity to look at the telephone to enter a phone number is a distraction. This suggests that another value is to minimize distraction in using the phone.

Another value for a wireless phone might be that it has voice mail. You should ask "Why does this matter?" and the response may be "for convenience." This would suggest that there might be other important aspects of convenience. One value might be that the ringing of the phone should not interrupt some event. An implication is, of course, that a design feature that can switch off the ringer might be desirable. On the other hand, since one may not want to miss phone calls, it might be useful to offer a vibration alert for an incoming call and have caller identification, so one could see if one wishes to answer it.

The intent of the interview process is to come up with as complete a list of customer values to define quality as one can. Then one goes through a logical process of examining these values to suggest possible design features that influence quality. This process hopefully creates a rich (i.e., large and diverse) set of potential design alternatives to choose from. The design process of selecting the design alternative to develop is separate and should follow only after a rich set of alternatives is established. If the creative process of coming up with designs is combined with the evaluative process of eliminating the less desirable ones, the process of creating alternatives is stymied.

This paper develops an explicit approach, grounded in common sense, to elicit values from customers and use them to create design alternatives. Two illustrations are first presented to provide a background for the general approach and procedures that follow. Section 7.2 presents a case involving cellular telephones, and Section 7.3 presents a case involving wireless communication plans. With these cases as background, Section 7.4 then presents the procedures for eliciting, understanding, and organizing customer values, while Section 7.5 presents the procedures for creating innovative design alternatives based on those values. Conclusions follow in Section 7.6.

7.2 CELLULAR TELEPHONES—A CASE STUDY

The cellular telephone market is dynamic and competitive. It is a fast-changing field with new designs being introduced regularly. By eliciting and structuring customer values, one can provide useful insights to guide the process of creating potentially successful designs.

In early 2000, I elicited values of six very experienced cellular telephone customers. These six were extremely knowledgeable about the desires of cell phone customers in general. At the time of the elicitations, they were the founders, chief technical officer and sales staff of IceWireless.com, a small Internet firm that provided small- and medium-sized companies with a software product to help each of their employees select a cell phone and wireless plan consistent with the company's policies and individual needs. At the time, I was the vice president of decision sciences at IceWireless.

Separate discussions of 30 minutes to an hour were held with each individual. I first asked each individual to write down everything that prospective customers might value about a cellular phone. When finished, usually after about 10 minutes, I used the initial responses to expand their list of values and to better understand each stated value. For each stated value, such as button size, form factor, durability, popularity and number of characteristics displayed, I probed the individual's thinking with questions such as: "What do you mean by this?" "Why is it important?" "How might it be measured?" "How might you achieve it?" The responses often suggested additional values that were subsequently probed in detail. The result of each discussion was a list of all the values that the individual could think of that might be relevant to a customer wanting a cellular telephone.

I then created a combined list of values. The individuals' lists naturally had much in common, but each individual also had some values that were not on other lists. The next step was to organize the combined list of values into categories (i.e., major values) and to identify the means-ends relationships among them. This facilitates the identification of possibly missing values and enhances the foundation to stimulate the identification of creative design alternatives.

The major values of cellular phones are shown in Fig. 7.2. It distinguishes between the values corresponding to customer objectives and the design objectives that were depicted in the general model of Fig. 7.1. Each of the major values in Fig. 7.2 is specified in much more detail in Table 7.1, which lists component values.

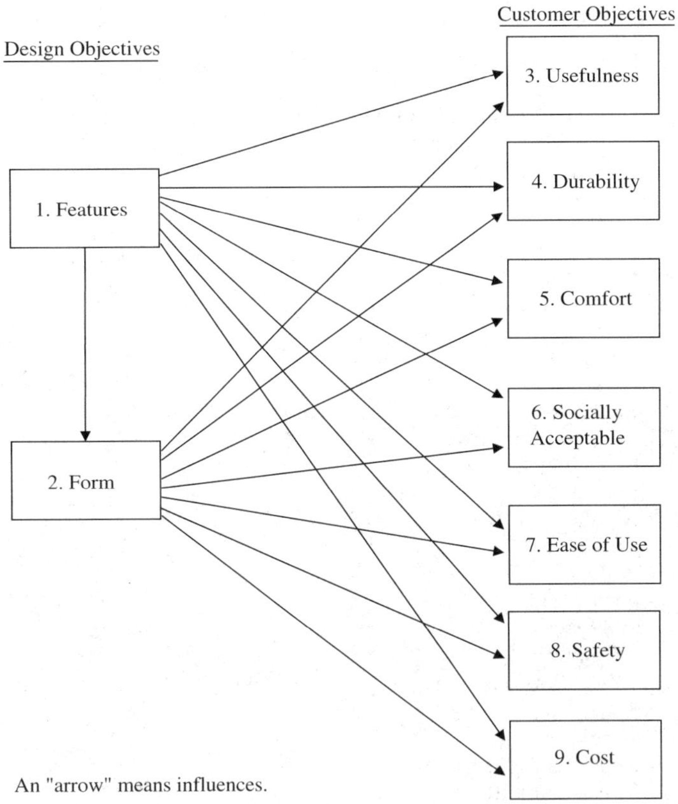

FIG. 7.2 RELATIONSHIP AMONG MAJOR VALUES FOR CELLULAR TELEPHONES

It is the information in Fig. 7.2 and Table 7.1 that provides the basis for creating design alternatives.

7.2.1 Creating Cellular Telephone Design Alternatives

A potentially better design is one that achieves at least one of the values in Table 7.1 better than existing alternatives. Hence, to stimulate the creation of design alternatives, for each value we ask, "How can we better achieve this?" As simple as this sounds, it is often subtle to implement in practice and, of course, getting the set of values corresponding to Table 7.1 isn't necessarily easy. Let us illustrate the creative process with some examples.

The value "durable—not easy to break" clearly suggests a range of design options to build a telephone out of stronger materials. These stronger materials may of course affect the weight of the telephone and its cost. All of this is important at the evaluation stage of potential designs, but here we are trying to generate creative design alternatives.

Consider the value under "usefulness—enhance voice communication" that refers to storing recent incoming phone numbers. One may ask why this is important. Some customers may state that it is important to have numbers available for return calls. This would suggest a design alternative that kept track of incoming phone numbers. As most phone calls are likely from friends and associates, a device that keeps track only of phone numbers not already in the directory might be smaller and lighter than a device that kept track of all recently used phone numbers.

Ask why vibration alert under the "usefulness" value is important and we find that one value is not to disturb people in situations such as concerts or business meetings. We can ask whether there are other ways to signal incoming calls that do not disturb others. One way might be to have a light on a pen or finger ring that would signal the call. Another question might pursue other situations where it is important not to disturb people. We have all been in an airport lounge or elevator where someone is speaking loudly and seemingly unaware that he or she is disturbing others. This suggests design alternatives that allow the person to talk less loudly and yet be heard. For those who don't perceive that they are disturbing others, a sophisticated phone could signal the speaker with a beep when the decibel level got higher than some level that was chosen by the user.

The e-mail value under "usefulness" implies that the feature of a screen is needed and suggests that the screen size is important. Bigger screens may increase the size of the telephone, but are better for e-mail. In pursuing why bigger screens are better, one reason is that it is easier to read the text. This suggests design alternatives that provide larger text on a smaller screen and allow the user to adjust the text size.

Consider the feature of button size. Large buttons farther apart from one another facilitate ease of use, whereas smaller buttons placed closer together allow one to have a smaller telephone, which is easier to carry. Accounting for both concerns, you could design a phone with six larger buttons, each button used for two numbers. Push the first button on the top to indicate 1; push it on the bottom to indicate 2. Alternatively, one might ask, "Why have buttons at all?" since they are means for ease of use and size of the cellular telephone. One could have voice input for telephone numbers and eliminate the buttons altogether, or just have a couple of buttons programmed for special purposes.

Customer values concerning comfort and social acceptability suggest potentially useful research. Research on comfort would investigate what feels good and appropriately fits the hand and face of different classes of potential customers. This research could directly be used to guide the design of alternatives. Regarding social acceptability, research could focus on classes of potential customers, such as lawyers, and pursue their complete set of values. This might lead to a telephone that could better meet their specific needs. For instance, given that many lawyers bill their time in segments involving minutes, a telephone that kept track of the talking time with specified phone numbers might be useful for billing purposes.

Consider "usefulness" at a high level. One might focus on the basic reason for a telephone, namely to talk to another person, and delete many of the other potential features, such as text communication, personal organization, Internet use and games. At the extreme, one might have a telephone similar to an old home telephone. You could make a call or receive a call when you are available, and that is it. Such a phone might be cheaper than existing models and much simpler to operate.

Values concerning "convenience" and "safety" are relevant when using cellular telephones in emergency situations. Some people may only be interested in a cell phone for such purposes. A design could allow only outgoing phone calls, or only outgoing calls to some numbers. Indeed, one could create a simple phone with, for instance, five buttons that corresponded to five important emergency numbers. For special circumstances, such as a two-week hike in the wilderness, one might create a disposable cellular telephone analogous to the disposable cameras that are regularly used for special purposes.

Many cellular customers would like to manage (i.e., minimize) their monthly bills. The value of "usefulness—facilitates cost management" suggests many design alternatives. If certain features of your cellular telephone plan were programmed into your phone, it could indicate the cost of a call and its cost components just after completing it. If programmed, it could indicate the full cost of a proposed call before placing it. Then a user could begin to

TABLE 7.1 THE CUSTOMER VALUES FOR CELLULAR TELEPHONES

1. *Features*
    - Has a screen
    - Has easily readable text
    - Has a memory
    - Has a directory
    - Can connect to a computer
    - Is data capable
    - Size
    - Weight
    - Talk time
    - Stand by time
    - Battery life
    - Button size
    - Mode (single, double or triple band)
    - Has flip top
    - Screen size
    - Memory size
    - Directory size
    - Number of characters displayed on screen
    - Has a working antenna

2. *Form* (These are often referred to as form factors.)
    - Is fashionable
    - Is slick
    - Is thin
    - Is shiny
    - Comes with colored face plates
    - Looks good

3. *Usefulness* (This concerns what functions you want your wireless telephone to be able to perform. The use of having a telephone conversation is assumed and not included.)
    - Enhance Voice Communication
        - Voice mail
        - Two-way radio
        - Allows group calls
        - Has caller ID
        - Has vibration alert
        - Has a speaker phone
        - Has speed dialing
        - Can adjust volume dynamically
        - Has a phone directory
        - Stores recently used phone numbers
        - Stores recent incoming phone numbers
        - Missed-call indicator
        - Has voice recorder
    - Enhance Text Communication
        - E-mail
        - Has alphanumeric paging
    - Facilitate Cost Management
        - Indicate cost of completed call
        - Monitor monthly usage (i.e., minutes in different cost categories)
    - Enhance Personal Organization
        - Personal calendar
        - Put address information in PDA
        - Provide reminders to users
        - Indicates time
        - Has an alarm
        - Has a clock
        - Has a calculator
    - Provide for Internet Use
        - Internet access
        - Web access
    - Has games

4. *Durability* (This is mainly an item for people in professions like construction.)
    - Be rugged
    - Be reliable
    - Not easy to break

5. *Comfort*
    - Feels good
    - Fits face
    - Fits hand

6. *Socially Acceptable*
    - Popular
    - Consistent with your profession
    - Consistent with your position
    - Consistent with your peers

7. *Ease of Use*
    - Simple to program
    - Easy to maintain the telephone (i.e., recharging battery)
    - Easy to use the telephone
        - Regular use
        - Outgoing calls only
        - Emergency calls
        - Special occasions
        - Easy dialing
    - Easy Access
        - Belt clip
        - Fits in pocket
    - Hard to Lose

8. *Safety*
    - While driving a vehicle
    - From regular use
    - In emergencies

9. *Cost*
    - Cost of phone
    - Cost of accessories

internalize the costs of calls. Also, the phone could keep track of monthly minutes used and/or minutes left in time periods that had additional expenses after the plan minutes (e.g., 300 prime-time minutes per month) were used.

7.3 WIRELESS COMMUNICATION PLANS—A CASE STUDY

To use a cellular phone, a customer must select a company and a plan for telephone service. The plan specifies the services provided and the price. It is a design decision that leads to each plan. We are concerned with the process of creating potential alternative plans to consider in that design decision.

Wireless communication plans are generically different from cellular telephones in several respects. First, a plan is a service (i.e., an intangible product), whereas cellular phones are tangible physical products. Second, the customer purchases the cellular telephone, but typically signs up for a plan. Third, the customer then owns the telephone, but uses the plan. Even with these differences, the same concepts to stimulate design alternatives for cellular telephones are useful for stimulating design alternatives for wireless communication plans.

In the same time period that I elicited values for cellular telephones, I assessed customer values for wireless communication plans from the same six individuals. The same process as described in Section 7.2 was followed.

The major values of a communication plan are shown in Fig. 7.3, which distinguishes between values relevant to design objectives and customer objectives. Component values of those major values are listed in Table 7.2. The structure in this figure and table indicates two interrelated decision contexts concerning the quality of a wireless communication plan: One involves building the network to support wireless communications. Decisions about the network affect what communication plans are technically feasible, the quality that customers receive and the price that they pay for using those plans. The other decision context concerns the quality of non-telephone service provided in conjunction with various plans. Decisions about these services affect customers' choices about whether to sign up for a plan as well as the company's bottom line.

7.3.1 Creating Wireless Communication Plan Alternatives

Using the values listed in Table 7.2, we can stimulate the creation of numerous potential alternative plans. This is illustrated with several examples.

Consider the value "coverage." For a customer to use a cellular phone in a particular area, the company needs adequate capacity for the network in that area. Decisions about capacity concern the design of the network and not the design of specific plans directly. Related to coverage in an area is the issue of blocked calls,

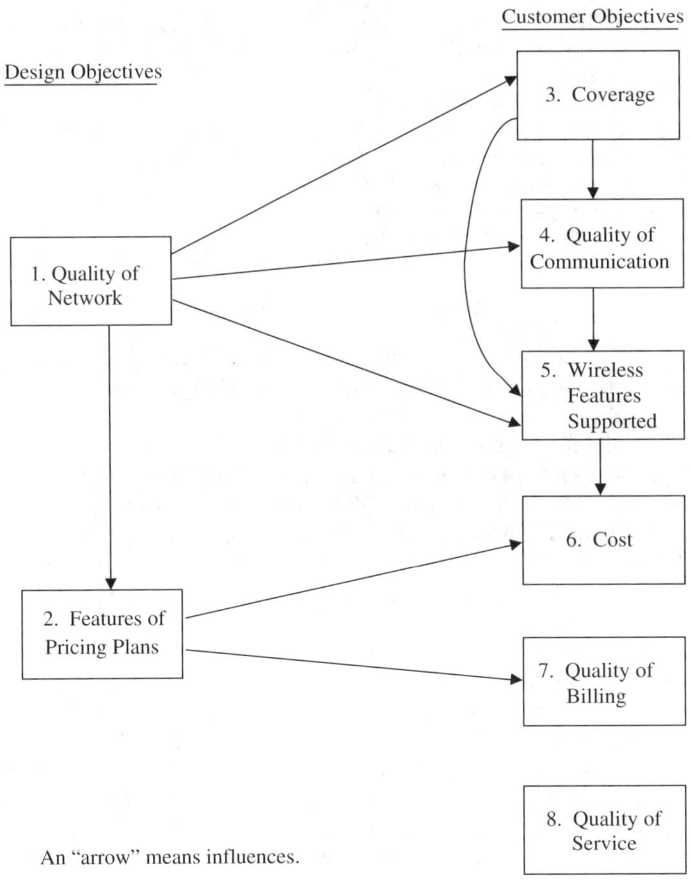

FIG. 7.3 RELATIONSHIP AMONG MAJOR VALUES FOR WIRELESS COMMUNICATIONS PLANS

described under the "quality of communication" value. Blocked calls result from high demand beyond the ability of the network to provide for them. A simple analysis may indicate that the major blockage problems occur between the hours of 5 and 7 p.m. To reduce peak-load telephone traffic and thus blocked calls, a design feature of plans might include peak-load pricing: cheaper rates during off-peak hours and/or higher rates from 5 to 7 p.m. Another alternative might try to promote short calls during that time period. For example, there could be a surcharge for each call over five minutes in high-capacity areas during peak hours.

Concerning "quality of billing," different customers may want their bills organized in different ways. A business person may have one cellular telephone for both business and personal use. It may be helpful to have the bill sorted by a predetermined list of phone numbers of business clients, personal friends and other. Then only the category "other" would need to be examined for billing purposes, which may save the customer time and effort.

With the complexity of all of the features of pricing plans, it is often difficult to decide on the best plan and to understand the complete bill each month. To simplify, a new plan could eliminate all special features and offer unlimited service in the United States for a fixed price of say $150 per month. A different type of alternative would be to put several existing plans in a "basket" plan. Each month, the company would determine which of the plans in the basket would lead to a customer paying the lowest price and then bill them using that plan. This would alleviate the anxiety of individuals in choosing a plan and reduce the irritation of paying for something that they didn't get if they underused the prescribed service, or paying very high prices if they used the service more than they had intended.

Consider the objective of the company to maximize profits. Components of this are to minimize billing expenses and disputed call costs as well as to minimize uncollectable charges (i.e., customers that default). Associated with the $150 per month fixed price, one might simply provide a bill with no details of individual calls, which should reduce billing and dispute costs. Another potential alternative might be to require prepayment in exchange for an overall cheaper communication plan rate. This should reduce the default rate significantly and would also avoid the time, hassle and cost of pursing nonpayment by customers.

7.4 ELICITING AND ORGANIZING CUSTOMER VALUES

There are systematic procedures to elicit customer values and use them to create design alternatives (for example, see [7] and [11]). This and the following section outline the procedures developed for use in the cases discussed in Sections 7.2 and 7.3. Here, we present procedures to elicit and organize values by considering four interrelated issues:

- Who to gather customer values from
- How and how many individuals to involve
- What the substance of the interaction should be
- How to organize the resulting information

7.4.1 Who to Gather Customer Values From

To gather customer values, the general principle is to ask people knowledgeable about customer values. If customer values are provided by many people, each need not be knowledgeable about all customers or all values of some customers.

For existing products, the obvious people knowledgeable about customer values are customers. If you can question customers about

TABLE 7.2 THE CUSTOMER VALUES FOR A WIRELESS COMMUNICATIONS PLAN

1. *Quality of Network* (There are relationships among listed values. For instance, the number and location of towers affects the capacity of and dead spots on the network.)
   - Number of towers
   - Location of towers
   - Capacity of the network
   - Dead spots on the network

2. *Features of Pricing Plans* (Features refer to all of the items that can have an effect on the overall cost of the wireless communications plan.)
   - Minutes included in access fee
   - Additional costs for peak-time minutes
   - Additional costs for off-peak minutes
   - Roaming charges
   - Long-distance costs
   - Incoming minute charges
   - Round-up policy for length of calls
   - Pooled minutes
   - Shared minutes
   - Corporate discounts
   - Volume discounts
   - Parameters (local, regional or national plan)
   - Protected usage (not easy to misuse)
   - Contract length
   - Cost of changing plan

3. *Coverage*
   - Cover personal usage area
   - Where individual lives
   - Where individual works
   - Between individual's workplace and home
   - An individual's building
   - Areas traveled in by individual

4. *Quality of Communication* (The quality of communication felt by an individual is mainly a result of the quality of the network and coverage.)
   - Sound clarity
   - Blocked calls
   - Blurred calls
   - Dropped calls

5. *Wireless Features Supported* (These pertain to the functions that can be performed via the wireless telephone using the wireless communications network.)
   - Voice mail
   - E-mail
   - Internet access
   - Caller ID
   - Paging
   - Digital and analog applications
   - Two-way communication

6. *Cost* (These costs are those that relate to the customer's usage.)
   - Monthly cost of wireless communication usage (averaged over some appropriate period of time)

7. *Quality of Billing* (The bill should be functional for the company and categorize various costs in any way that's useful to the company.)
   - Ability to read bill
   - Ability to comprehend bill
   - Aggregate billing for employees in a company
   - Breakdown in billing
     - By users
     - By cost center
     - By region
     - By use (e.g., e-mail versus telephone communication)

8. *Quality of Service* (It should be noted that aspects of billing could be considered quality of service, but I dealt with it separately above.)
   - Minimize time to order
   - Minimize time to set up communications (i.e., have your wireless communications ready for use)
   - Reorder ease
   - Ease in changing the plan
   - Provides desired electronic reports

their values, this is very useful. For these products, asking prospective customers about their values may provide values different from those of existing customers. If they had the same values, they could have become customers. For products that do not exist now, there are no current customers, so potential customers should be interviewed. For advanced technological products, von Hippel [25] pioneered the idea of using "lead users" of the product.

There are groups of individuals other than customers and prospective customers with very useful knowledge about customer values. These include people in the businesses that make and sell the general product of interest. Such people are in sales, marketing, management and engineering. Individuals from each group may have a different perspective, which is useful for developing a comprehensive list of customer values.

7.4.2 How and How Many Individuals to Involve

Deciding how and how many people to involve in providing customer values are strongly related. The "how" part always involves asking individuals to develop a list of customer values and then asking them to expand their lists. The process can be carried out with or without a facilitator and done either individually or in groups. The intent is always to help each individual to develop a written list of all his or her knowledge about customer values.

Except for the fact that personal interviews take more time and are more expensive, the ideal is for a facilitator to interact personally with each individual. The facilitator can deeply probe an individual's knowledge and do the work of writing it down. This frees the individual to just think. The substance of such an interview, discussed in the next subsection, provides the model that less-intensive approaches try to follow.

When personally interacting with a group, the facilitator asks many questions to help each individual separately record his or her ideas about customer values in written form. If one does not directly interact with individuals, paper or electronic questionnaires can guide participating individuals in providing a written list of customer values. On the Internet especially, the questionnaire can be dynamic in pursuing the thinking of a participant based on his or her previous responses.

The number of people to involve in providing customer values depends on the time and money available, the usefulness of the information and how the individuals are interviewed. When the lists of values being provided by additional individuals do not include any new customer values, enough people have been interviewed.

In general, it is useful to interview at least 5 and up to 50 individuals to begin to understand the range of customer values [6]. This group should include people with potentially different perspectives to enhance the likelihood that a combined list of values will cover the full range of values.

With knowledge of this combined list, you can conduct any subsequent discussions with groups more intelligently. You can also design written and Internet questionnaires to productively gather more information about customer values. With an Internet questionnaire, you can ask a very large number of individuals about customer values and automatically update the combined list as new values are provided.

7.4.3 The Process of Gathering Customer Values

Generating the initial values from individuals is a creative process, as you are going from nothing to something. The general idea is to help an individual to think hard and express everything in his or her mind that matters about the product. You first explain that you want a list of everything that they care about regarding the potential product of interest (e.g., a cellular telephone or wireless communication plan). You begin by simply asking them what it is they value or want or don't want in this context. After they have initially exhausted their thoughts, you begin to probe broader and deeper.

There are numerous devices from the marketing literature [3, 5, 24] and the decision literature [11] to facilitate thinking more broadly. If the individuals currently have the product, you ask them about problems and shortcomings they have experienced or features that they might like to have. You might ask individuals to identify as many situations as possible where they might use the product. For each situation, ask them what is important about that use. You may ask them to consider specific alternatives, hypothetical or real, and ask what is good or bad about each. Any questions that stimulate thoughts of the individual about product values are useful.

The process of deepening our understanding of one's values involves inquiring why the individual cares about each item on the list, and how one can influence the achievement of each item. Asking why provides reasoning for a means to an ends relationship. Asking how provides the reasoning for an ends to a means. With a cellular phone, an individual may say that easy-to-use buttons are valued. Asking why leads to the response that it reduces errors in dialing and the attention needed to correctly dial. Asking why reducing errors matters leads to avoiding unnecessary costs and wasting time. Asking why these matter, the individual may simply say, "because they are some of the things that I care about". This suggests that the latter two values are fundamental customer values in this situation. Asking how one can influence the value of easy to use buttons, the individual may state, "make the buttons bigger and further apart". Each of these values suggests potential design alternatives.

The process described above is for a facilitator interviewing individuals one at a time. When a facilitator interacts with a group, it is not possible to go into the same level of depth. You try to provide personal attention to push deeper thinking of individuals without loosing interest of other group members. With questionnaires, because it is easier to involve large numbers of individuals, you may identify some completely missing values that may provide insights for creating products. In such cases, it may be useful to discuss these new values in subsequent personal interactions with the same or other individuals to increase your understanding of these values.

7.4.4 Organizing Customer Values

Once you have obtained lists of customer values from several individuals, the lists should be combined. This is a straightforward process: First, put all items on any individual list on a common list. Then eliminate duplicate values. If the same words are used for a value, this is trivial. If the words are similar, such as "large buttons" and "big buttons," then select one word and combine these. In more difficult cases, you might need to decide if "readable type" and "large type" mean the same thing. In this case, I'd reason that large type is a means to readable type and keep them both on the list. Finally, combine at the detailed level. For values like "ease of use" or "simplicity," keep them separate at this stage, as they can be aggregated later if appropriate. For stimulating creative designs, potential redundant stimulants are not a shortcoming.

The combined list of values will contain items in many different forms. Some might be considered criteria, interests, measures, alternatives, aspirations, concerns, goals or objectives. The list will include nouns, verbs and adjectives. To better understand the list of values and to enhance its usefulness, it is important to develop consistency. This is done by converting each item on the list into an objective. An objective is something that is desired and can be stated using a verb and a noun. For instance, if "phone number storage" is on the list of values, the corresponding objective might be "maximize size of phone directory." If "keep costs under $200" is on someone's list, this might be converted to "minimize cellular telephone cost." To reduce clutter, several verbs that are obvious were deleted from Fig. 7.2 and 7.3 and Tables 7.1 and 7.2.

It is useful to understand the relationships among different customer values. Specifically, one cares about means-ends relationships [16]. Examining the full list of values will help identify many of the means-ends relationships. Others can be made apparent by asking how and why questions for each of the objectives now on the list. At this stage, we would expect that most responses to these how and why questions would lead to other values already on the master list. If not, they should be added.

It is often useful to aggregate closely related values by making them components of a major value. The cases illustrated in Sections 7.2 and 7.3 used such an aggregation. For instance, major values for cellular telephones included usefulness, cost and ease of use. When there are many detailed values, as there were in these cases, it is difficult to see the overall picture if all means-ends relationships are illustrated. Demonstrating the relationships among aggregated major values can help one understand the entire value structure. This provides a better foundation for creating potential design alternatives.

7.5 CREATING DESIGN ALTERNATIVES

Using values (e.g., wants and needs) to create design alternatives is generally accepted as a useful thing to do. But exactly how should you use those values? The cases discussed in Sections 7.2 and 7.3 illustrated the use of values to identify several possible design alternatives. From these, it is useful to learn the general principles used in the creation process. Examining the illustrated cases suggested the general procedures, which are organized into the five categories listed in Table 7.3. Alternatives created in

TABLE 7.3 GENERAL PROCEDURES FOR STIMULATING THE CREATION OF DESIGN ALTERNATIVES USING VALUES

1. Use Values Directly
   - Use individual values
   - Use combinations of values
   - Use company values

2. Use Means-Ends Relationships
   - Pursue means to ends value relationships
   - Pursue ends to means value relationships

3. Tailor Alternatives to Individuals' Values
   - Segmentation
   - Personalization

4. Combine alternatives
   - Combine features of different products
   - Allow choice of product after use

5. Expand Range of Product Alternatives
   - Segment by stressing only some customer values
   - Create a new product dimension that customers value

Sections 7.2 and 7.3 are used below to better describe each general procedure. Many of the examples that illustrate one procedure might also be considered to illustrate another procedure to create design alternatives. Such redundancy is not a shortcoming of the creation process, as the purpose is to create as many good potential alternatives as possible.

7.5.1 Use Values Directly

A straightforward way to create potential design alternatives is to use the individual values of customers. A simple case regarding telephones concerns the value of having e-mail. The designs in this case are simply to have it or not. Regarding the single value concerning button size, there is a continuum of potential button sizes that can be considered for design alternatives. There is also a continuum of the distance between buttons that could be considered and a continuum of the button height that can be considered.

An example concerning combinations of values relates to storing numbers of recent incoming calls. Because of other values concerning the size, weight and cost of the telephone, it might make sense simply to store only numbers that were not in that telephone file already. Regarding the use of major values, one customer value concerns the cost of the plan. An alternative might be to provide a plan for $150 a month that covers all use within the United States.

One company objective of cellular plans is to maximize profit. Aspects that contribute to profit by decreasing costs involve printing and sending detailed bills and having to write off customers who don't pay. Design alternatives that involve prepayment and little detail on the bill are examples of design alternatives based on this company objective.

7.5.2 Use Means-Ends Relationships

The usefulness value and the desire to have e-mail lead to the design values of having a large screen size and easily readable text. These values, which are a means to usefulness, suggest a design alternative of larger text. Indeed, a dial could allow the user to vary the text size depending on circumstances.

One can pursue the ends values of any stated customer value. An example concerns the desire to have a vibrating alert, which eventually leads to the desire not to disturb others as an end. Examining other situations where people might be disturbed involves cellular phone speakers in crowded quarters and circumstances where quiet is desired. A design alternative that indicates when the speaker is talking above a certain decibel level was developed from this value.

7.5.3 Tailor Alternatives to an Individual's Values

By examining sets of values, one can find grounds for segmentation in creating potential winning designs. For instance, certain classes of prospective cellular phone users might want them only for emergencies or for special occasions, like vacations. This led us to the ideas of very simple telephones with only five buttons for emergency uses that would be associated with a cheaper price and cheaper service plan. It also led to the idea of a disposable cellular phone, similar to a disposable camera, that might be used only for special occasions.

Personalization is difficult for tangible products, but less so for service products. Using the value of an individual who might want a specific type of bill for a wireless phone service, the suggestion of a bill that distinguished groups of telephone numbers into a business category, a personal category and others is an example of a personalized product that could be developed.

7.5.4 Combine Alternatives

Combining alternatives can often create another alternative. One way is to combine features of different products. Another is to allow the customer to use a general product and then choose the best one. Both phones for emergency use only and disposable cellular phones were discussed above. One could obviously combine these into a disposable emergency phone. Risky endeavors of different kinds from remote adventure travel to a two-week stay in a hospital would be situations where such a phone may be useful. In the former case, a global positioning system that automatically communicated the location of the caller might be included in one design alternative.

With service products, it may be useful to design a combined product that allows the customer to choose the eventual product only after use. For instance, many wireless communication companies have numerous plans, but it's very difficult for an individual to decide which one is the best for his or her use. A combined alternative is a basket plan that works as follows: Each month, each of the plans in the basket would be used to calculate the price an individual would pay were that plan in effect. Then, the price charged would simply be the minimum of those monthly costs calculated from the plans in the basket.

7.5.5 Expand Range of Product Alternatives

One can stress some customer values at the expense of others to create alternatives. The new alternative might have great appeal to a segment of the potential customers [13]. For example, consider the ease-of-use values for cellular telephones. Ease of use clearly means different things to different people. For some people, all the features included on most phones simply make them difficult to use. For such individuals, a simple cellular phone that works in a manner similar to that of the standard telephone used in a home might be desirable. You could answer it if you were there and the phone rang, and you could call someone. Otherwise, you wouldn't use it. Such a phone would be different from many cellular telephones and would distinguish it on the dimension of ease of use.

If you can create a new product feature that has value to some customers, it might be extremely useful for selling your product. For instance, suppose a cellular phone was automatically set up to also ring on your residence or office phone, or at other locations that you might be at. This would provide the potential to always be in contact via telephone. For some this might be a nightmare, but for others it could be very desirable. If one could preprogram a cellular phone such that this simultaneous ringing occurred only for a predetermined set of incoming phone numbers, it might become a much more desirable feature. For instance, if one had a relatively incapacitated relative, they might have the confidence that they could always reach someone if necessary, and that might be very important.

7.5.6 Process Suggestions

A common way that the procedures described above might be used is within a design team. The "science" of the process was discussed, but there is "art" to the process as well. A few suggestions may be helpful.

The general guideline is that you want each team member to think individually and develop his or her own ideas initially. Later, these can be discussed, combined and used to stimulate additional thinking. Each team member should first expand the set of customer values. Then he or she should create a list of alternatives using any of the general procedures described above.

Two big pitfalls to avoid are evaluating alternatives and focusing on the small picture. The intent is to create alternatives. Any evaluation should be a separate process that comes later. If individuals begin to evaluate alternatives prematurely, it will severely stifle the creative process. One can also be bogged down in a single objective like "button size": Are small buttons better because they allow for a smaller and lighter phone or are large buttons better because they are easier to use and avoid misdialing? Attempting to resolve such issues is part of evaluation and discussion. Such details also inhibit creativity. Just continue to focus on creating potential phones that are small, light, easy to use, and let you dial accurately while in the creative process.

CONCLUSIONS

The intent of this paper is to suggest a sound practical approach to stimulate the development of design alternatives. If you ask the question, "Why do I care about the design of a product?" the answer must be "Because I want a high-quality product." The notion of "quality" is one of value. The purpose of design is therefore to increase value. Hence, to guide the design process, it makes sense to begin with the values that you hope to achieve.

This paper presents and illustrates procedures to elicit values for potential products from individuals and then use these values to stimulate the creation of alternatives. The intent of the illustrations is to indicate that this is not a theoretical approach, but an extremely practical approach. In stimulating creativity, it is not complex mathematical or scientific skills that are required. Rather, it is the willingness to systematically apply common sense and pursue thoroughness in expressing values. The technical skills simply involve making and organizing lists of values.

Once you have the complete list of values, we suggest many different procedures to use these values to create alternatives. In this process, there has to be that spark of insight for the "aha" always present in creative processes. So if you still need that creative spark, what is so special about this approach? The difference is, the creative spark does not start from nothing. It starts from the list of stated values, and the jump from there to a conceptual product design is not as great as the jump from no organized structure of what matters in a particular design situation to a proposed design alternative. Also, since the set of values lays out the landscape of all that is valued regarding a potential design, we have a more complete space to stimulate our thoughts. This should stimulate a larger set of potential design alternatives from which to choose. It's simply a truism that if you have a much richer set of alternatives from which to choose, it is likely that some of these are much better than the best alternatives in a small set of alternatives.

An interesting question is whether the described studies were used. They definitely were used to create products, but not exactly the products described. As mentioned at the beginning of 7.2, IceWireless was a small Internet firm whose business was to help individuals in companies select a cell phone and wireless plan. Our (i.e., IceWireless) software products were decision models that allowed individuals to compare potential products in terms of the set of objectives they felt were important. These software products—one for cell phones and one for wireless plans—let individuals select the relevant objectives from the lists in Tables 7.1 and 7.2. They also selected the set of alternatives that had any appeal, and our decision models then helped them systematically zero in on better choices.

REFERENCES

1. Ackoff, R. L., 1978. *The Art of Problem Solving*, John Wiley & Sons, Inc., New York, NY.
2. Adams, J. L., 1979. *Conceptual Blockbusting: A Guide to Better Ideas*, W.W. Norton & Company, New York, NY.
3. Dahan, E. and Hasuer, J. R., 2002. "The Virtual Customer," *J. of Prod. Innovation Mgmt.*, Vol. 19, pp. 332–353.
4. Green, P. E., Krieger, A. M. and Wind, Y., 2001. "Thirty Years of Conjoint Analysis: Reflections and Prospects," *Interfaces*, 31(3), pp. S56–S73.
5. Green, P. E. and Srinivasan, V., 1978. "Conjoint Analysis in Consumer Research: Issues and Outlook," *J. of Consumer Res.*, 5(2), pp. 103–123.
6. Griffin, A. and Hauser, J. R., 1993. "The Voice of the Customer," *Marketing Sci.*, 12(1), pp. 1–27.
7. Hauser, J. R. and Clausing, D. P., 1988. "The House of Quality," *Harvard Bus. Rev.*, 66(3), pp. 63–73.
8. Hazelrigg, G. A., 1998. "A Framework for Decision-Based Engineering Design," *J. of Mech. Des.*, Vol. 120, pp. 653–658.
9. Hazelrigg, G. A., 1999. "An Axiomatic Framework for Engineering Design," *J. of Mech. Des.*, Vol. 121, pp. 342–347.
10. Jungermann, H., von Ulardt, I. and Hausmann, L., 1983. "The Role of the Goal for Generating Actions," *Analyzing and Aiding Decision Processes*, P. Humphreys, O. Svenson and A. Vari, eds., Amsterdam, North Holland.
11. Keeney, R.L., 1992. *Value-Focused Thinking*, Harvard University Press, Cambridge, MA.
12. Keller, L.R. and Ho, J. L., 1988. "Decision Problem Structuring: Generating Options," *IEEE Trans. on Sys., Man, and Cybermetrics*, Vol. 18, pp. 715–728.
13. Kim, W. C. and Mauborgne, R., 1997. "Value Innovation: The Strategic Logic of High Growth," *Harvard Bus. Rev.*, January-February, pp. 103–112.
14. Krishnan, V. and Ulrich, K. T., 2001. "Product Development Decisions, A Review of Literature," *Mgmt. Sci.*, Vol. 47, pp. 1–21.
15. Lilien, G. L. and Rangaswamy, A., 1998. *Marketing Engineering: Computer Assisted Analysis and Planning*, Prentice Hall, Englewood Cliffs, NJ.
16. Newell, A. and Simon, H. A., 1972. *Human Problem Solving*, Prentice Hall, Englewood Cliffs, NJ.
17. Pitz, G. F., Sachs, N. T. and Heerboth, T., 1980. "Procedures for Eliciting Choices in the Analysis of Individual Decisions," *Org. Behavior and Human Performance*, Vol. 26, pp. 396–408.

18. Rouse, W. B. and Cody, W. J., 1989. "A Theory Based Approach to Supporting Design Decision-Making and Problem Solving," *Information and Decision Tech.*, Vol. 15, pp. 291–306.
19. Shah, J., 1998. "Experimental Investigation of Progressive Idea Generation Techniques in Engineering Design," *Proc., ASME Des. Theory and Methodology Conf.*
20. Thurston, D. L., 2001. "Real and Misconceived Limitations to Decision-Based Design with Utility Analysis," *J. of Mech. Des.*, 123(June), pp. 176–182.
21. Thurston, D. L. and Nogal, A., 2001. "Meta-Level Strategies for Reformulation of Evaluation Function During Iterative Design," *J. of Engrg. Des.*, 12(2), pp. 93–115.
22. Tribus, M., 1969. *Rational Descriptions, Decisions, and Designs*, Pergamon Press, Elmsford, NY.
23. Ulwick, A. W., 2002. "Turn Customer Input into Innovation," *Harvard Bus.* Rev., January, pp. 5–11.
24. Urban, G. L. and Hauser, J. R., 1992. *Design and Marketing of New Products, 2nd Ed.*, Prentice Hall, Englewood Cliffs, NJ.
25. von Hippel, E., 1986. "Lead Users: A Source of Novel Product Concepts," *Mgmt. Sci.*, Vol. 32, pp. 791–805.
26. von Hippel, E., 2001. "User Toolkits for Innovation," *J. of Prod. Innovation Mgmt.*, Vol. 18, pp. 247–257.
27. von Hippel, E. and Katz, R., 2002. "Shifting Innovation to Users via Toolkits," *Mgmt. Sci.*, Vol. 48, pp. 821–833.
28. Yang, J. B. and Sen, P., 1997. "Multiple Attribute Design Evaluation of Complex Engineering Products Using Evidential Reasoning Approach," *J. of Eng. Des.*, Vol. 8, pp. 211–230.

CHAPTER

8

Generating Design Alternatives Across Abstraction Levels

William H. Wood and Hui Dong

8.1 GENERATING ALTERNATIVES IN DECISION-BASED DESIGN

Design is often cast as a process of generating and testing. The evaluation of potential design options effectively steers the subsequent generation of new designs. In turn, the options generated help to shape the evaluations used to choose from among them. This coupling between generation and evaluation is a vital part of the design process and is perhaps the most significant way in which decision-based design can be differentiated from decision-making: Designers are charged not only with selecting the best design options, but also with generating the options from which they must choose.

Human designers use abstraction throughout the design process, guiding the design through a series of representations: the social constructs of the customer/environment from which the design need springs, textual/graphical descriptions of this environment, "black box" functional descriptions of how a design will interact with the environment, detailed descriptions of interacting components that accomplish this interaction, spatial layout of these components, detailed geometry of connections among them, manufacturing/assembly plans for each part, etc. Designers develop and evaluate design alternatives at each of these levels of abstraction before moving to the next; DBD can help to formalize the decision-making within and across abstraction levels, at some points removing candidates from the set of options, at others selecting the best one.

In this chapter we lay out a methodology for generating design alternatives across these design abstraction levels and build on decision-making concepts from prior chapters to help control this process: probabilistic design modeling, value functions, expected value, decision-making under uncertainty and information value theory. Each of these is reframed for the task of conceptual design. In addition, a new concept–design freedom–is introduced to help designers evaluate the impact of their decisions on the diversity of the design candidate set. DBD over multiple abstraction levels is not a simple process of generate and select; it is a set of cascading decisions that refine both the abstract design and the initial requirements used to evaluate it.

8.1.1 Design Generation

As a human process, conceptual design emphasizes on generating a large number of potential solutions—establishing the design space. Brainstorming, a common method for generating design concepts, emphasizes quantity over quality, recommending that judgment of quality be (at least initially) divorced from the generation process in early design. In order to generate as many potential designs as possible, it can be useful to allow infeasible designs—relaxing enforcement of the laws of physics is not only useful in concept generation, it is a brainstorming method unto itself! As designs are developed further, some initially infeasible designs can be brought into line, others might open conceptual doors to feasible concepts and others might simply be discarded.

Individual designers and design teams strive for quantity in the design space, but are limited both by time and their collective experience; computational synthesis can support this process by collecting large amounts of experience and providing rigor in the generation of designs. Choosing the appropriate degree of rigor in synthesis requires a fundamental trade-off between completeness (the ability to generate all possible designs) and soundness (the ability to generate only feasible designs). Early design generation naturally tends toward completeness, sacrificing soundness to ensure that the entirety of the design space be explored. At some point in the design process, soundness must prevail so that the designs that are generated represent real possibilities—design decision-making can easily be distorted by infeasible designs (i.e., designs that don't obey the laws of physics often appear to perform better than those that do). So as we select representations for designs, early, high-level representations should emphasize completeness while later, low-level representations stress soundness. Abstraction mechanisms must be put in place to support the transition among design representations and the synthesis modes that operate on them.

8.1.2 Abstraction

Human designers use various levels of abstraction throughout the design process. In part this is due to the mismatch between design goals expressed in a human context and the concrete artifacts introduced into that context to address the needs of the customer. To ensure completeness, high-level designs are generally abstractions—*function* representing the aspects of artifact *behavior* intended to address the design need. Whereas function is a human construct, behavior is manifested in the physical world. But even here, behavioral models exist at varying levels of abstraction—the ideal speed-torque curve of a DC motor gives way to more realistic models that bring in brush friction, torque ripple, rotor inertia, heat conduction/convection, etc. Finally, behavior comprises individual parts that must be assembled into systems—brushes, plate laminations, bushings/bearings, windings, magnets, etc. Working from the bottom up: physical objects generate behavior that can be modeled; this behavior is generalized into useful relationships among design attributes; these relationships are abstracted into functionality; these functions transform the physical world in ways useful to design customers. However, abstractions useful to humans may not be particularly useful in a computational environment where reasoning processes are vastly different. Humans have difficulty managing large amounts of data in parallel, a strength of computers. This could mean fine-tuning human-designer abstraction for application to computational synthesis, or could mean discarding or adding abstractions where the mismatch between human and computational capabilities is large.

Perhaps the most widely recognized systematic method for design is that proposed by Pahl and Beitz [1]. First, design requirements are separated into those related to performance and those related to function. Functional modeling is separated into two components: *flows*—energy, material and information; and *functions*—transformations of those flows to produce the desired effects. These are related to each other through the *function-structure* boxes containing functions connected to each other by directed lines carrying the flows. High-level functionality is expressed as a "black box" in which all system flows pass through a single box containing all of the desired functions. This black box is then decomposed by separating out flow paths and creating a network of simpler, interacting functions. As decomposition progresses, functions reach a point of specificity at which they suggest physical realizations, called solution principles. These solution principles are then composed into a system and that system's behavior evaluated. At each point in the decomposition, multiple design options are generated.

Functional topology generally determines the idealized behavior of a system; this topology must be realized physically. High-level material and mechanical energy flows have position, direction and orientation that must match system-functional requirements. These spatial relationships and the internal transformations they require help define the physical layout of the functional topology in the configuration design stage. Solution principles along with this physical layout then help to define the connections required to link functions to each other and/or to a common structural backbone (referred to as the functional "ground"). These connections, along with material and manufacturing concerns help determine the shape, processing and assembly of each part.

These transitions—function to solution principle, functional topology to layout, layout to detail—each entail a change of representation and reasoning method. Computational synthesis in such a framework requires not only developing design primitives and composition rules within each level, but establishing methods for translating the results of one level into goals for the next. We now use an example in mechatronic design to demonstrate multi-abstraction computational synthesis.

8.2 EXAMPLE: FORCE-FEEDBACK MOUSE

The basic functional requirements for the design example are situated in the environment of computer use. It is suggested that the user interface of computers might be improved by introducing force feedback into the mouse-pointer interface to allow users to "feel" graphical features as they move the mouse pointer across the screen. Figure 8.1 shows both the black box function structure defined for this design and a rough set of physical requirements on the size and shape of the device, as well as the position and orientation of mechanical energy flows entering and leaving it. Note that even this high level of abstraction embodies a significant number of design decisions: input and output will be manual, the device will map a horizontal surface to the screen, the device will sit on a user-supplied surface, etc.

8.2.1 Customer Needs and Functional Design

Customer needs can be divided roughly into two components: what the design must do (its *function*) and how well, in the eye of the customers, it must do it (its *performance*). The process of developing design requirements from each is iterative and coupled with one another. From the standpoint of computational synthesis, functional requirements provide the guidance necessary to instantiate the generation process, and performance requirements feed into the decision process of determining which generated solutions are worth pursuing in greater detail. We will leave the interplay between function and performance for the latter half of this chapter, where we introduce decision-based methods for controlling computational synthesis. For now, we will focus mainly on abstraction and the generation of design alternatives.

Pahl and Beitz [1] recommend a process of abstraction on the functional side, attempting to identify the core functionality that is required: "customers don't really *need* a drill, they *need* holes." After abstracting the problem, function-based design proceeds with the development of a black box function structure which models the main input and output flows to the system and describes the general transformations that must occur. Our force-feedback computer input device shown in Fig. 8.1 captures motion/force from a user's hand and transforms it into a displacement signal, while information about required force is transformed into force transmit back to the user. Part of generating a force is making it react against a reference surface. Finally, the hand is supported ("clicking" functions are ignored for clarity). Functional design must proceed through several stages of decomposition in which specific solutions are generated to provide these overall system

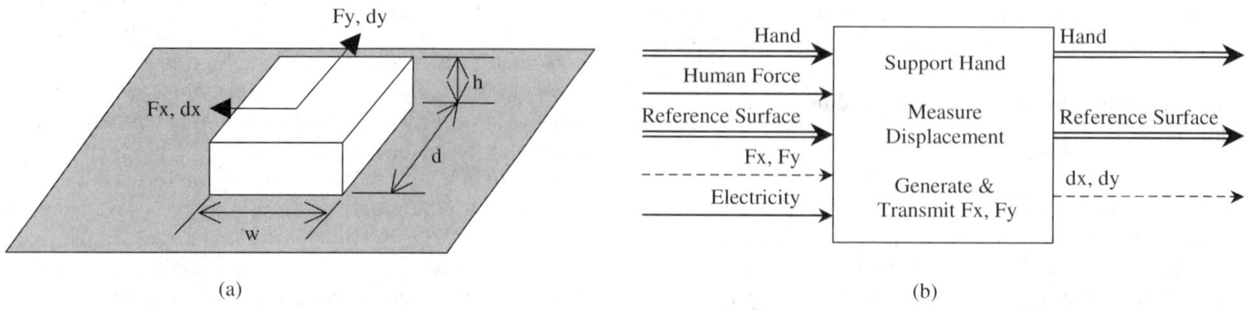

FIG. 8.1 SPATIAL AND FUNCTIONAL REQUIREMENTS FOR THE FORCE-FEEDBACK MOUSE EXAMPLE

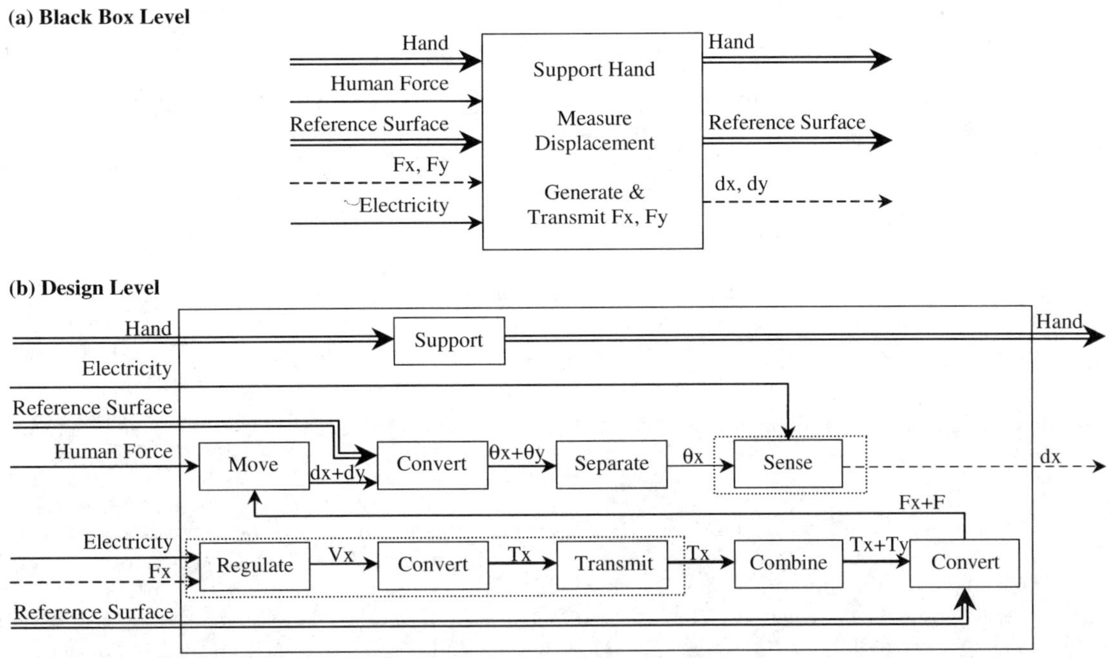

FIG. 8.2 TWO LEVELS OF ABSTRACTION FOR THE FUNCTION OF A FORCE FEEDBACK MOUSE; NOTE: BOXED SECTIONS ARE REPEATED FOR X AND Y AXES

needs. This process is very open-ended: a force-feedback device could transform any number of user motions: arm, hand, finger, etc. Figure 8.2 shows a decomposition for an input device based on a computer mouse; Fig. 8.3 presents the "sense" function of Fig. 8.2 generated by reverse-engineering (see panel for a discussion of reverse engineering) an actual computer mouse.

The lack of rigor in conceptual design challenges computational synthesis. Function-based design as defined by Pahl and Beitz allows the freedom of expression needed in early design, constraining the types of information captured in a functional decomposition (flows and transformations) but not its expression. Computerizing functional information for capture and reuse requires regularizing its expression in some way. Toward this end, [2, 3] propose the "functional basis" a hierarchical controlled vocabulary for both function and flow. Verma and Wood [4] further refine flow representation, augmenting the modeling of flow in key engineering aspects to improve design reuse. They find that augmenting material flows with models for size, shape and basic mechanical behavior markedly improves the ability of a computer system to find designs similar to those in the current context.

Design Primitives With a defined grammar (function-structures) and vocabulary (the functional basis), we could set about generating designs, but this would be like putting together sentences from English grammar rules and a dictionary. Clearly, we could generate all possible expressions (the process is complete) but most of the expressions generated would be gibberish (the process is unsound in terms of communicating ideas). Mitchell [5] establishes the fundamental trade-off: bias in a representation provides greater reasoning power within that representation at the cost of expressiveness (essentially, greater bias promotes soundness at the cost of completeness). In the case of textual expression, biasing the representation toward actual words and a framework for using them results in better performance than the proverbial "million monkeys typing." Further biasing the representation toward phrases that have actually been used in the past might produce even better performance in terms of generating meaningful new expressions. For function generation, we apply even greater bias toward functions and groups of functions that have been useful in the past. To do this, function-structure building blocks (functions and flows combined together—analogous to sentences or phrases in text) are extracted from existing products through reverse engineering as well as from catalogs of standard components, primarily sensors and actuators.

Geitka et al. [6] find abstraction in the modeling of function problematic: The expression of function does not map easily from high-to low-level representations of the same function. Figure 8.3 shows two different ways of accomplishing the same function. In Fig. 8.3(a), an off-the-shelf encoder is modeled; in 8.3(b) the same functions are accomplished by a customized set of parts in a computer mouse: angular displacement is transmitted into the function through a part with a molded-in encoder wheel; this

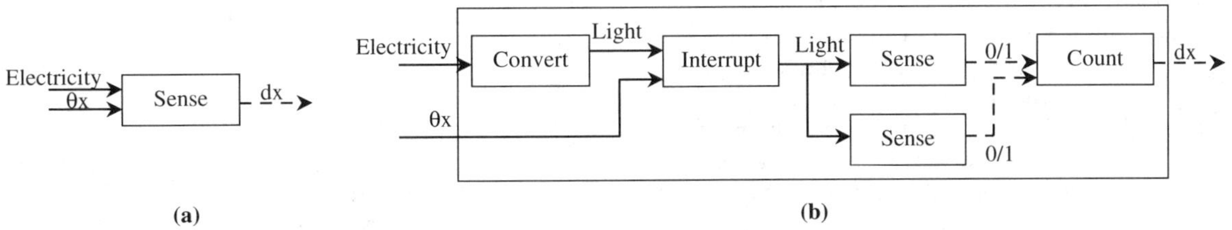

FIG. 8.3 FUNCTION STRUCTURES FOR THE "SENSE" FUNCTION IN FIGURE 8.2(B): (A) ABSTRACT FUNCTION; (B) ACTUAL FUNCTIONAL UNIT IN THE BALL MOUSE, REVERSE ENGINEERED AT THE "PARTS" LEVEL

The Reverse Engineering Process

Because it plays such a basic role in identifying building blocks for the computational synthesis methods presented, a brief supplementary discussion of reverse engineering is appropriate. Consistency of representation is important, ensuring that multiple "reverse engineers" produce similar decompositions for the same product. Toward ensuring consistency, we require that products be disassembled into their distinct parts. Parts are relatively easy to identify—it is straightforward to determine which flows enter and leave a part—and each part also represents a single manufacturing process (or, in the case of an off-the-shelf component, represents a make/buy decision).

For each product, the main input and output flows are identified. The product is disassembled along these flow paths and parts identified. For each part, input and output flows are identified and the part functions induced (a single part may produce multiple functions). The primary focus is on flow and function related to the overall function of the system. Functions that primarily deal with manufacturing are excluded for clarity of decomposition.

In addition to functional decomposition, each product is examined for the presence of standard mechanical solution principles (e.g., the gear systems, linkages, etc.). Salient features of these are captured into the database: maximum force/power/speed of operation, number of parts, bounding box size, etc.

Finally, for each part the material, manufacturing process and its primary and secondary process axes are captured. Connections between parts are identified as the interfaces that produce the degree of freedom required by the overall function. For each connection, the connection matrix and connection-type matrix are stored along with manufacturing information like size, assembly axes (relative to the manufacturing process axes), number of toleranced features and part count (small parts required for assembly/disassembly are folded into the connection definition).

This reverse engineering process produces information useful at all levels of abstraction. In addition, the representations are all indexed to the basic parts identified in the process, so information across abstraction levels can be related to one another readily. Connections can inherit force/power from the solution principle level, relationships between function and size or part count can be induced at high levels of abstraction, etc.

encoder wheel interrupts light passing from an emitter (which uses electricity to generate the light) to a pair of detectors that together detect magnitude and direction of displacement. From the standpoint of design, these two structures represent the same fundamental function; however, to the computer this similarity is hidden in their representations. Such mismatches in functional representation can stem not only from different implementations of the same function, but also from human interpretations of functionality in the reverse-engineering process. Verma and Wood [7] find that functional decompositions of the same artifact vary considerably across reverse engineers unless a rigorous process of part-by-part disassembly and functional decomposition is followed. This focus on parts turns out to be a windfall: It highlights the make/buy decision, which is a critical one in the design process. Still, the mismatch between Fig. 8.3(a) and (b) must be overcome.

Verma and Wood [4] explore an array of methods for aggregating functions to produce high-level representations from reverse-engineering function-structures. Using retrieval performance as a criterion, they find the best performance for a method that focuses on collapsing function-structures along "internal" flows: those flows that are produced and consumed within the system. For example, Fig. 8.4 shows the aggregation of Fig. 8.3(b) into higher-level structures. On the surface, the loss of function information in this aggregation would seem disastrous. However, naming functions in the first place can be problematic; while it is easy to identify and classify the flows in a product and to identify the points where they interact with one another, there remains ambiguity in how to label that interaction. For this reason, the functional basis provides sets of function synonyms to be mapped onto a common functional term, but the definition of function may be even more subjective than just the choice of the term: does a gearbox amplify torque or reduce speed? With function still somewhat ambiguous even in a restricted vocabulary, repetition in reverse engineering is derived from capturing flow information accurately. Thus, the extensions to flow representation in the functional basis intended to enhance design reuse do double duty in replacing function in the aggregated function-structures.

Given sets of decompositions done at low levels of abstraction and methods for aggregating these into more abstract representations, the final choice is the abstractions to be stored. To favor completeness at this highest level of abstraction, it is useful to have a relatively liberal mechanism for retrieving function from the reverse-engineering case base. We use a matching algorithm similar to that used in text retrieval, favoring functions whose input and output flows match the functional query (i.e., the black box function-structure) exactly but also identifying partial matches, ranking them in order of strength of agreement. By storing each reverse-engineered product first as a collection of low-level functions and then creating and storing function aggregates by repeatedly collapsing the function structure along internally generated

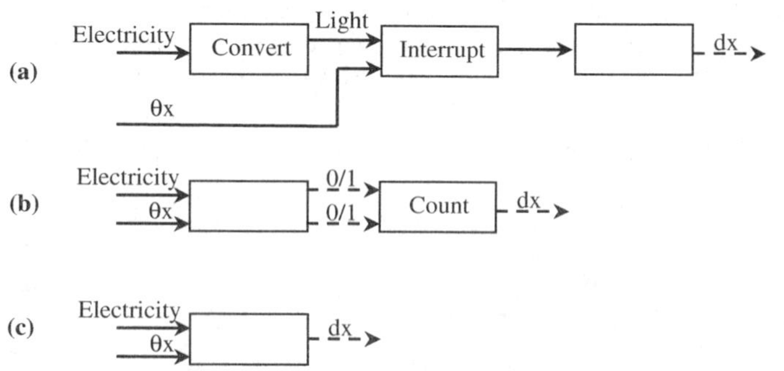

FIG. 8.4 SEVERAL STAGES OF AGGREGATION FOR THE REVERSE ENGINEERING-LEVEL DECOMPOSITION OF FIGURE 8.3(B)

flows, we allow the matching process to automatically find the appropriate abstraction level for a given query input.

Composition Function structures allow virtually any interpretation of function to be represented, but the lack of rigor in the representation hinders logic-based reasoning [8]. This is to be expected: functions structures allow completeness of the design search and this completeness comes at the cost of soundness. For the design primitives, we regain soundness by relying on case experience to bias the search toward pre-existing functions. Because function-structures are stored at multiple levels of abstraction, design queries identify precomposed fragments rather than having to compose functionality from base-level primitives. Still, these higher-level functional fragments must be composed into overall solutions. With flows as the main connection mechanism in function-structures, connecting fragments on common flow paths is the obvious choice. Two options exist: insertion of fragments into the series of a flow or branching the flow to create a new flow parallel to the existing flow. For completeness, both these methods must be applied toward "knitting" together different function fragments along flows. Of some help in composing energy flows is the formalism of "through" and "across" variable definitions. If energy is represented as a through variable (i.e., displacement, current, flow, etc.), then insertion should take place in series; if the energy is defined as across (i.e., force, voltage, pressure, etc.), then parallel combination makes the most sense.

Example Figure 8.2(a) shows the black box function-structure for the force-feedback mouse used to query the case base. The resulting "design" level function-structure of Fig. 8.2(b) is drawn from a combination of a ball-type computer mouse (for most of the structure) with a power screwdriver (for converting electrical energy into force reacted by the hand and a reference surface). This is one of many such possibilities (other computer mouse decompositions include optical or trackball types), and is itself a set of possibilities defined at several levels of abstraction. Composition in this case interrupts the mechanical energy flow path from the computer mouse to insert force/torque transmission functions.

8.2.2 Configuration

Function-level designs are passed to the configuration level where their general structure undergoes refinement to both flesh out the functional topology and satisfy the spatial constraints placed on that functionality. These constraints include flow type (rotation or translation) and flow vector (position and orientation) along with additional performance aspects (power, force, speed, etc.). In addition, the overall system must satisfy overall size and/or shape constraints.

Design Primitives Three separate classes of design primitives are used for configuration design. The first class focuses on the transformation from one class of flow to another, or from one type of energy flow to another. For example, the encoder of Fig. 8.3 transforms an energy flow into information. Electric motors transform electrical energy into mechanical energy. Transduction of a flow across domains generally relies on a physical phenomenon; catalogs of such transducers are readily formed and their performance-based selection (e.g., size and shape relative to mechanical flow direction, power, accuracy, etc.) is driven by rational, decision-based methods [9].

The second class of primitive deals with the physical nature of mechanical energy flows, transforming flow across subdomains (e.g., rotation to translation), changing flow orientation (*xx*-rotation to *yy*-rotation), changing flow magnitude (increase rotational speed) or changing flow position (*xx*-rotation at point 1 to *xx*-rotation at point 2). Toward capturing these transformations, we build on existing representations: Kota et al. [10–12] draw from a catalog of mechanical components, selecting among them based on changes of axis (*x*/*y*/*z*/skew) and motion type (translation/rotation) as well as additional characteristics like linearity, reversibility and continuity; Chakrabarti and Bligh [13–15] use a similar component catalog, but focus more on the spatial relationships among mechanical flows (*x*/*y*/*z* position, required axis offsets, etc.); and finally CADET [16] provides qualitative models among spatially defined flow variables (*x*/*y*/*z* position, velocity, etc.). Table 8.1 shows parallel representations for some common components.

TABLE 8.1 CONFIGURATION DESIGN BUILDING BLOCKS

Functional Unit	Kota et al.	Charkrabarti and Bligh	CADET
Key	In: $[T_x\ T_y\ T_z\ R_x\ R_y\ R_z]$ Out: $[T_x\ T_y\ T_z\ R_x\ R_y\ R_z]$ C: $(cont\ \ lin\ \ rev\ \ [I/O])$	Axes z x y Rotation ⇛ Translation → Inline o Offset o–o	$A = B^+$: A is an increasing function of B
Slider-Crank 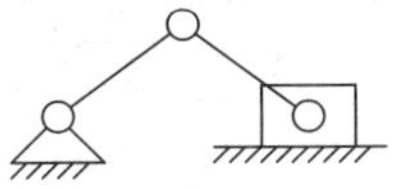	In: [0 0 0 1 0 0] Out: [0 1 0 0 0 0] C: $\left(1\ 0\ 1\ \begin{bmatrix}0&0&0\\0&0&0\\1&1&0\end{bmatrix}\right)$	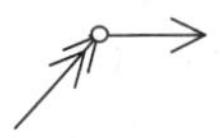	$T_y = R_x^+$
Lead Screw 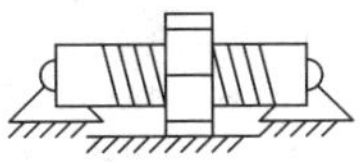	In: [0 0 0 1 0 0] Out: [1 0 0 0 0 0] C: $\left(1\ 0\ 0\ \begin{bmatrix}0&0&0\\0&0&0\\0&0&1\end{bmatrix}\right)$		$T_x = R_x^+$
Rack & Pinion 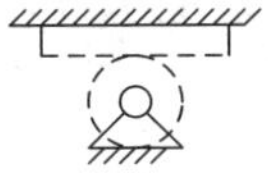	In: [0 0 0 1 0 0] Out: [0 1 0 0 0 0] C: $\left(1\ 1\ 1\ \begin{bmatrix}0&0&0\\0&0&0\\0&0&1\end{bmatrix}\right)$	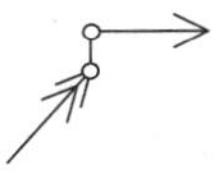	$T_y = R_x^+$

The third class of primitive builds on the flexibility of the CADET representation, connecting the mechanical-energy based representations that provide soundness in composing mechanical flows to less formal functionality, again derived from case experience. These connections, while less formal than those used to compose mechanical components, provide a basic mechanism for joining mechanical energy flows to material flows.

Composition For physical phenomena, selection is the primary reasoning mode. This selection does, however, anchor the composition of mechanical flow elements both spatially and in the orientation and magnitude of the mechanical flows they create or consume. For mechanical flow, primitives are composed first by identifying components whose inputs match unresolved system outputs and whose outputs match unresolved required inputs; those components that match both are favored over those matching just an input or output. As at the functional level, dealing with multiple inputs and outputs creates distinct possibilities in composition: Components inserted along different flow paths can be combined one after the other in a serial fashion, or can work in parallel to achieve the desired functionality. For each topology of oriented components, second-order effects like offset or reversibility help to select from among the candidates. Sets of solutions are built progressively until all flows are accounted for. Finally, rough models are created to ensure basic feasibility; infeasible designs are corrected if possible (e.g., by inserting additional elements) and the set finalized.

Example First, transducers that can generate the desired position information from the mechanical energy flows are selected. Two such devices (encoders and direct optical sensors) are highlighted, although the total number of possibilities is much larger. The encoders require a rotational input; the optical sensors determine position directly from the motion of the device relative to the reference surface. At the same time, devices to transform electrical energy into mechanical energy are selected from catalogs. Electric motors of two types (torque motors and voice-coil motors) satisfy space and power requirements; each produces a rotational torque (input function-structures passed down from more abstract designs have biased the solution toward rotational motors) that the system must transform into forces in the x and y directions.

Mechanical composition builds from these initial inputs that are not yet oriented spatially. Designs are generated from the basic components: Table 8.1 shows component types that transform rotation into translation, the simplest machines that satisfy the given functional requirements. Because the system has multiple inputs and outputs, four possible designs for the force flow path are illustrated in Fig. 8.5 (many more designs are possible).

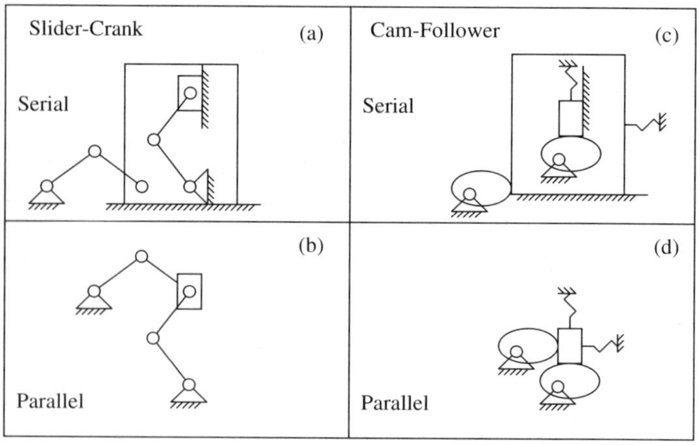

FIG. 8.5 SERIAL AND PARALLEL COMBINATIONS OF X AND Y SLIDER-CRANKS AND CAM-FOLLOWERS

8.2.3 Detail Design

Configuration design relies primarily on catalogs to map from function-structure to a functional topology and spatial layout of components. In this process, two types of connection are assumed: connections between functional elements along a flow path and connections between the functional element and the reference structure from which it acts (we call this the mechanical "ground"). The details of these connections have a large impact on the manufacturability of the system: connections represent a large proportion of the parts in a design; the interfaces between components also produce the majority of tolerance specifications. Both parts and tolerances are significant factors in manufacturing – each part in a design must be formed, handled, inventoried, assembled, etc. Boothroyd and Dewhurst [17] emphasize part count reduction as a major thrust of design for manufacturability. Fagade and Kazmer [18] find that the number of toleranced features in a part is a major determining factor in tooling costs for injection molding; similar relationships can be safely inferred for other casting processes as well. Ground connections play a major role in manufacturing: the ground itself can account for a large number of toleranced features [19], often making it the most expensive component in a product. This expense can yield dividends in ease of assembly. A well-designed ground (or set of grounds) can support assembly by eliminating fasteners and promoting simple, tip-down assembly motions. Because connections drive cost on both the forming and assembly sides of manufacturing, computational synthesis is focused on developing connection details for each configuration. In addition to helping establish manufactured cost, these connection designs also contribute to more accurate behavioral models, perhaps capturing friction or tolerance information.

Design Primitives Roth [20] provides a representation for connections among components that integrates well with the Cartesian representation used in configuration synthesis. Contact matrices are defined to establish the degrees-of-freedom associated with each connection; contact types parameterize these matrices, filling in details of the contacts (i.e., same material, form fit, elastic, friction, gravity constraint, etc.). As with function-level design, reverse engineering provides actual instantiations of these contact types for use as primitives. The first two columns in Table 8.2 show connection designs for a single-degree-of-freedom rotational joint and their corresponding contact-type matrices. In addition to contact and contact-type matrices, information like size, power transmit, number of toleranced features, etc. are captured for each connection and can be readily composed into part- and system-level models.

Composition Composition is a matter of grafting connections onto configured components. To do this, components defined at the configuration level of abstraction are defined in terms of individual parts and connections among them. These connections are defined in terms of the degrees-of-freedom necessary for function. Configuration-level components are composed by grafting connection details onto individual parts. To help ensure manufacturability of the resulting components, material, process, process direction (primary and secondary) and assembly direction are captured in the case base for each "side" of a connection (whether it is

TABLE 8.2 DETAILED DESIGN BUILDING BLOCKS

Single DOF Joint Connections

$$\textit{Contact Matrix} = \begin{bmatrix} 1 & 1 & 0 & 0 \\ 1 & 1 & 1 & 1 \\ 1 & 1 & 1 & 1 \end{bmatrix} \quad \left(\textit{Legend}: \begin{bmatrix} x+ & x- & xx+ & xx- \\ y+ & y- & yy+ & yy- \\ z+ & z- & zz+ & zz- \end{bmatrix} \right)$$

Contact-Type	Form	# Parts	Process Axis: Part	Process Axis: Connect / Ground	Assy Axis
$\begin{bmatrix} s & s & E & E \\ s & s & s & s \\ s & s & s & s \end{bmatrix}$		1	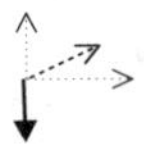	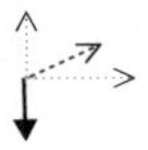	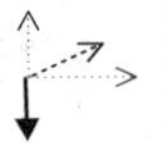
$\begin{bmatrix} f & f & r & r \\ f & f & f & f \\ f & f & f & f \end{bmatrix}$	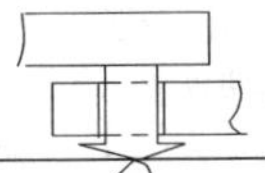	2	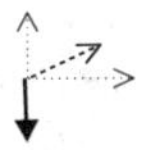	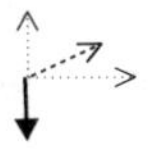	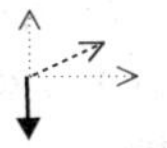
$\begin{bmatrix} f & f & r & r \\ f & f & f & f \\ f & f & f & f \end{bmatrix}$	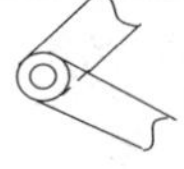	5	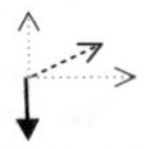	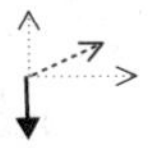	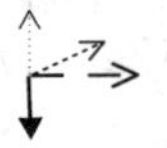

part-part or part-ground). Composition favors grafting connections of similar material and process directions to the base parts; maintaining a common assembly direction for each part is an additional consideration.

The above discussion lays out the general transition from customer requirement to functional concept to configuration to detail design. The process is supported initially through the use of cases derived from reverse engineering. These cases instantiate not only functional designs but also suggest components and mechanisms useful in the past for accomplishing that function. The result is a "sketch" of a design but, because reasoning about abstract function is imprecise, the concepts generated may not be feasible. Thus, while the system can suggest design alternatives likely to satisfy functional requirements, human designers must still be involved in selecting from among those alternatives. The set of alternatives presented to the designer can be further narrowed through additional semi-automated design processes that can predict the performance of design concepts. Based on functional and spatial requirements, configuration design selects machine components from catalogs to compose possible solutions. Detail design takes these "bread-board" configurations and attaches to each component the connections (again drawn from real-world cases) necessary to integrate it into the system. At each level of this process, the goal of computational synthesis is to produce a large set of design possibilities, balancing completeness in the search of the design space against soundness in the generated designs. At each stage, performance goals can be used to narrow down the set of alternatives to a size more manageable for human designers.

8.3 DECISION-BASED SYNTHESIS

To summarize we have:

(1) A methodology for modeling the design space based on design instances.
(2) A computational synthesis process in which both functional design and detail design draw heavily on design instances and in which configuration design is based on catalogs of standard components.
(3) Overall performance goals that can be determined through the design space model.
(4) Functional goals including desired functionality and input/output flow type, position and orientation.

From this basis, we must create not only an effective design but also an efficient process for arriving at that design. This requires an interplay between generation and evaluation—estimates of design performance at high levels of abstractiown can identify the most promising candidates for further development and determine aspects of the uncertain requirements whose resolution would help make this identification task easier. This process must be done at each of the abstraction levels defined in the computational synthesis process, carefully narrowing the set of designs to limit search effort while still maintaining the potential of the set of concepts to address still uncertain design performance goals.

8.3.1 The Uncertain Nature of Design

In a "typical" DBD formulation, there exists a set of alternatives from which the designer must choose, along with models for predicting the performance of those alternatives. When performance models are deterministic, optimization methods explore a multidimensional parameter space that defines all possible designs (i.e., an infinite option set), navigating toward areas of improved performance. Where performance models are uncertain (typically due to noise superimposed on the design vector), robust design modifies the basic optimization apparatus both in the evaluation of objective functions and the handling of design constraints. As uncertainty in the evaluation grows (due to uncertain performance estimation, uncertain value functions or accounting for risk tolerance), formal decision methods are applied to manage the uncertainty; their higher cost tends to mean exploring fewer design alternatives.

Both performance and value uncertainty are at their highest in conceptual design. Abstract designs have no distinct physical instantiation – they represent a range of possible artifacts. The closer one is to a physical instantiation of a design, the better one can predict its behavior; the closer one can come to introducing a physical artifact into the targeted environment, the better one can map behavior into value. So, in general, we expect design evaluations to improve both in accuracy (required for selecting among diverse, heterogeneous design concepts) and in precision (for selecting the best among very similar, homogenous design concepts) as we move from abstract concepts to physical artifacts. Because designers cannot afford to create physical artifacts for each design concept or to present all possible designs to all possible customers for evaluation, design must take place in the presence of a large amount of initial uncertainty.

8.3.2 Foundations of Decision-Based Conceptual Design

Decision-based conceptual design has been developed to manage uncertainty in the early phases of design. It is based on several key components: a probabilistic modeling method, expected value decision-making, information value theory and communication theory. Each of these components will be laid out individually.

8.3.3 Design Space Modeling

We seek to model the relationships among various attributes of a design as a multidimensional space. Unlike optimization methods in which the design space is usually limited to attributes under the direct control of the designer, dimensions in this space can be any quantifiable attribute of a design. The reason for confounding

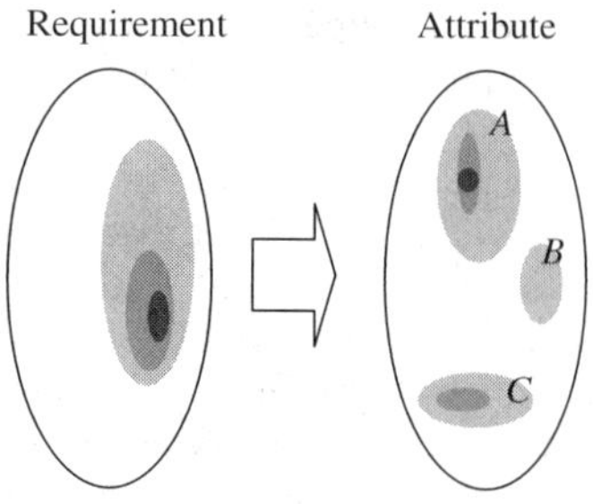

FIG. 8.6 MAPPING REQUIREMENTS INTO ATTRIBUTES

"design" and "performance" attributes stems from a design model proposed by Yoshikawa: General Design Theory [21]. In ideal design (one component of the theory), the design space is split into two: aspects directly under the control of the designer are contained in the "attribute" space; aspects representing design performance inhabit the "function" space. Design is defined as a mapping from the function space to the attribute space (this is the inverse of engineering science, where models predict performance based on known attributes). Requirements on performance are established, first as ranges (i.e., uncertain requirements). As Figure 8.6 shows, these ranges map onto multiple design concepts (A, B and C define the neighborhoods of high-level design concepts). As requirements are refined (increasingly smaller neighborhoods shown as progressively darker ellipses), the number of concepts they imply decreases until only one remains. Two main challenges confront someone who attempts to actually perform ideal design: (1) one must know all of the inverse mappings between requirement and attribute; and (2) one must know all possible requirements (and/or designs).

Confounding design and performance attributes eliminates the need to know all mappings—they are embodied in the design instances. Knowing all designs is still a problem; generalizing over a set of known designs can help to "fill in the blanks" between designs but cannot represent unknown designs (although the computational synthesis process presents a compromise, extrapolating new designs by interpolating performance from known design fragments). Using a probabilistic framework means that we are no longer tied to the restrictions of mathematical functions, as Fig. 8.6 demonstrates; even though there are likely functions mapping from attribute to performance, no inverse function exists—the same requirement maps to multiple designs. A probability-density function (PDF) readily models multimodal relationships such as those that map requirements onto designs in the ideal design. Because we want to exploit data from actual designs (computational synthesis draws heavily on experience rather than theory), the form of the PDF is one that can incorporate actual performance and attribute data drawn from design fragments. In addition, the model must be able to augment these raw data vectors with information generated from analysis models. The design vectors themselves must allow both discrete and continuous design attributes. Finally, as design classes are eliminated the design space becomes more homogeneous (designs that are more similar can be compared more directly: "apples vs. oranges" gives way to "Macintosh vs. Granny Smith"), the model must adapt to changing design descriptions (i.e., longer design attribute vectors).

While there are potentially many ways of satisfying these varied requirements, the most straightforward starts with the set of known design instances and builds a probability-density function from them:

$$p(\mathbf{x}) = \frac{1}{(2\pi\sigma^2)^{d/2}} \sum_{i=1}^{n} e^{\frac{-(\mathbf{x}-\mathbf{x}_i)(\mathbf{x}-\mathbf{x}_i)^T}{2\sigma^2}} \qquad \text{Eq. (8.1a)}$$

$$\sigma = b n^{-1/(d+4)}, \; b \in [0.01, 0.35] \qquad \text{Eq. (8.1b)}$$

where $\mathbf{x}$ = design description vector whose probability is sought; $\mathbf{x}_i$ = set of n design instances used as the experience base; d = dimensionality of the design description; and b = a smoothing factor whose indicated range is suggested in [22]. This model centers a Gaussian PDF on each known design point; the probability of any point in design space is simply the sum of its probability with respect to each of these Gaussian "islands." The model generalizes between the samples, does not assume any functional form for the relationships among attributes, and is easily extended by increasing the length of the vector describing each design instance (perhaps by applying analysis models to each design instance or by responding to an increase in the homogeneity of the design space—design instances that have more descriptive attributes in common).

Figure 8.7 shows the PDF generated by the above model for a function level design of the force-feedback mouse. It models two basic types of design: one based on crank-sliders for generating output force, the other using rack and pinions. Note that the generated PDF relates two performance factors for which there are no analytical models: size (bounding box volume) and number of parts for the design. The design instances underlying the model are taken from the set of function-level skeleton designs for each type (composed from experience across multiple combinations of function fragments from the reverse engineering database).

8.3.4 Making Decisions in the Design Space

With the design space model relating the salient aspects of a design to one another, the process of design is one of refinement: reducing the uncertainty of a design attribute. It is convenient to separate the reduction of uncertainty into two classes, based on the nature of the attribute being constrained. For discrete variables, the act of constraining an attribute deterministically removes part of the design space, making a "hard" partition between designs

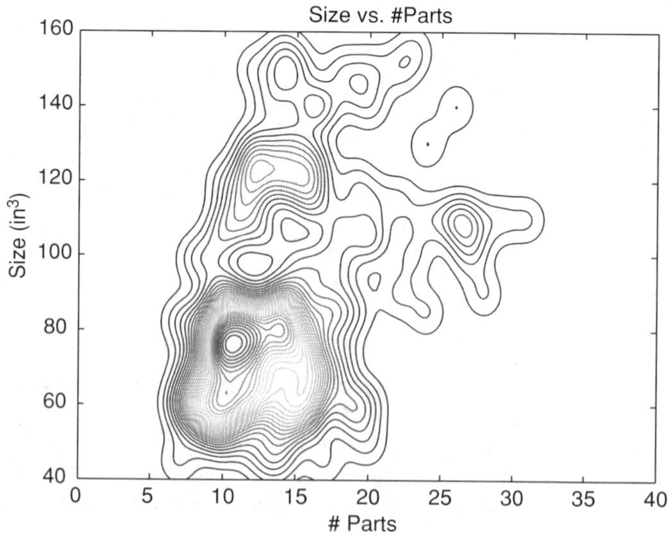

FIG. 8.7 DESIGN SPACE MODEL OF SIZE AND NUMBER OF PARTS

still under consideration and those discarded. (The careful reader might note that discrete variables are not specifically managed in the probabilistic design space model; rather than create a hybrid model for discrete and continuous variables, discrete variables are handled as a special case as we condition the joint; more information follows.) Due to generalization about design instances in the design space model, constraining continuous attributes still allows infeasible portions of the space near the constraint to influence the feasible space (e.g., even if the designer has constrained motor power to 50 watts, both 45 and 55 watt motors might help establish relationships among attributes for the desired motors). Thus, constraints on continuous variables make "soft" partitions in the design space. It is possible to "harden" these constraints simply by removing infeasible design instances from the design instance set.

The two different modes of design space partitioning lend themselves to two different aspects of decision theory. Hard partitions create distinct subspaces in the design space that can be evaluated using expected value decision-making, where a value function can be integrated over the probability space of each possible partition. Barring any further requirement refinement, the partition with the greatest expected value should be selected:

$$E(obj|dec_i, \mathbf{c}, \mathbf{v}) = \int_{\Omega_v} obj(dec_i, \mathbf{c}, \mathbf{v}) p(\mathbf{c}, \mathbf{v}) d\mathbf{v} \qquad \text{Eq.(8.2)}$$

where the expected value for the design objective, *obj*, is calculated for a possible design decision, dec_i, under both deterministic and uncertain constraints (**c** and **v**, respectively), simply by integrating objective function values for that decision over the uncertainty in the problem. Figure 8.8 shows two separate decisions in the force-feedback mouse example: selecting crank-slider or rack and pinion mechanisms and encoder or 2-D optical position sensing. Again, the criteria for selection are design size and the number of parts; vertical bars show the expected value of each selection criterion for each design type. These expected values are overlaid on the joint PDF of Fig. 8.7, conditioned for each design type and criterion. In selecting the mechanism, the crank-slider is superior in terms of size but the two are pretty much a wash in terms of the number of parts; resolving the large amount of modeling uncertainty might make this distinction more clear-cut for all decisions.

Of course, early in the design process there are many avenues for refining design requirements: revising performance expectations, examining the balance among competing design objectives, etc. Because constraints applied to the design condition the joint PDF, the design model produces marginal probability densities for each design attribute. Marginal probabilities express uncertainty in the problem that could be reduced through the introduction of a constraint (generally reducing the range of attribute values). Treated as "natural" constraints with uncertainty, these and other uncertain constraints can be analyzed using information value theory. For simplicity, we use the expected value of perfect information (EVPI):

$$EVPI(v_j) = \int_{\Omega_{v_j}} \left\{ \begin{matrix} \max_i E(obj|dec_i, \mathbf{c}, v_j) - \\ E(obj \mid dec^*, \mathbf{c}, v_j) \end{matrix} \right\} p(v_j \mid \mathbf{c}) dv_j \qquad \text{Eq. (8.3)}$$

Each uncertain attribute, v_j, is examined across its range; the best decision for a given value for the attribute is compared to the best decision minus that information and scaled by the marginal probability of that attribute value. Summing this over the range gives an upper bound on the expected value of determining that attribute exactly (i.e., picking a value for it at random from the given marginal probability density). Constraining attributes with a high EVPI helps make for better decisions among the "hard" partitions under consideration.

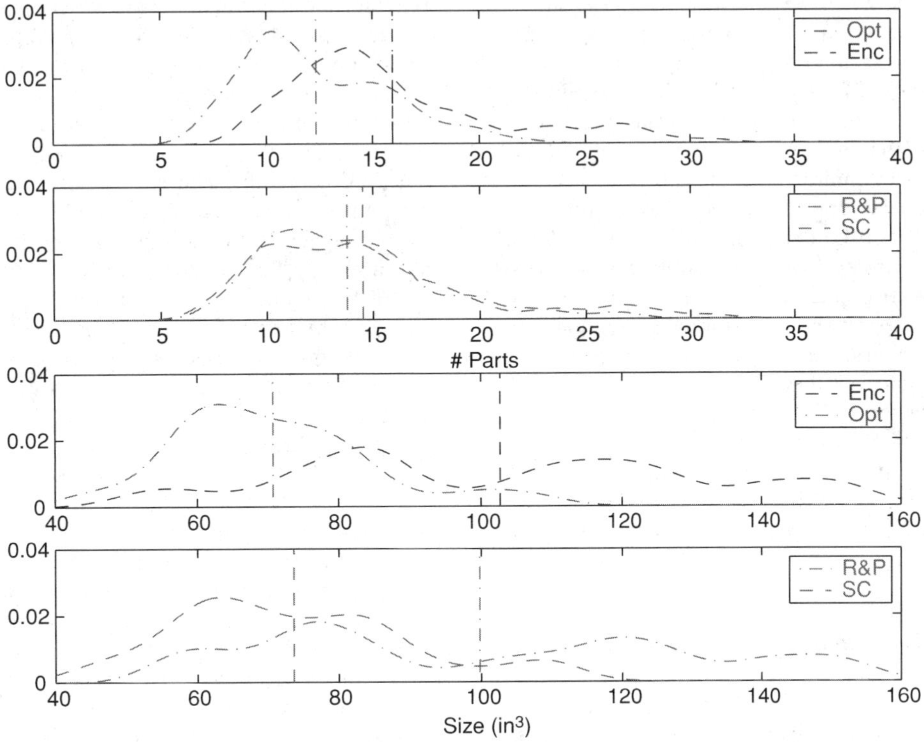

FIGURE 8.8 EVDM TO MINIMIZE NUMBER OF PARTS AND SIZE

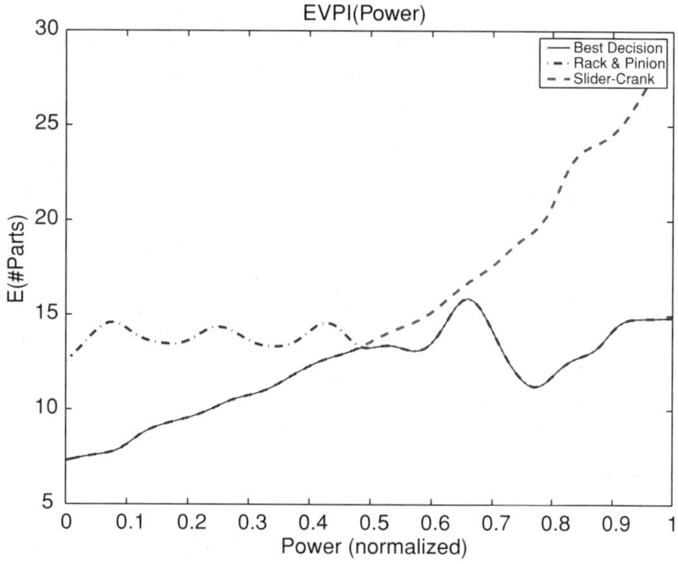

FIG. 8.9 PART COUNT VS. POWER

Figure 8.9 demonstrates how design value can change with further knowledge of a design attribute not currently part of the evaluation model. In this case, the expected number of parts for a rack and pinion system remains relatively constant with power, while parts for a crank-slider system vary strongly with power. For low power designs, crank-sliders are better; at a higher power level, the designer might instead choose a rack and pinion (the dark line shows the best decision for each level of power). Knowing power can help make the mechanism decision more clear-cut; making a better decision increases the expected value of the design.

8.3.5 Design Freedom

Decision-based conceptual design applies two central components of decision theory: expected value decision theory to evaluate deterministic partitions in the design space; and the expected value of perfect information to identify constraints on design attributes that amplify differences among the deterministic partitions. Both components seek to improve the *value* of a design by making *commitments* that reduce initial uncertainty in the design space. As long as a commitment changes the value of the design, decision methods can help identify the best commitments to make and the best point in the design process to make them. Because the design space is highly coupled, commitment along one dimension of a design often implies commitments along others. For example, deciding on a low-mass motor for an application may imply that the motor will operate at a high speed. If speed is not part of the design value function, the commitment to high-speed motors would not be reflected in a change in the expected value of the design. Further, if the favorable low-mass motor types do not change with operating speed, then the EVPI for speed would be zero. Still, a commitment has been made that might have implications: speed could change propeller efficiency or span, high speeds might require a gearbox or might generate more noise or vibration.

Because of the abstract, evolving nature of design concepts, the ramifications of implicit commitments can escape modeling. For sure, some commitments might not be important—it could be that low-mass motors all happen to be painted blue. Would a designer care about that? The key is to identify implicit commitments as they occur so that the designer can actively consider them (by including them in the value function or constraining them).

The first step is identifying implicit commitments as they are made. For this, we turn to the probabilistic design space and take some guidance from communication theory. In a probabilistic space, decisions produce constraints on some of the design attributes, conditioning the joint probability density. In response to this, both the overall joint PDF and the marginal probability density for "free" attributes (i.e., those that are not subject to an explicit, designer-specified constraint) change. In essence, the constraints are "communicating" through the joint PDF to individual design attributes, the design space model "telling" each of them how to respond to the constraint. While it may be difficult to track small, local changes in the marginal PDF for an attribute, it might be worthwhile to track the overall level of commitment in that attribute. As in the explicit decisions derived from decision theory, commitment in this case means constraining an attribute, thereby reducing its uncertainty. An unbiased measure of uncertainty of a random variable, z, is Shannon's entropy [23]

$$H(z) = \int_{\Omega_z} \log[p(z)dz \qquad \text{Eq. (8.4)}$$

We make a couple of basic changes to this formulation to adapt it to our modeling methods. First, because the probability function for the joint is not easily integrated Eq. (8.5) transforms Eq. (8.4) into a discrete sum over samples of the PDF instead of an integral. Second, because the PDF is only defined on $[0,1]^n$, entropy is scaled by a uniform density in that range (an infinite, uniform PDF in the continuous form leads to an infinite entropy). Because the result is not, strictly speaking, entropy, we will call it design freedom

$$DF(z) = -\sum_{i=1}^{m} \frac{P(z_i \mid c,u)\log[P(z_i \mid c,u)]}{\log(m)\sum_{i=1}^{m} P(z_i \mid c,u)} \qquad \text{Eq. (8.5)}$$

where m is number of samples of z used to calculate the design freedom. Design freedom ranges from one (a uniform PDF) to zero (a deterministic value for an attribute). It may be calculated across any number of design dimensions, but we will generally talk about freedom along a single design attribute.

In terms of detecting design commitment, design freedom for each design attribute can be calculated before and after the design joint PDF is conditioned by a decision/constraint. When design freedom for an attribute is reduced by a decision, this reduction can be flagged to the designer, who might then inspect the new marginal PDF for the attribute and decide how to proceed. If the loss of freedom is due to something of design importance (e.g., shifting to high-speed motors), then the designer might modify the current design requirements—applying a constraint to the attribute in question or modifying the design value function to include the attribute.

Design freedom also helps extend decision-based conceptual design into the target-setting process. DBD often emphasizes the design value function, selecting from among pre-defined design alternatives using utility or demand to determine the best one(s). As we move further upstream in the design process, constraints or targets that aid design generation become important. Of course constraints can be incorporated into value functions as penalties, but such penalties tend to distort the value function or put requirements on its form [24]. Two primary classes of methods deal directly with targets: goal programming and set-based design. In goal programming, distinct targets representing performance along multiple design attributes are set. The designer is charged with meeting each target in order of priority, until a target cannot be met. The designer must then examine the unmet targets and decide how much performance to give up (by relaxing higher-priority targets) in order to meet them. Conversely, if all targets can be met, the designer must determine how to "spend" the remaining design freedom.

Set-based design is a compromise between the two goal-programming scenarios. Setting goal ranges from which the final targets will be chosen, set-based design requires designers to produce solutions that satisfy all possible design targets. Applied in a manufacturing environment, this process reserves design freedom so that designs for short lead-time parts can be adapted to as-manufactured, long lead-time parts. As a more general design methodology, set-based design emphasizes that the development of design flexibility and the delay of commitment to final design targets provides the highly valuable "wiggle room" that a design needs to respond to unforeseen issues. This is behind the second "Toyota Paradox" [25], wherein delaying design commitment leads to better designs with lower design effort.

Figure 8.10 shows design freedom versus part count for the detail-level design of the force feed-back power transmission. In this case a new design attribute—joint stiffness—is introduced into the system as different detail design instances are created. The plot shows that for both low- and high-part counts, joint stiffness freedom is low; this implies that setting targets within either of these portions of the design evaluation may restrict the stiffness of the joints. For the mouse design, this is significant—if the joints are stiff, then the mouse will tend to move by itself to some neutral position rather than staying where the user has placed it. The decrease in design freedom for stiffness prompts the designer to examine the decision implied by a constraint, in this case rejecting designs toward the "optimal" end of the manufacturing range because of performance issues.

Design freedom is useful for the target setting process. Figure 8.11 shows design freedom for electric motors versus motor mass. Given a power requirement, a designer might want to minimize the mass of a motor, but there are many other issues involved in a design besides just mass. This figure shows design freedom for several other design attributes both continuous (performance) attributes like speed, torque, length and diameter, as well as discrete (classification) attributes like magnet type, commutation and frame type. While a designer is certainly free to select the lightest possible motor, it might be wise to relax this target a bit to increase

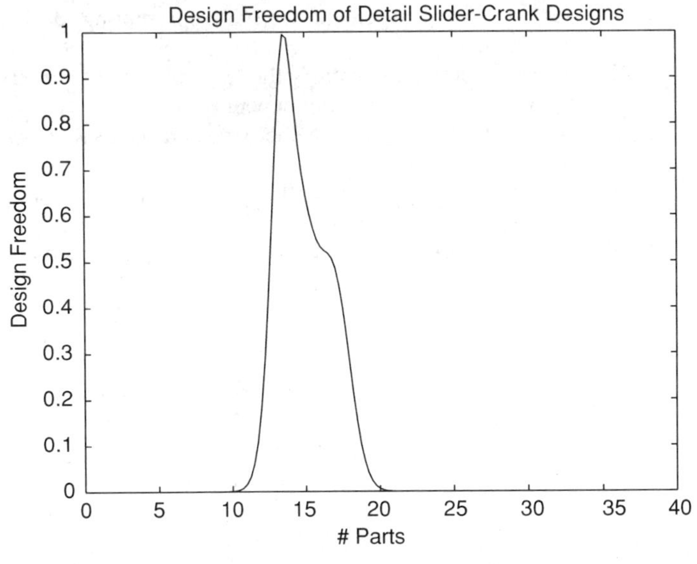

FIG. 8.10 DESIGN FREEDOM OF JOINT STIFFNESS

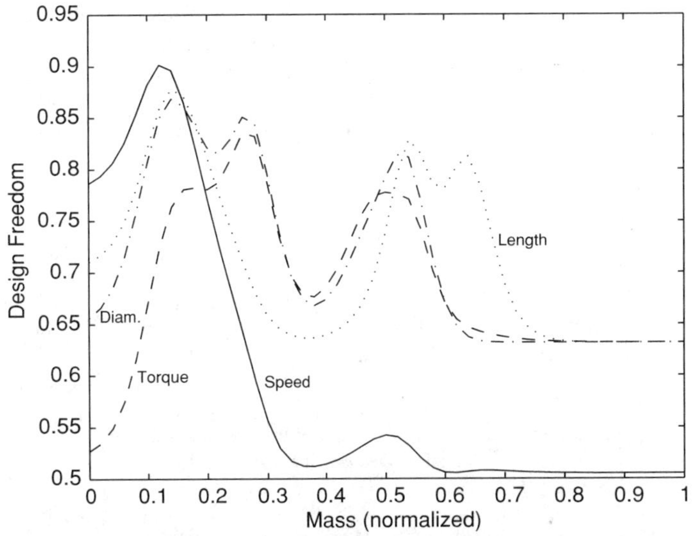

FIG. 8.10 DESIGN FREEDOM VS. MASS FOR VARIOUS ASPECTS OF AN ELECTRIC MOTOR

the diversity of the design space. In a goal-programming approach, one might set targets according to priority, reserving some amount of design freedom for remaining targets along the way. In a set-based design framework, one might simply identify a range of targets that includes high-performance, low-diversity designs as well as lower-performance, high-diversity ones. Design freedom helps to support the target-setting process in both frameworks, helping designers to evolve requirements, "spending" design freedom to improve the design.

8.4 SUMMARY

Decision-making in design takes place throughout the design process. In conceptual design, there is often too much uncertainty about the design (due to abstraction) and the requirements (because not enough is known about the design space) to allow the identification of the best design, so decisions largely involve figuring out the options to exclude from consideration. As the design process progresses and designs become more concrete, estimates of their performance increase in accuracy, leading to consideration of a smaller set of design concepts. But until design details are defined, the final trade-offs among the various dimensions of performance cannot be fully understood.

Generating design alternatives is an expensive part of the design process. Computational synthesis can help to reduce some of this expense by automating design generation. As with any automation process there are trade-offs. To generate the widest possible set of design alternatives, the synthesis process must start at high levels of abstraction. Function-level designs afford the completeness that we seek from computational synthesis, but at the cost of soundness. Biasing the generation of function-level designs by using fragments derived from reverse engineering helps ensure that the resulting designs are near-feasible. While human intervention could prune the infeasible designs, an alternative strategy is the automated generation of lower-level designs. Nonfunctional aspects of performance (e.g., size, part count, etc.) can be assessed at the configuration- and detail-level of design and fed back to the function level. This exploration can not only reduce the effort of human designers in evaluating the set of computer-generated designs, but can also help identify critical aspects of performance about which the current design evaluation model is ambiguous. Together, the combination of multi-abstraction computational synthesis and DBD provide a foundation for ideal design as defined by Yoshikawa: more "design" can be done in the performance space as we afford designers the ability to co-evolve design and requirement.

REFERENCES

1. Pahl, G. and Beitz, W., 1988. *Engineering Design- A Systematic Approach*, Springer-Verlag.
2. Hirtz, J. M., Stone, R. B. et al., 2001. "A Functional Basis for Engineering Design: Reconciling and Evolving Previous Efforts," *Res. in Engg. Des.*, 13(2), pp. 65–82.
3. Stone, R. B. and Wood, K. L., 2000. "Development of Functional Basis of Design," *J.* of *Mech. Des.*, 122(4), pp. 359–370.
4. Verma, M. and Wood, W., 2003. "Functional Modeling: Toward a Common Language for Design and Reverse Engineering," Proc., *2003 ASME Int. Conf. on Des. Theory and Methodology*, ASME, New York, NY.
5. Mitchell, T. M., 1990. "The Need for Biases in Learning Generalizations," *Readings in Machine Learning*, T. Dietterich, ed., Morgan Kauffman.
6. Gietka, P., Verma, M. et al., 2002. "Functional Modeling, Reverse Engineering, and Design Reuse," *Proc. 14th Int. Conf. on Des. Theory and Methodology*, ASME, New York, NY.
7. Verma, M. and Wood, W., 2004. "Toward Case-Based Functional Design: Matching Reverse Engineering Practice with the Design Process," *Design Studies*.
8. Segre, A., 1987. "On the Operationality/Generality Trade-off in Explanation-Based Learning," *Proc. 10th Int. Joint Conf. on Artificial Intelligence (IJCAI)*.
9. Wood, W. H. and Agogino, A. M., 2004. "Decision-Based Conceptual Design: Modeling and Navigating Heterogeneous Design Spaces," *J.* of *Mech. Des.*, 126(6).
10. Kota, S. and Erdman, A. G., 1997. "Motion Control in Product Design," *Mech. Engrg.* August, pp. 74–76.
11. Kota, S. and Chiou, S.-J., 1992. "Conceptual Design of Mechanisms Based on Computational Synthesis and Simulation of Kinematic Building Blocks," *Res. in Engrg. Des.*, 4(2), pp. 75–87.
12. Chiou, S.-J. and Kota, S., 1999. "Automated Conceptual Design of Mechanisms," *Mechanism and Machine Theory*, Vol. 34, pp. 467–495.
13. Chakrabarti, A. and Bligh, T. P., 2001. "A Scheme for Functional Reasoning in Conceptual Design," *Des. Studies*, Vol. 22, pp. 493–517.
14. Chakrabarti, A. and Bligh, T. P., 1994. "An Approach to Functional Synthesis of Solutions in Mechanical Conceptual Design. Part I: Introduction and Knowledge Representation," *Res. in Engg. Des.*, vol. 6, pp. 127-141.
15. Chakrabarti, A. and Bligh, T. P., 1996. "Approach to Functional Synthesis of Solutions in Mechanical Conceptual Design. Part II: Kind Synthesis," *Res. in Engrg. Des.*, 8(1), pp. 52–62.
16. Navinchandra, D., 1988. "Behavioral Synthesis in CADET, a Case-Based Design Tool," *Proc., DARPA Workshop on Case-Based Reasoning*, Morgan-Kaufman.
17. Boothroyd, G. and Dewhurst, P., 1989. *Product Design for Assembly*, Boothroyd Dewhurst Inc., Wakefield, RI.
18. Fagade, A. and Kazmer, D., 1999. "Optimal Component Consolidation in Molded Product Design," *Proc., 1999 ASME Des. for Manufacture Conf.*, ASME, New York, NY.
19. Verma, M. and Wood, W. H., 2001. "Form Follows Function: Case-Based Learning Over Product Evolution," *Proc., ASME DETC '01: Des. for Manufacture, Conf.*, ASME, New York, NY.
20. Roth, K.,1987. "Design Models and Design Catalogs," *Proc., Int. Conf. on Engg. Des.* (ICED '87), pp. 60–66.
21. Yoshikawa, H., 1981. "General Design Theory and a CAD System," *Proc., Man-Machine Communications in CAD/CAM, IFIP WG 5.2–5.3 Working Conf.* (Computer Aided Design / Computer Aided Manufacturing).
22. Specht, D., 1988. "Probabilistic Neural Networks for Classification, Mapping, or Associative Memory," *Proc., IEEE Int. Conf. on Neural Networks*, IEEE.
23. Shannon, C. E., 1948. "A Mathematical Theory of Communication," *Bell System Tech. J.* Vol. 27, pp. 379–423 and 623–656.
24. Scott, M. J. and Antonsson, E. K., 1999. "Aggregation Functions for Engineering Design Trade-offs," *Fuzzy Sets and Systems*, 99(3), pp. 253–264.
25. Ward, A., Liker, J. et al., 1995. "The Second Toyota Paradox: How Delaying Decision Can Make Better Cars Faster," *Sloan Magmt. Rev.*, Spring, pp. 43–61.

PROBLEMS

8.1 Select a small, mechanically oriented product that you can take apart (old, broken toys or electromechanical systems are often good subjects for study):

a. Try to identify the flows that enter and leave the system as well as the main functions it performs. Put these into a black-box functional decomposition.
b. Create a lower-level functional model by tracing each flow through the system and defining functions that transform the flow.
c. Take the system apart. For each part, identify the flows that enter and leave as well as the function of the part. Draw a detailed function-structure.
d. Identify the introduced flows, collapse the function-structure along them. Repeat this process, compare the resulting function-structures to the black-box and lower-level function-structures.

8.2 For a component of interest (electric motor, gearbox, etc.), find Web-based catalogs that capture a wide variety of performance.

a. Extract design instance data from the catalogs and use Eq. (8.1) to generate a pdf representing design performance.
b. Create a function to quantify the value of various aspects of performance for this component type.
c. Use Eq. (8.2) to calculate the expected value for each component type.
d. Use Eq. (8.3) to calculate EVPI for each uncertain design constraint.
e. Use Eq. (8.5) to calculate design freedom for unmodeled design attributes.

SECTION

4

DEMAND MODELING

INTRODUCTION

When viewing engineering design as an enterprise-wide decision-making activity, the design objective could be to maximize profit for private companies, or to maximize social welfare for government agencies. Each of these is a well-defined criterion in economic theory. While such a single-criterion approach to decision-making can overcome the limitations of using multi-attribute methods, one challenge of taking such an approach is to establish models for studying the economic impact of design changes. Among the economic performance metrics, *demand* plays a critical role in assessing the economic benefit as it contributes to the computation of both revenue and life-cycle cost. The interaction between supply and demand (market equilibrium) in economic theory implies the importance of considering both the producer's interests (supply) and the customer's desires (demand) in engineering design. Consequently, a reliable product demand model is needed to provide estimations of demand as a function of product attributes, including economic attributes such as price. In decision-based design, demand modeling is expected to facilitate engineering decision-making by providing the link between engineering attributes that are of interest to design engineers and those product attributes that are of interest to customers.

The challenge of demand modeling is to capture the variability among individuals (i.e., heterogeneity) and their individual choice behavior to avoid the paradox associated with aggregating preferences of a group of customers. As the competitors' product attributes and prices will have a direct impact on market share, it is paramount to consider competing products when forming a demand model. Besides, various sources of uncertainty exist and need to be quantified in demand modeling.

This section brings together different demand modeling approaches from the economics, transportation and engineering design communities. In Chapter 9, the essential methods developed in transportation economics and travel demand analysis are presented. Chapter 9 provides a solid foundation of discrete choice analysis (DCA), a disaggregated probabilistic demand modeling approach for demand analysis. In Chapter 10, the use of DCA techniques for engineering design problems is proposed and practical issues in demand modeling for designing complex engineering systems are studied. In addition to a walk-through example that provides the details of implementing the DCA technique, an engineering design case study is used in Chapter 10 to demonstrate the role of demand modeling in a DBD framework. In Chapter 11, an alternative approach to demanding modeling, called the S-model, is presented as a part of a new strategy of product planning. A formalism is established to link the planning process with marketing research, new technology development, manufacturing and financial forecasts.

While this section presents different approaches to demand modeling, it should be noted that other techniques also exist. For example, the choice-based conjoint analysis approach is not presented in this section, but it has been used in market research and applied to engineering design problems. Readers are encouraged to compare different demand modeling techniques by examining the pros and cons of each technique and to identify its applicability. For instance, the S-model approach uses a linear model, which means it is limited to small differences in demand between competing products. On the other hand, the S-model approach can be implemented with a small set of closed-form equations and with much fewer computational resources. Also note that while the presented approaches are generic, the examples of demand models used in this section are for a small number of industries (e.g., automobile industry). Extensions and modifications are needed when applying these approaches to problems in different industries.

CHAPTER

9

Fundamentals of Economic Demand Modeling: Lessons From Travel Demand Analysis

Kenneth A. Small

NOMENCLATURE

cdf = cumulative distribution function
D^i = alternative-specific dummy variable for alternative i
d_{jn} = choice variable (=1 if decision-maker n chooses alternative j)
E = expectation
G = function for generating GEV models of discrete choice
GEV = generalized extreme value
I_r = inclusive value for alternative group r
IID = identically and independently distributed
J = number of dependent variables (aggregate models) or alternatives (disaggregate models)
J_r = number of alternatives in alternative group r
L = leisure
$L(\cdot)$ = log-likelihood function
log = natural logarithm
N = number of people; number of vehicles in queue
n = indexes single individual consumer
P = choice probability
P_L = probability of being late
R = number of "replications" (random draws) in simulated probability; no. of groups in nested logit
s_n = vector of socioeconomic or other characteristics of decision-maker n
SDE = schedule delay early = work start time minus early actual arrival time
SDL = schedule delay late = late actual arrival time minus work start time
T = time spent in activities; travel time (usually in-vehicle) if used as scalar without sub- or superscripts
T^0 = out-of-vehicle travel time
T_F = free-flow travel time
T_R = random component of travel time
T_w = time spent at work (in value-of-time analysis)
t = time of arrival
t^* = desired time of arrival
t_d = departure time
U = utility function
V = indirect utility function
v_R = value of reliability
v_T = value of time (usually in-vehicle time)
W = welfare measure
w = wage rate (in travel-demand analysis)
X = generalized consumption good (numeraire good)
x = consumption vector
Y = unearned income
y = generalized argument for function G generating GEV models
z = independent variables for travel-demand models
α_i = alternative-specific constant for alternative i in discrete-choice indirect utility function; value of travel time in reliability analysis
β = parameter vector in discrete-choice indirect utility function (in travel-demand analysis); value of early-arrival time (in reliability analysis)
γ = value of late-arrival time (in reliability analysis)
γ_i = coefficient of an independent variable interacted with an alternative-specific constant for alternative i in discrete-choice utility function
ε_i = stochastic term for alternative i in discrete-choice indirect utility function
θ = value of late arrival (in reliability analysis)
λ = marginal utility of income
μ = scale parameter for probability density function (in discrete-choice analysis)
ρ = parameter of GEV functions (in discrete-choice analysis)

INTRODUCTION

In order to design facilities or products, one must know how and under what circumstances they will be used. In order to design them cost-effectively, one must also know how specific features are valued by users. These requirements can be met by applying tools of economic demand analysis.

This chapter illustrates the use of economic demand tools by providing a detailed account of their use in the field of urban transportation, where many of them were in fact first developed. Transportation applications include many design decisions, often at the level of an entire system such as a public mass-transit network. The design elements that can be addressed using transportation demand analysis include speed, frequency, reliability, comfort and ability to match people's desired schedules.

The chapter begins (Section 9.1) with a conventional aggregate approach to economic demand, and then moves to disaggregate models (Section 9.2), also known as "behavioral" because they depict decision-making processes by individual consumers. Some specific examples (Section 9.3) and more advanced topics (Section 9.4) are then discussed. Finally, Section 9.5 analyzes two behavioral results of travel-demand studies that are especially important for design, namely travelers' willingness to pay for travel-time savings and improved reliability.

9.1 AGGREGATE MODELS

In standard economic analysis of consumer demand, the aggregate demand for some product is explained as a function of variables that describe the product or its consumers. For example, total transit demand in a city might be related to the amounts of residential and industrial development, the average transit fare, the costs of alternative modes, some simple measures of service quality and average income.

In one type of study, known as *cross-sectional*, one examines influences on travel behavior by looking at variations across cities or across locations within a city. An example is the share of various travel modes by Kain and Liu [1] in Santiago, Chile. The share is measured for each of 34 districts and its logarithm is regressed on variables such as travel time, transit availability and household income.

Sometimes there are many cases of zero reported trips by a given mode between a pair of zones, making ordinary regression analysis invalid. This illustrates a pervasive feature of travel-demand analysis: many of the variables to be explained have a limited range. For this reason, travel-demand researchers have contributed importantly to the development of techniques to handle limited dependent variables [2]. We note here one such technique that is applicable to aggregate data.

Suppose the dependent variable of a model can logically take values only within a certain range. For example, if the dependent variable x is the modal share of transit, it must lie between zero and one. Instead of explaining x directly, we can explain the logistic transformation of x as follows:

$$\log\left(\frac{x}{1-x}\right)=\beta' z+\varepsilon \qquad \text{Eq. (9.1)}$$

where β = a vector of parameters; z = a vector of independent variables, and ε = an error term with infinite range. Equivalently,

$$x=\frac{\exp(\beta' z+\varepsilon)}{1+\exp(\beta' z+\varepsilon)} \qquad \text{Eq. (9.2)}$$

This is an *aggregate logit* model for a single dependent variable.

In many applications, several dependent variables x_i are related to each other, each associated with particular values z_i of some independent variables. For example, x_i might be the share of trips made by mode i, and z_i a vector of service characteristics of mode i. A simple extension of Eq. (9.2) ensures that the shares sum to one:

$$x_i=\frac{\exp(\beta' z_i+\varepsilon_i)}{\sum_{J=1}^{J}\exp(\beta' z_j+\varepsilon_j)} \qquad \text{Eq. (9.3)}$$

where J = number of modes.

In another type of study, known as *time-series*, one looks instead at variations over time within a single area. Several studies have examined transit ridership using data over time from a single metropolitan area or even a single transit corridor—for example, that of Gómez-Ibáñez [3] for Boston. Time-series studies are sensitive to the tendency for unobserved influences to persist over time, a situation known as autocorrelation in the error term. One may also postulate "inertia" by including among the explanatory variables one or more lagged values of the variable being explained. For example, [4], using U.S. nationwide data, considers the possibility that once people have established the travel patterns resulting in a particular level of vehicle-miles traveled, they change them only gradually if conditions such as fuel prices suddenly change. From the coefficients on the lagged dependent variables, one can ascertain the difference between short- and long-run responses.

It is common to combine cross-sectional and time-series variations, so that individual consumers analysis units are observed repeatedly over time. The resulting data are called *cross-sectional time-series* or *longitudinal* [5]. For example, [6] analyzes ridership data from 118 commuter-rail stations in metropolitan Philadelphia over the years 1978–91 to ascertain the effects of level of service and of demographics on rail ridership. Studies using panel data need to account for the fact that, even aside from autocorrelation, the error terms for observations from the same location at different points in time cannot plausibly be assumed to be independent. Neglecting this fact will result in unnecessarily imprecise and possibly biased estimates. Several approaches are available to account for this panel structure, the most popular being to estimate a "fixed effects" model in which a separate constant is estimated for every location.

9.2 DISAGGREGATE MODELS

An alternative approach, known as *disaggregate* or *behavioral* travel-demand modeling, is now far more common for travel demand research. Made possible by micro data (data on individual consumers), this approach explains behavior directly at the level of a person, household or firm. When survey data are available, disaggregate models are statistically more efficient in using such data because they take account of every observed choice rather than just aggregate shares; this enables them to take advantage of variation in behavior across individuals that may be correlated with variation in individual conditions, whereas such variations are obscured in aggregate statistics. Disaggregate models are also based on a more satisfactory microeconomic theory of demand. Most such models analyze choices among discrete rather than continuous alternatives and so are called *discrete-choice models*. Train [7] provides a thorough treatment.

9.2.1 Basic Discrete-Choice Models

The most widely used theoretical foundation for these models is the additive random-utility model in [8]. Suppose a consumer n facing discrete alternatives j = 1,…, J chooses the one that maximizes utility as given by

$$U_{jn}=V(z_{jn},s_n,\beta)+\varepsilon_{jn} \qquad \text{Eq. (9.4)}$$

where $V(\cdot)$ = a function known as the *systematic utility*; z_{jn} = a vector of attributes of the alternatives as they apply to this consumer; s_n = a vector of characteristics of the consumer (effectively allowing different utility structures for different groups of consumers); β = a vector of unknown parameters; and ε_{jn} = an unobservable component of utility that captures idiosyncratic preferences. U_{jn} and $V(\cdot)$ implicitly incorporate a budget constraint, and thus are functions of income and prices as well as product quantities and attributes; in economics terminology, such a utility function is called *indirect* to distinguish it from the direct or primal dependence of preferences on those quantities and attributes. U_{jn} and $V(\cdot)$ are also conditional on choice *j*. For these reasons they are known as *conditional indirect utility* functions.

The choice is probabilistic because the measured variables do not include everything relevant to the individual's decision. This fact is represented by the random terms ε_{jn}. Once a functional form for *V* is specified, the model becomes complete by specifying a joint cumulative distribution function (CDF) for these random terms, $F(\varepsilon_{1n} ..., \varepsilon_{Jn})$. Denoting $V(z_{jn}, s_n, \beta)$ by V_{jn}, the choice probability for alternative *i* is then

$$P_{in} = \Pr[U_{in} > U_{jn} \quad \text{for all} \quad j \neq i] \qquad \text{Eq. (9.5a)}$$

$$P_{in} = \Pr[\varepsilon_{jn} - \varepsilon_{in} < V_{in} - V_{jn} \quad \text{for all} \quad j \neq i] \quad \text{Eq. (9.5b)}$$

$$P_{in} = \int_{-\infty}^{\infty} F_i\left(V_{in} - V_{1n} + \varepsilon_{in}, \ldots, V_{in} - V_{Jn} + \varepsilon_{in}\right) d\varepsilon_{in} \quad \text{Eq. (9.5c)}$$

where F_i = partial derivative of *F* with respect to its *i*th argument. [F_i is thus the probability density function of ε_{in} conditional on the inequalities in Eq. (9.5b).]

Suppose $F(\cdot)$ is multivariate normal. Then Eq. (9.5) is the *multinomial probit* model with general covariance structure. However, neither *F* nor F_i can be expressed in closed form; instead, Eq. (9.5) is usually written as a $(J-1)$-dimensional integral of the normal density function. In the special case where the random terms are identically and independently distributed (IID) with the univariate normal distribution, *F* is the product of *J* univariate normal CDFs, and we have the *IID probit* model, which still requires computation of a $(J-1)$-dimensional integral. For example, in the IID probit model for binary choice (*J* = 2), Eq. (9.5) becomes

$$P_{1n} = \Phi\left(\frac{V_{1n} - V_{2n}}{\sigma}\right) \qquad \text{Eq. (9.6)}$$

where Φ = cumulative standard normal distribution function (a 1-D integral); and σ = standard deviation of $\varepsilon_{1n} - \varepsilon_{2n}$. In Eq. (9.6), σ cannot be distinguished empirically from the scale of utility, which is arbitrary; for example, doubling σ has the same effect as doubling both V_1 and V_2. Hence it is conventional to normalize by setting $\sigma = 1$.

The *logit* model (also known as multinomial logit or conditional logit) arises when the *J* random terms are IID with the extreme-value distribution (also known as Gumbel, Weibull or double-exponential). This distribution is defined by

$$\Pr[\varepsilon_{jn} < x] = \exp(-e^{-\mu x}) \qquad \text{Eq. (9.7)}$$

for all real numbers *x*, where μ = a scale parameter. Here the convention is to normalize by setting $\mu = 1$. With this normalization, McFadden [8] shows that the resulting probabilities calculated from Eq. (9.5) have the logit form:

$$P_{in} = \frac{\exp(V_{in})}{\sum_{j=1}^{J} \exp(V_{jn})} \qquad \text{Eq. (9.8)}$$

This formula is easily seen to have the celebrated and restrictive property of *independence from irrelevant alternatives*: namely, that the odds ratio (P_{in}/P_{jn}) depends on the utilities V_{in} and V_{jn} but not on the utilities for any other alternatives. This property implies, for example, that adding a new alternative *k* (equivalent to increasing its systematic utility V_{kn} from $-\infty$ to some finite value) will not affect the relative proportions of people using previously existing alternatives. It also implies that for a given alternative *k*, the cross-elasticities $\partial \log P_{jn} / \partial \log V_{kn}$ are identical for all $j \neq k$: hence if the attractiveness of alternative *k* is increased, the probabilities of all the other alternatives $j \neq k$ will be reduced by identical percentages. The binary form of Eq. (9.8) is: $P_{in} = \{1 + \exp[-(V_{1n} - V_{2n})]\}^{-1}$.

It is really the IID assumption (identically and independently distributed error terms) that is restrictive, whether or not it entails independence of irrelevant alternatives. Hence there is no basis for the widespread belief that IID probit is more general than logit. In fact, the logit and IID probit models have been found empirically to give virtually identical results when normalized comparably [9].[1] Furthermore, both probit and logit may be generalized by defining non-IID distributions. In the probit case the generalization uses the multivariate normal distribution, whereas in the logit case it can take a number of forms, to be discussed in Section 9.4.

As for the functional form of *V*, by far the most common is linear in unknown parameters β. Note that this form can easily be made nonlinear in *variables* just by specifying new variables equal to nonlinear functions of the original ones. For example, the utility on mode *i* of a traveler *n* with wage w_n facing travel costs c_{in} and times T_{in} could be:

$$V_{in} = \beta_1 \cdot (c_{in} / w_n) + \beta_2 T_{in} + \beta_{in} T_{in}^2 \qquad \text{Eq. (9.9)}$$

This is nonlinear in travel time and in wage rate. If we redefine z_{in} as the vector of all such combinations of the original variables [z_{in} and s_n in Eq. (9.4)], the linear-in-parameters specification is simply written as

$$V_{in} = \beta' z_{in} \qquad \text{Eq. (9.10)}$$

where β' = transpose of column vector β.

[1]Comparable normalization is accomplished by dividing the logit coefficients by $\pi/\sqrt{3}$ in order to give the utilities the same standard deviations in the two models. In both models, the choice probabilities depend on $(\beta/\sigma_\varepsilon)$, where σ_ε^2 is the variance of each of the random terms ε_{in}. In the case of probit, the variance of $\varepsilon_{1n} - \varepsilon_{2n}$, $2\sigma_\varepsilon^2$, is set to one by the conventional normalization; hence $\sigma_\varepsilon^{PROBIT} = 1/\sqrt{2}$. In the case of logit, the normalization $\mu = 1$ in Eq. (9.7) implies that ε_{in} has standard deviation $\sigma_\varepsilon^{LOGIT} = \pi/\sqrt{6}$ [10, p. 60]. Hence to make logit and IID probit comparable, the logit coefficients must be divided by $\sigma_\varepsilon^{LOGIT} / \sigma_\varepsilon^{PROBIT} = \pi/\sqrt{3} = 1.814$.

9.2.2 Estimation

For a given model, data on actual choices, along with traits z_{jn}, can be used to estimate the parameter vector β in Eq. (9.10) and to carry out statistical tests of the assumed error distribution and the assumed functional form of V. Parameters are usually estimated by maximizing the log-likelihood function:

$$L(\beta) = \sum_{N=1}^{N}\sum_{i=1}^{J} d_{in} \log P_{in}(\beta) \qquad \text{Eq. (9.11)}$$

where N = sample size. In this equation, d_{in} = choice variable, defined as 1 if consumer n chooses alternative i and 0 otherwise; and $P_{in}(\beta)$ = choice probability.

A correction to Eq. (9.11) is available for choice-based samples, i.e., those in which the sampling frequencies depend on the choices made. (For example, transportation mode choice might be estimated from data arising from roadside surveys and surveys taken on transit vehicles.) The correction simply multiplies each term in the outer summation by the inverse of the sampling probability for that sample member [11].

One of the major attractions of logit is the computational simplicity of its log-likelihood function, due to taking the logarithm of the numerator in Eq. (9.8). With V linear in β, the logit log-likelihood function is globally concave in β, so finding a local maximum ensures finding the global maximum. Fast computer routines to do this are widely available.

It is possible that the likelihood function is unbounded in one of the coefficients, making it impossible to maximize. This happens if one includes a variable that is a perfect predictor of choice within the sample. For example, suppose one is predicting car ownership (yes or no) and wants to include among variables s_n in Eq. (9.4) a dummy variable for high income. If it happens that within the sample everyone with high income owns a car, the likelihood function increases without limit in the coefficient of this dummy variable. We might solve the problem by re-specifying the model with more broadly defined income groups or more narrowly defined alternatives. Alternatively, we could postulate a *linear probability model*, in which probability rather than utility is a linear function of coefficients; this model has certain statistical disadvantages but is simple and may be adequate with large samples.

9.2.3 Data

Some of the most important variables for travel demand modeling are determined endogenously within a larger system of which the demand model is just one component. With aggregate data, the endogeneity of travel characteristics is an important issue for obtaining valid statistical estimates. Fortunately, endogeneity can usually be ignored when using disaggregate data because, from the point of view of the individual consumer, the travel environment does not depend appreciably on that one individual's choices.

Nevertheless, measuring the values of attributes z_{in}, which typically vary by alternative, is more difficult than it may first appear. How does one know the traits that a traveler would have encountered on an alternative that was not in fact used? One possibility is to use objective estimates, such as the *engineering values* produced by network models of the transportation system. Another is to use *reported values* obtained directly from survey respondents. Each is subject to problems. Reported values measure people's perceptions of travel conditions, which, even for alternatives they choose regularly, may differ from the measures employed in policy analysis or forecasting. Worse still, reported values may be systematically biased so as to justify the choice, thereby exaggerating the advantages of the alternative chosen and the disadvantages of other alternatives.

The data described thus far measure information about *revealed preferences* (RP), those reflected in actual choices. There is growing interest in using *stated preference* (SP) data, based on responses to hypothetical situations [12]. SP data permit more control over the ranges of and correlations among the independent variables, and they can also elicit information about potential travel options not available now. How accurately they described what people really do is still an open question. This is a very common dilemma in studies intended for use in engineering design, which have no choice but to rely on SP data if they concern product characteristics not available in actual situations.

It is possible to combine data from both revealed and stated preferences in a single estimation procedure in order to take advantage of the strengths of each [13]. As long as observations are independent of each other, the log-likelihood functions simply add. To prevent SP survey bias from contaminating inferences from RP, estimating certain parameters separately in the two portions of the data is recommended: namely, the scale factors μ for the two parts of the sample (with one but not both normalized), any alternative-specific constants, and any critical behavioral coefficients that may differ. The log-likelihood function Eq. (9.11) then breaks into two terms—one for RP observations and one for SP observations—with appropriate constraints among the coefficients in the two parts and with one part containing a relative scale factor to be estimated.

9.2.4 Interpreting Coefficient Estimates

When interpreting empirical results, it is useful to note that a change in $\beta' z_{in}$ in Eq. (9.10) by an amount of ±1 increases or decreases the relative odds of alternative i, compared to other alternatives, by a factor exp(1) = 2.72. Thus a quick gauge of the behavioral significance of any particular variable can be obtained by considering the size of typical variations in that variable, multiplying it by the relevant coefficient, and comparing with 1.0.

The parameter vector may contain *alternative-specific constants* for one or more alternatives i. That is, the systematic utility may be of the form

$$V_{in} = \alpha_i + \beta' z_{in} \qquad \text{Eq. (9.12)}$$

Since only utility differences matter, at least one of the alternative-specific constants must be normalized (usually to zero); that alternative then serves as a "base alternative" for comparisons. Of course, using alternative-specific constants makes it impossible to forecast the result of adding a new alternative unless there is some basis for a guess as to what its alternative-specific constant would be.

Equation (9.12) is really a special case of Eq. (9.10) in which one or more of the variables z are *alternative-specific dummy variables*, D^k, defined by D^k_{jn} = 1 if $j = k$ and 0 otherwise (for each j = 1,..., J). (Such a variable does not depend on n.) In this notation, parameter α_i in Eq. (9.12) is viewed as the coefficient of variable D^i included among the z variables in Eq. (9.10). Such dummy variables can also be interacted with (i.e., multiplied by) any other variable, making it possible for the latter variable to affect utility in a different way for each alternative. All such variables and interactions may be included in z, and their coefficients in β, thus allowing Eq. (9.10) still to represent the linear-in-parameters specification.

The most economically meaningful quantities obtained from estimating a discrete-choice model are often ratios of coefficients. By interacting the variables of interest with socioeconomic characteristics or alternative-specific constants, these ratios can be specified quite flexibly so as to vary in a manner thought to be a priori plausible. A particularly important example in transportation is the ratio of coefficients of time and money, often called the *value of travel-time savings*, or *value of time* for short. It represents the monetary value that the traveler places on an incremental time saving. Similarly, a per-unit value can be placed on any product attribute that consumers care about: for example, interior capacity of a vehicle, throughput rate of a communications device or resolution of a visual display.

The value of time in the model Eq. (9.9) is

$$(v_T)_{in} \equiv -\left(\frac{dc_{in}}{dT_{in}}\right)_{V_{in}} \equiv \frac{\partial V_{in} / \partial T_{in}}{\partial V_{in} / \partial c_{in}} = \left(\frac{\beta_2 + 2\beta_3 T_{in}}{\beta_1}\right) \cdot w_n \qquad \text{Eq. (9.13)}$$

which varies across individuals since it depends on w_n and T_{in}.

As a more complex example, suppose we extend Eq. (9.9) by adding alternative-specific dummies, both separately (with coefficients α_i) and interacted with travel time (with coefficients γ_i):

$$V_{in} = \alpha_1 + \beta_1 \cdot (c_{in} / w_n) + \beta_2 T_{in} + \beta_3 T_{in}^2 + \gamma_i T_{in} \qquad \text{Eq. (9.14)}$$

where one of the α_i and one of the γ_i are normalized to zero. This yields the following value of time applicable when individual n chooses alternative i:

$$(v_T)_{in} = \left(\frac{\beta_2 + 2\beta_3 T_{in} + \gamma_i}{\beta_1}\right) \cdot w_n \qquad \text{Eq. (9.15)}$$

Now the value of time varies across modes even with identical travel times, due to the presence of γ_i. In the same way, the value consumers place on a specified increase in resolution of a visual display could depend on the income (or any other characteristic) of the individual and on the particular model or display type selected.

Confidence bounds for a ratio of coefficients can be estimated by standard approximations for transformations of normal variates.[2] Or they can be estimated using a Monte Carlo procedure: take repeated random draws from the distribution of β (which is estimated along with β itself), and then examine the resulting values of the ratio in question. The empirical distribution of these values is an estimate of the actual distribution of the ratio, and one can describe it in any number of ways, including its standard deviation. As another example, the 5th and 95th percentile values of those values define a 90 percent confidence interval for β. See [7, Chapter 9] for how to take such random draws.

[2]Letting $v_T = \beta_2/\beta_1$, the standard deviation σ_v of v_T obeys the intuitive formula: $(\sigma_v/v_T)^2 \cong (\sigma_1/\beta_1)^2 + (\sigma_1/\beta_1)^2 - 2\sigma_{12}/(\beta_1\beta_2)$, where σ_1 and σ_2 are the standard deviations of β_1 and β_2, respectively, and σ_{12} is their covariance.

9.2.5 Randomness, Scale of Utility, Measures of Benefit and Forecasting

The variance of the random utility term in Eq. (9.4) reflects randomness in the behavior of individuals or, more likely, heterogeneity among observationally identical individuals. Hence it plays a key role in determining how sensitive travel behavior is to observable quantities such as price, service quality and demographic traits. Little randomness implies a nearly deterministic model, one in which behavior suddenly changes at some crucial switching point (for example, when transit service becomes as fast as a car). Conversely, if there is a lot of randomness, behavior changes only gradually as the values of independent variables are varied.

When the variance of the random component is normalized, however, the degree of randomness becomes represented by the inverse of the scale of the systematic utility function. For example, in the logit model Eq. (9.8), suppose systematic utility is linear in parameter vector β as in Eq. (9.10). If all the elements of β are small in magnitude, the corresponding variables have little effect on probabilities so choices are dominated by randomness. If the elements of β are large, most of the variation in choice behavior is explained by variation in the observable variables. Randomness in individual behavior can also be viewed as producing variety in aggregate behavior.

It is sometimes useful to have a measure of the overall desirability of the choice set being offered to a consumer. Such a measure must account for both the utility of the individual choices being offered and the variety of choices offered. The value of variety is directly related to randomness because both arise from unobserved idiosyncrasies in preferences. If choice were deterministic, the consumer would care only about the traits of the best alternative; improving or offering inferior alternatives would have no value. But with random utilities, there is some chance that an alternative with a low value of V_{in} will nevertheless be chosen; so it is desirable for such an alternative to be offered and to be made as attractive as possible. A natural measure of the desirability of choice set J is the expected maximum utility of that set, which for the logit model has the convenient form:

$$E \max_j (V_j + \varepsilon_j) = \mu^{-1} \log \sum_{j=1}^{J} \exp(\mu V_j) + \gamma \qquad \text{Eq. (9.16)}$$

where $\gamma = 0.5772$ is Euler's constant (it accounts for the nonzero mean of the error terms ε_j in the standard normalization). (Here we have retained the parameter μ, rather than normalizing it, to make clear how randomness affects expected utility.) When the amount of randomness is small (large μ), the summation on the right-hand side is dominated by its largest term (let's denote its index by j^*); expected utility is then approximately $\rho \cdot \log[\exp(V_{j^*}/\rho)] = V_{j^*}$, the utility of the dominating alternative. When randomness dominates (small μ), all terms contribute more or less equally (let's denote their average utility value by V); then expected utility is approximately $\mu^{-1} \cdot \log[J \cdot \exp(\mu V)] = V + \mu^{-1} \cdot \log(J)$, which is the average utility plus a term reflecting the desirability of having many choices.

Expected utility is, naturally enough, directly related to measures of consumer welfare. Small and Rosen [14] show that, in the absence of income effects, changes in aggregate consumer surplus (the area to the left of the demand curve and above the current price) are appropriate measures of welfare even when the demand curve is generated by a set of individuals making discrete choices. For a set of individuals n characterized by systematic utilities V_{jn}, changes in consumer surplus are proportional to changes in this expected maximum utility. The proportionality constant is the inverse of λ_n, the marginal utility of income. Thus a useful welfare measure for such a set of individuals, with normalization $\mu = 1$, is:

$$W = \frac{1}{\lambda_n} \log \sum_{j=1}^{J} \exp(V_{jn}) \qquad \text{Eq. (9.17)}$$

(The constant γ drops out of welfare comparisons so it is omitted here.) Because portions of the utility V_i that are common to all alternatives cannot be estimated from the choice model, λ_n cannot be estimated directly. However, typically it can be determined from Roy's Identity:

$$\lambda_n = -\frac{1}{x_{in}} \cdot \frac{\partial V_{in}}{\partial c_{in}} \qquad \text{Eq. (9.18)}$$

where x_{in} = consumption of good i conditional on choosing it from among the discrete alternatives. In the case of commuting-mode choice, for example, x_{in} is just the individual's number of work trips per year (assuming income and hence welfare are measured in annual units).

Once we have estimated a disaggregate travel-demand model, we face the question of how to predict aggregate quantities such as total transit ridership or total travel flows between zones. Ben-Akiva and Lerman [15, Chapter 6] discuss several methods. The most straightforward and common is *sample enumeration*. A sample of consumers is drawn, each assumed to represent a subpopulation with identical observable characteristics. (The estimation sample itself may satisfy this criterion and hence be usable as an enumeration sample.) Each individual's choice probabilities, computed using the estimated parameters, predict the shares of that subpopulation choosing the various alternatives. These predictions can then simply be added, weighting each sample member according to the corresponding subpopulation size. Standard deviations of forecast values can be estimated by Monte Carlo simulation methods.

9.2.6 Ordered and Rank-Ordered Models

Sometimes there is a natural ordering to the alternatives that can be exploited to guide specification. For example, suppose one wants to explain a household's choice from among owning no vehicle, one vehicle, or two or more vehicles. It is perhaps plausible that there is a single index of propensity to own many vehicles, and that this index is determined in part by observable variables like household size and employment status.

In such a case, an *ordered response* model might be assumed. In this model, the choice of individual n is determined by the size of a "latent variable" $y_n^* = \beta' z_n + \varepsilon_n$ with choice j occurring if this latent variable falls in a particular interval $[\mu_{j-1}, \mu_j]$ of the real line, where $\mu_0 = -\infty$ and $\mu_J = \infty$. The interval boundaries $\mu_1, \dots, \mu_{J-1}$ are estimated along with β, except that one of them can be normalized arbitrarily if $\beta' z_n$ contains a constant term. The probability of choice j is then

$$P_{jn} = \Pr[\mu_{j-1} < \beta' z_n + \varepsilon_n < \mu_j] = F(\mu_j - \beta' z_n) - F(\mu_{j-1} - \beta' z_n) \qquad \text{Eq. (9.19)}$$

where $F(\cdot)$ = cumulative distribution function assumed for ε_n. In the *ordered probit* model $F(\cdot)$ is standard normal, while in the *ordered logit* model it is logistic, i.e., $F(x) = [1 + \exp(-x)]^{-1}$. Note that all the variables in this model are characteristics of individuals, not of the alternatives, and thus if the latter information is available this model cannot easily take advantage of it.

In some cases the alternatives are integers indicating the number of times some random event occurs. An example would be the number of trips per month by a given household to a particular destination. For such cases, a set of models based on Poisson and negative binomial regressions is available [16]. In other cases, information is available not only on the most preferred alternative, but on the individual's ranking of other alternatives. Efficient use can be made of such data through the *rank-ordered logit* model, also called "expanded logit" [17].

9.3 EXAMPLES OF DISAGGREGATE MODELS

Discrete-choice models have been estimated for nearly every conceivable travel situation. In this section we present two examples.

9.3.1 Mode Choice

A series of models explaining choices of automobile ownership and commuting mode in the San Francisco Bay area were developed as part of planning for the Bay Area Rapid Transit System, which opened in 1975. One of the simplest models explains only the choice from among four modes: (1) auto alone; (2) bus with walk access; (3) bus with auto access; and (4) carpool (two or more occupants). The model's parameters are estimated from a sample of 771 commuters to San Francisco or Oakland who were surveyed prior to the opening of the Bay Area Rapid Transit system.[3]

Mode choice is explained by three independent variables and three alternative-specific constants. The three variables are: c_{in}/w_n, the round-trip variable cost (in \$ U.S.) of mode i for traveler n, divided by the traveler's post-tax wage rate (in \$ per minute); T_{in}, the in-vehicle travel time (in minutes); and T_{in}^o the out-of-vehicle travel time, including walking, waiting and transferring. Cost c_{in} includes parking, tolls, gasoline and maintenance. The estimated utility function is:

$$V = \underset{(0.0054)}{-0.0412 \cdot c/w} - \underset{(0.0072)}{0.0201 \cdot T} - \underset{(0.0070)}{0.0531 \cdot T^o} - \underset{(0.26)}{0.89 \cdot D^1} - \underset{(0.24)}{1.78 \cdot D^3} - \underset{(0.25)}{2.15 \cdot D} \qquad \text{Eq. (9.20)}$$

where the subscripts denoting mode and individual have been omitted, and standard errors of coefficient estimates are given in parentheses. Variables D^j are alternative-specific dummies.

This utility function is a simplification of Eq. (9.14) (with $\beta^3 = \gamma_i = 0$), except that travel time is broken into two components, T and T^o. Adapting Eq. (9.15), we see that the "value of time" for each of these two components is proportional to the post-tax wage rate: specifically, the estimated values of in-vehicle and out-of-vehicle time are 49 percent and 129 percent, respectively, of the after-tax wage. The negative alternative-specific constants indicate that the hypothetical traveler facing equal times and operating costs by all four modes will prefer bus with walk access (mode 2, the base mode); this is probably because each of the other three modes requires owning an automobile, which entails fixed costs not included in variable c. The strongly negative constants for bus with auto access (mode 3) and carpool (mode 4) probably reflect

[3]This is the "naive model" reported by McFadden et al. [18, pp. 121–123].

unmeasured inconvenience associated with getting from car to bus stop and with arranging carpools.

The model's fit could undoubtedly be greatly improved by including automobile ownership, perhaps interacted with ($D^1 + D^3 + D^4$) to indicate a common effect on modes that use an automobile. However, there is good reason to exclude it because it is endogenous—people choosing one of those modes for other reasons are likely to buy an extra car as a result. This, in fact, is demonstrated by the more complete model of Train [19], which considers both choices simultaneously. The way to interpret Eq. (9.20), then, is as a "reduced-form" model that implicitly incorporates the automobile ownership decision. It is thus applicable to a time frame long enough for automobile ownership to adjust to changes in the variables included in the model.

9.3.2 Choice of Free or Express Lanes

Lam and Small [20] analyze data from commuters who have the option of paying to travel in a set of express lanes on a very congested freeway. The data set contains cross-sectional variation in the cost of choosing the express lanes because the toll depends on time of day and on car occupancy, both of which differ across respondents. Travel time also varies by time of day, fortunately in a manner not too highly correlated with the toll. The authors construct a measure of the unreliability of travel time by obtaining data on travel times across many different days, all at the same time of day. After some experimentation, they choose the median travel time (across days) as the best measure of travel time, and the difference between 90th and 50th percentile travel times (also across days) as the best measure of unreliability. This latter choice is based on the idea that people are more averse to unexpected delays than to unexpected early arrivals.

The model explains a pair of related choices: (1) whether to acquire a transponder (required to ever use the express lanes); and (2) which lanes to take on the day in question. A natural way to view these choices is as a hierarchical set, in which the transponder choice is governed partly by the size of the perceived benefits of being able to use it to travel in the express lanes. As we will see in the next section, a model known as "nested logit" has been developed precisely for this type of situation, and indeed Lam and Small estimate such a model. As it happens, though, they obtain virtually identical results with a simpler "joint logit" model in which there are three alternatives: (1) no transponder; (2) have a transponder but travel in the free lanes on the day in question; and (3) have a transponder and travel in the express lanes on the day in question. The results of this model are[4]:

$$
\begin{aligned}
V = &-0.862 \cdot D^{\text{Tag}} + 0.023 \cdot Inc \cdot D^{\text{Tag}} - 0.766 \cdot ForLang \cdot D^{\text{Tag}} - 0.789 \cdot D^3 \\
&\quad (0.411) \qquad\quad (0.0058) \qquad\qquad\quad (0.412) \qquad\qquad\qquad (0.853) \\
&- 0.357c - 0.109 \cdot T - 0.159 \cdot R + 0.074 \cdot Male \cdot R + \text{(other terms)} \\
&\quad (0.138) \quad (0.056) \qquad (0.048) \qquad (0.046)
\end{aligned}
$$

Eq. (9.21)

Here $D^{\text{tag}} \equiv D^2 + D^3$ = a composite alternative-specific dummy variable for those choices involving a transponder, or "toll tag"; its negative coefficient presumably reflects the hassle and cost of obtaining one. Getting a transponder is apparently more attractive to people with high annual incomes (*Inc*, in $1,000s per year) and less attractive to those speaking a foreign language (dummy variable *ForLang*). The statistical insignificance of the coefficient of D^3, an alternative-specific dummy for using the express lanes, suggests that the most important explanatory factors are explicitly included in the model.

The coefficients on per-person cost c, median travel time T and unreliability R can be used to compute dollar values of time and reliability. Here we focus on two aspects of the resulting valuations: First, reliability is highly valued, achieving coefficients of similar magnitudes as travel time. Second, men seem to care less about reliability than women; their value is only 53 percent as high as women's, according to the estimates of the coefficient of unreliability (-0.159 for women, $-0.159 + 0.074 = -0.085$ for men). (A qualification to this is that the difference, i.e., the coefficient of $Male \cdot R$, is not quite statistically significant at a 10-percent significance level.) Several studies of this particular toll facility have found women noticeably more likely to use the express lanes than men, and this formulation provides tentative evidence that the reason is a greater aversion to the unreliability of the free lanes.

9.4 ADVANCED DISCRETE-CHOICE MODELING

9.4.1 Generalized Extreme Value Models

Often it is implausible that the additive random utility components ε_j be independent of each other, especially if important variables are omitted from the model's specification. This will make either logit or IID probit predict poorly.

A simple example is mode choice among automobile, bus transit and rail transit. The two public-transit modes are likely to have many unmeasured attributes in common. Suppose a traveler initially has available only auto ($j = 1$) and bus ($j = 2$), with equal systematic utilities V_j so that the choice probabilities are each one-half. Now suppose we want to predict the effects of adding a type of rail service ($j = 3$) whose measurable characteristics are identical to those for bus service. The IID models (without alternative-specific constants) would predict that all three modes would then have choice probabilities of one-third, whereas in reality the probability of choosing auto would most likely remain near one-half while the two transit modes divide the rest of the probability equally between them. The argument is even stronger if we imagine instead that the newly added mode is simply a bus of a different color: this is the famous "red bus, blue bus" example.

The probit model generalizes naturally, as already noted, by allowing the distribution function in Eq. (9.5) to be multivariate normal with an arbitrary variance-covariance matrix. It must be remembered that not all the elements of this matrix can be distinguished (*identified*, in econometric terminology) because, as noted, only the $(J - 1)$ utility differences affect behavior.[5]

The logit model generalizes in a comparable manner, as shown by McFadden [21, 22]. The distribution function is postulated to be *generalized extreme value* (GEV), given by: $F(\varepsilon_1,\ldots,\varepsilon_J) = \exp[-G(e^{-\varepsilon_1},\ldots,e^{-\varepsilon_J})]$, where G = a function satisfying

[4]This is a partial listing of the coefficients in Lam and Small [20, Table 11, Model 4b], with coefficients of T and R divided by 1.37 to adjust travel-time measurements to the time of the survey, as described on their p. 234 and Table 11, note a. Standard errors are in parentheses.

[5]The variance-covariance matrix of these utility differences has $(J - 1)^2$ elements and is symmetric. Hence there are only $J(J - 1)/2$ identifiable elements of the original variance-covariance matrix, less one for utility-scale normalization [23].

certain technical conditions. Logit is the special case $G(y_1,\ldots, y_J) = y_1 + \ldots + y_J$.

The best known GEV model, other than logit itself, is *nested logit*, also called *structured logit* or *tree logit*. In this model, certain groups of alternatives are postulated to have correlated random terms. This is accomplished by grouping the corresponding alternatives in G in a manner we can illustrate using the auto-bus-rail example, with auto as the first alternative:

$$G(y_1, y_2, y_3) = y_1 + (y_2^{1/\rho} + y_3^{1/\rho})^{\rho} \qquad \text{Eq. (9.22)}$$

Here ρ = a parameter between 0 and 1 that indicates the degree of dissimilarity between bus and rail; more precisely, $\sqrt{1-\rho^2}$ is the correlation between ε_1 and ε_2 [24]. The choice probability for this example may be written:

$$P_i = P(B_{r(i)}) \cdot P(i \mid B_{r(i)}) \qquad \text{Eq. (9.23)}$$

$$P(B_r) = \frac{\exp(\rho \cdot I_r)}{\sum_{s=1}^{2} \exp(\rho \cdot I_s)} \qquad \text{Eq. (9.24)}$$

$$P(i \mid B_r) = \frac{\exp(V_i / \rho)}{\sum_{j \in B_r} \exp(V_j / \rho)} \qquad \text{Eq. (9.25)}$$

where $B_1 = \{1\}$ and $B_2 = \{2, 3\}$ = a partition of the choice set into groups; $r(i)$ = indexes the group containing alternative i; and I_r = *inclusive value* of set B_r, defined as the logarithm of the denominator of Eq. (9.25):

$$I_r = \log \sum_{j \in B_r} \exp(V_j / \rho) \qquad \text{Eq. (9.26)}$$

When $\rho = 1$ in this model, ε_2 and ε_3 are independent and we have the logit model. As $\rho \downarrow 0$, ε_2 and ε_3 become perfectly correlated and we have an extreme form of the "red bus, blue bus" example, in which auto is pitted against the better (as measured by V_i) of the two transit alternatives; in this case $\rho I_1 \rightarrow V_1$ and $\rho I_2 \rightarrow \max\{V_2, V_3\}$.

The model just described can be generalized to any partition $\{B_r, r = 1,\ldots, R\}$ of alternatives, and each group B_r can have its own parameter ρ_r in Eqs. (9.22) to (9.26), leading to the form:

$$G(y_1,\ldots,y_J) = \sum_r \left(\sum_{j \in B_r} y_j^{1/\rho_r} \right)^{\rho_r} \qquad \text{Eq. (9.27)}$$

This is the general two-level nested logit model. It has choice probabilities Eqs. (9.23) to (9.26), except that the index s in the denominator of Eq. (9.24) now runs from 1 to R. The welfare measure for the two-level nested logit model is:

$$W = \frac{1}{\lambda} \log \sum_r \exp(\rho_r \cdot I_r) \qquad \text{Eq. (9.28)}$$

where again λ = marginal utility of income.

In nested logit, $\{B_r\}$ is an exhaustive partition of the choice set into mutually exclusive subsets. Therefore Eq. (9.25) is a true conditional probability, and the model can be estimated sequentially: First estimate the parameters (β / ρ) from Eq. (9.25), use them to form the inclusive values Eq. (9.26) and then estimate ρ from Eq. (9.24). Each estimation step uses an ordinary logit log-likelihood function, so it can be carried out with a logit algorithm. However, this sequential method is not statistically efficient and is rarely used today. Several studies show that maximum-likelihood estimation gives more accurate results [25].

A different direction for generalizing the logit model is to maintain independence between error terms while allowing each one to have a unique variance. This is the heteroscedastic extreme value model of Bhat [26]; it is a random-utility model but not in the GEV class, and its probabilities cannot be written in closed form so they require numerical integration. Other extensions of the logit model are described by Koppelman and Sethi [27].

9.4.2 Combined Discrete and Continuous Choice

In many situations, the choice among discrete alternatives is made simultaneously with some related continuous quantity. For example, a household's choice of type of automobile to own is closely intertwined with its choice of how much to drive. Estimating equations to explain usage, conditional on ownership, creates a *sample selection bias* [28]: For example, people who drive a lot are likely to select themselves into the category of owners of nice cars, so we could inadvertently overstate the independent effect of nice cars on driving. A variety of methods are available to remove this bias, as described in Train [29, Chapter 5] and Washington *et al.* [16, Chapter 12].

More elaborate systems of equations can be handled with the tools of *structural equations modeling*. These methods are quite flexible and allow one to try out different patterns of mutual causality, testing for the presence of particular causal links. They are often used when large data sets are available describing mutually related choices. Golob [30] provides a review.

9.4.3 Panel Data

Just as with aggregate data, data from individual respondents can be collected repeatedly over time. A good example is the Dutch Mobility Panel, in which travel-diary information was obtained from the same individuals (with some attrition and replacement) at 10 different times over the years 1984–89. The resulting data have been widely used to analyze time lags and other dynamic aspects of travel behavior [31].

The methods described earlier for aggregate longitudinal data are applicable to disaggregate data as well. In addition, attrition becomes a statistical issue: over time, some respondents will be lost from the sample and the reasons need not be independent of the behavior being investigated. The solution is to create an explicit model of what causes an individual to leave the sample, and to estimate it simultaneously with the choice process being considered. Pendyala and Kitamura [32] and Brownstone and Chu [33] analyze the issues involved.

9.4.4 Random Parameters and Mixed Logit

In the random utility model of Eqs. (9.4) and (9.5), randomness in individual behavior is limited to an additive error term in the utility function. Other parameters, and their functions, are deterministic: that is, variation in them is due only to observed variables. Thus, for example, the value of time defined by Eq. (9.13) varies with observed travel time and wage rate but otherwise is the same for everyone.

Experience has shown, however, that parameters of critical interest to transportation policy vary among individuals for reasons that we do not observe. Such reasons could be missing socioeconomic characteristics, personality, special features of the travel environment and data errors. These, of course, are the same reasons for the inclusion of the additive error term in utility function Eq. (9.4). So the question is, Why not also include randomness in the other parameters?

The only reason is tractability, and that has largely been overcome by advances in computing power. Consider first how one could allow a single parameter in the logit model to vary randomly across individuals. Suppose we specify a distribution, such as normal with unknown mean and variance, for the parameter in question. The overall probability is then determined by embedding the integral in Eq. (9.5c) within another integral over the density function of that distribution. This simple idea has been generalized to allow for general forms of randomness in many parameters, including alternative-specific constants, leading to a many-dimensional integral. Nevertheless, the model is tractable because the outer integration (over the distribution defining random parameters) can be performed using simulation methods based on random draws, while the inner integration (that over the remaining additive errors ε_{jn}) is unnecessary because, conditional on the values of random parameters, it yields the logit formula Eq. (9.8). The model is called *mixed logit* because the combined error term has a distribution that is a mixture of the extreme value distribution with the distribution of the random parameters.

Writing this out explicitly, the choice probability conditional on random parameters is

$$P_{in|\beta} = \frac{\exp(\beta' z_{in})}{\sum_j \exp(\beta' z_{jn})} \qquad \text{Eq. (9.29)}$$

Let $f(\beta|\Theta)$ = density function defining the distribution of random parameters, which depends on some unknown "meta-parameters" Θ (such as means and variances of β). The unconditional choice probability is then simply the multidimensional integral:

$$P_{in} = \int P_{in|\beta} \cdot f(\beta \mid \Theta) d\beta \qquad \text{Eq. (9.30)}$$

Integration by simulation consists of taking R random draws β^r, $r = 1,\ldots, R$, from distribution $f(\beta|\Theta)$, calculating $P_{in|\beta}$ each time and averaging over the resulting values: $P_{in}^{sim} = (1/R)\Sigma_{r=1}^{R} P_{in|\beta}^{r}$. Doing so requires, of course, assuming some trial value of Θ, just as calculating the usual logit probability requires assuming some trial value of β. Under reasonable conditions, maximizing the likelihood function defined by this simulated probability yields statistically consistent estimates of the meta-parameters Θ. Details are provided by Train [7].

Brownstone and Train [34] demonstrate how one can shape the model to capture anticipated patterns by specifying which parameters are random and what form their distribution takes–in particular, whether some of them are correlated with each other.[6] In their application, consumers state their willingness to purchase various makes and models of cars, each specified to be powered by one of four fuel types: gasoline (G), natural gas (N), methanol (M), or electricity (E). Respondents were asked to choose from among hypothetical vehicles with specified characteristics. A partial listing of estimation results is as follows: $V = -0.264 \cdot [p/\ln(inc)] + 0.517 \cdot range + (1.43+7.45\phi_1) \cdot size + (1.70+5.99\phi_2) \cdot luggage + 2.46\phi_3 \cdot nonE + 1.07\phi_4 \cdot nonN$ + (other terms), where p (vehicle price) and *inc* (income) are in thousands of dollars; the *range* between refueling (or recharging) is in hundreds of miles; *luggage* is luggage space relative to a comparably sized gasoline vehicle; non$E$ is a dummy variable for cars running on fuel that must be purchased outside the home (in contrast to electric cars); non$N$ is a dummy for cars running on fuel stored at atmospheric pressure (in contrast to natural gas); and $\phi_1 - \phi_4$ are independent random variables with the standard normal distribution. All parameters shown above are estimated with enough precision to easily pass tests of statistical significance.

This model provides for observed heterogeneity in the effect of price on utility, since it varies with *inc*. It provides for random coefficients on size and *luggage*, and for random constants as defined by *nonE* and *nonN*. This can be understood by examining the results term by term.

The terms in parentheses involving ϕ_1 and ϕ_2 represent the random coefficients. The coefficient of *size* is random with mean 1.43 and standard deviation 7.45. Similarly, the coefficient of *luggage* has mean 1.70 and standard deviation 5.99. These estimates indicate a wide variation in people's evaluation of these characteristics. For example, it implies that many people (namely, those for whom $\phi_2 < -1.70/5.99$) actually prefer less luggage space; presumably they do so because a smaller luggage compartment allows more interior room for the same size of vehicle. Similarly, preference for vehicle size ranges from negative (perhaps due to easier parking for small cars) to substantially positive.

The terms involving ϕ_3 and ϕ_4 represent random alternative-specific constants with a particular correlation pattern, predicated on the assumption that groups of alternatives share common features for which people have idiosyncratic preferences—very similar to the rationale for nested logit. Each of the dummy variables *nonE* and *nonN* is simply a sum of alternative-specific constants for those car models falling into a particular group. The two groups overlap: any gasoline-powered or methanol-powered car falls into both. If the coefficients of ϕ_3 and ϕ_4 had turned out to be negligible, then these terms would play no role and we would have the usual logit probability conditional on the values of ϕ_1 and ϕ_2. But the coefficients are not negligible, so each produces a correlation among utilities for those alternative in the corresponding group. For example, all cars that are not electric share a random utility component $2.46\phi_3$, which has standard deviation 2.46 (since ϕ_3 has standard deviation one by definition). Thus the combined additive random term in utility (including the random constants), $\varepsilon_{in} + 2.46\phi_{3n} \cdot nonE_i + 1.07\phi_{4n} \cdot nonN_i$, exhibits correlation across those alternatives i representing cars that are not electric. A similar argument applies to ϕ_4, which produces correlation across those alternatives representing cars that are not natural gas. Those alternatives falling into both *nonE* and *nonN* are even more highly correlated with each other. Note that because the distributions of ϕ_3 and ϕ_4 are centered at zero, this combined random term does not imply any overall average preference for or against various types of vehicles; such absolute preferences are in fact included in *other terms*.

The lesson from this example is that mixed logit can be used not only to specify unobserved randomness in the coefficients of certain variables, but also to mimic the kinds of correlation patterns among the random constants for which the GEV model was developed. Indeed, McFadden and Train [36] show that it can closely approximate virtually any choice model based on random utility.

---

[6] The following simplified explanation is adapted from Small and Winston [35].

9.5 VALUE OF TIME AND RELIABILITY

Among the most important quantities inferred from travel demand studies are the monetary values that people place on saving various forms of travel time or improving the predictability of travel time. The first, loosely known as the *value of time* (VOT), is a key parameter in cost-benefit analyses that measures the benefits brought about by transportation policies or projects. The second, the *value of reliability* (VOR), also appears important, but accurate measurement is a science in its infancy. The benefits or losses due to changes in time and reliability are normally captured as part of consumer surplus, for example that given by Eq. (9.17), as long as they are part of the demand model. However, it is often enlightening to separate them explicitly.

9.5.1 Value of Time

The most natural definition of value of time is in terms of compensating variation. The value of saving a given amount and type of travel time by a particular person is the amount that person could pay, after receiving the saving, and be just as well off as before. This amount, divided by the time saving, is that person's average value of time saved for that particular change. Aggregating over a class of people yields the *average value of time* for those people in that situation. The limit of this average value, as the time saving shrinks to zero, is called the *marginal value of time*, or just "value of time"; by definition, it is independent of the amount of time saving. It was defined empirically in Eq. (9.13).

Value of time may depend on many aspects of the trip maker and of the trip itself. To name just a few, it depends on trip purpose (e.g., work or recreation), demographic and socioeconomic characteristics, time of day, physical or psychological amenities available during travel and the total duration of the trip. There are two main approaches to specifying a travel-demand model so as to measure such variations. One is known as *market segmentation*: the sample is divided according to criteria such as income and type of household, and a separate model is estimated for each segment. This has the advantage of imposing no potentially erroneous constraints, but the disadvantage of requiring many parameters to be estimated, with no guarantee that these estimates will follow a reasonable pattern. The second approach uses theoretical reasoning to postulate a functional form for utility that determines how VOT varies. This approach often builds on a framework by Becker [37], in which utility is maximized subject to a time constraint. Becker's theory has been elaborated in many directions, most of which predict some relationship between value of time and the wage rate. For example, the theory of Oort [38] predicts that the value of time will exceed the wage rate if time spent at work is enjoyed relative to that spent traveling, and fall short of it if the opposite is true. Thus the value of time, even for nonwork trips, depends on conditions of the person's job.

These theories can provide guidance about how to specify the systematic utilities V_k in a discrete choice model. Suppose, for example, one believes that work is disliked (relative to travel), with its relative marginal disutility a fixed fraction of the wage rate. Then the value of time is a fraction of the wage rate as, for example, in specification Eq. (9.9) with $\beta_3 = 0$. Alternatively, one might think that work enjoyment varies nonlinearly with the observed wage rate—perhaps negatively due to wage differentials that compensate for working conditions, or perhaps positively due to employers' responses to an income-elastic demand for job amenities. Then the value of time is a nonlinear function of the wage rate, which could suggest using Eq. (9.9) with a nonzero term, β_3.

9.5.2 Value of Reliability

It is well known that uncertainty in travel time, which may result from congestion or poor adherence to transit schedules, is a major perceived cost of travel. A parallel with other types of products is fairly obvious: uncertainty in how well a product will perform the desired function will reduce its value to the user.

How can reliability be captured in a theoretical model of travel? Adapting Noland and Small [39], we can begin with a model of trip-scheduling choice, in which the trip cost depends on the degree of adherence to a desired time of arrival at work. Define *schedule delay*, S_D, as the difference (in minutes, rounded to the nearest five minutes) between the arrival time represented by a given alternative and the official work start time t^*. Define "schedule delay late" as $SDL = \text{Max}\{S_D,0\}$ and "schedule delay early" as $SDE = \text{Max}\{-S_D,0\}$. Define a "late dummy," DL, equal to one for the on-time and all later alternatives and equal to 0 for the early alternatives. Define T as the travel time (in minutes) encountered at each alternative. Suppose, then, that the trip cost is a linear function of these variables:

$$C(t_d,T_r) = \alpha \cdot T + \beta \cdot SDE + \gamma \cdot SDL + \theta \cdot DL \quad \text{Eq. (9.31)}$$

where $\alpha \equiv v_T/60$ is the per-minute value of travel time; β and γ = per-minute costs of early and late arrival; and θ = a fixed cost of arriving late. The functional notation $C(t_d,T_r)$ is to remind us that each of the components of the trip cost depends on the departure time, t_d, and a random (unpredictable) component of travel time, $T_r \geq 0$. Our objective is to measure the increase in expected cost C due to the dispersion in T_r, given that t_d is subject to choice by the traveler. Letting C^* denote this expected cost after the user chooses t_d optimally, we have

$$C^* = \underset{t_d}{\text{Min}} E[C(t_d,t_r)] = \underset{t_d}{\text{Min}}[\alpha \cdot E(T) + \beta \cdot E(SDE) + \gamma \cdot E(SDL) + \theta \cdot P_L] \quad \text{Eq. (9.32)}$$

where E = an expected value taken over the distribution of T_r; and $P_L \equiv E(DL)$ = probability of being late. This equation can form the basis for specifying the reliability term in a model like Eq. (9.21).

To focus just on reliability, let's ignore congestion for now by assuming that $E(T)$ is independent of departure time. Remarkably, the optimal value of t_d then does not depend on the distribution of T_r, provided that its probability density is finite everywhere. To find this optimal departure time, let $f(T_r)$ be the distribution function, and let T_f be travel time when $T_r = 0$. The next-to-last term in the square brackets of Eq. (9.32) can then be written as:

$$\gamma \cdot E(SDL) = \gamma \cdot E(t_d + T_r - \tilde{t} \mid T_r > \tilde{t} - t_d)$$
$$= \gamma \cdot \int_{\tilde{t}-t_d}^{\infty} (t_d + T_r - \tilde{t}) \cdot f(T_r)dT_r$$

where $\tilde{t} \equiv t^* - T_f$ time the traveler would depart if T_r were equal to zero with certainty. Differentiating yields:

$$\frac{d}{dt_d}\gamma \cdot E(SDL) = 0 + \gamma \cdot \int_{\tilde{t}-t_d}^{\infty} \left[\frac{d}{dt_d}(t_d + T_r - \tilde{t}) \cdot f(T_r)\right]dT_r = \gamma P_L^*$$

where P_L^* = optimal value of the probability of being late.[7] Similarly, differentiating the term involving β in Eq. (9.32) yields $-\beta\cdot(1-P_L^*)$. Finally, differentiating the last term yields $-\theta f^0$, where $f^0 \equiv f(\tilde{t}-t_d^*)$ is the probability density at the point where the traveler is neither early nor late. Combining all three terms and setting them equal to zero gives the first-order condition for optimal departure time:

$$P_L^* = \frac{\beta+\theta f^0}{\beta+\gamma} \qquad \text{Eq. (9.33)}$$

In general this does not yield a closed-form solution for t_d^* because f^0 depends on t_d^*. However, in the special case $\theta = 0$, it yields $P_L^* = \beta/(\beta+\gamma)$, a very intuitive rule for setting departure time that is noted by Bates et al. [40, p. 202]. The rule balances the aversions to early and late arrival.

The cost function itself has been derived in closed form for two cases: a uniform distribution and an exponential distribution for T_r. In the case of a uniform distribution with range b, Eq. (9.33) again simplifies to a closed form: $P_L^* = [\beta+(\theta/b)]/[\beta+\gamma]$. The value of C^* in this case is given by Noland and Small [39] and Bates et al. [40]. In the special case $\theta = 0$, it is equal to the cost of expected travel time, $\alpha\cdot E(T)$, plus the following cost of unreliability:

$$v_R = \left(\frac{\beta\gamma}{\beta+\gamma}\right)\cdot\frac{b}{2} \qquad \text{Eq. (9.34)}$$

The quantity in parentheses is a composite measure of the unit costs of scheduling mismatch, which plays a central role in the cost functions considered in the next chapter. Thus Eq. (9.34) indicates that reliability cost derives from the combination of costly scheduling mismatches and dispersion in travel time.

More generally, we can see from Eq. (9.32) that whatever the form of the distribution of uncertain travel time, the expected trip cost will increase with dispersion in that distribution. Furthermore, if $\gamma > \beta$ and/or if θ is large, both of which are confirmed by the empirical findings of Small [41], expected cost will be especially sensitive to the possibility of values of T_r high enough to make the traveler late even though t_d is chosen optimally. Therefore, the cost of unreliability depends especially on the upper tail of the distribution of uncertain travel times. This property was used in creating the reliability variable in the study by Lam and Small [20] described earlier.

In a similar manner, the reliability of a product design may need to be measured primarily by one part of the distribution of random events associated with the product's functioning. If a boat rudder bends under certain wave conditions, this may reduce its efficiency, with some minor loss of value; whereas if it bends so far as to break, the loss is much greater.

9.5.3 Empirical Results

Research has generated an enormous literature on empirical estimates of value of time, and a much smaller one on the value of reliability. Here we rely mainly on reviews of this literature by others.

Reviewing studies for the U.K., Wardman [42, Table 6] finds an average VOT of £3.58/hour in late 1994 prices, which is 52 percent of the corresponding wage rate.[8] Gunn [43] find that Dutch values used by planners in the late 1980s track British results (by household income) quite well; however, he also finds a substantial unexplained downward shift in the profile for 1997, possibly resulting from better in-vehicle amenities. Transport Canada [44] and U.S. Department of Transportation [45] recommend using a VOT for personal travel by automobile equal to 50 percent of the gross wage rate. A French review by the Commissariat Général du Plan [46, p. 42] finds VOT to be 59 percent of the wage on average for urban trips. Finally, a Japanese review suggests using 2,333 yen/hour for weekday automobile travel in 1999, which was 84 percent of the wage rate.[9]

There is considerable evidence that VOT rises with income but less than proportionally. The easiest way to summarize this issue is in an elasticity of VOT with respect to income. Wardman [47], using a formal meta-analysis, finds elasticity to be 0.72 when income is measured as gross domestic product per capita. Wardman's [48] meta-analysis focuses on how VOT depends on various trip attributes. There is a small positive relationship (elasticity 0.13) with trip distance, a 16 percent differential between commuting and leisure trips, and considerable differences across modes, with bus riders having a lower-than-average value and rail riders a higher-than-average value—possibly due to self-selection by speed. Most important, walking and waiting time are valued much higher than in-vehicle time—a universal finding conventionally summarized as 2 to $2\frac{1}{2}$ times as high, although Wardman finds them to be only 1.6 times as high.

One unsettled methodological issue is an apparent tendency for SP data to yield considerably smaller values of time than RP data. Brownstone and Small [49] find that SP results for VOT are one-third to one-half the corresponding RP results. One possible explanation for this difference is hinted at by the finding from other studies that people overestimate the actual time savings from the toll roads by roughly a factor of two; thus when answering SP survey questions, they may indicate a per-minute willingness to pay for *perceived* time savings that is lower than their willingness to pay for *actual* time savings. If one wants to use a VOT for purposes of policy analysis, one needs it to correspond to actual travel time since that is typically the variable considered in the analysis. Therefore, if RP and SP values differ when both are accurately measured, it is the RP values that are relevant for most purposes.

From this evidence, it appears that the value of time for personal journeys is almost always between 20 and 90 percent of the gross wage rate, most often averaging close to 50 percent. Although it varies somewhat less than proportionally with income, it is close enough to proportional to make its expression as a fraction of the wage rate a good approximation and more useful than its expression as an absolute amount. There is universal agreement that VOT is much higher for travel while on business, generally recommended to be set at 100 percent of total compensation including benefits. The value of walking and waiting time for transit trips is probably 1.6 to 2.0 times that of in-vehicle time, not counting some

[7] The term "0" in this equation arises from differentiating the lower limit of integration: $-[d(\tilde{t}-t_d)/dt_d]\cdot[(t_d+T_r-\tilde{t})\cdot f(T_r)]_{T_r=\tilde{t}-t_d} = 1\cdot 0 = 0$.

[8] Mean gross hourly earnings for the U.K. were £6.79 and £7.07/hour in spring 1994 and 1995, respectively. Source: National Statistics Online [50, Table 38].

[9] Japan Research Institute Study Group on Road Investment Evaluation [51, Table 3-2-2], using car occupancy of 1.44 (p. 52). Average wage rate is calculated as cash earnings divided by hours worked, from [52].

context-specific disutility of having to transfer from one vehicle to another.

There has been far less empirical research on VOR. Most of it has been based on SP data for at least two reasons: (1) if is difficult to measure unreliability in actual situations; and (2) unreliability tends to be correlated with travel time itself. However, a few recent studies, including [20], have had some success with RP data. Brownstone and Small [49] review several such studies in which unreliability is defined as the difference between the 90th and 50th percentile of the travel-time distribution across days, or some similar measure; in those studies, VOR tends to be of about the same magnitude as VOT. One of those studies, using data from the high-occupancy toll (HOT) lane on State Route 91 in the Los Angeles region, finds that roughly two-thirds of the advantage of the HOT lane to the average traveler is due to its lower travel time and one-third is due to its higher reliability.[10] In prospective studies of a possible £4 cordon toll for Central London, May, Coombe and Gilliam [53] estimate that reliability would account for 23 percent of the benefits to car users.

9.6 CONCLUSIONS

The methods discussed here have spread far beyond transportation to applications in labor economics, industrial organization and many other fields. The field of marketing has taken them up with special vigor, adapting and refining them to match the kinds of data often elicited in marketing surveys. Some of the refinements involve more sophisticated models, sometimes made feasible by large volumes of data. Others involve SP methodology, which is prevalent in marketing studies. Researchers have paid considerable attention to using information on the demand for product characteristics to forecast the reaction to new products.

In these and other ways, methods from travel demand analysis can bring to bear information on how consumers value the characteristics under consideration in design problems, as well as how the demand for products will depend on those design decision. There is ample room for specialists in design to both use and contribute to the tools described here.

REFERENCES

1. Kain, J. F. and Liu, Z., 2002. "Efficiency and Locational Consequences of Government Transport Policies and Spending in Chile," *Chile: Political Economy of Urban Development*, E. L. Glaeser and J. R. Meyer, eds., pp. 105–195.
2. McFadden, D., 2001. "Economic Choices," *Am. Eco. Rev.*, Vol. 91, pp. 351–378.
3. Gómez–Ibáñez, J. A., 1996. "Big-City Transit Ridership, Deficits and Politics: Avoiding Reality in Boston," *J. of the Am. Planning Assoc.*, Vol. 62, pp. 30–50.
4. Greene, D. L., 1992. "Vehicle Use and Fuel Economy: How Big is the Rebound Effect?" *Energy J.*, Vol. 13, pp. 117–143.
5. Kitamura, R., 2000. "Longitudinal Methods," *Handbook of Transport Modelling*, D. Hensher and K. Button, eds., Pergamon, Elsevier Science, Amsterdam, The Netherlands, pp. 113–129.
6. Voith, R., 1997. "Fares, Service Levels, and Demographics: What Determines Commuter Rail Ridership in the Long Run?" *J. of Urban Eco.*, Vol. 41, pp. 176–197.
7. Train, K., 2003. *Discrete Choice Methods With Simulation*, Cambridge University Press, Cambridge, UK.
8. McFadden, D., 1973. "Conditional Logit Analysis of Qualitative Choice Behavior," *Frontiers in Econometrics*, P. Zarembka, ed., Academic Press, New York, NY, pp. 105–142.
9. Horowitz, J. L., 1980. "The Accuracy of the Multinomial Logit Model as an Approximation to the Multinomial Probit Model of Travel Demand," *Transportation Res. Part B*, Vol. 14, pp. 331–341.
10. Hastings, N. A. J. and Peacock, J. B., 1975. *Statistical Distributions: A Handbook for Students and Practitioners*, Butterworth, London, U.K.
11. Manski, C. F. and Lerman, S. R., 1977. "The Estimation of Choice Probabilities From Choice Based Samples," *Econometrica*, Vol. 45, pp. 1977–1988.
12. Hensher, D. A., 1994. "Stated Preference Analysis of Travel Choices: The State of Practice," *Transportation*, Vol. 21, pp. 107–133.
13. Louviere, J. J. and Hensher, D. A., 2001. "Combining Sources of Preference Data," *Travel Behaviour Research: The Leading Edge*, D. A. Hensher, ed., Pergamon, Oxford, pp. 125–144.
14. Small, K. A. and Rosen, H. S., 1981. "Applied Welfare Economics With Discrete Choice Models," *Econometrica*, Vol. 49, pp. 105–130.
15. Ben-Akiva, M. and Lerman, S. R., 1985. *Discrete Choice Analysis: Theory and Application to Travel Demand*, MIT Press, Cambridge, MA.
16. Washington, S. P., Karlaftis, M. G. and Mannering, F. L., 2003. *Statistical and Econometric Methods for Transportation Data Analysis*, Chapman and Hall, Boca Raton, FL.
17. Hausman, J. A. and Ruud, P. A., 1987. "Specifying and Testing Econometric Models for Rank-Ordered Data," *J. of Econometrics*, Vol. 34, pp. 83–104.
18. McFadden, D., Talvitie, A. P. et al. 1977, "*Demand Model Estimation and Validation. Urban Travel Demand Forecasting Project.*" *Special Rep. UCB–ITS–SR–77–9, Phase I Final Rep. Ser., Vol. V*, University of California Institute of Transportation Studies, Berkeley, CA.
19. Train, K. 1980. "A Structured Logit Model of Auto Ownership and Mode Choice," *Rev. of Eco. Studies*, Vol. 47, pp. 357–370.
20. Lam, T. C. and Small, K. A., 2001. "The Value of Time and Reliability: Measurement from a Value Pricing Experiment," *Transportation Res. Part E*, Vol. 37, pp. 231–251.
21. McFadden, D., 1978. "Modelling the Choice of Residential Location," *Spatial Interaction Theory and Planning Models*, A. Karlqvist, L. Lundqvist, F. Snickars and J. W. Weibull, eds., North-Holland, Amsterdam, The Netherlands, pp. 75–96.
22. McFadden, D., 1981. "Econometric Models of Probabilistic Choice," *Structural Analysis of Discrete Data with Econometric Applications*, C. F. Manski and D. McFadden, eds., MIT Press, Cambridge, MA., pp. 198–272.
23. Bunch, D. S., 1991. "Estimability in the Multinomial Probit Model," *Transportation Res. Part B*, Vol. 25, pp. 1–12.
24. Daganzo, C. F. and Kusnic, M., 1993. "Two Properties of the Nested Logit Model," *Transportation Sci.*, Vol. 27, pp. 395–400.
25. Brownstone, D. and Small, K. A., 1989. "Efficient Estimation of Nested Logit Models," *J. Bus. and Eco. Statistics*, Vol. 7, pp. 67–74.
26. Bhat, C., 1995. "A Heteroscedastic Extreme Value Model of Intercity Travel Mode Choice," *Transportation Res. Part B*, Vol. 29, pp. 471–483.
27. Koppelman, F. S. and Sethi, V., 2000. "Closed–Form Discrete–Choice Models," *Handbook of Transport Modelling*, D. Hensher and K. Button, eds., Pergamon, Elsevier Science, Amsterdam, The Netherlands, pp. 211–227.
28. Heckman, J. J., 1979. "Sample Selection Bias as a Specification Error," *Econometrica*, Vol. 47, pp. 153–162.
29. Train, K., 1986. *Qualitative Choice Analysis: Theory, Econometrics, and an Application to Automobile Demand*, MIT Press, Cambridge, MA.
30. Golob, T. F., 2003. "Structural Equation Modeling for Travel Behavior Research," *Transportation Res. Part B*, Vol. 37, pp. 1–25.
31. Van Wissen, L. J. G. and Meurs, H. J., 1989. "The Dutch Mobility Panel: Experiences and Evaluation," *Transportation*, Vol. 16, pp. 99–119.
32. Pendyala, R. M. and Kitamura, R., 1997. "Weighting Methods for Attrition in Choice-Based Panels," *Panels for Transportation Planning: Methods and Applications*, T. F. Golob,R. Kitamura and L. Long, eds., Kluwer, Dordrecht, The Netherlands, pp. 233–257.

[10]An updated version of that study is [54].

33. Brownstone, D. and Chu, X., 1997. "Multiply-Imputed Sampling Weights for Consistent Inference with Panel Attrition," *Panels for Transportation Planning: Methods and Applications*, T. F. Golob, R. Kitamura and L. Long, eds., Kluwer, Dordrecht, The Netherlands, pp. 259–273.
34. Brownstone, D. and Train, K., 1999. "Forecasting New Product Penetration With Flexible Substitution Patterns," *J. of Econometrics*, Vol. 89, pp. 109–129.
35. Small, K. A. and Winston, C., 1999. "The Demand for Transportation: Models and Applications," *Transportation Policy and Economics: A Handbook in Honor of John R. Meyer*, J. A. Gómez-Ibáñez, W. Tye and C. Winston, eds., Brookings Institution, Washington, D.C., pp. 11–55.
36. McFadden, D. and Train, K., 2000, "Mixed MNL Models for Discrete Response," *J. of Appl. Econometrics*, Vol. 15, pp. 447–470.
37. Becker, G. S., 1965. "A Theory of the Allocation of Time," *Eco. J.*, Vol. 75, pp. 493–517.
38. Oort, C. J., 1969. "The Evaluation of Travelling Time," *J. of Transport Eco. and Policy*, Vol. 3, pp. 279–286.
39. Noland, R. B. and Small, K. A., 1995. "Travel-Time Uncertainty, Departure Time Choice, and the Cost of Morning Commutes," *Transportation Res. Rec.*, Vol. 1493, pp. 150–158.
40. Bates, J., Polak, J., Jones, P. and Cook, A., 2001. "The Valuation of Reliability for Personal Travel," *Transportation Res. E: Logistics and Transportation Rev.*, Vol. 37, pp. 191–229.
41. Small, K. A., 1982. "The Scheduling of Consumer Activities: Work Trips," *Ame. Eco. Rev.*, Vol. 72, pp. 467–479.
42. Wardman, M., 1998. "The Value of Travel Time: A Review of British Evidence," *J. of Transport Eco. and Policy*, Vol. 32, pp. 285–316.
43. Gunn, H. 2001. "Spatial and Temporal Transferability of Relationships between Travel Demand, Trip Cost and Travel Time," *Transportation Res. Part E*, Vol. 37, pp. 163–189.
44. Transport Canada, 1994. *Guide to Benefit-Cost Analysis in Transport Canada, http://www.tc.gc.ca/finance/BCA/en/TOC_e.htm,* Ottawa accessed December 30, 2004.
45. U.S. Department of Transportation, 1997. *The Value of Travel Time: Departmental Guidance for Conducting Economic Evaluations*, Washington, D.C.
46. Commissariat Général du Plan, 2001. *Transports: Choix des Investissements et coût des nuisances (Transportation: Choice of Investments and the Cost of Nuisances)*, Paris, June.
47. Wardman, M., 2004. "Public Transport Values of Time," *Transport Policy*, Vol. 11, pp. 363–377.
48. Wardman, M., 2001. "A Review of British Evidence on Time and Service Quality Valuations," *Transportation Res. Part E*, Vol. 37, pp. 107–128.
49. Brownstone, D. and Small, K. A., 2005. "Valuing Time and Reliability: Assessing the Evidence from Road Pricing Demonstrations," *Transportation Res. Part A*, Vol. 39, pp. 279–293.
50. U.K. National Statistics Online, 2004. *Labour Force Survey (LFS) Historical Quarterly Supplement, http://www.statistics.gov.uk/STATBASE/Expodata/Spreadsheets/D7938.xls*, accessed December 18, 2004.
51. Japan Research Institute Study Group on Road Investment Evaluation, 2000. *Guidelines for the Evaluation of Road Investment Projects*, Tokyo, Japan.
52. Japan Ministry of Health, Labour and Welfare, 1999. *Final Report of Monthly Labour Survey: July 1999,http://www.mhlw.go.jp/english/database/db-l/*, accessed December 30, 2004.
53. May, A. D., Coombe, D. and Gilliam, C., 1996. "The London Congestion Charging Research Programme. 3: The Assessment Methods," *Traffic Engrg. and Control*, Vol. 37, pp. 277–282.
54. Small, K. A., Winston, C. and Yan, J., 2005. "Uncovering the Distribution of Motorists' Preferences for Travel Time and Reliability: Implications for Road Pricing," *Econometrica*, Vol. 73, pp. 1367–1382.

ACKNOWLEDGMENT

This chapter is adapted from Chapter 2 of *Urban Transportation Economics*, by K. A. Small and E. Verhoef (2nd Ed. Routledge, forthcoming 2007). The work has benefited from past or recent comments by Alex Anas, Richard Arnott, David Brownstone, Marc Gaudry, Amihai Glazer, David Hensher, Sergio Jara–Díaz, Charles Lave, Kenneth Train and Clifford Winston. All responsibility for accuracy and interpretation lies with the author.

PROBLEMS

9.1. Suppose the choice between travel by automobile (alternative 1) and bus (alternative 2) is determined by the following logit model:

$$P_2 = \frac{1}{1+\exp\left[(\beta_1 / w)\cdot(c_1 - c_2)+\beta_2\cdot(T_1 - T_2)\right]}$$

where w = wage rate; and c_i and T_i = cost and time of using alternative i.

a. If c_2 is varied, show that the probability of choosing bus varies according to the derivative:

$$\frac{dP_2}{dc_2} = (\beta_1 / w)\cdot P_2\cdot(1-P_2)$$

b. Write the above formula as a price elasticity of demand for bus travel, assuming combined travel by automobile and bus is fixed. You may assume that everybody has the same wage w and bus fare c_2.

Note: If a demand function is $Q = Q(c)$ where c is price, the price elasticity of demand is defined as $(c/Q)\cdot dQ/dc$.

c. Derive the cross-price elasticity of demand for bus travel: that is, the elasticity of P_2 with respect to the cost of automobile travel, c_1.

d. For a *small* increase in cost c_1 of automobile travel in this model, say from c_1 to $c_1 + \Delta c_1$, the expected loss in consumer surplus is just $P_1\Delta c_1$. Show that this is identical to the change in the welfare measure given by the following equation:

$$W = \frac{1}{\lambda}\left[\exp(V_1)+\exp(V_2)\right]$$

where $V_i = (\beta_1/w)\cdot c_i + \beta_2\cdot T_i$.

9.2.

Suppose utility is

$$U_i = \beta' z_i + \varepsilon_i \qquad i = 1, 2$$

and stochastic terms ε_i are identically and independently distributed (IID) with the extreme value distribution, whose cumulative distribution function (CDF) is:

$$F(\varepsilon) = \exp[-e^{-(\varepsilon-a)/b}]$$

a. Compute the mean and variance of the distribution.

Hint: $\int_0^\infty e^{-x}\log(x)dx = -\gamma;\qquad \int_0^\infty e^{-x}\left[\log(x)\right]^2 dx = (\pi^2/6)+\gamma^2$

where γ = Euler's constant = $\lim_{p\to\infty}\left[\sum_{n=1}^{p}(1/n) - \log(p)\right] = 0.5772157$

b. Show that the CDF of $(\varepsilon_2 - \varepsilon_1)$ is logistic, i.e.,

$$F(x) \equiv \Pr(\varepsilon_2 - \varepsilon_1 \le x) = \frac{1}{1+\exp(\beta' z_2 / b)}$$

c. Show that the probability of choosing alternative 1 is logit, i.e.,

$$P_1 = \frac{\exp(\beta' z_1 / b)}{\exp(\beta' z_1 / b) + \exp(\beta' z_2 / b)}$$

d. Explain from this formula why we can normalize a and b, for example $a = 0$, $b = 1$.

e. Show that as z_1 and z_2 vary, they affect P_1 only through their *difference* $(z_1 - z_2)$.

9.3.

This problem is about aggregating alternatives, and the properties of the resulting aggregate sets. *Note: This problem does not require any integration, providing you make use of results in the chapter!*

Let $U_i = V_i + \varepsilon_i$ for $i = 1,\ldots, 4$, where V_i is a known constant and ε_i are IID error terms with extreme value distribution (normalized as usual with location parameter 0 and scale parameter 1).

(a) Write the CDF of U_i.

(b) Define $U_a = \max\{U_1, U_2\}$. Use the answer to part (a) to write the CDF of U_a; and show that it is an extreme value distribution with a nonzero location parameter. Use this result to transform U_a to a random variable that has a normalized extreme value distribution.

(c) Define similarly $U_b = \max\{U_3, U_4\}$, and transform it just as you did U_a. Use the resulting CDFs to derive the probability that $U_a > U_b$. (*Hint: Make use of the fact that if two random variables are IID with the normalized extreme value distribution, their difference has a logistic distribution, as shown in problem 9.2b.*)

9.4.

How would you expect the value of time of a person who is not in the labor force to depend on the wage rate or work enjoyment of that person's employed spouse?

CHAPTER

10

DISCRETE CHOICE DEMAND MODELING FOR DECISION-BASED DESIGN

Henk Jan Wassenaar, Deepak Kumar, and Wei Chen

NOMENCLATURE

A	customer's product selection attributes
C	total product cost
CA	Conjoint Analysis
DBD	decision-based design
DCA	discrete choice analysis
E	engineering design (ED) attributes
$E(U)$	expected value of enterprise utility
IIA	independence of irrelevant alternatives
IID	independent and identically distributed
MDA	multiple discriminant analysis
MDS	multidimensional scaling
MNL	multinomial logit
P	product price
Q	product demand
RP	revealed preference
S	customer demographic attributes
SP	stated preference
t	time interval for which demand/market share is to be predicted
U	enterprise utility
u_{in}	true utility of alternative i by customer n
V	selection criterion used by the enterprise (e.g., profit, market share, revenues, etc.)
W_{in}	deterministic part of the utility of alternative i by customer n
X	design options
Y	exogenous variables (represent sources of uncertainty in the market)
ε_{in}	random unobservable part of the utility of alternative i by customer n

10.1 INTRODUCTION

Decision-Based Design (DBD) is emerging as a rigorous approach to engineering design that recognizes the substantial role that decisions play in design and in other engineering activities, which are largely characterized by ambiguity, uncertainty, risk and trade-offs [1–6]. The DBD optimization seeks to maximize the *utility* of a designed artifact while considering the interests of both the producer and the end-users [1, 6]. Although there is great consensus that for a profit-driven company, the *utility* of a product should be a measure of the profit[1] it brings, there is concern over using profit as the single criterion in DBD because of the belief that profit seems too difficult to model. One difficulty in modeling the profit is the construction of a reliable product demand model that is critical for assessing the revenue, the total product cost and eventually the profit.

In market research, there exist a number of approaches in demand modeling that explore the relationship between customer choice and product characteristics (attributes). Various analytical methods such as multiple discriminant analysis [7], factor analysis [8], multi-dimensional scaling [9], conjoint analysis [10–12] and Discrete Choice Analysis [32] have been used. They can be classified according to the type of data used (stated versus actual choice), type of model used (deterministic versus probabilistic) and the inclusion or noninclusion of customer heterogeneity. Even though demand modeling techniques exist in market research, they do not address the specific needs of engineering design, in particular for engineering decision-making.

Efforts have been made in the design community in recent years to extend the demand modeling techniques and incorporate customer preference information in product design [13–20]. Among them, the Comparing Multi-attribute Utility Values Approach from Li and Azarm [15] is a deterministic demand modeling approach, which estimates the demand by comparing deterministic multi-attribute utility values obtained through conjoint analysis. They also proposed a Customer Expected Utility Approach [16], which accounts for a range of attribute levels within which customers make purchase decisions and takes care of designers' preferences and uncertainty in achieving a desired attribute level. In recent years, the disaggregated probabilistic choice modeling approach in enterprise-driven engineering design applications has been employed [17–20]. Michalek et al. proposed a choice-based conjoint analysis approach within the multinomial logit (MNL) framework to analyze stated preference (SP) data. In this chapter, we illustrate how disaggregated probabilistic demand models based on discrete choice

[1]Profit is a result of accounting practices, which need not be related to engineering design, such as depreciation. Therefore, profit implies not revenue, i.e., the difference between revenue and expenditure. The net revenue can be discounted to present value.

analysis (DCA) can be incorporated in a decision-based design (DBD) framework for making rational product design decisions.

There is a detailed introduction of DCA in Chapter 9. DCA offers certain advantages over other demand modeling techniques. For instance, DCA-based demand models account for uncertainty in modeling by using probabilistic choice models and as well as the data of individuals instead of group averages, which enables a more accurate capturing of the variation in characteristics of individuals (i.e., heterogeneity) as detailed in Chapter 9 and avoids the paradox associated with aggregating the preferences of a group of customers. In addition to the fundamental principles of DCA, we provide here guidelines to apply the DCA approach to facilitate engineering decision-making, especially in the design of complex engineering systems. The mapping of customer desires to design attributes related to engineering analyses is discussed and a demand modeling procedure is developed to enable designers to focus the demand survey on specific features of the product.

The organization of the chapter is as follows. A discussion on the proposed DBD framework and the background of DCA are provided in Section 10.2. Sections 10.3 lays out the detailed sequence of steps for the implementation of DCA for DBD. The section also presents an approach to selecting the form of the customer utility function, used in the demand model, to enhance the predictive accuracy. A walk-through of a typical DCA implementation is shown in Section 10.4. The proposed approach is demonstrated in Section 10.5, using a vehicle engine case study, developed in collaboration with the market research firm J.D. Power & Associates and the Ford Motor Company.

10.2 TECHNICAL BACKGROUND

10.2.1 The Decision-Based Design (DBD) Framework

The flowchart of the proposed DBD framework (Fig. 10.1) is an enhancement of the DBD framework proposed by Hazelrigg [1]. DCA is presented in our DBD framework as a systematic approach to establish the relationship between the customer's product selection (CPS) attributes $\boldsymbol{A}$, price P, the socioeconomic and demographic background attributes $\boldsymbol{S}$ of the market population, time t and the demand Q. As presented in Chapter 9, DCA is a statistical technique, which identifies patterns in choices customers make between competing products and predicts the probability that an alternative is selected from a set of choice alternatives. In this chapter, the probability of selecting a particular alternative is extended to predict the probable market share and demand of a design option.

The arrows in the flowchart indicate the existence of relationships between the different entities (parameters) in DBD. The arrows do not necessarily coincide with the sequence of implementing DBD, part of which is detailed in Section 10.3 regarding demand modeling. We discern two different types of attributes in our approach, namely the engineering design (ED) attributes $\boldsymbol{E}$ and the customer's product selection attributes $\boldsymbol{A}$. Attributes $\boldsymbol{A}$ are product features and financial attributes (such as service and warranty) that a customer typically considers when purchasing the product. Attributes $\boldsymbol{E}$ are any quantifiable product properties that are used in the engineering product development process. The relationship between design options $\boldsymbol{X}$ and engineering design attributes $\boldsymbol{E}$ are determined through engineering analysis.

Alternative product designs, characterized by discrete or continuous design options $\boldsymbol{X}$ are determined during the "alternative generation" stage. It should be noted that design options $\boldsymbol{X}$ may include both engineering (product and process) design options and enterprise planning options, such as warranty options and annual percentage rate (APR) of auto loan, etc.; both influence the customer's product selection attributes $\boldsymbol{A}$. Engineering design attributes $\boldsymbol{E}$, apart from including the quantifications of some of the attributes $\boldsymbol{A}$, also include design attributes that are only of interest to design engineers. These attributes may act as physical constraints in DBD optimization, e.g., material stress, for instance, should not exceed the maximum allowable stress. Other engineering design attributes such as the product's weight will impact the total product cost C.

The total product cost C in the diagram accounts for all costs that occur during a product's life cycle, including the expenses for product development, manufacturing, overhead, storage, sales cost including distribution and marketing, warranty, liability, disposal,

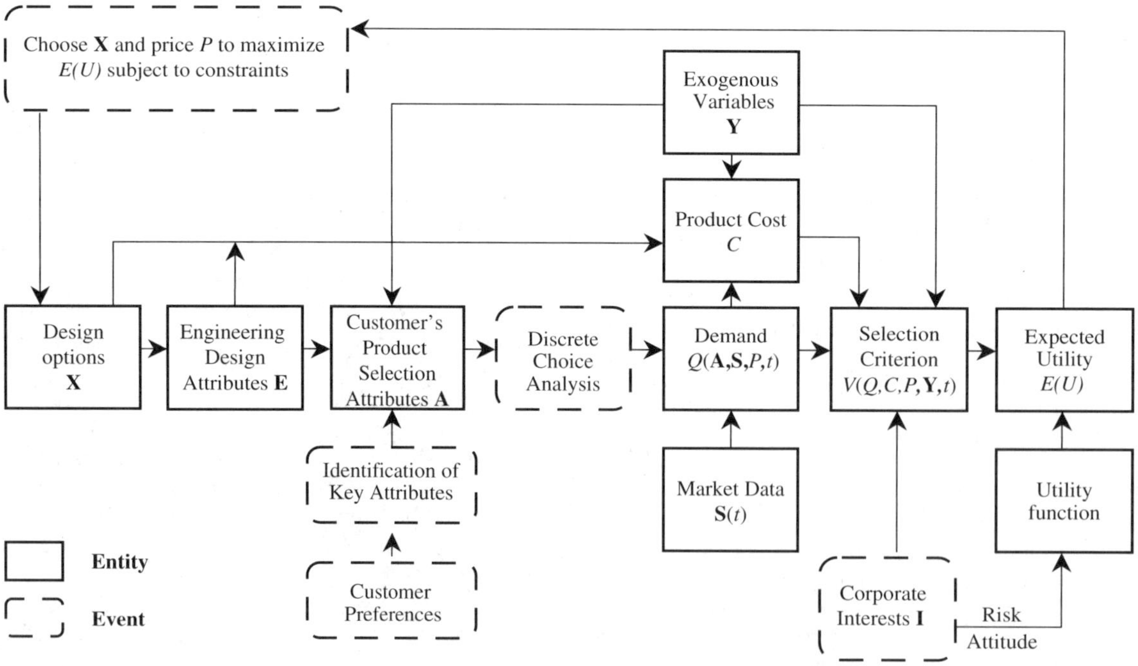

FIG. 10.1 DECISION-BASED DESIGN FLOWCHART

taxes, incentives, etc. Total product cost is impacted by the design options $\boldsymbol{X}$, exogenous variables $\boldsymbol{Y}$, engineering design attributes $\boldsymbol{E}$ and product demand (quantity) Q. Exogenous variables are uncertain parameters beyond the control of the design engineer (e.g., climate, legislation, demographics, financial markets, market trends). Total product cost can be estimated by determining the cost as a function of the characteristics of existing similar products such as cost per part or per unit of weight [21–23]. Since the product may be produced over several years, future costs of labor, capital, natural resources and supplies should be estimated, along with the availability of these production factors.

Under the DBD framework, a selection criterion V is needed to facilitate a valid comparison between design alternatives and to determine which alternative should be preferred. The net present value of profit is used as the selection criterion to avoid subjective trade-offs and problems of using multi-attribute utility associated with group decision-making [24,25]. The time t is considered when discounting V to the net present value. Owing to uncertainties in the calculations of $\boldsymbol{E}$, $\boldsymbol{A}$, C, Q and $\boldsymbol{Y}$, the resulting selection criterion V is a distribution of values. Therefore, the (expected) net present value of the product designs cannot be compared directly. For example, it is likely that one prefers a lottery with equal chance of receiving \$400 or \$600 over a lottery with an equal chance of receiving \$0 or \$1,000 even though the expected outcome for both lotteries is \$500. By assessing the risk attitude of the decision-maker the distribution of V is transformed into the expected utility $E(U)$, which is an integration of the utility function $U(V)$ and the distribution of V, i.e., $f(V)$. $U(V)$ expresses the decision-maker's risk attitude and could be assessed with von Neumann and Morgenstern lotteries [1].

The flowchart in Fig. 10.1 coincides with an optimization loop that identifies the best design option to maximize the expected utility. The optimal product design is determined by choosing both the design options $\boldsymbol{X}$ and the price P, such that the expected utility E(U) of the selection criterion is maximized while satisfying the constraints. It should be stressed that rigorous decision-making only allows constraints that are logically or physically necessary to be active at the selection of the preferred alternative. Otherwise, potentially valuable design alternatives could be accidentally excluded.

10.2.2 Background of Discrete Choice Analysis

A brief synopsis of DCA is provided in this section; a detailed explanation can be found in Chapter 9. DCA is based on probabilistic choice models, which have origins in mathematical psychology [26–29] and were developed in parallel by economists and cognitive psychologists. DCA identifies patterns in choices customers make between competing products and generates *the probability* that an option is chosen. Disaggregate approaches to demand modeling, such as DCA, use data of individual customers as opposed to aggregate approaches, which use group averages. Disaggregate demand models model the market share of each alternative as a function of the characteristics of the alternatives and sociodemographic attributes of the group of customers considered in the data set. Disaggregate approaches explain why an individual makes a particular choice given her/his circumstances, and, therefore, is better able to reflect changes in choice behavior due to changes in individual characteristics and the attributes of alternatives. Also, unlike aggregate models, disaggregate models are known to obtain unbiased coefficient estimates. Among the other advantages of DCA are: more freedom when formulating the survey questions; fewer problems with the degrees-of-freedom; a more natural task for the survey respondent; and the ability to handle more product attributes.

A quantitative process based on *multinomial analysis* is used in this chapter to create the demand model. The concept of random utility is adopted by assuming that the individual's true utility u consists of a deterministic part W and a random disturbance ε [see Eq.(10.1)]. The deterministic part utility W can be parameterized as a function of observable independent variables (product selection attributes $\boldsymbol{A}$, socioeconomic and demographic attributes $\boldsymbol{S}$ and price P), and unknown β-coefficients, which can be estimated by observing the choices respondents make (revealed or stated) and thus represent the respondent's taste, see Eq. (10.2). There is no functional form imposed on the utility function W, which is usually assumed to have a linear additive form in order to simplify computation as well as to enable easier interpretation of the choice models.

$$u_{in} = W_{in} + \varepsilon_{in} \qquad \text{Eq. (10.1)}$$

$$W_{in} = f(A_i, P_i, S_n; \beta_n) \qquad \text{Eq.(10.2)}$$

Alternative specific constants are part of the utility function corresponding to the expectation of the random disturbance ε, and thus representing preferences that are inherent and independent of specific attribute values, toward the alternatives [30]. Alternative specific constants are the equivalent of the intercept used in linear regression. An example of an alternative specific effect is brand image, which may affect the customer's utility beyond what can be explained by the product and customer attributes alone. The β-coefficients and utility functions are indicated with the subscript n, representing the nth respondent; the index i refers to the ith choice alternative. The probability that alternative 1 is chosen from a choice set containing two alternatives (binary choice) depends on the probability that the utility of alternative 1 exceeds the utility of alternative 2 or, alternatively, on the *probability that the difference between the disturbances does not exceed the difference of the deterministic parts of the utility*, i.e.,

$$\Pr(1)[1,2] = \Pr(W_{1n} + \varepsilon_{1n} \geq W_{2n} + \varepsilon_{2n}) \qquad \text{Eq. (10.3a)}$$

$$\Pr(1)[1,2] = \Pr(\varepsilon_{2n} - \varepsilon_{1n} \leq W_{1n} - W_{2n}) \qquad \text{Eq.(10.3b)}$$

The binary probit choice model [31] is presented in Eq. (10.4), where $\Phi()$ = standard normal distribution function.

$$\Pr_n(1) \mid [1,2] = \Pr_n(u_{1n} \geq u_{2n}) = \Phi\left(W_{1n} - W_{2n}\right) \qquad \text{Eq.(10.4)}$$

When the random disturbance is assumed normal, the normal distribution can be approximated with a logistical distribution, which can be evaluated in a closed format. Multinomial probit analysis assumes a multivariate normal distribution of the random disturbance ε, which allows complete flexibility of the variance-covariance matrix of the error terms. However, probit is computationally burdensome as it requires integration of the multivariate normal distribution. Eq. (10.5) shows the choice probability of the binary logit model, where $\Pr_n(1)$ = probability that respondent n chooses alternative 1 over alternative 2.

$$\Pr_n(1)\,|\,[1,2] = \Pr_n(u_{1n} \geq u_{2n}) = \frac{1}{1+e^{-\mu(W_{1n}-W_{2n})}} \qquad \text{Eq. (10.5a)}$$

$$\Pr_n(1)\,|\,[1,2] = \Pr_n(u_{1n} \geq u_{2n}) = \frac{e^{\mu W_{1n}}}{e^{\mu W_{1n}} + e^{\mu W_{2n}}} \qquad \text{Eq. (10.5b)}$$

The binary logistical distribution function of the difference of the (deterministic) utilities $W_{1n} - W_{2n}$ is depicted in Fig. 10.2. Note that the predicted choice probability does not reach unity or zero. The binomial logit model is extended to the multinomial logit (MNL) model that predicts the probability that alternative i is chosen by the nth respondent from among J competing products.

$$\Pr_n(i) = \frac{e^{W_{in}}}{\sum_{l=1}^{J} e^{W_{ln}}} \qquad \text{Eq. (10.6)}$$

The logit model [32,33] assumes that the error terms are independently and identically distributed (IID) across choice alternatives and observations (respondent choices). In other words, it pre-assumes that each alternative has the same unobserved error part ε in the utility [Eq.(10.1)]. This leads to the well-known independence of irrelevant alternatives (IIA) property, which assumes that when a customer is choosing between any pair of alternatives, this choice is independent of the remaining available alternatives. Therefore, in a logit model, changing the attributes of one alternative affects all other alternatives similarly. This allows for the addition or removal of an alternative to/from the choice set without affecting the structure or parameters of the model, enabling faster and easier computation of choice probabilities. But it also gives rise to the famous blue bus, red bus paradox [32]. Estimation techniques such as the maximum log-likelihood method can be used to determine the β-coefficients in Eq. (10.2), such that the predicted choice probabilities of the model match the observed choices as closely as possible. The total demand for a particular design is the summation of segment's population share of the total population [32].

The advantages of using the DCA technique for demand modeling in engineering design can be summarized as: (1) The method does not involve any ranking, weighting or normalization, thus avoiding the paradox associated with many multicriteria approaches. (2) Probabilistic choice addresses the uncertainties associated with unobserved taste variations, unobserved attributes and model deficiencies. (3) Competing products are considered, enabling analysis of market impact and competitive actions through "what if" scenarios. (4) The choice alternatives do not necessarily share the same set of attributes or attribute levels (required for conjoint analysis), expanding market testing possibilities and leaving more freedom to the marketing engineer. (5) The customer survey embedded in DCA resembles purchasing behavior more closely, reducing respondent errors and enabling the analysis of more attributes.

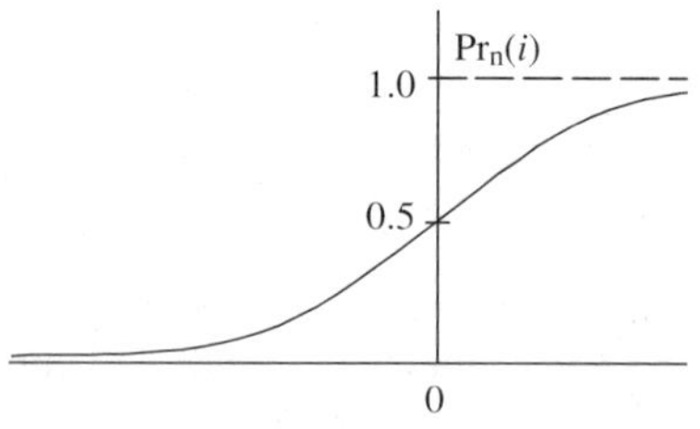

FIG. 10.2 CUMULATIVE DISTRIBUTION FUNCTION OF THE LOGIT DISTRIBUTION

10.3 IMPLEMENTING DCA FOR DEMAND MODELING IN ENGINEERING DESIGN

To facilitate engineering decision-making, a demand model is expected to relate the market demand to engineering measures of product attributes that can be used to guide product design decision-making. In this section, we focus on the procedure for implementing DCA for product demand modeling and discuss the potential issues involved in each phase. Our discussion follows the sequence of the four major phases for implementing DCA:

- Phase I Identify customer's product selection attributes ***A***, engineering design attributes ***E***, the range of price P and survey choice set (**attributes and choice set identification**)
- Phase II Collect quantitative choice data of proposed designs versus alternative choice options and record customers' socioeconomic and demographic background ***S*** (**data collection**)
- Phase III Create a model for demand estimation based on the probability of choice (**modeling**)
- Phase IV Use the demand model for market share and demand estimation (**demand estimation**)

10.3.1 Phase I: Attributes and Choice Set Identification

A useful demand model requires that there exists a causal relationship between the attributes and customers' purchase decisions. There are several methods available to assess what customers desire, what product attributes customers consider [34]: and what competing alternatives should be considered in a discrete choice survey. Focus groups [35] can be used for both existing products and products that are completely new (e.g., innovative design).

Through surveys, the identified customer desires can be clustered together into categories such as cost, performance, safety, operability, comfort, style, convenience, etc. These groups can be considered as *top-level customer desires.* The next step is to identify the customer's product selection attributes ***A*** that contribute to each customer desire. This involves translating the language of customer desires (e.g., good engine sound quality) into language that engineers can use in product development, i.e., identify suitable units of measurement for each customer desire. This transformation is very important in order to use the demand model for engineering decision-making. This task consists of cooperation between market researchers and engineers, and perhaps consultations with customers to verify the correct understanding of the customer desires. It implies that a design engineer must develop a preliminary understanding of the design and how the design can fulfill the customer's desires. Identification of the product attributes for some customer desires is straightforward, e.g., miles per gallon for fuel economy in vehicle design. For other customer desires this can be quite complicated, e.g., vehicle style or engine sound quality. It is possible that multiple design attributes need to be used to capture the properties of a customer desire, while one design attribute can impact multiple customer desires.

Figure 10.3 demonstrates how the top-level customer desires are mapped to specific customer desires (in customer language), to customer's product selection attributes ***A***, and then to engineering design attributes ***E***. Establishing such a mapping relationship is especially important in the design of complex engineering systems. The number of levels involved can be more than illustrated.

From a market analysis point of view, the input ***A*** of a demand model could be attributes with physical units (e.g., fuel economy) or without (e.g., level of comfort). However, to assist engineering decision-making, attributes ***A*** related to engineering performance need to be converted to quantifiable attributes ***E***. The set of engineering design attributes ***E***, apart from including the quantifications of some of the attributes ***A***, also include attributes that are only of interest to design engineers, e.g., stress level of a structure. These attributes might be introduced as intermediate variables or variables that impose physical restrictions on the design or impact the total product cost *C*. On the other hand, some of the non-performance-related attributes ***A*** are not influenced by the engineering design attributes ***E***, but by financial attributes. Therefore, ***A*** *and* ***E*** *can be viewed as two sets that share a number of common elements.*

To integrate the demand model into the DBD framework (see Fig. 10.1), engineering analysis (modeling) needs to be carried further to establish the relationship between design options ***X*** and attributes ***A***. As an example of mapping customer desires to a specific design, we show at the right side of Figure 10.3 that "noise while idling" can be considered as an attribute (***A***) that belongs to the group of "performance" under "product benefit," while radiated sound and engine mount vibration can be considered as attributes ***E*** for measuring the engine sound while idling. When using stated preference for demand modeling, attributes ***A*** are often used directly in a survey. On the other hand, when using revealed preference for demand modeling, quantitative engineering design attributes ***E*** could be used as exlanatory (i.e., independent) variables of the demand model to facilitate engineering decision-making. In the latter case, the demand model is expressed as $Q(\boldsymbol{E}, \boldsymbol{S}, P, t)$.

Engineering models can relate radiated sound to other engineering design attributes such as main bearing clearance and crankshaft stiffness. Finally, crankshaft stiffness can be modeled as a function of the design options, such as crankshaft material, pin diameter and cheek thickness.

10.3.2 Phase II: Data Collection

The choice data for the demand model is collected in the second phase of implementing DCA. There are two ways of collecting choice data: stated choice and revealed choice. *Revealed choice* concerns actual behavior that can be observed in real choice situations. *Stated choice* concerns controlled choice experiments that ask the respondents to state their purchase *intent*.

With stated choice, the survey respondent is asked to pick an alternative from a choice set in a process similar to real purchase decisions. An example of a choice set is presented in Table 10.1. A choice set contains a number of competing alternatives: a "survey alternative" (i.e., a new product or the alternative with the improved design), one or more competing alternatives from competitors and sometimes a "no choice" alternative (i.e., not to purchase anything). The alternatives are described by product selection attributes (***A***), including important business aspects such as price and warranty. The choice sets can be generated using design of experiment techniques. The survey results (choice data) are recorded, along with the respondent's customer background (***S***) such as age, income, product usage, etc.

Both stated choice and revealed choice have advantages and disadvantages [36]. Limitations of revealed choice are that it is not always clear what choice alternatives were available to the customer at the time of purchase. Failure to identify the customer's actual choice set may lead to biased results. Similarly, it may be difficult to identify all attributes ***A*** that are considered by the customer, which also may lead to biased results. Additionally, there can be a mismatch between the actual level of ***A*** and the level as *perceived* by the customer. Stated choice on the other hand is a controlled choice experiment. Unlike with revealed choice,

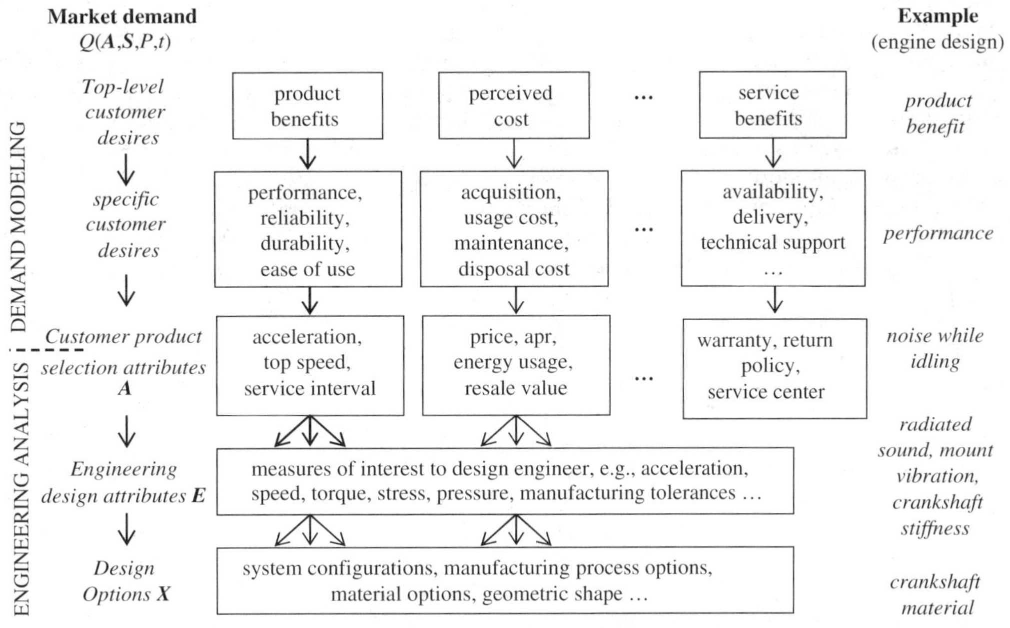

FIG. 10.3 MAPPING TOP-LEVEL CUSTOMER DESIRES TO DESIGN OPTIONS

TABLE 10.1 EXAMPLE OF A CHOICE SET FOR CELL PHONE DESIGN

Choice Set # 31	Survey Alternative (Cellular Phone)	Competing Product A (Cellular Phone)	Competing Product B (Cordless Phone)	None of These
A_1 Price	\$100	\$120	\$25	
A_2 Weight	3 ounces	3.5 ounces	6 ounces	
A_3 Battery life	10 hours	8 hours	N/A	
A_4 Internet Access	yes	yes	N/A	
A_5 ...	—	—	—	
Indicate whether this is the product you want to buy and how many				

alternatives, the attributes A and the attribute levels are controlled by the researcher and explicitly known to the respondent. However, a limitation of stated choice is that respondents don't need to commit to their choices (e.g., pay the purchase price), which can result in a mismatch between what respondents say they will do and the purchases they make in real life. Additionally, the respondent may not have thought of some of the attributes or competing products used in the choice, or may consider different attributes or competing products in a real purchase situation. Besides, not every competing product may be available to the respondent in a real purchase situation. Generally, revealed choice is used when similar products or services exist, e.g., when redesigning a power tool, while stated choice is used for innovative new designs, product features or services that do not yet exist.

To obtain accurate demand predictions it is necessary to get a representative sample of the market population. Typically, sampling design involves defining the target market population, the target market population size, etc., as well as a definition of the sampling unit, i.e., a customer or a product purchase. Random sampling cannot adequately capture the choice behavior of a very small population subgroup. This issue can be addressed using stratified random sampling [32], which divides the market population into mutually exclusive and exhaustive segments. Random samples are then drawn from each market segment. A demand model for each market segment can be constructed to predict each market segment's demand, which can then be properly weighted to arrive at an unbiased estimate for the total market demand.

DCA approaches like logit and probit assume that the data is normally distributed. Large deviations from normality and outliers may result in biased coefficient estimates. Therefore, before proceeding to fitting a demand model it is necessary to inspect the data for large deviations from the normal distribution like skewness, kurtosis, bimodality, etc., which may impair coefficient estimation. Most statistical software packages provide tests for normality, like the Shapiro-Wilks test. Inspecting the distribution or scatter plots directly is another possibility. Large deviations from normality can be reduced by removing outliers or through transformation, e.g., by using the log-transform to change a positively skewed distribution of a variable to a normal distribution.

10.3.3 Phase III: Modeling

Phase III is a quantitative process to generate the demand model. Based on the data, modeling techniques such as logit, as introduced in Chapter 9, are used to create a choice model that can predict the choices individual customers make and forecast the market demand for a designed artifact.

An accurate demand model is essential for the proposed engineering design framework shown in Figure 10.1. The predictive capability of a demand model depends largely on the attributes considered and the form of the utility function used in it. An important step in Phase III is to predetermine the functional form of the utility function W, shown in Eq. (10.2). It is common to initially assume a linear shape of the customer utility function and then to test different functional shapes (e.g., quadratic, exponential) for improvement of the model fit. However, any changes made to the customer utility function W should be supported by sound econometric reasoning (i.e., causal) to avoid overfitting the data and to obtain a model that not only fits the sample data well, but generalizes well to other data (e.g., the market population) for accurate predictive capability. One technique for assessing generalization capability is cross-validation, presented in Section 10.5.4. The predictions of a DCA model appear to be highly sensitive to changes in attribute values [37]. Part of this oversensitivity may be caused by the use of linear utility functions in the demand model. In reality, the relationship between an attribute and its utility is unlikely to be linear (consider diminishing marginal utility), thus a linear additive treatment of attributes may be too simplistic for engineering design. We propose using the Kano method [38] to facilitate the identification of the appropriate functional relationship between customer choice and product performance.

The Kano method, introduced in the late 1970s by Dr. Noriaki Kano of Tokyo Rika University, provides an approach to determine the generalized shape of the relation between product performance and customer satisfaction by classifying the customer attributes into three distinctive categories: *must-be, basic* and *excitive* (see Figure 10.4); (note: these terms may be named differently in various references). The three categories are described briefly; details regarding the classification process can be found in literature. *Must-be* attributes are expected by the customer and only cause

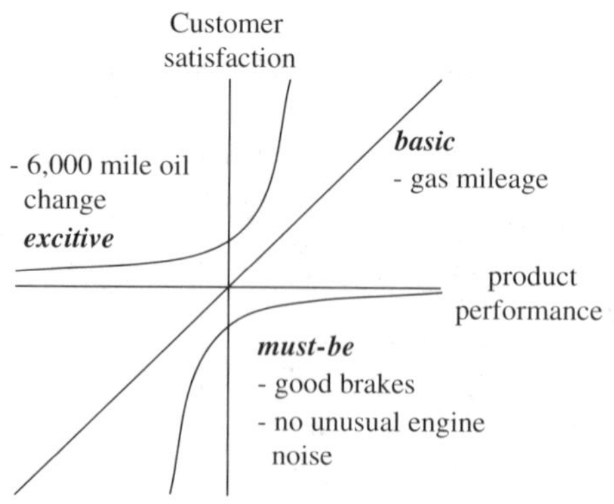

FIG. 10.4 KANO DIAGRAM OF CUSTOMER SATISFACTION

dissatisfaction if not met, e.g., unusual engine noise. Customer satisfaction never rises above neutral no matter how good the engine sounds; however, the consumer will be dissatisfied if an unusual engine noise occurs. Improving the performance of *basic* attributes increases satisfaction proportionally (i.e., linear), e.g., gas mileage (unless gas mileage is really bad). The *excitive* attributes increase satisfaction significantly for the reason that the customer does not expect them. For instance, the oil-change interval, a (unexpected) long interval, may be expected to significantly increase satisfaction. Attributes are thought to move over time from *excitive* to basic to must-be. For example, cup holders were once *excitive* when first introduced but are now expected and their absence can lead to great dissatisfaction.

We believe that the Kano method only allows a qualitative assessment of product attributes, i.e., the shape of the curves. However, it can be used to provide guidance to the demand modeling specialist in capturing the true customer behavior, improving explanatory power and predictive accuracy of demand models. Based on the Kano diagram we can assume a quadratic or a logistic function form for the *excitive* and *must-be* attributes in the choice model's customer utility function. This approach is expected to better capture the underlying behavior of consumers as opposed to randomly trying different functional shapes without proper econometric reasoning.

Representing the customer product selection attributes (***A***) using a sufficient number of quantifiable engineering design attributes ***E*** in a choice model is often desirable in facilitating engineering decision-making. For instance, to capture the sound quality (an attribute ***A***) as experienced by the vehicle occupants, the engineering design attributes ***E*** could include: noise level, harmonics and frequency. When these attributes are included, the demand model can be used to guide engineering decision-making related to air intake design, engine configuration, firing order, exhaust design, engine mount design, noise insulation, etc. However, while including more explanatory variables (attributes) may improve the model fit as additional variables help explain more data, using too many explanatory variables may lead to problems of overfitting. Two criteria can be used for comparing alternative model fits and for determining whether including additional explanatory variables is useful. They are the Akaike Information Criterion (AIC) [Eq. (10.7)] and the Bayesian Information Criterion (BIC) [39] [Eq. (10.8)]. Both criteria penalize models for having too many explanatory variables,

$$AIC = -2L + 2p \qquad \text{Eq. (10.7)}$$

$$BIC = -2L + p\ln(n) \qquad \text{Eq. (10.8)}$$

where L = log-likelihood; p number of explanatory variables; and n = number of observations (sample size). For both criteria, the best-fitting model is the model with the lowest score. A difference of six points on the BIC scale indicates strong evidence that the model with the lower value should be preferred [40]. Another issue that may arise when using large numbers of explanatory variables is collinearity, that is, some explanatory variables may be explained by combinations of other explanatory variables. Factor analysis [40] or latent variable modeling [41] could be used to combine explanatory variables that are correlated to each other into a fewer number of factors. Another solution approach is to constrain the (beta) coefficients of collinear variables in the utility function.

10.3.4 Phase IV: Demand Estimation

The choice model obtained through Phases I to III can be used to predict the choice probabilities for each alternative in the choice set given a customer's background (***S***) and descriptions of the product selection attributes that describe the choice alternatives. The logit demand model equation, Eq. (10.9), can be used to estimate demand based on sample enumeration using random samples of the market population N. Index i = choice alternative; and n= sampled individual.

$$Q(i) = \sum_{n}^{N} \Pr_n(i) = \sum_{n}^{N} \frac{e^{W_{in}}}{\sum_{k=1}^{J} e^{W_{kn}}} \qquad \text{Eq. (10.9)}$$

The accuracy of demand prediction can be improved by estimating choice models specifically for each market segment to account for systematic variations of taste parameters (β coefficients) among population subgroups. Ultimately, one can assume taste parameters that are log-normal distributed across the market population [32]. Including customer specific data in the customer background ***S*** can improve the accuracy of the demand predictions. For example, when one is estimating the demand in the passenger vehicle market, one can think of annual mileage driven, type of usage (commuting/recreational), etc. Such data can be recorded for each respondent when collecting the customer data and incorporated in the demand model.

A different approach for estimating the market demand is to use the choice model to predict the average choice probabilities (i.e., market shares) of the market population. In that case a separate specialized model can be used to estimate the total market sales volume. An advantage of this approach is that a separate model for predicting the market sales volume may be more accurate by accounting for economic growth, seasonal effects, market trends, etc., potentially leading to more accurate demand predictions.

10.4 WALK-THROUGH OF A TYPICAL MNL MODEL ESTIMATION

In this example, we illustrate how multinomial logit analysis can be used to create a demand estimation model for an academic power saw design scenario. First, product sales data are recast into data that can be used for demand modeling. This is followed by a discussion of the model estimation process as well as illustration of the estimation of several demand models with different utility function structures; this section also provides details on the statistical and behavioral measures used to evaluate the relative merits of demand models.

10.4.1 Constructing the Choice Set

We assume there are three (3) competing power saw alternatives in the market, each characterized by different levels of customer product selection attributes (speed and maintenance interval*, defined as ***A*** in the DBD framework) and the price P. Power saw 1 is the high-price, high-speed alternative, but requires more frequent maintenance. Saw 2 is the medium-price, medium-speed and low-maintenance alternative, while saw 3 is the low-speed, low-price and medium-maintenance alternative. For illustrative purposes, we examine a small sample data set representing the revealed preference of 15 customers who buy these saws from different vendors. Only normalized data has been used for convenience of computation and

*Defined as the time interval between successive maintenance visits.

TABLE 10.2 CUSTOMER SALES DATA

Customer No.	Income	Vendor	Alternative Chosen
1	0.44	A	2
2	0.62	B	3
3	0.81	C	1
4	0.32	D	3
5	0.78	E	2
6	1.00	F	1
7	0.84	G	1
8	0.39	H	2
9	0.55	I	3
10	0.62	J	3
11	0.66	K	1
12	0.50	L	3
13	0.43	M	1
14	0.76	N	1
15	0.32	O	3

interpretation, although normalization is not a requirement. Table 10.2 shows the sales data, along with the customer's income, which is the demographic attribute $\boldsymbol{S}$ considered in this example. Having demographic information related to the customer's age, income, education, etc. is useful in explaining the heterogeneity in customer choices and also helps a company design its products to target different market segments .

Table 10.3 shows the same three alternatives being sold at different prices by different vendors. Possible reasons of difference in prices could be due to different marketing strategies, different geographic locations, etc.

The data in Tables 10.2 and 10.3 are combined and transformed into a format that can be readily used for MNL analysis in Table 10.4. In Table 10.3, there are three rows of data for each customer, one for each choice alternative; each row in the data set contains the demographic attribute $\boldsymbol{S}$ of the individual customers, the customer's product selection attributes $\boldsymbol{A}$ that describe the alternative, price P, and the customer's observed choice (recorded in Table 10.2). Note that customer choice is treated as a binary variable in MNL analysis (Table 10.4). For example, customer 1 chose power saw alternative 2, which is indicated by a nonzero entry in the "*Choice*" column and the row corresponding to customer 1 and alternative 2. A few assumptions are typically made for the MNL analysis. One assumption is the Independence of Irrelevant Alternatives assumption property (see Section 10.2.2). Another important assumption is that the customers are fully aware of the product's attributes and make rational choices based on this knowledge. It is also assumed that customers did indeed consider all the three available alternatives before making their choices.

10.4.2 Walk-Through of the Demand Modeling

Several software tools are available for estimating MNL models, e.g., LINDEP, SAS, STATA, etc.—most software tools can use maximization of the log likelihood function as the optimization criterion for the model estimation. The results presented here are obtained from STATA. Typically, developing a satisfactory demand model involves estimating models of increasing complicated specification. That is, one has to progressively increase the number of variables in the utility function of the demand model [Eq. (10.2)] until obtaining a model that not only has excellent statistical goodness of fit, but also explains customer behavior in a manner consistent with our understanding of the market. The first step may involve building a zero-model,

TABLE 10.3 VENDOR PRICING INFORMATION

Vendor	Alternative	Price
	1	0.97
A	2	0.73
	3	0.63
	1	1
B	2	0.72
	3	0.55
	1	0.95
C	2	0.75
	3	0.6
	1	0.93
D	2	0.75
	3	0.6
	1	0.98
E	2	0.71
	3	0.56
	1	0.95
F	2	0.71
	3	0.58
	1	0.95
G	2	0.81
	3	0.61
	1	0.93
H	2	0.77
	3	0.57
	1	0.96
I	2	0.8
	3	0.58
	1	1
J	2	0.79
	3	0.59
	1	0.96
K	2	0.77
	3	0.59
	1	0.93
L	2	0.77
	3	0.6
	1	0.9
M	2	0.74
	3	0.63
	1	0.94
N	2	0.73
	3	0.64
	1	0.96
O	2	0.75
	3	0.61

which is also known as the *equally likely* model, named so, because this model does not have any parameters in the utility function that determines a customer's choice and assigns equal probability to each of the choice alternatives. In our case, the zero-model would assign a choice probability of one-third to each of the three power saws, i.e.;

$$\begin{aligned} \text{For } 1 \le n \le 15, \quad & \Pr_n(1)\left[1, 2, 3\right] = 1/3 \\ & \Pr_n(2)\left[1, 2, 3\right] = 1/3 \\ & \Pr_n(3)\left[1, 2, 3\right] = 1/3 \end{aligned} \qquad \text{Eq. (10.10)}$$

Here $\Pr_n(1)$ [1, 2, 3] represents the probability of customer n choosing alternative 1, when asked to choose from the alternative set {1, 2, 3}. The zero model is generally used as a reference to compare the goodness of fit of other models. But the zero model is not used for prediction purposes since it does not consider

TABLE 10.4 REVEALED PREFERENCE DATA USED FOR THE ANALYSIS

Customer No.	Alternative ID	Choice	Speed	Price	Maintenance Interval	Income
1	1	0	1	0.97	0.64	0.44
1	2	1	0.71	0.73	1	0.44
1	3	0	0.67	0.63	0.89	0.44
2	1	0	1	1	0.64	0.62
2	2	0	0.71	0.72	1	0.62
2	3	1	0.67	0.55	0.89	0.62
3	1	1	1	0.95	0.64	0.81
3	2	0	0.71	0.75	1	0.81
3	3	0	0.67	0.6	0.89	0.81
4	1	0	1	0.93	0.64	0.32
4	2	0	0.71	0.75	1	0.32
4	3	1	0.67	0.6	0.89	0.32
5	1	0	1	0.98	0.64	0.78
5	2	1	0.71	0.71	1	0.78
5	3	0	0.67	0.56	0.89	0.78
6	1	1	1	0.95	0.64	1
6	2	0	0.71	0.71	1	1
6	3	0	0.67	0.58	0.89	1
7	1	1	1	0.95	0.64	0.84
7	2	0	0.71	0.81	1	0.84
7	3	0	0.67	0.61	0.89	0.84
8	1	0	1	0.93	0.64	0.39
8	2	1	0.71	0.77	1	0.39
8	3	0	0.67	0.57	0.89	0.39
9	1	0	1	0.96	0.64	0.55
9	2	0	0.71	0.8	1	0.55
9	3	1	0.67	0.58	0.89	0.55
10	1	0	1	1	0.64	0.62
10	2	0	0.71	0.79	1	0.62
10	3	1	0.67	0.59	0.89	0.62
11	1	1	1	0.96	0.64	0.66
11	2	0	0.71	0.77	1	0.66
11	3	0	0.67	0.59	0.89	0.66
12	1	0	1	0.93	0.64	0.5
12	2	0	0.71	0.77	1	0.5
12	3	1	0.67	0.6	0.89	0.5
13	1	1	1	0.9	0.64	0.43
13	2	0	0.71	0.74	1	0.43
13	3	0	0.67	0.63	0.89	0.43
14	1	1	1	0.94	0.64	0.76
14	2	0	0.71	0.73	1	0.76
14	3	0	0.67	0.64	0.89	0.76
15	1	0	1	0.96	0.64	0.32
15	2	0	0.71	0.75	1	0.32
15	3	1	0.67	0.61	0.89	0.32

the impact of product attributes and customers' demographic attributes). Note that the market share predictions (obtained by aggregating the choice probabilities for each alternative across all individuals) from this model are $\{1/3, 1/3, 1/3\}$ for alternatives 1, 2 and 3, respectively, which do not match well with the observed market shares [i.e., {0.4, 0.2, 0.4}].

The estimation of the zero model is usually followed by the estimation of a model that has only constants in the utility function. A *constants-only* model has only alternative specific constants (ASC) [32] but no other explanatory variables like ***A*** and ***S*** in the utility function. ASCs are used to estimate the utility biases [ε in Eq. 10.2] due to excluded variables. The ASC corresponding to one of the alternatives is set to zero and the constants corresponding to the other alternatives are evaluated with respect to that reference (zero) alternative. For our data set, the *constants-only* model would carry two constants, e.g., β_{01} and β_{02}, for alternative 1 and 2, respectively. The ASC corresponding to alternative 3 (i.e. β_{03}) is then set to zero. As a result, the deterministic part of the utility function for each alternative would look as below:

For $1 \leq n \leq 15$,

$$W_{1n} = \beta_{01} \quad \text{Eq. (10.11a)}$$

$$W_{2n} = \beta_{02} \quad \text{Eq. (10.11b)}$$

$$W_{3n} = \beta_{03} (=0) \quad \text{Eq. (10.11c)}$$

The STATA output for this model is shown in Fig. 10.4. The output contains information on the iteration history of the log-likelihood values, number of observations in the data set (i.e., 45 with three observations for each customer in the sample data set). The output

```
Iteration 0:   log-likelihood =    -16.1443
Iteration 1:   log-likelihood = -15.827368
Iteration 2:   log-likelihood = -15.823803
Iteration 3:   log-likelihood = -15.823803

Conditional (fixed-effects) logistic regression   Number of obs   =        45
                                                  LR chi2(2)      =      1.31
                                                  Prob > chi2     =    0.5192
Log-likelihood = -15.823803                       Pseudo R2       =    0.0398

------------------------------------------------------------------------------
      Chosen |      Coef.   Std. Err.      z    P>|z|     [95% Conf. Interval]
-------------+----------------------------------------------------------------
       ASC_1 |   6.81e-17   .5773503     0.00   1.000    -1.131586    1.131586
       ASC_2 |  -.6931472   .7071067    -0.98   0.327    -2.079051    .6927566
------------------------------------------------------------------------------
```

FIG. 10.5 STATA OUTPUT FOR THE CONSTANTS-ONLY MODEL

also includes statistical goodness-of-fit values like the pseudo R-square value, i.e., ρ_0, the log-likelihood ratio with respect to the zero model as defined in Appendix 10 A.1 and values related to the chi-square test, explained in Appendix 10 A.2. The model, as expected, has a higher log-likelihood value (−15.823) than the zero model (−16.479). The ρ_0 value is 0.0398. which is low and indicates that the model is not much better than the zero-model. The output "`Prob >chi2`" entry in Figure 10.5 is the probability of significance with which the zero-model can be rejected in favor of the constants-only model, using the chi-square test. "`LR chi2(0)`" is the left-hand side of the chi-square test. The chi-square test shows that the zero model can be rejected in favor of the *constants-only* model with a probability of (1 − 0.5192) = 48.08%, which is low and reinforces the conclusion that the *constants-only* model does not explain much more variance in the data than the zero model. The ASC corresponding to alternative 1 is estimated as zero, which implies that it is equal to the ASC corresponding to alternative 3. The confidence intervals and the statistical significance of the estimators, computed based on the standard errors for these estimators, show that the coefficients are not significantly different from zero since the 95 percent confidence intervals for both coefficients β_{01} and β_{02} (i.e., ASC_1 and ASC_2 in the STATA output) include zero. Statistical significance of the different estimators becomes more relevant in models with a more detailed specification, (i.e., models with more variables in the utility function). Explanatory variables are usually retained in the utility function, if the signs and magnitudes of the estimators are satisfactory even though they may not be statistically significant.

Based on the estimations of the utility function coefficients in the STATA output, the choice probabilities can be calculated as shown in Eq. (10.12).

For $1 \le n \le 15$

$$W_{1n} = \beta_{01} = 0 \qquad \text{Eq. (10.12a)}$$

$$W_{2n} = \beta_{02} = -0.6932 \qquad \text{Eq. (10.12b)}$$

$$W_{2n} = \beta_{03} = 0 \qquad \text{Eq. (10.12c)}$$

$$\Pr_n(1)[1,2,3] = \frac{e^{W_{1n}}}{\left(e^{W_{1n}} + e^{W_{1n}} + e^{W_{1n}}\right)} = \frac{e^{-0.0932}}{\left(1 + e^{-0.6932} + 1\right)} = 0.4 \qquad \text{Eq. (10.12d)}$$

$$\Pr_n(2)[1,2,3] = \frac{e^{W_{1n}}}{\left(e^{W_{1n}} + e^{W_{1n}} + e^{W_{1n}}\right)} = \frac{1}{\left(1 + e^{-0.6932} + 1\right)} = 0.2 \qquad \text{Eq. (10.12e)}$$

$$\Pr_n(3)[1,2,3] = \frac{e^{W_{1n}}}{\left(e^{W_{1n}} + e^{W_{1n}} + e^{W_{1n}}\right)} = \frac{1}{\left(1 + e^{-0.6932} + 1\right)} = 0.4$$

Eq. (10.12f)

The utility function values, as well as the choice probabilities, are identical across all customers in the *constants-only* model. Therefore, the predicted market shares from the model are identical to the individual choice probabilities. Note that the predicted market share values match exactly with the observed market shares for this model. This result is expected since it is well known that any model, which has a *full set*[2] of ASC (like the *constants-only* model presented here), will always produce an exact match between predicted market shares (aggregated choice probabilities) and observed market shares [32]; any difference between the two is only due to numerical (computational) error.

Finally, a model that includes the customer's product-selection attributes ***A*** (speed and maintenance interval), price, *P*, and the demographic characteristics ***S*** (customer income) is estimated, assuming a linear form of the utility function. All demographic attributes are included as alternative specific variables (ASV) due to the nature of the MNL model. The coefficient of the income ASV for alternative 1 is set to zero and serves as the reference. The form of the deterministic part of the utility function is shown in Eq. (10.13):

For $1 \le n \le 15$

$$W_{1n} = \beta_{\text{speed}}[X_{\text{speed}(1)}] + \beta_{\text{price}}[X_{\text{price}(1)}] + \beta_{\text{maintenance}} t[X_{\text{maintenance}(1)}]$$

Eq. (10.13a)

$$W_{2n} = \beta_{\text{speed}}[X_{\text{speed}(2)}] + \beta_{\text{price}}[X_{\text{price}(2)}] + \beta_{\text{maintenance}}[X_{\text{maintenance}(2)}] + \beta_{\text{income}(2)}[S_{\text{income}(n,2)}]$$

Eq. (10.13b)

$$W_{3n} = \beta_{\text{speed}}[X_{\text{speed}(3)}] + \beta_{\text{price}}[X_{\text{price}(3)}] + \beta_{\text{maintenance}}[X_{\text{maintenance}(3)}] + \beta_{\text{income}(3)}[S_{\text{income}(n,3)}]$$

Eq. (10.13c)

where $X_{\text{price}(j)}$, $X_{\text{speed}(j)}$ and $X_{\text{maintenance}(j)}$ = price, speed and maintenance interval of alternative *j*, respectively; and $S_{\text{income}(n,j)}$ = income

[2]Any model considering choice data involving *J* alternatives is said to have a *full set* of alternative specific constants, if it has *(J–1)* alternative specific constants.

of customer n, used as ASV for alternative j. Note that the β-coefficients of the product attributes (speed, price and maintenance interval) are identical across all alternatives and all customers in the above utility functions. However, the coefficients for the alternative specific income variables do vary across alternatives. The results of the model estimation in STATA are shown in Fig. 10.6.

The signs of the coefficients in the utility function (as shown in the STATA output) indicate that customers prefer higher speeds, lower prices and higher maintenance intervals, which corresponds with our understanding of the market. Since the data in this example is normalized, the magnitudes of the coefficients also indicate the relative importance of the product attributes to the customers. The results indicate that customers view price as the most important factor and view speed as slightly less important; the maintenance interval of the product is considered least important. The coefficients of the demographic variables have to be interpreted in conjunction with background knowledge about the product. It is known that alternative 1 is the most expensive and alternative 3 is the least expensive. The income variables in the utility function have to be interpreted in that context. The negative signs of *income_2* and *income_3* indicate that customers with higher incomes view alternative 2 and alternative 3 as less desirable than alternative 1. Also, the larger magnitude of the coefficient for *income_3* indicates that customers with higher incomes would view alternative 3 (the low-price, low-speed alternative) less desirable than alternative 2. These results are reasonable and are consistent with our expectations, and therefore the model can be regarded favorably.

The log-likelihood (–7.8035) and pseudo R-square ρ_0 (0.5265) values are much higher when compared to the zero model and the *constants-only* model. Also, the chi-square test indicates that the zero model can be rejected in favor of the linear model, with a very high degree of statistical significance (1.0–0.0039)=99.61%. It can be shown that the *linear* model is superior to the *constants-only* model in a similar fashion. However, some of the coefficients in the model are not statistically significant at the 95% level. But since these coefficients are consistent with our expectation, the variables are retained in the model.

Sample calculations of the utility functions and choice probabilities are provided for customer 3 in Eq. (10.14). The computations show that the predicted choice probability for alternative 1 is the highest. This agrees well with the actual choice of the customer recorded in Table 10.4. The comparison between the actual and predicted choice is provided for all customers in Table 10.5, showing that in most cases, the alternative with the highest choice probability (as predicted by the model) is the one chosen by the customer.

For $n = 3$,

$$W_{1n} = 47.09(1) - 55.95(0.95) + 28.01(0.64) = 11.86 \quad \text{Eq. (10.14a)}$$

$$W_{2n} = 47.09(0.71) - 55.95(0.75) + 28.01(1) - 13.67(0.44) = 8.42 \quad \text{Eq. (10.14b)}$$

$$W_{3n} = 47.09(0.67) - 55.95(0.60) + 28.01(0.89) - 19.66(0.44) = 6.98 \quad \text{Eq. (10.14c)}$$

For $n = 3$

$$\Pr_n(1)[1,2,3] = \frac{e^{W_{1n}}}{(e^{W_{1n}} + e^{W_{2n}} + e^{W_{3n}})} = \frac{e^{11.86}}{(e^{11.86} + e^{8.42} + e^{6.98})} = 0.96 \quad \text{Eq. (10.14d)}$$

$$\Pr_n(2)[1,2,3] = \frac{e^{W_{2n}}}{(e^{W_{1n}} + e^{W_{2n}} + e^{W_{3n}})} = \frac{e^{8.42}}{(e^{11.86} + e^{8.42} + e^{6.98})} = 0.03 \quad \text{Eq. (10.14e)}$$

$$\Pr_n(3)[1,2,3] = \frac{e^{W_{2n}}}{(e^{W_{1n}} + e^{W_{2n}} + e^{W_{3n}})} = \frac{e^{6.98}}{(e^{11.86} + e^{8.42} + e^{6.98})} = 0.01 \quad \text{Eq. (10.14f)}$$

As noted before, aggregated predictions of choice probability, which translate to market share predictions, tend to agree with the actual market share values for unbiased models like the *constants-only* and the *linear* models presented here. However, the predictive capability of a demand model is better expressed when the model is tested on data that has not been used to estimate the model. The engine design example in Section 10.5 illustrates the use of cross-validation for this purpose.

```
Iteration 0:   log-likelihood = -14.359848
Iteration 1:   log-likelihood = -8.9144052
Iteration 2:   log-likelihood = -7.9954175
Iteration 3:   log-likelihood = -7.8154834
Iteration 4:   log-likelihood = -7.8035352
Iteration 5:   log-likelihood = -7.8034579

Conditional (fixed-effects) logistic regression   Number of obs   =        45
                                                  LR chi2(5)      =     17.35
                                                  Prob > chi2     =    0.0039
Log-likelihood = -7.8034579                       Pseudo R2       =    0.5265

------------------------------------------------------------------------------
      Choice |      Coef.   Std. Err.      z    P>|z|     [95% Conf. Interval]
-------------+----------------------------------------------------------------
       speed |   47.08898   36.66279     1.28   0.199    -24.76876    118.9467
       price |  -55.95013   24.87006    -2.25   0.024    -104.6945   -7.205713
maintenanc~n |    28.0128   26.03458     1.08   0.282    -23.01405    79.03964
    income_2 |  -13.66506    7.74274    -1.76   0.078    -28.84055    1.510429
    income_3 |  -19.66365   9.744276    -2.02   0.044    -38.76208   -.5652178
------------------------------------------------------------------------------
```

FIG. 10.6 STATA OUTPUT FOR THE LINEAR MODEL

TABLE 10.5 COMPARISON OF ACTUAL AND PREDICTED INDIVIDUAL CHOICE

Case No.	Alt ID	Chosen	Predicted Choice
1	1	0	0.018
1	2	1	0.866
1	3	0	0.115
2	1	0	0.008
2	2	0	0.304
2	3	1	0.687
3	1	1	0.962
3	2	0	0.031
3	3	0	0.007
4	1	0	0.021
4	2	0	0.178
4	3	1	0.800
5	1	0	0.243
5	2	1	0.591
5	3	0	0.166
6	1	1	0.977
6	2	0	0.022
6	3	0	0.001
7	1	1	0.997
7	2	0	0.001
7	3	0	0.002
8	1	0	0.019
8	2	1	0.020
8	3	0	0.961
9	1	0	0.128
9	2	0	0.015
9	3	1	0.857
10	1	0	0.092
10	2	0	0.069
10	3	1	0.838
11	1	1	0.631
11	2	0	0.090
11	3	0	0.279
12	1	0	0.429
12	2	0	0.101
12	3	1	0.470
13	1	1	0.567
13	2	0	0.347
13	3	0	0.086
14	1	1	0.899
14	2	0	0.100
14	3	0	0.001
15	1	0	0.006
15	2	0	0.279
15	3	1	0.715

10.5 INDUSTRIAL EXAMPLE

In this section we present an implementation of the DCA demand modeling approach to constructing a vehicle demand model with emphasis on evaluating engine design changes in a DBD model. The demand model developed in this case study can be used to assess the impact of engine design changes on vehicle demand, facilitating the evaluation of engine design and making proper trade-offs between performance and cost. Twelve vehicles (7 models, 12 trims) are considered in the demand model representing the midsize vehicle segment, which includes vehicles like the Ford Taurus, Toyota Camry, and the Honda Accord. All data illustrated are normalized to protect proprietary rights of the providers. Our implementation is subject to the assumption that customers only consider these 12 vehicle trims when purchasing a vehicle. The demand model developed is a static model, i.e., demand changes over time are not considered. In "what if" studies and DBD optimization, we assume that only the design of one vehicle changes at a time, while the other vehicle designs are kept the same.

10.5.1 Vehicle Demand Modeling: Attributes and Choice Set Identification

Based on J.D. Power's vehicle quality survey (VQS), we identify five groups of top-level customer desires related to vehicle choice at the vehicle system level. These are: engine/transmission performance, comfort and convenience, ride and handling performance, product image and price/cost (see Table 10.6). For reasons of simplicity, customer desires related to sound system, seats and style are not considered. In Section 10.3.1 we detailed the process of translating customer desires into customer product selection attributes $\boldsymbol{A}$ and then to quantifiable engineering design attributes $\boldsymbol{E}$ that are meaningful to both the demand-modeling specialist and to a design engineer. Specific customer desires can be identified for each top-level, vehicle-system customer desire. The attributes considered in our model are presented in Table 10.7, which shows a representative mapping of top-level customer desires to engineering design attributes. Let's take engine and transmission performance as an example of this mapping process. The specific customer desires include performance during rapid acceleration, passing power at highway speeds as well as a pleasant sound while idling at full throttle acceleration and low vibration levels. Interaction between engineering experts at the Ford Motor Company and market research specialists from J.D. Power helped identify the engineering design attributes corresponding to their specific customer desires. Another important activity of design engineers is linking the attributes $\boldsymbol{E}$ to the design options $\boldsymbol{X}$. The design options in this vehicle engine design case study are represented by the different settings of attribute levels in Table 10.7.

10.5.2 Vehicle Demand Modeling: Data Collection

The demand model is created using revealed choice data at the respondent level provided by J.D. Power. The data consist of 2,552 observed individual vehicle purchases (of the seven vehicles — 12 trims considered in this case study) of the year 2000 vehicle market in the U.S., including respondents' background. The values of the customer product selection attributes related to the general vehicle descriptions of the 12 discrete choices, such as weight, fuel economy and legroom are obtained from Ward's Automotive. The values of other attributes such as ride, handling, noise and vibration are provided by Ford. A representative part of the choice set input data table for one customer is presented in Table 10.7.

For each respondent there are 12 rows of data in the database, one for each choice alternative—each row containing the customer background, the vehicle attributes and the respondent's observed choice (real purchase). The customer choice is treated as a binary variable, and in this particular case the customer selected vehicle 2. In total, the database contains 30,624 observations (2,552 respondents * 12 vehicles). The correlation of a number of attributes of this data is presented in Table 10.8.

The following conclusions can be deducted from Table 10.9. (Note, the variables gender and USA/import of Table 10.9 are binary variables that is, female = 1, and import = 1, otherwise 0.) For instance, the negative sign of the correlations related to gender for wheelbase, vehicle width, and vehicle length indicates that women apparently buy smaller vehicles. The negative coefficient (−0.220) for USA/import indicates that older consumers tend

TABLE 10.6 PRODUCT SELECTION ATTRIBUTES STRUCTURE FOR VEHICLE ENGINE DESIGN EXAMPLE

Top-Level Customer Desires	Specific Customer Desires	Attributes
Engine and transmission Performance	Performance	Horsepower
		Torque
		Low-end torque
		Displacement
		Type (I4/V6)
	Noise	Noise at 2,000 rpm (highway)
		Noise at 4,000 rpm (accelerating)
		Noise at rpm of max Hp (db)
	Vibration	Overall vibration level
		Vibration @2,000 rpm (highway)
		Vibration @4,000 rpm (accelerating)
Comfort and convenience	Comfort	Front legroom
		Front headroom
		Rear legroom
		Rear headroom
	Convenience	Trunk space
		Range between fuel stops
Ride and handling	Handling	Roll Gradient (deg/g)
	Steering	SWA0. @5 g (deg) Window
		Rolling parking efforts
		Static parking efforts
	Ride	Choppiness (M/sec^2/minute)
Product image	Brand	Vehicle make
	Origin	USA/import
	Reliability	IQS (initial quality index)
	Durability	VDI (vehicle dependability index)
	Vehicle size	Vehicle mass
		Vehicle width
		Vehicle length
Product cost	Acquisition cost	MSRP price
		Rebate
	Usage cost	APR
		Fuel economy
		Resale index

to prefer domestic vehicles. The negative coefficient for rebate and USA/Import (−0.869) shows that imports are generally sold with smaller rebates. The correlation between customer background (gender, age, and income) and customer product selection attributes appears to be very weak, which is desirable. Highly correlated variables are prone to being collinear, giving problems with estimating the demand model coefficients. Further, high correlation between the dependent variable (in this case the vehicle choice) and independent explanatory variables (i.e., design attributes and customer demographic attributes) implies that few variables are sufficient to predict vehicle choice, limiting the use of many explanatory variables (engineering design attributes) in the demand model, which are required for engineering design decision-making.

TABLE 10.7 PARTIAL DEMAND MODEL INPUT DATA TABLE (NORMALIZED)

			Customer Background			Attributes			
Customer ID	Vehicle ID	Observed Choice	Gender	Age	Income	Msrp Price	Horse power	Torque	Fuel Economy
1	1	0	0	27	5	1.07	1.13	1.09	0.96
1	2	1	0	27	5	0.87	0.89	0.85	1.15
1	3	0	0	27	5	1.15	1.09	1.02	0.98
1	4	0	0	27	5	1.02	1.06	1.02	0.90
1	5	0	0	27	5	1.05	1.08	1.12	0.98
1	6	0	0	27	5	0.89	0.77	0.82	1.12
1	7	0	0	27	5	0.96	1.04	0.94	1.00
1	8	0	0	27	5	0.89	0.93	0.97	1.00
1	9	0	0	27	5	1.07	1.02	1.10	1.00
1	10	0	0	27	5	1.03	0.92	0.98	1.02
1	11	0	0	27	5	1.11	1.23	1.16	0.98
1	12	0	0	27	5	0.89	0.83	0.94	0.94

10.5.3 Vehicle Demand Modeling – Multinomial Logit

In this case study we use STATA to estimate the choice model. A linear customer utility function shape is initially considered for the utility function used in the logit choice model (Eqn. 10.2) and dividing up the market population in different segments is not considered. Over 200 customer utility functions with different combinations of linear and interaction items were examined, illustrating the effort typically involved in developing a demand model. Eventually a model using 38 explanatory variable items (including attribute interactions) was selected based on the Bayesian Information Criterion score (BIC) (see description in Section 10.3.3).

The observed and estimated market shares for the 12 vehicle models of the final demand model are shown in Table 10.9. It shows that the observed choice rate/market shares and the market shares as predicted by the model match quite well as would be expected for a model with a full set of alternative specific constants and in-sample prediction. The MS_R2, i.e., the R-square error measure of the observed market shares *versus* predicted market shares for this model is 0.995851. As proposed earlier in Section 10.3.3, the Kano method is used to further improve the predictive accuracy by identifying appropriate shapes for the customer utility function of the choice model [Eq. (10.2)]. According to Kano study results at Ford, all attributes should be considered as *basic* (i.e., linear) except for *fuel economy beyond 27.5 mpg*, which can be classified as *excitive*. The econometric reasoning for this is as follows: Fuel economy is considered *basic* if the fuel mileage is near what is expected for the vehicle's class—in this case the midsize market segment.

But when the fuel mileage is significantly higher than its competitors, then it becomes a distinguishing feature, e.g., "I bought this vehicle because of its remarkable fuel economy." We test a quadratic function shape for the key customer attributes "fuel economy" and "range between fuel stops" in the customer utility function of the demand model. The BIC score shown in Table 10.10 indicates that the demand model using the utility function shape as assessed by the Kano method provides a better fit for the collected data given that the BIC score improved by more than six points.

TABLE 10.8 PARTIAL CORRELATION MATRIX OF VEHICLE ATTRIBUTES AND CUSTOMER BACKGROUND

	Gender	Age	Income	USA/ Import
Gender	1	—	—	—
Age	−0.192	1	—	—
Income	−0.074	−0.176	1	—
USA/import	0.150	−0.220	0.087	1
Msrp_price	0.006	−0.041	0.141	0.183
Rebate	−0.101	0.256	−0.141	−0.869
Apr	−0.072	0.173	−0.017	−0.425
Resale index	0.178	−0.215	0.031	0.869
VDI (dependability)	−0.117	0.036	0.024	−0.746
IQS (initial quality)	−0.162	0.187	−0.059	−0.928
Horsepower/mass	−0.011	−0.104	0.180	0.212
Torque/mass	−0.051	−0.005	0.148	0.013
Low-end torque/mass	−0.087	0.036	0.120	−0.255
Fuel economy	0.127	−0.047	−0.102	0.444
Fuel range	0.138	−0.063	−0.045	0.680
Wheel base	−0.106	0.076	0.050	−0.667
Vehicle width	−0.119	0.157	−0.066	−0.918
Vehicle length	−0.149	0.154	-0.038	−0.907
Front headroom	−0.013	−0.103	0.145	0.290
Front legroom	0.072	−0.094	0.116	0.762
Rear headroom	−0.162	0.132	0.053	−0.695
Rear legroom	−0.140	0.157	0.013	−0.731
Trunk space	−0.132	0.139	0.004	−0.844

TABLE 10.9 OBSERVED AND ESTIMATED MARKET SHARES FOR VEHICLE DEMAND MODEL

Vehicle ID.	Choice Rate (#)	Market Shares	
		Observed	Estimated
1	251	0.098354	0.098814
2	190	0.074451	0.074544
3	335	0.131270	0.130938
4	220	0.086207	0.086117
5	231	0.090517	0.090972
6	192	0.075235	0.075440
7	199	0.077978	0.077447
8	167	0.065439	0.064866
9	67	0.026254	0.027256
10	435	0.170455	0.170324
11	213	0.083464	0.083507
12	52	0.020376	0.019776

10.5.4 Cross-Validation of Demand Model

The predictive capability of a demand model cannot be assessed using in-sample data, i.e., the data that is used to estimate the demand model but has to be carried out through model validation. Due to the limited scope of our study, we won't use the current market demand data to validate the demand model. The approach we take for validating the final vehicle demand model is through the technique of cross-validation [42], which does not require the collection of additional data. The data set consisting of 2,552 individuals is divided into five subsets of approximately equal size using random sampling. The model is fitted to the combined data of four out of the five data sets. The fitted model is then used to predict the choice for the remaining (out-of-sample) choice set and the R-square value for the market shares, which is used as error measure, is calculated. This procedure is repeated five-fold, every time using a different data set from the five data sets for prediction and error measure calculation. The R-square value of the (in-sample) demand model fitted on the full data set is 0.99. The R-square value decreased to an average 0.92 for the five cross-validation tests, which is still an acceptable value. The cross-validation helps us build more confidence in using the proposed DCA approach to demand modeling and demand prediction. It also shows that the accuracy of the obtained demand model is satisfactory.

10.5.5 Market Share Prediction and "What If" Scenarios

The impact of attribute level changes (which reflect engineering design changes) on the vehicle market shares can be predicted by updating the vehicle descriptions and recalculating the predicted

TABLE 10.10 COMPARISON BETWEEN LINEAR AND QUADRATIC CUSTOMER UTILITY FUNCTION FITS

Statistical Metrics	Kano (quadratic)	Regular (linear)
MS_R2	0.998293	0.995984
Maximum likelihood	−5,820.69	−5,831.48
BIC	11,930.61	11,941.85

TABLE 10.11 RESULTS OF "WHAT IF" SCENARIOS

	Market Shares (%)			
Vehicle ID	Base	Scenario 1	Scenario 2	Scenario 3
1	9.84	9.81	9.41	9.38
2	7.45	7.47	7.18	7.15
3	13.13	12.91	12.42	12.37
4	8.62	8.53	8.21	8.18
5	**9.05**	**8.81**	**12.15**	**12.08**
6	7.52	7.37	7.12	7.08
7	7.80	7.63	7.38	7.34
8	6.54	6.45	6.20	6.17
9	2.63	2.71	2.62	2.60
10	17.05	17.09	16.49	16.41
11	**8.35**	**9.25**	**8.92**	**8.87**
12	**2.04**	**1.95**	**1.89**	**2.36**

choice probabilities for each individual. To illustrate how the demand model can be used to study the impact of design changes and possible actions of competitors, we consider the following "*what if*" scenarios. Vehicles 11 and 12 are two trims of one vehicle model from the same manufacturer; one of them is a basic version, while the other is a more powerful luxury version. We assume that the manufacturer decides to improve the fuel efficiency of the base model (vehicle 11) with 10%; the impact on the market shares is shown in Table 10.11 under the heading "Scenario 1." The model results show that increasing the fuel efficiency of vehicle 11 increases its market share from 8.35% to 9.25%, but it also shows that vehicle 12's market share is negatively affected. This negative impact of feature upgrades of a product on other members of the same manufacturer is known in marketing literature as "cannibalism." It implies that *the product being designed should not be considered in isolation*. Scenario 2 shows the impact on the market shares if the producer of Vehicle 5 decides to introduce a rebate of $500 to boost its market share. Finally, Scenario 3 shows the impact of increasing vehicle 12's engine power with 5%.

In addition to the market share, the feasibility or the desirability of design changes depends on the impact on profit, which necessitates the consideration of the cost of such changes. This is considered in the DBD design alternative selection example in the next section.

10.5.6 Decision-Based Design for Vehicle Engine Alternative Selection

We integrate the vehicle demand model with a cost model into a DBD model (see its framework in Figure 10.1). The DBD model is used to select the best engine design from five different engine design configurations considered for Vehicle 11. To simplify matters, the design options are represented by setting of the attribute values rather than the design options themselves. The cost model considers the impact of the performance improvements related to power, torque and low-end torque on the total cost. Low-end torque is the maximum torque an engine produces at approximately 2,000 rpm; it is important for accelerating to pass a vehicle when driving at highway speed.

The five alternative engine designs for Vehicle 11 are presented in Table 10.12. Engine Design 1 offers increased power, torque and low-end torque with 3% and a price increase of 5% relative to the performance of the existing engine used in vehicle 11. Engine Design 2 is similar in performance to Engine Design 1 but is sold at the base price. Engine Design 3 offers a 3% power and 5% price increase relative to the base model, while the performance of Engine Design 4 is the same as Engine Design 3, but sold at the base price. A fifth engine design alternative (Engine Design 5) is added by considering reusing an existing engine design for Vehicle 11 of a different vehicle model, which is less powerful but enables a reduction in price of 5% when compared with the base model. The market size M of the 12 midsize vehicles is estimated at 1,000,000 vehicles annually. Uncertainty is introduced by assuming a normal distribution of the market size with a standard deviation of 50,000 vehicles. To facilitate the consideration of the impact of engine changes of Vehicle 11 on Vehicle 12 and on the same manufacturer's profit, we assume that Vehicle 12 contributes $1,100 per vehicle to the profit.

TABLE 10.12 DESIGN ALTERNATIVES FOR DECISION-BASED DESIGN CASE STUDY

	Design Alternative (Vehicle 11) (% change attribute level)				
Attribute Design #	Design 1	Design 2	Design 3	Design 4	Design 5
Price	5	0	5	0	−5
HP	3	3	3	3	0
Torque	3	3	0	0	−10
Low-end Torque	3	3	0	0	210

The manufacturer's expected utility is obtained by assuming a risk-averse attitude, which is obtained by taking the log of the profit. The DBD optimization problem, shown in Figure 10.7, is formulated as follows: *given* the vehicle demand model (Section 10.5.3) and the decision-maker's risk attitude, *maximize* the expected utility of profit *with respect to* price, horsepower, torque and low-end torque.

The market share impact (% change) for the 12 vehicles and the impact on the profit (in millions of dollars) of the manufacturer of Vehicle 11 and Vehicle 12 together with the expected utility for the five design alternatives (Vehicle 11) are presented in Table 10.13. For example, it is noted that under design alternative 1, increasing the horsepower, torque and low-end torque with 3% and price with 5% leads to a 9.7% market share gain for Vehicle 11 and a drop in Vehicle 12's market share with 3.8%. When considering the (maximum of) expected utility of the five design alternatives, it appears that design alternative 4, consisting of a 3% torque increase while leaving the price unchanged, should be preferred. It should be noted that even though the DBD model is used to select the best design among a set of discrete alternatives in this study, the DBD model can be used to select the best alternative among a range of continuous decision variables via optimization.

10.6 CONCLUSION

In this chapter, DCA is established as a systematic procedure to estimate demand and the guidelines for implementing the discrete choice demand modeling approach in the context of DBD are provided. The transformation of top-level customer desire groups to customer desires, and further into quantifiable engineering design attributes, is introduced to bridge the gap between market analysis and engineering. As such, the customer' product selection attributes form the link between the design options and demand (and consequently profit), thus facilitating engineering design decision-making. The Kano method is used to provide econometric

GIVEN
Market size M — 1000,000 vehicles annually
Standard deviation σ_M — 50,000
Customer-driven design attributes A
Demand model Q
The demand model is obtained using the multinomial logit technique to fit the discrete choice survey data
Cost model C
Determines the relationship between A and C
Corporate interests I
None other than the single selection criterion, V
Single criterion V
Net revenue — $V = Q\,P - C$
Utility function $U(V)$
$U(V) = \log(V)$
Market Data S (Socioeconomic and demographic attributes)
Data related to gender, age and income

FIND
Key customer attributes A and price P

MAXIMIZE
Expected utility of the net present value of profit V

FIG. 10.7 VEHICLE ENGINE DBD DESCRIPTION

justification for selecting the shape of the customer utility function, which better captures the underlying purchase behavior and enhances the predictive capability of demand models. The proposed approach is demonstrated using an illustrative walk-through example and a (passenger) vehicle engine design problem as a case study, developed in collaboration with the market research firm J.D. Power and Associates and the Ford Motor Company. The estimated demand model is shown to be satisfactory through cross-validation.

It should be noted that in contrast to some existing design approaches that construct a single utility function for a group of customers, the proposed DBD approach optimizes a single-criterion utility function that is related to the profit of a product. As a part of the profit estimation, the demand modeling based on DCA predicts the choice for each individual customer and finally sums up the choice probabilities across individual decision-makers (customers) to arrive at the market share of different products, thus avoiding the paradox associated with aggregating the utility or preference of a group of customers.

TABLE 10.13 MARKET SHARE IMPACT (% CHANGE), PROFIT ($ MILLION) AND EXPECTED UTILITY FOR CASE STUDY

	Design Alternative				
Vehicle ID	1	2	3	4	5
1	– 0.4	– 0.6	0.1	– 0.1	0.2
2	– 0.8	– 0.9	– 0.3	– 0.5	– 0.1
3	– 1.1	– 1.3	– 0.6	– 0.9	– 0.5
4	– 1.0	– 1.1	– 0.5	– 0.7	– 0.3
5	– 0.3	– 0.5	0.1	– 0.1	0.4
6	– 0.6	– 0.7	– 0.1	– 0.4	0.1
7	– 1.5	– 1.7	– 1.1	– 1.3	– 0.8
8	– 1.8	– 1.9	– 1.3	– 1.5	– 1.0
9	2.9	2.7	3.4	3.1	3.7
10	– 1.0	– 1.1	– 0.5	– 0.7	– 0.3
11	9.7	11.4	4.4	7.0	2.0
12	– 3.8	– 3.9	– 3.4	– 3.6	– 3.0
Expected impact on profit	77.77	77.00	87.60	89.10	31.01
Expected utility	90.84	90.78	91.43	91.52	86.24

The demand modeling approach presented here as part of the DBD framework can be expected to facilitate the communication and collaboration of a company's employees in engineering, marketing and management. The application of the methodologies developed in this work can contribute to the development of more competitive products because in the approach presented here, products will be improved in a systematic way, considering not only the engineering requirements, but the business interests, customers' preferences, competitors' products and market conditions.

Our proposal to employ the Kano method to select and econometrically justify the customer utility function shape is a first step in improving the predictive capabilities of the proposed demand modeling approach. Another approach that can be adapted to enhance the capturing of the customer's perception of the customer's product selection attributes is through consideration of the unobservable top-level customer desires in the customer utility function using latent variables. In addition, the impact of marketing incentives, distribution and competition needs to be addressed within the DBD framework.

ACKNOWLEDGMENTS

We would like to thank the specialists at J.D. Power & Associates and Ford Motor Company for their thoughtful contributions and their efforts to gather data for our vehicle demand model. We also thank J.D. Power & Associates for providing the opportunity to work with vehicle demand modeling experts during an internship in summer 2002. The support from NSF grant DMII DMI--0335880 are acknowledged.

REFERENCES

1. Hazelrigg, G.A., 1998. "A Framework for Decision Based Engineering Design," *ASME J. of Mech. Des.*, Vol. 120, pp. 653–658.

2. Gu, X., Renaud, J.E., Ashe, L.M., Batill, S.M., Budhiraja, A S. and Krajewski, L.J., 2002. "Decision-Based Collaborative Optimization," *ASME J. of Mech. Des.*, 124(1), pp. 1–13.
3. Tappeta, R.V. and Renaud, J.E., 2001. "Interactive Multiobjective Optimization Design Strategy for Decision Based Design," *ASME J. of Mech Des.*, 123(2), pp. 205–215.
4. Wan, J. and Krishnamurty, S., 2001. "Learning-Based Preference Modeling in Engineering Design Decision-Making," *ASME J. of Mech Des.*, 123(2), pp. 191–198.
5. Thurston, D.L., 2001. "Real and Misconceived Limitations to Decision Based Design With Utility Analysis," *ASME J. of Mech. Des.*, 123(2), pp.176–186.
6. Wassenaar, H.J. and Chen, W., 2003. "An Approach to Decision Based Design with Discrete Choice Analysis for Demand Modeling," *ASME J. of Mech. Des.*, 125(3), pp. 490–497.
7. Johnson, R. M., 1971. *Multiple Discriminant Analysis: Marketing Research Applications in Multivariate Methods for Market and Survey Research*, J., Seth, ed., pp. 65–82.
8. Green, P.E. and Tull, D.S. 1988. *Research for Marketing Decisions*, Englewood Cliffs.
9. Green, P. E. and Carmone, F. J., 1970. *Multidimensional Scaling and Related Techniques in Marketing Analysis*, Allyn & Bacon, Boston, MA.
10. Green, P.E. and Wind, Y., 1975. "New Ways to Measure Consumer Judgments," *Harvard Bus. Rev.*
11. Green, P.E. and Srinivasan, V., 1978. "Conjoint Analysis in Consumer Research: Issues and Outlook," *J. of Consumer Res.*, Vol. 5.
12. Green, P.E. and Srinivasan, V., 1990. "Conjoint Analysis in Marketing: New Developments with Implications for Research and Practice," *J. of Marketing.*
13. Cook, H. E., 1997. *Product Management: Value, Quality, Cost, Price, Profit, and Organization*, Chapman & Hall, London, UK.
14. Donndelinger, J. and Cook, H.E., 1997. "Methods for Analyzing the Value of Automobiles," *SAE Paper 970762. Society of Automotive Engineers, Inc*, Warrendale, PA.
15. Li, H. and Azarm, S. 2000. "Product Design Selection under Uncertainty and with Competitive Advantage," *ASME Des. Tech. Conf., DETC2000/DAC-14234.* Baltimore, MD.
16. Besharati, B., Azarm, S. and Farhang-Mehr, A., 2002. "A Customer-Based Expected Utility for Product Design Selection," *Proc., ASME Des. Engrg. Tech. Conf.*, Montreal, Canada.
17. Wassenaar, H.J. 2003. "*An Approach to Decision-Based Design*," Ph.D. dissertation, University of Illinois, Chicago.
18. Wassenaar, H.J., Chen, W., Cheng, J. and Sudjianto, A., 2004. "An Integrated Latent Variable Modeling Approach for enhancing Product Demand Modeling," *Proc., DETC 2004 ASME Des. Engrg. Tech. Conf.*, Salt Lake City, UT.
19. Wassenaar, H.J. Chen, W., Cheng, J., and Sudjianto, A., 2005. "Enhancing Discrete Choice Demand Modeling for Decision-Based Design," *ASME J. of Mech. Des.* 127(4), *in press.*
20. Michalek, J., Feinberg, F. and Papalambros, P. Y., 2004. "Linking Marketing and Engineering Product Design Decisions via Analytical Target Cascading," *J. of Prod. Innovation Mgmt: Special Issue on Des. and Marketing in New Product Development*, in press.
21. Mileham, A.R., Currie, G.C., Miles, A. W. and Bradford, D.T., 1993. "A Parametric Approach to Cost Estimating at the Conceptual Stage of Design," *J. of Engrg. Des.*, 4 (2), pp. 117–125.
22. Matthews, L.M., 1983. *Estimating Manufacturing Costs: A Practical Guide for Managers and Estimators*, McGraw-Hill Book Co., New York, NY.
23. Stewart, R.D., 1982. *Cost Estimating*, John Wiley & Sons, New York, NY.
24. Arrow, K.J., 1963. *Social Choice and Individual Values*, John Wiley & Sons, Inc., New York, NY.
25. Arrow, K.J. and Raynaud, H., 1986. *Social Choice and Multicriterion Decision-Making*, Massachusetts Institute of Technology, Boston, MA.
26. Thurstone, L., 1927. "A Law of Comparative Judgment," *Psych. Rev.*, Vol. 34, pp. 273–286.
27. Luce, R., 1959. *Individual Choice Behavior: A Theoretical Analysis, John Wiley & Sons, Inc.*, New York, NY.
28. Marschak, J., 1960. "Binary Choice Constraints on Random Utility Indicators," *Proc., Stanford Symp. on Math. Methods in the Soc. Sci.*, K. Arrow, ed., Stanford University Press, Stanford, CA.
29. Tversky, A., 1972. "Elimination by Aspects: A Theory of Choice," *Psych. Rev.*, Vol. 79, p.281–299.
30. Bierlaire, M., Lotan, T. and Toint, P., 1997. "On the Overspecification of Multinomial and Nested Logit Models Due to Alternative Specific Constants," *Transportation Sci.*, 31 (4).
31. Daganzo, C., 1979. "*Multinomial Probit, the Theory and its Application to Demand Forecasting*," Academic Press Inc., New York, NY.
32. Ben-Akiva, M. and Lerman, S.R., 1985. *Discrete Choice Analysis*, The MIT Press, Cambridge, MA.
33. Hensher, D. A. and Johnson, L.W., 1981. *Applied Discrete Choice Modeling*, Halsted Press, New York, NY.
34. Otto, K.N. and Wood, K., 2001. *Product Design: Techniques in Reverse Engineering and New Product Development*, Prentice Hall, Upper Saddle River, NJ.
35. Krueger, R.A., 1994. *Focus Groups: A Practical Guide for Applied Research*, 2nd Ed., Sage Publications, Thousand Oaks, CA.
36. Louviere, J.J., Hensher, D.A. and Swait, J.D., 2000. *Stated Choice Methods, Analysis and Application*, Vol. 24, J.F. Hair, ed., Cambridge University Press.
37. Ben-Akiva, M., Walker, J., Bernardino, A.T., Gopinath, D.A., Morikawa, T. and Polydoropoulou, A., 2002. "Integration of Choice and Latent Variable Models, *Perpetual Motion: Travel Behaviour Research Opportunities and Application Challenges,* E. Mahmassani, ed., Elsevier Science, Chapter 21, pp. 431–470.
38. Shiba, S., Graham, A. and Walden, D., 1993. *New American TQM: Four Practical Revolutions in Management*, Productivity Press, Cambridge, MA.
39. Hastie, T., Tibshirani, R. and Friedman J., 2001. *The Elements of Statistical Learning, Springer.*
40. Raftery, A., 1995. "Bayesian Model Selection in Social Research," *Soc. Methodology.*
41. Loehlin, J.C., 1998. *Latent Variable Models, an Introduction to Factor, Path, and Structural Analysis,* 3rd Ed., L. Erlbaum Associates, Mahwah, NJ.
42. Breiman, L. and Spector, P., 1992. "Submodel Selection and Evaluation in Regression: The X-Random Case," *Int. Statistical Rev.*, Vol. 60, pp. 291–319.
43. Saari, D.G., 2000. "Mathematical Structure of Voting Paradoxes. I: Pairwise Vote. II; Positional Voting," *Eco. Theory,* Vol. 15, pp.1–103.

PROBLEMS

10.1 Consider 100 customers: 35 own a bike, 20 own a bike and a car and 45 own a car. Then the following statements are true: The majority of the customers own a bike; the majority of the customers own a car. However, the conclusion that the majority of consumers own a bike and a car is false. Use this example to show why *a group of decision-makers cannot be represented by an imaginary average individual.* Discuss advantages and disadvantages of aggregate level and individual level choice modeling.

10.2 Arrow [24] showed that preferences cannot be combined to form a group preference. Demand modeling doesn't combine preferences but predicts choices customers make. By adding choices across all individuals, an indication is obtained of how many of a particular product can be sold and at what price. Please take a look at the following table that is adapted from Saari [43] to show that the number of alternatives included in the choice set could affect demand modeling. For example, if the choice set consists of alternatives {A, B, C, D} then 9 consumers buy alternative A, 8 purchase alternative B, and so on. The group preference is thus A f B f C f D by 9:8:7:6.

a. As a demand model specialist you decide what and how many alternatives to include in the demand model's choice set. Suppose that for some reason alternative D is not included in the choice set. Determine the new group preference using the preferences provided in Table 10.14.
b. Does the paradox you noticed in question (a) pose a problem for demand modeling? Can a choice model capture the customer's true preference structure? Provide your answer for both stated preference and revealed preference. How do you think this issue can be mitigated? Consider that not all products may be available to a consumer at all times. Note: Demand modeling is a descriptive (not normative) modeling approach.
c. To what demand modeling techniques does this paradox of dropping alternatives in question (a) apply? Discuss aggregate *versus* disaggregate models and binary choice models *versus* multinomial choice models.
d. Discuss the impact of the paradox noticed in question (a) on design alternative selection. Will the demand modeling specialist's choice of alternatives considered in the choice set affect the design selection, thus affecting rational decision-making?

10.3 In the walk-through example in Section 10.4, alternative specific constants (ASC) and alternative specific variables (ASV) were used in demand model estimation.

a. Show that the coefficients for demographic variables (e.g., customer age, income, education, etc.) cannot be estimated unless they are included as ASVs.
b. Estimation of models with ASCs and/or ASVs involves setting at least one of the coefficients to zero, and the alternative to that ASV or ASC is taken as reference. Does changing the reference alternative change the choice probabilities?
c. In the walk-through, the coefficients of the alternative specific income variables were interpreted with respect to a reference alternative. How would the values of these coefficients change (i) if the reference alternative were changed to 3, instead of 1; (ii) if the reference alternative were changed to 2, instead of 1.

10.4

a. Only linear terms were used in the utility function in the walk-through example in section 10.4. However, in most real cases, it's a good idea to explore the interactions between various explanatory variables. One possible interaction that is generally used is the one between product price and customer income. Provide a brief write-up on how this can be done for the example shown in Section 10.4, within the MNL framework. Estimate an MNL model with these interaction term(s) and interpret the results. Are the results for statistical goodness-of-fit and coefficient estimates consistent with your understanding of the problem?
b. In economic modeling, diminishing returns implies that the additional utility of each additional unit of a product decreases with increasing quantities of that product, i.e., the utility function shape is concave. The relevance to demand model estimation is that customer preference for a particular product attribute (expressed as the b-coefficient of that attribute in the utility function) may not remain uniform throughout the range of that attribute. Suggest ways to incorporate this nonlinearity in customer preferences, in the utility function.

10.5

a. In Section 10.4, it was claimed that the *linear* model was superior to the *constants-only* model. Do the two models have a *restricted-unrestricted* relationship (see Appendix 10 A.2.)? What statistical evidence would you use to suggest that one of the models is better than the other? If one of the models is indeed better than the other, what is the statistical significance at which the inferior model can be rejected?
b. In Table 10.15, comment on whether the models being compared have a restricted-unrestricted relationship. Based on your assessment, choose the appropriate statistical test to compare and evaluate the statistical level of significance at which one of the models can be rejected in favor of the other.

10.6 Write a computer program that can estimate a binary demand model. Use the computer code for an example application (e.g., the work-through problem). Provide the output of coefficient estimates and the maximum likelihood value. Compare your program results with those obtained using a commercial package like STATA.

a. Write a computer program for binary logit demand modeling
b. Write a computer program binary probit demand modeling

TABLE 10.14 DROPPING ALTERNATIVES FROM THE CHOICE SET

No. of Consumers	Preference	No. of Consumers	Preference
3	$A \succ C \succ D \succ B$	2	$C \succ B \succ D \succ A$
6	$A \succ D \succ C \succ B$	5	$C \succ D \succ B \succ A$
3	$B \succ C \succ D \succ A$	2	$D \succ B \succ C \succ A$
5	$B \succ D \succ C \succ A$	4	$D \succ C \succ B \succ A$

TABLE 10.15 COMPARISON OF LOG-LIKELIHOOD ESTIMATES

Goodness-of-fit estimates*	(1) price (2) reliability (3) performance	(1) Price (2) Reliability (3) Performance (4) Order-to-ship time (5) Customer service
Log-likelihood of the estimated model $LL(\beta)$	−5,056	−5,040
Log-likelihood of the zero-model $LL(0)$	−5,565	−5,565

* see Appendix 10A

APPENDIX 10 ASTATISTICAL GOODNESS-OF-FIT MEASURES

10 A.1 Pseudo R-Square Values

The statistical goodness-of-fit of MNL models is evaluated using maximum likelihood estimates and pseudo R-square (ρ^2) values, a performance measure evaluated with the zero-model and the constants-only model as references.

$$\rho_0^2 = 1 - \frac{LL(\beta)}{LL(0)} \qquad \text{Eq.(10.15a)}$$

$$\rho_c^2 = 1 - \frac{LL(\beta)}{LL(\beta_c)} \qquad \text{Eq.(10.15b)}$$

In Eq. (10.15), ρ_0^2 and ρ_c^2 = pseudo R-square estimates, evaluated with respect to the zero-model and the *constants-only* model, respectively. $LL(0)$ and $LL(\beta_c)$ represent the log-likelihood estimates for the zero-model and the *constants-only* models, while $LL(\beta)$ represents the log-likelihood estimate for the model being evaluated. The zero-model is a model that has no parameters, i.e., the individual is assumed to have equal probability of choosing any of the alternatives in the choice set available to him. The *constants-only* model includes only a full set of constants, i.e., alternative specific constants (ASC) corresponding to each of the alternatives, with one of the alternatives chosen as the reference alternative. From the above relationships, it is easy to see that only an *ideal model* would have $\rho_0^2 = 1$. In such a case, the log-likelihood would be zero and the actual and predicted choices would match perfectly.

10 A.2 Chi-Square Test

Another test used to compare different choice models is the chi-square test. In this test, models that have a *restricted-unrestricted* relationship with each other can be compared. Two statistical models are said to have a *restricted-unrestricted* relationship when the explanatory variables of one of the models (called the *restricted* model and represented by (β_r) form a proper subset* of set of the explanatory variables of the other model (called the *unrestricted* model and represented by (β_u). In short, the *unrestricted* model has all the explanatory variables that are included in the restricted model plus a few more. The *restricted* model can be rejected in favor of the *unrestricted* model if the following relationship is satisfied:

$$-2[LL(\beta_r)-LL(\beta_u)] > \chi^2_{NR} \qquad \text{Eq. (10.16)}$$

In Eq. (10.16), $LL(\beta_r)$ and $LL(\beta_u)$ = log-likelihood estimates for the two models being considered; and $X^2{}_{NR}$ = chi-square value corresponding to NR degrees-of-freedom. NR is the number of additional explanatory variables that the *unrestricted* model has, compared with the *restricted* model. However, usually a modified form of the above test is used, in which the *restricted* model is the zero-model and the *unrestricted* model is the model in question.

10 A.3 Non-Nested Test

Finally, we need a test that can compare different models that do not necessarily have have a *restricted-unrestricted* relationship. While this test can be used to compare models that have such a relationship, it is more useful in cases where such a relationship does not exist. The non-nested test is used when considering any two models with different log-likelihood values, to evaluate the significance of rejecting the model with the lower likelihood value. The test is presented in Eq. (10.17):

$$\textit{Significance of rejection} = \Phi\left(\left[-\left[-2(\rho_H^2 - \rho_L^2) \times LL(0)\right] + (K_H + K_L)\right]^{1/2}\right) \qquad \text{Eq. (10.17)}$$

where subscript H = model with the higher likelihood value; subscript L = model with the lower likelihood value; ($K_H - K_L$) = difference in the number of explanatory variables between the two models; and $\Phi(.)$ = standard normal distribution.

*A set S_2 is a proper subset of another set S_1 if every element in S_2 is in S_1 has some elements that are not in S_2

CHAPTER

11

THE ROLE OF DEMAND MODELING IN PRODUCT PLANNING

H. E. Cook

NOMENCLATURE

A_i = annual cash flow for product i
A_i^* = annual cash flow *target* for product i
C = variable cost
C_i = variable cost of product i
CTV = critical to value
D_T^* = average annual demand over N products
dBA = noise level in decibels on A scale
D_i = annual demand for product i
D_i^* = annual demand target for product i
D_{Max} = maximum possible demand
DP = demand price (analysis)
D_T^* = total annual demand target for the segment
$D_{T,0}$ = total annual demand for identical products at cartel point
DV = direct value (method)
E_2 = price elasticity of average demand
EEV = expected economic value
f_{Alt} = fraction of buyers selecting the alternative product
$F_{Cost,i}$ = annual fixed cost for product i
$F^*_{Cost,i}$ = annual fixed cost target for product i
f_i = market share of product i
f_{Opt} = fraction of buyers selecting the option
G = annual net value to society of product
$g_{0,i}$ = baseline size of attribute i
$g_{C,i}$ = critical size of attribute i
g_i = size of attribute i
$g_{I,i}$ = ideal size of attribute i
IIA = independence of irrelevant alternatives (axiom)
K = negative slope of the demand curve at cartel point
LIB = larger is better
M_i = annual investment for product i
$M^*_{T,i}$ = investment plus interest target paid as mortgage over time horizon
mvA = minivan A
mvB = minivan B
N = number of products being considered
NIB = nominal is best
$\bar{P}$ = mean price of N products
P_0 = price (baseline) of N identical products at cartel point
P_{Bal} = balanced price for bargaining
P_C = Cournot-Nash price
$P_{C,i}$ = Cournot-Nash price forecast for product i
P_i = price of product i
P_N = neutral price for the alternative product
P_{Opt} = price of optional attribute or feature
SIB = smaller is better
U_i = utility computed from logit model for product i
V = total value of a product
$\bar{V}$ = mean value of N products
V_0 = value (baseline) of N identical products at cartel point
V_{Alt} = value of alternative product
VDI = J.D. Power and Associates Vehicle Dependability Index Associates
V_i = value to customer of product i
V_{Opt} = value of option
Y = time horizon for the product
β = price coefficient in logit model
β_{Opt} = price coefficient for option
$v(g_i)$ = normalized value coefficient at attribute size g_i

11.1 INTRODUCTION: PRODUCT PLANNING FOR HYPER-COMPETITIVE MARKETS

The global economy has materialized into a double-edged sword for manufacturing firms because the economic opportunities offered by new markets have been offset by a large increase in the number of strong competitors. For example, the number of major automotive companies in the United States was three before globalization; now it is seven or more and growing. But it is not just automotive. Globalization has put acute pressure on firms across all industries to continuously reduce costs while simultaneously speeding up the development of innovative, new products.

The job of the product planner has become particularly challenging. Not only does the rate of product improvement have to be increased while driving costs down, but planners must discover how to use limited resources to select, develop and implement new technologies in a way that best serves the diverse needs and tastes of customers around the globe. Moreover, these are customers who can now choose from a wide array of strong, competing alternatives. Thus, there is a pressing need to make the product planning process more effective. The solution recommended here is to put product planning on a firmer, more scientific foundation. The major challenge is to have a practical methodology that balances rigor with simplicity and transparency so that it will have wide appeal and usage across the diverse disciplines within a manufacturing firm.

The goal of any product planning methodology should be to identify, in a timely manner, the technologies and attributes for new products that offer the best improvement in the firm's bottom line. *A critical element in generating sound financial forecasts is having a trustworthy and insightful algorithm for forecasting demand, which is used in arriving at the price of the product and in forecasting cash flow.* The overall planning methodology should be science-based, which means that the quantitative forecasts of demand and cash flow made during the product development process are compared with actual outcomes once the product is in production. Shortcomings found in the process and model must be identified and improved upon, if not eliminated. The purpose of this chapter is to explore the foundations of a product planning methodology based on a model for product demand that is simple yet rigorous in the limit of small departures in the values and prices of the products competing in a segment [1, 2]. Alternate approaches to the product planning/design problem, wide ranging in assumptions and complexity, can be found in Chapters 16 through 20. The nature of the problem at hand and the experience of the user will dictate which approach is favored.

11.1.1 A Product Planning Template

Because every product planning problem is unstructured at the outset, the first step in establishing a plan is to structure it using a solution template. The template used here, Fig. 11.1 [3], has two loops. The loop on the right starts with customer needs, which are translated into key, measurable system-level attributes that are critical to customer value (CTV attributes). Market research and economic calculations are used to quantify how changes in the CTV attributes affect the value of the product to the customer. Changes in the CTV attributes can also affect cost (variable, fixed and investment). Value, cost and the pace of innovation (the speed at which the loops are navigated) represent a set of fundamental financial metrics [4] that determine the bottom-line financial metrics of price, demand, cash flow.

The loop on the left traces how societal needs and values are impacted by the externalities associated with the design, manufacture, use and disposal of products. Externalities are controlled largely by governmental regulations. Adherence to regulations also affects costs, the waypoint where the customer and the societal loops connect.

The two merged loops then flow through the bottom-line metrics of price and demand arriving at cash flow. Thus the needs of the three stakeholders, the customer, society and the firm, are coupled. The firm must not only satisfy customer needs and societal values, it must satisfy the needs of a sufficient number of customers *better* than competition to provide the cash flow required to develop the future products necessary for staying in business. The risks to the manufacturer in planning a new product are very large. Disruptions in material availability can cause an increase in variable costs. A key supplier may go into bankruptcy. Consumer tastes can change between the time the marketing research was done and when the product was produced. But the greatest risk of all is not being able to make a reliable assessment of the improvements that competitors will make in their products.

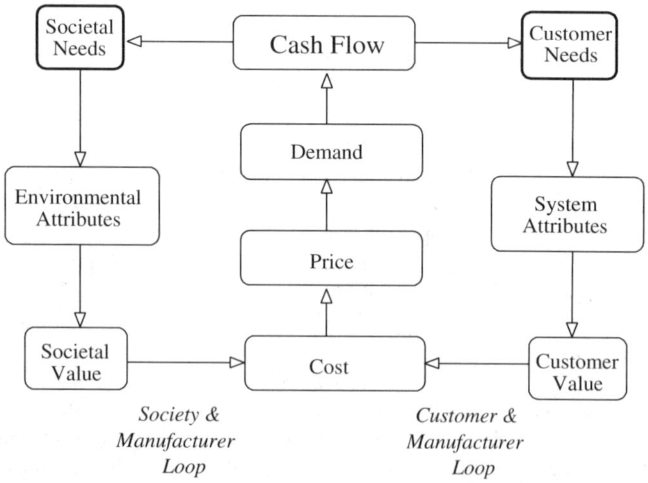

FIG. 11.1 THE S-MODEL TEMPLATE FOR STRUCTURING THE PLANNING OF A NEW PRODUCT CONSISTS OF A CUSTOMER & MANUFACTURER LOOP AND A SOCIETY & MANUFACTURER LOOP (SOURCE: [3]; REPRODUCED WITH PERMISSION OF SPRINGER-VERLAG LONDON LTD.)

11.2 ANALYSIS OF A SIMPLE MARKET TRANSACTION

The connections between the fundamental and bottom-line metrics shown in Fig. 11.1 are explored in the simple market transaction shown in Fig. 11.2, which involves three agents: the seller, the buyer and the rest of society. The net value to the buyer is value, V, minus price, P, and the net value to the seller is price minus variable cost, C. For the transaction to take place, both the buyer and the seller need to sense that they will receive a positive gain in net value. If the price is not posted, they can bargain—the seller seeking a high price and the buyer a low price. With equal bargaining power, they are likely to arrive at a balanced price, P_{Bal}, for the transaction that results in sharing of net value:

$$P_{\text{Bal}} - C = V - P_{\text{Bal}} \qquad \text{Eq. (11.1)}$$

The solution of Eq. (11.1) for P_{Bal} is given by:

$$P_{\text{Bal}} = \frac{V + C}{2} \qquad \text{Eq. (11.2)}$$

This price is the average of the sum of the customer's value and the seller's value, which is equal to the variable cost, C, of the product. If V is less than C, it follows from Eq. (11.2) that P_{Bal} will be less than cost. Thus, for the transaction to take place profitably for the seller, value must be greater than variable cost,

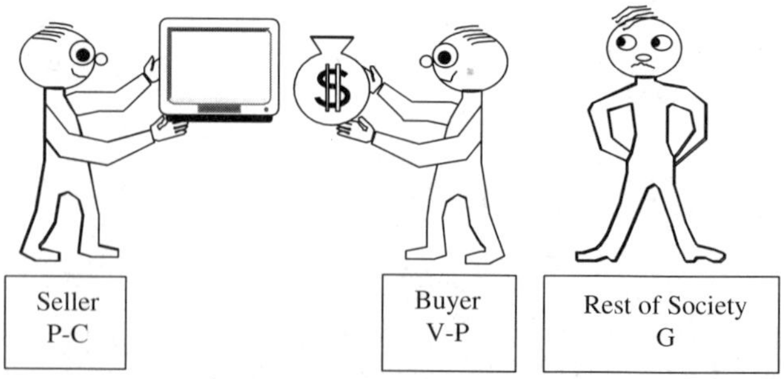

FIG. 11.2 SCHEMATIC OF NET VALUE CHANGES FOR A SIMPLE TRANSACTION. (SOURCE: [2] WITH PERMISSION FROM SPRINGER SCIENCE + BUSINESS MEDIA)

$V > C$. In other words, *the value of the product to the buyer must be greater than the value to the seller for a sale to take place.*

The person on the for right in Fig. 11.2 is assessing the net value, G, to society of the externalities associated with the product. For the transaction to be beneficial to society as a whole, the sum of the three net values for the simple transaction under review must be greater than zero:

$$V - C + G > 0 \qquad \text{Eq. (11.3)}$$

This relationship can also be written as $V > (C–G)$. As G is often negative, societal concerns place an additional condition on the size of value relative to cost. The externalities for new product development are managed, in large part, by strict adherence to governmental regulations. Environmentally conscious buyers will see added value in products that are deemed environmentally friendly.

Demand for a product is the aggregate *desired* rate of purchase. It will equal the actual sales rate if there is a product available for every buyer who has the funds and desire to purchase it. When demand starts to exceed supply, it is not uncommon for the seller to increase price, which helps keep sales in line with demand. Demand is considered as being "pent up" (greater than sales) when there are more potential buyers willing to pay the asking price than products available.

11.3 DEMAND IN A MARKET SEGMENT HAVING *N* COMPETITORS

11.3.1 Linear Model

Products such as cars, trucks, construction equipment, aircraft and laptop computers compete most often within well-defined product segments. Family sedans, for example, can be segmented into entry-level (small), lower-middle, upper-middle, large and luxury. Although there is competition between segments, most is within a segment. If the simplifying assumption is made that a product competes only with the other N–1 products in its segment and that all have similar but different values and prices, the aggregate, annual demand for each product i can be written as a function of the values and prices of each of the N products:

$$D_i = D_i(V_1, V_2, \ldots, V_N;\, P_1, P_2, \ldots, P_N) \qquad \text{Eq. (11.4)}$$

If demand is assumed to be analytic in the values and prices and expanded as a Taylor series up to and including linear terms, Eq. (11.4) becomes a linear set of N simultaneous equations of the form [1]:

$$D_i = K\left[V_i - P_i - \frac{1}{N}\sum_{j\neq i}(V_j - P_j)\right] = K\left[\left(\frac{N+1}{N}\right)(V_i - P_i) - (\bar{V} - \bar{P})\right]$$

Eq. (11.5)

The terms $\bar{V}$ and $\bar{P}$ = average value and average price, respectively. The expansion is about a so-called cartel reference point. The term "cartel" is used to define the reference point for the Taylor expansion because the products at this point are assumed to be identical in attributes, value, price and demand. At the cartel point, total demand, value and price are defined as $D_{T,0}$, V_0 and, P_0, respectively. These parameters are identified with and computed from the reference state for the problem at hand. For example, planners for a new minivan could use the current year as the reference state. The cartel parameters $D_{T,0}$, V_0 and P_0 would be set equal to the total demand, *average* value and *average* price, respectively, of the minivans in the baseline market. When both sides of Eq. (11.5) are summed from $i = 1 \rightarrow N$ over the reference state, the constant K, which is the negative slope of the total demand curve at the cartel point, is found to be:

$$K = \frac{D_{T,0}}{V_0 - P_0} = \frac{D_{T,0}E_2}{P_0} \qquad \text{Eq. (11.6)}$$

The term E_2 defined as $-[\delta\bar{D}/D_0]/[\delta P/P_0]$ and equal to $P_0/(V_0 - P_0)$ = price elasticity of baseline average demand D_0 (= $D_{T,0}/N$) when each of the competitors change price by the same amount, δP. The average value V_0 represents a marginal quantity in that it will be a function of price if demand is not linear in price. Total demand for an arbitrary state is given by:

$$D_T = K(\bar{V} - \bar{P}) \qquad \text{Eq. (11.7)}$$

This expression follows on summing both sides of Eq. (11.5) from i equal 1 to N. The expression for market share, f_i, obtained from Eqs. (11.5) and (11.7) is given by:

$$f_i = \frac{N+1}{N}\left(\frac{V_i - P_i}{\bar{V} - \bar{P}}\right) - 1 \qquad \text{Eq. (11.8)}$$

When $N = 1$, it follows from Eq. (11.8) that $f_i = 1$, as it should. It also follows from Eq. (11.6) and the definition for E_2 that the average values and prices are related by:

$$\bar{V} = \bar{P}\left(\frac{1 + E_2}{E_2}\right) \qquad \text{Eq. (11.9)}$$

When the elasticity E_2 has been measured using econometric methods, average value can be estimated using Eq. (11.9).

If demand is not linear with price, Eqs. (11.5), (11.7) and (11.8) are valid only in a region about the cartel point. This can be formally described by adding a term for model error, e_m, to the right-hand side (RHS) of Eq. (11.5), which goes to zero as the cartel point is approached. Also if Eq. (11.5) is used to forecast the demand for a new product, the uncertainties in the values and prices on the RHS generate additional error. A major source of uncertainty in a forecast will be the values and prices of the N–1 products competing against product i in the future.

There is also uncertainty in forecasting the value and price of product i. Assume, for example, that product i is an automobile and the new design has a new exterior body style, a reduction in interior noise level by 10% and an increase in acceleration performance by 15%. There is uncertainty in the additional amount that potential buyers will be willing to pay for the changes in these CTV attributes. Moreover, not all potential buyers will value the attribute changes the same. There may be other CTV attributes not identified (unobserved) in the planning process that change. Such oversights will lead to additional error. In spite of these uncertainties, the feedback gained over time from comparing forecasts with actual outcomes will identify the shortcomings and improper assumptions, which can then be corrected to sharpen the model.

11.3.2 Pricing

The linear demand model is helpful in working through the nontrivial problem in a transparent manner of how best to price a product in a competitive market. Each competitor would ideally like to price its product to maximize its profit. If a cartel were legal, the members could cooperate and set prices openly in an optimal manner. The same result can be obtained approximately and legally when one of the companies becomes a price leader, with the others following along. However, such arrangements often fall apart as there is usually a gain, albeit short-term, for one competitor to break away from the pattern and reduce the price somewhat.

The model used here for pricing in a highly competitive market is based upon the assumption that each competitor sets its price to maximize annual cash flow, A_i (or profit), believing, *incorrectly*, that the other N–1 competitors will not change their prices. Arguments of this type were first employed in 1838 by Cournot and later by Bertrand. This results in N simultaneous equations of the form:

$$\frac{\partial A_i}{\partial P_i} = 0 \qquad \text{Eq. (11.10)}$$

Annual cash flow for product i is given by:

$$A_i = D_i[P_i - C_i] - F_{\text{Cost},i} - M_i \qquad \text{Eq. (11.11)}$$

in which $F_{\text{Cost},i}$ = annual fixed cost; and M_i = annual investment. The set of equations represented by Eq. (11.10) is evaluated with cash flow given by Eq. (11.11). When starting with the intended price optimization given by Eq. (11.10), one might expect to arrive at a Bertrand pricing model. Instead, the resulting prices represent a Cournot or Cournot-Nash equilibrium price for $i = 1 \rightarrow N$ competing products differentiated in value, cost and price [5]:

$$P_{C,i} = \frac{(N^2 + 2N)C_i + (N^2 + N + 1)V_i + N\left[\sum_{j \neq i}(C_j - V_j)\right]}{2N^2 + 3N + 1} \qquad \text{Eq. (11.12)}$$

The Cournot-Nash expression for N products not differentiated in value, cost and price has been given by Pashigian [6]. The price given by Eq. (11.12) for undifferentiated duopoly products is not equal to variable cost as in the well-known Bertrand duopoly model, but equal to the Cournot duopoly price.

Price is reduced for the oligopoly condition ($N > 1$) because the bargaining power of the seller is weakened by having N–1 competitors vying for the same customers. The price for a monopoly ($N = 1$) from Eq. (11.12) is $(V + C)/2$, which is the same result found for the price in Eq. (11.2) when the buyer and seller bargained with equal strength.

It follows that a model of cost plus a percentage model for pricing likely does not apply to competitive markets. The above model for price equal to a linear combination of value and cost is far different. Software is a good example of a product that is priced in relation to its value as variable cost is almost zero.

Equation (11.12) should not be taken to suggest that firms sharpen their pencils and take partial derivatives before setting prices. This expression represents the outcome of a behavioral model for highly competitive markets prone to price wars [5, 7]. The Cournot-Nash prices represent the set of prices where a price war should end, in theory, as there is no short-term gain for any single firm to reduce price further. If the firms chose to cooperate and if all had the same values and costs, their prices would again be $(V + C)/2$. However, in a highly competitive market, the prices predicted for a price war scenario are likely to be more representative than those for cooperative behavior.

11.3.3 Logit and Probit Models for Demand

Although the applicability of the simple linear model is being stressed here, nonlinear models should be used for analyzing the outcomes of market research surveys because the wide ranges of price and market share found in surveys are often outside the limits of validity for the linear model. The so-called multinomial logit model for aggregate demand (referred to here simply as the logit model) is given by [8]:

$$D_i = D_T \frac{\exp(U_i)}{\sum_{j=1}^{N} \exp(U_j)} \qquad \text{Eq. (11.13)}$$

In Eq. (11.13), U_j = utility of product j. This expression is constructed by aggregating the probability functions given by the logit model for individuals whose utilities are taken to be a random number having a deterministic part plus a random error term that is independently and identically Gumbel-distributed. On the other hand, if the error terms are assumed to be normally distributed, then the multinomial probit (MNP) model is obtained [9]. Although theoretically more satisfying due to the Central Limit Theorem, the MNP model does not provide a closed-form solution and the simpler logit form given by Eq. (11.13) is more widely used. Recent computational advances may change this [10]. However, when only paired comparisons are being evaluated in a survey, the binomial probit can be used with ease to evaluate the outcomes.

When Eq. (11.13) is divided through by total demand, the expression for market share is found to be:

$$f_i = \frac{\exp(U_i)}{\sum_{j=1}^{N} \exp(U_j)} \qquad \text{Eq. (11.14)}$$

It follows that the ratio of the two demands is the same as the ratio of f_i and f_j:

$$\frac{D_i}{D_j} = \frac{f_i}{f_j} = \exp(U_i - U_j) \qquad \text{Eq. (11.15)}$$

When the logarithm of the demand ratio is taken and evaluated in the region close to the cartel point, $Ln(D_i / D_j)$ can be replaced by $(D_i - D_j)/\bar{D}$. Thus, as the cartel point is approached in terms of prices and values, the difference between the two utilities approaches the demand difference divided by $\bar{D}$ $(= D_T / N)$:

$$U_i - U_j \rightarrow \frac{D_i - D_j}{\bar{D}} \quad \text{Eq. (11.16)}$$

Because both the logit and linear models are analytic, the logit model in the region near the cartel point must, therefore, be equal to the linear model. Thus from Eqs. (11.16) and (11.5), the difference between the two utilities is given by [11]:

$$U_i - U_j = \frac{(N+1)E_2}{\bar{P}_0}[(V_i - V_j) - (P_i - P_j)] \quad \text{Eq. (11.17)}$$

Often a linear model of the form:

$$U_i - U_j = \beta[(V_i - V_j) - (P_i - P_j)] \quad \text{Eq. (11.18)}$$

is used to represent the utility differences over a wide range of values and prices. Comparison with Eq. (11.17) shows that the theoretical expression for the price coefficient β is given by [11]:

$$\beta = \frac{(N+1)E_2}{P_0} = \frac{N+1}{V_o - P_0} \quad \text{Eq. (11.19)}$$

It follows from Eq. (11.19) that the independence of irrelevant alternatives (IIA) axiom for utility differences does not hold for the logit model because *β is a function of the number of alternatives (competitors), N.*

11.4 MODEL VALIDATION

When the demand for a monopoly is plotted on the x-axis versus price on the y-axis, the linear model given by Eq. (11.5) for $N = 1$ predicts that the line should shift upward by $\Delta P = \Delta V$ if value is increased by ΔV and that price should approach value as demand approaches zero. These predictions were tested using an experimental market for lottery tickets with known payoffs [3]. One ticket had a 50% probability of paying \$100 and the other had an 80% chance of paying \$100, the expected economic values (EEVs) of the two tickets being \$50 and \$80, respectively. A total of 78 respondents were surveyed. The survey form for each lottery ticket asked the respondent to decide between buying or not buying the ticket over a range of prices from \$0 to \$95 in \$5 increments. The linear demand model can be written as:

$$D = K(V - P) = D_{Max}\left(1 - \frac{P}{V}\right) \quad \text{Eq. (11.20)}$$

which follows because maximum demand, $D_{Max} = KV$, is obtained when price equals zero. Thus, the fraction of *maximum* demand is given by:

$$f = \frac{D}{D_{Max}} = \left(1 - \frac{P}{V}\right) \quad \text{Eq. (11.21)}$$

Their results are replotted here as Fig. 11.3 and the values computed from the price intercepts are listed in Table 11.1 for two ranges of data. The first range includes all of the data and the second covers f from 3% to 97% for both tickets as it is a stretch to apply the linear model over the entire range.

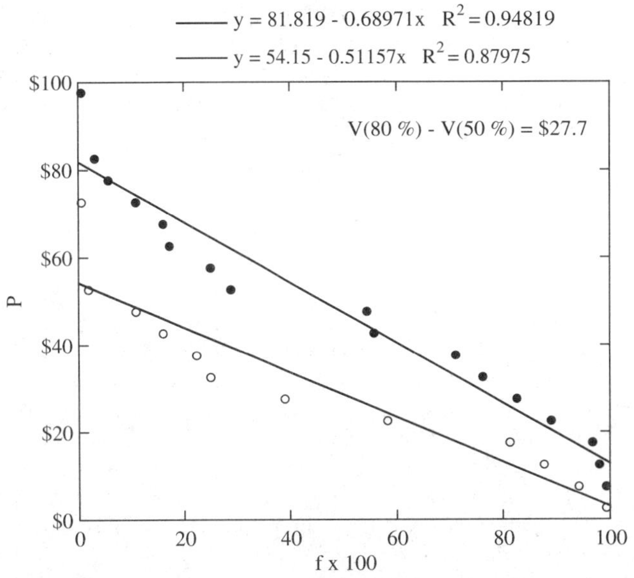

FIG. 11.3 PLOT OF PRICE VERSUS PERCENT OF RESPONDENTS SELECTING LOTTERY TICKET (SOURCE: [3]; REPRODUCED WITH PERMISSION OF SPRINGER-VERLAG LONDON LTD.)

Price divided by the EEVs of the respective tickets is plotted over the reduced range in Fig. 11.4. The value differences and the standard errors are listed in Table 11.2. The observed difference in value between the two tickets is within two standard errors (SE) of the difference of \$30 between the two EEVs. The shorter range shows a smaller SE as should be expected. Figure 11.3 shows that a small fraction of respondents were risk takers in that they would pay more than the EEVs of the tickets. The results shown in Figs. 11.3 and 11.4 and Tables 11.1 and 11.2 lend strong support to the value and price relationships in the demand model given by Eq. (11.5).

Figure 11.5 shows the normal probability plot associated with a BNP model using the full range of data, the prices again being divided by their respective EEVs. As the distributions are assumed normal, demand never goes to zero and never reaches D_{Max}. The vertical line drawn through $P / EEV = 1$ shows empirically that the respective EEVs are reached at the 5% level on the upper tail of the distribution. A similar result is found if the logit function $Ln[f / (1-f)]$ is plotted versus P / EEV.

The R^2 values in Figs. 11.4 and Fig. 11.5 give a slight edge in fidelity to the BNP model over the linear model. The strength of

TABLE 11.1 THE SEPARATE VALUES AND STANDARD ERRORS (IN \$) COMPUTED FROM THE LINEAR MODEL FOR THE 80% AND 50% CHANCES OF WINNING \$100 (THE COMPUTATIONS ARE COMPARED FOR THE TWO RANGES OF *F*)

	50% Ticket		80% Ticket	
	0% to 100%	3% to 97%	0% to 100%	3% to 97%
Value	54.2	47.2	81.8	77.7
SE	3.4	2.0	2.5	1.8

TABLE 11.2 VALUE DIFFERENCES AND THEIR STANDARD ERRORS (IN $) BETWEEN THE 80% AND 50% CHANCES OF WINING $100 AS COMPUTED FROM THE LINEAR MODEL (THE COMPUTATIONS ARE COMPARED FOR TWO RANGES)

	0% to 100%	3% to 97%
Value Difference	27.7	30.6
SE	4.3	2.7

the linear model is that it gives better insight into the outcomes of the experiment. This includes the predictions that demand should approach zero as price approaches the EEV and that plots of P / EEV versus *f* should overlay according to Eq. (11.21). Moreover, the linear model was a good representation of thse data over a significant range of *f*.

It is interesting that the values of 0.598 and 0.577 for the average prices divided by their respective EEVs shown in normal probability plots in Fig. 11.5 are actually higher than 0.5, which is the theoretical amount that a monopoly should charge. Assume that the cost of the lottery ticket was zero. (This would be the case if the seller skipped town after selling the tickets and did not pay off the winners.) For this case, the optimal price for the monopolist is where cash flow, *A*, is at a maximum. Normalized cash flow before payout is given by $fP = A / D_{Max}$ for the 80% lottery, which is shown in Fig. 11.6 as a function of price for both the probit model (full range) and the linear model (3% to 97% range). The peak in cash flow for the probit model is at $37.5. The linear model has its maximum very close to EEV/2 of $40, the linear pricing model for a monopoly assuming zero variable cost. For the lottery to be profitable to the seller after the payout of the winnings, the price of a ticket needs to be in excess of the EEV, which is equal to the average cost of a ticket to the seller. The buyers would be the risk takers in the probit model distribution in Fig. 11.6, willing to pay more than the EEV of $80.

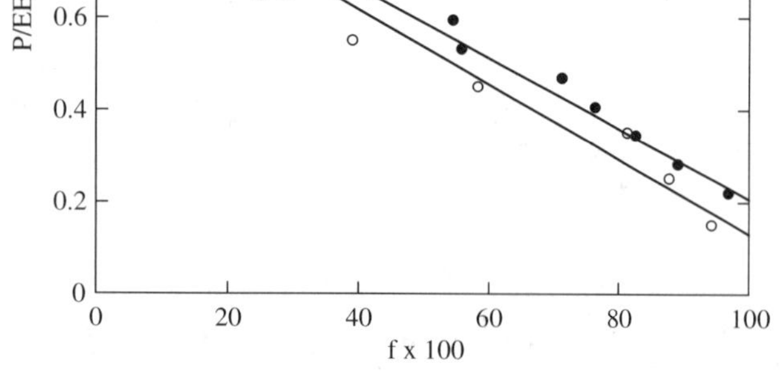

FIG. 11.4 PLOT OF PRICE DIVIDED BY THE EXPECTED ECONOMIC VALUE OF THE LOTTERY TICKET VERSUS PERCENT OF RESPONDENTS SELECTING TICKET FOR *f* X 100 RANGING BETWEEN 3% AND 97% (SOURCE: [3]; REPRODUCED WITH PERMISSION OF SPRINGER-VERLAG LONDON LTD.)

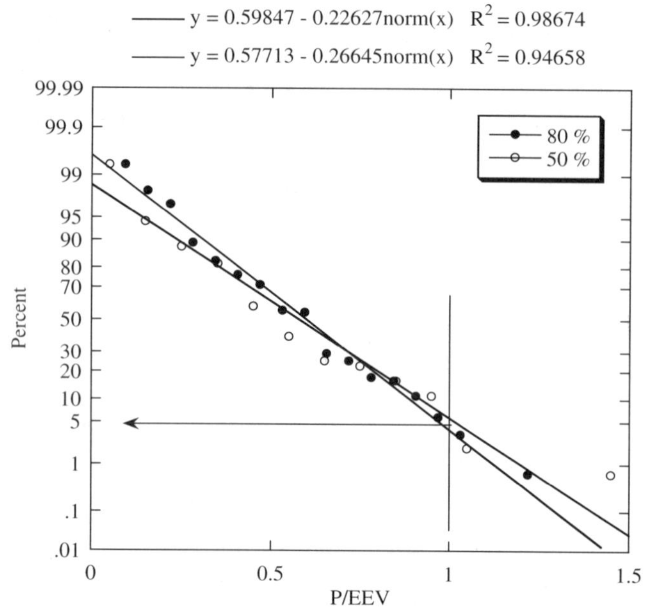

FIG. 11.5 NORMAL PROBABILITY PLOT OF PERCENTAGE OF RESPONDENTS SELECTING LOTTERY TICKET AS A FUNCTION OF PRICE DIVIDED BY THE EXPECTED ECONOMIC VALUE OF THE TICKET (SOURCE: [3]; REPRODUCED WITH PERMISSION OF SPRINGER-VERLAG LONDON LTD.)

11.5 SIMULATING THE GLOBAL MARKETPLACE

As shown by the simulated demand for the lottery tickets, the linear model provided an easy-to-interpret, approximate model for demand behavior over a reasonable range of *f*. Thus, the pricing

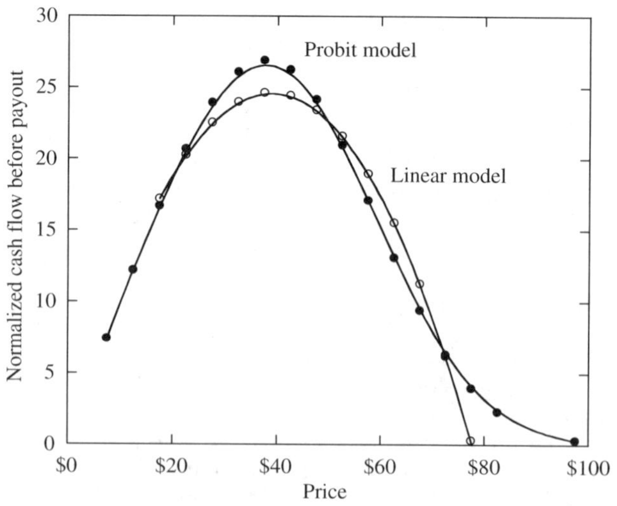

FIG. 11.6 PLOT OF *FP* AS A FUNCTION OF PRICE FOR THE PROBIT AND LINEAR MODELS

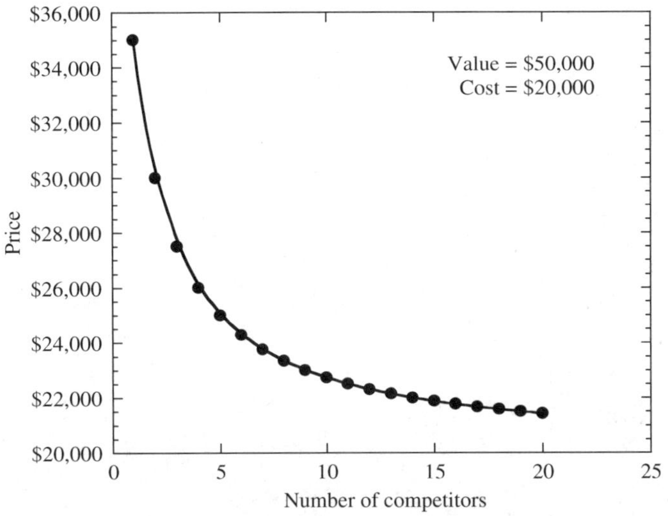

FIG. 11.7 PRICE ACCORDING TO THE COURNOT-NASH MODEL AS A FUNCTION OF THE NUMBER OF COMPETITORS, *N*

model given by Eq. (11.12) should provide both qualitative and semi-quantitative insight into the demand and pricing behavior for the global marketplace. The relationship between the Cournot price and N is shown in Fig. 11.7. The products are assumed to be identical for simplicity, having value and variable cost of $50,000 and $20,000, respectively. The baseline annual demand for the $N = 1$ monopoly condition was taken to be 600,000 units, and the annual fixed cost per manufacturer was assumed to be $500 million. Although these numbers have no basis in fact, they are not outlandish for the upper middle segment in the U.S. family sedan market.

Price is seen to fall dramatically from $35,000 to $25,000 as N increases from 1 to 5. Adding insult to injury, average demand per manufacturer also declines as shown in Fig. 11.8. However, total demand grows because of the major price reductions being made as N increases as shown by Fig. 11.7. The combined impact of lower prices and lower demand per competitor leads to a dramatic reduction in cash flow per competitor as shown in Fig. 11.9.

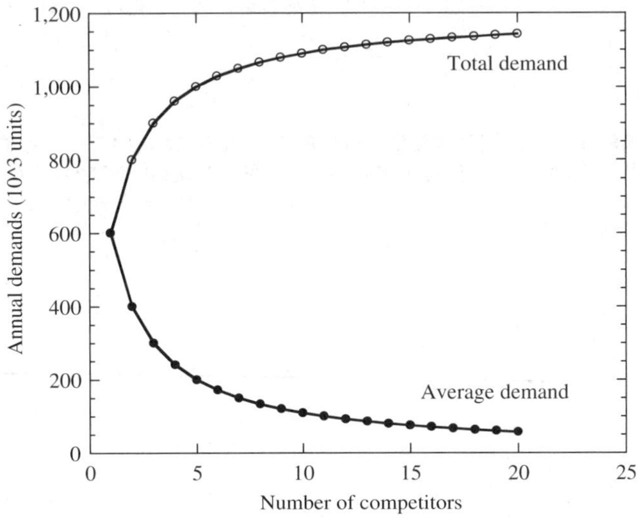

FIG. 11.8 AVERAGE ANNUAL DEMAND AND TOTAL AVERAGE DEMAND AS A FUNCTION OF THE NUMBER OF COMPETITORS, *N*

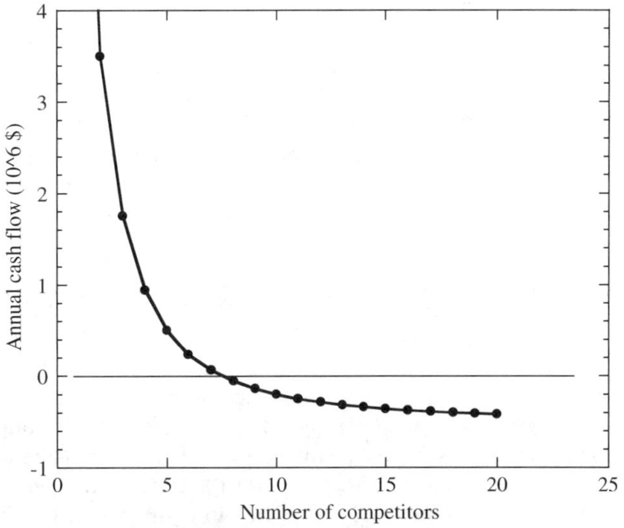

FIG. 11.9 ANNUAL CASH FLOW AS A FUNCTION OF THE NUMBER OF COMPETITORS, *N*

Cash flow is seen to be negative for $N > 7$. Thus for the parameters chosen, the market will not profitably support more than seven products.

The assumption that the N products are identical, of course, will not hold for a real market. As the number of competitors increase, the weaker products that carry higher costs but lower values will have negative cash flow at lower levels of N. If the manufacturers of such products cannot turn this around, it is only a matter of time before such products drop from the segment. Thus, the number of competing products in a market segment is expected to grow as globalization materializes but then drop back as the weaker products fail. The silver lining of the intense competition is that customers benefit because it leads to a higher pace of product innovation and lower prices.

11.6 SCIENCE-BASED PRODUCT PLANNING

11.6.1 Financial Target Setting

The major steps in the product planning process are highlighted in Fig. 11.10. Financial planning begins with the needs of the firm represented by cash flow, which is shown top center in the solution template, Fig. 11.1, used here to structure and guide the planning process. Initial bottom-line targets are set for cash flow, demand and price. The degree of stretch represented by these targets should

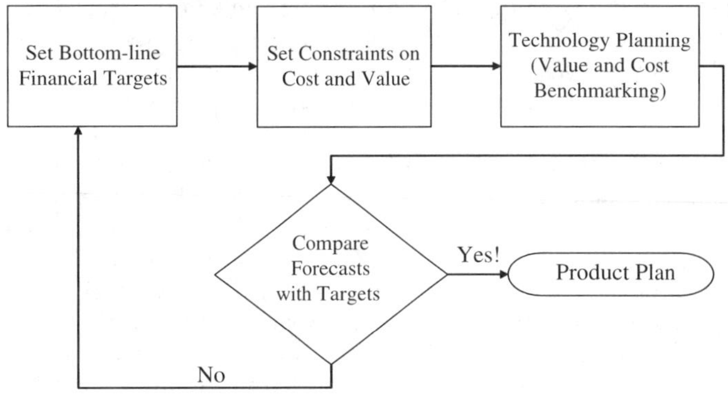

FIG. 11.10 PLANNING FLOWCHART

be challenging but not unreasonable. If they do border on being dreamlike, it will quickly become obvious as the planning process moves forward and the initial targets are adjusted.

11.6.2 Cost and Value Constraints

The bottom-line targets place constraints on the fundamental metrics, the second step in Fig. 11.10. The variable cost constraint obtained from Eq. (11.11) is well-known and can be expressed as:

$$C_i < P_i^* - \left(\frac{A_i^* + F_{\text{Cost},i}^* + M_{T,i}^* / Y}{D_i^*} \right) \quad \text{Eq. (11.22)}$$

where P_i^*, A_i^*, $F_{\text{Cost},i}^*$, and D_i^* = planned price, cash flow, fixed cost and demand *targets*, respectively, for product i; $M_{T,i}^*$ = planned total investment plus interest, which is paid off as a mortgage with equal annual payments of $M_{T,i}^* / Y$ over the time period; and Y = number of years the product is expected to be in production. The constraint on value is computed from Eq. (11.8) and is given by:

$$V_i > P_i^* + \frac{N}{K(N+1)}\left[D_i^* + D_T^*\right] \quad \text{Eq. (11.23)}$$

In writing Eq. (11.23), the result that the forecast total demand, D_T^*, is equal to $K(\bar{V} - \bar{P})$ was used. It follows from Eqs. (11.22) and (11.23) that cost must be less than price and value must be greater than price. For example, if the demand target is set very high, this lessens the variable cost task but exacerbates the value task. If a high target is set for annual cash flow, the cost target will likely not be met unless the demand target is also high. The constraints are thus highly coupled.

11.6.3 Technology Planning

11.6.3.1 DP Analysis for Value Trends The third step, technology planning, involves value and cost benchmarking, which is a dual exercise of: (1) identifying the new technologies that can improve value and reduce cost; and (2) assessing what leading competitors are likely to do with their products. A study of historical trajectories in demand (sales) and price can indicate by extrapolation where the overall market and individual competitors may be headed. For greater insight, planners can *convert observed demand and price trends into revealed value trends by replacing the inequality sign in Eq. (11.23) with an equal sign and eliminating the superscripts on the price and demand variables.* This procedure has been called demand/price analysis for value trends, or simply DP analysis.

The value trends from 1996 through 2001 (in year 2001 dollars) are shown in Fig. 11.11 for $N = 5$ sedans in the upper-middle segment. The lines are drawn simply to aid the eye. The elasticity E_2 was set equal to one for the computations. The individual values are shown in Table 11.3. The prices used in the computations were actual transaction prices and the demands were set equal to retail sales (total sales less fleet sales). Prices and retail sales are proprietary and thus not shown. The larger open circles are used to identify a major redesign. They are seen to bump up value, at least temporarily. However, all of the vehicles enjoyed a rise in value from 1996 to 1998 whether or not a major redesign was made. Apart from the boost of the major redesign of vehicle C in 2000 (and a change in brand name), values in this important segment have trended down since the 1998 model year.

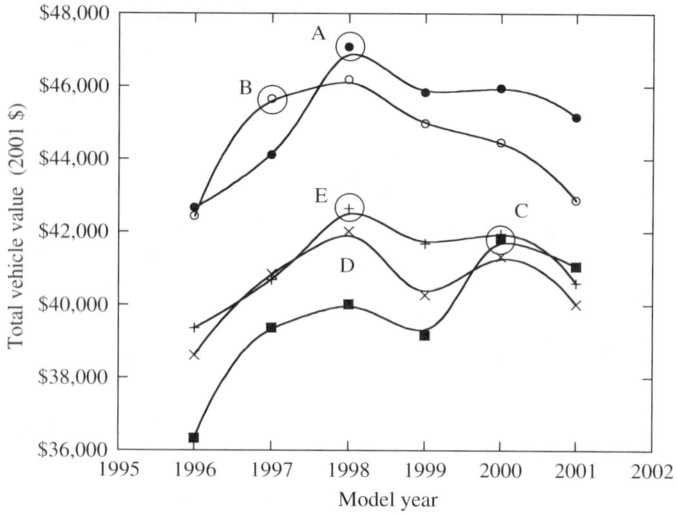

FIG. 11.11 PLOT OF TOTAL VALUE IN 2001 $ FOR FIVE VEHICLES COMPETING IN THE UPPER-MIDDLE SEGMENT ACCORDING TO THE LINEAR MODEL (CIRCLES DENOTE MAJOR REDESIGNS; LINES ARE DRAWN TO AID THE EYE. TRANSACTION PRICES AND RETAIL DEMAND USED IN COMPUTING VALUES WERE PROVIDED BY POWER INFORMATION NETWORK, LLC, AN AFFILIATE OF J.D. POWER AND ASSOCIATES)

Another observation is that vehicles A and B are roughly $2,000 to $4,000 higher in value than vehicles C, D and E. Differences in vehicle durability/dependability may be largely responsible. This is supported by Fig. 11.12, which is a plot of vehicle value versus the J.D. Power and Associates Vehicle Dependability Index (VDI) reported in 2002 for 1998 model year vehicles. The index is representative of problems per 100 vehicles over the first three years of operation. It is likely proportional to the total number of repairs over the lifetime of use, which buyers use as one factor in making their next purchase decision. The average lifetime of number of repairs per vehicle (repairs over seven to 10 years) should be more than two times VDI/100. The value for R^2 in Fig. 11.12 suggests that VDI explains about 80% of the variation. Other factors such as styling, brand loyalty, performance, etc. also play an important role.

TABLE 11.3 TOTAL VEHICLE VALUES COMPUTED FOR FIVE FAMILY SEDANS IN THE UPPER-MIDDLE SEGMENT FROM THEIR DEMAND AND PRICE TRENDS

Model	Total Vehicle Value (2001 $)				
	Brand				
Year	A	B	C	D	E
1996	$42,661	$42,422	$36,327	$38,616	$39,360
1997	$44,106	$45,640	$39,374	$40,846	$40,686
1998	$47,069	$46,171	$40,022	$42,008	$42,641
1999	$45,815	$44,977	$39,173	$40,266	$41,684
2000	$45,939	$44,458	$41,824	$41,322	$41,952
2001	$45,151	$42,873	$41,059	$40,015	$40,597

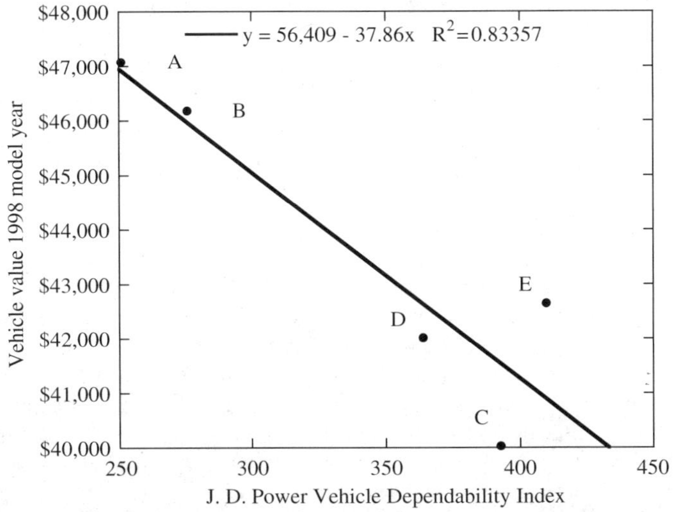

FIG. 11.12 VEHICLE VALUE COMPUTED FOR 1998 MODEL YEAR VERSUS J.D. POWER AND ASSOCIATES 2002 VEHICLE DEPENDABILITY INDICES FOR BRANDS BUILT IN THE 1998 MODEL YEAR

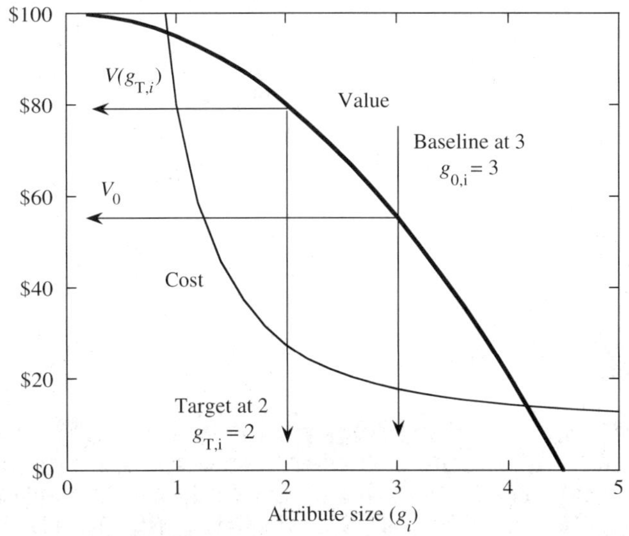

FIG. 11.14 SIMULATED VALUE AND COST CURVES FOR AN SIB ATTRIBUTE USED TO DETERMINE THE DETERMINATION OF THE OPTIMAL SIZE OF THE ATTRIBUTE

The value trends from DP Analysis for two minivans are shown in Fig. 11.13, mvA being the market leader during this time period and mvB being a serious challenger [3]. The value trends for the other minivans are omitted for simplicity of presentation. The initial minivan offering by mvB was developed by the firm's truck division and, not surprisingly, the outcome was a smaller version of a standard truck van with reduced towing capacity and reduced interior volume.

This entry missed the market established by the manufacturer of mvA, which was a replacement for and an improvement over the station wagon. It had a good view of the road, a car-like ride and the roominess offered by front-wheel drive. In fact, minivans made by the manufacturer of mvA were referred to as "tall cars" within the company. Starting in the latter part of the 1994 model year, the manufacturer of mvB released an all new minivan with a new brand name having the right attributes for a direct attack on mvA, including front-wheel drive, added room and fresh, aerodynamic styling. With these changes, mvB became the value leader.

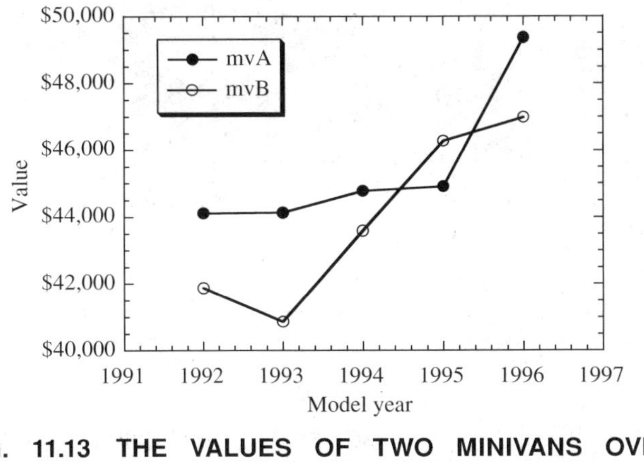

FIG. 11.13 THE VALUES OF TWO MINIVANS OVER TIME COMPUTED USING EQ. (2.15). (SOURCE: [3]; REPRODUCED WITH PERMISSION OF SPRINGER-VERLAG LONDON LTD.)

The competitive response by mvA in the 1996 model year included the addition of a rear sliding door on the driver's side, which was a unique feature at the time. A marketing research study [3] found that the door added over $1,200 in value, which is more than one-third of the value increment for mvA from the 1995 to the 1996 model year. Faced with a sizeable value disadvantage, the manufacturer of mvB initiated an early, costly structural redesign to accommodate the second sliding door.

The trends shown in Figs. 11.11 and 11.13 are for total value. The next step in the planning process is to consider each of the CTV attributes individually and then in concert to see how new technologies can be used to meet the value and cost targets. For example, two continuous CTV attributes for a laptop computer are its weight and the time to do a standard computational exercise. Both are smaller is better (SIB) attributes in that value should improve as each is reduced. Of course, such improvements will often entail a cost penalty. A hypothetical SIB value curve is shown in Fig. 11.14 along with its cost. Both are shown as a continuous function of the attribute g_i. However, real cost curves are highly dependent upon the manufacturing processes involved and may at best only be piecewise continuous.

The attribute where value minus variable cost is greatest locates the target specification, assuming that investment and fixed cost are not a function of g_i. The current baseline specification, $g_{0,i}$, equal to 3 represents the target for the existing but now outdated technology. The total value of the product at $g_{0,i}$ is given by the baseline value, V_0. The cost curve for the new technology shown in Fig. 11.14 results in a new target specification, $g_{T,i}$, which generates a higher total value equal to $V(g_{T,i})$. In contrast to the cost curve, the value curve represents a state function of the CTV attribute in that it is not expected to change much over time after it is normalized by dividing through by the baseline value, V_0.

Value curves are mirror images of the "cost of inferior quality" as defined by Taguchi and Wu [12] in their seminal description of robust design. Following the nomenclature used by Taguchi for loss functions, there are three basic types of value curves: SIB, nominal is best (NIB) and larger is better (LIB). The reciprocal of the LIB attribute is generally taken, which converts it into an SIB attribute.

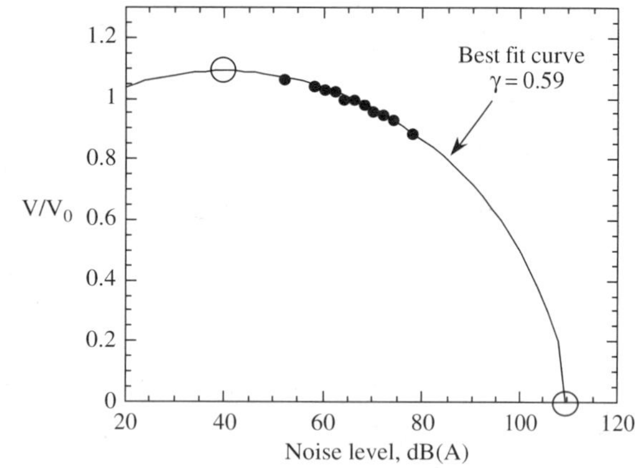

FIG. 11.15 THE EXPONENTIALLY WEIGHTED THREE-POINT VALUE CURVE FOR INTERIOR NOISE AT 70 MPH (THE CIRCLED POINTS AT 40 AND 110 DBA ARE ESTIMATES OF IDEAL AND CRITICAL POINTS, RESPECTIVELY, TAKEN FROM PUBLISHED HUMAN FACTOR STUDIES; REPRINTED WITH PERMISSION FROM *SAE PAPER NO. 980621* © 1998 SAE INTERNATIONAL)

Value curves can be fit empirically to an exponentially weighted parabola of the form:

$$\frac{V(g_i)}{V_0} = v(g_i) = \left[\frac{(g_{I,i} - g_{C,i})^2 - (g_{I,i} - g_i)^2}{(g_{I,i} - g_{C,i})^2 - (g_{I,i} - g_{0,i})^2} \right]^{\gamma} \qquad \text{Eq. (11.24)}$$

The terms $g_{I,i}$ and $g_{C,i}$ = ideal and critical specifications, respectively, for attribute i. Value is a maximum at the ideal specifications and zero at the critical specification. This expression was used by Pozar and Cook [13] to draw the curve shown in Fig. 11.15 through the points for value as a function of the interior noise of a luxury car at cruising speeds. The exponential weighting factor, γ, for the best fit was found to be 0.59.

11.6.3.2 DV Method of Marketing Research The points shown as filled circles in Fig. 11.15 were obtained using the direct value (DV) method of marketing research [14]. (The value for the second sliding rear door on the minivan discussed earlier was also obtained using the DV method.) The points at 40 and 110 dBA shown by the large circles were obtained from human factor studies. The parameter, v(gi), is the dimensionless value coefficient for attribute gi. The DV method has been used with random samples and with convenience samples of respondents to evaluate the value of a variety of attributes and options, including the interior noise in a vehicle [13], reliability [14], fuel economy [15], acceleration performance [15], four-wheel drive [16], forest product features [17], brand names [14], aircraft performance [18], minivan options [19, 20], farm equipment [21], interior seating capacity [22] and options for the Ford Mustang [23].

The DV method is a variant of the stated choice survey [24], the major difference being the use of paired comparisons in which an alternative is evaluated over a range of prices relative to a fixed, well-defined baseline product at a fixed price. In this manner, a neutral price, P_N, for the alternative can be obtained by interpolation where respondents, as an aggregate, are indifferent to the choice between the baseline and the alternative. In other words, the aggregate demands for the two are equal at the neutral price. It follows from Eq. (11.5) for the linear model and from Eqs. (11.15) and (11.17) for the logit model that the value, V_{Alt}, of the alternative less the value of the baseline, V_0, is equal to the neutral price minus the baseline price, P_0.

$$V_{\text{Alt}} - V_0 = P_N - P_0 \qquad \text{Eq. (11.25)}$$

An important feature of the DV survey versus a regular choice survey is that the value differences are computed simply from the difference in the two prices in Eq. (11.25). There are no empirical coefficients that need to be evaluated. It can also be shown that Eq. (11.25) follows from the probit model.

The dependence of the price coefficient in Eq. (11.19) on N needs to be accounted for in analyzing the outcomes of choice surveys that do not find a neutral price. In a stated choice survey, N represents the number of alternatives in a given choice set. For example, in a DV survey N is always two, a baseline and an alternative having one or more attribute changes from the baseline. In a revealed choice survey, N should be set equal to the average number of alternatives actually considered by a buyer in making a purchase decision. For example, buyers of Mustang vehicles considered 4.5 other vehicles, making $N = 5.5$ [23].

In the examination of the value of vehicle interior noise, respondents sat in front of a computer screen, Fig. 11.16, and listened to the noise levels of the baseline and the alternative through headphones. They could listen to one and then the other sound level until they were satisfied before making a selection, but they were not told what the noise levels were. The baseline noise level of 66 dBA and baseline price of $40,000 were fixed in keeping with the findings of Prospect Theory for making consistent paired comparisons [25].

The opening screen shows both vehicles at the same price. If the noise level of the alternative were significantly lower than 66 dBA, most respondents would select it. The computer would then automatically increase the price of the alternative. This process was repeated until the respondent switched to the baseline. If the noise level of the alternative was higher than 66 dBA, the respondent would likely choose the baseline and the computer would then reduce the price of the alternative until the respondent switched to

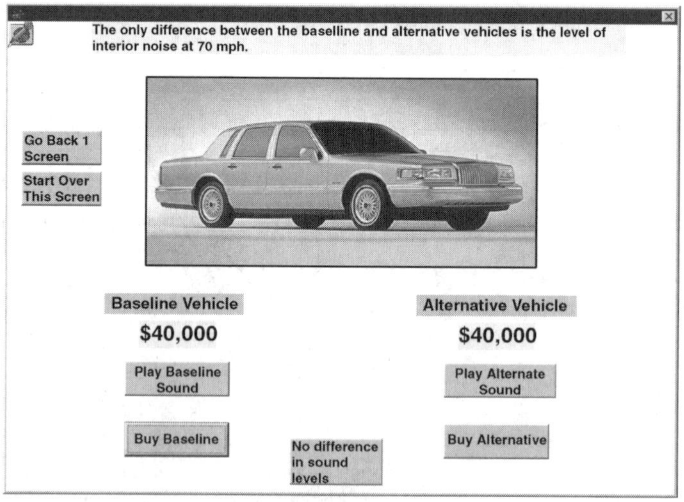

FIG. 11.16 COMPUTER SCREEN USED IN THE DV SURVEY FOR THE VALUE OF INTERIOR NOISE (REPRINTED WITH PERMISSION FROM *SAE PAPER NO. 980621* © 1998 SAE INTERNATIONAL)

the alternative. From Eqs. (11.15) and (11.18) for the logit model, the logarithm of f_{Opt} / f_0, where f_{Opt} is the fraction selecting the option (alternate) and $f_0 = 1 - f_{Opt}$ is the fraction selecting the baseline, is given by:

$$\text{Ln}\left(\frac{f_{Opt}}{1-f_{Opt}}\right) = \beta_{Opt}\left[V_{Opt} - P_{Opt}\right] \qquad \text{Eq. (11.26)}$$

The term V_{opt} = added value of the option; and P_{Opt} = price of the product with the option minus the price of the baseline without it. The price coefficient β_{Opt} for the option will not, in general, be the same as the price coefficient β for the baseline product. The logit model is used because of the large range in f_{Opt} usually found when using the DV survey.

The logit plot for the alternative at 62 dBA is shown in Fig. 11.17. The neutral price less the baseline price is seen to be $1,895, which is taken as the incremental value improvement for reducing the noise level in the luxury car from 66 to 62 dBA. As already stated, the values plotted in Fig. 11.15 have been normalized to form the function $v(g) = V(g) / V_0$. In doing so, the normalized curve determined for a luxury sedan can be used to generate, at least approximately, the value curve for a sedan in another segment having a different baseline values, V'_0, using the relation $V'(g) \cong v(g)V'_0$.

Two key features of the DV survey are: (1) the ability to tune the price range for the value of the alternative of interest using written or computer surveys; and (2) the minimization of cognitive stress as respondents evaluate only one alternative at a time in making a choice. As already stated, DV surveys measure the value of a single alternative relative to a fixed baseline. A *third* choice of "not buy" in addition to the baseline and the alternative is not only unnecessary with the DV method, it is not permitted. (Simulations of the demand curves for the lotteries discussed earlier with the DV method were generated using "not buy" as the baseline, which represented the purchase of a lottery ticket at a price of 0$ having 0% chance of paying off.) If several attributes are to be bundled, they, in concert, form a single alternative. If there is a concern that two attributes have a strong interaction, a DV survey is designed for each of the two attributes separately and a third is designed for the combination.

The DV survey also includes demographic questions such as gender, occupation, household income, education level and age. With this information, the value improvement for the attribute change or the feature can be computed for each of the consumer segments that make up the buyers in the product segment. When the logit model is used, the values for each demographic group should be determined from plots similar to Fig. 11.17. If a demand forecast is to be made for a fixed price, the demand for each group should be determined from its respective values. The demands need to be weighted by the fraction of the potential buyers represented by the group. The results can then be summed to arrive at the total demand forecast, provided that the demographics are segmented so that a given respondent is included in just one group.

A given product feature may not be seen as an economic good by all of the respondents. An excellent example of this was found in a DV survey of Mustang buyers [23]. Roughly one-half would pay for an automatic transmission and the other half would have to be paid to take it, resulting in an average value of approximately zero. The procedure for assessing the value of a feature for the fraction of buyers that sees an attribute as an economic good has been described by McConville and Cook [23]. When opinion is

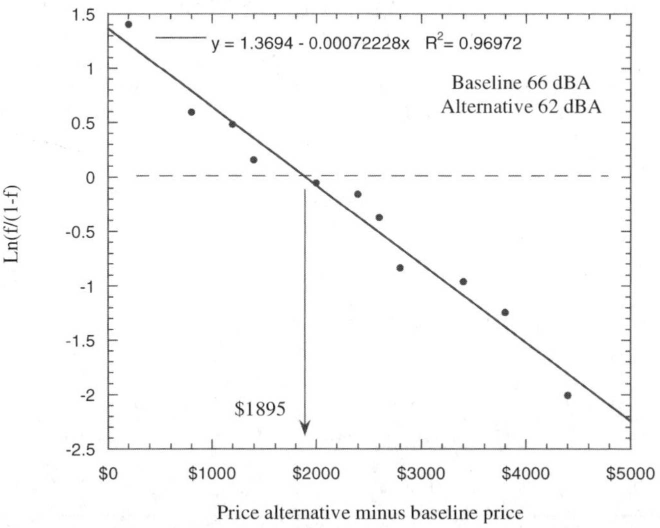

FIG. 11.17 LOGIT PLOT FOR ASSESSING VALUE OF 66 VERSUS 62 DBA (REPRINTED WITH PERMISSION FROM *SAE PAPER NO. 980621* © 1998 SAE INTERNATIONAL)

strongly split on whether a feature is an economic good, it should be offered as an option as opposed to standard equipment.

If respondents judge options independently when making purchase decisions, then the DV survey findings for the options given by plots of the type shown in Fig. 11.17 can be used to assess the predicted take rate given the price of the option on the x-axis. The outcomes of this approach are shown in Table 11.4 for options evaluated for the 1995 model year Mustang using a DV survey [23]. Except where noted in Table 11.4, reasonable agreement is found between the predicted and actual take rate fractions of these options by the respondents to the survey. The V8 and convertible options could not be purchased individually as noted at the bottom of Table 11.4. Bundling results in a package of higher value and higher price. However, in Table 11.4 all of the values are those obtained from the survey for the single option; they do not include the unknown added value of the other bundled options. This is likely the major cause of the difference between the predicted and actual take rates of the V8 and convertible. The leather seats option was not available on the base model, which reduced its availability. Another source of error is that the list prices used [25] may not be the same as the actual transaction prices.

The neutral price for an option is not necessarily the optimal price for an option. The price that maximizes cash flow depends on how the availability of the option affects the initial buy decision. If the option influences the initial buy decision, it should be priced lower than if the option decision is made after the initial buy decision. The reason is that such an option influences total demand for the product if it influences the initial buy decision. It does not if made afterward.

For example, customers rated the availability of the second sliding door of the minivans from Chrysler Corporation highly in regard to their purchase decision [19, 20]. In the 1996 model year the door was priced less than one-half its value [3] and soon made part of the vehicle's standard equipment. It is also important to note that an option may be profitable even though its variable cost is higher than the point estimate for its value given by Eq. (11.25). The reason is that the distribution of f_{Opt} or $\text{Ln}(f_{opt} / [1 - f_{opt}])$ for some options may be sufficiently flat with price to allow them be priced higher

TABLE 11.4 PREDICTED TAKE RATE FROM DV SURVEY USING LOGIT MODEL COMPARED TO ACTUAL TAKE RATE OF OPTIONS BY RESPONDENTS TO THE SURVEY (NOTE: THE V8 AND CONVERTIBLE OPTIONS WERE BUNDLED WITH OTHER FEATURES)

	$\text{Ln}(f_{opt}/[1-f_{opt}])$ vs. P_{Opt}			Price to Obtain	Single Option		
Option	**Slope**	**Intercept**	**Value**	**Option**	**Prediction**	**Actual**	**Notes**
V-8	–0.00145	3.99	$2,751	$3,575	0.23	0.55	A
Convertible	–0.00084	2.75	$3,290	$5,578	0.13	0.22	B
ABS V-6	–0.00342	3.11	$908	$565	0.76	0.60	—
ABS V-6 & V-8	–0.00342	3.11	$908	$565	0.76	0.78	—
Auto. Trans.	–0.00055	–0.03	$(58)	$790	0.39	0.48	—
AC	–0.00447	6.77	$1,517	$780	0.96	1.00	—
Leather seats	–0.00162	0.89	$549	$500	0.52	0.32	C

A: The V8 option is bundled with the GT model, which comes with four-way head restraint, power lumbar supports, GT suspension package, traction-lok axle, fog lamps, rear deck-lid spoiler, leather-wrapped steering wheel, all-season tires and alloy wheels.
B: The convertible comes bundled with power mirrors, power door locks and deck-lid release and power windows.
C. The actual take rate is reduced because the option was not available on the base vehicle.

than value and still generate enough demand to give an acceptable rate of return.

11.6.3.3 Other Tools for Estimating Value Market surveys are not the only technique that can be used to estimate the value of options and attributes. For example, Simek and Cook [22] estimated the value of interior roominess using a model of how the human body fits into the space provided in the vehicle. Economic computations can be used to evaluate the value of vehicle range and fuel economy [14, 15]. If there is no time to initiate and complete a marketing research study, an estimated value curve can be generated by using a jury evaluation to estimate the ideal and critical specifications as well as the weighting coefficient for the CTV attributes of interest. The weighting coefficient is estimated by a jury evaluation of the fraction of time the attribute is deemed important during use of the product.

11.6.3.4 Value of Multiple CTV Attributes An empirical expression given by [14]:

$$V(g_1, g_2, g_3, \ldots) = V_0\left[v(g_1), v(g_2), v(g_2), \ldots\right] \quad \text{Eq. (11.27)}$$

is used to compute the value of multiple attributes because it has the property that if any CTV attribute is at its critical specification, the value of the entire product is zero. For example, if a cell phone was perfect in every way except that it weighed more than anyone would want to carry or was so large that it was cumbersome to carry, it would essentially have zero value as a cell phone. If a car was perfect in every way, but had a turning radius so large that it could not turn at an intersection, it would have zero value. If a commercial plane had an interior noise level at 120 dBA that could not be changed, the major airlines would likely not buy it at any price. There can be other attributes or features of a product that are important to value but are optional, such as leather seats in a car or a color screen on a cell phone. These simply add to value.

Because the demand model given by Eq. (11.5) does not consider, for simplicity, competition between segments, limits need to be imposed when considering how much value and cost is to be added to a new product that is part of a larger product line that extends across multiple segments. Consider the segmentation of 1993 model year family sedans shown in Fig. 11.18 [5]. The open circles represent the coordinates of the average price and value for the segment, the average value being computed from Eq. (11.9) for $E_2 = 1$. Each segment has a value and price range. Cash flow is not necessarily highest for the product at the high value end of the segment. It is interesting to note, however, that almost all of the products at the high value ends of the segments had the highest reliability ratings [5].

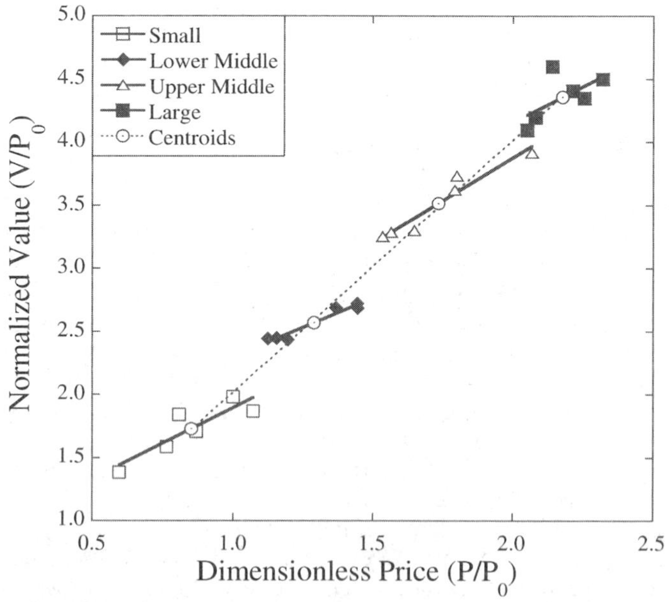

FIG. 11.18 NORMALIZED VALUE VERSUS NORMALIZED PRICE FOR FOUR FAMILY SEDAN SEGMENTS IN THE 1993 MODEL YEAR (REPRINTED WITH PERMISSION FROM *SAE PAPER 970765* © 1997 SAE INTERNATIONAL)

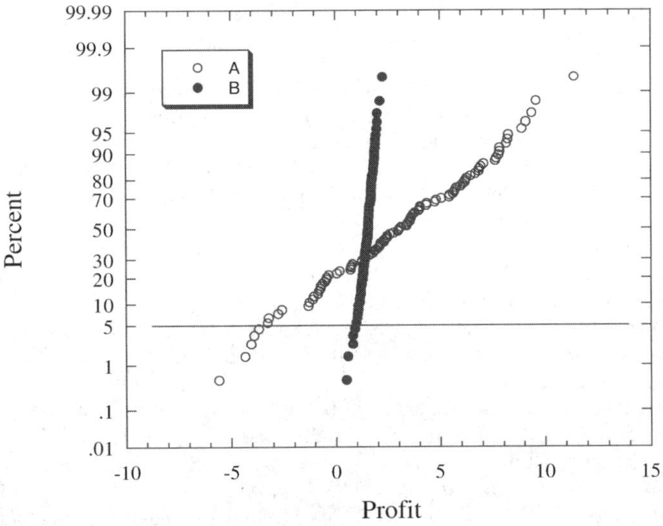

FIG. 11.19 THE SIMULATED CUMULATIVE DISTRIBUTIONS FOR TWO TECHNOLOGIES, A AND B, PLOTTED ON NORMAL PROBABILITY PAPER (THE HORIZONTAL LINE REPRESENTS THE 95% CONFIDENCE LEVEL.)

11.6.3.5 Prioritizing New Technologies for Implementation New technologies can be prioritized on the basis of their profitability forecasts [7, 23]. Each new technology, q, will likely impact several CTV attributes. Mutually exclusive technologies such as a diesel engine versus a spark-ignited engine versus a fuel-cell engine can be made by computing the value for each technology using Eq. (11.27), with point estimates for the CTV attributes. The total value and variable cost for each technology are used to compute their Cournot-Nash prices. These metrics along with the investment and fixed cost for each technology are then substituted into Eq. (11.11) for the cash flow estimates. At this juncture, it is sufficient to assume that competitors will do nothing. This will of course overestimate cash flows, but the relative rankings should be unaffected.

When technologies are not mutually exclusive, a design of experiments (DOE) approach can be taken [27]. The experiments can be run either by computer simulation or with prototypes. Each experimental trial will measure the mean and variance of each CTV attribute, variable cost, fixed cost and investment. The CTV attributes and variances for each trial are converted into value. Value and variable cost are then converted into a Cournot-Nash price [5]. These metrics can be used to estimate the cash flow for each trial. A regression analysis of the outcomes will show how each technology affects value, variable cost, price and cash flow. The technologies can then be ranked. All of the technologies showing positive cash flow can be chosen, if not mutually exclusive, *provided that the price of the product stays within the envelope for the segment and the required investment is within the amount available.*

The above processes for selecting new technologies are based upon point estimates for cash flow. However, the uncertainty in the cash flow also has to be taken under consideration as illustrated in Fig. 11.19. Distributions for cash flow as shown in Fig. 11.19 can be made by combining Monte Carlo simulations with the DOE process [27]. The mean for technology A is higher than for B but the 95% confidence level cash flow for A is negative; whereas, the 95% confidence level for B is positive and would likely be preferred by most planners.

11.7 SUMMARY

Excellence in product planning is critical for surviving the intense competition in the global marketplace. The first product planning step described here was to structure the problem around the fundamental metrics of the values to the customer and society, cost and the pace of innovation. Changes in these metrics are then used to forecast the bottom-line metrics of demand, price and cash flow. Both value and cost targets need to be set when planning a new product.

Particular attention was given to the linear demand model because it provides closed-form solutions for demand, price and cash flow. The linear model should be accurate when the value and price changes are small (this is often the case for products undergoing continuous improvement), assuming that demand is analytic in the values and prices of the competing products. When demand changes are large, there is a direct connection between the parameters in the linear model and the logit and probit models.

The DV method of marketing research was developed based upon the realization that customer value can be forecast simply from the difference between the neutral price and the baseline price obtained from market surveys. No empirical constants are involved. The neutral price is the price of the alternative that results in one-half of the respondents choosing the alternative and the other half choosing the baseline. The alternative can differ from the baseline in one or many CTV alternatives. The key feature of the D Value method is that it will almost always result in less cognitive stress to the respondent than the respective stated choice or conjoint survey.

REFERENCES

1. Cook, H. E. and Kolli, R. P., 1994. "Using Value Benchmarking to Plan and Price New Products and Processes," *Manufacturing Rev.*, **7**(2), pp. 134–247.
2. Cook, H. E., 1997. *Product Management*, Kluwer Academic (formerly Chapman & Hall), Amsterdam, The Netherlands, pp. 56–63.
3. Cook, H. E. and Wu, A. E., 2001. "On the Valuation of Goods and Selection of the Best Design Alternative," Res. Eng. Des., Vol. 13, pp. 42–54.
4. Cook, H. E. and DeVor, R. E., 1991. "On Competitive Manufacturing Enterprises. I: The S-Model and the Theory of Quality," Manufacturing Rev., 4(2), pp. 96–105.
5. Monroe, E., Silver, R. and Cook, H. E., 1997. "Value versus Price Segmentation of Family Automobiles," *SAE Paper 970765, Proc., SAE Int. Cong.*, Soc. of Automotive Engrs., Warrendale, PA.
6. Pashigian, B.P.,1995. *Price Theory and Applications*, McGraw-Hill, New York, NY, pp. 247–249.
7. Cook, H. E., 1997. Product Management, Kluwer Academic (formerly Chapman & Hall), Amsterdam, The Netherlands, pp. 66–71.
8. McFadden, D., 1974. "Conditional Logit Analysis of Qualitative Choice Behavior," Frontiers of Econometrics, P., Zarembka, ed. Academic Press, New York, NY, pp. 105–42.
9. Daganzo, C., 1980. Multinomial Probit, Academic Press, New York, NY.
10. Bolduc, D., 1999. "A Practical Technique to Estimate Multinomial Probit Models in Transportation," Trans. Res. Part B, Vol. 33, pp. 63–69.
11. Cook, H. E., 1997. Product Management, Kluwer Academic (formerly Chapman & Hall), Amsterdam, The Netherlands, p. 63.
12. Taguchi, G. and Wu, Y., 1980. Introduction to Off-line Quality Control, Central Japan quality Association, Nagoya, Japan.

13. Pozar, M. and Cook, H. E., 1997. "On Determining the Relationship Between Vehicle Value and Interior Noise," SAE Trans. J. of Passenger Cars, Vol. 106, pp. 391–401.
14. Donndelinger, J. D. and Cook, H. E., 1997. "Methods for Analyzing the Value of Automobiles." SAE Trans., J. of Passenger Cars, Vol. 106, 1263–1281.
15. McConville, G., and Cook, H. E. 1995. "Examining the Trade-Off between Automobile Acceleration Performance and Fuel Economy," *SAE Trans., J. of Mat. Manufacturing*, 105(37), pp. 37–45.
16. Monroe, E. and Cook, H. E. 1997. "Determining the Value of Vehicle Attributes Using a PC Based Tool," *SAE Paper 970764I, Proc., SAE Int., Cong.*, Soc. of Automotive Engrs., Warrendale, PA.
17. Bush, C. A., 1998. "*Comparison of Strategic Quality Deployment and Conjoint Analysis in Value Benchmarking*," M.S. thesis, Dept. of Mech. and Ind. Engrg., University of Illinois at Urbana-Champaign, IL.
18. LeBlanc, A., 2000. Personal communication, Pratt & Whitney, Canada. Inc.
19. Wu, A. E., 1998. "*Value Benchmarking the Minivan Segment*," M.S. thesis, Dept. of Mech. and Ind. Engrg., University of Illinois at Urbana-Champaign, IL.
20. Lee, M. D., 1998. "*Brands, Brand Management, and Vehicle Engineering*," M.S. thesis, Dept. of Mech. and Ind. Engrg., University of Illinois at Urbana-Champaign, IL.
21. Silver, R. L., 1996. "*Value Benchmarking to Improve Customer Satisfaction*," M.S. thesis, Dept. of Mech. and Ind. Engrg., University of Illinois at Urbana-Champaign, IL.
22. Simek, M. E. and Cook, H. E., 1996. "A Methodology for Estimating the Value of Interior Room in Automobiles," *SAE 1996 Trans., J. of Mat. & Manufacturing*, Vol. 105, pp. 13–26.
23. McConville, G. and Cook, H. E, 1997. "Evaluating Mail Surveys to Determine the Value of Vehicle Options," *SAE Trans., J. of Passenger Cars*, Vol. 106, pp. 1290–1297.
24. Louviere, J. J., Hensher, D. A. and Swait, J. D., 2000. *Stated Choice Methods Analysis and Applications*, Cambridge University Press, Cambridge, U.K.
25. Tversky, A. and Kahneman, D., 1991. "Loss Aversion in Riskless Choice: A Reference-Dependent Model," *Quarterly J. of Eco.*, pp. 1039–1061.
26. *Consumer Guide Auto '95*, 1995. Pub. Int., Ltd., 558(12), pp. 71–73.
27. Cook, H. E., 2005. *Design for Six Sigma as Strategic Experimentation*, Am. Soc. for Quality, Milwaukee, WI.

SECTION

5

VIEWS ON AGGREGATING PREFERENCES IN ENGINEERING DESIGN

INTRODUCTION

Decisions in design are made in many different ways. Nevertheless, the foundations for making design decisions are the *preferences of the decision-maker* that dictate the choice of one option over another. In design it is often difficult to develop a single criterion upon which to base decisions. In actuality, many design decisions are made based on preferences applied to multiple, potentially conflicting criteria (e.g., quality of materials and cost) or attributes of the product and its production. Therefore, the chapters in this section focus on approaches for multicriteria decision-making by a single decision-maker or a group of decision-makers that has reached considerable consensus.

Two general approaches to modeling and making multiple-criteria decisions are presented in this section. The first general approach is to aggregate preferences over multiple criteria into a single evaluation metric, while the second approach focuses on modeling and handling trade-offs between the criteria directly. Regardless of whether the first, second or a combination of the approaches is used, there are a set of fundamental issues common to this class of decisions that must be managed. These challenges form the heart of the chapters in this section and include:

- Mathematically aggregating diverse and conflicting criteria
- Quantitatively capturing a decision-maker's preferences over multiple criteria
- Providing an ordered scale (ordinal or cardinal) of decision options that can be used to select the best one

In this section, some basic multi-attribute decision principles are presented along with some effective decision support methods that provide insight into these issues.

In Chapter 12, a method for preference and criteria *aggregation* is discussed based on the foundations of utility theory. In Chapter 13, approaches for multicriteria decision-making are presented and compared, resulting in a valuable and practical approach to making selection decisions that uses pairwise *trade-off* comparisons. In Chapters 14 and 15, multicriteria *trade-offs* are used to help *aggregate* preferences in decision-making. In Chapter 14, an approach is presented that elicits decision-maker preferences between hypothetical alternatives using multicriteria *trade-offs* in order to properly *aggregate* the attributes into a single criterion. Some fundamental flaws of some other common approaches are also illustrated. In Chapter 15, an approach is presented that elicits preference information from a decision-maker on multiple criteria and then performs objective *trade-offs* using an *aggregation* scheme called Physical Programming.

While each of the methods discussed in this section cover a specific type of decision, such as an alternative selection decision or an alternative optimization decision, the principles presented are more general and apply to any decision in a design process that involves multiple criteria. The approach employed by a designer should be selected based on the features or characteristics of the specific decision problem at hand. These include: whether the decision is being made early or late in a design process, the availability and fidelity of quantifiable alternative information, and whether the decision is from a finite or infinite set of options.

The chapters in this section focus on preference aggregation in engineering design and are not intended to address the accounting for risk and uncertainty in multicriteria decision-making. Recall that basic principles for handling risk and uncertainty in decision-making were presented in Chapters 3 and 4 of Section 2. Additionally, in Chapters 17 and 18 of Section 6, some principles for handling uncertain parameters in predictive decision-making in an enterprise context will be detailed. The methods presented in Section 6 of this text focus on approaches for single-criterion decision-making, such as using profit or cost as the only objective function. Finally, the chapters in this section focus on decisions being made by a single decision-maker, or a group of decisions-makers that has reached considerable consensus. In Section 7 the decisions focused on will be those made by multiple decision-makers with conflicting objectives in decentralized or collaborative environments.

A note on the slight differences in terminology used by contributing authors is warranted. For instance, one author will use the term "multi-attribute decision-making," another author will use "multi-objective optimization" and yet another will use the term "multicriteria selection." These terms are used interchangeably in the design decision-making literature. They allude to the same set of fundamental issues surrounding the complexities in modeling and capturing preferences over multiple criteria in order to provide decision recommendations. This use of different terminologies, while from a scholarly perspective is not at all ideal, does provide the reader with an appreciation for the diversity of fields that contribute to multicriteria decision-making in engineering design. Being able to understand similar expressions will help develop "multilingual" skills to communicate with engineers and decision-makers from other disciplines (e.g., management, customers, manufacturing, sales, marketing).

CHAPTER

12

Multi-attribute Utility Analysis of Conflicting Preferences

Deborah L. Thurston

12.1 INTRODUCTION

This chapter addresses the aggregation of preferences for conflicting design objectives. It will clarify what a decision-based approach can and cannot contribute to the design process. A decision-based approach with multi-attribute utility analysis can be *directly* employed only for design evaluation and selection. However, by providing a logical structure for organizing and using all the information and analysis employed by designers, it can also contribute indirectly to all phases of design, including problem identification, creativity, synthesis, product development, experimentation and analysis. Specific problems that can be resolved with utility analysis are: lack of a systematic procedure for integrating preferences into the traditional design analytic framework, trade-off inaccuracies, inconsistencies and suboptimality. These problems result in a design process that sometimes takes too long (especially concurrent design), fails to address all interests early in the design process (or address then at all) and produces results that are not competitive in the marketplace. A constrained multi-attribute utility approach can help remedy these problems. The next section presents the formulation of the multi-attribute design optimization problem. Section 12.3 describes and resolves the most common misconceptions about multi-attribute utility analysis in design related to the independence conditions, the functional form, the distinction between trade-offs and constraints, subjectivity and group decision-making. Section 12.4 summarizes.

12.2 DESIGN DECISION PROBLEM FORMULATION

12.2.1 Pareto Optimality

First we describe the type of design problem addressed here, one involving trade-off decisions as shown in Fig. 12.1. The axes indicate the cost and weight of design alternatives. The alternatives that lie in the region above the Pareto optimal (PO) frontier [1] are those that are technically feasible, those in the region below are infeasible, and those that lie directly on the PO frontier are those where it is not possible to improve one attribute (such as weight) without worsening another (such as cost). Any alternative that lies directly on the PO frontier is superior to any that lies above it. If one begins in the region above the PO frontier, expert analytic design expertise can be employed to specify design changes that improve weight, cost or both. This drives the iterative design process toward the Pareto optimal frontier. Of course, these design changes should always be made. However, the designer is eventually thwarted in the attempt to simultaneously optimize all objectives, when the design decisions that further improve one objective worsen another. For example, substituting advanced polymer composite materials for steel improves weight but can worsen cost.

So, the problem we address is twofold: First, what design modifications will help one move toward the PO frontier? Second, what specific location on the frontier provides the *best combination* of cost and weight? Figure 12.1 indicates that the optimal solution lies on the greatest feasible iso-utility curve, where $U = 0.5$.

To move from the inferior or dominated design region to the PO frontier, designers typically begin with a house of quality (HOQ) matrix approach to illuminate the cause and effect relationships between performance, quality, cost and engineering decisions [2]. The matrix rows list all attributes of product performance x_i, such as cost, performance, weight, environmental impact, etc. The columns list all the engineering decision variables y_j that the *designer can directly control* in order to improve each attribute x_i. Typical decision variables y_j include material choice, geometric configuration, manufacturing process, assembly technology, etc. Design for "X" methods, where "X" might be quality, assembly, disassembly, life cycle, etc., can help guide the designer toward including decision variables that might otherwise remain unexplored.

The central problem is how to identify the design variable values that result in the optimal combination of attributes. What materials, product configuration, manufacturing processes and end-of-life strategies offer the best combination of cost, performance and environmental impact?

When the Pareto optimal frontier is reached or approximated, it is no longer possible to improve one objective without worsening another. Here we make a semantic distinction that is important to decision-based design (DBI); once we reach the Pareto optimal frontier, minimizing weight is no longer an *objective*. Instead, maximizing some *function of a combination of attributes* x_i, which includes weight, becomes the objective, as shown in Eq. (12.1)

$$\text{Maximize } f(x_1, x_2, \ldots, x_n) \qquad \text{Eq. (12.1)}$$

12.2.2 Constrained Multi-Attribute Utility Optimization Modeling

Engineering design is typically an iterative configure-evaluate-reconfigure process. So, too, is the process of preference modeling with multi-attribute utility analysis and optimization. The definition of each element influences the others, which in turn might

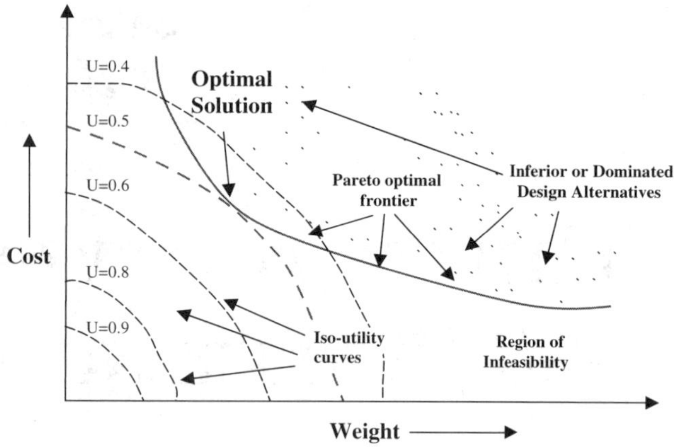

FIG. 12.1 DESIGN ALTERNATIVE SPACE

lead to redefining the first element. As the design artifact evolves and certain decisions are made (such as material choice), the model should be reformulated and solved again to accurately reflect the actual decision being made, as well as the problem constraints, as in [3]. Nonetheless, this section describes the general structure of the problem.

The four major elements in constrained multi-attribute optimization modeling are: a vector of product performance attributes $\mathbf{x} = (x_1, x_2, x_3, \ldots, x_n)$ that often conflict with one another (such as cost, quality and environmental impact), an objective utility function $f(\mathbf{x})$, which one is attempting to maximize, a vector of decision variables $\mathbf{y} = (y_1, y_2, y_3, \ldots, y_m)$ over which the designer has direct control (such as material choice, geometry, assembly process, etc.), and a vector of constraint functions $\mathbf{h}(\mathbf{y}) = \mathbf{x}$ that model the analytic cause and effect relationships between design decision variables $\mathbf{y}$ and the product performance vector $\mathbf{x}$. Table 12.1 summarizes model elements.

The first step is to reduce the very large number of product attributes $\mathbf{x}$ and engineering parameters $\mathbf{y}$ listed in the matrix to the subset upon which experimental and product development time are best spent [4]. A preliminary conception of the objective function can be used to guide this process, but its explicit definition is not necessary at this stage. The performance attribute vector $\mathbf{x} = (x_1, x_2, x_3, \ldots, x_n)$ is then defined in terms of the design decision variable vector $\mathbf{y} = (y_1, y_2, y_3, \ldots, y_m)$ as a vector of functions $\mathbf{h} = (h_1, h_2, h_3, \ldots, h_n)$, or $\mathbf{h}(\mathbf{y}) = \mathbf{x}$ shown in Eq. (12.2)

$$\begin{aligned} h_1(y_1, y_2, y_3, \ldots, y_m) &= x_1 \\ h_2(y_1, y_2, y_3, \ldots, y_m) &= x_2 \\ \vdots \quad & \quad \vdots \\ h_n(y_1, y_2, y_3, \ldots, y_m) &= x_n \end{aligned} \qquad \text{Eq. (12.2)}$$

These constraint equations are most often determined through traditional, expert design knowledge of the availability and behavior of materials, structures, mechanisms, kinematics, cost and quality estimation, etc. For example, activity-based cost estimation might be employed where x_1 is cost, statistical process control where x_2 is scrap rate, finite-element analysis where x_3 is deflection, etc. The cause and effect relationships between design decisions $\mathbf{y}$ and life-cycle environmental impacts are much more difficult to assess, but life-cycle analysis methods are continually improving. For a linear model, these constraint functions can be written as a matrix where a_{ij} are the coefficients of $\mathbf{h}(\mathbf{y}) = \mathbf{x}$, and a_{i0} are constants as shown in Eq. (12.3)

TABLE 12.1 ELEMENTS OF MULTI-ATTRIBUTE UTILITY OPTIMIZATION IN DESIGN

Model Element	Expression	Meaning
Design Attributes	$x = (x_1, x_2, x_3, \ldots, x_n)$, *for example:* x_1 = *cost* x_2 = *quality* x_3 = *environmental impact, etc.*	Elements of product performance that are both relevant and negotiable over a specific range.
Decision Variables	$y = (y_1, y_2, y_3, \ldots, y_m)$, *for example:* y_1 = *material choice* y_2 = *gauge* y_3 = *manufacturing settings, etc.*	Engineering parameters the designer can directly control in order to improve attributes $(x_1, x_2, x_3, \ldots, x_m)$.
Constraints	$h_i(y_1, y_2, \ldots, y_m) = x_i$ for i = 1, 2, . . . , n	Relationships between engineering parameters and product performance.
Cost	$x_{1,l} \leq h_1(y_1, y_2, y_3, \ldots, y_m) \leq x_{1,u}$	Cost (x_1) is a function of decision variables $(y_1, y_2, y_3, \ldots, y_m)$. Typical units are total amortized cost per unit product.
Quality	$x_{2,l} \leq h_2(y_1, y_2, y_3, \ldots, y_m) \leq x_{2,u}$	Quality (x_2) is a function of decision variables $(y_1, y_2, y_3, \ldots, y_m)$. Typical units are percent scrap rate, reliability, fit and finish, etc.
Environmental Impact	$x_{3,l} \leq h_3(y_1, y_2, y_3, \ldots, y_m) \leq x_{3,u}$	Environmental impact (x_3) is a function of decision variables $(y_1, y_2, y_3, \ldots, y_m)$. Typical units are pounds of waste, ecopoints, etc.
Objective Function	max $\mathbf{U}(\mathbf{y}) = \frac{1}{K}\left[\left(\prod_{i=1}^{n}\left\{Kk_iU_i[h_i(y_1, y_2, y_3, \ldots, y_m)]+1\right\}\right)-1\right]$	Utility maximization is determined through identification of the optimal set of design decision variables **y**.

$$\begin{bmatrix} a_{10} \\ a_{20} \\ a_{30} \\ \\ \\ a_{n0} \end{bmatrix} + \begin{bmatrix} a_{11} & a_{12} & a_{13} & a_{14\ldots} & a_{1m} \\ a_{21} & a_{22} & a_{23} & a_{24\ldots} & a_{2m} \\ a_{31} & a_{32} & a_{33} & a_{34\ldots} & a_{3m} \\ \\ \\ a_{n1} & a_{n2} & a_{n3} & a_{n4\ldots} & a_{nm} \end{bmatrix} \begin{bmatrix} y_1 \\ y_2 \\ y_3 \\ \\ \\ y_m \end{bmatrix} = \begin{bmatrix} x_1 \\ x_2 \\ x_3 \\ \\ \\ x_n \end{bmatrix}$$

Eq. (12.3)

The constraints shown in Eq. (12.2) must be satisfied simultaneously, so only certain combinations of attributes $\mathbf{x} = (x_1, x_2, x_3, \ldots, x_n)$ are technically feasible. For example, it is not possible to improve x_1 (say, stiffness) by increasing y_1 (thickness of a component) without simultaneously worsening x_2 (weight), all else equal.

Now the central problem is how to identify which feasible combination of attributes $\mathbf{x}$ is best. Traditional design objective functions minimize a single attribute, such as weight, or perhaps cost. However, our problem is how to decide which combination of several attributes $\mathbf{x} = (x_1, x_2, x_3, \ldots, x_n)$ best suits our purpose. Since $\mathbf{x} = \mathbf{h}(\mathbf{y})$, our objective function is of the form:

$$f(x_1, x_2, x_3, \ldots, x_n) = f[(h_1(y), h_2(y), h_3(y), \ldots, h_n(y)] = g(\mathbf{y}) \qquad \text{Eq. (12.4)}$$

The form of $g(\mathbf{y})$ should reflect preferences for the conflicting attributes $\mathbf{x}$. The simplest form is a linear weighted average shown in Eq. (12.5), but this approach has been demonstrated to be highly unreliable for design problems [4]. The reason is that the arbitrary assessment of "relative importance" employed to assign the weighting factors w_i can be systematically biased by irrelevant factors. The perceived relative importance (and thus the willingness to make trade-offs) often does not remain static throughout the real design space [4]. Thus, relative weighting factor approaches can lead to designs that do not best satisfy the decision-maker's preferences.

$$g(\mathbf{y}) = \sum_{i=1}^{n} w_i[h_i(\mathbf{y})] \qquad \text{Eq. (12.5)}$$

Instead, we recommend a multi-attribute utility function formulation as shown in Eq. (12.6). Multi-attribute utility analysis is a rigorously normative methodology based on the axioms of utility theory [5]. The scaling factors k_i more accurately reflect the designer's willingness to make trade-offs among the attributes over their entire range of feasibility. The n single attribute utility functions $U_i[h_i(\mathbf{y})]$ can reflect the decision-maker's nonlinear valuation of each attribute i and his or her attitude toward risk; they are assessed using the lottery method [5] and presented for design in [4]. With careful problem definition, they can often take the form of a well-behaved exponential function $U_i(x_i) = b_i + a_i\exp(-c_ix_i)$. Coefficient c_i reflects the degree of risk aversion and a_i and b_i are calculated such that $U_i(x_i)$ is scaled from 0 to 1.

The multiplicative form in Eq. (12.6) can be employed only after conditions of preferential and utility independence are verified. The goal is to determine the set of design decision variables $\mathbf{y}$ that maximize overall utility. It is very important to note that the attribute scaling constants k_i should *not* be viewed as "weights" reflecting the "relative importance" of each attribute. In contrast, they help provide a more accurate assessment of the decision-maker's willingness to make trade-offs as it changes throughout the design space.

In addition, inequality constraint Eqs. (12.8) and (12.9) define the range over which the designer is both able and willing to make trade-offs among the attributes $\mathbf{x}$. Where less is preferred to more (such as cost or environmental impact) the upper bound is defined as the "worst" that the decision-maker is willing to tolerate (*not* the worst possible). The lower bound is defined as an optimistic, yet realistic, "best" from the viewpoint of technical feasibility. This is the range of technical and preferential negotiability. Additional equality or inequality constraints Eqs. (12.10) and (12.11) may be necessary, and the specific form of Eqs. (12.6) to (12.11) will depend on the problem.

$$\text{Max } U(\mathbf{y}) = \frac{1}{K}\left[\left(\prod_{i=1}^{n}\left\{Kk_iU_i[h_i(y_1, y_2, y_3, \ldots, y_m)] + 1\right\}\right) - 1\right]$$

Eq. (12.6)

Where: $h_i(y_1, y_2, y_3, \ldots, y_m) = x_{i,}$ for $i = 1, 2, \ldots, n$ Eq. (12.7)

Subject to:

$$h_i(y_1, y_2, y_3, \ldots, y_m) \geq x_{i,l} \text{ for } i = 1, 2, \ldots, n \qquad \text{Eq. (12.8)}$$

$$h_i(y_1, y_2, y_3, \ldots, y_m) \leq x_{i,u} \text{ for } i = 1, 2, \ldots, n \qquad \text{Eq. (12.9)}$$

and perhaps

$$q_k(y_1, y_2, y_3, \ldots, y_m) = 0 \text{ for } k = 1, 2, \ldots, p \qquad \text{Eq. (12.10)}$$

$$h_{n+j}(y_1, y_2, y_3, \ldots, y_m) \geq 0 \text{ for } j = 1, 2, \ldots, r \qquad \text{Eq. (12.11)}$$

where K= normalizing parameter, calculated from

$$1 + K = \prod_{i=1}^{n}(1 + Kk_i) \qquad \text{Eq. (12.12)}$$

12.2.3 Computational Issues

While the nonlinear objective function in Eq. (12.6) appears to present some computational complexity, several factors facilitate its solution. First, many design problems eventually distill to a relatively small number of incommensurate, conflicting objectives after elements of the attribute set $(x_1, x_2, x_3, \ldots, x_n)$ and their ranges have been defined in such a way that preferential and utility independence conditions are satisfied. If satisfied, these conditions facilitate straightforward identification of single-attribute utility functions $U_i(x_i)$, and indicate the multiplicative form shown in Eq. (12.6). For example, ergonomics might appear in the HOQ, but demand functions (if known) can directly convert ergonomic performance to expected profits. In fact, some design theory researchers argue that expected profit alone is the only accurate metric for design evaluation. Second, with careful definition of attributes and their ranges, single-attribute utility functions can often take the form of the well-behaved exponential function $U_i(x_i) = b_i + a_i \exp(-c_ix_i)$. The third factor that reduces computational complexity is that several elements of the design decision variable set $(y_1, y_2, y_3, \ldots, y_m)$ are often discrete or even binary variables, such as material choice, enabling exhaustive enumeration techniques. All of the utility optimization problems performed in the Decision Systems Laboratory at the University of Illinois have been readily solved on a personal computer using a spreadsheet add-in solver that employs simplex and branch and bound for linear and integer problems, and a generalized reduced gradient algorithm for nonlinear problems.

12.2.4 Turnbuckle Example

As a simple, brief example of the model described above, Fig. 12.2 shows a turnbuckle configuration that was generated as a result of the "creative" or "synthesis" stage of design. Utility analysis cannot be directly employed during this stage, but it can enable the designer to think in terms of function rather than form, freeing the designer from cognitive biases such as framing and anchoring, as illustrated for the design of a chair in [4].

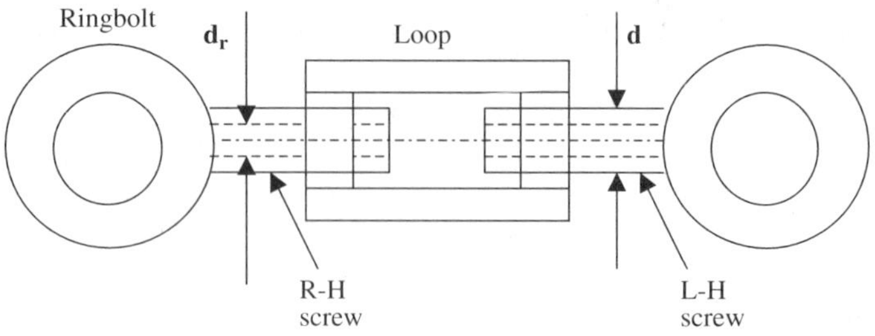

FIG. 12.2 CREATIVITY GENERATES TURNBUCKLE CONFIGURATION

For the turnbuckle problem developed in Thurston and Locascio [6], a two-stage design decision analysis revealed that after the best material was identified, the relevant attributes in the utility function were weight x_1, cost x_2 and fatigue strength x_3. Traditional engineering design analysis was employed to derive the constraint relationships $h(d) = x_i$ between the design decision variable (diameter d in mm) and the resulting weight x_1 and fatigue performance x_3, shown in Figures 12.3 and 12.4. Figure 12.5 shows the most direct contribution of utility analysis; identification of the design decision variable (diameter d) that results in the best combination of weight and fatigue performance.

12.2.5 Multi-Attribute Utility Formulation Summary

Other chapters in this book present alternative evaluation metrics. However, multi-attribute utility analysis is the tool best suited for making normative trade-off decisions that exhibit one or both of the following features: nonlinearity of preference over an attribute range; a willingness to make trade-offs that can change

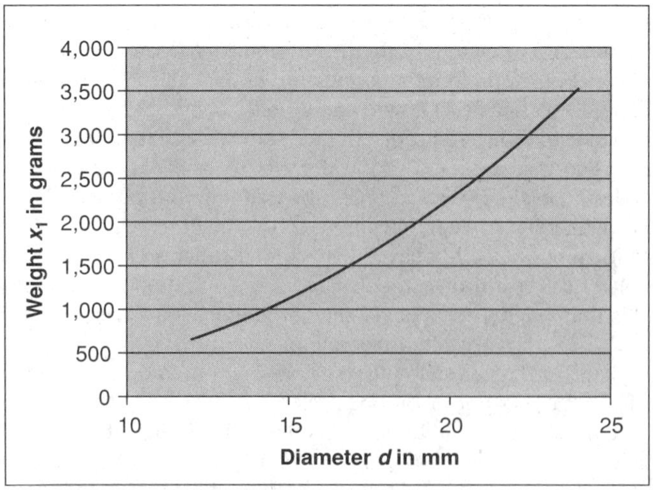

FIG. 12.3 ANALYSIS PREDICTS HOW WEIGHT WORSENS AS DIAMETER INCREASES

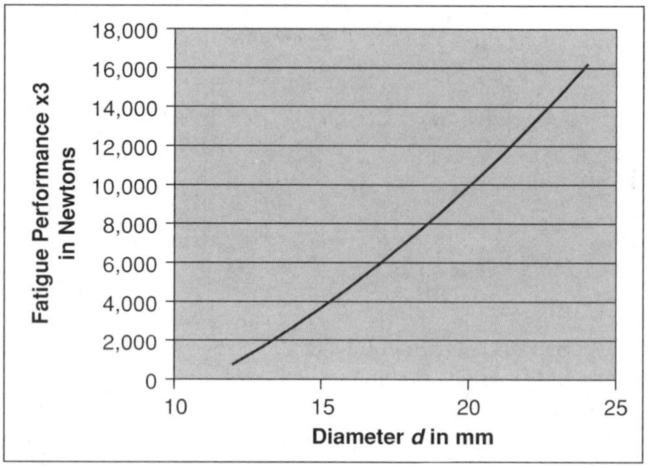

FIG. 12.4 ANALYSIS PREDICTS HOW FATIGUE PERFORMANCE IMPROVES AS DIAMETER INCREASES

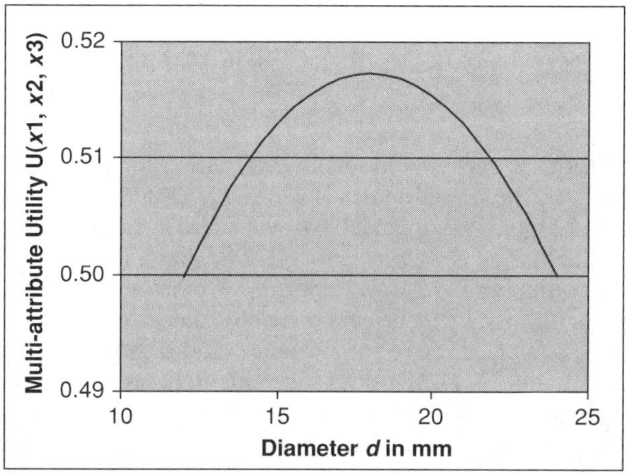

FIG. 12.5 UTILITY FUNCTION IDENTIFIES DIAMETER RESULTING IN BEST COMBINATION OF WEIGHT AND FATIGUE RESISTANCE

as product development progresses; and uncertainty which affects desirability (where that uncertainty can be modeled probabilistically). Multi-attribute utility analysis has been employed successfully to identify the design with the optimal set of trade-offs for a bumper beam [7] and for automotive body panels [8]. Thurston and Liu [9] and Tian et al. [10] demonstrate how utility analysis can be employed to reflect the effect of uncertainty. Nogal et al. [3] present a set of rules for using utility analysis to streamline the iterative design process.

12.3 MISCONCEPTIONS ABOUT MULTI-ATTRIBUTE UTILITY ANALYSIS

This section describes and resolves the most common misconceptions about multi-attribute utility analysis in design. These include misconceptions about the independence conditions, the form of the utility function, scaling and range definition, the purpose of normative subjective preference modeling and group decision-making. More detail can be found in Thurston [11].

12.3.1 Misconception: The Independence Conditions in Utility Analysis are Rarely Satisfied

The most widespread misconception relates to the independence conditions, specifically the mistaken belief that the independence conditions of utility analysis are not valid for design since the attributes are interdependent. Nam Suh [12] proposed axioms of design, one of which is independence. Suh proposed that the designer attempt to develop a configuration in which he or she can control (and thus improve) each important attribute without affecting (or worsening) another. Ideally, this decreases the number of objectives that will be in conflict, enabling designers to optimize individual attributes through control of a smaller number of engineering design decision variables. This can save significant time and effort during the design process. In the framework

$$\begin{bmatrix} a_{10} \\ a_{20} \\ a_{30} \\ a_{40} \\ a_{50} \end{bmatrix} + \begin{bmatrix} a_{11} & 0 & 0 & 0 & 0 & 0 & 0 \\ 0 & a_{22} & a_{23} & 0 & 0 & 0 & 0 \\ 0 & 0 & 0 & a_{34} & a_{35} & 0 & 0 \\ 0 & 0 & 0 & 0 & 0 & a_{46} & 0 \\ 0 & 0 & 0 & 0 & 0 & 0 & a_{57} \end{bmatrix} \begin{bmatrix} y_1 \\ y_2 \\ y_3 \\ y_4 \\ y_5 \\ y_6 \\ y_7 \end{bmatrix} = \begin{bmatrix} x_1 \\ x_2 \\ x_3 \\ x_4 \\ x_5 \end{bmatrix} \qquad \text{Eq. (12.13)}$$

presented above, the set of constraints for a 5 attribute, 7 decision variable multi-attribute optimization problem might look as shown in Equation (12.13) if Suh's independence axiom is obeyed.

Note that the designer can seek to improve attribute x_2 by making modifying design decision variables of the physical design artifact y_2 and y_3 without affecting any of the other attributes. Thus, this design is configured in such a way that Suh's independence ideal is achieved. In contrast, the two independence conditions of utility analysis have nothing to do with the physical design artifact, but rather with *preferences* for attribute combinations. Their main purpose is to make the job of utility assessment easier. Specifically, if preferential and utility independence conditions are satisfied, then the decision-maker's preferences need be assessed once and only once over each single-attribute range, independent of the other attributes. Loosely stated, preferential independence means that rank ordering of preferences for one attribute does not depend on the levels of the other attributes, when the other attributes levels are held constant. The test to determine if x_1 is preferentially independent of x_2 is shown below, where A and B refer to particular levels of each attribute:

If (x_{1A}, x_{2A}) is preferred to (x_{1B}, x_{2A})
and (x_{1A}, x_{2B}) is preferred to (x_{1B}, x_{2B})
then x_1 is preferentially independent of x_2

A design example might be where x_1 is weight and x_2 is cost:
If (50 pounds, \$100) is preferred to (40 pounds, \$100)
and (50 pounds, \$80) is preferred to (40 pounds, \$80)
then weight x_1 is preferentially independent of cost x_2

The test to determine whether x_1 is utility independent of x_2, or $x_1 UI\, x_2$, is shown in Fig. 12.6.

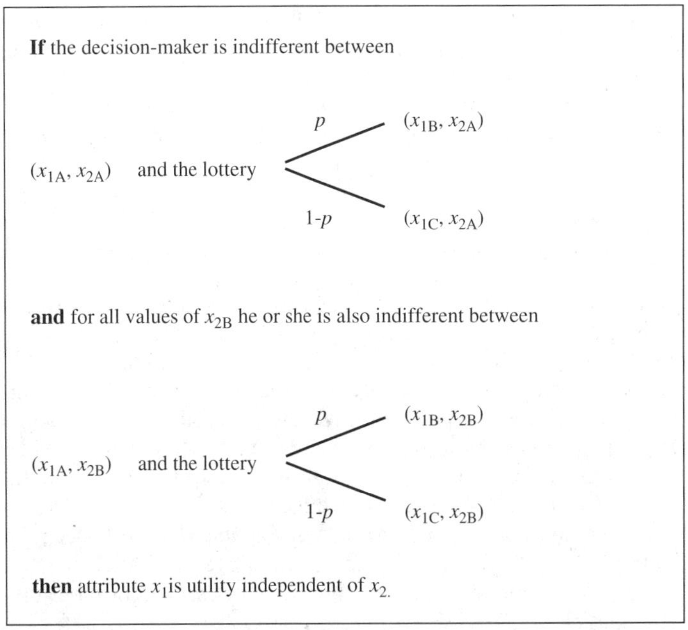

FIG. 12.6 TEST TO DETERMINE IF X_1 IS UTILITY INDEPENDENT OF X_2

Generally speaking, when x_1 is utility independent (UI) of x_2, it means that the degree of risk aversion for x_1 remains constant regardless of the level of x_2. It is important to note that even when the independence conditions do not hold, it might still be possible to assess a multi-attribute utility function, but the combinatorial explosion of required preference statements and the complexity of the resulting functional form would often prove intractable. For example, when preferences for the attribute weight are *not* preferentially independent of cost, the decision-maker might prefer a low-weight alternative to a high-weight alternative when all else is equal, including low cost, but when cost is high, would prefer the high-weight alternative. When preferences for weight are preferentially independent but *not* utility independent of cost, the decision-maker would consistently prefer the low-weight alternative (when all else is equal) at any level of cost because of preferential independence. However, since weight is not utility independent of cost, the decision-maker might be risk averse with respect to weight when cost is low, but risk neutral or risk seeking when cost is high. Such utility functions would be extremely time-consuming to assess, and their functional form would be very difficult to determine with accuracy. Even for the relatively simple two-attribute case where x_1 is utility independent of x_2, but x_2 is not utility independent of x_1, multiple assessments of $U(x_1, x_2)$ would be required over the full range of x_1 since the degree of risk aversion changes over this range. It should be noted that neither preferential nor utility independence conditions possess reflexive properties. For example, both (x_1 UI x_2) and (x_2 UI x_1) must be tested. In contrast, when preferences for x_1 and x_2 are both preferentially independent and utility independent, the only assessments required are one single-attribute utility function for x_1 and one single-attribute utility function for x_2 to determine risk aversion, and two lottery question assessments of k_1 and k_2 to determine willingness to make trade-offs.

In design, when these conditions do not hold it is often because the defined attributes can be directly substituted for one another, from a functional viewpoint. When this occurs, the recommended procedure is to redefine the attributes, their units and/or their ranges in such a way that the independence conditions hold; for example, by aggregating two or more attributes that are directly substitutable into one. For example, the attributes "material cost," "energy cost" and "labor cost" would likely not be mutually utility independent, since the designer's degree of risk aversion over material cost would likely depend on whether labor costs were known with certainty to be very high or very low. Then, it would be appropriate to aggregate all three into one attribute of "piece cost." The same holds true for the monotonicity axiom. To facilitate utility assessment, one should define attributes and their units such that preferences are monotonically increasing (more is always preferred to less) or monotonically decreasing (less is always preferred to more) over the tolerable range. If these conditions for formulating the utility function are violated, it does not imply that the basic concepts of utility analysis are invalid, but rather makes the assessment much more cumbersome.

In summary, Suh's independence axiom is an ideal that is rarely fully achieved in real engineering design, while the independence conditions of utility analysis are much more commonly satisfied. Suh's independence axiom states that the design should be configured in such a way as to minimize the physical interdependence of attributes on one another. If one can fully obey this axiom to the extreme, then one can simultaneously optimize each objective. Then by definition, the objectives are not in conflict. If on the other hand this is not possible, unavoidable trade-offs between conflicting objectives must be made, and preferential and utility independence conditions simply facilitate assessment of the trade-off function.

12.3.2 Misconception: The Form of Multi-Attribute Utility Function Is Chosen

This misconception is the mistaken belief that the mathematical form of the evaluation function, even a multi-attribute utility function, is arbitrarily chosen or selected. This misconception most likely arose because one can find a wide variety of arbitrarily defined evaluation metrics in the design literature. Instead, the correct functional form is determined only after testing to determine which independence conditions hold. Keeney and Raiffa [5] demonstrate that if and only if preferential and utility independence conditions are satisfied, then the appropriate functional form is the multiplicative, as shown in Eq. (12.14):

$$U(\mathbf{x}) = \frac{1}{K}\left(\left\{\prod [Kk_iU_i(x_i)+1]\right\}-1\right) \qquad \text{Eq. (12.14)}$$

where $U(\mathbf{x})$ = overall utility characterized by attribute vector $\mathbf{x} = (x_1, \ldots, x_n)$, scaled from 0 to 1; x_i = performance level of attribute i; $U_i(x_i)$ = single-attribute utility function for attribute i, scaled from 0 to 1; $i = 1, 2, \ldots, n$ attributes; k_i = single-attribute scaling constant; and K = normalizing constant so that $U(\mathbf{x})$ scales 0 to 1, derived from:

$$1 + K = \prod (1 + Kk_i) \qquad \text{Eq. (12.15)}$$

Although it appears to be simpler, the additive form shown in Eq. (12.16) is actually much more restrictive, and is a special case of the multiplicative form;

$$U(\mathbf{x}) = \sum k_iU_i(x_i) \qquad \text{Eq. (12.16)}$$

It is valid only if preferential, utility *and* the additive independence condition shown in Fig. 12.7 are satisfied [13]. This special case of the multiplicative form rarely appears in practice, when

$$\sum k_i = 1 \qquad \text{Eq. (12.17)}$$

It is important to note the analyst cannot arbitrarily require that k_i's sum to 1 as in Eq. (12.17); their value is determined by the decision-maker's response to the lottery questions shown in Figure

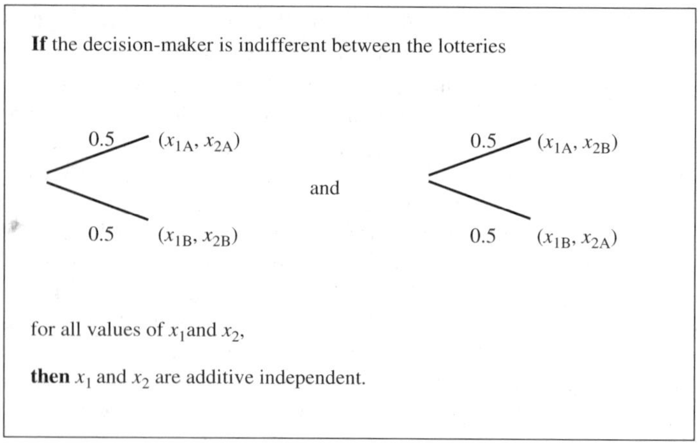

FIG. 12.7 TEST OF ADDITIVE INDEPENDENCE CONDITION BETWEEN X_1 AND X_2

12.7. If their sum is very close to 1 and the test shown in Figure 12.7 is satisfied, only then is the additive form accurate.

12.3.3 Misconception: Designers Don't Make Trade-Offs

The is a failure to distinguish the hard constraints imposed by absolute performance requirements and the range over which the designer does, in fact, make trade-offs. A closely related issue is definition of the attribute ranges used during the assessment procedure. The endpoints are correctly defined as the range beyond which the decision-maker is no longer able and willing to make trade-offs. The "worst" endpoint is arbitrarily assigned a single-attribute utility of 0 on a scale of 0 to 1, but this does not mean that at that level the attribute under question, say, weight, has no utility to the decision-maker. It simply means that that particular weight is the worst that the decision-maker is willing to consider. At the extreme, a design alternative that exhibits the worst levels of all attributes $\mathbf{x}_w$ is assigned overall utility $U(\mathbf{x}) = 0$, but still *by definition* satisfies all required design specifications and constraints.

12.3.4 Misconception: Why Model "Subjective" Preferences, Which Might Be Irrational?

"But it's all subjective, and depends on whose utility function you assess" is meant as a criticism that utility analysis does not yield one universal "answer" that is correct for all decision-makers all the time. In traditional design analysis, there is in fact one correct answer to questions such as "How much will this beam deflect under this load?" But answering such questions is not the goal of utility analysis. Instead, the type of question it seeks to answer is, "What design gives me the best combination of deflection, weight and cost, when I'm willing to pay $X to reduce weight by one pound but less than $X for the second pound, or when I am uncertain what the final cost will actually be?" Utility analysis answers this question by determining a mathematical model of a particular decision-maker's preferences, especially nonlinear preferences under uncertainty. Subjective preferences, by definition, can vary from decision-maker to decision-maker, and depend on many factors, including the current competitive position of the firm, the market position of the particular product under design, etc. Utility analysis of material selection for sailboat masts has been performed, resulting in different choices depending on whether the end-user is a cruiser or a racer [14]. Further, an individual's utility function can and often should change over time in response to external factors.

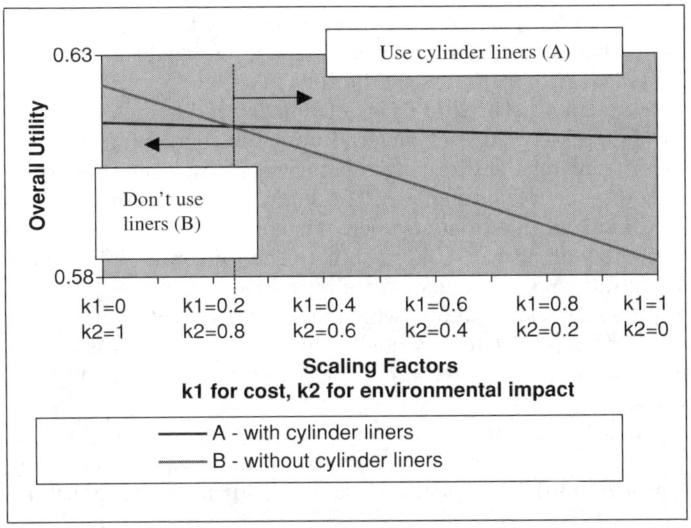

FIG. 12.8 SENSITIVITY ANALYSIS OF UTILITY TO SCALING FACTOR VALUES FOR COST (k_1) AND ENVIRONMENT (k_2)

However, one should be able to validate a properly assessed utility function through repeated assessment within a short time frame.

Krill and Thurston [15] analyzed design decisions for remanufacturing for engine blocks. Remanufacturing offers the potential for simultaneously recovering the economic value of manufactured components, and improving the environment. Design for remanufacturing aims to make remanufacturing less expensive and/or increase the proportion of components that can be remanufactured. For example, sacrificial components such as cylinder liners can be used to protect key parts from wear. But some design for remanufacturing decisions can increase original production costs and create their own environmental impact. For example, the addition of cylinder liners to protect against wear creates several additional steps during original production and remanufacturing. This results in a slightly higher environmental impact. So trade-offs are involved. Krill and Thurston [15] demonstrate that remanufacturing lowers overall costs when two life cycles are considered, that sacrificial cylinder liners should be employed for small (2 liter) engines and that their superiority increases with multiple remanufacturing cycles. Figure 12.8 shows sensitivity analysis on the scaling factors, which reflect the decision-maker's subjective willingness to make trade-offs. The dashed line denotes the point at which the decision-maker's preference switches between alternatives A and B. For purposes of illustration, we assume here that $k1$ and $k2$ = 1. When the scaling factor for cost ($k1$) is greater than 0.25, alternative A (use cylinder liners) is preferred. If the scaling factor for the cost is less than 0.25, alternative B (don't use cylinder liners) is preferred. Our hypothesis would be that the scaling factors for most manufacturers would fall within the "use cylinder liners" region.

In addition, some misinterpret the term "subjective" to mean "anything goes," and are concerned that one might model an irrational or inconsistent decision-makers' preferences. Quite the opposite is true. As a normative theory, a properly carried out utility analysis helps decision-makers avoid irrationality and inconsistencies.

12.3.5 Misconception: Group Design Cannot be Aided by Utility Analysis

While it is true that utility analysis does not resolve the central dilemma posed by Arrow's impossibility theorem [16], it can contribute significantly to the problem of how to aggregate the conflicting preferences of individuals in a group. From a modeling perspective, the group decision-making problem is of the same structure as the preference aggregation problem posed in Eq. (12.4); one merely substitutes conflicting attributes x_i with the conflicting interests of individuals. The central difficulty is how to define the function $f(\mathbf{x})$ that combines these conflicting interests in such a way that is best for the group as a whole. If one assumes the simple weighted average of Eq. (12.5), the element that poses difficulty is the "weighting factors" w_i for each individual in the group. Design is most often a team effort, requiring specialists in materials, structures, electronics, manufacturing, marketing, etc. Again, each specialist employs his or her knowledge to develop a configuration that optimizes the objective or objectives relevant to them. The problem then becomes one of a group searching for the best design, where group members might have competing objectives or priorities. The "group" also might be defined as comprising separate market segments for a product. From a decision analytic viewpoint, team design is technically not the same as the "social choice" problem as addressed in the classical literature briefly summarized below; issues of fairness or equitable distribution of welfare among team members are not relevant as long as one can assume that team members share a common corporate objective function. While one could argue that design teams share the common goal of developing the best product possible, one must also acknowledge that there are competing forces at work within many corporations. For example, while all team members agree that minimizing cost and maximizing quality are both ideal goals, marketing might place a higher priority on cost, while manufacturing might place a higher priority on quality. Also, the effect of exposure to unavoidable risk and uncertainty might vary dramatically between team members. Addressing these types of differences between team members' preference functions is the central problem of group decision-making in design.

One must first define a decision criterion. A variety of voting schemes are often employed, such as majority, runoff voting or a series of pairwise comparison votes. Unfortunately, Arrow's impossibility theorem [16] showed that all group decision-making procedures are flawed in the sense that: (1) the result depends on the decision procedure or voting scheme employed; and (2) no one procedure or voting scheme can be identified as best. As a result, it is not possible to construct the definitive, normative group utility function. Hazelrigg [17] describes the implications for design, which are that total quality management or quality function deployment can lead to erroneous results.

Arrow [16] further showed that there is no procedure for combining individual rankings that does not directly or indirectly include interpersonal comparisons of preference. No methods exist for accurately comparing subjective preferences between individuals. Disturbingly, Kirkwood [18] showed that strictly efficient methods that have Pareto optimality or maximization of total welfare as the sole objective are incompatible with attempts to address equity. In other words, when one attempts to make the distribution of welfare (or level of satisfaction) among individuals more even or "fair," then one *must* sacrifice total group welfare (or in the case of design, overall design worth).

Utility analysis does not resolve these central problems of classical group decision-making. However, team design decisions must be made, and utility analysis does provide an organized, structured framework, which disaggregates the problem into smaller, easier to resolve decision problems. There are two approaches. The first is to assess utility functions on the group as a whole, posing the lottery questions to the entire design team simultaneously, allowing for discussion and debate on which the team can (hopefully) easily reach consensus. The second approach avoids assignment of individual weighting factors through careful definition of "individuals" such that each is perceived by all to be of relatively "equal importance." For example, although the entire design team might comprise eight individuals, only one would represent materials, one would represent manufacturing, one marketing and so on.

12.4 SUMMARY AND CONCLUSIONS

This chapter has presented a framework for a mathematical model for aggregating (often) conflicting preferences in design. It has described the limitations to utility analysis in design, and several misconceived limitations. Table 12.2 summarizes the real limitations and the potential benefits of DBD. The first and second real limitations are that it cannot be directly employed to generate a set of feasible alternative design configurations, nor to perform what is commonly referred to as design analysis [defining $\mathbf{h}(\mathbf{y}) = \mathbf{x}$]. A properly assessed utility function does free the creative designer to think in terms of function rather than form, but the design knowledge of material properties, structural analysis, stress/strain relationships, manufacturing processes, etc. must originate from the design engineer. This knowledge is required to reach the Pareto optimal frontier, where it is not possible to improve one attribute without worsening another. The third real limitation is that utility analysis cannot directly resolve the central problems of group decision-making, which are that different voting methods yield different results (and no one voting method is clearly superior), interpersonal comparison of utilities and optimality versus equity.

These limitations aside, utility analysis can contribute significantly to the design process by providing a formal, structured way in which to model subjective trade-offs, particularly those that are

TABLE 12.2 LIMITATIONS AND BENEFITS OF DECISION-BASED DESIGN WITH UTILITY ANALYSIS

Design Stage	Limitations of DBD With Utility Analysis	Benefits of DBD With Utility Analysis
Customer/designer need expressed	Cannot create or influence customer preferences, other than to reveal inconsistencies of prior choices	Separates true objectives from superfluous Defines true trade-off range Avoid biases, inconsistencies, paradoxes in customer preferences
Creative synthesis of alternatives	Cannot replace creativity Cannot replace engineering expertise	Frees designer to think in terms of function rather than form Defines initial filter for feasible material, configuration, manufacturing options based on attribute and range definition
Analysis	Cannot define analytic constraint equations (strength of materials, kinematics, structural analysis, etc.)	Indicates which analytic equations are relevant, based on attributes and range Indicates where experimentation or other effort is worthwhile to improve analytic model
Trade-off evaluation	Cannot determine which trade-offs are technically feasible (must be done through analysis) Cannot define Pareto Optimal frontier Cannot provide solution procedures, optimization algorithms	Rank orders preliminary alternatives Identifies alternatives worth further analysis Determines which trade-offs are desirable Focuses effort where payoff is greatest Defines objective function for optimal solution
Decision-making under uncertainty	Cannot remove uncertainty	Provides method for modeling uncertainty Includes effect of uncertainty on rank order of alternatives Avoids irrationality under uncertainty Determines when it is worthwhile to gather more information vs. when to act, via expected value of information
Team design	Does not resolve Arrow's Impossibility Theorem	Provides framework for obtaining preference information from individuals and/or group Communicates preference information to team members Breaks decision problem into components on which consensus can be reached

nonlinear and/or that must be made under uncertainty. The contribution is to help identify the design that provides the optimal set of conflicting objectives for an individual designer. For group decision-making or team design, utility analysis decomposes the decision problem into smaller subproblems, through which individuals can express and resolve conflicting preferences using methods described in the previous section.

Several misconceived limitations to utility analysis in design have been described here. The most important is confusion about the distinction between independence of attributes and independence of various aspects of preference. The former relates to the unavoidable cause and effect relationships between design decisions and objectives exhibited in the design configuration. If achieved, attribute independence frees the designer to improve each objective without worsening another. In contrast, the independence conditions employed in utility analysis relate to cause and effect relationships between preferences over each objective. If achieved, they simply facilitate the task of assessing the multi-attribute utility function. If not, the simplest course of action is to reformulate the attribute set (definition, units used, range, etc.) so that they are achieved. In general, the other common misconceptions center on misinterpretation of terminology. For example, "subjective" preferences are those that are specific to a particular decision-maker, rather than those that are arbitrary, inconsistent or irreproducible.

Utility analysis cannot be the only analytic tool employed in design. It cannot contribute much to the creative or configuration phase, except to free the designer to think in terms of function rather than form. It cannot tell the designer which raw material options are available, nor the beam cross section required to bear a particular load. Neither can it fully resolve the problem of defining the optimal group decision, one which has long plagued economists. Like many useful analytic tools, it can be used naively or incorrectly, and there are special cases that yield inconsistent or nonsensical results.

However, design theory and methodology is an arena worthy of endeavor because traditional design processes sometimes take too long, result in products that are too costly, are difficult to manufacture, are of poor quality, don't satisfy customer needs, impact the environment adversely, and provide design teams with only ad hoc methods for communicating and resolving conflicting preferences. Utility analysis can help remedy these problems by quickly focusing creative and analytic efforts on decisions that affect important design functions, by identifying the best trade-offs (particularly under uncertainty) and by disaggregating design team decision problems into subproblems on which consensus can be reached. So while decision theory by itself does not constitute a theory of design, integrating it throughout several design phases, including configuration and analysis, can improve both the process and the product of design.

ACKNOWLEDGMENT

The author would like to thank the many colleagues within the design theory and methodology community with whom she has enjoyed fruitful discussions over the years, and also the National Science Foundation grant DMI 0217491.

REFERENCES

1. Pareto, V., 1971. *Manual of Political Economy,* Translation of the French edition (1927) by A.S. Schwier, London-Basingslohe, The McMillan Press Ltd.
2. Hauser, J. R. and D. Clausing, 1988. "The House of Quality," *Harvard Bus. Rev.* 66, (3).
3. Nogal, A.M., Thurston, D.L. and Tian, Y.T., 1994. "Meta-Level Reasoning in the Iterative Design Process," *Proc., ASME Winter Annual Meeting on Des. Automation.*
4. Thurston, D. L., 1991. "A Formal Method for Subjective Design Evaluation With Multiple Attributes," *Res. in Engrg. Des.*, 3, (2).
5. Keeney, R.L. and Raiffa, H., 1993. *Decisions With Multiple Objectives: Preferences and Value Tradeoffs*, Cambridge University Press; first published by Wiley and Sons, 1976.
6. Thurston, D.L. and Locascio, A., 1994. "Decision Theory for Design Economics," *The Engrg. Economist*, 40 (1).
7. Thurston, D.L., Carnahan, J.V. and Liu, T., 1994. "Optimization of Design Utility," *ASME J. of Mech. Des.* 116 (3).
8. Thurston, D.L. and Essington, S., 1993. "A Tool for Optimal Manufacturing Design Decisions," *Manufacturing Rev.* 6 (1).
9. Thurston, D. L. and Liu, T., 1991. "Design Evaluation of Multiple Attribute Under Uncertainty," *Sys. Automation: Res. and Appl.* 1 (2).
10. Tian, Y.Q., Thurston, D.L. and Carnahan, J.V., 1994 "Incorporating End-Users' Attitudes Towards Uncertainty Into an Expert System," *ASME J. of Mech. Des.* 116 (2).
11. Thurston, D.L., 2001. "Real and Misconceived Limitations to Decision Based Design With Utility Analysis," *ASME J. of Mech. Des.*, 123 (2), pp. 176-186.
12. Suh, N., 1988 *The Principals of Design*, Oxford University Press, Oxford, U.K.
13. Fishburn, P.C., 1970 *Utility Theory for Decision Making*, Wiley, New York, NY.
14. Thurston, D.L., and Crawford, C.A, 1994. "A Method for Integrating End-User Preferences for Design Evaluation in Rule-Based Systems," *ASME J. of Mech. Des.*, 116 (2).
15. Krill, M. and Thurston, D. 2005. "Remanufacturing: Effects of Sacrificial Cylinder Liners," *ASME J. of Manufacturing Sci. and Engrg.*, 127 (3).
16. Arrow, K.J., 1951. *Social Choice and Individual Values, 2nd Ed.*, John Wiley and Sons, New York, NY.
17. Hazelrigg, G., 1960. "The Implications of Arrow's Impossibility Theorem on Approaches to Optimal Design," *ASME J. of Mech. Des.*, 118 (2).
18. Kirkwood, C.W., 1979. "Pareto Optimality and Equity in Social Decision Analysis," *IEEE Trans. on Sys., Man and Cybernetics*, SMC-9 (2).

PROBLEMS

12.1 You are trying to formulate a design decision problem to maximize multi-attribute utility. First you must determine whether the attributes you have defined satisfy the preferential and utility independence conditions. Attribute X = profit, ranging from –5 to +30 million dollars and attribute Y = environmental impact on a scale of 1 to 10, where 1 is best and 10 is worst.

a. Show the results of a test that indicates that your preferences for X are not preferentially independent of Y.
b. Show the results of a test that indicates that your preferences for X are not utility independent of Y.

12.2 You are performing a utility assessment in order to help a designer make decisions that will affect the cost of a new vehicle. It is not going well. She has exhibited the Allais paradox on lotteries over x = $dollars ranging from $10,000 to $100,000.

a. What are three possible implications for your utility assessment?
b. Show three things you might do to solve the problem. Feel free to make (and describe) any assumptions necessary to illustrate your point.

12.3 You are trying to assess a designer's multi-attribute utility function for cost and environmental impact. For the attributes

X = cost, ranging from \$20,000 to \$100,000 per year, and Y = environmental impact, ranging on a unitless scale from 10 to 90, where 10 is best and 90 is worst.

a. Show the results of a test that indicates that $k_x = 0.8$.
b. Show the results of a test that indicates that $k_y = 0.3$.

12.4 A customer's preferences for widgets have been assessed for x = performance and y = cost in dollars. $U(x) = 1 - e^{-x/25}$, $U(y) = 1 - y/100$, where both x and y range from 0 to 100, and $k_x + k_y = 1$. The preferential, utility and additive independence conditions have each been tested and satisfied. The alternatives are Widget A with performance $x = 10$ and cost y = \$20; Widget B with performance $x = 50$ and cost y = \$60; and Widget C with performance x = 90 and cost y = \$80.
On a plot where the x-axis is k_x and the y-axis is $U(x,y)$, clearly indicate the range over which each widget is preferred.

12.5 A customer's preferences for widgets have been assessed for x = performance and y = cost in dollars. Given $U(x) = 1\text{-}e^{x/25}$, $U(y) = y/100$, where both x and y range from 0 to 100, and $k_x = 0.8$ and $k_y = 0.2$:

a. Sketch indifference curves where $U(x,y) = 0.2$, $U(x,y) = 0.5$ and $U(x,y) = 0.8$.
b. The best widget currently on the market has performance of $x = 10$ and $y = 30$. Your new, improved widget design offers performance of $x = 40$. What is the most the customer would be willing to pay for your widget?

True or False

12.6 T or F The utility independence condition is not satisfied in engineering design problems when the designer cannot improve one attribute without worsening another.

12.7 T or F Assessing utilities using the lottery method is only appropriate when the decision problem involves uncertainty.

12.8 T or F For an engineering design problem, a correctly assessed utility function can be used to determine which trade-offs are technically feasible.

12.9 T or F The scaling constants k_i reflect the decision-maker's subjective attitude toward risk.

12.10 T or F The scaling constants k_i reflect the decision-maker's subjective willingness to make trade-offs.

12.11 T or F The scaling constants k_i reflect the decision-maker's subjective utility derived from attribute i.

12.12 T or F The decision analyst should choose the additive form of the multi-attribute utility function in order to simplify the calculations.

CHAPTER

13

ON THE LEGITIMACY OF PAIRWISE COMPARISONS

Clive L. Dym, William H. Wood, and Michael J. Scott

13.1 INTRODUCTION

People routinely compare similar instantiations of objects or classes of objects in which they are interested, whether they are digital cameras, colleges or potential spouses. Designers are routinely charged with, first, *generating* or *creating* a set of alternatives, and, second, choosing one of them as the most preferred. Thus, designers routinely rank objectives, design attributes and designs using decision-centric methods and techniques, many of which are variants of the standard pair-by-pair *pairwise comparison* [1–8]. The common underlying concept in these methods, now all embodied within the rubric of *decision-based design* (DBD), is that *some part of* design thinking can be represented and modeled as a decision-making process that is aimed at addressing the need for a *rational* way to choose among alternatives. But, *which* part of the design process—or perhaps *how much* of the design process—is decision-making? We attempt to address such questions based very much on a prior analysis [9], some of which was also detailed in [10], and augmented in part by a recent inquiry into design thinking, teaching and learning [11].

13.1.1 Decision-Making in Design

Hazelrigg has argued [12] that "... a mathematics of design is needed ... based on the recognition that engineering design is a decision-intensive process and adapting theories from other fields such as economics and decision theory." Moreover, Hazelrigg observed that the conventional engineering approach utilizes scientific deterministic models, thus yielding a limited set of decisions. He extended his argument by leveraging decision theory to construct a set of axioms for designing and to formulate two theorems that could be applied to construct statistical models that would account for uncertainty, risk, information, preferences and external factors such as competition—the elements of decision theory [13]. This approach arguably results in numerous decisions, only one of which would be optimal. Hazelrigg concluded that the axiomatic approach yields a more accurate representation and produces results having a higher probability of winning in a competitive situation.

No one can doubt that designers must—and do—make decisions. However, one question has emerged in recent years about how design and decision-making relate [14, 15]: Can designers *legitimately* make design decisions using the methods outlined above to inform their own preferences and aggregate those of others?

The relevance of applying existing decision-centric views to evaluate and choose among alternative design concepts was demonstrated by Dieter [1] when he constructed a decision-matrix to determine the intrinsic worth of outcomes associated with competing design concepts. Dieter's method was based on utility theory and formalized the development of values in decision-making. The "Pugh selection chart" and similar decision matrices have since become well established [3–7].

Another decision-centric approach was set forth by Radford and Gero [8], who used (deterministic) optimization—in contrast with Hazelrigg's probabilistic model—to account for ambiguity. Radford and Gero stressed that goals are essential to design, forcing decisions as to how they should be achieved. They also argued that exploring the relationship between design decisions and performance of the resulting solutions is fundamental to design, using optimization as the mechanism to introduce goal-seeking directly into the process of exploring the design space.

While these and other attempts at improving decision support within design were evolving and becoming part of the designer's tool kit, the role of decision-making in design became somewhat controversial because questions remain unanswered about the validity of such an identification because some of the underlying premises of decision theory do not seem to be appropriate models for completely describing design processes. For example, as detailed in sections 13.1.2 and 13.5, Arrow's Impossibility Theorem seems an unduly restrictive model for describing how designers actually work when they make design decisions. In addition, DBD assumes *that alternative design concepts and choices have already been generated* and can be represented in forms to which DBD can be applied. DBD cannot account for how concepts and alternatives are generated, nor does it suggest a process for doing so.

Some decision theorists acknowledge these limitations by recognizing that decision analysis can only be practiced after a certain point. Howard asked [16]; "Is decision analysis too narrow for the richness of the human decision?" He then argued that "framing" and "creating alternatives" should be addressed before decision analysis techniques are applied to ensure that "we are working on the right problem." Howard [16] also observed that "Framing is the most difficult part of the decision analysis process; it seems to require an understanding that is uniquely human. Framing poses the greatest challenge to the automation of decision analysis."

Finally, the role of decision-making in design derives in large part from the *seeming* success that "neoclassical" (e.g., decision theory) economics has had in mathematically modeling how rational human beings would behave by making rational choices, based on their own knowledge of their preferences or utilities, as well as of the alternatives and choices available to them. However, a new approach to modeling human decision-making—*called behavioral economics*—has emerged because economists have begun to recognize that it might be more worthwhile to *model how decision-makers actually behave*, and how that behavior influences the decisions thus reached, rather than focusing on what economists think "rational" decision-makers should do [17]. Thus, behavioral economics departs from the neoclassical about rational behavior. It is built on H. A. Simon's notion of *bounded rationality* that recognizes that we do not have mathematical models of human cognitive behavior—which behavior is far too rich and too complex to be subsumed within the confines of neoclassical economics [18–20]. (It is also interesting to note that research in behavioral economics has moved beyond traditional statistical and econometric tests of neoclassical economics to include experiments and surveys about the actual decision-making processes of individuals [17].)

13.1.2 Establishing Rankings With Pairwise Comparisons

In recent years, questions have been raised about the means by which designers establish rankings of alternatives, with a special focus on how pairwise comparisons are performed to assembling information on rankings. In *pairwise comparisons*, the elements in a set (i.e., the objectives, design attributes or designs) are ranked two at a time, on a pair-by-pair basis, until all of the permutations have been exhausted. Points are awarded to the winner of each comparison. (As both described and practiced, the number of points awarded in pairwise comparisons is often nonuniform and subjectively or arbitrarily weighted. But it is quite important to award the points in pairwise comparisons in multiples of fixed increments.) Then the points awarded to each element in the set are summed and the rankings are obtained by ordering the elements according to points accumulated.

This methodology has been criticized from two standpoints. In the first, Hazelrigg [14, 15] argues that processes for aggregating pairwise comparisons are subject to *Arrow's Impossibility Theorem* [21], which is a proof that a "perfect" or "fair" voting procedure cannot be developed whenever there are more than two candidates or elements that are to be chosen. Arrow's theorem has been stated in many ways, including the formulation due to Scott and Antonsson [22] that conforms to Arrow's original and which we follow here. Thus, a voting procedure can be characterized as fair *only* if five axioms are obeyed:

(1) The *unrestricted* axiom: All conceivable rankings registered by individual voters are actually possible;
(2) The *no imposed orders* or *citizen's sovereignty* axiom: There is no pair *A*, *B* for which it is impossible for the group to select one over the other.
(3) The *no dictator* axiom: The system does not allow one voter to impose his or her ranking as the group's aggregate ranking.
(4) The *positive response* axiom: If a set of orders ranks *A* before *B*, and a second set of orders is identical to the first except that individuals who ranked *B* over *A* are allowed to switch, then *A* is still preferred to *B* in the second set of orders. This axiom is an ordinal version of *monotonicity*.
(5) The *independence of irrelevant alternatives* (IIA) axiom: If the aggregate ranking would choose *A* over *B* when *C* is not considered, then it will not choose *B* over *A* when *C* is considered.

(As noted, this statement conforms to Arrow's original presentation. Arrow subsequently showed [21] that axioms 2 and 4 could be replaced by the *Pareto Condition,* which states that if everyone ranks *A* over *B*, the societal ranking has *A* ranked above *B*. Arrow's presentation also formally states that both individual and group rankings are *weak orders*, that is, they are transitive orderings that include all alternatives and allow for indifference among alternatives.)

Arrow proved that at *least one of these properties must be violated* for problems of reasonable size (at least three voters expressing only ordinal preferences among more than two alternatives). Hazelrigg stated Arrow's theorem informally by saying that "in general, we cannot write a utility function for a group" (p. 165 of [14]). It is worth noting that a consistent social choice (voting) procedure can be achieved by violating any one of the five (or four) conditions. Indeed, the questions we address in this paper are; "Which axioms are violated by designers as they make sensible choices?" *and* "What are the consequences of these "violations"?" Dictatorship (axiom 3) and violations of the Pareto condition (axioms 2 and 4) are intuitively offensive. Further, Scott and Antonsson [22] argue that engineering approaches that use quantified performance rankings do not violate axiom 5, since comparison of two alternatives on the basis of measurable criteria is independent of the performance of other alternatives; however, they often violate axiom 1, since many theoretically possible orders are not admissible in practice, as many engineering criteria must be of the less-is-better, more-is-better or nominal-is-best varieties [22].

The mathematician Saari [23, 24] notes that some *voting procedures* based on pairwise comparisons are faulty in that they can produce ranking results that offend our intuitive sense of a reasonable outcome. Saari further claims that virtually any final ranking can be arrived at by correct specification of the voting procedure. Saari also suggests that among pairwise comparison procedures, the Borda count most "respects the data" in that it avoids the counterintuitive results that can arise with other methods. Indeed, Saari notes (in [24]) that the Borda count "*never elects the candidate which loses all pairwise elections ... always ranks a candidate who wins all pairwise comparisons above the candidate who loses all such comparisons.*"

In the case of the Borda count, the fifth Arrow axiom, the IIA, is violated. What does this mean for design? In principle, and in the conceptual design phase where these questions are often most relevant, the possible space of design options is infinite. The designer must at some point stop generating and start choosing from among the alternatives, just as the customer must at some point stop evaluating and start choosing (and buying!) from among the available attributes and design choices. Simon's articulated principle of *bounded rationality* [18–20] suggests that the designer must find a way to limit that set of design alternatives, to make it a finite, relatively small set of options. We cannot afford to decide to generate all of the options because we would then never get beyond generating options. As a process, design requires that we generate design alternatives and select one (or more) of them. We may eliminate options because they don't meet some goals or criteria or are otherwise seen as poor designs. Given these two bases for selection, how important is IIA? Does it matter if we violate IIA? Are we likely to erroneously remove promising designs early in the process? Is our design process flawed because of these removed designs?

The violation of IIA leads to the possibility of *rank reversals*, that is, changes in order among *n* alternatives that may occur when one alternative is dropped from a once-ranked set before a second ranking of the remaining *n*–1 alternatives (see Section 13.3). The elimination of designs or candidates *can* change the tabulated rankings of those designs or candidates that remain under consideration. The determination of which design is "best" or which candidate is "preferred most" may well be sensitive to the set of designs considered.

Now, it is thought that these rank reversals occur because of a loss of information that occurs when an alternative is dropped or removed from the once-ranked set [24]. In addition, rank reversals occur when there are *Condorcet cycles* in the voting patterns: $[A \succ B \succ C,\ B \succ C \succ A,\ C \succ A \succ B\ ]$. When aggregated over all voters and alternatives, these cycles cancel each other out because each option has the same Borda count. When one of the alternatives is removed, this cycle no longer cancels. Thus, removing *C* from the above cycle unbalances the Borda count between *A* and *B*, resulting in a unit gain for *A* that is propagated to the final ranking results.

Paralleling the role of the Borda count in voting procedures, the PCC is the most consistent pairwise procedure to apply when making design choices. Both implementations are better than "drop and revote," whether viewed from the standpoint of bounded rationality embedded in Simon's concept of *satisficing* [20] or from Saari's analysis of voting procedures [23]: Both say we should *consider all of the information we have*. We may not attain perfect rationality and complete knowledge, but we should proceed with the best available knowledge. Design iterates between generating options and selecting among them, with the richness of information increasing as the process proceeds. At each stage, design selection tools must operate at an appropriate information level—as more information is developed, more complex tools can be applied: decision and information value theory, demand modeling, etc. While these tools can overcome the IIA violations inherent to the Borda count, they do so at a cost. Selection actions could be delayed until design information is rich enough to apply techniques that won't violate IIA, but this would commit the designer to the added expense of further developing poor designs. Rather than "drop and revote," design is more akin to sequential runoff elections in which the (design) candidates continue to "debate" throughout the design selection process.

In the end, no selection method can overcome poor design option generation. However, the act of selection helps to clarify the design task. From a practical standpoint, both designers and teachers of design have found that pairwise comparisons appear to work well by focusing their attention, bringing order to large numbers of seemingly disparate objectives, attributes or data points. In addition, these rankings often produce good designs, which at least suggests that the process has a high probability of producing good outcomes.

We are interested in enabling and contributing to a positive discussion of improving methods of design decision-making. In this spirit, we describe here a way to use pairwise comparisons in a structured approach that produces results that are identical to the accepted vote-counting standard, the Borda count. The method is a structured extension of pairwise comparisons to a *pairwise comparison chart* (PCC) or matrix (pp. 60–66 of [7]). We show that the PCC produces consistent results quickly and efficiently, and that these results are identical with results produced by a Borda count. We illustrate this with examples that have been used to show the inconsistencies produced by pairwise comparisons [15, 25], and we provide a formal proof of the equivalence of the PCC and the Borda count.

13.2 PAIRWISE COMPARISONS AND BORDA COUNTS

We begin with an example that highlights some of the problems of (non-Borda count) pairwise comparison procedures. It also suggests the equivalence of the Borda count with a structured PCC.

Twelve (12) voters are asked to rank order three candidates: *A*, *B* and *C*. In doing so, the 12 voters have, collectively, produced the following sets of orderings:

1 preferred $A \succ B \succ C$ 4 preferred $B \succ C \succ A$ Eq. (13.1a)

4 preferred $A \succ C \succ B$ 3 preferred $C \succ B \succ A$ Eq. (13.1b)

Saari [25] has shown that pairwise comparisons other than the Borda count can lead to inconsistent results for this case. For example, in a widely used plurality voting process called *the best of the best*, *A* gets 5 first-place votes, while *B* and *C* each get 4 and 3, respectively. Thus, *A* is a clear winner. On the other hand, in an "antiplurality" procedure characterized as *avoid the worst of the worst*, *C* gets only 1 last-place vote, while *A* and *B* get 7 and 4, respectively. Thus, under these rules, *C* could be regarded as the winner. In an iterative process based on *the best of the best*, if *C* were eliminated for coming in last, then a comparison of the remaining pair *A* and *B* quickly shows that *B* is the winner:

1 preferred $A \succ B$ 4 preferred $B \succ A$ Eq. (13.2a)

4 preferred $A \succ B$ 3 preferred $B \succ A$ Eq. (13.2b)

On the other hand, a Borda count produces a clear result. The Borda count procedure assigns numerical ratings separated by a common constant to each element in the list. Thus, sets such as (3, 2, 1), (2, 1, 0) and (10, 5, 0) could be used to award points to rank a three-element list. If we use (2, 1, 0) for the rankings presented in Eq. (13.1), we find total vote counts of (*A*: 2 + 8 + 0 + 0 = 10), (*B*: 1 + 0 + 8 + 3 = 12) and (*C*: 0 + 4 + 4 + 6 = 14), which clearly shows that *C* is the winner. Furthermore, if *A* is eliminated and *C* is compared only to *B* in a second Borda count:

1 preferred $B \succ C$ 4 preferred $B \succ C$ Eq. (13.3a)

4 preferred $C \succ B$ 3 preferred $C \succ B$ Eq. (13.3b)

Candidate *C* remains the winner, as it also would here by a simple vote count. It must be remarked that this consistency cannot be guaranteed, as the Borda count violates the IIA axiom.

TABLE 13.1 A PAIRWISE COMPARISON CHART (PCC) FOR THE BALLOTS CAST BY 12 VOTERS CHOOSING AMONG THE CANDIDATES *A*, *B* AND *C* [SEE EQ. (1)]

Win/Lose	*A*	*B*	*C*	Sum/Win
A	—	1+4+0+0	1+4+0+0	**10**
B	0+0+4+3	—	1+0+4+0	**12**
C	0+0+4+3	0+4+0+3	—	**14**
Sum/lose	**14**	**12**	**10**	**—**

TABLE 13.2 A REDUCED PAIRWISE COMPARISON CHART (PCC) FOR THE VOTING PROBLEM OF TABLE 13.1 AFTER THE "LOSER" *A* IN THE FIRST RANKING IS REMOVED FROM CONSIDERATION

Win/Lose	*B*	*C*	Sum/Win
B	—	1+0+4+0	5
C	0+4+0+3	—	7
Sum/lose	7	5	—

We now make the same comparisons in a PCC matrix, as illustrated in Table 13.1. As noted above, a point is awarded to the winner in each pairwise comparison, and then the points earned by each alternative are summed. In the PCC of Table 13.1, points are awarded row-by-row, proceeding along each row while comparing the row element to each column alternative in an individual pairwise comparison. This PCC result shows that the rank ordering of preferred candidates is entirely consistent with the Borda results just obtained:

$$C \succ B \succ A \qquad \text{Eq. (13.4)}$$

Note that the PCC matrix exhibits a special kind of symmetry, as does the ordering in the "Win" column (largest number of points) and the "Lose" row (smallest number of points): the sum of corresponding off-diagonal elements, $X_{ij} + X_{ji}$, is a constant equal to the number of comparison sets.

We have noted that a principal complaint about some pairwise comparisons is that they lead to rank reversals when the field of candidate elements is reduced by removing the lowest-ranked element between orderings. (Strictly speaking, rank reversal can occur when any alternative is removed. In fact, and as we note further in Section 13.3, examples can be constructed to achieve a specific rank reversal outcome. Such examples usually include a dominated option that is not the worst. Also, rank reversals are possible if new alternatives are *added*.) Practical experience suggests that the PCC generally preserves the original rankings if one alternative is dropped. If element *A* is removed above and a two-element runoff is conducted for *B* and *C*, we find the results given in Table 13.2. Hence, once again we find:

$$C \succ B \qquad \text{Eq. (13.5)}$$

The results in inequality Eq. (13.5) clearly preserve the ordering of inequality (13.4), that is, no rank reversal is obtained as a result of applying the PCC approach. In those instances where some rank reversal does occur, it is often among lower-ranked elements where the information is strongly influenced by the removed element (see section 13.5).

13.3 THE PCC IMPLEMENTS THE BORDA COUNT

We now prove that the PCC is an implementation of the Borda count or, in other words, that *they are equivalent*. In both procedures, the number of times that an element outranks another in pair-by-pair comparisons is tallied to determine a final overall ranking. More formally, we prove that these methods are identical, always producing the same rank order for a given collection of individual orderings.

13.3.1 Preliminaries

We start by supposing that a set of n alternatives

$$\{A_1, A_2, \ldots, A_n\} \qquad \text{Eq. (13.6)}$$

is ranked individually m times. Each rank order R_i takes the form

$$A_{i_1} \succ A_{i_2} \succ \ldots \succ A_{i_n} \qquad \text{Eq. (13.7)}$$

where $A \succ B$ indicates that A outranks or is ranked ahead of B. Each rank order R_i can then be expressed as a permutation σ_i of (1, 2,..., n):

$$\sigma_i = (i_1, i_2, \ldots, i_3) \qquad \text{Eq. (13.8)}$$

Let σ_{ij} be the jth entry of σ_i, so $\sigma_{ij} = i_j$. Let $\sigma_i(k)$ be the index of the entry with value k in σ_i (for k = 1,2,..., n). Then:

$$\sigma_i(\sigma_{ij}) = j \qquad \text{Eq. (13.9)}$$

Then $\sigma_i(k)$ is equal to the ordinal position that alternative A_k holds in the rank order σ_i. To take an example with $n = 3$, if R_i expresses the ranking

$$A_3 \succ A_1 \succ A_2 \qquad \text{Eq. (13.10)}$$

that is, if $\sigma_i = (3,1,2)$, then $\sigma_i(1) = 2$, $\sigma_i(2) = 3$, and $\sigma_i(3) = 1$.

13.3.2 Borda Count Sums

In a Borda count, each alternative A_k is assigned a number of points for each individual rank order R_i depending on its place in that order, and then the numbers are summed. Although there is an infinite number of equivalent numbering schemes, the canonical scheme—which may be used without loss of generality—assigns $(n - \sigma_i(k))$ points to alternative A_k from rank order R_i. For example, the rank ordering in Eq. (10) assigns two points to alternative A_3, one point to A_1, and no points to A_2. The Borda sum for the alternative A_k is obtained by summing over all individual orders R_i:

$$A_k^B = \sum_{i=1}^{m} \left(n - \sigma_i(k)\right) \qquad \text{Eq. (13.11)}$$

13.3.3 Pairwise Comparison Chart (PCC) Sums

To generate the kth row of the PCC, for each $j \neq k$ we count the number of permutations σ_i for which $\sigma_i(k) < \sigma_i(j)$, assigning one point for each such σ_i. (Notice that $\sigma_i(k) < \sigma_i(j)$ if and only if A_k outranks A_j in R_i.) For any σ_i, if $\sigma_i(k) < n$, then one point will be added to the A_k row in each of the columns $A_{\sigma_i(k)+1}, \ldots, A_n$. If $\sigma_i(k) = n$, no points are added to that row. Thus, the total points added to the A_k row as a result of R_i is $(n - \sigma_i(k))$. The grand total for A_k in the "sum/win" column is simply

$$A_k^{PCC:W} = \sum_{i=1}^{m} (n - \sigma_i(k)) \qquad \text{Eq. (13.12)}$$

which is exactly equal to the Borda sum given in Eq. (13.11). Therefore, the two methods are equivalent: The PCC is thus either an alternate representation of or a simple method for obtaining a Borda count (or vice versa).

Note that the sum for the "sum/lose" row in the PCC is just

$$A_k^{PCC:L} = m(n-1) - \sum_{i=1}^{m} (n - \sigma_i(k)) \qquad \text{Eq. (13.13)}$$

Therefore, the information contained in the "sum/lose" row is immediately available if the Borda count is known.

13.4 PAIRWISE COMPARISONS AND RANK REVERSALS

Rank reversals can occur when alternatives are dropped and the PCC procedure is repeated. Consider the following example. Thirty (30) designers (or consumers) are asked to rank-order five designs, A, B, C, D and E, as a result of which they produce the following sets of orderings:

$$10 \text{ preferred } A \succ B \succ C \succ D \succ E \quad \text{Eq. (13.14a)}$$

$$10 \text{ preferred } B \succ C \succ D \succ E \succ A \quad \text{Eq. (13.14b)}$$

$$10 \text{ preferred } C \succ D \succ E \succ A \succ B \quad \text{Eq. (13.14c)}$$

Here, too, the procedure chosen to rank order these five designs can decidedly influence or alter the results. For example, all of the designers ranked C and D ahead of E in the above tally. Nonetheless, if the following sequence of pairwise comparisons is undertaken, an inconsistent result obtains:

$$C \text{ vs } D \rightarrow C;\ C \text{ vs } B \rightarrow B;\ B \text{ vs } A \rightarrow A;\ A \text{ vs } E \rightarrow E \quad \text{Eq. (13.15)}$$

If we construct a PCC matrix for this five-design example, we find the results shown in Table 13.3, and they clearly indicate the order of preferred designs to be

$$C \succ B \succ D \succ A \succ E \quad \text{Eq. (13.16)}$$

If the same data are subjected to a Borda count, using the weights (4, 3, 2, 1, 0) for the place rankings, we then find the results displayed in Table 13.4. When we compare these results to the PCC results shown in Table 13.3, we see that the PCC has achieved the same Borda count results, albeit in a slightly different fashion.

What happens if we drop the lowest-ranked design and redo our assessment of alternatives? Here design E is least preferred, and we find the results shown in Table 13.5 if it is dropped. These results show a rank ordering of

$$C \succ B \succ A \succ D \quad \text{Eq. (13.17)}$$

Rank order is preserved here for the two top designs, C and B, while the last two change places. Why does this happen? Quite simply, because of the relative narrowness of the gap between A and D when compared to the gap between A and E, the two lowest ranked in the first application of the PCC in this example.

TABLE 13.4 THE BORDA COUNT FOR THE DATA IN TABLE 13.3 USING THE WEIGHT SET (4, 3, 2, 1, 0)

Element	Points
A	40 + 0 + 10 = 50
B	30 + 40 + 0 = 70
C	20 + 30 + 40 = 90
D	10 + 20 + 30 = 60
E	0 + 10 + 20 = 30

It is also useful to "reverse engineer" this example. Evidently it was constructed by taking a Condorcet cycle [$A \succ B \succ C$, $B \succ C \succ A$, $C \succ A \succ B$] and replacing C with an ordered set ($C \succ D \succ E$) that introduces two dominated (by C) options that are irrelevant by inspection. Removing only E produced a minor rank reversal of the last two alternatives, A and D. Removing only D, the third best option, produces the same result among A, B and C as removing E, although without creating a rank reversal. Removing both D and E produces a tie among A, B and C.

In a design context, assuming that designs D and E are always inferior to design C, they would seem to be dominated members of the same basic design family. Thus, in order to avoid these (minor) rank reversals, it is important to group designs into similar families, pick the best and then use PCCs to rank the best across families. In other words, we need to be sure that we are not evaluating inferior alternatives from one class of design along with the best options from that class and from other classes. This suggests that PCCs should be applied hierarchically to avoid artificial spacing in the Borda count among design alternatives. In early design, it is too costly to acquire quantitative measures of performance that can indicate how much better one alternative is than another. By grouping alternatives into families, we can lessen the chance that alternatives that are actually quite close to each other in performance will appear far apart due to the sheer number of alternatives deemed to fall in the middle.

It is also worth noting that rank reversals of any two alternatives can be "cooked" by adding enough irrelevant alternatives to a Borda count. This follows directly from the fact that the Borda count depends upon the number of alternatives *between* two alternatives, as does its PCC equivalent. Consider the following case.

TABLE 13.3 A PAIRWISE COMPARISON CHART (PCC) FOR THE PERSONAL VOTES CAST BY 30 DESIGNERS CHOOSING AMONG FIVE CANDIDATES *A*, *B*, *C*, *D* AND *E* [SEE EQS. (13.14a-13.14c)]

Win/Lose	A	B	C	D	E	Sum/Win
A	—	10 + 0 + 10	10 + 0 + 0	10 + 0 + 0	10 + 0 + 0	**50**
B	0 + 10 + 0	—	10 + 10 + 0	10 + 10 + 0	10 + 10 + 0	**70**
C	0 + 10 + 10	0 + 0 + 10	—	10 + 10 + 10	10 + 10 + 10	**90**
D	0 + 10 + 10	0 + 0 + 10	0 + 0 + 0	—	10 + 10 + 10	**60**
E	0 + 10 + 10	0 + 0 + 10	0 + 0 + 0	0 + 0 + 0	—	**30**
Sum/lose	**70**	**50**	**30**	**60**	**90**	**—**

TABLE 13.5 A REDUCED PAIRWISE COMPARISON CHART (PCC) WHEREIN THE "LOSER" *E* IN THE FIRST RANKING IN TABLES 13.3 AND 13.4 IS REMOVED FROM CONSIDERATION (NOTE: RANK ORDER IS PRESERVED FOR THE TOP TWO DESIGNS)

Win/Lose	*A*	*B*	*C*	*D*	Sum/Win
A	—	10 + 0 + 10	10 + 0 + 0	10 + 0 + 0	**40**
B	0 + 10 + 0	—	10 + 10 + 0	10 + 10 + 0	**50**
C	0 + 10 + 10	0 + 0 + 10	—	10 + 10 + 10	**60**
D	0 + 10 + 10	0 + 0 + 10	0 + 0 + 0	—	**30**
Sum/lose	**50**	**40**	**30**	**60**	**—**

There are $n + 1$ alternatives and $m + 1$ voters. Alternative A is ranked first (n points) and alternative B last (0 points) by m voters, while the remaining voter casts B as second-to-last (1 point) and A as last (0 points). Thus, A has $m \times n$ points and B has 1. It is clear that it doesn't really matter what the absolute rankings are: A has gotten n more points than B from m voters and B has gotten 1 *more* than A on the last criterion—as far apart as the two alternatives can be without having A dominate B. Suppose new alternatives are added. Any new alternative that is either better than both A and B or worse than both will not affect the ranking of A and B. However, if a new alternative falls between A and B, the relative ranking will change. Therefore, if we find $m \times n$ new alternatives that are more or less preferred than both A and B by the original m voters that favor A, but that fall between B and A for the last voter, we can change the aggregated scores to $m \times n$ for A and $(m \times n) + 1$ for B. Thus, again, we have changed the aggregate scores by (artificially) introducing a large number ($m \times n$) of irrelevant "ringers."

Perhaps one of the main points of all of the above discussion is that the tool that should be used to rank or to calculate aggregate demand depends on how much data is available, with what granularity and on how much the data gatherers are prepared to spend. Pairwise comparisons are cheap and require little detailed knowledge, and are thus invaluable in conceptual design. Focusing on the best candidates or exemplars in a set introduces a certain granularity in the data, which can help avoid IIA-induced rank reversals. Alternatives that fit an existing group don't earn a separate, distinguishable space in the PCC, and the spacing between different alternatives is less likely to be "padded" by alternatives that are actually quite close in performance.

13.5 RESPECT THE DATA

We now present an example that shows how pairwise ranking that does not consider other alternatives can lead to a result exactly opposite to a Borda count, which does consider other alternatives. It also indicates that attempting to select a single best alternative may be the wrong approach.

One hundred (100) customers are "surveyed on their preferences" with respect to five mutually exclusive design alternatives, A, B, C, D and E [15]. The survey reports that "45 customers prefer A, 25 prefer B, 17 prefer C, 13 prefer D and no one prefers E." These data seem to indicate that A is the preferred choice, and that E is entirely "off the table."

However, as reported, these results assume either that the customers are asked to list only one choice or, if asked to rank order all five designs, that only their first choices are abstracted from their rank orderings. Suppose that the 100 customers were asked for rankings and that those rankings are [15]:

$$45 \text{ preferred } A \succ E \succ D \succ C \succ B \qquad \text{Eq. (13.18a)}$$

$$25 \text{ preferred } B \succ E \succ D \succ C \succ A \qquad \text{Eq. (13.18b)}$$

$$17 \text{ preferred } C \succ E \succ D \succ B \succ A \qquad \text{Eq. (13.18c)}$$

$$13 \text{ preferred } D \succ E \succ C \succ B \succ A \qquad \text{Eq. (13.18d)}$$

Again, the procedure used to choose among the rank orderings of these five designs can decidedly influence or alter the results. For example, if A and B are compared as a (single) pair, B beats A by a margin of 55 to 45. And, continuing a sequence of pairwise comparisons, we can find that:

$$A \text{ vs } B \to B;\ B \text{ vs } C \to C;\ C \text{ vs } D \to D;\ D \text{ vs } E \to E \qquad \text{Eq. (13.19)}$$

Equation (13.19) provides an entirely different outcome, one that is not at all apparent from the vote count originally reported. How do we sort out this apparent conflict?

We resolve this dilemma by constructing a PCC matrix for this five-product example, as shown in Table 13.6, and whose results clearly indicate the order of preferred designs to be:

$$E \succ D \succ A \succ C \succ B \qquad \text{Eq. (13.20)}$$

A Borda count of the same data [of Eq. (13.18)], using the weights (4, 3, 2, 1, 0) for the place rankings, confirms the PCC results, with the Borda count numbers being identical to those in the "win" column of the PCC in Table 13.6, that is:

$$E(300) \succ D(226) \succ A(180) \succ C(164) \succ B(130) \qquad \text{Eq. (13.21)}$$

In this case, removing B and revoting generates a relatively unimportant rank reversal between A and C, thus demonstrating the meaning of IIA and showing that dropping information can have consequences.

This example is one where the "best option" as revealed by the PCC/Borda count is not the most preferred by anyone. Is the PCC lying to us? In a real market situation, where all five options are available, none of the surveyed customers would buy E. Two explanations for this survey come to mind: First, this data could have been collected across too broad a spectrum of customers in a segmented market in which design E is something of a "common denominator"; the other four designs respond better to four disparate market "niches." Under this explanation, there is really no

TABLE 13.6 A PAIRWISE COMPARISON CHART (PCC) FOR THE PREFERENCES EXPRESSED IN A SURVEY OF 100 CUSTOMERS CHOOSING AMONG FIVE CANDIDATE PRODUCTS A, B, C, D AND E [SEE EQ. (18)]

Win/Lose	*A*	*B*	*C*	*D*	*E*	Sum/Win
A	—	45 + 0 + 0 + 0	45 + 0 + 0 + 0	45 + 0 + 0 + 0	45 + 0 + 0 + 0	**180**
B	0 + 25 + 17 + 13	—	0 + 25 + 0 + 0	0 + 25 + 0 + 0	0 + 25 + 0 + 0	**130**
C	0 + 25 + 17 + 13	45 + 0 + 17 + 13	—	0 + 0 + 17 + 0	0 + 0 + 17 + 0	**164**
D	0 + 25 + 17 + 13	45 + 0 + 17 + 13	45 + 25 + 0 + 13	—	0 + 0 + 0 + 13	**226**
E	0 + 25 + 17 + 13	45 + 0 + 17 + 13	45 + 25 + 0 + 13	45 + 25 + 17 + 0	—	**300**
Sum/lose	**220**	**270**	**236**	**174**	**100**	**—**

"best design," although *E* seems to be a good starting point from which to search. Unfortunately, there is also no identifiable "worst design," although one could also argue that *E* is the "worst."

A second explanation is that these designs are all extremely close to each other in performance, so that small variations in performance have translated into large differences in the PCC. If this is the case, a designer might try to generate new design options by better merging the apparent desires of consumers. Methods such as the *House of Quality* require that designs be ranked along several significant (and possibly linguistic or non-quantifiable) performance criteria [3, 5]. The goal in such a process shifts from selecting the "best" design to identifying the characteristics of a composite, winning design. Of course, there is no guarantee that such a winning composite design exists, but PCCs can help the ranking process that might lead to its generation.

Both of the above explanations point to the need to integrate the PCC into a hierarchy of design decision methods. Deciding just when the PCC should give way to more information-rich methods is perhaps the main problem in this task. The PCC shown in Table 13.6 shows strong support for option *E*, yet we have argued that more information should be developed before a design is selected. Inconclusive results generated by the PCC are generally easy to detect and can be corrected by moving to a more detailed selection method. While such graceful degradation of performance is typical of the PCC in practice, the above example, unfortunately, is of a case in which the PCC yields clear selection results at a point where more detailed selection procedures might be more appropriate.

13.6 ON PAIRWISE COMPARISONS AND MAKING DECISIONS

The structured PCC—an implementation of the Borda count—can support consistent decision-making and choice, notwithstanding concerns raised about pairwise comparisons and violations of Arrow's theorem. Rank reversals and other infelicities do result when "losing" alternatives are dropped from further consideration. But simulation suggests that such reversals are limited to alternatives that are nearly indistinguishable [26]. Pairwise comparisons that are properly aggregated in a PCC produce results that are identical to the Borda count, which in Saari's words [24] is a "unique positional procedure which should be trusted."

Practicing designers use the PCC and similar methods very early in the design process where rough ordinal rankings are used to bound the scope of further design work. The PCC is more of a discussion tool than a device intended to aggregate individual orderings of design team members into a "group" decision. Indeed, design students are routinely cautioned against overinterpreting or relying too heavily on small numerical differences. In political voting, we usually end up with only one winner, and any winner *must* be one of the entrants in the contest. In early design, it is perfectly fine to keep two or more winners around, and the ultimate winner often does not appear on the initial ballot. Indeed, it is often suggested that designers look at *all* of the design alternatives and try to incorporate the good points of each to create an improved, composite design. In this framework, the PCC is a useful aid for understanding the strengths and weaknesses of individual design alternatives. Still, pairwise comparison charts should be applied carefully and with restraint. As noted above, it is important to cluster similar choices and to perform the evaluations at comparable levels of detail.

In addition, given the subjective nature of these rankings, when we use such a ranking tool, we should ask *whose* values are being assessed. Marketing values are easily included in different rankings, as in product design, for example, where a design team might need to know whether it's "better" for a product to be cheaper or lighter. On the other hand, there might be deeper issues involved that, in some cases, may touch upon the fundamental values of both clients and designers. For example, suppose two competing companies, GRAFT and BJIC, are trying to rank order design objectives for a new beverage container. We show the PCCs for the GRAFT- and BJIC-based design teams in Tables 13.7. It is clear from these two charts and the scores in their right-hand columns that the GRAFT designers were far more interested in a container that would generate a strong brand identity and be easy to distribute than in it being environmentally benign or having appeal for parents. At BJIC, on the other hand, the environment and taste preservation ranked more highly, thus demonstrating that subjective values show up in PCCs and, eventually, in the marketplace!

It is also tempting to take our *ranked* or ordered objectives and put them on a *scale* so that we can manipulate the rankings in order to attach relative weights to goals or to do some other calculation. It would be nice to be able to answer questions such as: "*How much more* important is portability than cost in a ladder?" Or, in the case of a beverage container, "*How much more* important is environmental friendliness than durability?" A little more? A lot more? Ten times more? We can easily think of cases where one of the objectives is substantially more important than any of the others, such as safety compared to attractiveness or to cost in an air traffic control system, and other cases where the objectives are essentially very close to one another. However, and sadly, there is no mathematical foundation for normalizing

TABLE 13.7 USING PCCs TO RANK-ORDER DESIGN OBJECTIVES AT TWO DIFFERENT COMPANIES DESIGNING NEW BEVERAGE CONTAINERS (*AFTER* [7])

(A) Graft's Weighted Objectives

Goals	Environ Benign	Easy to Distribute	Preserve Taste	Appeals to Parents	Market Flexibility	Brand ID	Score
Environ. benign	••••	0	0	0	0	0	0
Easy to distribute	1	••••	1	1	1	0	4
Preserve taste	1	0	••••	0	0	0	1
Appeals to parents	1	0	1	••••	0	0	2
Market flexibility	1	0	1	1	••••	0	3
Brand ID	1	1	1	1	1	••••	5

(B) BJIC's Weighted Objectives

Goals	Environ. Benign	Easy to Distribute	Preserve Taste	Appeals to Parents	Market Flexibility	Brand ID	Score
Environ. benign	••••	1	1	1	1	1	5
Easy to distribute	0	••••	0	0	1	0	1
Preserve taste	0	1	••••	1	1	1	4
Appeals to parents	0	1	0	••••	1	1	3
Market flexibility	0	0	0	0	••••	0	0
Brand ID	0	1	0	0	1	••••	2

the rankings obtained with tools such as the PCC. The numbers obtained with a PCC are *approximate*, *subjective* views or judgments about relative importance. We must not inflate their importance by doing further calculations with them or by giving them unwarranted precision.

13.7 CONCLUSIONS

We have argued that design *is not* decision-making per se. The notion that design *is* decision-making succumbs to the same shortfalls that are giving birth to the emergence of behavioral economics and a decline in neoclassical economics. Simply put, decision theory does not offer us:

- Good models of how people compare and evaluate alternatives
- Usable models of how people actually make decisions
- Guidance on how to generate design alternatives.

On the other hand, decisions are an important part of design that must be made with as much care as possible. We have demonstrated that effective decision-making is possible in the practice of engineering design, notwithstanding concerns raised about pairwise comparisons and Arrow's Impossibility Theorem. The identification of the structured PCC as an implementation of the well-known Borda count and its application to oft-cited "pathological" examples suggests several ideas. First, the individual pairwise comparisons do not lead to erroneous results. Rather, rank reversals and other infelicities result from serial aggregation of pairwise comparisons when "losing" alternatives are dropped from further consideration. Pairwise comparisons that are properly aggregated in a PCC produce results that are identical to the Borda count, a "unique positional procedure which should be trusted" [24]. Indeed, our proof that the PCC is identical to the Borda count confirms that it compensates for and removes the same inherent cancellations.

It is important to recall that, in practice, the PCC and similar methods are used very early in the design process where rough ordinal rankings are used to bound the scope of further development work. The PCC is more of a discussion tool than a device intended to aggregate individual orderings of design team members into a "group" decision. Indeed, design students are routinely cautioned against overinterpreting or relying too heavily on small numerical differences (p. 113 of [7]). In voting, we usually end up with only one winner, and any winner *must* be one of the entrants in the contest. In early design, it is perfectly fine to keep two or more winners around, and the ultimate winner often does not appear on the initial ballot. Indeed, it is often suggested that designers look at *all* of the design alternatives and try to incorporate the good points [3, 5, 7, 27] of each to create an improved, composite design. In this framework, the PCC is a useful aid for understanding the strengths and weaknesses of individual design alternatives, holistically or along more detailed performance criteria.

PCCs can be used not only to rank designs, but to order design criteria by importance. This information helps structure other design selection methods (e.g., Pugh concept selection [3]), showing the design team where comparative differences among candidate designs are most important. This emphasis on *team* is significant. PCCs that implement the Borda count by having individuals vote in the pairwise comparisons are useful in the design process. However, they are most useful for encouraging student design teams to work on designs *as a team*. True collaboration takes place when team members must reach consensus on each comparison. The discussion necessary to reach this consensus helps foster the shared understanding that is so important for good design. This collaborative approach might not be relevant to a social choice framework. In design and design education, however, where we are encouraged (and able) to improve design alternatives midstream, fostering constructive discussion is a significant reason for using any structured design approach. Thus, the matrix format of the PCC is perhaps a more useful tool in design education and design practice for conveying the same results obtained with the Borda count implemented as a piece of formal mathematics.

ACKNOWLEDGMENTS

This chapter is an extended version of the paper [9] "Rank Ordering Engineering Designs: Pairwise Comparison Charts and Borda Counts," which appeared in *Research in Engineering Design*. It was extended with permission of Springer-Verlag. We are also grateful to Elsevier Academic Press for permission to use problems from Section 9.4 of [10], *Principles of Mathematical Modeling*.

REFERENCES

1. Dieter, G. E., 1983. *Engineering Design: A Materials and Process Approach*, McGraw-Hill, New York, NY.
2. Rowe, P. G., 1987. *Design Thinking*, MIT Press, Cambridge, MA.
3. Pugh, S., 1990. *Total Design: Integrated Methods for Successful Product Engineering*, Addison-Wesley, Wokingham, U.K.
4. Pugh, S., 1996. "Concept Selection: A Method that Works," *Creating Innovative Products Using Total Design,* D. Clausing and R. Andrade, (eds.) Addison-Wesley, Reading, MA.
5. Ulrich, K. T. and Eppinger, S. D., 2000. *Product Design and Development, 2nd Ed.*, McGraw-Hill, Boston, MA.
6. Otto, K. N. and Wood, K. L., 2001. *Product Design: Techniques in Reverse Engineering and New Product Development*, Prentice-Hall, Upper Saddle River, NJ.
7. Dym, C. L. and Little, P., 2004. *Engineering Design: A Project-Based Introduction, 2nd Ed.*, John Wiley, New York, NY.
8. Radford, A. D. and Gero, J. S., 1985. "Multicriteria Optimization in Architectural Design," *Design Optimization,* J. S. Gero, ed., Academic Press, Orlando, FL.
9. Dym, C. L., Wood, W. H. and Scott, M. J., 2002. "Rank Ordering Engineering Designs: Pairwise Comparison Charts and Borda Counts," *Res. in Engrg. Des.*, Vol. 13, pp. 236–242.
10. Dym, C. L., 2004. *Principles of Mathematical Modeling, 2nd Ed.* Elsevier Academic Press, New York, NY.
11. Dym, C. L., Agogino, A. M., Eris, O., Frey, D. D. and Leifer, L. J., 2005. "Engineering Design Thinking, Teaching, and Learning," *J. of Engrg. Edu.*
12. Hazelrigg, G. A., 1999. "An Axiomatic Framework for Engineering Design," *J. of Mech. Des.*, Vol. 121, pp. 342–347.
13. von Neumann, J. and Morgenstern, O., 1947. *Theory of Games and Economic Behavior,* Princeton University Press, Princeton, NJ.
14. Hazelrigg, G. A., 1996. *Systems Engineering: An Approach to Information-Based Design*, Prentice Hall, Upper Saddle River, NJ.
15. Hazelrigg, G. A., 2001. "Validation of Engineering Design Alternative Selection Methods," unpublished manuscript, courtesy of the author.
16. Howard, R. A., 1988. "Decision Analysis: Practice and Promise," *Mgmt. Sci.*, Vol. 34, pp. 679–695.
17. *www.sfb504.uni-mannheim.de/glossary/behave.htm* 2004. accessed August 10. 2004.
18. Simon, H. A., 1987(a). "Behavioral Economics," *The New Palgrave: A Dictionary of Economics*, J., Eatwell, M. Millgate and P. Newman, eds., Macmillan, London and Basingstoke, U.K.
19. Simon, H. A., 1987(b). "Bounded Rationality," *The New Palgrave: A Dictionary of Economics,* J., Eatwell, M. Millgate and P. Newman, eds., Macmillan, London and Basingstoke, U.K.
20. Simon, H. A., 1996. *The Sciences of the Artificial, 3rd Ed.*, MIT Press, Boston, MA.
21. Arrow, K. J., 1951. *Social Choice and Individual Values, 1st Ed.*, John Wiley, New York, NY.
22. Scott, M. J. and Antonsson, E. K., 1999. "Arrow's Theorem and Engineering Decision Making," *Res. in Engrg. Des.*, Vol. 11, pp. 218–228.
23. Saari, D. G., 1995. *Basic Geometry of Voting,* Springer-Verlag, New York, NY.
24. Saari, D. G., 2001. "Bad Decisions: Experimental Error or Faulty Decision Procedures," unpublished manuscript, courtesy of the author.
25. Saari, D. G., 2001. *Decisions and Elections: Explaining the Unexpected*, Cambridge University Press, New York, NY.
26. Scott, M. J. and Zivkovic, I., 2003. "On Rank Reversals in the Borda Count," *Proc., 2003 ASME Des. Engrg. Tech. Conf.*, Chicago, IL, p. 378.
27. Pahl, G. and Beitz, W., 1996. *Engineering Design: A Systematic Approach*, Springer-Verlag, London, U.K.

PROBLEMS

13.1 Are there election procedures that violated Arrow's third axiom that you would find offensive? Explain your answer.

13.2 Would an election procedure that violated the Pareto condition, Arrow's fourth axiom, be offensive to you? Explain your answer.

13.3 Engineering designers often use quantified performance rankings to compare alternatives on the basis of measurable criteria. If this comparison were done on a pairwise basis, would it violate Arrow's fourth axiom? Explain your answer.

13.4 Defend or refute the proposition that ranking criteria that are of the less-is-better, more-is-better or nominal-is-best varieties will violate Arrow's first axiom. (*Hint*: Are all theoretically possible orders admissible in practice?)

13.5 Verify the ordering of the five alternatives displayed in Eq. (13.16) by performing the appropriate individual pair-by-pair comparisons.

13.6 Construct a PCC of the data presented in Eq. (13.18) and confirm the Borda count results given in Eq. (13.21).

13.7 Using the weights (4, 3, 2, 1, 0), perform a Borda count of the preferences expressed in Eq. (13.18) and confirm the results obtained in Eq. (13.21) and in the previous problem.

CHAPTER

14

MULTI-ATTRIBUTE DECISION-MAKING USING HYPOTHETICAL EQUIVALENTS AND INEQUIVALENTS

Tung-King See, Ashwin Gurnani, and Kemper Lewis

In this chapter, the problem of selecting from among a set of alternatives using multiple, potentially conflicting criteria is presented. The theoretical and practical flaws with a number of commonly employed methods in engineering design are first presented by demonstrating their strengths and weaknesses using an aircraft selection problem. With the same aircraft example, this chapter presents the concept of hypothetical equivalents and inequivalents method (HEIM) that utilizes the strength of those commonly employed methods to make a selection decision in multi-attribute decision-making. Finally, the visualization techniques, coupled with an indifferent point analysis, are then used to understand the robustness of the solution and determine the appropriate additional constraints to identify a single robust optimal solution.

14.1 INTRODUCTION

There are always trade-offs in decision-making. We have to pay more for better quality, carry around a heavier laptop if we want a larger display or wait longer in a line for increased airport security. More specifically, in engineering design, we can be certain that there is no one alternative that is best in every dimension. Therefore, how to make the "best" decision when choosing from among a set of alternatives in a design process has been a common problem in research and application in engineering design. When the decision is multi-attribute in nature, common challenges include aggregating the criteria, rating the alternatives, weighting the attributes and modeling strength of preferences in the attributes. In recent years, decision-based design (DBD) has proposed that decisions such as these are a fundamental construct engineering design [1–3].

In general, the multi-attribute decision problem can be formulated as follows:

Choose an alternative i;

maximize $$V_i = \sum_{j=1}^{n} w_j r_{ij} \quad \text{Eq. (14.1a)}$$

subject to $$\sum_{j=1}^{n} w_j = 1 \quad \text{Eq. (14.1b)}$$

where V = value function for alternative i, w = weight for attribute j; and r = normalized score for alternative i on attribute j. There are many ways to implement and solve this formulation. Most methods focus on formulating the attribute weights w_j and/or the alternative scores r_{ij} indirectly or directly from the decision-maker's preferences. For instance, for a set of vehicle alternatives whose attributes include miles-per-gallon (mpg), the mpg rating for one of the vehicles would simply be the vehicle's mpg value, normalized between 0 and 1 using the highest and lowest mpg values of all the candidate vehicles. In new product development, a common challenge in a design process is how to capture the preferences of the end-users while also reflecting the interests of the designer(s) and producer(s). Typically, preferences of end-users are multidimensional and multi-attribute in nature. If companies fail to satisfy the preferences of the end-user, the product's potential in the marketplace will be severely limited. For example, the Ford Motor Company selected and introduced the Edsel and lost more than $100 million. General Motors was forced to abandon its Wankel Rotary Engine after more than $100 million had been invested in the project [4]. At some point in Ford's and GM's design process, the decision of selecting these concepts was deemed to be sound and effective. However, good decisions that are successful have also been made. For example, Southwest Airline's decision to only select the 737 aircraft for the entire fleet was excellent, as it lowered the maintenance and training costs. While the specific process used by these companies to make these selection decisions is not in the scope of this chapter, it hypothesizes that perhaps the process being used to make selection decisions impacts the outcome more than the information used in the decision. In fact, studies have shown this to be true, as when the number of alternatives approaches seven, the process used to make the decision influences the outcome 97% of the time [5]. In addition, it is difficult to evaluate the value of a decision based on the outcome itself. Rather, the process being used should be used as the evaluation and validation standard [6]. In this chapter, an attempt is made to demonstrate the effect of a decision process on the outcome and present a method that facilitates the practical selection from among a set of alternatives using theoretically sound decision theory principles.

In the next sections, a simple example is used to present the strengths and weaknesses of common decision-making processes: pairwise comparison, ranking, rating/normalization, strength of preferences and the weighted sum method. Then, the method of hypothetical equivalents and inequivalents (HEIM) is presented. The latter half of the chapter investigates the robustness of the solution using visualization.

14.2 MULTI-ATTRIBUTE DECISION METHODS

In this section, a number of common approaches are used to solve the following multi-attribute decision problem. For illustration purposes, suppose a fictional airline carrier, Jetair, is planning to establish an air fleet to serve the routes on major cities among Asia Pacific countries and the United States. Jetair has decided to purchase only one type of aircraft for its entire fleet to reduce operating cost, similar to the strategy used by Southwest Airlines and Jetblue Airways [7]. At this point, Jetair has identified four possible choices that meet Jetair's requirements and budget constraints: Boeing 777–200 (long range), Boeing 747–200, Airbus 330–200 and Airbus 340–200. After reflecting upon the appeal of each of the four aircraft, Jetair has identified three key attributes:

(1) The number of passengers the plane can hold, which obviously reflects revenue for each flight.
(2) The cruise range, where a longer cruise range will provide passengers with nonstop service.
(3) The cruise speed, where a faster cruise speed means shorter times needed for each flight. Potentially, this could increase the frequency of turnaround times.

In Table 5.1, the data of the three attributes for the four aircrafts [8–9] are given. This problem is simplistic and is not meant to be realistic about how airliners choose which aircraft to purchase. It is meant to illustrate the practical and theoretical advantages and disadvantages when using common decision-making methods to make selection decisions from among a set of alternatives in a multicriteria environment.

14.2.1 Pairwise Comparisons

Jetair first uses a pairwise comparison to make its decision, first comparing B777 with B747 attribute by attribute, and then choosing the aircraft that "wins" on the most attributes. This process is repeated, taking the "winner" of the previous comparison and comparing it with the next alternative. This process is similar to any kind of tournament approach to determine the winner from among many competitors. More generally, the pairwise comparison method takes two alternatives at a time and compares them to each other. A pairwise approach is used in the analytic hierarchy process (AHP) to find relative importances among attributes [10]. Adaptations of AHP and other pairwise methods are widely used to obtain relative attribute importances [11], to select from competing alternatives [12], as well as to aggregate individual preferences [13, 14].

Ordinal-scale comparison is used in this problem. Thus, the B747 is better than the B777 because B747 has a faster maximum speed and greater passenger capacity. So, the B747 is then compared to the A330 and is preferred because of a longer cruise range and greater passenger capacity. However, the A340 is preferred over the B747 because of its greater speed and cruise range. Thus, Jetair concludes that the A340 is the superior aircraft for its needs. However, if Jetair compares the A340 with the B777, B777 is the preferred aircraft. Thus, Jetair's decision process will produce the following rankings, where "$\succ$" indicates "preferred to":

$$B747 \succ B777 \succ A340 \succ B747$$

TABLE 14.1 ATTRIBUTE DATA FOR AIRCRAFT ALTERNATIVES

	Attribute		
Aircraft	Speed (Mach)	Max. Cruise Range (nmi)	No. of Passengers
B777–200	0.84	8,820	301
B747–200	0.85	6,900	366
A330–200	0.85	6,650	253
A340–200	0.86	8,000	239

which is a set of intransitive preferences that will lead to decision cycling [15]. There are two fundamental flaws in this method:

- It ignores strength of preference: suppose aircraft E is just a little better than aircraft F on two out of three attributes, but much worse on the third attribute. Clearly, most airliners would disregard aircraft E, but pairwise comparisons ignore this information.
- This procedure ignores the relative important of the attributes: in AHP, pairwise comparisons are used to find relative importances, but then the problems with pairwise comparisons to choose from among alternatives only increase.

Further details regarding the theoretical problems with pairwise comparisons can be found in [5, 16, 17]. In the next section, a ranking method is used to make the same decision.

14.2.2 Ranking of Alternatives

Rankings are commonly used to rank order a set of alternatives. *U.S. News and World Report* annually ranks colleges based upon a number of attributes [18]. The NCAA athletic polls are based on a ranking system. Compared with pairwise methods, ranking methods are slightly more elaborate. However, ranking methods still make limiting assumptions and are limited in their applicability to engineering design.

Suppose Jetair uses the data from Table 14.1 and ranks the alternatives with respect to each attribute. Jetair assigns four points for the top-ranked alternative for a given attribute, three points for second, two points for third and one point for the worst. For a tie, Jetair averages the points. Table 14.2 presents the results of this procedure.

The preferred aircraft using this method is B747 with 8.5 points; while B777 and A340 follow closely behind it with 8 points; A330 is clearly a *noncontender*. Noncontenders are alternatives that are *equal to or worse than* at least one other alternative with respect to every attribute. Therefore, the A330 alternative can be dropped from consideration, since it should never be picked. Making the rational decision to drop the A330 from contention, the resulting rankings are shown in Table 14.3.

As shown in Table 14.3, all three alternatives are tied. There is no clear preferred aircraft. This outcome has demonstrated that

TABLE 14.2 RESULTS OF RANKING PROCEDURE

	Attribute			
Aircraft	Speed (Mach)	Max Range (nmi)	No. of Passengers	Total Score
B777–200	1	4	3	8
B747–200	2.5	2	4	8.5
A330–200	2.5	1	2	5.5
A340–200	4	3	1	8

TABLE 14.3 RANKINGS WITHOUT THE A330

Aircraft	Attribute: Speed (Mach)	Max. Range (nmi)	No. of Passengers	Total Score
B777–200	1	3	2	6
B747–200	2	1	3	6
A340–200	3	2	1	6

TABLE 14.4 NORMALIZED ALTERNATIVE SCORES

Aircraft	Attribute: Speed (Mach)	Max Range (nmi)	No. of Passengers	Total Score
B777–200	0	100	48.8	148.8
B747–200	50	11.5	100	161.5
A330–200	50	0	11	61.0
A340–200	100	62.2	0	162.2

the ranking procedure has violated the independence of irrelevant alternatives (IIA) principle, which states that "*the option chosen should not be influenced by irrelevant alternatives or clear non-contenders*" [19]. If a noncontender exists, it would never be rational to choose this alternative.

Further, although it is not shown here, noncontenders can be chosen to make *any* of the alternatives (except A330) win. Ranking methods, while violating the IIA principle, also assume linear preference strengths. That is, the difference between first and second place is the same as the difference between fourth and fifth and so on. In the next section, a rating procedure is used for the same problem.

14.2.3 Normalization Rating

When aggregating attributes that have different units of measure, normalization is a common way to eliminate dimensions from the problem. Since in the problem of Table 14.1 the dimensions for all three attributes are different, normalization could certainly convert these attributes into a dimensionless scale, so they can be aggregated.

Assume a simple linear method to normalize the aircraft attribute data into a scale from 0 to 100, where 0 is assigned to the worst value and 100 to the best value, as shown in the following:

Speed (mach):	0.84 = 0 points
	0.86 = 100 points
Range (nmi):	6,650 = 0 points
	8,820 = 100 points
No. of passengers	239 passengers = 0 points
	366 passengers =100 points

The intermediate values for each attribute are calculated using linear interpolation. Table 14.4 shows the normalized scale for the example.

It is now possible to sum the individual ratings for each alternative since all the attributes are on the same scale. By doing this, A340 is determined to be the preferred aircraft.

As opposed to a ranking procedure, normalization rating does satisfy the IIA principle because the noncontenders do not affect the relative scores. However, these normalized values depend on the relative position of the attributes value within the range of values. The lack of a rigorous method to determine the normalizing range leads to paradoxes [3]. Further, this procedure still neglects the strength of preference within each attribute. Ignoring the strength of preferences can lead to a result that does not reflect the decision-maker(s) preferences. In addition, relative importances of the attributes are not used. While weights could certainly be assigned to each attribute (in Table 14.4 it is assumed that all the weights are equal) and then used to determine the final score, this creates further complications as shown in the next section.

14.2.4 Strength of Preferences and Weighted Sums

Using a linear preference scale may not truly reflect a decision-maker's preferences. Jetair would be better off using a nonlinear strength of preference representation, better reflecting its true preferences. In this chapter, simple assumptions are made for illustration purposes. For the cruise speed, assume that an increase from 0.85 to 0.86 is preferred to an increase from 0.84 to 0.85. For the aircraft range, assume an increase from 6,500 to 7,000 nmi is preferred over an increase from 8,000 to 9,000 nmi (because if the cruise range is less than 7,000 nmi, the aircraft may have to make multiple stops for refueling). For the number of passengers assume that an increase from 290 to 340 is slightly preferred over an increase from 240 to 290. There are a number of ways to assess the strength of preferences, including utility theory methods [3, 20, 21]. These strength of preferences are shown, in Figure 14.1(*a*), (*b*) and (*c*), respectively.

Table 14.5 shows the numerical values for each attribute according to these strength of preference functions as well as the aggregation of scores for each alternative. Here, B747 is the winner with 185 points and is followed closely by A340 with 180 points.

Even though using strength of preferences more accurately represents decision-makers' preferences and does not violate the IIA principle, determining the relative importance of the attributes is largely an arbitrary process. This arbitrary process can create a number of complications in multi-attribute decision-making and optimization [22–25], some of which are discussed here.

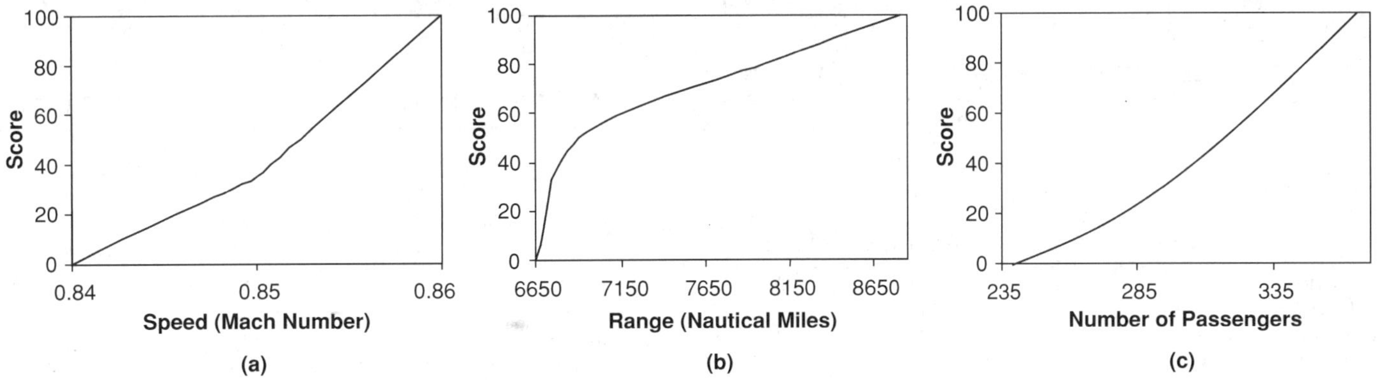

FIG. 14.1 STRENGTH OF PREFERENCE FOR (A) CRUISE SPEED; (B) RANGE; AND (C) NO. OF PASSENGERS

TABLE 14.5 STRENGTH OF PREFERENCE ASSESSMENTS

Aircraft	Attribute: Speed (Mach)	Max Range (nmi)	No. of Passengers	Total Score
B777–200	0	100	35	135
B747–200	35	50	100	185
A330–200	35	0	5	40
A340–200	100	80	0	180

First, suppose that Jetair has decided that cruise range is the most important attribute, followed by the numbers of passengers and then the speed. Therefore, Jetair has decided to use the weights, 0.1, 0.6, and 0.3, respectively, for speed, range and passengers. Using these weights and Eq. (14.1), the B777 aircraft is determined to be the winner, as shown in Table 14.6. Note that the preference strengths shown in Table 14.5 are also used here.

Suppose that some time later (maybe even after the first decision has been made), Jetair has decided that the number of passengers is the most important attribute and not the cruise range, or decided to use a moderate set of weights. Undeniably, a different set of weights lead to different preferred aircraft as shown in Table 14.7.

As shown in Tables 14.6 and 14.7, different sets of weight can lead to very different results. This dependence on a largely arbitrary assessment of weights that can fluctuate is the primary drawback of using any method where weights are chosen not using strict decision theory principles [26]. In the next section, a more rigorous method, called hypothetical equivalents, to find a theoretically correct set of weights based upon a decision-maker's stated preferences is discussed. This method is applied to the aircraft selection problem.

14.2.5 Hypothetical Equivalents

The hypothetical equivalents approach determines the attribute weights using a set of preferences rather than selecting weights arbitrarily based on intuition or experience. While first encountered in the management literature [27], in this chapter it is developed and expanded for design decisions. The approach is based on developing a set of *hypothetical alternatives* that the decision-maker is *indifferent between*. In other words, it is based on identifying hypothetical alternatives that have equal value to the decision-maker. These indifference points are then used to analytically solve for the theoretically correct set of attribute weights. The approach is best illustrated through use of an example: Suppose that Jetair felt uncomfortable assessing weights directly, and therefore, it started

TABLE 14.6 RESULTS FOR WEIGHTS OF [0.1, 0.6, 0.3]

Aircraft	Attribute & Weights: 0.1 Speed (Mach)	0.6 Max. Range (nmi)	0.3 No. of Passengers	Value
B777–200	0	100	35	70.5
B747–200	35	50	100	63.5
A330–200	35	0	5	5
A340–200	100	80	0	58

TABLE 14.7 RESULTS FOR VARIOUS WEIGHT COMBINATIONS

Attribute Weights (Speed, Range, No. of Passengers)	Preferred Aircraft
(0.2, 0.2, 0.6)	B747
(0.3, 0.4, 0.3)	A340

by considering a number of hypothetical choices. These hypothetical choices can be developed by the decision-maker in order to meet the indifference requirement and are shown in Table 14.8 for this problem.

Assume that Jetair is indifferent between aircraft *A* and *B*. That is both aircraft are equivalent to them and it would not matter which one they chose. Based on the strength of preferences that are used in Section 14.2.4, aircraft *A* is at the bottom of the range on both speed and range, but at the top in terms of number of passengers. Aircraft *B* is at the bottom on range and number of passengers, but at the top in terms of speed.

Therefore, by saying they are indifferent between aircraft *A* and aircraft *B*, the total value (represented by the total score in Table 14.9) must be equal, which gives Eq. (14.2):

$$w_1 = w_3 \qquad \text{Eq.(14.2)}$$

Since there are three attributes, three weights must be solved for. This requires three equations, Eq. (14.2) being one of them. Another equation is generated from the fact that the weights are normalized and sum to one:

$$w_1 + w_2 + w_3 = 1 \qquad \text{Eq. (14.3)}$$

Therefore, one more indifference point must be found in order to generate the third equation. Assume that Jetair is indifferent between aircrafts *C* and *D*. Using the strength of preferences in Section 14.2.4, the total scores for each aircraft are shown in Table 14.9. This indifference point results in the following equation:

$$100w_1 + 50w_2 + 100w_3 = 100w_2,$$

or,

$$2(w_1 + w_3) = w_2 \qquad \text{Eq. (14.4)}$$

Together, solving Eqs. (14.2), (14.3) and (14.4) give

$$w_1 = 1/6;\ w_2 = 2/3;\ \text{and } w_3 = 1/6$$

TABLE 14.8 HYPOTHETICAL AIRCRAFT CHOICES

Aircraft	Attribute & Weights: w_1 Speed (Mach)	w_2 Max. Range (nmi)	w_3 No. of Passengers
Aircraft A	0.84	6,650	300
Aircraft B	0.86	6,650	250
Aircraft C	0.84	8,820	250
Aircraft D	0.86	6,900	300

TABLE 14.9 NORMALIZED SCORES FOR HYPOTHETICAL AIRCRAFT

	Attribute & Weights			
Aircraft	w_1 Speed (Mach)	w_2 Max. Range (nmi)	w_3 No. of Passengers	Value
Aircraft A	0	0	100	$100w_3$
Aircraft B	100	0	0	$100w_1$
Aircraft C	0	100	0	$100w_2$
Aircraft D	100	50	100	$100w_1 + 50w_2 + 100w_3$

With these attribute weights, a weighted sum result using the strength of preferences is shown in Table 14.10. The preferred aircraft is B777.

The concept of indifferent points is also used in other decision-making contexts. In utility theory, one method to construct utility functions queries a decision-maker for his/her indifference point between whether or not to accept a guaranteed payoff or play a lottery for a chance at a potentially larger or smaller payoff [28]. In [29], indifference relationships are used to determine preferences that are then used to solve for weights and compensation strategies. While other work on indifference points uses lottery probabilities or preferences to find relative importances among attributes, hypothetical alternatives utilize product alternatives and their attributes directly. However, finding hypothetical equivalents that are exactly of *equivalent* value to a decision-maker, or "indifference point," can be a challenging and time-consuming task [30], specifically in the context of constructing utility functions. Therefore, the hypothetical equivalents method is expanded to a more general approach that is easier to apply to complex decisions, called HEIM, which is explained in the next section.

14.3 AN APPROACH TO DECISION-MAKING USING HYPOTHETICAL EQUIVALENTS AND INEQUIVALENTS

HEIM has been developed to elicit stated preferences from a decision-maker regarding a set of hypothetical alternatives in order to access attributes' importance as well as to determine the weights directly from a decision-maker's stated preferences [31]. While integrating the concept of hypothetical equivalents, HEIM also accommodates inequivalents in the form of stated preferences,

TABLE 14.10 RESULTS USING HYPOTHETICAL EQUIVALENTS

	Attribute & Weights			
Aircraft	1/6 Speed (Mach)	2/3 Max. Range (nmi)	1/6 No. of Passengers	Value
B777–200	0	100	35	72.5
B747–200	35	50	100	55.8
A330–200	35	0	5	6.7
A340–200	100	80	0	70.0

such as "I *prefer* hypothetical alternative *A over B*." When a preference is stated, by either equivalence or inequivalence, a constraint is formulated and an optimization problem is constructed to solve for the attribute weights. The weights are solved by formulating the following optimization problem,

$$\text{Minimize} \qquad f(x) = \left(1 - \sum_{i=1}^{n} w_i\right) \qquad \text{Eq. (14.5)}$$

$$\text{subject to} \qquad h(x) = 0 \qquad g(x) \leq 0$$

where, the objective function ensures that the sum of the weights is equal to one; $\boldsymbol{X}$ = vector of attribute weights; n = number of attributes, and w_i = weight of attribute i. The constraints are based on a set of stated preferences from the decision-maker. The equality constraints are developed based on the stated preference of "I prefer alternatives *A* and *B* equally." In other words, the value of these alternatives is equal, giving the following equation:

$$V(A) = V(B) \text{ or } V(A) - V(B) = 0 \qquad \text{Eq. (14.6)}$$

The value of an alternative (alternative *A* in this case) is given as:

$$V(A) = \sum_{i=1}^{n} w_i r_{Ai} \qquad \text{Eq. (14.7)}$$

where r_{Ai} = rating of alternative A on attribute i. The inequality constraints are developed based on the stated preference of "I prefer *A* over *B*." In other words, the value of alternative *A* is more than alternative *B*, as shown in the following:

$$V(A) > V(B) \text{ or } V(B) - V(A) + \delta \leq 0 \qquad \text{Eq. (14.8)}$$

where δ = a small positive number to ensure the inequality of the values in Eq. (14.3). The value of an alternative is given by Eq. (14.1), as mentioned earlier.

In concept, the HEIM approach to decision-making is similar to the method described in [32], which is based on a least-distance approximation using pairwise preference information. However, in HEIM, the constraints are formed solely based on stated preferences from a decision-maker. The normalization constraint that requires the sum of the weights to be equal to one is converted into the objective function in Eq. (14.5). This allows the generation of multiple equivalent feasible solutions that are in turn used to refine the decision-maker's preferences to ensure a single, robust winning alternative. In HEIM, the distance or "slack" variables introduced in [32] are not utilized, simplifying the problem formulation and its solution. The formulation given in [32] is also generated for problems where a set of pairwise preferences is not transitive. This chapter focuses on transitive sets of preferences. Also, note that even though an additive model as shown in Eq. (14.7) is used in this chapter, more general utility functions models can also be used in HEIM.

While HEIM has been shown to avoid the theoretical pitfalls of the common decision-making processes as discussed in Section 14.2, there are still significant research issues associated with applying the method to many types of multi-attribute decisions in design.

The following sections systematically demonstrate how HEIM is used to solve a multi-attribute decision problem by using the same aircraft example from Section 14.2. In Section 14.4, the uniqueness and robustness of the solution is investigated.

14.3.1 Identify the Attributes

The first step is to identify the attributes that are relevant and important in the decision problem. This is because HEIM is not able to identify the absence of an important attribute. Techniques such as factor analysis [4] or value-focused thinking [33] can be used to identify the important/key attributes, reduce the attribute space, or eliminate unimportant or irrelevant variables/attributes. If an unimportant attribute is included in the process, HEIM will indicate the attribute's limited role with a low weighing factor through the sequence of stated preferences over the hypothetical alternatives (e.g., the hypothetical alternatives that score well in important attributes will be preferred over those alternatives that score well in unimportant attributes). Also, by having unimportant attributes in the problem, the computational time of the method will increase. Therefore, identifying the key attributes is important to reduce the computational effort. Section 14.2 has identified the three attributes to be speed, maximum cruise range and number of passengers.

14.3.2 Determine the Strength of Preference Within Each Attribute

As discussed in Section 14.2.4, assessing a decision-maker's true strength of preferences with respect to a given attribute is necessary to develop accurate decision models and make effective decisions. These strength of preference functions are based on the ranges of each attribute in the decision problem. If another alternative is added to the decision problem that has an attribute value outside of the current range of attribute values, then the strength of preference functions must be formulated and normalized again. For instance, in Fig. 14.1(b), the lowest and highest cruise ranges, 6,650 and 8,820 nmi, are used to formulate the preference score. If another alternative with a cruise range lower than 6,650 nmi or higher than 8,820 nmi is added to the decision problem, the strength of the preference function must be reformulated using the new upper and lower cruise ranges. In this section, the strength of preferences as shown in Fig. 14.1 is used.

14.3.3 Set up Hypothetical Alternatives

In order to use HEIM, setting up the hypothetical alternatives is the next important step. This step is important to sample the design space where $\sum_{j=1}^{n} w_j = 1$. The purpose of this step is to establish a set of hypothetical alternatives that a designer feels indifferent between or that a designer can differentiate if one alternative is preferred over the other. This is done so that the preference structure can be modeled using not only equality equations, but also inequality equations. Therefore, the set of preference weights (design variables) can then be solved by using optimization techniques.

In [30], the hypothetical alternatives were developed by simply mixing the upper and lower bounds of each attribute in different combinations. However, a more systematic approach is needed to develop the hypothetical alternative so as to efficiently sample the design space. In this chapter, fractional factorial experimental design [34] is used. Other effective experimental designs such as central composite design [34] and D-optimal [35] designs could also be used. A 3^{3-1} fractional factorial design is used with three levels for each of the three attributes (the 0, 50 and 100 score levels from the strength of preference curves in Figure 14.1). Table 14.11 shows the resulting experimental design and hypothetical alternatives with their corresponding attribute values.

TABLE 14.11 HYPOTHETICAL ALTERNATIVES

Alternative	Speed	Range	Passengers
A	0.84	6,650	235
B	0.85	6,900	366
C	0.86	8,820	320
D	0.84	6,900	320
E	0.85	8,820	235
F	0.86	6,650	366
G	0.84	8,820	366
H	0.85	6,650	320
I	0.86	6,900	235

14.3.4 Normalize the Scale and Calculate the Value for Each Alternative

Normalization is required to eliminate the dimensions from the problem. However, normalization can be carried out only after the preference strengths have been determined in order to avoid the flaws of assuming a linear preference structure. In addition, the values of each alternative as a function of the attribute weights are also calculated and are used in the optimization problem in the next section. The normalized scores and value equations for the hypothetical alternatives are shown in Table 14.12.

14.3.5 Gathering the Preference Structure

Before applying the optimization technique for HEIM, the preference structure is identified based on the hypothetical alternatives in Table 14.14 to provide constraints to the optimization problem.

Assume that Jetair feels rating nine alternatives at once is difficult and, therefore, it rates three alternatives at a time. For the first three alternatives, Jetair has the preference structure as $C \succ B \succ A$, where $\succ$ indicates "preferred to." From this first set of preferences, two nonredundant constraints can be generated, $C \succ B$ and $B \succ A$. By using the values shown Table 14.12, the constraints can be written as

$$G_1 = -0.5w_1 - 0.5w_2 + 0.5w_3 + \delta \leq 0 \quad \text{Eq. (14.9a)}$$

$$G_2 = -0.5w_1 - 0.5w_2 - w_3 + \delta \leq 0 \quad \text{Eq. (14.9b)}$$

where when δ is 0.001, it is sufficient to ensure the inequality of the value. For the remaining two sets of alternatives, the preference structures by Jetair are $F \succ E \succ D$ and $G \succ I \succ H$. This

TABLE 14.12 NORMALIZED SCORE FOR HYPOTHETICAL ALTERNATIVES

	Attribute & Weights			
Alternative	Speed (w_1)	Max. Range (w_2)	No. of Passengers (w_3)	Total Values
A	0	0	0	0
B	0.5	0.5	1	$0.5w_1 + 0.5w_2 + w_3$
C	1	1	0.5	$w_1 + w_2 + 0.5w_3$
D	0	0.5	0.5	$0.5w_2 + 0.5w_3$
E	0.5	1	0	$0.5w_1 + w_2$
F	1	0	1	$w_1 + w_3$
G	0	1	1	$w_2 + w_3$
H	0.5	0	0.5	$0.5w_1 + 0.5w_3$
I	1	0.5	0	$w_1 + 0.5w_2$

comparison has provided six constraints, which can be the formulation of the optimization problem in the next step.

14.3.6 Formulate the Preference Structure as an Optimization Problem

Therefore, the complete optimization problem for this example is shown in Eq. (14.10).

$$\text{Min } F = \left[1 - (w_1 + w_2 + w_3)\right]^2$$

subject to

$$\begin{aligned} G_1 &= -0.5w_1 - 0.5w_2 + 0.5w_3 + \delta \leq 0 \\ G_2 &= -0.5w_1 - 0.5w_2 - w_3 + \delta \leq 0 \\ G_3 &= -0.5w_1 + w_2 - w_3 + \delta \leq 0 \\ G_4 &= -0.5w_1 - 0.5w_2 + 0.5w_3 + \delta \leq 0 \\ G_5 &= w_1 - 0.5w_2 - w_3 + \leq 0 \\ G_6 &= -0.5w_1 - 0.5w_2 + 0.5w_3 + \leq 0 \end{aligned} \qquad \text{Eq. (14.10)}$$

Side constraints: $0 \leq w_i \leq 1$

Note that G_4 and G_6 = redundant constraints (they are the same as G_1). In the computational stage, these two redundant constraints are not included.

14.3.7 Solve for the Preference Weights

Solution for the preference weights can be obtained using any optimization technique. However, since the constraints are linear, sequential linear programming (SLP) or generalized reduced gradient (GRG) methods work well [36]. Using SLP, and given a single starting point, one feasible solution set of weights is [0.33, 0.33, 0.33].

14.3.8 Make Decision

With the attribute weights from the preceding section, [0.33, 0.33, 0.33], a weighted sum result is shown in the first value column of Table 14.13. The preferred aircraft is B747. Since it is assumed that a linear combination of attributes represents the value of an alternative [Eq. (14.1)], and because the domain of choices is discrete, many of the noted pitfalls of weighted-sum approaches are avoided [22–25]. In other words, new alternatives are not searched for and developed outside of those in Table 14.13. However, the sensitivity of the best alternative to changes in the weights is important, as the following discussion illustrates.

Because the weights were found using their sum as an objective function, there may be many possible sets of weights whose sum equals one and that satisfy the constraints from the stated preferences. Using another starting point to solve the optimization problem using SLP, a different set of weights is found [0.4,0.3,0.3]. The modified weighted sum result for this set of weights is also shown in the second value column of Table 14.13.

As seen in Table 14.13, the A340 aircraft is now the winning alternative with the highest score. This indicates that by using the preference structure and resulting constraints shown in Eq. (14.10), more than one winning alternative can be found. This is obviously not a desirable state. Since the winning alternative is not robust (it can change depending on the starting point of the optimization problem solution), it would indicate a need to investigate the presence of multiple solutions of Eq. (14.10). In fact, it would indicate that Eq. (14.10) is an underconstrained problem. If more constraints were added, perhaps the robustness of the solution would increase and the winning alternative would not change across multiple sets of feasible weights. This is precisely the issue that will be investigated in the next section using visualization techniques and indifference point analyses.

14.4 DETERMINATION OF A SINGLE ROBUST SOLUTION

In the preceding section, it was shown that by using a different starting point for the optimization problem, different weight values were obtained that were both optimal (sum to one) and feasible (satisfy the preference constraints). Additionally, different weight values resulted in different alternatives emerging as the overall choice, indicating a need to investigate the issue of the robustness of the winning alternative with respect to changes in the weight values.

As it is possible for multiple alternatives to be the preferred solution to the selection problem, it is desirable that the HEIM method be able to identify one, robust solution. The classical definition of "robust" is a solution that is insensitive to variations in control and noise factors [37]. The term "robust" in the context of this chapter refers to a preferred alternative that is insensitive to different sets of feasible weights. From Table 14.13 of Section 14.3.8, it is obvious that the winning alternative is *not robust* since a change in the weight values changes the winning alternative.

Since the aircraft example has only three attributes, the design space can be represented by the three weights and visualized using the OpenGL Programming API [38]. The different attribute weights are represented along the three axes, using the normalized attribute scale. Next, a large number of weight sets that satisfy the various constraints and sum to one are randomly generated and plotted in Fig. 14.2. The different winning alternatives corresponding to the different weight values are shown in different colors along with the plane representing the sum of weights equal to one. The region where the B747 wins is shown with gray points, while the region where the A340 aircraft wins is shown in black. The points

TABLE 14.13 DIFFERENT RESULTS USING HEIM

	Attribute & Weights				
Aircraft	Speed (w_1)	Max. Range (w_2)	No. of Passengers (w_3)	Value with (w_1, w_2, w_3) = (0.33,0.33,0.33)	Value with (w_1, w_2, w_3) = (0.4,0.3,0.3)
B777–200	0	100	35	44.6	40.5
B747–200	35	50	100	61.1	59.0
A330–200	35	0	5	13.2	15.5
A340–200	100	80	0	59.4	64.0

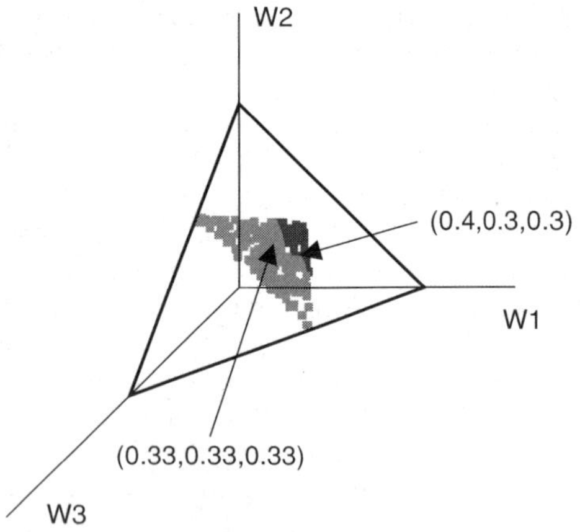

FIG. 14.2 FINAL FEASIBLE SPACE INCLUDING ALL CONSTRAINTS SPECIFIED IN EQ. (14.6)

corresponding to the weights given in Table 14.16 are also shown on the figure. It is obvious that the problem can result in any one of the two alternatives emerging as the winner, based on the chosen starting point for the solution of the optimization problem.

From Figure 14.1, it is concluded that the feasible region would require more constraints to have a single winning alternative region. In order to determine the additional constraints necessary, it is necessary to determine the line separating the region of gray and black points in Fig. 14.2. If a mathematical representation of this line can be determined and converted into a preference constraint, then one side of the line could be deemed infeasible, eliminating either the gray or black regions from consideration. This dividing line is the line of indifference between the gray and black regions because any combination of weight values on this line will give the same overall score for both alternatives. In order to determine the indifference line equation, the value functions for B747 and A340 aircrafts from Figure 14.2 are equated. The value functions for the two alternatives are: $V(\text{B747}) = 0.35w_1 + 0.5w_2 + w_3$; $V(\text{A340}) = w_1 + 0.8w_2$.

Therefore:

$$V(\text{B747}) = V(\text{A340})$$

$$0.35w_1 + 0.5w_2 + w_3 = w_1 + 0.8w_2 \qquad \text{Eq. (14.11)}$$

$$-0.65w_1 - 0.3w_2 + w_3 = 0$$

As mentioned earlier, hypothetical alternatives are used to elicit stated preferences without biasing the decision-maker toward one particular alternative. Having the decision-maker state his/her preferences directly over the actual winning alternatives goes against the ideology of HEIM. Therefore, using Eq. (14.11), two new hypothetical alternatives are constructed over which the decision-maker can then state his/her preferences.

To create new hypothetical alternatives, the terms in Eq. (14.11) are rearranged and the preference curves of Figure 14.1 are used to unnormalize the normalized attribute ratings. Rearranging Eq. (14.11), which is

$$0.35w_1 + 0.7w_2 + w_3 = w_1 + w_2 \qquad \text{Eq. (14.12)}$$

It is important to note that Eq. (14.12) is just one possible rearrangement. The right- and left-hand side of Eq. (14.12) are two value functions that correspond to two different hypothetical alternatives. Using the strength of preference curves of Figure 14.1, the two hypothetical alternatives are unnormalized and presented in Table 14.14.

TABLE 14.14 NEW HYPOTHETICAL ALTERNATIVES FOR ROBUSTNESS

Alternative	Speed	Range	Passengers
J	0.847	7,620	366
K	0.86	8,820	235

Now, in order to achieve a robust winning alternative, the decision-maker states his/her preference over the hypothetical alternatives *J* and *K*. If the decision-maker states a preference of *J* over *K*, then

$$\text{Alternative } J > \text{Alternative } K$$

$$0.35w_1 + 0.7w_2 + w_3 > w_1 + w_2 \qquad \text{Eq. (14.13)}$$

Equation (14.13) provides the extra constraint needed to achieve a single robust winner. This constraint is incorporated into the design space, and the result is shown in Fig. 14.3(*a*). As seen in Fig. 14.3(*a*), the feasible region is now only populated with gray points, representing the B747 aircraft as being the robust winning alternative. On the other hand, if the decision-maker reversed his/her preferences over the new hypothetical alternatives, then

$$\text{Alternative } J < \text{Alternative } K$$

$$0.35w_1 + 0.7w_2 + w_3 < w_1 + w_2 \qquad \text{Eq. (14.14)}$$

and the feasible space is populated solely with black points as shown in Fig. 14.3(*b*), representing the A340 aircraft as being the robust winning alternative. Thus, a single winning alternative is obtained in either case, even though multiple weight values result from the solution of the initial optimization problem of HEIM. A more formal presentation of this extension to HEIM is presented in [39], where the necessary steps are outlined to ensure a robust winning alternative for problems with any number of attributes.

14.5 CONCLUSIONS

In this chapter, an approach to decision-making using the concepts of hypothetical equivalents and inequivalents is presented.

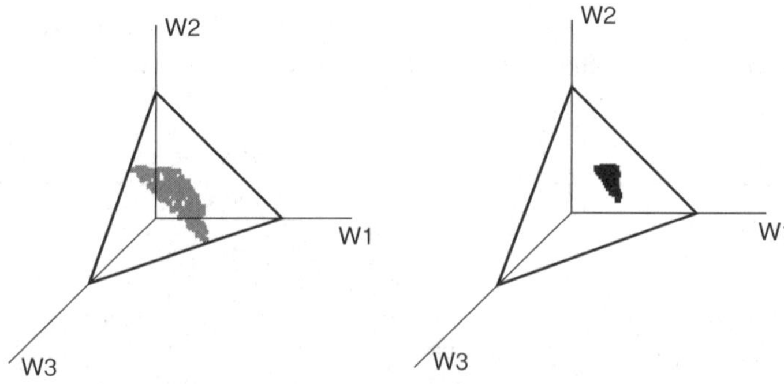

FIG. 14.3 NEW FEASIBLE REGIONS INCORPORATING: (A) EQ. (14.13); AND (B) EQ. (14.14)

The developments presented here are generally applicable to decision situations where one decision-maker is making the decision. The method is mathematically rigorous in that it assesses the true decision-maker's stated preferences on a number of hypothetical alternative choices and solves for a set of attribute weights that accurately represent the preferences. If only hypothetical equivalents are used, the solution is found by solving a set of simultaneous equations. If hypothetical inequivalents are used with or without equivalents, then optimization techniques are used to solve for the attribute weights. The set of attribute weights accurately represent the stated preferences of the decision-maker, and are more theoretically sound and practically representative of actual preferences than methods that simply assign weights, try various weight combinations or use a standard default of assuming all weights to be equal. This chapter also investigated the presence of multiple solutions in HEIM and their impact on the alternative chosen. Then, an approach is formulated to determine a single robust winning alternative by generating hypothetical alternatives based on equating the value functions of multiple winning alternatives. This approach ensures that enough preference constraints are elicited to identify one preferred alternative across the entire feasible region.

ACKNOWLEDGMENTS

We would like to thank the National Science Foundation, grant No. DMII-9875706, for its support of this research.

REFERENCES

1. Chen, W., Lewis, K.E. and Schmidt, L., 2000. "Decision-Based Design: An Emerging Design Perspective," *Engrg. Valuation & Cost Analysis, Special Ed. on "Decision Based Design: Status & Promise*, **3**(1), pp. 57–66.
2. Hazelrigg, G.A., 1998. "A Framework for Decision-Based Engineering Design," *ASME J. of Mech. Des.*, Vol. 120, pp. 653–658.
3. Wassenaar, H. J. and Chen, W., 2003. "An Approach to Decision-Based Design With Discrete Choice Analysis for Demand Modeling," *ASME J. of Mech. Des.*, 125(3), pp. 490–497.
4. Urban, G. L. and Hauser J. R., 1993. *Design and Marketing Of New Products, 2nd Ed.*, Prentice Hall, pp. 1–16.
5. Saari, D.G., 2000. "Mathematical Structure of Voting Paradoxes. I: Pairwise Vote. II: Positional Voting," *Eco. Theory*, Vol. 15, pp. 1–103.
6. Matheson, D. and Matheson, J., 1998. *The Smart Organization*, Harvard Business School Press, Boston, MA.
7. Jetblue Airway, 2001. http://www.jetblue.com.
8. Airbus, 2001. "A330/A340 Family," http://www.airbus.com.
9. Boeing, 2001. "Commercial Airplane Info," http://www.boeing.com/commercial/flash.html.
10. Saaty, T. L., 1980. *The Analytic Hierarchy Process*, McGraw-Hill.
11. Fukuda, S. and Matsura, Y., 1993. "Prioritizing the Customer's Requirements by AHP for Concurrent Design," *Des. for Manufacturability*, Vol. 52, pp. 13–19.
12. Davis, L. and Williams, G., 1994. "Evaluating and Selecting Simulation Software Using the Analytic Hierarchy Process," *Integrated Manufacturing Sys.*, **5**(1), pp. 23–32.
13. Basak, I. and Saaty, T.L., 1993. "Group Decision Making Using the Analytic Hierarchy Process," *Math. and Computer Modeling*, **17**(4–5), pp. 101–110.
14. Hamalainen, R. P. and Ganesh, L.S., 1994. "Group Preference Aggregation Methods Employed in AHP: An Evaluation and an Intrinsic Process for Deriving Members' Weightages," *Euro. J. of Operational Res.*, 79(2), pp. 249–265.
15. Arrow, K. J., 1951. *Social Choice and Individual Values*, Wiley & Sons, New York, NY.
16. Barzilai J., Cook, W. D. and Golany, B., 1992. "The Analytic Hierarchy Process: Structure of the Problem and Its Solutions," *Systems and Management Science by Extremal Methods*, F.Y. Phillips and J.J. Rousseau, eds., Kluwer Academic Publishers, pp. 361–371.
17. Barzilai, J. and Golany, B., 1990. "Deriving Weights from Pairwise Comparison Matrices: The Additive Case," *Operations Res. Letters*, Vol. **96**, pp. 407–410.
18. U.S. News and World Report, 2003. "Graduate School Rankings," http://www.usnews.com/usnews/edu/grad/rankings/rankindex.htm.
19. Peter, H. and Wakker, P., 1991. "Independence of Irrelevant Alternatives and Revealed Group Preferences," *Econometrica J.*, **59**(6), pp. 1787–1801.
20. Callaghan, A. and Lewis, K., 2000. "A 2-Phase Aspiration-Level and Utility Theory Approach to Large Scale Design," Proc., *ASME Des. Automation Conf., DETC00/DTM-14569*, ASME, New York, NY.
21. Thurston, D. L., 1991. "A Formal Method for Subjective Design Evaluation with Multiple Attributes," *Res. in Engrg. Des.*, Vol. 3, pp. 105–122.
22. Messac, A., Sundararaj, J. G., Tappeta, R. V. and Renaud, J. E., 2000. "Ability of Objective Functions to Generate Points on Non-Convex Pareto Frontiers," *AIAA J.*, **38**(6), pp. 1084–1091.
23. Chen. W., Wiecek, M. and Zhang, J., 1999. "Quality Utility: A Compromise Programming Approach to Robust Design," *ASME J. of Mech. Des.*, **121**(2), pp. 179–187.
24. Dennis, J.E. and Das, I., 1997. "A Closer Look at Drawbacks of Minimizing Weighted Sums of Objective for Pareto Set Generation in Multicriteria Optimization Problems," *Struc. Optimization*, **14**(1), pp. 63–69.
25. Zhang, J., Chen, W. and Wiecek, M., 2000. "Local Approximation of the Efficient Frontier in Robust Design," *ASME J. of Mech. Des.*, **122**(2), pp. 232–236.
26. Watson, S. R. and Freeling, A. N. S., 1982. "Assessing Attribute Weights," *Omega*, **10**(6), pp. 582–583.
27. Wu, G., 1996. "Exercises on Tradeoffs and Conflicting Objectives," *Harvard Bus. School Case Studies*, Vol. 9, pp. 396–307.
28. Keeney, R. L. and Raiffa, H., 1993. *Decisions with Multiple Objectives: Preferences and Value Tradeoffs*, Cambridge University Press.
29. Scott, M. J. and Antonsson, E. K., 2000. "Using Indifference Points in Engineering Decisions," *Proc., 12th ASME Des. Theory and Methodology Conf., DETC2000/DTM-14559*. ASME, New York, NY.
30. Thurston, D. L., 2001. "Real and Misconceived Limitations to Decision Based Design with Utility Analysis," *ASME J. of Mech. Des.*, **123**(2), pp. 176–182.
31. See, T. K. and Lewis, K., 2002. "Multi-Attribute Decision Making Using Hypothetical Equivalents," *Proc., ASME Des. Tech. Conf., Des. Automation Conf., DETC02/DAC-02030*, ASME, New York, NY.
32. Yu, P.-L., 1985. Multiple-criteria Decision-Making: Concepts, Techniques and Extensions, Plenum Press, Chapter 6, pp. 113–161.
33. Keeney, R. L., 1996. *Value-Focused Thinking: A Path to Creative Decision Making*, Harvard University Press.
34. Montgomery, D. C., 1997. *Design and Analysis of Experiments, 4th Ed., John Wiley & Sons*, New York, NY.
35. Atkinson, A. C. and Donev, A. N., 1992. *Optimum Experimental Designs*, Oxford University Press.
36. Vanderplaats, G. N., 1999. *Numerical Optimization Techniques for Engineering Design, 3rd Ed.*, Vanderplaats Research & Development, Inc.
37. Phadke, M. S., 1989. *Quality Engineering Using Robust Design*, Prentice Hall.
38. Neider, J., Davis, T. and Woo, M., 1994. *OpenGL Programming Guide, Release 1*, Addison-Wesley.
39. Gurnani, A. P., See, T. K. and Lewis, K., 2003. "An Approach to Robust Multi-Attribute Concept Selection," *Proc., ASME Des. Tech. Conf., Des. Automation Conf., DETC03/DAC-48707*, ASME, New York, NY.

PROBLEMS

14.1 Eight compact sedan alternatives are presented as shown in Table 14.15. The objective is to choose the winning alternative based on your preferences. Provide support for this decision in the form of a set of steps, explanations and a formal decision matrix. The following steps shall be carried out:

- Develop strength of preferences for each attribute scale
- Normalize the scale for each attribute based on your strength of preferences
- Determine and assign the relative weight of each attribute
- Multiply the relative weight times the rating (normalized scale) and sum for each alternative
- Choose the alternative with the highest score.

Some things that people tend to overlook or forget:

- Strength of preferences can be linear or nonlinear and need to reflect your preferences for each attribute
- Make clear how you determined the relative importance of the attributes.

14.2 Using Table 14.15 shown in the previous problem, carry out HEIM process to determine the winner for the compact car. The following steps shall be carried out:

- Develop strength of preferences for each attribute scale (which has been done in the previous problem)
- Set up the hypothetical alternatives
- Normalize the scale and calculate the value for both actual and hypothetical alternatives
- Gather the preference structure by comparing the hypothetical alternatives
- Formulate the preference structure as an optimization problem
- Solve for the attribute relative weights

Compare the attribute weights solved by HEIM with the attributes weights in problem 14.1.

- Are they different? Do you have different winners?
- Are the ranking of the attributes the same even though the weights value might be different?

14.3 Determine whether the optimization formulation in problem 14.2 gives a robust solution. If not, would more comparisons of hypothetical alternatives make it possible to narrow down the design space?

Or would the process described in Section 14.4 produce a single robust solution?

TABLE 14.15 ATTRIBUTE DATA FOR AUTOMOBILE ALTERNATIVES

	Attributes and Relative Weights				
Automobile	w_1 **Engine**	w_2 **Howrsepower**	w_3 **MPG**	w_4 **Price**	w_5 **Acceleration (0–60 mph)**
Car #1	2.0	145	36	$13,425	8.6
Car #2	1.7	127	38	$13,470	10.5
Car #3	2.0	140	33	$11,995	9.9
Car #4	1.8	130	40	$13,065	9.5
Car #5	2.0	132	36	$12,917	10.0
Car #6	2.0	130	31	$13,315	10.4
Car #7	2.2	140	33	$13,884	7.9
Car #8	2.0	135	33	$12,781	9.8

CHAPTER

15

MULTIOBJECTIVE DECISION-MAKING USING PHYSICAL PROGRAMMING

Achille Messac

15.1 INTRODUCTION

Engineering design is an iterative process, where several design alternatives are analyzed to obtain the design that satisfactorily performs a desired task. Traditionally, design approaches have relied more on the intuition and past experience of the designer and less on sound scientific or engineering principles to perform this iterative process of analyzing and choosing the best design.

In recent years, the field of engineering design has witnessed a significant evolution, promoted largely by an exponential growth in the computational resources available to a designer. Moreover, with increasing global competition, designs are required not only to be functional, but also cost-effective, efficient and innovative. With numerous factors judging the desirability of a design, the designer cannot solely rely on the traditional design approach of *manually* choosing the best design.

The advances in computer technology and the high performance needs of the aerospace industry, coupled with increasing global competition, have fueled the development of the field of *optimal design*. The optimal design approach provides the necessary mathematical and analytical tools required to systematically and rapidly examine various design alternatives, and select what some may consider the *best* design. Optimal design approaches typically use computer-based numerical optimization techniques to maximize or minimize a measure of the system performance, subject to certain design constraints.

Most realistic engineering design problems are *multiobjective* in nature, where a designer is interested in simultaneously optimizing two or more objectives. A class of multiobjective problem formulation methods in the literature combines the multiple objectives into a single objective function, also known as the *aggregate objective function* (AOF). One of the several challenges in the area of multiobjective optimization is to properly formulate an AOF that satisfactorily models the designer's preferences. In this chapter, we describe the challenges associated with aggregating preferences in multiobjective problems. More specifically, we study the *Physical Programming* method, which provides a framework to effectively incorporate the designer's preferences into the AOF.

The material presented in this chapter is organized into 10 sections. Section 15.2 consists of a review of the basic terminology in optimization. Multiobjective optimization is introduced in Section 15.3, where we discuss the importance and the challenges associated with multiobjective problem formulation. Section 15.4 outlines three methods to formulate and solve multiobjective problems. Readers familiar with the area of multiobjective optimization can skip this section and go to Section 15.5, which introduces the Physical Programming method. Section 15.6 describes the Linear Physical Programming method, followed by Section 15.7, which presents a description of Nonlinear Physical Programming. In Section 15.8, we present interesting comparisons between Goal Programming and Physical Programming. Section 15.9 illustrates the Linear Physical Programming method through an example. Section 15.10 summarizes the material presented in this chapter. The Appendix contains a sample Matlab code for implementing the Linear Physical Programming algorithm.

15.2 BASICS OF OPTIMIZATION

15.2.1 Optimization Terminology

The formulation of a design optimization problem usually requires translation of a verbal description of the design specifications into a mathematical form [1]. The mathematical form consists of three components: the *design variables*, the *objective function* and the *design constraints*.

In an optimization problem, the designer is usually interested in finding the design that maximizes or minimizes a certain measure of the system performance, called the *objective function* (the terms "objective function", "design objective", "design criterion" and "design metric" are unfortunately used interchangeably in the literature), subject to certain conditions, called the *design constraints*. The objective function and the constraints are functions of quantities called the *design variables*. These are quantities whose values can be changed during optimization to improve the design performance. A design that satisfies all the constraints is called a *feasible design*. A feasible design that optimizes the objective function is called the *optimum design*, which is expected to be the solution of interest to the designer. A typical optimization problem formulation is given as

$$\min_x \quad J(x) \qquad \text{Eq. (15.1)}$$

subject to

$$g(x) \leq 0 \qquad \text{Eq. (15.2)}$$

$$h(x) = 0 \qquad \text{Eq. (15.3)}$$

$$x_{\min} \leq x \leq x_{\max} \qquad \text{Eq. (15.4)}$$

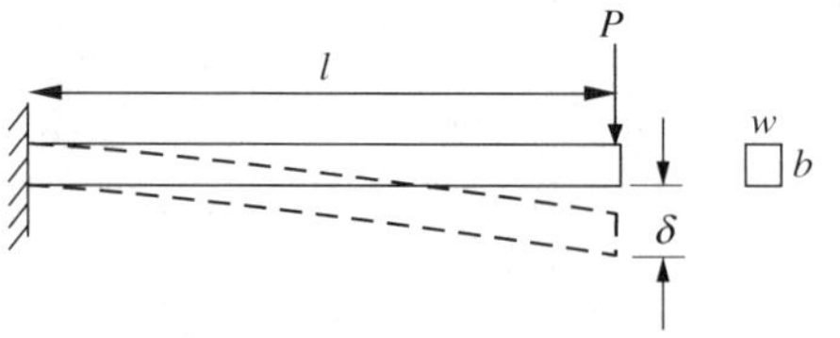

FIG. 15.1 BEAM EXAMPLE

where x = design variable; and the function $J(x)$ = objective function that is minimized. The constraint given by Eq. (15.2) is called an *inequality constraint*. The constraint given by Eq. (15.3) is called an *equality constraint*. The constraint given by Eq. (15.4) is called a *side constraint*. The quantities x_{min} and x_{max} are the minimum and the maximum acceptable values, respectively, for the design variable, x. Note that the quantities x, g and h could be vectors.

15.2.2 Example

Let us illustrate the optimization problem formulation stated above with the help of an example. Consider the task of designing a cantilever beam with a rectangular cross section, subjected to the load P, as shown in Fig. 15.1. We are interested in finding the cross-sectional dimensions, b and w, of the lightest beam that can safely support the load P. We also specify the maximum and the minimum acceptable values for b and w as b_{max}, b_{min}, w_{max} and w_{min}.

This beam design can be posed as an optimization problem, where the design variables are the cross-sectional dimensions, b and w, of the beam and the objective function is the mass of the beam. The constraints can be specified as: (1) the maximum bending stress in the beam should not exceed the maximum allowable stress; and (2) the cross-sectional dimensions should be within the specified limits. The optimization problem can be stated as

$$\min_{b,w} \quad M = \rho bwl \qquad \text{Eq. (15.5)}$$

subject to

$$S < S_{max} \qquad \text{Eq. (15.6)}$$

$$b_{min} \le b \le b_{max} \qquad \text{Eq. (15.7)}$$

$$w_{min} \le w \le w_{max} \qquad \text{Eq. (15.8)}$$

where M = mass of the beam; ρ density of the material of the beam; $S = 6PL/(wb^2)$ = stress, which is induced at the root of the beam due to the load P; and S_{max} = maximum allowable stress.

The above-stated optimization problem can be solved to obtain the optimum values of b and w that result in the minimum mass of the beam, while not violating the constraints. Assume the following design parameters: P = 600 kN, L = 2 m, Young's Modulus E = 200 GPa, ρ = 7,800 kg/m^3, S_{max} = 160 MPa, b_{min} = 0.1 m, b_{max} = 0.8 m, w_{min} = 0.1 m and w_{max} = 0.5 m. Solving the optimization problem in Matlab using the command *fmincon* yields an optimum solution of w = 0.1000 m, b = 0.6708 m and the optimum mass, M = 1,046.5 kg. This optimum solution represents the lightest beam design that can safely support the load P. Note that here we have neglected all issues of uncertainty and manufacturing imperfections, which should impact design decisions in practice.

15.2.3 Classification of Optimization Problems

Optimization problems can be classified into several categories, based on the nature of the design variables, the objective function and the constraint functions. The following are a few important classes of optimization problems.

(1) **Constrained versus Unconstrained:** If an optimization problem has no constraints, it is classified as an *unconstrained* optimization problem. If there are constraints (equality, inequality or side constraints) in the problem, then it is a *constrained* optimization problem.
(2) **Linear versus Nonlinear:** If the objective function and the constraints are linear functions of the design variables, then the optimization problem is a *linear programming*, or a linear optimization, problem. However, if either the constraints or the objective function, or both, are nonlinear, then the optimization problem is called a *nonlinear optimization* problem.
(3) **Continuous versus Discrete:** If the design variables are continuous, the problem is said to be a *continuous* optimization problem. On the other hand, if any design variable can only assume discrete values (for example, the number of rivets in a joint), then the problem is called a *discrete* optimization problem.
(4) **Deterministic versus Stochastic:** If there is no uncertainty modeled in the design variables of an optimization problem, then it is said to be a *deterministic* problem. If the randomness or the uncertainty present in the design variables is modeled in the optimization problem, it is said to be a *stochastic* optimization problem (note that we could also discuss other forms of uncertainty, such as modeling uncertainty).
(5) **Single Objective versus Multiobjective:** As the name suggests, single-objective problems consist of only one objective function to be optimized. Multiobjective problems consist of two or more objectives to be optimized simultaneously.

Several numerical optimization algorithms are available to solve the above types of problems. A detailed discussion of the methods can be found in most optimization books [1, 2, 3]. Our emphasis in this chapter is primarily on multiobjective optimization problems, which are prevalent in engineering design, and particularly, our focus is on means of aggregating preferences between multiple objectives.

15.3 MULTIOBJECTIVE OPTIMIZATION

Most realistic engineering design problems are multiobjective in nature, where the designer is interested in simultaneously optimizing multiple objectives. These multiple objectives are usually conflicting in nature (for example, minimize cost and maximize productivity). The designer is required to resolve the trade-offs between these competing objectives. Multiobjective optimization can be used as an important decision-making tool to resolve such conflicts in engineering design. A typical multiobjective optimization formulation is given as

$$\min_{x} \quad \{\mu_1(x), \ldots, \mu_n(x)\}^T \qquad \text{Eq. (15.9)}$$

subject to

$$g(x) \le 0 \qquad \text{Eq. (15.10)}$$

$$h(x) = 0 \qquad \text{Eq. (15.11)}$$

$$x_{\min} \le x \le x_{\max} \qquad \text{Eq. (15.12)}$$

where $\mu_i(x)$ = ith objective function to be minimized; and n = number of objectives.

In the beam example discussed in the previous section, assume now that the designer wishes to minimize the deflection of the beam, δ (see Fig. 15.1), and the mass of the beam. We now have a biobjective optimization problem, with the two design objectives being the mass and the deflection of the beam.

Notice that the two design objectives are conflicting in nature. As the mass *decreases*, the deflection tends to *increase*. Conversely, as the deflection *decreases*, the mass tends to *increase*. Ideally, the designer would be interested in obtaining an optimum that minimizes the mass and the deflection *simultaneously*. This, in practice, is not possible because of the conflicting nature of the objectives. Instead, the designer could obtain a solution that achieves a compromise, or a trade-off, between the objectives. A so-called *Pareto optimal solution* is one that achieves such a compromise.

One of the interesting features of multiobjective optimization problems is that the Pareto optimal solution is generally not unique. There exists a set, called the *Pareto optimal set*, which represents a complete representation of the set of solutions for a multiobjective problem. Pareto optimal solutions are those for which any improvement in one objective results in the worsening of at least one other objective.

If we plot the entire Pareto set in the design objective space (a plot with the design objectives plotted along each axis), the set of points obtained is called a *Pareto frontier*. This is widely used in multiobjective decision-making problems to study trade-offs between objectives. Under some multiobjective optimization approaches, the complete Pareto frontier is generated first. The Pareto solution that possesses the most desirable trade-off properties is then chosen as the final design.

Figure 15.2 shows an example of a Pareto frontier for a generic biobjective problem [let $n = 2$ in Eq. (15.9)] in the design objective space. The constraints in Eqs. (15.10), (15.11) and (15.12) define the feasible design space. The point A_1 is obtained by minimizing objective 1 alone, and the point A_2 is obtained by minimizing objective 2 alone.

A *dominated* design point, C (see Fig. 15.2) is one for which there exists at least one feasible design point that is better than C in all design objectives. For example, the design point C is dominated by the design point D, as shown in Fig. 15.2. From the definitions of a Pareto optimal point and a dominated design point, observe that there does not exist any feasible design point better than a Pareto optimal point in all design objectives. Therefore, Pareto optimal points are also known as *non-dominated points*.

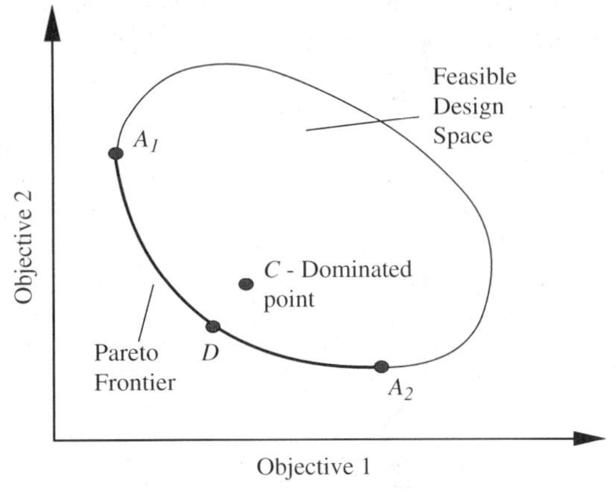

FIG. 15.2 PARETO FRONTIER IN THE DESIGN OBJECTIVE SPACE

15.4 FORMULATION OF AGGREGATE OBJECTIVE FUNCTIONS

Formulating a multiobjective optimization problem is generally a challenging task. A multiobjective problem is usually posed as a single objective problem by combining all the objectives, and thereby forming an *aggregate objective function* (AOF). The importance of proper formulation of the AOF should be understood: the optimum solution will only be as effective as the AOF [4].

Multiobjective design optimization typically consists of the following three phases [4]: (1) modeling the physical system in terms of design objectives, design variables and constraints; (2) combining the design objectives using the designer's preferences to form an AOF; and (3) optimizing the AOF to obtain the most preferred solution.

Robust computational and analytical tools are available for the first and the third phases, i.e., the modeling and optimization phases. However, the second phase, which involves formulation of the AOF, is not an easy task. This complexity arises because it is not intuitively clear how to combine the individual objectives such that the resulting AOF is indeed a mathematical representation of the designer's specifications and preferences.

Moreover, all the design objectives in a multiobjective problem may not be of equal importance to the designer. The designer might have relative *preferences* among the objectives. Consider the beam example: If minimizing the mass of the beam is more important to the designer when compared to minimizing the deflection, the designer is said to express an *inter-criteria preference*, or a preference among several objectives. Another type of preference the designer could express is an *intra-criterion preference*, or a preference within an objective. For example, in the beam design problem, a mass of 2,000 kg might be more desirable to the designer than a mass of 2,500 kg. The challenge in formulating an AOF lies in translating the designer's (often subjective) preferences, both intra-criterion and inter-criteria, into a mathematical form.

In this section, we describe some of the popular AOF formulation techniques.

15.4.1 Weighted Sum Method

The *weighted sum* method is one that is most widely used for multiobjective optimization. As its name suggests, the AOF in this method is a weighted sum of the individual objectives. The designer chooses numerical weights for each objective. These weights are expected to reflect the relative importance of each objective, i.e., an objective of higher importance is generally given a higher weight, after appropriate scaling.

For a generic biobjective optimization problem, the AOF by the weighted sum approach is given as

$$J_{ws} = w_1\mu_1 + w_2\mu_2 \qquad \text{Eq. (15.13)}$$

where w_1 and w_2 = weights reflecting the relative importance of each objective. For example, if we specify $w_1 = 0$ and $w_2 = 1$ in Eq. (15.13), it implies that we are interested in minimizing objective μ_2 alone. Similarly, if we specify $w_1 = 1$ and $w_2 = 0$ in Eq. (15.13), it implies that we are interested in minimizing objective μ_1 alone.

The weighted sum method relies on the designer to choose the weights that correctly reflect his/her relative preferences among objectives. Often, it is not very clear how to choose these weights, since they are not physically meaningful. Let us say that the designer is interested in finding the optimum values of b and w that result in a mass of approximately 1,200 kg and a deflection of approximately 0.2 mm. Then, it is not clear what set of weights the designer needs to use to formulate the AOF in Eq. (15.13). This ambiguity often results in many iterations in the correct choice of weights.

In addition, the weighted sum method does not provide the means to effectively specify intra-criterion preferences. Typically, the designer would wish to express a higher preference for an objective value that is desirable when compared to an objective value that is not desirable. For example, in the beam design example, a mass of 2,000 kg may be more desirable to the designer than a mass of 2,500 kg. The weighted sum method does not effectively model such intra-criterion preferences into the AOF, since each design objective is assigned only a single weight irrespective of the designer's ranges of desirability. These notions are discussed in more detail later.

Obtaining the Pareto Frontier Using the Weighted Sum Method The weighted sum method can also be used to obtain the Pareto frontier of a multiobjective problem. Consider the case of a generic biobjective problem given in Eq. (15.13). If we specify $w_1 = 0$ and $w_2 = 1$ in Eq. (15.13), we obtain the point A_2 in Fig. 15.2. Similarly, the point A_1 is obtained by setting $w_1 = 1$ and $w_2 = 0$ in Eq. (15.13). The Pareto points between A_1 and A_2 can be obtained by choosing different relative preferences (or weights) for each objective. By sequentially varying the weights, say between zero and one, one can obtain different Pareto solutions for the problem.

The task of choosing weights and obtaining a Pareto frontier is even more challenging in the presence of design objectives of disparate magnitudes. For example, in the beam problem the mass of the beam is several orders of magnitude larger than the deflection of the beam. The multiobjective formulation is given as:

$$\min_{b,w} \quad J_{\text{beam}} = w_1(M) + w_2(\delta) \qquad \text{Eq. (15.14)}$$

subject to

$$S < S_{\max} \qquad \text{Eq. (15.15)}$$

$$b_{\min} \le b \le b_{\max} \qquad \text{Eq. (15.16)}$$

$$w_{\min} \le w \le w_{\max} \qquad \text{Eq. (15.17)}$$

If we use the weighted sum method to solve this problem, and vary the weights evenly between zero and one, we obtain a very poor representation of the Pareto frontier. Figure 15.3 shows: (1) the actual Pareto frontier; and (2) the *two* points obtained through a careless application of the weighted sum method.

One possible technique to obtain a good representation of the Pareto frontier in problems with scaling issues is to choose weights so as to compensate for the difference in the magnitudes of the objectives. For example, in the beam problem the weight for deflection can be chosen to be several orders of magnitude higher than the weight for mass. Equivalently, the design objectives can first be normalized or scaled by dividing them by a typical or good value of the respective objective.

Moreover, the weighted sum method often cannot yield all the Pareto solutions, especially those lying on the non-convex regions of the Pareto frontier. Let us now recall the definition of a non-convex set. In Fig. 15.2, the feasible design space is convex, since a line segment joining any two points in the feasible design set lies inside the set. The Pareto frontier is also convex, since a line segment joining any two points on the Pareto curve lies entirely in the feasible space. On the other hand, consider the feasible design space in Fig. 15.4(a). The feasible design space is non-convex. Also, note that the Pareto frontier, shown by the thick line, is non-convex.

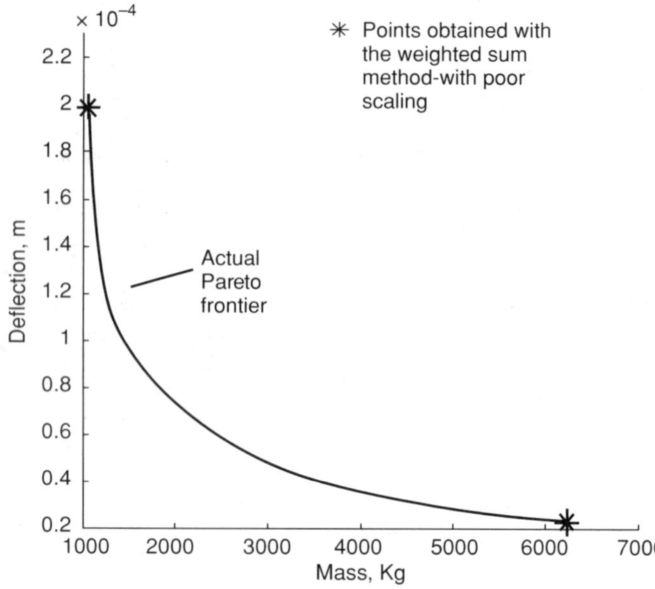

FIG. 15.3 BEAM EXAMPLE—PARETO FRONTIER USING THE WEIGHTED SUM METHOD

Equation (15.13) indicates that the AOF in the weighted sum method is a linear function of the objectives. The contours of the AOF are straight lines in Fig. 15.4(a), and therefore cannot capture the solutions lying on the non-convex regions of the Pareto frontier. For example, in Fig. 15.4(a), the weighted sum algorithm will not yield the point P_3 as a Pareto point. This is the case because it is always possible to further decrease the objective function beyond the point P_3.

15.4.2 Compromise Programming

The *compromise programming* method also involves weights with which the designer specifies preferences among the design objectives. The AOF for this method is a simple extension of the weighted sum method, and is given as:

$$J_{cp} = w_1\mu_1^m + w_2\mu_2^m \qquad \text{Eq. (15.18)}$$

where m = an even integer. However, important distinctions apply. Note that the AOF in this case is not a linear function of the objectives, as in the case of the weighted sum method. The advantage of such an objective function is that it can reach into the non-convex regions of the Pareto frontier [see Fig. 15.4(b)]. It can yield the Pareto points lying on the non-convex Pareto regions, such as P_3, unlike the weighted sum approach. The proper value of m is usually dictated by the nonlinearity present in the problem at hand. A more comprehensive examination of issues related to choosing m is provided in [5, 6].

The *weighted square sum* method is a special case of the compromise programming method, where m = 2. The AOF for the weighted square sum method can be given as:

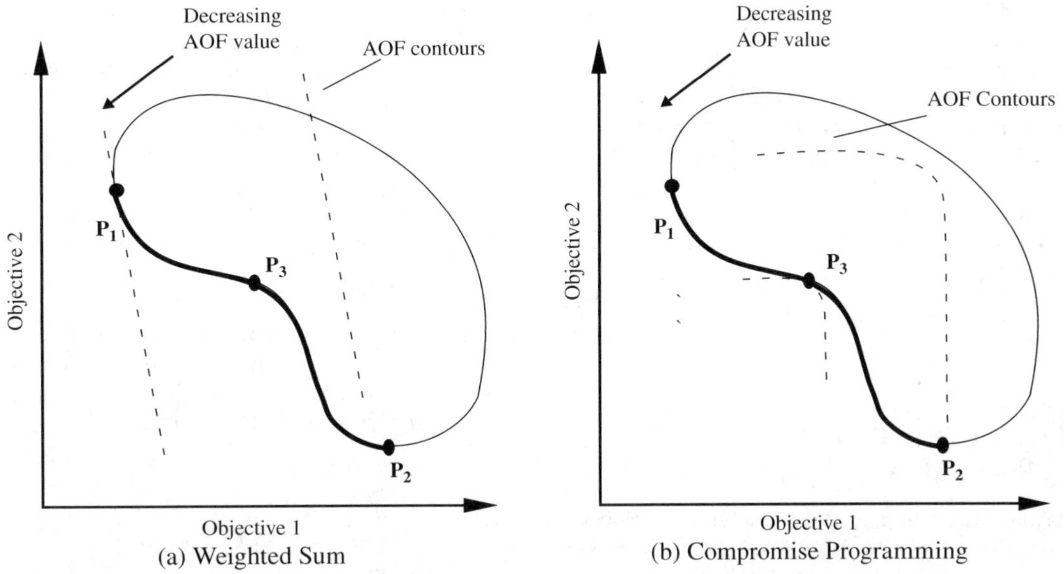

FIG. 15.4 WEIGHTED SUM VERSUS COMPROMISE PROGRAMMING

$$J_{wss} = w_1\mu_1^2 + w_2\mu_2^2 \qquad \text{Eq. (15.19)}$$

The following equation presents a slight variation in the problem formulation:

$$J = w_1(\mu_1 - G_1)^2 + w_2(\mu_2 - G_2)^2 \qquad \text{Eq. (15.20)}$$

The term $(\mu_1 - G_1)^2$ in the objective function ensures that the optimum value of μ_1 approaches a desired target value, G_1, instead of zero.

Even though compromise programming yields solutions on the non-convex regions of the Pareto frontier, the designer is still required to choose weights to indicate the relative preferences among the design objectives. The ambiguity associated with the choice of weights in the weighted sum and the compromise programming methods is a significant drawback of both methods.

15.4.3 Goal Programming

Goal Programming (GP) was first developed by Charnes and Cooper [7, 8], and later extended by Ijiri [9], Lee [10] and Ignizio [11]. It is a well-known approach for solving multiobjective optimization problems.

Goal programming requires setting a target or a goal for each objective, as opposed to a simple minimization or maximization. For each objective, the designer specifies a single desired target value. Also, the designer specifies two weights for each objective, w_{GP}^+ and w_{GP}^-, which indicate the penalties for deviating from the target value on either side (see Fig. 15.5). The basic principle behind the GP approach is to minimize the deviation of each design objective from its corresponding target value, which is defined by a *deviational variable*.

In Fig. 15.5, the concept of deviational variables is illustrated. The objective function value, μ_i, is represented on the horizontal axis. The function that we minimize for each objective, also known as the *preference function*, z_i, is represented on the vertical axis. Figure 15.5 shows that the preference function is zero if the design objective is at the target. On either side of the target, the preference function is linear. Note that the slopes of the preference function on either side of the target value can be different, based on the preferences of the designer. This feature is not provided in the weight-based approaches described so far.

For a generic biobjective optimization problem, the GP problem can be stated as:

$$\min_{d_{GP1}^+, d_{GP2}^+, d_{GP1}^-, d_{GP2}^-} J_{gp} = w_{GP1}^+ d_{GP1}^+ + w_{GP1}^- d_{GP1}^- + w_{GP2}^+ d_{GP2}^+ + w_{GP2}^- d_{GP2}^- \qquad \text{Eq. (15.21)}$$

subject to

$$\mu_1 - d_{GP1}^+ \le \alpha_1 \qquad \text{Eq. (15.22)}$$

$$\mu_2 - d_{GP2}^+ \le \alpha_2 \qquad \text{Eq. (15.23)}$$

$$\mu_1 + d_{GP1}^- \ge \alpha_1 \qquad \text{Eq. (15.24)}$$

$$\mu_2 + d_{GP2}^- \ge \alpha_2 \qquad \text{Eq. (15.25)}$$

$$d_{GP1}^+, d_{GP2}^+, d_{GP1}^-, d_{GP2}^- \ge 0 \qquad \text{Eq. (15.26)}$$

where $d_{GP1}^+, d_{GP2}^+, d_{GP1}^-$ and d_{GP2}^- = deviational variables to be minimized; μ_1 and μ_2 = two design objectives; α_1 and α_2 = targets (goals) to be attained for each objective; and $w_{GP1}^+, w_{GP2}^+, w_{GP1}^-$ and

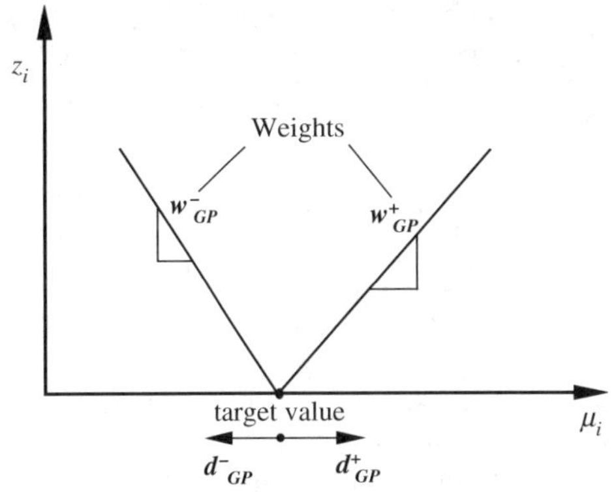

FIG. 15.5 ILLUSTRATION OF THE PREFERENCE FUNCTION IN GOAL PROGRAMMING

w_{GP2}^{-} = slopes of the preference functions for the two objectives to be chosen by the designer.

The AOF of the GP method (Eq. 15.21) is an improvement over the AOF of the weighted sum method (Eq. 15.13) because the designer can specify two different weights for each objective, one on each side of the target value. The GP approach provides flexibility to the designer to specify intra-criterion preferences through these two weights. However, as we explored in Section 15.4.1, the process of choosing weights is not an easy task. The GP approach suffers from the drawback that the designer is required to choose a special set of weights to reflect his/her preferences.

15.5 INTRODUCING PHYSICAL PROGRAMMING

All the methods that we discussed so far for solving multiobjective problems require the designer to specify numerical weights in order to fully define the AOF. This process is usually ambiguous. For example, consider the following: (1) How can the designer specify weights in weight based approaches? (2) Do the weights reflect the designer's preferences accurately? If the designer chooses to increase the importance of a particular objective, by how much should he/she increase the weight? Is 25% adequate? Or is 200% adequate? (3) Does the AOF denote a true mathematical representation of the designer's preferences? (4) How does the designer choose the weights w_{GP}^{+} and w_{GP}^{-} in the goal programming formulation?

Keeping in mind the above-raised questions, we can observe that the problem of determining "good weights" can be difficult and dubious. Due to this ambiguity, the weight selection process is often a computational bottleneck in large-scale design optimization problems. The above discussion paves the way for a multiobjective problem formulation framework that alleviates the above mentioned ambiguities—*Physical Programming.*

The Physical Programming (PP) approach was developed by Messac [12]. It systematically develops an AOF that effectively reflects the designer's wishes. It provides a more natural problem formulation framework, which affords substantial flexibility to the designer. This approach eliminates the need for iterative weight setting, which alleviates the above-discussed ambiguities. Instead of choosing weights, the designer chooses ranges of desirability for each objective. The PP method formulates the AOF from these ranges of desirability, while yielding some interesting and useful properties for the AOF.

Let us now examine the PP application procedure in greater detail. We first study Linear Physical Programming (LPP) in detail, and then proceed to describe Nonlinear Physical Programming (NPP).

15.6 LINEAR PHYSICAL PROGRAMMING

15.6.1 Classification of Preferences

Using PP, the designer can express preferences for each design objective with more flexibility, as opposed to specifying *maximize, minimize, greater than, less than* or *equal to,* which are the only choices available in conventional optimization approaches. Using the PP approach, a designer can express preferences with respect to each design objective using four different classes.

Figure 15.6 illustrates the four classes available in LPP. A generic design objective, μ_i, is represented on the horizontal axis; and the function to be minimized for that objective, z_i, hereby called the *preference function* or the *class function*, is represented on the vertical axis (compare Fig. 15.6 with the preference function of the goal programming approach in Fig. 15.5). Each class consists of two subclasses, *hard* and *soft*, referring to the sharpness of the preference. These subclasses are also illustrated in Fig. 15.6, and are characterized as follows:

(1) Soft Classes:

a. Class 1S: Smaller-is-better (minimization)
b. Class 2S: Larger-is-better (maximization)
c. Class 3S: Value-is-better
d. Class 4S: Range-is-better

(2) Hard Classes:

a. Class 1H: Must be smaller
b. Class 2H: Must be larger
c. Class 3H: Must be equal
d. Class 4H: Must be in range

For example, in the beam problem, the design objectives mass and deflection fall under the Class 1S. Physical Programming offers a flexible lexicon to express ranges of desirability for both hard and soft classes. The lexicon consists of six ranges of desirability for classes 1S and 2S, 10 ranges for the class 3S and 11 ranges for the class 4S.

15.6.2 Physical Programming Lexicon

Let us examine the different ranges of desirability under LPP, with which a designer can express his/her preferences. To illustrate, consider the case of class 1S shown in Fig. 15.7. The ranges of desirability are defined as follows, in the order of decreasing preference:

(1) Ideal Range ($\mu_i \le t_{i1}^{+}$) (Range 1): A range over which every value of the criterion is *ideal*, or the most desirable possible (for example, in the beam problem, the ideal range for the mass of the beam could be specified as $M \le 2{,}000$ kg). Any two points in this range are of equal value to the designer (see discussion in [13]).
(2) Desirable Range ($t_{i1}^{+} \le \mu_i \le t_{i2}^{+}$) (Range 2): An acceptable range that is desirable (for example, the desirable range for the mass of the beam could be specified as 2,000 kg $\le M \le$ 3,000 kg).
(3) Tolerable Range ($t_{i2}^{+} \le \mu_i \le t_{i3}^{+}$) (Range 3): This is an acceptable, tolerable range (for example, 3,000 kg $\le M \le$ 3,500 kg could be specified as the tolerable range for the mass of the beam).
(4) Undesirable Range ($t_{i3}^{+} \le \mu_i \le t_{i4}^{+}$) (Range 4): A range that, while acceptable, is undesirable (for example, the undesirable range for the mass of the beam could be specified as 3,500 kg $\le M \le$ 4,000 kg).
(5) Highly Undesirable Range ($t_{i4}^{+} \le \mu_i \le t_{i5}^{+}$) (Range 5): A range that, while still acceptable, is highly undesirable (for example, 4,000 kg $\le M \le$ 4,500 kg could be specified as the highly undesirable range for the mass of the beam).
(6) Unacceptable Range ($\mu_i \ge t_{i5}^{+}$) (Range 6): The range of values that the design objective must not take (the range $M \ge$ 4,500 kg could be specified as the unacceptable range for the mass of the beam).

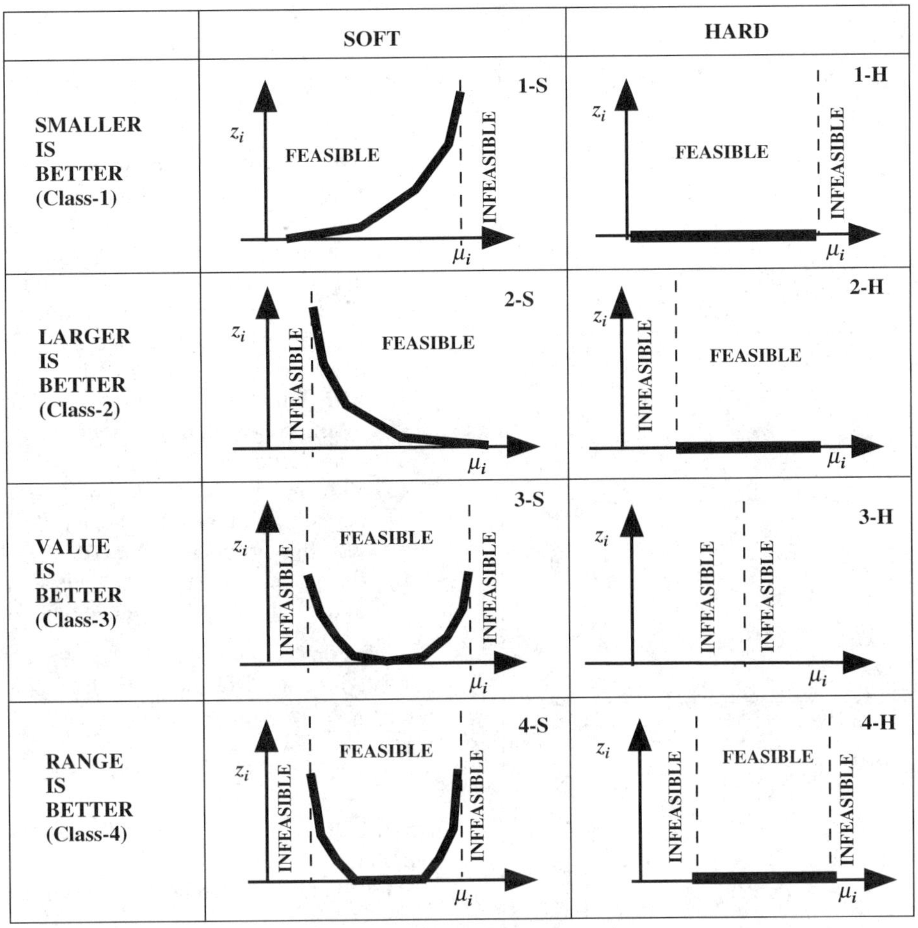

FIG. 15.6 CLASSIFICATION OF DESIGN OBJECTIVES IN LPP

The parameters (t_{i1}^+) through (t_{i5}^+) defined above for soft classes are physically meaningful constants that are specified by the designer to quantify the preferences associated with the ith design objective [for example, the set of values specified above for the mass of the beam in kg (2,000, 3,000, 3,500, 4,000, 4,500)]. The class functions shown in Fig. 15.7 provide the designer the means to express ranges of desirability for a given design objective.

In the case of hard classes, only two ranges are defined, *acceptable* and *unacceptable*. All soft class functions become constituent components of the AOF to be minimized, and all the hard class functions simply appear as constraints in the LPP problem formulation.

The preference functions map the design objectives, such as mass and deflection, into nondimensional, strictly positive real numbers. This mapping, in effect, transforms disparate design objectives with different physical meanings onto a dimensionless scale through a unimodal convex function. The preference functions are piecewise linear and convex in the LPP method, as seen in Fig. 15.7 (recall that a function is unimodal in an interval $a \leq x \leq b$ if and only if it is monotonic on either side of the single optimum point x^* in the interval [2]).

15.6.3 Intra-Criterion and Inter-Criteria Preferences—OVO Rule

Once the designer specifies the ranges of desirability for each design objective using the above-stated PP lexicon, the intra-criterion preference statement is complete [13]. In order to completely formulate the multiobjective optimization problem, the designer also needs to specify the inter-criteria preferences. The PP method operates within an inter-criteria heuristic rule, called the *one versus others* (OVO) *rule*. The inter-criteria preference for each soft criterion, μ_i, is defined as follows. Consider the following options:

(1) Full improvement of μ_i across a given range, and
(2) Full reduction of all the other criteria across the next better range.

Then, the PP method formulates the AOF such that option 1 is preferred over option 2. The OVO rule has a built-in preemptive nature by which the worst criterion tends to be minimized first. For example, consider a multiobjective problem with 10 objectives. According to the OVO rule, it is preferable for a single objective to improve over a full tolerable range, than it is for the remaining nine to improve over the full desirable range. In the next subsection, we explain how the OVO rule is implemented in the LPP method.

15.6.4 Definition of the LPP Class Function

As mentioned in Section 15.6.2, the class function maps the design objectives into nondimensional, strictly positive real numbers that reflect the designer's preferences. In order to be able to do so, the class function, z_i, is required to possess the following properties:

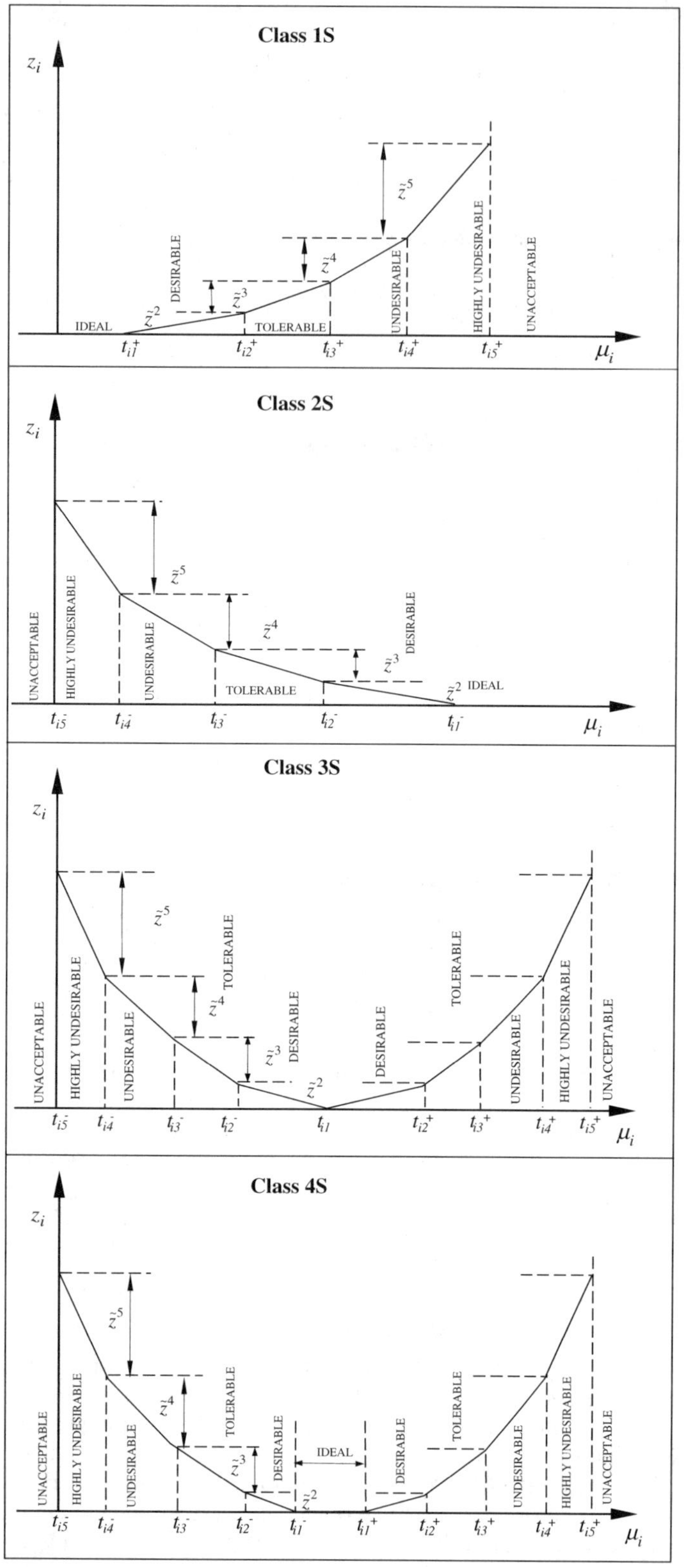

FIG. 15.7 RANGES OF PREFERENCES FOR SOFT CLASSES IN LPP

(1) A lower value of the class function is preferred over a higher value thereof (see Fig. 15.7). Irrespective of the class of a criterion (1S, 2S, 3S or 4S), the ideal value of the criterion always corresponds to the minimum value of the class function, which is zero.

(2) A class function is positive ($z_i \geq 0$).

(3) A class function is continuous, piecewise linear and convex.

(4) The value of the class function at a given range limit, $z_i(t_{is}^+)$ is always fixed (see Fig. 15.7). From criterion to criterion, only the location of the limits (t_{is}^+) changes, but not the corresponding z_i values. Because of this property, as one travels across all the criteria and observes a given range type, the change in the class function value, z_i, will always be of the same magnitude (see Fig. 15.7). This property of the class function results in a normalizing effect, which eliminates numerical conditioning problems that arise because of improper scaling among design objectives of disparate magnitudes.

(5) The magnitude of the class function's vertical excursion across any range must satisfy the OVO rule [we shall represent this property in Eq. (15.29)]. Observe in Fig. 15.7 that the value of $\tilde{z}^2$ (desirable) is less than that of $\tilde{z}^5$ (highly undesirable). This is in keeping with the OVO rule.

Based on the above properties, we now present the mathematical relations used in the LPP algorithm. From property 4 given above, we can write the relation:

$$z^s = z_i(t_{is}^+) = z_i(t_{is}^-); \quad \forall i; \quad (2 \leq s \leq 5); \quad z^1 = 0 \quad \text{Eq. (15.27)}$$

where s and i = a generic range intersection and the soft criterion number, respectively.

The change in z_i across the s th range is given by

$$\tilde{z}^s = z^s - z^{s-1}; \quad (2 \leq s \leq 5); \quad z^1 = 0 \qquad \text{Eq. (15.28)}$$

The OVO rule is enforced by:

$$\tilde{z}^s > (n_{sc} - 1)\tilde{z}^{s-1}; \quad (3 \leq s \leq 5); \quad (n_{sc} > 1) \quad \text{Eq. (15.29)}$$

or

$$\tilde{z}^s = \beta(n_{sc} - 1)\tilde{z}^{s-1}; \quad (3 \leq s \leq 5); \quad (n_{sc} > 1); \quad \beta > 1 \quad \text{Eq. (15.30)}$$

where n_{sc} = number of soft criteria; and β = convexity parameter. In order to use Eq. (15.30), the value of $\tilde{z}^2$ needs to be specified. We can assume $\tilde{z}^2$ to be equal to a small positive number (say 0.1) in practice. Note that Eq. (15.30) does not guarantee convexity of the class function, because the convexity also depends on the targets chosen by the decision-maker.

Let us now present the relations that specifically enforce convexity of the class function. We define the following quantities:

$$\tilde{t}_{is}^+ = t_{is}^+ - t_{i(s-1)}^+; \qquad (2 \leq s \leq 5) \qquad \text{Eq. (15.31)}$$

$$\tilde{t}_{is}^- = t_{is}^- - t_{i(s-1)}^-; \qquad (2 \leq s \leq 5) \qquad \text{Eq. (15.32)}$$

Note that the above equations define the length of the sth range of the ith criterion. Using the above definition, the magnitude of the slope of the class function is given by:

$$w_{is}^+ = \frac{\tilde{z}^s}{\tilde{t}_{is}^+}; \qquad (2 \leq s \leq 5) \qquad \text{Eq. (15.33)}$$

$$\tilde{w}_{is}^{-} = \left|\frac{\tilde{z}^{s}}{\tilde{t}_{is}^{-}}\right|; \qquad (2 \le s \le 5) \qquad \text{Eq. (15.34)}$$

Note that these slopes change from range to range and from criterion to criterion. The convexity requirement can be enforced by using the relation:

$$\tilde{w}_{\min} = \min_{i,s}(\tilde{w}_{is}^{+}, \tilde{w}_{is}^{-}) > 0; \qquad (2 \le s \le 5) \qquad \text{Eq. (15.35)}$$

where

$$\tilde{w}_{is}^{+} = w_{is}^{+} - w_{i(s-1)}^{+}; \qquad (2 \le s \le 5) \qquad \text{Eq. (15.36)}$$

$$\tilde{w}_{is}^{-} = w_{is}^{-} - w_{i(s-1)}^{-}; \qquad (2 \le s \le 5) \qquad \text{Eq. (15.37)}$$

$$w_{i1}^{-} = w_{i1}^{+} = 0; \qquad (2 \le s \le 5) \qquad \text{Eq. (15.38)}$$

The quantities $\tilde{w}_{is}^{+}$ and $\tilde{w}_{is}^{-}$ = slope increments of the class function between the different ranges of desirability. Equation (15.35) implies that if all the incremental weights are positive, the class function (which is piecewise linear) will be convex. Let us now proceed to discuss the algorithm that can be used to define the class function using the equations given in this subsection.

15.6.5 LPP Weight Algorithm

The LPP weight algorithm is given below.

(1) Initialize: $\beta = 1.1$; $w_{i1}^{-} = w_{i1}^{+} = 0$; $\tilde{z}^{s}$ = small positive number, say 0.1. $i = 0$; $s = 1$; n_{sc} = number of soft criteria.
(2) Set $i = i + 1$.
(3) Set $s = s + 1$.
(4) Evaluate in the same order: $\tilde{z}^{s}$, $\tilde{t}_{is}^{+}$, $\tilde{t}_{is}^{-}$, w_{is}^{+}, w_{is}^{-}, $\tilde{w}_{is}^{+}$, $\tilde{w}_{is}^{-}$ and $\tilde{w}_{min}$.
(5) If $\tilde{w}_{min}$ is less than some chosen small positive value (say 0.01), increase β: Set $i = 0$, $s = 1$ and go to step 2.
(6) If $s \ne 5$, go to step 3.
(7) If $i = n_{sc}$, terminate. Otherwise, go to step 2.

A Matlab code that uses this algorithm to compute weights, given the preference values for each criterion, is given in the Appendix. Once we obtain the weights from the above algorithm, we can define the piecewise linear class function for each criterion.

Note that the formulation of the LPP problem involves the presence of numerous weights because of the piecewise linear nature of the class function. Important, however, is the fact that the work of the designer is actually much simpler, as all these weights are automatically evaluated by the LPP weight algorithm. Keeping this in mind, we define the AOF using *deviational variables*, denoted by d_{is}^{-} and d_{is}^{+}. In the LPP method, a deviational variable is defined as the deviation of the ith design criterion from its sth range intersection. The class function for soft classes can then be defined in terms of the deviational variables as

$$z_i = \sum_{s=2}^{5}\{\tilde{w}_{is}^{-}d_{is}^{-} + \tilde{w}_{is}^{+}d_{is}^{+}\} \qquad \text{Eq. (15.39)}$$

15.6.6 LPP Problem Model

So far we have learned some important concepts in LPP. Let us now use these concepts to define the LPP problem. The LPP application procedure consists of four distinct steps.

(1) Specify the class type for each design objective (1S-4H).
(2) Provide the ranges of desirability (t_{is}^{+}, or t_{is}^{-}, or both) for each class (see Fig. 15.7). The designer specifies five limits for classes 1S or 2S, nine limits for class 3S and 10 limits for class 4S. For hard classes, the designer specifies one limit for classes 1H, 2H and 3H, and two limits for 4H (see Fig. 15.6).
(3) Use the LPP weight algorithm to obtain the incremental weights, $\tilde{w}_{is}^{+}$ and $\tilde{w}_{is}^{-}$. Note that the designer does not need to explicitly define the class function z_i.
(4) Solve the following linear programming problem.

$$\min_{d_{is}^{-},d_{is}^{+},x} J = \sum_{i=1}^{n_{sc}}\left[\sum_{s=2}^{5}\{\tilde{w}_{is}^{-}d_{is}^{-} + \tilde{w}_{is}^{+}d_{is}^{+}\}\right] \qquad \text{Eq. (15.40)}$$

subject to

$$\mu_i - d_{is}^{+} \le t_{i(s-1)}^{+}; \quad d_{is}^{+} \ge 0; \quad \mu_i \le t_{i5}^{+} \quad (1S,3S,4S) \qquad \text{Eq. (15.41)}$$

$$\mu_i + d_{is}^{-} \ge t_{i(s-1)}^{-}; \quad d_{is}^{-} \ge 0; \quad \mu_i \ge t_{i5}^{-} \quad (2S,3S,4S) \qquad \text{Eq. (15.42)}$$

$$\mu_j(x) \le t_{j,max} \qquad 1H \qquad \text{Eq. (15.43)}$$

$$\mu_j(x) \ge t_{j,min} \qquad 2H \qquad \text{Eq. (15.44)}$$

$$\mu_j(x) = t_{j,val} \qquad 3H \qquad \text{Eq. (15.45)}$$

$$t_{j,\,min} \le \mu_j(x) \le t_{j,\,max} \qquad 4H \qquad \text{Eq. (15.46)}$$

$$x_{min} \le x \le x_{max} \qquad \text{Eq. (15.47)}$$

where $i = \{1, 2, ..., n_{sc}\}$, $s = \{2, ..., 5\}$, $j = \{1, 2, ..., n_{hc}\}$; n_{hc} = number of hard classes; x = design variable vector; and $\mu_i = \mu_i(x)$.

The above formulation concludes our presentation of the LPP method. Let us now proceed to describe the NPP method.

15.7 NONLINEAR PHYSICAL PROGRAMMING

The NPP method can be advantageous when compared to the LPP method in solving optimization problems. When all the constraints and objective functions are linear in terms of the design variables, the LPP method is indeed the one of choice. However, when the constraints or the objective functions are nonlinear, the LPP method should be avoided. The piecewise linear nature of the class function in LPP may lead to computational difficulties because of the discontinuities in the class function derivatives at the intersection of the range limits. The NPP method alleviates this difficulty by defining a class function that is smooth across all range limits. However, it operates fully in the nonlinear programming domain, and as such, is computationally less efficient.

In this section, we provide a brief discussion of the NPP method. Interested readers can refer to [12] for a more detailed description. Let us begin our discussion of NPP by first identifying the similarities and differences between LPP and NPP.

15.7.1 LPP versus NPP: Similarities and Differences

There are a few similarities and differences between the LPP and the NPP methods. The following are the similarities:

(1) The class and the subclass definitions are the same in LPP and NPP.
(2) The PP lexicon and the classification of preferences are the

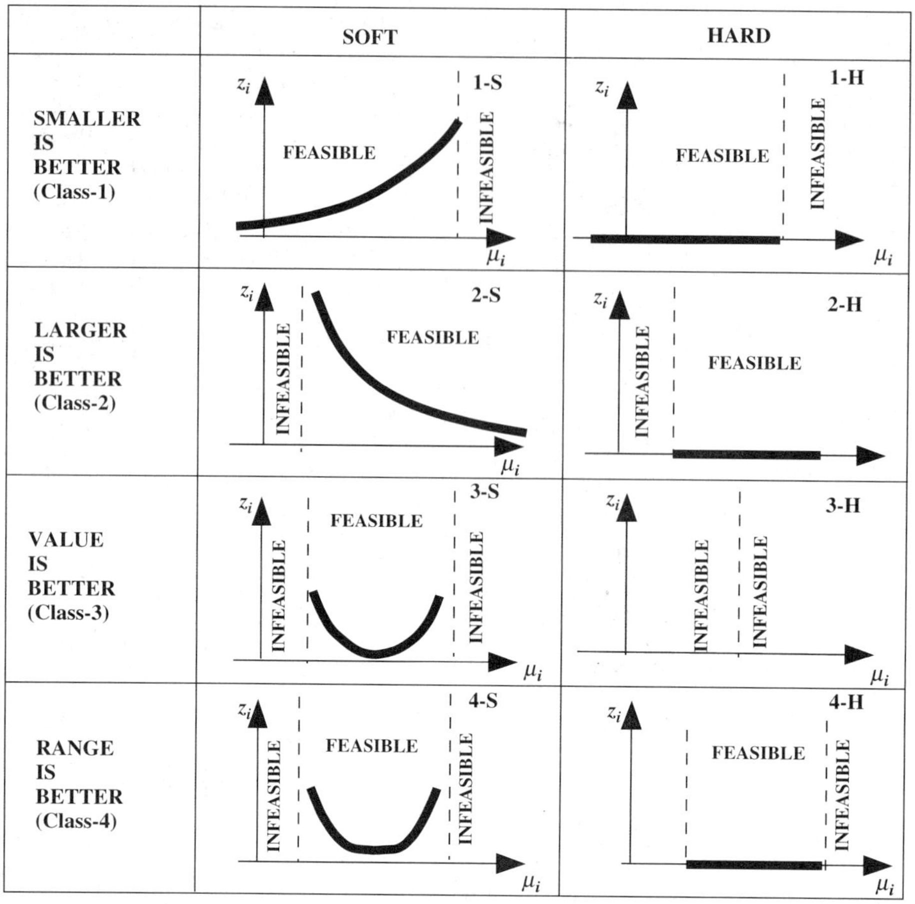

FIG. 15.8 CLASSIFICATION OF DESIGN OBJECTIVES IN NPP

same for NPP and LPP, with one exception: the analog of *ideal* (LPP) is *highly desirable* (NPP). Figures 15.8 and 15.9 show the classification of design objectives and the ranges of preferences for soft classes in NPP.

(3) The OVO rule is defined in the same manner in LPP and NPP.

How then are LPP and NPP different from each other? Compare the class function plot in Fig. 15.9 with that in Fig. 15.7. In the case of NPP, the class functions are not piecewise linear. In fact, they are nonlinear and smooth.

Specifically, the class functions in NPP are defined using a special class of *splines*. A detailed discussion on the mathematical development of these splines can be found in [12]. More flexible and powerful splines were developed later. Here, we present a summary of the mathematical background for NPP.

15.7.2 Definition of the NPP Class Function

A suitable class function in NPP must possess the following properties.

(1) All soft class functions must

a. Be strictly positive
b. Have continuous first derivatives
c. Have strictly positive second derivatives (implying convexity of the class function)

(2) All the above-defined properties must hold for any practical choice of range limits.

Observe the class functions for NPP given in Fig. 15.9. The class function in the highly desirable range is defined by a decaying exponential function, while in all the other ranges, the class functions are defined by spline segments [12]. A complete description of the class function properties and definition is provided in [12].

15.7.3 NPP Problem Model

Having defined the class function, we now use the following steps to generate an NPP problem model:

(1) Specify the class type for each design objective (1S–4H)
(2) Provide the ranges of desirability for each design objective (see Fig. 15.9)
(3) Solve the constrained nonlinear minimization problem that is given by

$$\min_{x} \quad J = \log_{10}\left[\frac{1}{n_{sc}}\sum_{i=1}^{n_{sc}} z_i[\mu_i(x)]\right] \quad \text{(for soft classes)} \qquad \text{Eq. (15.48)}$$

subject to

$$\mu_i(x) \le t_{i5}^{+} \quad \text{(for class 1S objectives)} \qquad \text{Eq. (15.49)}$$

$$\mu_i(x) \ge t_{i5}^{-} \quad \text{(for class 2S objectives)} \qquad \text{Eq. (15.50)}$$

$$t_{i5}^{-} \le \mu_i(x) \le t_{i5}^{+} \quad \text{(for class 3S, 4S objectives)} \qquad \text{Eq. (15.51)}$$

$$\mu_j(x) \le t_{i,max} \quad \text{(for class 1H objectives)} \qquad \text{Eq. (15.52)}$$

$$\mu_j(x) \ge t_{i,min} \quad \text{(for class 2H objectives)} \qquad \text{Eq. (15.53)}$$

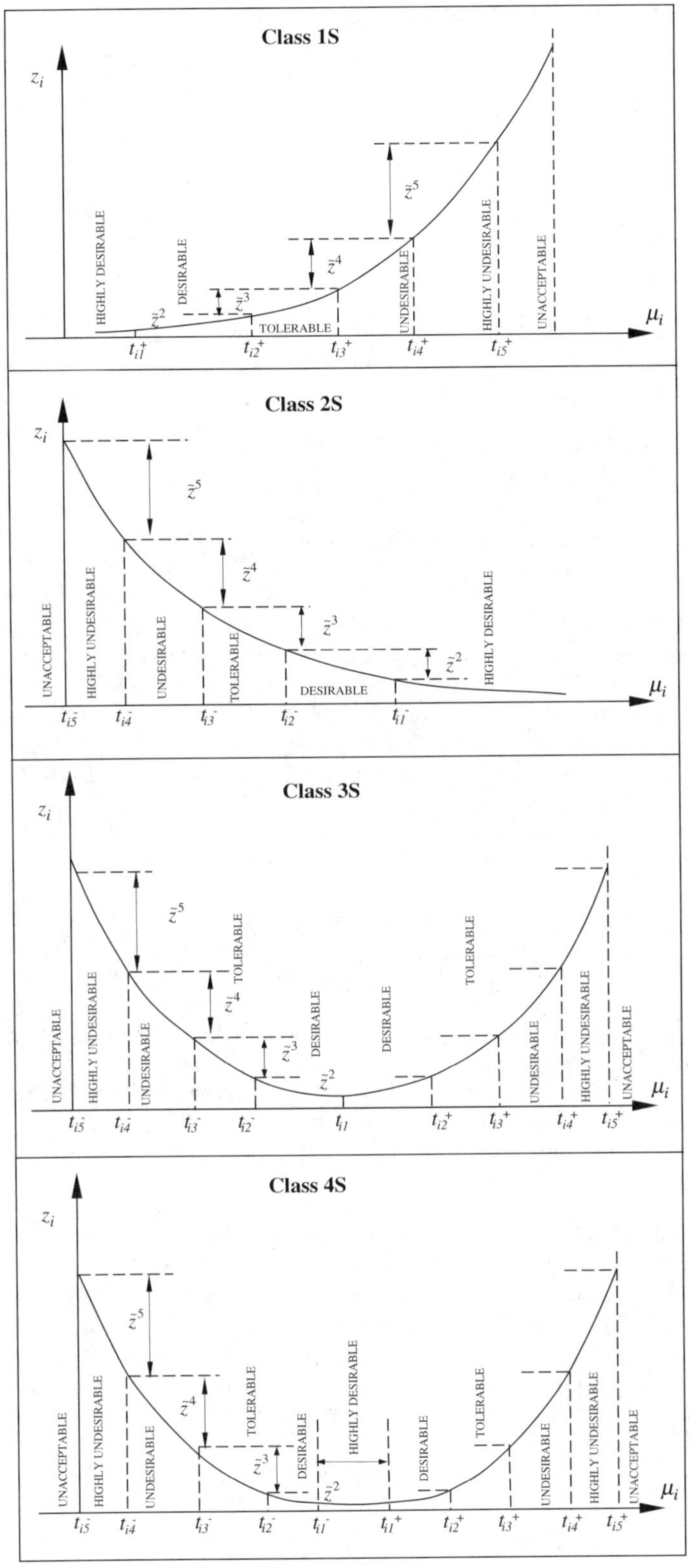

FIG. 15.9 RANGES OF PREFERENCES FOR SOFT CLASSES IN NPP

$$\mu_j(x) = t_{i,val} \quad \text{(for class 3H objectives)} \qquad \text{Eq. (15.54)}$$

$$t_{j,\,min} \le \mu_j(x) \le t_{j,\,max} \quad \text{(for class 4H objectives)} \qquad \text{Eq. (15.55)}$$

$$x_{j,\min} \le x_j \le x_{j,\max} \quad \text{(for design variables)} \qquad \text{Eq. (15.56)}$$

where $t_{i,\,min}$, $t_{i,\,max}$, and $t_{i,\,val}$ = specified preferences values for the ith hard objective; $x_{j,\,min}$ and $x_{j,\,max}$ = minimum and maximum values, respectively, for x_j; ranges of desirability, t^+_{i5} and t^-_{i5}, are provided by the designer; and n_{sc} = number of soft objectives. In the above formulation, observe that hard classes are treated as constraints and soft classes become part of the objective function.

15.8 COMPARISON OF THE GP AND LPP METHODS

In this section, we compare the flexibility offered by the LPP method to that offered by the GP method. Figure 15.10 shows the flexibility offered to the designer by each of the above methods. The GP method offers limited flexibility with the option of choosing two weights and a target value for each objective. As discussed in Section 15.4.3, it is not intuitively clear how to choose the appropriate values of the weights that reflect one's preferences with respect to each design objective.

The LPP method, on the other hand, lets the designer choose up to 10 physically meaningful target values or ranges of desirability for each design objective. The LPP method defines the class function for each objective, and completely eliminates the need for the designer to deal with weights.

Figure 15.11 shows the behavior of the AOF for the GP and the LPP methods [see Eqs. (15.21) and (15.40)], respectively. The XY plane of each figure shows the contour plots of the AOF for each method. In the typical GP form, we have two-sided goal criteria, yielding an intersection of four planes. Also observe that the contour plots of the GP AOF are quadrilaterals. In LPP, the AOF surface is obtained by the intersection of 81 planes (for the 4-S

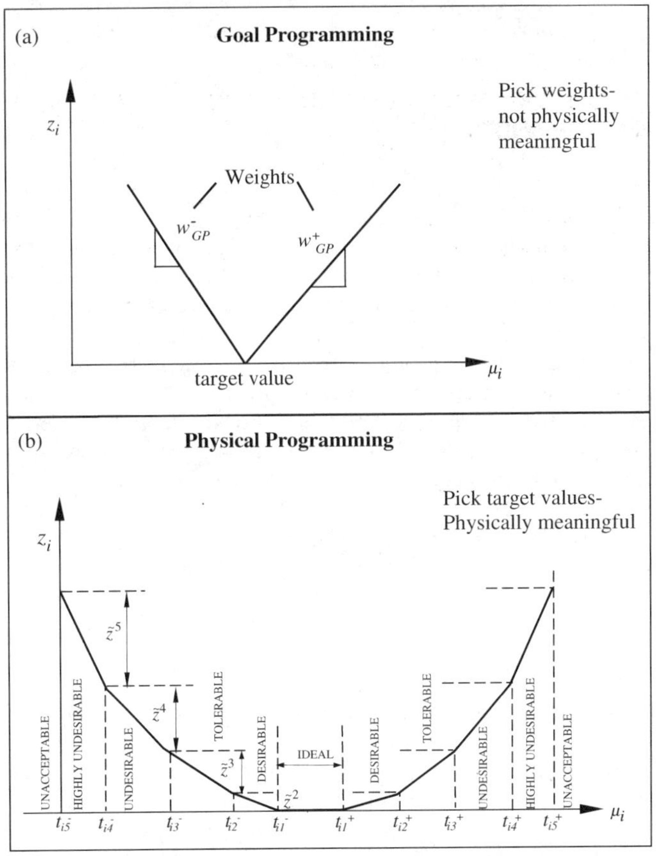

FIG. 15.10 COMPARISON OF FLEXIBILITY OF GP AND LPP

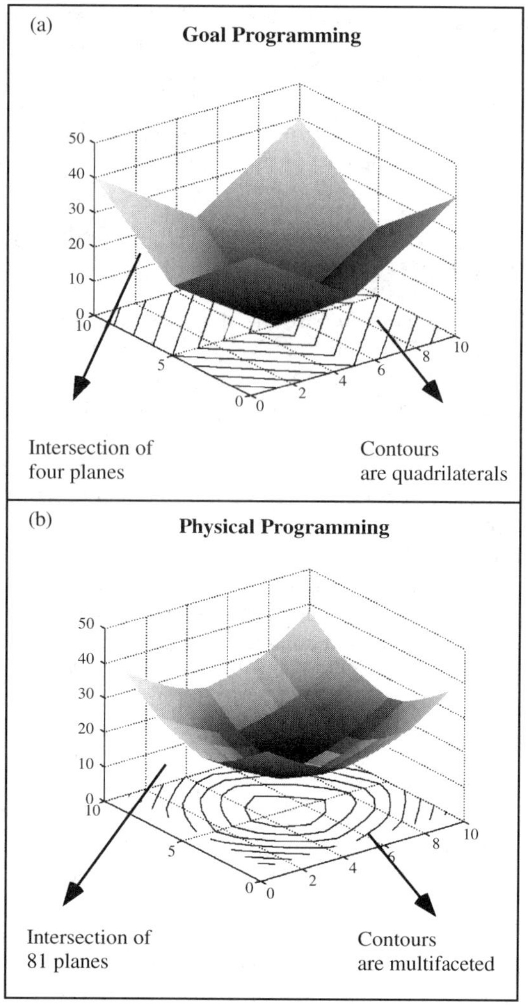

FIG. 15.11 THREE-DIMENSIONAL VISUALIZATION OF THE AOF – GP AND LPP

criterion), which reflects a more realistic preference. Observe the multifaceted contours of the AOF for the LPP method. Note that, even when one should use a multisided GP function, the designer is still left with the prohibitive task of choosing a large number of weights under GP.

The LPP method provides the designer with the flexibility of effectively specifying ranges of preferences (such as ideal, desirable, tolerable), in contrast with the GP method. The effectiveness of LPP comes from the judiciously defined class function, which tailors itself to the complex nature of the designer's choices.

Let us solve an example problem using the GP and the LPP methods, and then compare the results obtained.

15.9 EXAMPLE

A company manufactures two products, A and B. The ideal production levels per month for A and B are 25 units and 10 units, respectively. The profit per unit sold for A and B are \$12K and \$10K, respectively. Under these conditions, the monthly profit is \$400K. The company needs to make a profit of at least \$580K to stay in business. The designer has certain preferences for the

TABLE 15.1 PREFERENCE RANGES FOR μ_1 AND μ_2

Preference Level	μ_1	μ_2
Ideal	<25	<10
Desirable	25–31	10–18
Tolerable	31–36	18–26
Undesirable	36–44	26–33
Highly undesirable	44–50	33–40
Unacceptable	>50	>40

production levels for A and B, given in Table 15.1. Let us define μ_1 and μ_2 as the two design criteria, which denote the production levels of products A and B, respectively. The profit constraint function is given as:

$$12\mu_1 + 10\mu_2 \geq 580 \qquad \text{Eq. (15.57)}$$

The GP formulation for this problem is given as

$$\min_{\mu_1, \mu_2, d^+_{GP1}, d^+_{GP2}} \left[w^+_{GP1} d^+_{GP1} + w^+_{GP2} d^+_{GP2} \right] \qquad \text{Eq. (15.58)}$$

subject to

$$\mu_1 - d^+_{GP1} \leq 25 \qquad \text{Eq. (15.59)}$$

$$\mu_2 - d^+_{GP2} \leq 10 \qquad \text{Eq. (15.60)}$$

$$12\mu_1 + 10\mu_2 \geq 580 \qquad \text{Eq. (15.61)}$$

$$\mu_1 \leq 50 \qquad \text{Eq. (15.62)}$$

$$\mu_2 \leq 40 \qquad \text{Eq. (15.63)}$$

$$d^+_{GP1}, d^+_{GP2}, \mu_1, \mu_2 \geq 0 \qquad \text{Eq. (15.64)}$$

The slopes of the preference functions for the GP formulation are specified by w^+_{GP1} and w^+_{GP2}. The target for μ_1 is 25 and that of μ_2 is 10. The results obtained using GP are shown in Fig. 15.12(a), (b) and (c). The three solutions obtained with GP are for the cases where the ratio of slopes w^+_{GP1} / w^+_{GP2} is less than, equal to and greater than 12/10 = 1.2.

In Fig. 15.12, the shaded area represents the feasible region and the solid dots represent the optimum solutions. The solution when $w^+_{GP1} / w^+_{GP2} < 1.2$ is the point P = (40, 10) in Fig. 15.12(a). The solution when $w^+_{GP1} / w^+_{GP2} > 1.2$ is the point Q = (25, 28) in Fig. 15.12(c). In Fig. 15.12(b), when $w^+_{GP1} / w^+_{GP2} = 1.2$, the slope of the objective function given in Eq. (15.59) is equal to that of the constraint given in Eq. (15.62). There are infinitely many solutions along the straight line segment shown by the thick line in Fig. 15.12(b).

Let us now examine how LPP can be used to solve this problem.

From the values of preferences given in Table 15.1, we note that μ_1 and μ_2 belong to the class 1S. The LPP model is formulated using the linear programming model given in Section 15.6.6, Eq. (15.40). The solution obtained is R = (31, 20.8), as shown in Fig. 15.12(d).

Compare the solutions P and Q obtained by the GP method, and the solution R obtained by the LPP method. The solution obtained with GP highly depends on the weights chosen for each objective. For the point P = (40, 10) [see Fig. 15.12(a)], μ_1 lies in the undesirable range and μ_2 lies in the desirable range. For the point Q = (25, 28) [see Fig. 15.12(c)], μ_1 lies in the desirable range and μ_2 lies in the

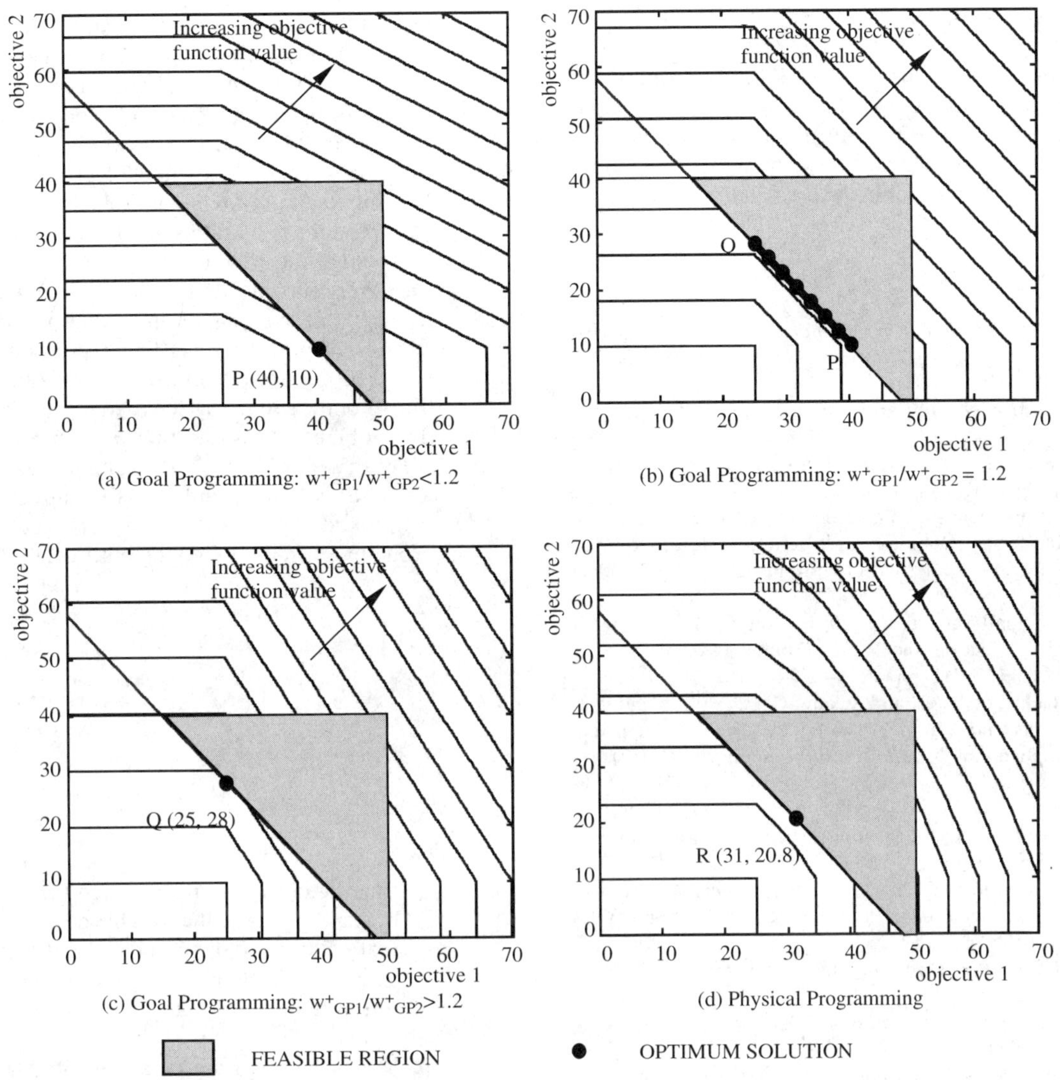

FIG. 15.12 EXAMPLE: SOLUTION OBTAINED USING GP AND LPP

undesirable range. The solutions P and Q lie in the undesirable ranges because the GP problem formulation does not fully model the designer's preferences given in Table 15.1. The LPP method, on the other hand, utilizes all the information provided by the designer in Table 15.1 to formulate the problem. With the LPP method, observe that the optimum point R = (31, 20.8) [Fig. 15.12(d)] lies on the desirable/tolerable boundary for μ_1 and within the tolerable range for μ_2.

Also, observe the contours of the LPP AOF versus the GP AOF in Fig. 15.12. The shape and the number of sides of these contours are significantly different. These observations can be better understood from Fig. 15.11, where a 3-D representation is shown.

15.10 SUMMARY

Multiobjective optimization is a useful tool for the design of large-scale multidisciplinary systems. Most numerical optimization algorithms are developed for application to single-objective problems. In order to pose the multiobjective problem in a single-objective framework, the designer needs to effectively *aggregate* the criteria into a single AOF. In doing so, he/she has to model the intra-criterion and inter-criteria preferences into the AOF. There are several methods available in the literature to aggregate preferences. In this chapter, we discussed some popular methods available to formulate a multiobjective optimization problem. We discussed their relative advantages and shortcomings.

We studied the PP framework for AOF formulation. The PP method provides a framework to unambiguously incorporate the designer's preferences into the AOF. The PP method precludes the need for the designer to specify physically meaningless weights. The PP algorithm generates the weights of the class function based on the designer's preferences, allowing the designer to focus on specifying physically meaningful preference values for each objective. This renders this method unique, and provides an effective framework for multiobjective decision-making.

We also discussed the LPP and the NPP approaches in this chapter. The PP method has been applied to a wide variety of applications, such as product design, multiobjective robust design, production planning and aircraft structure optimization. Interested readers are invited to visit www.rpi.edu/~messac to access more publications on Physical Programming.

ACKNOWLEDGMENTS

I would like to express my special thanks to my doctoral student Sirisha Rangavajhala for her substantial and extensive help in the creation of this chapter. Her thoughtful contribution has been invaluable and is truly appreciated. Many thanks also go to my doctoral student Anoop Mullur for his valued contributions.

REFERENCES

1. Arora, J. A., 1989. *Introduction to Optimal Design*, Mc Graw Hill, Inc.
2. Reklaitis, G. V., Ravindran, A. and Ragsdell, K. M., 1983. *Engineering Optimization: Methods and Applications*, John Wiley and Sons, New York, NY.
3. Vanderplaats, G. N., TK. *Numerical Optimization Techniques for Engineering Design: With Applications*, 3rd Ed., Vanderplaats Research and Development, Inc, Colorado Springs, CO.
4. Messac, A., 2000. "From Dubious Construction of Objective Functions to the Application of Physical Programming",. *AIAA J.*, **38**(1), pp. 155–163.
5. Messac, A. and Ismail-Yahaya, A., 2001. "Required Relationship Between Objective Function and Pareto Frontier Orders: Practical Implications," *AIAA J.*, **39**(11), pp. 2168– 2174.
6. Messac, A., Melachrinoudis, E. and Sukam, C. P., 2000. "Aggregate Objective Functions and Pareto Frontiers: Required Relationships and Practical Implications," *Optimization and Engrg. J.*, **1**(2), pp. 171–188.
7. Charnes, A., Cooper, W. W. and Ferguson, R. O., 1955. "Optimal Estimation of Executive Compensation by Linear Programming," *Mgmt. Sci.*, **1**(2), pp. 138–151.
8. Charnes, A. and Cooper, W. W., 1961. *Management Models and Industrial Applications of Linear Programming, Vol. 1*, John Wiley and Sons, New York, NY.
9. Ijiri, Y., 1965. *Management Goals and Accounting for Control*, Rand McNally, Chicago, IL.
10. Lee, S. M., 1972. *Goal Programming for Decision Analysis*, Auerbach Publishers, Philadelphia, PA.
11. Ignizio, J. P., 1976. *Goal Programming and Extensions*, Springer-Verlag, Berlin, Germany.
12. Messac, A., 1996. "Physical Programming: Effective optimization for design". *AIAA Journal*, **34**(1), p. 149.
13. Messac, A., Gupta, S. and Akbulut, B., 1996. "Linear Physical Programming: A New Approach to Multiple Objective Optimization," *Trans. on Operational Res.*, Vol. **8**, pp. 39–59.

PROBLEMS

Note: The problems given here assume the use of Matlab. You may also use any other software of your choice.

15.1 You are a water-storage-tank builder. You have a special order to build the tank shown in Fig. 15.13. The cost of building the tank (per unit volume) is \$ p. You are asked to build a tank with the largest possible capacity, but the total cost of building it should not exceed \$ a. Due to space restrictions, the width, w, of the tank is required to be equal to half the height, h, of the tank, and the depth, d, of the tank cannot exceed 3 feet. The thickness, t, of the tank is allowed to lie between 0.3 feet and 0.5 feet.
Formulate an optimization problem by defining the design variables, the constraints and the objective function.

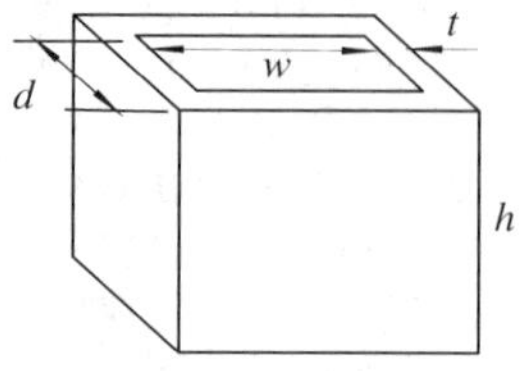

(Note: All Dimensions are in feet)

FIG. 15.13 PROBLEM 1: WATER TANK

15.2 Download the Matlab optimization toolbox tutorial from the Web site of Mathworks Inc., www.mathworks.com. Explore the commands in Matlab to solve the following classes of problems: (1) linear programming; (2) unconstrained nonlinear optimization; and (3) constrained nonlinear optimization. Discuss your findings.

15.3 In problem 1, assume that p = \$5 and a = \$500. Classify this problem (for example, as linear/nonlinear, etc.). Solve the problem using an appropriate Matlab command.

15.4 You are given the following optimization problem.

$$\min_{x} \quad 8x_1 + 10x_2 + 4 \qquad \text{Eq. (15.65)}$$

subject to

$$x_1 - x_2 \geq -4 \qquad \text{Eq. (15.66)}$$

$$x_1 + x_2 \leq 6 \qquad \text{Eq. (15.67)}$$

$$x_1, x_2 \geq 0 \qquad \text{Eq. (15.68)}$$

a. Plot the objective function and the constraint equations by hand. Identify the feasible design space and the optimal solution. Plot the objective function and the constraints with Matlab and compare your plots. Categorize this problem into one of the classes discussed in the chapter.
b. Use an appropriate command in Matlab to solve this particular class of problems, and compare your results from part (a).
c. Now, let us remove all the constraints from the above problem. By looking at our plots, comment on what the new optimum will be. What can you say about unconstrained optimization in this class of problems?

15.5 Explain the concept of Pareto optimality in your own words with the help of an engineering example of your choice. Clearly state all the assumptions made.

15.6 You are given the following biobjective optimization problem.

$$\min_{x} \quad \mu_1 = x^2; \quad \mu_2 = (x-4)^2 \qquad \text{Eq. (15.69)}$$

subject to

$$-5 \leq x \leq 5 \qquad \text{Eq. (15.70)}$$

a. Provide the multiobjective problem formulation using the weighted sum method.
b. Use the weighted sum method to generate the Pareto frontier.
c. Comment on the performance of the method. (Hint: Write a program in Matlab, which sequentially varies weights; each set of weights will yield a Pareto point.)

15.7

a. Repeat parts (a) and (b) in the above problem for the following problem in Eqs. (15.72) and (15.73). Did the weighted sum method perform satisfactorily? Explain.

$$\min_{x} \quad \mu_1 = \sin^2(x); \ \ \mu_2 = 1 - \sin^7(x) \qquad \text{Eq. (15.71)}$$

subject to

$$0.5326 \le x \le 1.2532 \qquad \text{Eq. (15.72)}$$

b. For the above problem, use the compromise programming method to obtain the Pareto frontier. Choose an appropriate exponent for compromise programming. How do the results compare with those of the weighted sum method?

15.8 Consider the beam example used in the chapter (see Fig. 15.1).

a. Generate the single-objective optimization results given in Section 15.2.
b. Duplicate the results that are generated by the weighted sum method shown in Fig. 15.3.
c. Now, choose the weights appropriately such that the difference in the magnitudes of the two weighted objectives is compensated. Generate the entire Pareto frontier shown by the thick line in Fig. 15.3 using appropriate weights.

15.9 You are a manufacturer of steel bolts. You are concerned about the following production parameters, and would like to perform multiobjective optimization: (1) cost per bolt; and (2) annual production volume. Use the PP method to perform the following tasks:

a. Choose an appropriate class function for each criterion.
b. Define some reasonable ranges of desirability (assume values of your choice; use your engineering judgement).
c. Sketch a figure analogous to Fig. 15.7 for your case.

15.10 You are given the following ranges of desirability for two criteria: t_1= [2 4 6 8 10] and t_2 = [20 40 60 80 100]. Identify the classes to which these criteria belong. Using the LPP weight generation code given in the Appendix, compute the weights of the preference functions. Prepare a flowchart of the LPP weight algorithm.

15.11 Read the 2000 paper by Messac, A.: "From Dubious Construction of Objective Functions to the Application of Physical Programming" (AIAA J. 38, (1), pp. 155–163; available at www.rpi.edu/~messac). Prepare a two-page summary in your own words.

APPENDIX

Matlab Code for the LPP weight algorithm:

```
% This code is used to compute weights for
% LPP.
% Inputs to the function are:
% nsc = no. of objectives
% t = preference matrix: eg. [10 20 30 40
%     50; 10 20 30 40 50];
% type1 = "1S" or "2S"

function [weights, z1] = lppw(t, nsc, type1);

% The following parameters are initialized.
%Note that "b" here stands for "beta"
b=1.1;
zbar(2)=0.1;
min1=0.0001;
min2=-0.01;
i=1;

%w(i, j) denotes the weight for the jth
%class of the ith criterion.
% Similar notation is used for all the
%other variables.
for i=1:nsc

    w(i,1)=0
end

while(i<=2)
    for s=2:5
        if s>2

    zbar(s)=(nsc-1)*(zbar(s-1))*b
    end

    tbar(i,s)=t(i,s)-t(i,s-1)
    w(i,s)=(zbar(s))/(tbar(i,s))
    w _ tilda(i,s)=w(i,s)-w(i,s-1)]

% Finding wmin
    if type1= "1S"

        if s==2

         wmin1=[w _ tilda(i,s)]      %Eq.(15.35)

        else
            if [w _ tilda(i,s)]<wmin1
                wmin1=[w _ tilda(i,s)]
           end
        end
      else
        if s==2
           wmin2=(w _ tilda(i,s))
        else
          if (w _ tilda(i,s))>wmin2
             wmin2=(w _ tilda(i,s))

           end
        end
      end
    end

% Checking for tolerance, incrementing b
    if type1= = "1S"
           if wmin1>min1
           % if wmin1 is greater than some
           %small positive value,
           % go to the next objective,
           %increment i
               i=i+1
           else
```

```
        %otherwise, repeat the proce
        %dure for all criteria
              i=1;
        %increment b
              b=b+0.5
        end

else
        if wmin2<min2
              i=i+1;
        else
              i=1;
              b=b+0.5;
                    end
              end
        end
        z(1)=0;
        %Finding the class function value
         for s=2:5
              z(s)=zbar(s)+z(s-1);
        end
        z1=z;
        weights=abs(w _ tilda);
```

SECTION

6

MAKING PRODUCT DESIGN DECISIONS IN AN ENTERPRISE CONTEXT

INTRODUCTION

The design and development of a product involve decision-makers from all activity areas in any enterprise and have an impact on all aspects of its operation. The word *enterprise* is used in this chapter to mean the operation of the business entity as a whole, not just a single product line or functional division. Design decisions, for all but the simplest of products, will always involve trade-offs between engineering performance criteria (i.e., technical objectives). However, there are also many competing enterprise performance objectives that bear consideration during design decision-making as well. Examples of enterprise-level objectives include the development of reliable demand models to estimate the market strength a product area has (the topic of Section 4), the growth and acquisition of proprietary technical knowledge, labor costs for manufacturing and assembly, long-term corporate strategies for growth and environmental regulations imposed by states and countries. These enterprise-level performance objectives are as dynamic a set as any, and they influence design and development.

While Section 5 presents multicriteria approaches that aggregate the preference of decision-makers and focus on trade-offs between technical objectives, the first three chapters of this section (Chapters 16, 17 and 18) promote a top-down, single-criterion approach to design decision-making that focuses on the economic benefits accrued by a product's design. The adopted perspective is maximizing economic benefits of product design decisions as they contribute to enterprise-wide profit-making activities. The methods presented in Chapters 16 and 18 show quantitatively the single-criterion optimization based decision-making models for hierarchical and non-hierarchical complex systems, respectively. In Chapter 17 the principles of economics and finance relevant to product development are introduced and used with engineering technical performance attributes to define a design decision model for the product development firm to determine an optimal level of product differentiation. Integrating the effects of these decisions on required investments and, subsequently, overall business profit reveals that engineering design decisions ultimately impact consumer choice behavior and production costs. Thus, one of the themes of these first several chapters is that design decisions on product attributes must be considered in concert with all other decisions in the business enterprise of the larger organization.

Work in the next two chapters of Section 6 broaden the decision-making perspective to include observed evolution and behavior in actual product development organizations. Chapter 19 provides a bridge between methods for design decision-making tasks and a broader view of design as part of a larger, enterprise-wide, decision-making system. The vehicle development process provides the context for the discussion of design decision-making in Chapter 19 and demonstrates the connection between decision support tools and decision-making tasks. Chapter 20 presents an approach for viewing design decision-making as a decision production system, a process characterized by information flow among decision-makers.

The chapters in this Section display differences between the use of mathematical optimization models to make product design decisions and a reliance on a systems-level model of decision-makers who are informed by the results of decision-support tools. In Chapters 16, 17 and 18 it is implied that product development decisions can be made through mathematical modeling. Later chapters in the section describe product design decision-making as relying on a systems-level model of decision-makers. These are not inconsistent positions. In Chapters 19 and 20 the profitability objective is not ignored. Rather, because other aspects of the design process aren't easily modeled in profitability optimization approaches, the profitability problem is, in practice, decomposed into a series of interrelated decisions, as described in these chapters.

While developing decision-making approaches that model all the inherent uncertainty and risk in design is beyond the scope of any single section of this text, it should be noted that several chapters in Section 6 do provide illustrations for adding uncertainty into decision problem formulation. The first three chapters use a single-criterion utility optimization approach to optimize the expected utility, which captures the uncertain outcome of the profitability objective and decision-maker's risk attitude toward this uncertainty.

CHAPTER

16

DECISION-BASED COLLABORATIVE OPTIMIZATION OF MULTIDISCIPLINARY SYSTEMS

John E. Renaud and Xiaoyu (Stacey) Gu

NOMENCLATURE

$\boldsymbol{a}$ = vector of attributes
C_F = fixed cost
C_T = total cost
C_u = unit cost
C_V = variable cost
c_i = demand coefficient for attribute i
c_{price} = demand coefficient for price
$\boldsymbol{d}$ = vector of compatibility (or discrepancy) constraints
d_i = compatibility constraint in discipline i
$E(?)$ = expected value
F = objective function in the optimization problem
$\boldsymbol{g}$ = vector of constraints
$\boldsymbol{g}_i$ = vector of constraints in discipline i
NR = net revenue
n_{cost} = number of cost-related variables
n_a = number of attributes
P = price
PNR = present value of net revenue
n_{ss} = number of subspaces (disciplines)
$\boldsymbol{p}$ = vector of parameters
$(p_v)_i$ = variable cost parameter associated with cost-related
= variable i
q = demand
q_B = baseline demand
u = utility
$\boldsymbol{x}$ = vector of design variables
$\boldsymbol{x}_{aux}$ = vector of auxiliary design variables
$(\boldsymbol{x}_{aux})_{ij}$ = vector of auxiliary design variables corresponding to $\boldsymbol{y}_{ij}$
$\boldsymbol{x}_i$ = vector of local design variables for discipline i
$\boldsymbol{x}_{sh}$ = vector of shared design variables
$(\boldsymbol{x}_{sh})_i$ = vector of shared design variables used in discipline i
$\boldsymbol{x}_{ssi}$ = vector of subspace design variables in discipline i
$(\boldsymbol{x}^*_{ss})_i$ = vector of optimal target values in discipline i
$\boldsymbol{x}_{sys}$ = vector of system-level design variables
$(x^{o}_{sys})_i$ = vector of targets (i.e., system-level design variables)
= sent down to discipline i
y = vector of state variables
y_i = vector of state variables calculated as output from discipline i
y_{ij} = vector of coupling variables calculated as output from discipline i and used as input to discipline j
$(?)^{o}$ = system-level target variable
$(?)^{*}$ = variable at the optimum design

INTRODUCTION

Increasing attention has been paid to the notion that engineering design is a decision-making process [7, 13, 1]. This notion is consistent with the definition of decision as a choice from among a set of options and as an irrevocable allocation of resources. The approach of decision-based design (DBD) is built upon this notion. Rooted in more than 200 years of research in the field of decision science, economics, operations research and other disciplines, DBD provides a rigorous foundation for design, which enables engineers to identify the best trade-off and focus on where the payoffs are greatest.

Engineering design involves the generation of design alternatives or options and the selection of the best one. Since the number of possible design options is practically infinite for most products, human judgment is needed to decide which options to include in the consideration of alternative designs and which to neglect. Moreover, an appropriate value measure has to be determined in order to compare and rank-order design options. It is impossible to know exactly how a particular design alternative will perform before it is built. However, the product cannot be built until it is selected. Evidently engineers have to make their selection a priori, without full knowledge of the consequence of this certain selection. Thus design is always a matter of normative decision-making under uncertainty and risk.

Optimization techniques have been widely applied to select the preferred design options from the set of alternatives taken into consideration without having to explicitly evaluate all possible alternatives in the set. There exists a close relationship between decision-making and optimization. In general, decision-making has three elements: generation or identification of options, assignment of expectations on each option, and the application of preferences to determine the preferred choice. An optimization problem involves the maximization or minimization of an objective function or functional $F(\boldsymbol{x})$ in the feasible region of design variables $\boldsymbol{x}$. A careful comparison between decision-making and optimization will reveal that options, expectations and preferences are all present in optimization. The option space is equivalent to the set of permissible values of $\boldsymbol{x}$ in the feasible region. The expectation of any given $\boldsymbol{x}$ is assigned by $F(\boldsymbol{x})$, and the preference is stated that more is better (maximization) or less is better (minimization). Thus optimization can be used to capture the properties of decision-making. This recognition allows the application of rigorous decision theory to the case of optimization.

16.1 DECISION-BASED DESIGN FRAMEWORK

Application of DBD within an optimization domain requires practitioners to formulate valid objective functions for proper decision-making. Most of the research in the field of optimization has been focused on the solution of the optimization problem such as development, improvement and implementation of search methods to locate the optimum, while little attention has been paid to optimization problem formulation. In fact, the issue of problem formulation is of the same significance, if not more, as the issue of finding the optimal solution. Any solution obtained using any search method is no better than the objective function chosen for the optimization. If an irrelevant objective is used, the solution is equally irrelevant [9, 11]. Therefore, a primary concern in DBD is the development of a mathematically sound objective function. Recognizing that design is a decision-making process, it is imperative to construct an objective function that is able to capture the preferences of rational decision-makers in system design problems involving risk. The DBD framework of Hazelrigg [9], [12] (Fig. 16.1) provides a basis for exploring this issue.

16.1.1 The Rule of Rational Decisions

The aforementioned DBD framework implements the concept of rational decisions. Rational decisions follow the rule that the preferred decision is the option whose expectation has the highest value. In a normative approach, decisions involve options, expectations and values. An expectation combines the possible outcomes of an alternative with probabilities of occurrence of each possible outcome. Note that in general expectations are not equivalent to outcomes. An outcome refers to what actually happens after a decision is made to select a certain option, while an expectation refers to what is expected to happen, based on available knowledge, as the result of a decision before it is made. In other words, expectations are associated with what *will* happen; therefore, expectations relate to the future. In the process of engineering design, it is practically impossible to predict the future with precision and certainty. The outcomes of most options in the design option space cannot be determined with certainty prior to the decision to select one option. Hence expectations are always probabilistic. Engineers are forced to make decisions under uncertainty and risk.

Unlike problem-solving, decision-making cannot be conducted in the absence of human values. The purpose of values in decision-making is to rank-order alternatives according to the preference of decision-makers. In the case of optimization, engineers seek a design that maximizes value, and the objective function serves as a numerical value function to automate the process of rank ordering. By means of an objective function, a real scalar is assigned to each design alternative in accordance with the decision-maker's preference. In this sense an objective function serves as a utility function in the context of economics. Note that utilities are determined by preferences and so is the objective function. A necessary condition for the existence of an objective function in DBD is that decision-makers or design engineers are rational individuals whose preferences and indifferences between all pairs of outcomes in the design space exists and comprises a transitive set.

16.1.2 Decision-Making/Optimization Under Uncertainty

Due to the nature of engineering design, expectations on design alternatives can never be determined with certainty. Particular care should be taken in the formulation of an objective function when risk is present. It is imperative that the objective function (or utility function in the context of economics) be valid under conditions of uncertainty and risk. The von Neumann-Morgenstern (vN-M) utility is such a value measure. Built upon the notion of von Neumann Morgenstern lottery and six rigorous axioms [18], the normative framework for decision-making under risk leads to a simple but profound result—the so-called expected utility theorem: "The utility of a lottery is the sum of the utilities of all possible outcomes of the lottery weighted by their probabilities of occurrence." It follows that the preferred choice from among a set of risky options is the option with the highest expected utility.

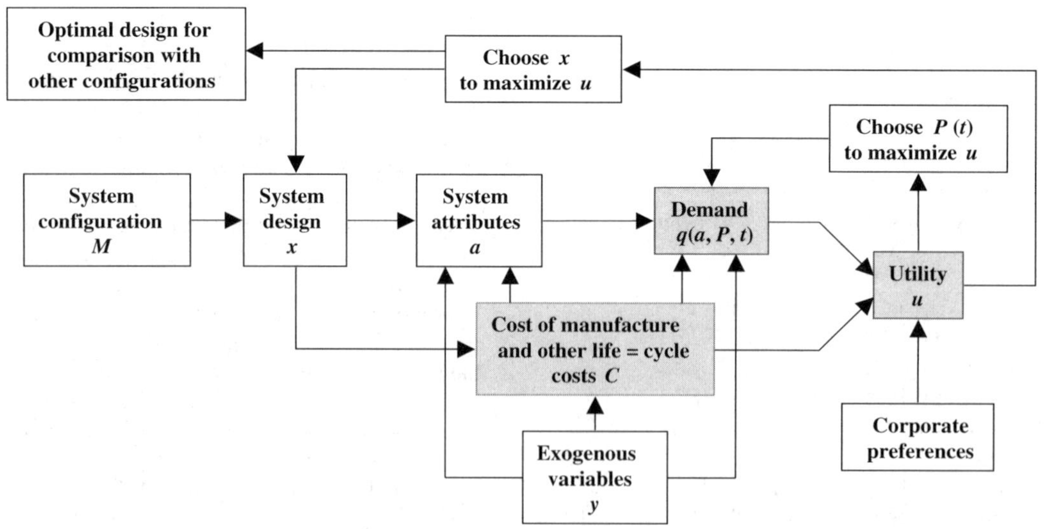

FIG. 16.1 DECISION-BASED DESIGN FRAMEWORK [HAZELRIGG, 1996(b)]

In the case of optimization, the axioms of vN-M utility should be adhered to and the objective function should entail the assignment of a vN-M utility to each design alternative under consideration.

16.1.3 Characteristics of DBD Framework

The normative framework of DBD implements the concept of rational decisions. It facilitates vN-M utility as a measure of value against which design alternatives can be compared and optimal designs sought.

The DBD framework may view the objective of systems design as one of maximizing profit. Profit, which is also referred to as net revenue (*NR*), is revenue generated by the product less all costs generated by the product. Revenue is the sum of products sold times their prices; in other words, it can be calculated as the product of the demand q and the price P at which the product is sold. Costs are the sum of things bought multiplied by their prices. Total cost C_T consists of: cost of manufacture and all other life-cycle costs such as costs of research, design, maintenance, repair and much more. The calculation of *NR* can be summarized as:

$$NR = P.q - C_T \qquad \text{Eq. (16.1)}$$

To account for the time value of money since revenues and costs are spread over periods of time, a better index for profit would be the present value of net revenue (*PNR*), which can be obtained from properly discounting *NR* to the present and integrating it over time.

Recognizing that most products are designed to make money, a valid objective function for optimization or decision-making under uncertainty and risk can be established according to the rule of rational decisions: In the DBD process, the optimizer should seek to maximize the expected vN-M utility of the profit or *NR*.

The axioms of vN-M utility dictate that utilities are not arbitrarily selected. Rather, utilities must provide proper rank-ordering of design alternatives in the case of the vN-M lottery. In nondeterministic optimization, where uncertainty and risk cannot be neglected, the decision-maker's risk preference toward money must be taken into consideration. In general, the risk preference of the decision-maker toward money leads to a utility of money that has diminishing marginal value [3]. Thus profit or NR itself is not a valid utility under such circumstances.

16.2 MULTIDISCIPLINARY ENTERPRISE MODEL

Design is inherently a multidisciplinary process. Traditionally, multidisciplinary design has focused on disciplines within the field of engineering analysis, such as aerodynamics, solid mechanics, kinematics, control and many others. In the DBD approach, engineers are compelled to look at the design from a broader viewpoint. Indeed, DBD encompasses not only the engineering disciplines but also business disciplines, including economics, marketing, operation research and more. In addition to aiming to improve the performance of a design, engineers must be aware of the substantial impact of nonengineering disciplines on the goal of design. Effective communications between engineers and experts in the business field are vital to producing successful designs.

The DBD framework of Hazelrigg [9, 10, 12] combines both engineering and business performance simulations in a single-level, all-at-once optimization approach. In the current research the Hazelrigg framework (Fig. 16.1) has been decomposed into the multidisciplinary enterprise model shown in Fig. 16.2. The decomposed system consists of two major elements: the engineering disciplines and the business discipline. The work in engineering disciplines focuses on predicting the performance of the product for different design configurations, as well as satisfying performance targets set in the business discipline (i.e., management). The role of the business discipline centers is to provide targets for performance improvements in order to

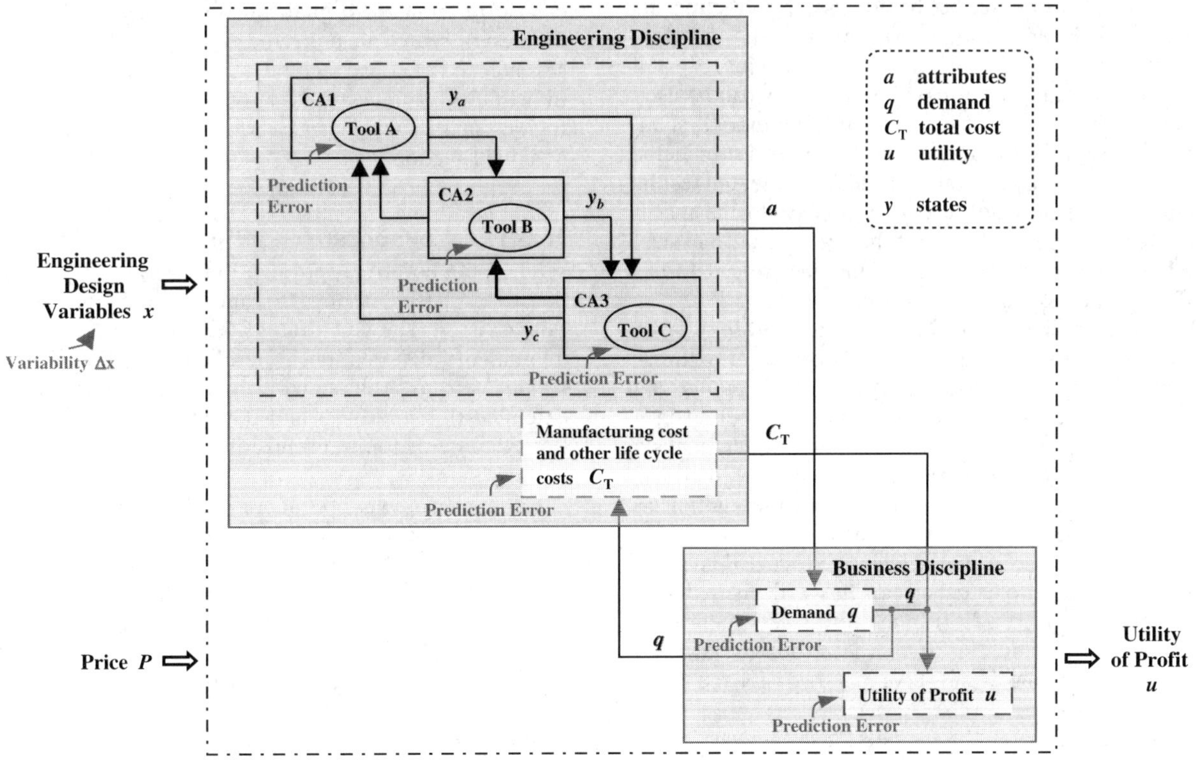

FIG. 16.2 MULTIDISCIPLINARY ENTERPRISE MODEL

yield higher profit. These two organizations are coupled through attributes $\boldsymbol{a}$, total cost C_T and demand q.

Attributes $\boldsymbol{a}$ refers to the performances of a product. These performances directly influence customer demand. Examples of attributes include speed, acceleration, comfort, quality, reliability, safety, etc. Prediction of attributes $\boldsymbol{a}$ can be obtained from the engineering disciplines. Such information will be used as input into the business discipline to estimate the demand q for the product, which will then be used to determine the profit or net revenue. In order to compute profit, an estimate of the total cost C_T is required. C_T is again computed in the engineering disciplines. Since total cost is influenced by the number of products manufactured and/or sold, estimation of demand q is called for to feed back into the engineering disciplines. In this chapter it is assumed that the demand for the product, the amount of the product manufactured and the amount of the product sold are equal.

The prediction of product attributes $\boldsymbol{a}$ and total cost C_T is conducted in the engineering disciplines. The engineering disciplines module is partitioned into a module of product performance and a module of manufacturing and life-cycle costs. The set of variables, which uniquely defines a specific design, are referred to as engineering design variables $\boldsymbol{x}$. Engineers usually have control over these variables. The estimation of demand q, NR and, ultimately, the expected utility of net revenue, are managed in the business discipline. The price of the product not only directly affects the amount of profit or net revenue [see Eq. (16.1)], it is also an important factor driving the demand q, which is also affected by the product attributes $\boldsymbol{a}$. Mathematically speaking, q can be modeled as a function of attributes $\boldsymbol{a}$ and the price, P.

$$q = \bar{q}\,(\boldsymbol{a}, P) \qquad \text{Eq. (16.2)}$$

Note that the decision-maker is free to choose the price therefore, price should be treated as a design variable.

In the context of multidisciplinary design, the term "system analysis" (SA) is often used to describe the process of predicting the performance of an engineering artifact using numerical simulation, which is part of the multidisciplinary enterprise model investigated in this chapter. In fact, simulation-based design tools can be used to assist in the analyses within the entire multidisciplinary system. Under these circumstances, the analysis of the enterprise model is no different than an SA, only of a larger scale. The research results of multidisciplinary design can be readily applied to the multidisciplinary enterprise model. Note that the equations of physical laws involved in the system analysis are sometimes referred to as "constraints" in the field of decision science.

The performance predictions obtained from an SA are referred to as states $\boldsymbol{y}$ in the context of multidisciplinary design. If we think of the multidisciplinary enterprise model (Fig. 16.2) as one big SA, then the attributes $\boldsymbol{a}$, total cost C_T, demand q and net revenue NR are components of the system states. In this chapter, states $\boldsymbol{y}$ are used to represent the engineering performance predictions, calculated in the product performance module of the multidisciplinary enterprise model. Therefore, not all of the states $\boldsymbol{y}$ are attributes $\boldsymbol{a}$. Only those states that influence product demand q will be included in the set of attributes $\boldsymbol{a}$. In the decision-based collaborative optimization (DBCO) framework, the system-level optimization treats attributes $\boldsymbol{a}$ as system-level design variables x_{sys}, which influence demand q and serve as targets for the discipline designers to satisfy.

Note that variability exists in the engineering design variables $\boldsymbol{x}$ and all simulation-based design tools employed in each discipline have prediction errors associated with them.

16.3 SYSTEM OPTIMIZATION

The multidisciplinary enterprise model discussed above belongs to the type of so-called non-hierarchic systems (or coupled/networked systems). Such systems, usually of a fairly large scale, are characterized by large numbers of design variables $\boldsymbol{x}$ and parameters $\boldsymbol{p}$, large numbers of requirements or constraints $\boldsymbol{g}$ and a high level of coupling between participating disciplines that are intrinsically linked to one another. Coupling results from the information exchange that is required within the SA. This occurs where the output of one discipline is required as input for another discipline and vice versa. Some of the mathematical terms, defined by Balling and Sobieszczanski-Sobieski [2], are briefly reviewed below and will be used from here on.

In a coupled system the design variable $\boldsymbol{x}$ scan be decomposed into the set of shared variables $\boldsymbol{x}_{sh}$ and n_{ss} sets of local variables $\boldsymbol{x}_i$, where i ranges from 1 to, n_{ss}, and n_{ss} is the total number of subspaces (or disciplines). The set of shared variables $\boldsymbol{x}_{sh}$ contains design variables that are needed by more than one discipline. The set of local variables $\boldsymbol{x}_i$ includes design variables associated with discipline i only. The set of $\boldsymbol{x}_{sh}$ and the sets of $\boldsymbol{x}_i$ are mutually exclusive subsets of the set of design variables $\boldsymbol{x}$. The term $(\boldsymbol{x}_{sh})_i$ is used to represent the shared design variables that are needed in discipline i.

The set of output from discipline i is denoted by $\boldsymbol{y}_i$. The set of system states $\boldsymbol{y}$ is composed of all the $\boldsymbol{y}_i$'s in the n_{ss} disciplines. The term $\boldsymbol{y}_{ij}$ is introduced to represent the output from discipline i which is also used as input in another discipline j. Note that $\boldsymbol{y}_{ij}$ are the subset of coupling variables in the set of $\boldsymbol{y}_i$.

The vector $\boldsymbol{g}_i$ contains the design constraints associated with discipline i. It is assumed that no constraint will stretch over more than one discipline. It is also assumed that each inequality constraint has been formulated such that zero is the permissible value, and the constraint is satisfied when it is greater than or equal to zero. In conventional design optimization, constraints can be formulated to guard against system failure or to restrict the design search to preferred regions of the design space. This second class of constraint, related to design preference, is not used in the DBCO framework developed here. In this chapter preferred regions of the design space are imposed implicitly through the demand function.

16.3.1 Collaborative Optimization (CO)

The collaborative optimization (CO) strategy was first proposed by Kroo et al. [15] and Balling and Sobieszczanski-Sobieski [2]. The CO strategy has been successfully applied to a number of different design problems, including the design of a single-stage-to-orbit launch vehicle [5]. Tappeta and Renaud [17] extended this approach and developed three different formulations to provide for multiobjective optimization of non-hierarchic systems.

CO is a two-level optimization method specifically created for large-scale distributed-analysis applications. The basic architecture of collaborative optimization [4] is shown in Fig. 16.3. The CO framework facilitates concurrent design at the discipline design level. To achieve the concurrency in the subspace level, auxiliary design variables $(\boldsymbol{x}_{aux})_{ij}$ are introduced as additional design variables to replace the coupling variables $\boldsymbol{y}_{ij}$ so that the analyses in disciplines i and j can be executed concurrently. Compatibility constraints (or discrepancy functions) $\boldsymbol{d}$ are added to ensure consistency such that $(\boldsymbol{x}_{aux})_{ij} = \boldsymbol{y}_{ij}$. Compatibility constraints are usually in the form of equality constraints.

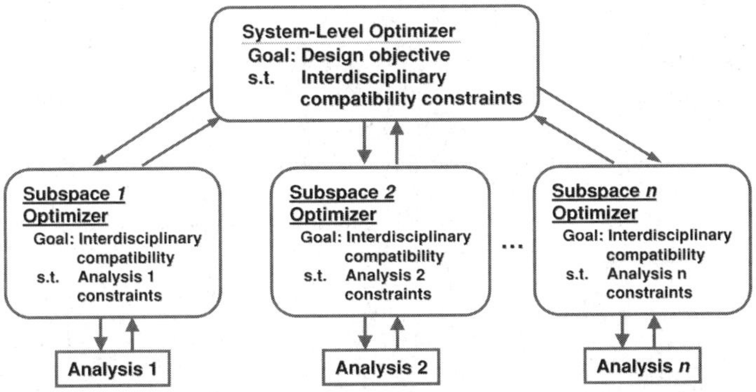

FIG. 16.3 BASIC COLLABORATIVE OPTIMIZATION ARCHITECTURE [4]

The system-level optimizer attempts to minimize a system-level objective function F while satisfying all the compatibility constraints $\boldsymbol{d}$. System-level design variables $\boldsymbol{x}_{sys}$ consist of not only the shared variables $\boldsymbol{x}_{sh}$, but also the auxiliary $\boldsymbol{x}_{aux}$ variables. These are specified by the system-level optimizer and are sent down to subspaces as targets to be matched. Each subspace, as a local optimizer, operates on its own set of design variables $\boldsymbol{x}_{ssi}$ with the goal of matching target values posed by the system level as well as satisfying local constraints $\boldsymbol{g}_i$. The matching can be attained by minimizing the discrepancy d_i between some of the local design variables and/or local states and their corresponding target values; in other words, the objective functions at subspace level are identical to the system-level (compatibility) constraints. This formulation allows the use of post-optimal sensitivities calculated at the subspace optimum to be used as the gradients of the system-level constraints. This important feature improves the overall efficiency of CO by eliminating the need to execute subspace analysis for the sole purpose of calculating system constraint gradients by finite differencing.

16.4 DECISON-BASED COLLABORATIVE OPTIMIZATION

16.4.1 Decision-Based CO (DBCO) Framework

In the DBCO framework developed in this chapter, the multidisciplinary enterprise model (Fig. 16.2) is decomposed along analysis boundaries into several subsystems. The CO method of is used to determine the optimal design of this complicated model, where an optimizer is integrated within each subsystem or discipline. The resulting DBCO framework rigorously simulates the existing relationship between business and engineering in multidisciplinary systems design, as shown in Fig. 16.4 A brief discussion about this framework is given below.

16.4.1.1 System-Level Optimization In the DBCO framework, the business decisions are made at the system level. As discussed before, the goal of DBD optimization is to maximize the expected utility of net revenue $\{E[u(NR)]\}$ of the engineering artifact being designed. The system-level optimizer in the decision-based collaborative optimization framework attempts to increase expected utility of net revenue while satisfying compatibility constraints $\boldsymbol{d}$. According to the analyses in the business discipline and the subspace optimization results, the system-level optimizer determines price P and establishes a set of performance targets for shared design variables $\boldsymbol{x}_{sh}$ and auxiliary design variables $\boldsymbol{x}_{aux}$, including demand q and total cost C_T. Attributes $\boldsymbol{a}$ are treated as auxiliary design variables that influence demand. Note that q and C_T are also among the auxiliary variables. Also note that since the design variable price P is only needed in the business discipline, it is not a target for any subspace.

The business discipline is operated on directly by the system-level optimizer. It is *not* further decomposed into demand and utility of profit subspaces. The reasons why the business discipline is dealt with in this manner are as follows: First, there is no two-way coupling between the analysis in the demand subspace and the utility of profit subspace. The demand q is fed-forward into the profit

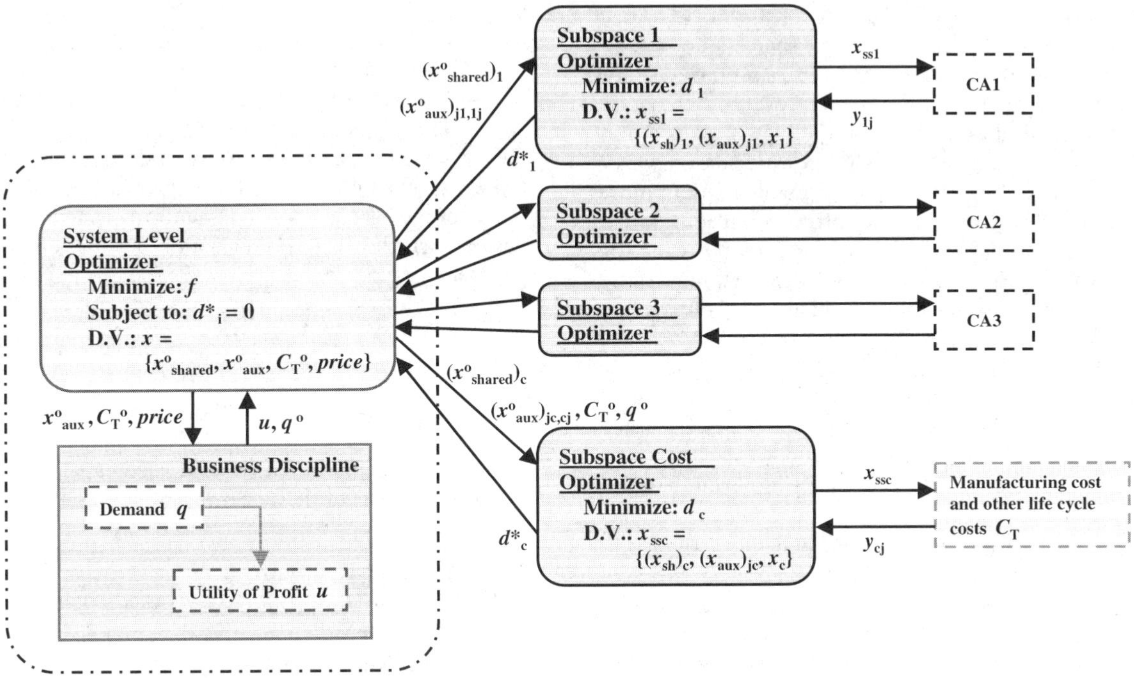

FIG. 16.4 DECISION-BASED COLLABORATIVE OPTIMIZATION

subspace. Second, the analyses involved in this discipline are relatively simple and straightforward [Eqs. (16.1 and 16.2)]. Consequently it is relatively easy to obtain sensitivity information with respect to the profit (or net revenue) or the utility of NR. In the cases where an analytical demand model and an analytical utility model are supplied, the sensitivity information is readily acquired through the application of the chain rule.

The system-level optimization problem in its standard form is given in Eq. (16.3):

$$\begin{aligned} &\underset{\text{w.r.t. } \boldsymbol{x}^{o}_{sys}}{\text{Minimize:}} && F = -E[u(NR)] \\ &\text{Subject to:} && d_i^* = 0 \quad i = 1,2,\ldots, n_{ss} \\ &&& (\boldsymbol{x}^{o}_{sys})_{min} \le \boldsymbol{x}^{o}_{sys} \le (\boldsymbol{x}^{o}_{sys})_{max} \\ &&& \boldsymbol{x}^{o}_{sys} = (\boldsymbol{x}^{o}_{sh}, \boldsymbol{x}^{o}_{aux}) \\ &&& P > 0 \end{aligned} \qquad \text{Eq. (16.3)}$$

Note that maximizing the expected utility $E[u(NR)]$ is equivalent to minimizing its negative value. Note also that NR is determined by the system-level design variable x_{sys} and price P. Since system design variables $\boldsymbol{x}_{sys}$ are posed as targets to the subspaces, the term $\boldsymbol{x}^{o}_{sys}$ is used in Eq. (16.3) so that they can be distinguished from subspace design variables in the subproblem formulations. The term d_i^* refers to the optimal value of the discrepancy function d_i obtained by the subproblems. The formulation of d_i is discussed in the next section.

16.4.1.2 Subspace-Level Optimization The subspace optimizer seeks to satisfy the targets sent down by the system-level optimizer and reports the discrepancy d_i^* back to the system level. Meanwhile the subspace optimizers are subject to local design constraints $\boldsymbol{g}_i$. In the field of engineering design, the design constraints normally guard against failure or restrict the search to a preferred region of the design space. One example of failure-related constraints is to require that "the axial load in a beam not exceed its buckling load." The statement that "the mass of the beam should be less than 7 kg" is an example of constraints based on preference. The use of constraints to restrict the search to the preferred region of design space is not recommended in the DBD approach: First, to impose a preferred space, the engineer must decide and quantify what level of behavior is unacceptable or undesirable (i.e., not preferable). This is a matter of decision-making and by imposing constraints of preferences, the designer is removing some degrees-of-freedom in the design process. The resulting system optimization may fail to identify the design with the best trade-off, especially when this constraint is active or near active at the optimal solution. In the example of the beam, a beam of mass greater than 7 kg is said to be unacceptable. But it is possible that a beam of 7.1 kg can support a much higher load than a beam of 7 kg. If the goal of the optimization is to find a light beam that can support a large load, the beam of 7.1 kg might be a better design. Yet if a constraint is set to ensure that the mass of the beam be no greater than 7 kg, the optimizer will not locate the beam of 7.1 kg, even though it may be more preferred.

Upon closer examination, undesirable behaviors (i.e., non-preferred region) are often undesirable because such behaviors lead to a decrease in the demand of the product and/or an increase in the cost of the product, which, in the end, results in a decrease in profit. In DBD the market place is used to determine the preferred region of the design space through demand and cost models, and thus constraints related to undesirability or preference are eliminated. Therefore, the local constraints in the DBCO framework tend to be those that guard against system failure. Other traditional engineering constraints related to consumer preference are eliminated from the subspace optimization and are instead incorporated in the demand model and/or cost model.

The subspace optimization problem for discipline 1 of Fig. 16.4 in its standard form is given below:

$$\begin{aligned} \underset{\text{w.r.t. } \boldsymbol{x}_{ss1}}{\text{Minimize:}} \quad d_1 = {} & [(\boldsymbol{x}_{sh})_1 - (\boldsymbol{x}^{o}_{sh})_1]^2 \\ & + \sum_{j=2}^{n_{ss}} [(\boldsymbol{x}_{aux})_{j1} - (\boldsymbol{x}^{o}_{aux})_{j1}]^2 \\ & + \sum_{k=2}^{n_{ss}} (\boldsymbol{y}_{1k} - (\boldsymbol{x}^{o}_{aux})_{1k})^2 \end{aligned} \qquad \text{Eq. (16.4a)}$$

$$\begin{aligned} \text{Subject to:} \quad & \boldsymbol{g}_1 \ge 0 \\ & (\boldsymbol{x}_{ss1})_{min} \le \boldsymbol{x}_{ss1} \le (\boldsymbol{x}_{ss1})_{max} \\ & \boldsymbol{x}_{ss1} = [(\boldsymbol{x}_{sh})_1, (\boldsymbol{x}_{aux})_{ji1}, \boldsymbol{x}_1] \end{aligned} \qquad \text{Eq. (16.4b)}$$

The subspace must satisfy local constraints $\boldsymbol{g}_i$ while attempting to minimize discrepancies in system-level targets. Note that attribute targets are imposed in the second and the third terms of the discrepancy function.

16.4.1.3 System-Level Constraints Gradient The gradient of system-level constraints plays an important role in forming search directions for the system-level optimization. As mentioned earlier, one important feature of CO is that post-optimality sensitivity analysis from the converged subspace optimization problem can be used to provide system-level derivatives for compatibility constraints [15]. As a result, both computational expense and numerical error are reduced. This is possible because the system-level design variables are treated as parameters (i.e., targets) in the subproblems. Note that for a certain discipline i, depending on the contributing analysis involved, not all of the system-level design variables $\boldsymbol{x}_{sys}$ are necessarily posed as targets to be matched. It is possible that only a subset of $\boldsymbol{x}_{sys}$, referred to as $(\boldsymbol{x}_{sys})_i$, is sent down as subspace i targets. Generally, all the subsets for n_{ss} subspaces are not mutually exclusive, i.e., their intersections exist. The set of system-level design variables $\boldsymbol{x}_{sys}$ is the union of all the subsets $(\boldsymbol{x}_{sys})_i$. The gradient of system level constraint d_i^* with respect to the subset $(\boldsymbol{x}_{sys})_i$ of the system-level design variables sent down as targets to discipline i is given below in Eq. (16.5). The gradient of system-level constraint d_i^* with respect to those system-level design variables that are not imposed as targets for discipline i is apparently zero.

$$\frac{\partial d_i^*}{\partial (\boldsymbol{x}^{o}_{sys})_i} = -2[(\boldsymbol{x}^*_{ss})_i - (\boldsymbol{x}^{o}_{sys})_i] \qquad \text{Eq. (16.5)}$$

The term $(\boldsymbol{x}^*_{ss})_i$ refers to the vector formed by the converged optimal values of local variables and states in discipline i at the end of subspace optimization. The elements in $(\boldsymbol{x}^*_{ss})_i$ are the optimal counterparts of the system-level targets $(\boldsymbol{x}_{sys})_i$.

16.4.2 Test Problem

A preliminary application of the DBCO framework has been tested on an aircraft concept sizing (ACS) problem. This problem was originally developed by the MDO research group at the University of Notre Dame[19]. It involves the preliminary sizing of a general aviation aircraft subject to certain performance constraints.

The design variables in this problem comprise variables relating to the geometry of the aircraft, propulsion and aerodynamic characteristics, as well as flight regime. Appropriate bounds are placed on all design variables. The problem also includes a number of parameters that are fixed during the design process to represent constraints on mission requirements, available technologies and aircraft class regulations.

The original problem has 10 design variables and five parameters. The design of the system is decomposed into six contributing analyses. This problem has been modified by Tappeta [16] to fit the framework of multi-objective coupled MDO systems. It is further modified in this chapter to be suitable for the DBD case. For comparison, the rest of this section gives a brief description of the modified ACS problem by Tappeta. It will be referred to as the ACS problem from here on. The DBD version of the ACS problem will be discussed in the following sections.

The ACS problem has three disciplines (see Fig. 16.5) aerodynamics (CA1), weight (CA2) and performance (CA3). It can be observed from the dependency diagram that the system has two feed-forwards and there are no feed-backs between disciplines. Table 16.1 lists the design variables and their bounds in the ACS problem. Table 16.2 lists the usage of design variables as inputs to each discipline. It can be seen that there are five shared design variables ($x_1 \sim x_4$ and x_7). Table 16.3 lists the parameters and their values. Table 16.4 lists the state variables and their relations with each discipline. Clearly there are two coupled states (y_2 and y_4). Table 16.5 contains all the relevant information for the ACS problem in the standard MDO notation introduced earlier.

The objective in the ACS problem is to determine the least gross take-off weight within the bounded design space subject to two performance constraints. The first constraint is that the aircraft range must be no less than a prescribed requirement, and the second constraint is that the stall speed must be no greater than a specified maximum stall speed. In standard form, the optimization problem is given below:

TABLE 16.1 LIST OF DESIGN VARIABLES IN ACS PROBLEM

	Name (Unit)	L	U
x_1	Aspect ratio of the wing	5	9
x_2	Wing area (ft^2)	100	300
x_3	Fuselage length (ft)	20	30
x_4	Fuselage diameter (ft)	4	5
x_5	Density of air at cruise altitude (slug/ft^3)	0.0017	0.002378
x_6	Cruise speed (ft/sec)	200	300
x_7	Fuel weight (lbs)	100	2,000

TABLE 16.2 INPUT DESIGN VARIABLES OF EACH DISCIPLINE IN ACS*

	CA1 (Aero.)	CA2 (Weight)	CA3 (Perf.)
x_1	√	√	—
x_2	√	√	√
x_3	√	√	—
x_4	√	√	—
x_5	—	√	—
x_6	—	√	—
x_7	—	√	√

*Shaded cells in the table indicate shared variables.

$$\text{Minimize: } F = \text{Weight} = y_4$$

$$\text{Subject to: } g_1 = 1 - \frac{y_6}{V_{\text{stall}_{\text{req}}}} \geq 0$$

$$g_2 = 1 - \frac{\text{Range}_{\text{req}}}{y_5} \geq 0$$

$$V_{\text{stall}_{\text{req}}} = 70 \text{ ft/sec}$$

$$\text{Range}_{\text{req}} = 560 \text{ miles} \qquad \text{Eq. (16.6)}$$

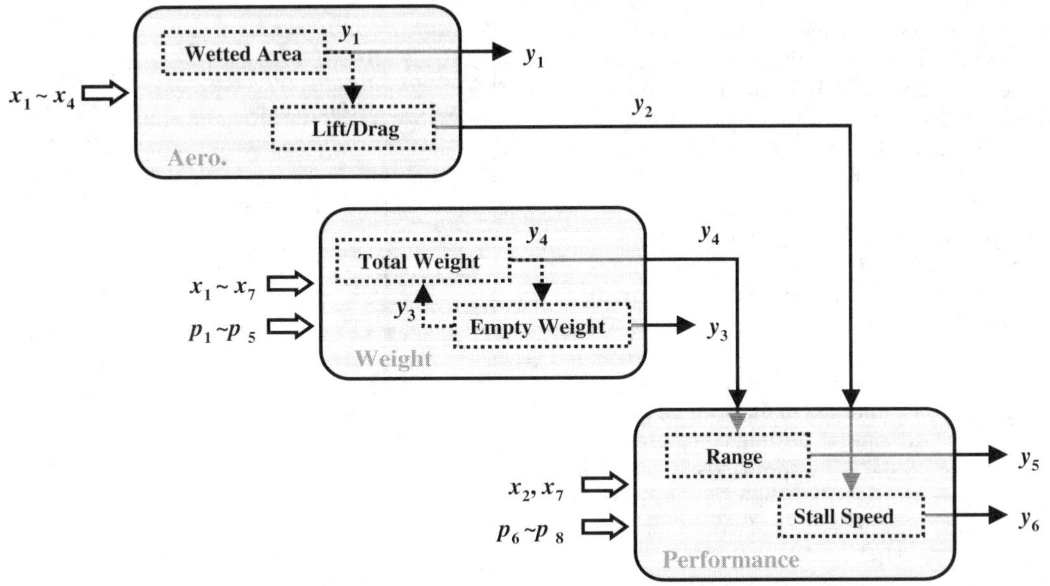

FIG. 16.5 AIRCRAFT CONCEPT SIZING PROBLEM

TABLE 16.3 LIST OF PARAMETERS IN ACS PROBLEM

	Name	Description	Value
p_1	N_{pass}	Number of passengers	2
p_2	N_{en}	Number of engines	1
p_3	W_{en}	Engine weight	197 (lbs)
p_4	W_{pay}	Payload weight	398 (lbs)
p_5	N_{zult}	Ultimate load factor	5.7
p_6	Eta	Propeller efficiency	.85
p_7	c	Specific fuel consumption	0.4495 (lbs/hr/hp)
p_8	$C_{l_{max}}$	Maximum lift coeff. of the wing	1.7

TABLE 16.4 LISTS OF STATES IN ACS PROBLEM*

	Name (Unit)	Output From	Input To
y_1	Total aircraft wetted area (ft^2)	CA1	—
y_2	Max lift-to-drag ratio	CA1	CA3
y_3	Empty weight (lbs)	CA2	—
y_4	Gross take-off weight (lbs)	CA2	CA3
y_5	Aircraft range (miles)	CA3	—
y_6	Stall speed (ft/sec)	CA3	—

*Shaded cells in the table indicate coupling states.

16.4.3 Demand Model and Cost Model

The DBD approach requires engineers to not only focus on the product performance, but also life-cycle costs as well as demand and the profit obtained over the life cycle of the product. Thus it is very important to construct a proper demand model and a proper cost model for the product. The authors are aware that the task of building such models is not an easy one, and engineers are generally not trained for this task. Since this chapter concentrates on the optimization aspect of DBD, it is reasonable to assume that other discipline experts have developed such demand and cost models and made them available to the optimizer.

In the case of the ACS problem, neither a demand model nor a cost model was available from the previous studies. In order to apply the DBCO framework of Fig. 16.4, a demand model and a cost model have been developed. These models are built in a way such that they agree with industry trends for this specific class of aircraft and that they lead to reasonable optimization behavior. Although they are by no means complete, they serve fairly well as concept models for the application of DBCO at the current stage. Figure 16.6 illustrates the demand and cost models of the ACS problem. Only the annual demand, annual cost and annual profit are considered in the current research.

16.4.3.1 Demand Model The first step in building the demand model is to identify the attributes that influence the demand q of this aircraft. The conventional optimization ACS problem [Eq. (16.6)] tries to minimize gross take-off weight (y_4) while satisfying two performance constraints: one on stall speed (y_5) and the other on aircraft range (y_6). Closer examination reveals that the objective function and two constraints, imposed in the original problem, are based on the designer's estimate of customer preference for weight, stall speed and range. In the DBD approach, it is more appropriate to treat these quantities (takeoff weight y_4, aircraft range y_5 and stall speed y_6) as attributes of demand. Hence there are no performance constraints in the DBD version of the ACS problem, and the goal of the optimization is to maximize profit.

It is also assumed that customers are interested in the cruise speed (x_6) of the aircraft as well as how much room they would have on the airplane. A new state variable, fuselage volume (y_7), is introduced to reflect the concern for passenger room on the aircraft. In all, there are five attributes of demand in the DBD version of the ACS problem: take-off weight y_4, aircraft range y_5, stall

TABLE 16.5 DESIGN VECTORS AND FUNCTIONS FOR ACS PROBLEM

	Vector or Function	Variables or Content
	$\boldsymbol{x}$	$[x_1, x_2, x_3, x_4, x_5, x_6, x_7]$
	$\boldsymbol{x}_{sh}$	$[x_1, x_2, x_3, x_4, x_7]$
	$\boldsymbol{x}_{aux}$	Goals for y_2, y_4
	$\boldsymbol{x}^{o}_{sys}$	$[x^o_1, x^o_2, x^o_3, x^o_4, x^o_7, y^o_2, y^o_4]$
	F	$F = y_4$
CA1	System targets to be matched	$(\boldsymbol{x}^o_{sys})_1 = [x^o_1, x^o_2, x^o_3, x^o_4, y^o_2]$
	$\boldsymbol{x}_1$	Empty set
	$(\boldsymbol{x}_{sh})_1$	$[x_1, x_2, x_3, x_4]$
	$(\boldsymbol{x}_{aux})_1$	Empty set
	$\boldsymbol{x}_{ss1}$	$[x_1, x_2, x_3, x_4]$
	$\boldsymbol{g}_1$	Empty set
	Analysis	$[y_1, y_2] = CA1[x_1, x_2, x_3, x_4]$
	Optimal target values	$(\boldsymbol{x}^*_{ss})_1 = [x^*_1, x^*_2, x^*_3, x^*_4, y^*_2]$
CA2	System targets to be matched	$(\boldsymbol{x}^o_{sys})_2 = [x^o_1, x^o_2, x^o_3, x^o_4, x^o_7, y^o_4]$
	$\boldsymbol{x}_2$	$[x_5, x_6]$
	$(\boldsymbol{x}_{sh})_2$	$[x_1, x_2, x_3, x_4, x_7]$
	$(\boldsymbol{x}_{aux})_2$	Empty set
	$\boldsymbol{x}_{ss2}$	$[x_1, x_2, x_3, x_4, x_5, x_6, x_7]$
	$\boldsymbol{g}_2$	Empty set
	Analysis	$[y_3, y_4] = CA2[x_1, x_2, x_3, x_4, x_5, x_6, x_7]$
	Optimal target values	$(\boldsymbol{x}^*_{ss})_2 = [x^*_1, x^*_2, x^*_3, x^*_4, x^*_7, y^*_4]$
CA3	System targets to be matched	$(\boldsymbol{x}^o_{sys})_3 = [x^o_2, x^o_7, y^o_2, y^o_4]$
	$\boldsymbol{x}_3$	Empty set
	$(\boldsymbol{x}_{sh})_3$	$[x_2, x_7]$
	$(\boldsymbol{x}_{aux})_3$	$[y_2, y_4]$
	$\boldsymbol{x}_{ss3}$	$[x_2, x_7, y_2, y_4]$
	$\boldsymbol{g}_3$	$[g_1, g_2]$
	Analysis	$[y_5, y_6] = CA3[x_2, x_7, y_2, y_4]$
	Optimal target values	$(\boldsymbol{x}^*_{ss})_3 = [x^*_2, x^*_7, y^*_2, y^*_4]$

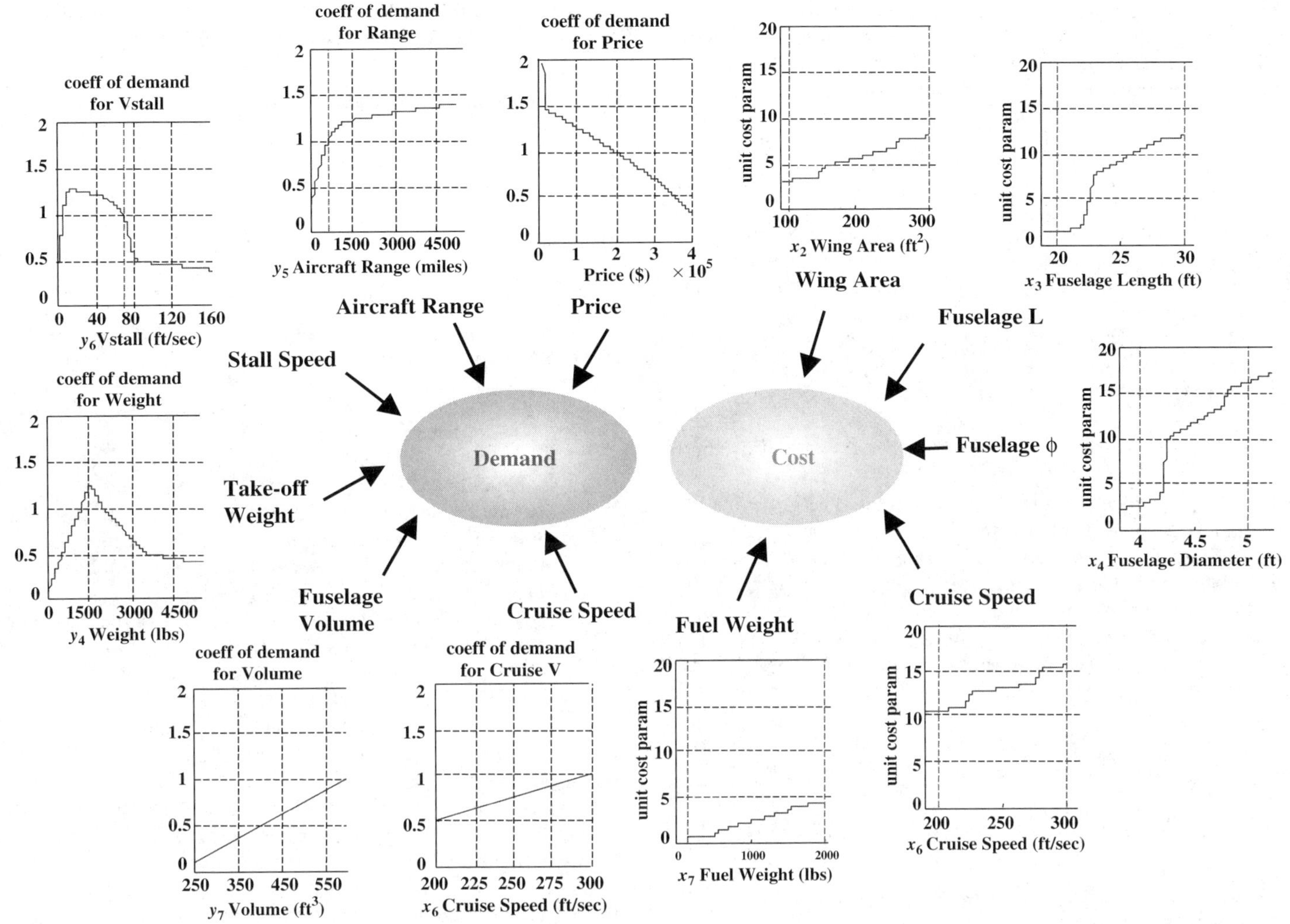

FIG. 16.6 CONCEPT DEMAND MODEL AND CONCEPT MODEL

speed y_6, fuselage volume y_7 and cruise speed x_6. Note that demand is also influenced by price P.

The demand model developed is a multiplicative model:

$$q = \bar{q}(\boldsymbol{a}, P) = q_B \left(\prod_{i=1}^{n_a} c_i \right) c_{\text{price}}$$

$$\text{where} \quad \boldsymbol{a} = \{y_4, y_5, y_6, y_7, x_6\}$$
$$n_a = 5$$
$$q_B = 1{,}200 \qquad \text{Eq. (16.7)}$$

The term q_B represents a baseline demand, which is set to 1,200. The number of attributes is denoted by n_a. The effect of each attribute on the final demand is reflected by the demand coefficient c_i. Similarly, the term c_{price} denotes the demand coefficient of price P. The final demand q is the product of all demand coefficients and the baseline demand q_B.

Demand coefficients for each attribute and price are developed by financial analysts and marketing personnel within the business discipline and vary with time. For the purpose of this study, they are assumed fixed with respect to time and are given in Fig. 16.6. The curves in Fig. 16.6 plot the coefficient of demand on the ordinate and the corresponding attribute (or price) on the abscissa:

(1) Gross take-off weight (y_4): The lower the take-off weight, the higher the demand, but an aircraft with a very light weight is not desired.
(2) Aircraft range (y_5): The longer the aircraft range, the higher the demand, but after the range reaches more than 600 miles, there is no significant increase in demand. This formulation is not unlike the original performance constraint g_2.[Eq. (16.6)], where the coefficient is set to 1 when aircraft range equals 560 miles.
(3) Stall speed (y_6) : The lower the stall speed, the higher the demand, but a near-zero stall speed is not necessary. This formulation is not unlike the original performance constraint g_1. [Eq. (16.6)], where the coefficient is set to 1 when stall speed equals 70 ft/sec.
(4) Fuselage volume (y_7): The larger the fuselage volume, the higher the demand.
(5) Cruise speed (x_6): The faster the cruise speed, the higher the demand.
(6) Price (P): The lower the price, the higher the demand; if the aircraft is sold for free (P=0), the demand approaches infinity.

16.4.3.2 Cost Model It is assumed that all costs (C_T) related to the production of the aircraft can be divided into two categories: fixed cost C_F and variable cost C_V [14]. The fixed cost C_F is the part of the total cost C_T that remains consistent regardless of changes in the amount of product produced, for example, the annual cost of renting or buying equipment or facilities. The variable cost C_V is the portion of the total cost C_T that varies directly with quantity of product produced, such as cost per unit for material and labor. If we assume the quantity of product produced and sold per year is equal to the demand q for the product per year, the total cost C_T per year is:

$$C_T = C_F + qC_V \quad \text{Eq. (16.8)}$$

It is assumed that the variable cost C_V in the ACS problem is dependent on five of the seven design variables, including wing area (x_2), fuselage length (x_3), fuselage diameter (x_4), cruise speed (x_6) and fuel weight (x_7). A variable cost parameter $(p_V)_i$ is assigned to each cost-related variable to represent the portion of variable cost (per unit) associated with each variable. The total variable cost per unit is the sum of all variable cost parameters:

$$C_V = \sum_{i=1}^{n_{cost}} (p_V)_i \quad \text{Eq. (16.9)}$$

where n_{cost} = number of cost-related variables. The guideline for assigning variable cost parameters is: the larger the variable, the higher the cost. The curves in Fig. 16.6, associated with each cost-related variable, plot the variable cost parameter (unit: \$10,000) on the ordinate and the corresponding cost-related variable on the abscissa. The step jumps in the curves represents the need to purchase (or rent) and/or install new equipment (or facilities) when the size of the aircraft exceeds existing production capabilities.

Substituting Eq. (16.9) into (16.8), the model of the total cost C_T is:

$$C_T = C_F + q\sum_{i=1}^{n_{cost}} (p_V)_i \quad \text{Eq. (16.10)}$$

and the unit cost C_u can be obtained by dividing both sides of Eq. (16.10) by demand q:

$$C_u = \frac{C_F}{q} + \sum_{i=1}^{n_{cost}} (p_V)_i \quad \text{Eq. (16.11)}$$

Note that the number of product produced (q) may have a discounting effect on the variable cost parameters p_V. For instance, usually the cost per unit for material will decrease when the total amount of material bought increases. Thus a q-discounting option has been included in the determination of variable cost parameter in the cost model.

16.4.3.3 Note The demand and cost models developed in this chapter are by no means complete. They are conceptual and rather simplistic. Future work on the modification of these models will include (and not be limited to) the following issues:

(1) Gross take-off weight (y_4): It has been pointed out to the authors that, to a customer, higher gross weight is actually desirable because it leads to longer aircraft range. Meanwhile, higher gross weight leads to higher manufacture cost. Therefore, a modified demand model would include the gross take-off weight as a slightly favorable feature (i.e., the higher the take-off weight, the higher the demand). On the other hand, a modified cost model may include the gross take-off weight as a strongly negative factor (i.e., the higher the take-off weight, the higher the cost).

(2) Aspect ratio of the wing (x_1): Increasing the aspect ratio will increase the wing structural weight, which will in turn lead to an increase in the aircraft gross take-off weight, thus adding to the total cost. A modified cost model would include aspect ratio as another negative factor.

(3) Price (P): It has been brought to the authors' attention that in the real world, due to maintenance requirements such as insurance and hanger, the demand will not approach infinity if the aircraft is given out for free. A modified demand model will address this issue by assigning a definite number to the demand coefficient for price when price is set to zero. This definite number will be associated with the maximum demand possible for this aircraft.

TABLE 16.6 INPUT DESIGN VARIABLES TO EACH DISCIPLINE IN ACS (DBD VERSION)*

	CA1 (Aero.)	CA2 (Weight)	CA3 (Perf.)	CA_c (Cost)	CA_b (Busin.)
x_1	√	√	—	—	—
x_2	√	√	√	√	—
x_3	√	√	—	√	—
x_4	√	√	—	√	—
x_5	—	√	—	—	—
x_6	—	√	—	√	√
x_7	—	√	√	√	—
P	—	—	—	—	√

*Shaded cells indicate shared variables.

16.4.4 DBCO Formulation

The DBCO framework has been applied to the DBD version of the ACS problem. This application is a preliminary study and focuses on the collaborative optimization feature of the DBCO framework. The issues of propagated uncertainty are neglected in this study. The utility of profit is assumed to be the profit itself. Hence, the objective of the resulting deterministic optimization is to maximize profit (or net revenue). During the optimization, the demand q is treated as a continuous variable, rather than an integer. At the end of the system optimization, q is rounded to the nearest integer.

Two additional disciplines are added to the DBD version of the ACS problem: cost (CA_c) and business (CA_b). Price P is a new design variable and a new state variable (fuselage volume y_7) is introduced. Table 16.6 provides the list of input design variables to each discipline in the DBD version of ACS problem. Clearly, design variable x_6 (cruise speed) enters the set of shared variables. Table 7 lists the states y, demand q, total cost C_T, and NR. It also depicts how they are related to each discipline. The set of coupling variables expands to include five additional members: y_5 (aircraft range), y_6 (stall speed), y_7 (fuselage volume), q (demand) and C_T (total cost). Table 16.8 contains all the design vector information

TABLE 16.7 LISTS OF STATES IN ACS PROBLEM (DBD VERSION)*

	Output From	Input To
y_1	CA1	—
y_2	CA1	CA3
y_3	CA2	—
y_4	CA2	CA3, CA_b
y_5	CA3	CA_b
y_6	CA3	CA_b
y_7	CA1	CA_b
q	CAb	CA_c
C_T	CAc	CA_b
NR	CAb	—

*Shaded calls indicate coupling states.

TABLE 16.8 DESIGN VECTORS FOR ACS PROBLEM (DBD VERSION)

Vector or Function		Variables or Content
	$\boldsymbol{x}$	$[x_1, x_2, x_3, x_4, x_5, x_6, x_7, P]$
	$\boldsymbol{x}_{sh}$	$[x_1, x_2, x_3, x_4, x_6, x_7]$
	$\boldsymbol{x}_{aux}$	Goals for $y_2, y_4, y_5, y_6, y_7, C_T, q$
	$\boldsymbol{x}^{o}_{sys}$	$[x_1^o, x_2^o, x_3^o, x_4^o, x_6^o, x_7^o, y_2^o, y_4^o, y_5^o, y_6^o, y_7^o, C_T^o, q^o]$
	F	$F = -NR$
CA1	System targets to be matched	$(\boldsymbol{x}^o_{sys})_1 = [x_1^o, x_2^o, x_3^o, x_4^o, y_2^o, y_7^o]$
	$\boldsymbol{x}_1$	Empty set
	$(\boldsymbol{x}_{sh})_1$	$[x_1, x_2, x_3, x_4]$
	$(\boldsymbol{x}_{aux})_1$	Empty set
	$\boldsymbol{x}_{ss1}$	$[x_1, x_2, x_3, x_4]$
	$\boldsymbol{g}_1$	Empty set
	Analysis	$[y_1, y_2, y_7] = CA1[x_1, x_2, x_3, x_4]$
	Optimal target values	$(\boldsymbol{x}^*_{ss})_1 = [x_1^*, x_2^*, x_3^*, x_4^*, y_2^*]$
CA2	System targets to be matched	$(\boldsymbol{x}^*_{sys})_2 = [x_1^o, x_2^o, x_3^o, x_4^o, x_6^o x_7^o, y_4^o]$
	$\boldsymbol{x}_2$	$[x_5]$
	$(\boldsymbol{x}_{sh})_2$	$[x_1, x_2, x_3, x_4, x_6, x_7]$
	$(\boldsymbol{x}_{aux})_2$	Empty set
	$\boldsymbol{x}_{ss2}$	$[x_1, x_2, x_3, x_4, x_5, x_6, x_7]$
	$\boldsymbol{g}_2$	Empty set
	Analysis	$[y_3, y_4] = CA2[x_1, x_2, x_3, x_4, x_5, x_6, x_7]$
	Optimal target values	$(\boldsymbol{x}^*_{ss})_2 = [x_1^*, x_2^*, x_3^*, x_4^*, x_6^*, x_7^*, y_4^*]$
CA3	System targets to be matched	$(\boldsymbol{x}^o_{sys})_3 = [x_2^o, x_7^o, y_2^o, y_4^o, y_5^o, y_6^o]$
	$\boldsymbol{x}_3$	Empty set
	$(\boldsymbol{x}_{sh})_3$	$[x_2, x_7]$
	$(\boldsymbol{x}_{aux})_3$	$[y_2, y_4]$
	$\boldsymbol{x}_{ss3}$	$[x_2, x_7, y_2, y_4]$
	$\boldsymbol{g}_3$	Empty set
	Analysis	$[y_5, y_6] = CA3[x_2, x_7, y_2, y_4]$
	Optimal target values	$(\boldsymbol{x}^*_{ss})_3 = [x_2^*, x_7^*, y_2^*, y_4^*, y_5^*, y_6^*]$
CA_c	System targets to be matched	$(\boldsymbol{x}^o_{sys})_c = [x_2^o, x_3^o, x_4^o, y_6^o, y_7^o, q^o, C_T^o]$
	$\boldsymbol{x}_c$	Empty set
	$(\boldsymbol{x}_{sh})_c$	$[x_2, x_3, x_4, x_6, x_7]$
	$(\boldsymbol{x}_{aux})_c$	$[q]$
	$\boldsymbol{x}_{ssc}$	$[x_2, x_3, x_4, x_6, x_7, q]$
	$\boldsymbol{g}_c$	Empty set
	Analysis	$C_T = CAc[x_2, x_3, x_4, x_6, x_7, q]$
	Optimal target values	$(\boldsymbol{x}^*_{sys})_c = [x_2^*, x_3^*, x_4^*, y_6^*, y_7^*, q^*, C_T^*]$

for the DBD version of subspaces 1 (aerodynamics), 2 (weight), 3 (performance) and c (cost) of the ACS problem in MDO standard notation. Table 16.9 contains the design vector information of subspace b (business), since it is operated directly by the system-level optimizer. The difference between the DBD version of ACS problem and the modified ACS problem by Tappeta [16] can be clearly observed by comparing Tables 16.6, 16.7, 16.8 and 16.9 with Tables 16.2, 16.4 and 16.5, respectively.

The system-level optimization problem, for this application, in its standard form is detailed in Eq. (16.12):

16.4.4.1 System-Level Optimization

$$\begin{aligned}
&\underset{\text{w.r.t. } \boldsymbol{x}^o_{sys}}{\text{Minimize:}} \quad F = -NR \\
&\text{Subject to:} \quad d_1^* = 0 \\
&\qquad d_2^* = 0 \\
&\qquad d_3^* = 0 \\
&\qquad d_c^* = 0 \\
&\qquad (\boldsymbol{x}^o_{sys})_{min} \le \boldsymbol{x}^o_{sys} \le (\boldsymbol{x}^o_{sys})_{max} \\
&\text{where} \quad \boldsymbol{x}^o_{sys} = [x_1^o, x_2^o, x_3^o, x_4^o, x_6^o, x_7^o, \\
&\qquad y_2^o, y_4^o, y_5^o, y_6^o, y_7^o, C_T^o, P] \\
&\qquad q^o = \overline{q}(y_4^o, y_5^o, y_6^o, y_7^o, x_6^o, P) \\
&\qquad NR = P \cdot q^o - C_T^o
\end{aligned} \qquad \text{Eq. (16.12)}$$

Note that the system-level optimizer calls the business discipline directly to obtain demand q^o and the system-level objective NR. There are 13 system-level design variables and four compatibility constraints that are evaluated in subspace 1, 2, 3 and c.

The subspace optimization problems for each discipline in their standard forms are given by Eqs. (16.13) to (16.16)

16.4.4.2 Subspace 1 (Aerodynamics) Optimization

$$\begin{aligned}
&\underset{\text{w.r.t. } \boldsymbol{x}_{ss1}}{\text{Minimize:}} \quad d_1 = (x_1 - x_1^o)^2 + (x_2 - x_2^o)^2 \\
&\qquad + (x_3 - x_3^o)^2 + (x_4 - x_4^o)^2 \\
&\qquad + (y_2 - y_2^o)^2 + (y_7 - y_7^o)^2 \\
&\text{Subject to:} \quad (\boldsymbol{x}_{ss1})_{min} \le \boldsymbol{x}_{ss1} \le (\boldsymbol{x}_{ss1})_{max} \\
&\text{where} \quad \boldsymbol{x}_{ss1} = [x_1, x_2, x_3, x_4] \\
&\qquad [y_1, y_2, y_7] = CA1[x_1, x_2, x_3, x_4]
\end{aligned} \qquad \text{Eq. (16.13)}$$

TABLE 16.9 DESIGN VECTORS IN THE SUBSPACE B (BUSINESS) FOR ACS PROBLEM (DBD VERSION)

CA_b	**Targets Sent Down to Other Subspaces**	$(\mathbf{x}^{o}_{sys})_b = [x^{o}_6, y^{o}_4, y^{o}_5, y^{o}_6, y^{o}_7, q^{o}, C^{o}_T]$
	$\boldsymbol{x}_b$	$[P]$
	$(\boldsymbol{x}_{sh})_b$	x_6
	$(\mathbf{x}_{aux})_b$	$[y_4, y_5, y_6, y_7, C_T]$
	$\boldsymbol{x}_{ssb}$	$[x^{o}_6, y^{o}_4, y^{o}_5, y^{o}_6, y^{o}_7, C^{o}_T, P]$
	$\boldsymbol{g}_b$	Empty set
	Analysis	$[q^{o}, NR] = CA_b[x^{o}_6, y^{o}_4, y^{o}_5, y^{o}_6, y^{o}_7, C^{o}_T, P]$

16.4.4.3 Subspace 2 (Weight) Optimization

$$\underset{\text{w.r.t. } \boldsymbol{x}_{ss2}}{\text{Minimize:}} \quad d_2 = (x_1 - x^{o}_1)^2 + (x_2 - x^{o}_2)^2 + (x_3 - x^{o}_3)^2 + (x_4 - x^{o}_4)^2 + (x_6 - x^{o}_6)^2 + (x_7 - x^{o}_7)^2 + (y_4 - y^{o}_4)^2$$

$$\text{Subject to:} \quad (\boldsymbol{x}_{ss2})_{\min} \le \boldsymbol{x}_{ss2} \le (\boldsymbol{x}_{ss2})_{\max}$$

$$\text{where} \quad \boldsymbol{x}_{ss2} = [x_1, x_2, x_3, x_4, x_5, x_6, x_7]$$

$$[y_3, y_4] = CA2[x_1, x_2, x_3, x_4, x_5, x_6, x_7] \qquad \text{Eq. (16.14)}$$

16.4.4.4 Subspace 3 (Performance) Optimization

$$\underset{\text{w.r.t. } \boldsymbol{x}_{ss3}}{\text{Minimize:}} \quad d_3 = (x_2 - x^{o}_2)^2 + (x_7 - x^{o}_7)^2 + (y_2 - y^{o}_2)^2 + (y_4 - y^{o}_4)^2 + (y_5 - y^{o}_5)^2 + (y_6 - y^{o}_6)^2$$

$$\text{Subject to:} \quad (\boldsymbol{x}_{ss3})_{\min} \le \boldsymbol{x}_{ss3} \le (\boldsymbol{x}_{ss3})_{\max}$$

$$\text{where} \quad \boldsymbol{x}_{ss3} = [x_2, x_7, y_2, y_4]$$

$$[y_5, y_6] = CA3[x_2, x_7, y_2, y_4] \qquad \text{Eq. (16.15)}$$

16.4.4.5 Subspace c (Cost) Optimization

$$\underset{\text{w.r.t. } \boldsymbol{x}_{ssc}}{\text{Minimize:}} \quad d_c = (x_2 - x^{o}_2)^2 + (x_3 - x^{o}_3)^2 + (x_4 - x^{o}_4)^2 + (x_6 - x^{o}_6)^2 + (x_7 - x^{o}_7)^2 + (q - q^{o})^2 + (C_T - C^{o}_T)^2$$

$$\text{Subject to:} \quad (\boldsymbol{x}_{ssc})_{\min} \le \boldsymbol{x}_{ssc} \le (\boldsymbol{x}_{ssc})_{\max}$$

$$\text{where} \quad \boldsymbol{x}_{ssc} = [x_2, x_3, x_4, x_6, x_7, q]$$

$$C_T = CAc[x_2, x_3, x_4, x_6, x_7, q] \qquad \text{Eq. (16.16)}$$

Note that other than the variable bounds, there are no local constraints for the subspace optimization problems.

16.4.5 Optimization Results and Discussion

A sequential quadratic programming (SQP) method was used for optimization in both the system-level and the subspace optimization. The SQP solver, *fmincon*, was obtained from the Matlab Optimization Toolbox. The program converged to an optimum after 37 system-level iterations. The optimal solution is listed in Table 16.10. Figure 16.7 shows the system-level optimization history of convergence of the system-level objective function (negative of profit, in subplot 8), the convergence history of the four compatibility (discrepancy) constraints (d^*_1, d^*_2, d^*_3 and d^*_c, in subplots 9 to 12), and the convergence history of the seven system-level design variables (cruise speed x^{o}_6, aircraft range y^{o}_5, stall speed y^{o}_6, fuselage volume y^{o}_7, price *P*, demand q^{o}, and total cost C^{o}_T in subplots 1 to 7). The abscissa of each subplot is the number of system-level iterations. Note that the value of profit (not the negative of profit) was plotted in subplot 8 for easy reading. For the same reason unit cost C^{o}_u was plotted instead of total cost C^{o}_T in subplot 7.

As can be seen from the convergence plots, the system-level optimizer tries to minimize both the negative of profit and the constraint violations simultaneously. At the beginning of the optimization, the system-level optimizer sets targets high for price, high for the levels of performance (to ensure high demand) and low for cost based on the results of the business analyses. However, these targets conflict with one another and lead to a large discrepancy at the subspace level. Thus the system-level optimizer, while trying to keep profit as high as possible, was forced to lower price, downgrade performance and tolerate higher cost so that the subspace discrepancy could be reduced. Gradually, the system-level optimizer found the best trade-off among the targets and reached a consistent optimal design. The optimization history observed in the ACS problem resembles the existing relationship between business and engineering in multidisciplinary systems design.

The demand model and the cost model play an important role in the DBD approach. In order to illustrate the influence of the demand and cost models, a conventional all-at-once optimization was performed according to the problem formulation in Eq. (16.6). The conventional optimum obtained is also listed in Table 16.10. Note that fuselage volume (y_7) at the conventional optimum is determined by the optimal fuselage length (x_3) and optimal fuselage diameter (x_4). It can be observed that the conventional optimum outperforms the DBCO optimal design on lower weight (y_3, y_4). However, it possess

TABLE 16.10 OPTIMAL SOLUTION FOR ACS PROBLEM

	Name (Unit)	DV Bounds	DBCO Optimum	Conven. Optimum
x_1	Aspect ratio of the wing	5–9	7.968	5
x_2	Wing area (ft^2)	100–300	230.3	176.53
x_3	Fuselage length (ft)	20–30	21.927	20
x_4	Fuselage diameter (ft)	4–5	4.1871	4
x_5	Density of air at cruise altitude (slug/ft^3)	0.0017 –0.002378	0.0023	0.0017
x_6	Cruise speed (ft/sec)	200–300	219.65	200
x_7	Fuel weight (lbs)	100–2000	231.22	142.86
y_1	Total aircraft wetted area (ft^2)	—	887.21	710.3
y_2	Max lift-to-drag ratio	—	14.273	10.971
y_3	Empty weight (lbs)	—	1556. 6	1,207.6
y_4	Gross take-off weight (lbs)	—	2185.9	1,748.4
y_5	Aircraft range (miles)	—	953.67	560
y_6	Stall speed (ft/sec)	—	68.525	70
y_7	Fuselage volume (ft^3)	—	301.92	251.33
P	Price ($)	—	3.56e5	(3.56e5)
q	Demand	—	87	(32)
C_T	Total cost ($)	—	2.02e7	(6.39e6)
C_U	Unit cost ($)	—	2.31e5	(2.02e5)
NR	Net revenue or profit ($)	—	10.9e6	(6.39e6)

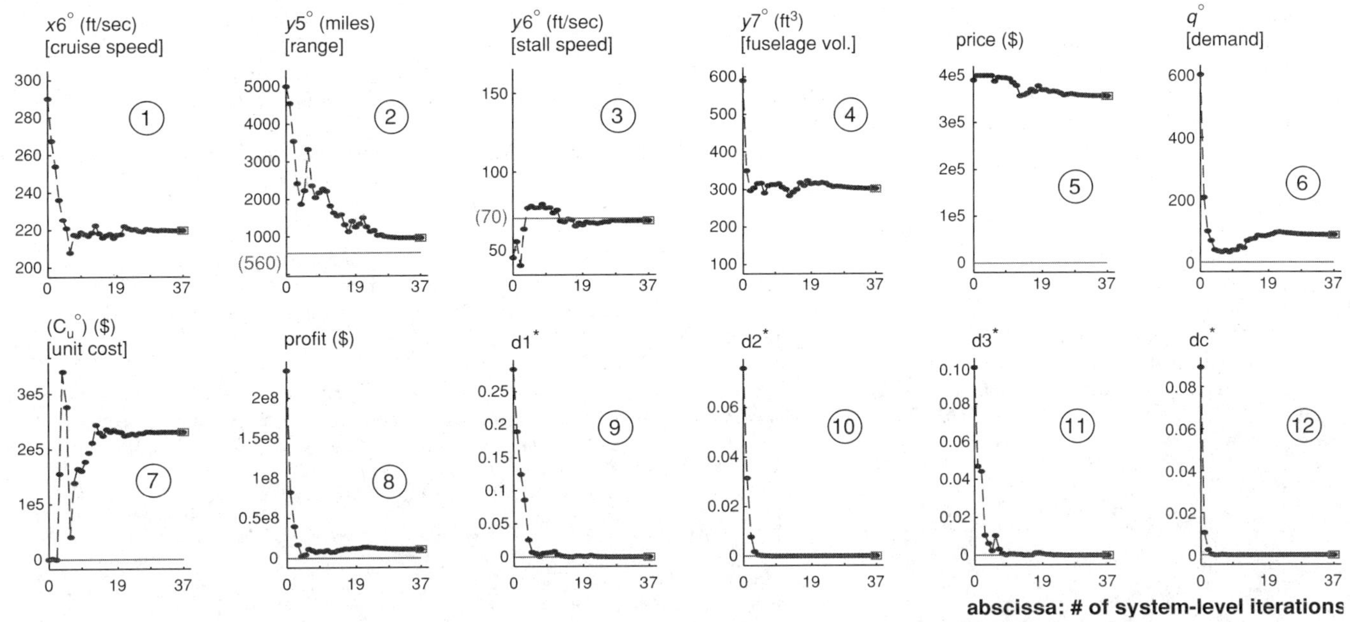

FIG. 16.7 SYSTEM-LEVEL CONVERGENCE PLOTS FOR ACS PROBLEM (DBD VERSION)

poor characteristics in many aspects, such as smaller aircraft range (y_5), higher stall speed (y_6) and smaller fuselage volume (y_7). Such an outcome is no surprise since the main concern of the conventional ACS problem is to minimize take-off weight, while the DBD approach takes into account other performance attributes because of the DBD objective of maximizing profit.

If we assume that the aircraft configuration at the conventional optimum design will be sold at the same price as the DBD optimum design, the demand, cost and profit of the conventional product can be obtained according to the demand model [Eq. (16.7)] and cost model [Eqs. (16.10) and (16.11)] developed earlier. These values are listed in Table 16.10 in parentheses because of the assumption. Notice that the unit cost of conventional optimal design is lower than the unit cost of the DBD optimal design. However, the poor performance attributes cause the demand for conventional optimal design to be much lower than the DBD optimal design. Hence the DBD optimal design leads to higher profit.

CONCLUSIONS

In this chapter a DBCO framework that incorporates the concepts of normative DBD and the strategies of CO has been developed. This bilevel nondeterministic optimization framework more accurately captures the existing relationship between business and engineering in multidisciplinary systems design. The business decisions are made at the system level, which result in a set of engineering performance targets that disciplinary engineering design teams seek to satisfy as part of subspace optimizations. The objective of the DBCO is to maximize the expected von Neumann-Morgenstern (vN-M) utility of the profit or *NR* of a product.

A preliminary application of this approach (deterministic case) has been conducted on a multidisciplinary test problem named the ACS test problem. Conceptual demand and cost models have been developed. The corresponding optimization results are discussed and compared with the conventional optimization solutions.

Future work is being targeted towards a nondeterministic implementation of the DBCO framework in which the issues of propagated uncertainty in such a bilevel optimization framework will be addressed [8]. A conceptual utility model for the *NR* will be adapted from the literature in the field of DBD. The conceptual demand and cost models developed in this chapter will be modified to better reflect the real world.

ACKNOWLEDGMENTS

This work is based on Ph. D. research conducted by Xiaoyu (Stacey) Gu at the University of Notre Dame. This multidisciplinary research effort was supported in part by the following grants: NSF grant DMI98-12857 and NASA grant NAG1-2240.

REFERENCES

1. Azarm, S. and Narayanan, S., 2000. "A Multiobjective Interactive Sequential Hybrid Optimization Technique for Design Decision Making," *Engrg. Optimization*, Vol. 32, pp. 485–500.
2. Balling, R.J. and Sobieszczanski-Sobieski, J., 1994. "Optimization of Coupled Systems: A Critical Overview of Approaches," *AIAA-94-4330-CP, Proc., 5th AIAA/NASA/USAF/ISSMO Symp. on Multidisciplinary Analysis and Optimization*, Panama City, FL.
3. Bernoulli, D., 1738. "Exposition of a New Theory of Risk Evaluation," *Econometirca*, 22 (January), 1954, pp. 22–36, translated by Louise Sommers.
4. Braun, R.D., Gage, P., Kroo, I. and Sobieski, I., 1996(a). "Implementation and Performance Issues in Collaborative Optimization," *AIAA-96-4017, Proc., 6th AIAA/NASA/USAF/ISSMO Symp. on Multidisciplinary Analysis and Optimization*, Bellevue, WA.
5. Braun, R.D., Kroo, I.M. and Moore, A.A., 1996(b). "Use of the Collaborative Optimization Architecture for Launch Vehicle Design," *AIAA-96-4018, Proc., 6th AIAA/NASA/USAF/ISSMO Symp. on Multidisciplinary Analysis and Optimization*, Bellevue, WA.
6. Chen, W., Schimidt, L., Lewis, K., 1998. Decision-Based Design Open Workshop, http://dbd.eng.buffalo.edu/.
7. Chen, W., Lewis, K. Schmidt, L., 2000. "Open workshop on Decision-Based Design: Origin, Status, Promise, and Future," *J. of Engrg. Valuation & Cost Analysis*, 3, (2), pp. 57–66, Reading, England.

8. Gu, X. and Renaud, J.E., 2001. "Implicit Uncertainty Propagation for Robust Collaborative Optimization," Proc., *2001 ASME Design Automation Conf.*, Pittsburgh, PA.
9. Hazelrigg, G.A., 1996(a). *Systems Engineering: An Approach to Information-Based Design*, Prentice Hall, Upper Saddle River, NJ.
10. Hazelrigg, G.A., 1996(b). "The Implications of Arrow's Impossibility Theorem on Approaches to Optimal Engineering Design," *J. of Mech. Des.*, 118, (June), pp. 161–164.
11. Hazelrigg, G.A., 1997. "On Irrationality in Engineering Design," *J. of Mech. Des.*, 119(June), pp. 194–196.
12. Hazelrigg, G.A., 1998. "A Framework for Decision-Based Engineering Design," *J. of Mech. Des.*, 120(4), pp. 653.
13. Kim, H.M., Michelena, N.F., Papalambros, P.Y. and Jiang, T., "Target Cascading in Optimal System Design," 2000. *DAC-14265, Proc., 2000 ASME Des. Automation Conf., Baltimore, MD.*
14. Krajewski, L.J. and Ritzman, L.P., 1999. *Operations Management: Strategy and Analysis*, 5th Ed., Addison-Wesley.
15. Kroo, I., Altus, S., Braun, R., Gage, P. and Sobieski, I., 1994. "Multidisciplinary Optimization Methods for Aircraft Preliminary Design," *AIAA-94-4325-CP, Proc., 5th AIAA/NASA/USAF/ISSMO Symp. on Multidisciplinary Analysis and Optimization*, Panama City, FL.
16. Tappeta, R.V., 1996. "An Investigation of Alternative Problem Formulations for Multidisciplinary Optimization," M.S. Thesis, University of Notre Dame, IN.
17. Tappeta, R.V. and Renaud, J.E., 1997. "Multiobjective Collaborative Optimization," *ASME J. of Mech. Des.*, 119(3), pp. 403–411.
18. Von Neumann, J. and Morgenstern, O., 1953. *The Theory of Games and Economic Behavior*, 3rd Ed., Princeton University, Princeton, NJ.
19. Wujek, B.A., Renaud, J.E., Batill, S.M., Johnson, E.W. and Brockman, J.B., 1996. "Design Flow Management and Multidisciplinary Design Optimization in Application to Aircraft Concept Sizing," *AIAA-96-0713, 34th Aerospace Sci. Meeting & Exhibit*, Reno, NV.

CHAPTER

17

A DESIGNER'S VIEW TO ECONOMICS AND FINANCE

Panos Y. Papalambros and Panayotis Georgiopoulos

NOMENCLATURE

C	= cost
D	= firm's debt
E	= firm's equity
E_p	= price elasticity of product demand
f	= analysis function
$\mathbf{g}$	= vector of inequality constraints
$\mathbf{h}$	= vector of equality constraints
H	= horizon of product life cycle
I	= amount of a financial investment
K	= capacity of production
M	= firm's market share
MR	= marginal revenue
MC	= marginal cost
P	= product price
PV	= present value of a future cash flow stream
Q	= industry's product demand
q	= firm's product demand
R	= revenue
r_f	= risk-free interest rate
r_m	= financial market return
t	= time
t_0	= commercialization time
U	= capacity utilization
WACC	= weighted average cost of capital
w	= criterion preference weight
x	= product decisions
x_b	= business decisions
x_e	= economic decisions
x_f	= financial decisions
z	= Wiener process
α	= observed product characteristics
β	= sensitivity of firm's stock to market return
δ	= mean utility level from consuming a product
θ	= intercept of the product demand curve
λ	= slope of the product demand curve
μ	= growth of product demand
ξ	= unobserved product characteristics
π	= profit
σ	= volatility of product demand
τ	= product

17.1 INTRODUCTION

In recent years, the engineering design community has expanded its quest to address analytically design intent, particularly as it is manifested within a producing enterprise; see, for example, [46], [45], [44], [35], [11], [15], [21], [22], [34], [33], [41], [53], [1, 2], [9], [19], [28, 29], [31, 32], [50], [51, 52], [16], [12], [36], [26], [17]. These investigators, while following different approaches, have shown that designers can and should not only balance technical trade-offs but also account more directly for the needs of users and producers, positioning engineering trade-offs in a societal context. In such a view, technical objectives not only compete amongst each other, but generate a long chain of effects that influence purchasing behavior, firm growth, regional economics and environmental policies.

In this article we review some basic concepts that are the building blocks for modeling economic and finance considerations in the context of product development. Along the way we will use case studies to demonstrate how to incorporate these concepts in design decision-making. We conclude with proposing a simple synthesis of engineering design, economics and finance in a single design optimization model.

17.2 USERS AND PRODUCERS

The product design decision-making process involves matching a customer's demand for differentiation with the firm's capacity to supply differentiation [18, p. 221]. Differentiation translates to uniqueness and involves all the activities the firm performs. The direct utility that consumers gain from a bundle of observable product attributes (i.e., tangible differentiation) in conjunction with socioeconomic, psychographic and demographic considerations (i.e., intangible differentiation) are driving purchasing behavior.

The ability of the firm to satisfy demand for differentiation profitably to a large extent depends on a set of product decisions $\boldsymbol{x}$ that encompass engineering design decisions $\boldsymbol{x}_d$ and business decisions $\boldsymbol{x}_b$. In turn business decisions $\boldsymbol{x}_b$ encompass economic decisions $\boldsymbol{x}_e$, such as product pricing and production output decisions, and financial decisions $\boldsymbol{x}_f$, such as product valuation and investment timing. The optimal combination of $\boldsymbol{x}_d$, $\boldsymbol{x}_e$ and $\boldsymbol{x}_f$ would lead to increased profitability π:

$$\pi(x), x = [x_d, x_b] \qquad \text{Eq. (17.1)}$$

First, the relevant mathematical models of economic decisions will be reviewed. Financial decision model review will follow. A case study will help us demonstrate the respective disciplinary models. This case study will serve as a platform to present the final holistic product development decision model.

17.3 ECONOMIC DECISIONS

The evaluation criterion of economic decisions is the summation of monetary costs and benefits at one instant of time. In this section we will focus on two decisions: product pricing and production output.

The demand curve is the main tool used in economics that links the quantity of a product the consumer is willing to buy to the price of that product. Assuming that the demand curve is downward sloping and linear, then the following equation models the relationship between price P and quantity Q [39, pp. 21, 31]:

$$Q = \theta - \lambda P \qquad \text{Eq. (17.2)}$$

where θ is the intercept; $\lambda = \Delta Q / \Delta P$ the slope of the demand curve; and λ is the change in quantity associated with a change in price.

But what does the slope of the demand curve represent? From calculus we know that $\lambda = \Delta Q / \Delta P$ captures the change in quantity demanded from a change in price. Dividing by the original level of quantity and price we derive a normalized quantity called the price elasticity of demand

$$E_P = \frac{\dfrac{\Delta Q}{Q}}{\dfrac{\Delta P}{P}} = \frac{\%\Delta Q}{\%\Delta P} \qquad \text{Eq. (17.3)}$$

Given that usually when we increase price the quantity falls, the price elasticity of demand is a negative number. This fall in demand is the consumer's response to a pricing decision set by the decision-maker. Therefore, it models consumer preferences toward a specific product attribute, which is price. If the price elasticity is less than one in magnitude the consumers are inelastic and a change in price will not substantially affect the quantity demanded. The opposite applies when the price elasticity is greater than one and customers are elastic with respect to changes in price.

In the absence of cost we can formulate the following unconstrained optimization problem;

$$\begin{array}{ll} \text{maximize} & \{\text{revenue}\} \\ \text{with respect to} & \{\text{quantity}\} \end{array} \qquad \text{Eq. (17.4)}$$

where revenue is the product of price and quantity. If we were to decide on a single price then the solution of Eq. (17.4) is the largest rectangle under the demand curve (see Fig. 17.1). In the presence of cost, Eq. (17.4) becomes

$$\begin{array}{ll} \text{maximize} & \{\text{revenue} - \text{cost}\} \\ \text{with respect to} & \{\text{quantity}\} \end{array} \qquad \text{Eq. (17.5)}$$

17.3.1 Design Scenario: Demand Curve Formulation

We consider decisions to be made by an automotive manufacturing firm that markets premium-compact (PC), among others,

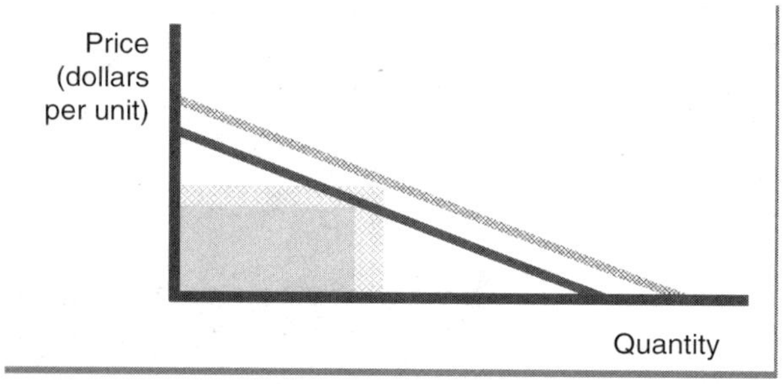

FIG. 17.1 THE SOLUTION OF EQ. (17.4) IS THE PRICE AND QUANTITY THAT MAXIMIZES THE SHADED RECTANGLE AREA UNDER THE DEMAND CURVE (A SHIFT IN THE DEMAND CURVE (EQ. 17.2) ASKS FOR NEW DECISIONS)

vehicles in the U.S. This market segmentation adopted in the study follows the J.D. Power classification for vehicles in the United States. The firm is assumed to operate in a mature industry where complementary assets [43], such as access to distribution channels, service networks, etc., are given. Finally, the decision-maker is assumed to be playing a game against nature, namely, the firm's strategy is affected by an exogenously generated random state, not by competitive interaction. The firm wishes to design new engines and transmissions for the PC segment. Representative of the PC product, the automatic transmission versions of the Chevrolet Cavalier LS Sedan has been selected and simulated using the Advanced Vehicle Simulator (ADVISOR) program [54].

The monthly profit is defined as

$$\pi = P \cdot q - C \qquad \text{Eq. (17.6)}$$

where P = price; q = quantity; and C = average total cost of producing a vehicle.

We draw the relationship between the price P of each product and the quantity q demanded from the observed demand for final goods. Knowing two different points on the demand curve we can use the arc elasticity of demand, see Eq. (17.3), which is defined as:

$$E_P = \frac{\Delta q}{\Delta P}\frac{\overline{P}}{\overline{q}} \qquad \text{Eq. (17.7)}$$

where $\overline{P}$, $\overline{q}$ = averages of the known prices and quantities at the two points.

In 2000 and 1999, General Motors PC vehicles (Chevrolet Cavalier and Prizm, Pontiac Sunfire, Saturn S-series) did not undergo a major design change. Using two pairs of data points ($P_{1/99}$, $q_{1/99}$), ($P_{1/00}$, $q_{1/00}$) (see Table 17.1) [49], we compute the price elasticity of the GM PC segment as equal to –4.9. The price elasticity of all U.S. automobiles has been found to be between –1 and –1.5 [22]. For individual models, price elasticities have been found to be between –3 and –6.7 [4]. For example in [4], the price elasticity

TABLE 17.1 PRICE AND QUANTITIES FOR THE PC SEGMENT

$P_{1/99}$	\$14,512
$q_{1/99}$	43,507
$P_{1/00}$	\$ 15,015 ('99 adj.)
$q_{1/00}$	36,755

of the Chevrolet Cavalier was found to be –6.4. It is reasonable to assume that the price elasticity of demand for all GM PC vehicles is less than the elasticity of an individual model in that segment.

Although there was no observed change in the quality of vehicles in the GMPC segment from 1999 to 2000, which could be a reason for a shift in the demand curve, other factors that affect demand may have taken place. To use the estimated elasticity, we assumed that between the two years there was no major change in consumers' income, product advertising, product information available to consumers, price and quality of substitutes and complementary goods, and population [10].

Using the average values of two data points ($\bar{P}$, $\bar{q}$), we derive the demand curve for the PC segment:

$$P = 17{,}753 - 0.075q \qquad \text{Eq. (17.8)}$$

Assuming a linear relationship between cost and output,

$$C = c_0 q \qquad \text{Eq. (17.9)}$$

the total profit for one time period is

$$\pi = \frac{\theta}{\lambda_P} q - \frac{1}{\lambda_P} q^2 - c_0 q \qquad \text{Eq. (17.10)}$$

Eq. (17.9) assumes that the marginal cost is constant, that is, for every unit increase in output, the average total cost increases by c_0 [40, p. 236], which is set at $13,500 for the PC segment. Eq. (17.9) also assumes that the firm is operating at its minimum efficient scale.

The economic profit is then

$$\pi = \frac{\theta}{\lambda_P} q - \frac{1}{\lambda_P} q^2 - c_0 q \qquad \text{Eq. (17.11)}$$

and the optimum is

$$q_t^* = \frac{\frac{\theta}{\lambda_P} - c_0}{\frac{2}{\lambda_P}} \qquad \text{Eq. (17.12)}$$

17.3.2 The Demand Curve of the Product Development Firm

Let us again review Eq. (17.2). At each period of a multiperiod decision time horizon the decision-maker is facing pricing and output decisions. This observation modifies Eq. (17.2) as follows:

$$q_t = \theta - \lambda P_t \qquad \text{Eq. (17.13)}$$

where t = period we are at. At time period t, the decision-maker will choose the price of the product P_t and the quantity produced q_t. This would essentially mean that economic decisions are moving along the demand curve. But given that there is a unique rectangle with maximum area under the demand curve, what is the motivation behind a time sequence of pricing and output decisions? The answer is that there are many factors that have changed Eq. (17.2). To name a few, since the previous period (t – 1) the income of the consumer may have contracted, the price and quality of substitutes may have risen and the population in the area under consideration may have decreased. All those factors would translate to a shift in the demand curve. This shift asks for new decisions. The decision-maker is updating her decisions to estimate the maximum area rectangle under the new demand curve (see Fig. 17.1). The shifts in the demand curve are represented as follows [39], [47, p. 290], [7]:

$$q_t = \theta_t - \lambda P_t \qquad \text{Eq. (17.14)}$$

where θ_t is now a time-dependent intercept that moves away from or toward the origin. The assumption here is that the slope of the demand curve λ is not changing as frequently as θ. Is it reasonable to assume that λ is not changing as frequently as θ? In the short run, the answer is positive. It takes time for people to change their consumer habits [39, p. 35]. However, one should pay special attention to the product under consideration and the elasticities of both supply and demand. In the chapters that follow, it is assumed that consumer behavior will not change during the life cycle of the product.

In the case where an evaluation of an investment opportunity is under consideration, the decision-maker is facing the ever-changing demand curve. How could she cope with this uncertainty? Fortunately, there is available theory that allows treatment of it. Using information available now, namely, information on past changes in θ and the deviation of those changes from their long-run trend, this source of uncertainty can be modeled. Still a critical piece will be missing, which is the future information that generates shocks on the demand of the product. Therefore, the uncertainty model should combine deterministic with stochastic information. The deterministic one should include past observations (old information) and the stochastic one will involve a simple random number generation (or Monte Carlo Simulation) simulating new information that is unknown to the decision-maker today.

Two possible ways to model the future values of θ are the additive and multiplicative models:

$$\theta(i+1) = a\theta(i) + u(i)$$

$$\theta(i+1) = u(i)\theta(i) \qquad \text{Eq. (17.15)}$$

where $u(i)$, $i = 0, 1, \ldots, N-1$ = random disturbances that cause the θ value to fluctuate in the i-period (i = 0, 1, 2, . . . ,H); and a is a constant. Once $u(0)$ is given, based on θ (0), which is the intercept of the demand curve today, progressively we can find future values. Therefore, values of θ depend only on the value at the most recent previous time and the random disturbance.

By taking the natural logarithm of both sides of the multiplicative model we have

$$\ln \theta(i+1) = \ln \theta(i) + \ln u(i) \Rightarrow$$

$$\Delta \ln \theta(t) = \ln u(i). \qquad \text{Eq. (17.16)}$$

We proceed by modeling the random variable $\ln u(i)$. We define the random-walk process z as:

$$z(t_{i+1}) = z(t_i) + \in(t_i)\sqrt{\Delta t},$$

$$t_{i+1} = t_i + \Delta t, \qquad \text{Eq. (17.17)}$$

$\in(t_i)$ is a normal random variable with mean 0 and variance 1. By taking the limit of the random walk process Eq. (17.17) at $\Delta t \to 0$, the Wiener process (or Brownian motion) is obtained [30, p. 306]:

$$\Delta z = \in(t)\sqrt{\Delta t} \qquad \text{Eq. (17.18)}$$

The generalized Wiener process is defined as

$$\Delta X(t) = a\Delta t + b\Delta z$$

$$\Delta z = \epsilon(t)\sqrt{\Delta t} \qquad \text{Eq. (17.19)}$$

where z is a Wiener process; a, b are constants; and $X(t)$ is a random variable.

The Ito process [30, p. 308]:

$$\Delta X(t) = a(\theta,t)\Delta t + b(\theta,t)\Delta z$$

$$\Delta z = \epsilon(t)\sqrt{\Delta t} \qquad \text{Eq. (17.20)}$$

is a generalization of Eq. (17.19). We assume that the random variable $\ln u(i)$ has expected value $\mathbb{E}[\ln u(i)] = v$ and variance σ^2. We can then describe $\ln u(i)$ as a generalized Wiener process:

$$\ln u(i) = v\Delta t + \sigma\Delta z, \qquad \text{Eq. (17.21)}$$

where z = a Wiener process. Then from Eqs. (17.16), (17.21):

$$\begin{aligned} \Delta \ln \theta(t) &= v\Delta t + \sigma\Delta z \Rightarrow \\ \frac{\Delta\theta(t)}{\theta(t)} &= \mu\Delta t + \sigma\Delta z \Rightarrow \\ \Delta\theta(t) &= (\mu\Delta t + \sigma\Delta z)\theta(t) \Rightarrow \\ \Delta\theta(t) &= \mu\theta(t)\Delta t + \sigma\theta(t)\Delta z \end{aligned} \qquad \text{Eq. (17.22)}$$

where $\mu = (1/2)\sigma^2 + v$ = a correction term necessary when one changes variables in Ito processes [30, p. 309]. Eq. (17.22) is termed the Geometric Brownian Motion. The full equation scheme is as follows:

$$\Delta\theta(t) = \mu\theta(t)\Delta t + \sigma\theta(t)\Delta z$$

$$\Delta z = \epsilon(t)\sqrt{\Delta t} \qquad \text{Eq. (17.23)}$$

$$\epsilon \sim N(0,1)$$

Historical data on θ allow the estimation of μ and σ. By generating random numbers $\epsilon(i)$ and having the value of θ today, we can estimate $\Delta\theta(t)$.

If one follows Eq. (17.14) as the representation of the demand curve, one assumes that all factors other than the price of the product are random disturbances. While many factors are indeed random, others are not. As stated earlier, in corporate strategy literature the product design decision-making process involves matching the customer's demand for differentiation with the firm's capacity to supply differentiation. Therefore, the supply of differentiation is a choice made by the firm. If the level of product differentiation at time t is different from that at time $(t-1)$ a shift in Eq. (17.14) will be realized. A simplified version of the Berry Levinsohn and Pakes (BLP) method [3, 4, 5, 37] will help us model the consumer demand for product differentiation.

The BLP method, as it is presented here, requires an a priori consumer market segmentation. Segmentation is different from differentiation [18, p. 220]. Segmentation is concerned with where the firm competes, e.g., consumer groups, and geographic regions. Differentiation is concerned with how the firm competes in the particular segment, making its product different from that of its competitors. An assumption being made in this chapter is that market segmentations at times t and $(t + 1)$ are the same. Although in the short run this assumption is most likely to hold, in the long run it does not. The value network of the consumer is dynamic. Static market segmentations, which are based on socioeconomic categories, have failed to address consumers' preferences in the past, see, e.g., the General Motors case in Grant [18, p. 227]. A change in market segmentation depends both on supply and demand. For example, the elasticities of supplying product reliability in the automotive industry were inelastic in the late 1980s. The demand for reliability asked for redesign of manufacturing processes not only by the automotive manufacturers, but by their suppliers as well. As a result Japanese automotive firms, which were relatively more elastic and thus had greater capacity to supply differentiation, met first the demand for differentiation both in the European and U.S. markets. To conclude, both market segment definition and the frequency of reviewing it are left at the discretion of the decision-maker.

Let us define a few terms: α are observed product characteristics, e.g., horsepower, fuel efficiency and price P; ξ are unobserved product characteristics, e.g, style, prestige, reputation and past experience; δ is the mean utility level (numerical measure of consumer preferences) obtained from consuming a product τ in the segment under consideration

$$\delta^\tau = \sum_{\kappa} \alpha_\kappa^\tau \phi_\kappa^\tau + \xi^\tau \Rightarrow \qquad \text{Eq. (17.24)}$$

$$\delta^\tau = -\mathrm{P}^\tau\phi_{\mathrm{P}}^\tau + \sum_{\kappa \neq P}^{K} \alpha_\kappa^\tau \phi_\kappa^\tau + \xi^\tau \quad \tau \in \{0,1,\dots,L\}, \ \ P,\kappa \in K$$

where ϕ is the market segment aggregate observed component of utility for the product characteristic κ. Let us assume that market share depends only on mean utility levels:

$$s^\tau = s^\tau(\delta), \ \ \tau \in \{1,\dots,L\} \qquad \text{Eq. (17.25)}$$

At the true values of utilities δ and market shares **s**, Eq. (17.25) must hold exactly. If Eq. (17.25) can be inverted to produce the vector $\delta = s^{-1}(\mathbf{s})$, then the observed market segment shares uniquely determine the means of consumer utility for each product τ. If M is the market segment size, that is, the number of consumers in the market, the output quantity sold is:

$$q^\tau = Ms^\tau(\alpha,\xi,\boldsymbol{P}) \qquad \text{Eq. (17.26)}$$

The consumer may choose to purchase an outside product τ = 0 instead of L products within the segment. The market share of product τ is then given by the logit formula

$$s^\tau(\delta) = \frac{e^{\delta^\tau}}{\sum_{\kappa=0}^{K} e^{\delta_\kappa^\tau}} \qquad \text{Eq. (17.27)}$$

with the mean utility of the outside product normalized to zero,

$$\ln s_\tau - \ln s_0 = \delta_j(\mathbf{s}) \equiv -P_\tau\phi_P + \sum_{\kappa \neq P}^{K} \alpha_\kappa^\tau \phi^\tau + \xi^\tau \qquad \text{Eq. (17.28)}$$

so δ_τ is uniquely identified directly from a simple algebraic calculation involving market shares. Thus the logit suggests a simple regression of differences in log market shares on $(\boldsymbol{\alpha}^\tau, P^\tau)$. From past observations on the number of firms within the market segment, ϕ can then be estimated. When Eq. (17.28) is solved, the κ product attribute sensitivity of demand would be equal to

($\partial Ms / \partial \phi_\kappa$). Having estimated the sensitivity, we can then estimate the κ product attribute elasticity of demand.

One needs to make a few back-of-the-envelope calculations before exercising Eq. (17.28). First a definition of market segment size M is needed. Often, publicly held firms are listing their sales in dollar values per segment. In this case M is equal to the sum of all sales in the market. If this is not the case, one can acquire volume data from market research firms. The estimation of the market size of the outside product requires some judgment. In the case where one wants to exercise Eq. (17.28) for compact cars, then it is reasonable to assume that the outside product consists of all other market segments and the market of used cars. It cannot be overstated that the calculation of Eq. (17.28) is very sensitive to the definition of the outside product.

In the case where one needs to take into account the price and attribute levels of a product's substitutes, Eq. (17.28) will give unreasonable crossprice elasticities. The cross-price elasticities represent percentage change in quantity demanded by one percent change in the price of competing products [40, p.32]. For the decision-maker interested in substitution patterns, the full version [4] of the BLP method addresses this problem.

We conclude this section by laying out the multidimensional demand curve of the firm for a given market segment, namely:

$$q_t = \theta_t - \lambda_p P_t + \boldsymbol{\lambda}_\alpha^T \boldsymbol{\alpha} \qquad \text{Eq. (17.29)}$$

Here Eq. (17.14) is augmented to include the impact of product characteristics on demand. Now we can assemble the entire demand model by including the calculation of θ from Eq. (17.23):

$$\begin{aligned} q_t &= \theta_t - \lambda_p P_t + \boldsymbol{\lambda}_\alpha^T \boldsymbol{\alpha} \\ \Delta\theta(t) &= \mu\theta(t)\Delta t + \sigma\theta(t)\Delta z \qquad \text{Eq. (17.30)} \\ \Delta z &= \epsilon(t)\sqrt{\Delta t} \end{aligned}$$

$$\epsilon \sim N(0,1)$$

Here are the necessary steps to construct the demand curve of the product development firm Eq. (17.30):

- The product market segment needs to be defined first. One can use publicly available definitions from market research firms.
- Observations of market shares of competing products and their respective levels of product attributes need to be collected. The time brackets of these observations require special attention. The decision-maker may be tempted to collect a large sample to improve the fidelity of descriptive statistics. However, technological changes may have improved the quality of the product, regardless of consumer preferences.
- Estimation of the market size and the size of the outside product comes next. Based on Eq. (17.28), product attributes sensitivities of demand $\boldsymbol{\lambda}_\alpha$ are then estimated.
- Having completed the study of the market segment we shift our focus to the firm itself. A time interval where the product attributes are fixed needs to be identified. For the defined time interval, observations of price and demand need to be collected.
- Based on these observations the drift μ and volatility σ need to be estimated. Given that θ follows a stochastic process, calculation of the (deterministic) μ and volatility need to be treated appropriately. Having estimated μ, σ, λ_α Eq. (17.30) can be now constructed. This is the product demand curve for a specific market segment and will be used for product decisions for the estimated life cycle.

The estimation of Eq. (17.30) is decomposable. It captures sensitivities of the consumer toward product attributes and product demand uncertainty. The latter allows us to generate random paths of the future, which allows the decision-maker to project future cash flows necessary for investment valuation.

17.3.3 Design Scenario: Single Vehicle Decision Model

The design of new engines and transmissions allows the firm to market a product with improved performance. From [4] we know that the miles-per-dollar and horsepower-to-vehicle-weight elasticities of demand for Chevrolet Cavalier are 0.52 and 0.42, respectively. That is, a 10% increase in miles-per-dollar and horsepower-to-vehicle-weight ratio will boost demand by 5.2% and 4.2%, respectively. We will use these elasticities as representative for the PC segment.

From Eq. (17.29) we define the demand curve as follows:

$$q = \theta - \lambda_P P + \lambda_{\frac{HP}{w}} \frac{HP}{w} + \lambda_{\frac{M}{\$}} M \qquad \text{Eq. (17.31)}$$

Using the average values of two data points ($\bar{P}, \bar{q}$), we derive the demand curve for the PC segment, extending Eq. (17.8):

$$P = 14{,}943 - 0.075q + 2{,}401\frac{HP}{w} + 805\frac{M}{\$} \qquad \text{Eq. (17.32)}$$

where HP is horsepower, w is weight in tens of pounds; and $M/\$$ is number of 10-mile increments one could travel for \$1.

The fuel economy ratings for a manufacturer's entire line of passenger cars must average at least 27.5 miles per gallon (mpg). Failure to comply with the corporate average fuel economy (CAFE) limit, L_i, results in a civil penalty of \$5 for each 0.1 mpg the manufacturer's fleet falls below the standard, multiplied by the number of vehicles it produces. For example, if a manufacturer produces 2 million cars in a particular model year, and its CAFE falls 0.5 mpg below the standard, it would be liable for a civil penalty of \$50 million. Specifically, for each vehicle, τ, the penalty (or credit) c_{CAFE} due to CAFE is:

$$\text{Cost}_{\text{CAFE}} = c_{\text{CAFE}}\, q \qquad \text{Eq. (17.33a)}$$

$$c_{\text{CAFE}} = \left[5 \times \frac{L - fe}{0.1}\right] \qquad \text{Eq. (17.33b)}$$

Fuel economy f_e is measured in miles per gallon M/G, and it is an engineering attribute computed in terms of the design decisions by the ADVISOR model [54].

The CAFE regulation can only hurt the firm's profits, not contribute to them. The PC segment allows the firm to get credit for less-fuel-efficient vehicles that yield higher profits. Therefore, each PC vehicle sold generates the opportunity for the firm to reap those profits. In a subsequent section we will study the impact of CAFE in a portfolio of two vehicles; one fuel-efficient and one fuel-inefficient.

The economic profit is:

$$\pi = \frac{\theta}{\lambda_P} q - \frac{1}{\lambda_P} q^2 + \frac{1}{\lambda_P}\left(\lambda_{\frac{HP}{w}} \frac{HP}{w} + \lambda_{\frac{M}{\$}} \frac{M}{\$}\right) q - c_0 q - c_{\text{CAFE}}\, q \qquad \text{Eq. (17.34)}$$

where CAFE is equal either CAFE penalty or contribution of credit. The optimum quantity for maximum profit is calculated to be:

$$q_t^* = \frac{\dfrac{\theta}{\lambda_P} + \dfrac{1}{\lambda_P}\left(\lambda_{\frac{HP}{w}}\dfrac{HP}{w} + \lambda_{\frac{M}{\$}}\dfrac{M}{\$}\right) - c_0 - c_{\text{CAFE}}}{\dfrac{2}{\lambda_P}} \quad \text{Eq. (17.35)}$$

We portray Eq. (17.35) in Fig. 17.2, where monthly profits are depicted on the acceptable values of engine size in kW and final drive ratio. Profits have been calculated based on Eq. (17.34). We can observe that the lower and upper engine size bounds yield the maximum profit for the firm. The lower engine size bound corresponds to a fuel-efficient vehicle, while the upper bound corresponds to a fuel-inefficient one.

Given that we have not yet accounted for technical requirements, some of the designs may be infeasible. Therefore, Eq. (17.35) was derived from an incomplete decision model.

We formulate now a design problem by taking into account the CAFE regulation of Eq. (17.33). Then we solve the problem again without Eq. (17.33) to understand the impact of government intervention in the supply of the final product.

The decision model is:

maximize π
with respect to: $\mathbf{x}$ = {engine size, final drive ratio}
subject to: fuel economy ≥ 27.3 mpg

$$t_{0-60} \leq 12.5 \text{ s}$$
$$t_{0-85} \leq 26.3 \text{ s}$$
$$t_{40-60} \leq 5.9 \text{ s} \quad \text{Eq. (17.36)}$$

(max acceleration) ≥ 13.0 ft/s²
(max speed) ≥ 97.3 mph
(max grade at 55 mph) ≥ 18.1%
(5-sec distance) ≥ 123.5 ft

This is the first step toward the synthesis model that will emerge in the conclusion of this chapter.

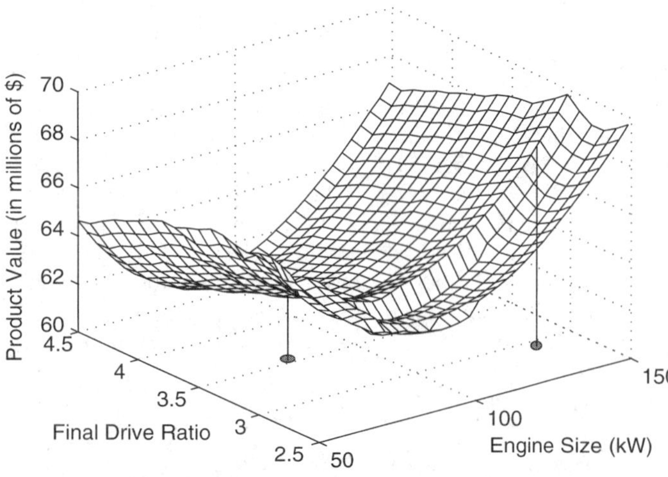

FIG. 17.2 OPTIMAL PROFITABILITY FOR DIFFERENT DESIGN DECISIONS (OPTIMAL DESIGNS LIE ON THE POINTS WHERE TECHNICAL CONSTRAINTS ARE ACTIVE)

TABLE 17.2 IMPACT OF GOVERNMENT REGULATIONS ON DESIGN DECISIONS AND PROFITABILITY

Variable	Solution With CAFE	Solution With no CAFE
Quantity	**29,121**	**29,994**
Price	**$15,670**	**$15,735**
Engine size	**97 (HP)**	**174 (HP)**
Final drive	**3.52**	**2.92**
(Fuel economy)	37.78(mpg)	27.30 (mpg)
(Acceleration 0 to 60)	12.42 (s)	8.02 (s)
(Acceleration 0 to 80)	26.27 (s)	15.91 (s)
(Acceleration 40 to 60)	5.66 (s)	3.82 (s)
(5-sec distance)	130 (ft)	182.55 (ft)
(Max acceleration)	16 (ft/s²)	16.24 (ft/s²)
(Max speed)	110.73 (mph)	137.93 (mph)
(Max grade at 55mph)	18.41 (%)	25.77 (%)
CAFE	($514)	n/a
Profit (Single period)	$63 M ($78M including CAFE)	$67M

We will solve the design model Eq. (17.36) twice. First by taking into account the cost of CAFE in computing profits π as stated in Eq. (17.34). The impact of regulation on the incentives of the firm to supply differentiation will be quantified by estimating profits without taking into account the CAFE cost (or credit) component [Eq. (17.33)].

The optimization algorithm employed to solve Eq. (17.36) is DIRECT [25]. In Table 17.2 the solution in each of the two cases is presented.

The two designs deviate substantially. In the case where CAFE is taken into account the firm has the incentive to design a fuel-efficient vehicle (left point in Fig. 17.2). The time 0 to 60 and 0 to 80 constraints are active. In the absence of regulations the fuel economy constraint becomes active. Under this scenario the vehicle has 80% increased horsepower (right point on Fig. 17.2). The firm realizes a 6% increase in profits. As the case study continues to evolve we will see that this increase is insignificant compared to the profits realized by supplying differentiation to the higher segment.

17.4 INVESTMENT DECISIONS

The evaluation criterion of investment decisions is the summation of monetary costs and benefits across time. The decision of interest is whether or not one should sacrifice current consumption for future consumption.

Let us assume that a university professor is considering investing $100,000 in a real estate property that will yield a return of $110,000 a year from now. However, she is also considering a trip to Jamaica after correcting the final exam two weeks from now. Assuming that the return of the real estate investment is riskless, there should be a financial institution willing to lend her money today. Assuming that the interest rate at the capital markets is 5%, the university professor could borrow \$110,000/1.05, which is equal to \$104,762. That is:

$$\text{present value} = \frac{\text{future value}}{1 + \text{interest rate}} \qquad \text{Eq. (17.37)}$$

Therefore, the net present value of the real estate investment is \$104,762 – \$100,000 = \$4,762. That is

$$\text{net present value} = \frac{\text{future value}}{1 + \text{interest rate}} - \text{investment} \qquad \text{Eq. (17.38)}$$

A year from now she will be able to pay off the loan from the return that the investment will yield.

Now, let us assume that a colleague of the university professor has access to this investment opportunity but has different preferences toward consumption. Unlike her, he prefers to travel to Michigan's Upper Peninsula after the exams and consume the accumulated wealth a year from now. The present value of the investment is \$110,000/1.05 = \$104,762. Although they have different preferences toward consumption, they both agree that the net present value of the investment is \$110,000/1.05 – \$100,000 = \$4,762.

The case where the two university professors are shareholders of the same firm is now considered. Let us assume that the required real estate investment is equal to \$10M. Regardless of their preferences, they would both authorize the management of the firm to invest in this opportunity. The net present value would now be equal to \$0.47M. If the two colleagues own 100 shares each, out of the total \$10M outstanding shares, then from this investment opportunity they would increase their wealth by 100 × \$0.47M/10M = \$4.7.

In the previous example we did valuation of the real investment opportunity under certainty. The risk of the firm not getting \$11M return was zero. Unfortunately, this is rarely the case. Investments in assets other than Treasury bills encompass many uncertainty factors that translate to risk. Relaxing the certainty assumption we demonstrate valuation under uncertainty.

Let us assume that the probability the publicly held firm will gain \$11M from the real estate investment is 80%. Many economics analysts have raised concerns regarding a possible burst in property prices. In that case the real estate is expected to have a value of \$9M. Let us assume that the probability of such an event to occur is 20%. The net present value of this investment is (\$11M × 0.8 + \$9*M* × 0.2)/1.05 – \$10M = \$95K. Therefore, it is advisable that the firm should undergo this investment.

So far, we made the assumption that the firm needs to make the investment decision today. Let us assume that the real estate company has enough opportunities available such that the firm can invest a year from now. Essentially, this relaxes the assumption of absence of flexibility in the decision-making process. The firm can adopt a "wait and see" approach. A year from now the decision-maker would be able to see if the economists were right in their projections regarding the assumed property-price bubble. The valuation question now turns to the following: How much is the option to invest a year from now worth to the firm? In the case where the analysts are right, then the net present value is (\$9M – \$10M)/1.05^2 = –\$900K. In the case where, say, the chairman of the Federal Reserve Board is right and property prices will indeed rise, then the net present value of the investment is \$11M – \$10M / 1.05^2 = \$900K. In both cases we have a three-period investment (today, period one and period two) and therefore we need to discount twice the expected profit to obtain the net present value. Taking into account the probability of each likely event, the value of this option is equal to 0.8 × \$900K + 0.2 × (–\$900K) = \$540 K. Given that the firm could exercise the option today and get \$95 K, the price of the option is equal to \$540K – \$95K = \$445K.

If the real estate company asked for a fee of more than \$445K to reserve the right for the publicly held firm to decide a year later whether to exercise the option to buy, then the firm should forego the investment opportunity.

Let us now assume that the university professor was hired by the real estate firm as a consultant. Her project is to find the net present value of an investment decision the firm is facing. In her personal investment decision we assumed that the discount factor or opportunity cost of capital is equal to 5% (equal to the rate of a Treasury bill). What is the appropriate discount rate for the firm's investment decision?

Let us assume that the stock of the firm is listed in the stock exchange. By using historical data of the past performance of the stock one can exercise the following regression:

$$r = a + \beta r_m + \epsilon \Rightarrow$$

$$\mathbb{E}(r) = a + \beta \mathbb{E}(r_m) \qquad \text{Eq. (17.39)}$$

where r is the return of the firm's stock; r_m is the return of the capital markets as proxied by the S&P500 index; ϵ is the error of the regression; a and β are constants; and $\mathbb{E}$ denotes expectation. The slope β models the sensitivity of the firm's stock return for 1% variation in the market return. This can be represented as follows:

$$\beta = \frac{\mathrm{Cov}(r, r_m)}{\mathrm{Var}(r_m)} \qquad \text{Eq. (17.40)}$$

where Cov stands to covariance; and Var for variance.

Based on certain assumptions that concern the behavior of the investors and the conditions for perfect and competitive capital markets [47, p. 44], it is assumed that any investor can invest a portion (β%) of her wealth at the S&P500 with $\mathbb{E}(r_m)$, so that its covariance with the capital markets is unity ($\beta = 1$). She can then borrow (1 – β%) at a "risk-free" rate r_f so that its covariance with the capital markets is 0 ($\beta = 0$). Her return would be:

$$\mathbb{E}(r) = \beta \mathbb{E}(r_m) + (1 - \beta) r_f \quad \text{or}$$

$$\mathbb{E}(r) = r_f + \beta\left[\mathbb{E}(r_m) - r_f\right] \qquad \text{Eq. (17.41)}$$

The intercept of Eq. (17.41) is the risk-free interest rate r_f and its slope the market risk premium [$\mathbb{E}(r_m) - r_f$]. Therefore, an investor should be willing to hold a stock with a particular β only if compensated by a return equal to Eq. (17.41).

In our example, the expected risk premium on the stock of the firm should be equal to

$$\mathbb{E}(r) - r_f = \beta\left[\mathbb{E}(r_m) - r_f\right] \quad \text{or} \qquad \text{Eq. (17.42a)}$$

$$\beta = \frac{\mathbb{E}(r) - r_f}{\mathbb{E}(r_m) - r_f} \qquad \text{Eq. (17.42b)}$$

which is called the capital asset pricing model (CAPM). The CAPM is essentially modeling how much more (or less) risky the stock of a firm is relative to the market. If the β of the firm's stock is two that means it is twice as risky as the expected market risk premium [$\mathbb{E}(r_m) - r_f$]. The β of the stock is calculated based on Eq. (17.39).

If the firm does not have any debt, then the appropriate discount factor for the investment that the firm is considering should be calculated based on Eq. (17.42). In the case where the firm does have debt, this makes the investment more risky. Assuming the firm has an amount of debt equal to D, and market capitalization (number of shares multiplied by the stock price) equal to E, where E stands for equity, then the total value of the firm is equal to ($D + E$). Then the weighted average cost of capital (WACC) is equal to:

$$\text{WACC} = \frac{D}{V} r_d + \frac{E}{V} r \qquad \text{Eq. (17.43)}$$

where r is the cost of equity calculated by Eq. (17.42); and r_d is the current borrowing rate of the firm. If the riskiness of the investment under consideration is that of the firm, then the university professor, and now consultant, should use Eq. (17.43) as the discount factor to evaluate this investment. Substituting Eq. (17.42) in Eq. (17.43):

$$\text{WACC} = \frac{D}{V} r_d + \frac{E}{V}(r_f + \beta[\mathbb{E}(r_m) - r_f]) \qquad \text{Eq. (17.44)}$$

This would be the situation we assume in the studies described in the following sections.

The investment decision that the firm is facing can be summarized in the following unconstrained optimization model

maximize	{net present value (future cash flows, investment cost, risk)}
with respect to	{invest now or, invest never or, invest later}. Eq. (17.45)

17.4.1 Design Scenario: Portfolio Decision Model

We now expand the design scenario to consider an automotive manufacturing firm that markets premium-compact (PC) and full-size sport utility (SUV) vehicles. This market segmentation follows the J.D. Power classification for vehicles in the United States.

The firm wishes to design new engines and transmissions for both PC and SUV segments. The PC and SUV segments are low and high profit margin segments, respectively. The Energy Policy and Conservation Act of 1975 required passenger car and light truck manufacturers to meet corporate average fuel economy (CAFE) standards applied on a fleet-wide basis. There are K units of monthly capacity currently in place for both segments, and so K is fixed, representing a capacity constraint. It is assumed that this capacity is not expandable. The decision-maker faces the following decisions: How should the units of capacity be allocated between the two segments in order to maximize the firm's value? What should the performance specifications for engines and transmissions be and how do these specifications affect the resource allocation decision? How much is this investment worth to the firm's stock owners?

We formulate the following decision model as:

$$\begin{aligned} &\text{maximize} && \text{NPV} \\ &\text{with respect to} && \{\mathbf{x}^\tau, \mathbf{q}^\tau\}\ \tau = 1, 2 \\ &\text{subject to} && \sum_{\tau=1}^{2} q^\tau = K \\ & && \sum_{\tau=1}^{2} \text{Cost}^\tau_{\text{CAFE}} \le 0 \qquad \text{Eq. (17.46)} \\ & && \mathbf{g}^\tau(\mathbf{x}^\tau) \le 0,\ \tau = 1,2 \end{aligned}$$

where NPV is the net present value; $\mathbf{q}$ is the vector of supply decisions (monthly production quantities q_1, q_2); and $\mathbf{x}^\tau$ is the vector of engineering design decisions (engine sizes and final drive ratios) for each vehicle. The equality is a production constraint that fixes the total available production, the first inequality is an enterprise constraint that will not allow CAFE penalties to be paid for the selected product mix, and the two vector inequalities are engineering constraints. We will use the index τ = 1, 2 for the two products, but we will also use the subscripts PC and SUV instead of $\tau = 1$ and $\tau = 2$, respectively, when convenient.

Net present value is the aggregation of future monthly cash flows or profits π^τ minus the investment cost I over the life H of product x with a weighted average cost of capital:

$$\text{NPV} = -I + \int_0^H \pi^\tau e^{-(WACC)t} dt \qquad \text{Eq. (17.47)}$$

The monthly profit is defined as

$$\pi^\tau = P^\tau q^\tau - C^\tau \qquad \text{Eq. (17.48)}$$

where P^τ is the price; q^τ is the quantity; and C^τ is the average total cost of producing vehicle τ.

We will first derive the model for the problem where instead of the NPV we maximize simply the monthly profits $\sum \pi^\tau$.

In Section 17.3, using price and quantity data points from years 2000 and 1999 we estimated the demand curve for the PC segment. For the GM SUV segment (Chevrolet Tahoe and Suburban, GMC Suburban/Yukon and Yukon XL) we used data points from the years 1999 and 1998 ($P_{1/98}$, $q_{1/98}$), ($P_{1/99}$, $q_{1/99}$) [49], where no major design change took place as well, finding the elasticity to be equal to –2.3 (see Table 17.3).

We assume that for each segment between the two years there was no major change in consumer's income, product advertising, product information available to consumers, price and quality of substitutes and complementary goods, and population [10]. We further assume that the two goods are independent, namely, a change in the price of the compact car has no effect on the quantity demanded for the SUV.

For demonstration purposes we assume that the horsepower to weight elasticity of demand of the traditional luxury segment is close to the SUV one. In [4], the Cadillac Seville horsepower-to-weight elasticity of demand is found to be 0.09. In this segment the miles per dollar elasticity of demand is found to be close to 0. This essentially means that the customer of that segment is satisfied with the current level of fuel economy performance.

The demand curves for both segments are:

$$P^{\text{PC}} = 14{,}943 - 0.075 q^{\text{PC}} + 2401 \frac{HP}{w} + 805 \frac{M}{\$}$$

$$P^{\text{SUV}} = 40{,}440 - 0.525 q^{\text{SUV}} + 2071 \frac{HP}{w} \qquad \text{Eq. (17.49)}$$

where HP = horsepower, w = weight in tens of pounds; and $M/\$$ = number of 10-mile increments one could travel for one dollar.

Assuming a linear relationship between cost and output,

$$C^\tau = c_0^\tau q^\tau \qquad \text{Eq. (17.50)}$$

TABLE 17.3 PRICE AND QUANTITIES FOR THE PC AND SUV SEGMENTS

PC	$P_{1/99}$ \$14,512	$q_{1/99}$ 43,507	$P_{1/00}$ \$15,015 ('99 adj.)	$q_{1/00}$ 36,775
SUV	$P_{1/98}$ \$28,628	$q_{1/98}$ 24,658	$P_{1/99}$ \$29,596 ('98 adj.)	$q_{1/99}$ 22,813

The economic profit becomes

$$\pi^\tau = \left(\frac{\theta^\tau}{\lambda_P^\tau} - \frac{1}{\lambda_P^\tau}q^\tau + \frac{1}{\lambda_P{}^\tau}\left(\boldsymbol{\lambda}_\alpha^\tau\right)^T \boldsymbol{\alpha}^\tau\right)q^\tau - c_0 q^\tau \qquad \text{Eq. (17.51)}$$

Eq. (17.50) assumes that the marginal cost is constant, that is, for every unit increase in output, the variable cost increases by c_0, which is set at \$13,500 and \$18,500[1] for the PC and SUV segments, respectively [40, p. 236]. We have assumed that the firm is operating at its minimum efficient scale [6, p. 73].

For the period 1992 to 2001, the CAFE penalty was nonpositive and approximately zero for DaimlerChrysler, Ford and General Motors [55]. This is represented as follows:

$$\sum_{\tau=1}^{2} \text{Cost}_{\text{CAFE}}^\tau \le 0 \qquad \text{Eq. (17.52)}$$

Note that the CAFE regulations can only hurt the firm's profits, not contribute to them. They function as a set of internal taxes (on fuel-inefficient vehicles) and subsidies (on fuel-effcient vehicles) within each firm [27].

Hence, the cost of each product, Eq. (17.50), is modified to be

$$C^\tau = c_0^\tau q^\tau + \text{Cost}_{\text{CAFE}}^\tau \qquad \text{Eq. (17.53a)}$$

$$C^\tau = \left(c_0^\tau + c_{\text{CAFE}}^\tau\right)q^\tau \qquad \text{Eq. (17.53b)}$$

$$C^\tau = \xi^\tau q^\tau \qquad \text{Eq. (17.53c)}$$

where, $\xi^\tau = c_0^\tau + c_{\text{CAFE}}^\tau$ and c_{CAFE}^τ is equal either to CAFE penalty or to contribution of credit to the portfolio. For example, if the PC vehicle generates \$3M of credit but the penalty incurred to the SUV is \$2.5M then Cost_{CAFE}^{PC} would be equal to \$2.5M. For a portfolio of n products the value of Cost_{CAFE}^τ is then redefined as:

$$c_{\text{CAFE}}^\tau q^\tau = \begin{cases} c_{\text{CAFE}}^\tau q^\tau, & \sum_1^n c_{\text{CAFE}}^\tau < 0 \\ c_{\text{CAFE}}^\tau + \dfrac{u(c_{\text{CAFE}}^\tau)c_{\text{CAFE}}^\tau q^\tau}{\sum_1^n u(c_{\text{CAFE}}^\tau)\left|c_{\text{CAFE}}^\tau\right| q^\tau} \sum_1^n c_{\text{CAFE}}^\tau q^\tau & \sum_1^n c_{\text{CAFE}}^\tau q^\tau < 0 \end{cases}$$

$$\text{Eq. (17.54)}$$

where $u(c_{CAFE}^\tau)$ = equal to 1 when c_{CAFE}^τ is negative and 0 when nonnegative.

As mentioned early in this section, the microeconomic model suggests that in each monthly period the firm should produce the quantity that maximizes total profit during that period:

$$\begin{aligned} &\text{maximize} && \sum_{\tau=1}^{2}\pi^\tau = \sum_{\tau=1}^{2}\left[\frac{\theta^\tau}{\lambda_P^\tau} - \frac{1}{\lambda_P^\tau}q^\tau + \frac{1}{\lambda_P^\tau}(\boldsymbol{\lambda}_\alpha^\tau)'\boldsymbol{\alpha}^\tau\right]q^\tau - \xi^\tau q^\tau \\ &\text{with respect to} && \{q^\tau\}, \tau = 1,2 \\ &\text{subject to} && \sum_{\tau=1}^{2} q^\tau = K \\ & && \sum_{\tau=1}^{2}\text{Cost}_{\text{CAFE}}^\tau = c_{\text{CAFE}}^1 q_1 + c_{\text{CAFE}}^2 q_2 \le 0. \end{aligned} \qquad \text{Eq. (17.55)}$$

For positive quantities of production q^τ, Eq. (17.55) can be solved analytically and the global optimum is

$$q^{PC*} = \max\left(\frac{\left(\frac{\theta^{\text{PC}}}{\lambda_P^{\text{PC}}} - \frac{\theta^{\text{SUV}}}{\lambda_P^{\text{SUV}}}\right) + \left(\frac{1}{\lambda_P^{\text{PC}}}\left(\lambda_{\alpha\text{PC}}^{\text{PC}}\right)^T \alpha^{\text{PC}} - \frac{1}{\lambda_P^{\text{SUV}}}\left(\boldsymbol{\lambda}_{\alpha\text{SUV}}^{\text{SUV}}\right)^T \boldsymbol{a}^{\text{SUV}}\right) + 2\frac{1}{\lambda_P^{\text{SUV}}}K - \left(\xi^{\text{PC}} - \xi^{\text{SUV}}\right)}{2\left(\frac{1}{\lambda_P^{\text{PC}}} + \frac{1}{\lambda_P^{\text{SUV}}}\right)}, \frac{Kc_{\text{CAFE}}^{\text{SUV}}}{c_{\text{CAFE}}^{\text{SUV}} - c_{\text{CAFE}}^{\text{PC}}}\right) \qquad \text{Eq. (17.56)}$$

$$q^{\text{SUV*}} = K - q^{\text{PC*}},$$

where

$$\frac{\left(\frac{\theta^{\text{PC}}}{\lambda_P^{\text{PC}}} - \frac{\theta^{\text{SUV}}}{\lambda_P^{\text{SUV}}}\right) + \left(\frac{1}{\lambda_P^{\text{PC}}}\left(\boldsymbol{\lambda}_{\alpha^{\text{PC}}}^{\text{PC}}\right)^T \boldsymbol{\alpha}^{\text{PC}} - \frac{1}{\lambda_P^{\text{SUV}}}\left(\boldsymbol{\lambda}_{\alpha^{\text{SUV}}}^{\text{SUV}}\right)^T \boldsymbol{\alpha}^{\text{SUV}}\right) + 2\frac{1}{\lambda_P^{\text{SUV}}}K - \left(\xi^{\text{PC}} - \xi^{\text{SUV}}\right)}{2\left(\frac{1}{\lambda_P^{\text{PC}}} + \frac{1}{\lambda_P^{\text{SUV}}}\right)}$$

is the interior optimum, and $Kc_{\text{CAFE}}^{\text{SUV}} / (c_{\text{CAFE}}^{\text{SUV}} - c_{\text{CAFE}}^{\text{PC}})$ is the boundary optimum.

From the derivation above we see that the enterprise-wide problem Eq. (17.46) can be partitioned to production and design problems (see also section Section 17.5). The production problem is Eq. (17.55), above, which can be solved separately for $\boldsymbol{q}^*$. With this partial optimization [38, p. 63] π^τ in Eq. (17.47) can be replaced by $\pi^\tau{}^*$ computed from Eq. (17.56), and the problem of Eq. (17.46) is reduced to Eq. (17.57) for fixed investment costs. Of course, $\boldsymbol{q}^*$ depends on the design variables via the quantities $\left(\boldsymbol{\lambda}_\alpha^\tau\right)^T \boldsymbol{\alpha}^\tau$ and c_{CAFE}^τ.

$$\begin{aligned} &\text{maximize} && \text{NVP}(\mathbf{x}) = -I + \int_0^H \sum_{\tau=1}^{2} \pi^{\tau*} e^{-(WACC)t} dt \\ &\text{with respect to} && \mathbf{x} \\ &\text{subject to} && \mathbf{g}^\tau(\mathbf{x})^\tau \le 0, \quad \tau = 1, 2 \end{aligned} \qquad \text{Eq. (17.57)}$$

The firm's product demand $\left(q_t^\tau\right)^D$ = product τ at time t is expressed as a product of the two sources of uncertainty defined above, namely, market product demand Q and market share M:

$$(q_t^\tau)^D = \begin{cases} 0 & 0 \le \text{month} \le 24 \\ Q_t^\tau M_t^\tau & 25 \le \text{month} \le 84 \end{cases} \qquad \text{Eq. (17.58)}$$

$(q_t^\tau)^D$ = a 1 × 84 vector and represents a random walk in the future. During the first 24 months of product development and production start-up time we have null sales.

[1]These are estimates from private d iscussions with automotive industry professionals.

If the actual demand exceeds the optimal capacity of the plant the firm will sell only at capacity. Thus, if the actual demand $(q_t^\tau)^D$ is less than the optimal capacity $q^{\tau *}$ of the plant, then the firm will supply only as much as the demand permits. That is, the actual supply $(q_t^\tau)^S$ is given by:

$$(q_t^\tau)^S = \begin{cases} (q_t^\tau)^D & \text{if}\,(q_t^\tau)^D < q^{\tau *} \\ q^{\tau} * & \text{otherwise} \end{cases} \qquad \text{Eq. (17.59)}$$

The assumption is that the firm does not possess flexibility in adjusting capacity. If that were not the case, then an appropriate theory of capital budgeting would have been needed to model the resource allocation decision.

We assume that future cash flows generated by a product's commercialization are only imperfectly predictable from the current observation. The probability distribution is determined by the present, but the actual path remains uncertain [13]. We consider product demand and the firm's market share as the two main sources of uncertainty.

To describe future product demand we assume that the automotive product demand Q follows Geometric Brownian motion. Seasonality [48] and life-cycle considerations [7] can be also taken into account.

$$\Delta Q_t = \mu Q_t \Delta t + \sigma Q_t \Delta z$$

$$\Delta z = \in \sqrt{\Delta t} \qquad \text{Eq. (17.60)}$$

$$\in \sim N(0,1)$$

Here $\mu \Delta t$ and $\sigma \sqrt{\Delta t}$ = expected value and the standard deviation, respectively, of $(\Delta Q_t / \Delta Q_t)$ in Δt. To simulate the path followed by Q we divide the life of the source of uncertainty into 73 monthly intervals from January 1995 until January 2001. The value of Q at time Δt (i.e., February 2001) is calculated from the initial value of Q (i.e., January 2001), the value at time $2\Delta t$ is calculated from the value at time Δt and so on [23]. One simulation trial involves constructing a complete path for Q using 73 random samples from a normal distribution. Data provided by J.D. Power & Associates for the period between January 1995 and January 2001 (see Table 17.4) have been employed for the estimation of the expected growth rate μ and volatility σ.

The Ornstein-Uhlenbeck or mean-reverting process is used to model market share uncertainty M.

$$\Delta M_t = \alpha\left(\bar{M}_t - M_t\right)\Delta t + \sigma' \Delta \qquad \text{Eq. (17.61a)}$$

$$\Delta z = \in \sqrt{\Delta t} \qquad \text{Eq. (17.61b)}$$

$$\in \sim N(0,\,1) \qquad \text{Eq. (17.61c)}$$

Here α is the speed of reversion; and $\bar{M}$ is the "normal" level of M, i.e., the level to which $\bar{M}$ tends to revert. By running the nonlinear regression [14]

$$\Delta M = p + bM_{t-1} + \in_t \qquad \text{Eq. (17.62)}$$

from data provided by J.D. Power & Associates (see Table 17.4) for the period between January 1995 and January 2001, we estimate:

TABLE 17.4 MARKET SHARE DATA OF A MAJOR U.S. AUTOMOTIVE MANUFACTURER FOR THE PREMIUM-COMPACT SEGMENT

Month	Units	Market Share
January 1995	135,928	26.25%
February 1995	154,139	26.19%
March 1995	176,348	30.82%
—	—	—
Jan-01	163,253	25.04%

$$\bar{M} = \hat{p} / \hat{b}$$

$$\hat{\alpha} = -\log(1+\hat{b}) \qquad \text{Eq. (17.63)}$$

$$\hat{\sigma}' = \hat{\sigma}'_\varepsilon \sqrt{\frac{\log(1+\hat{b})}{(1+\hat{b})^2 - 1}}$$

where $\hat{\sigma}_\epsilon$ = standard error of the regression.

As we noted in the presentation of Eq. (17.23) in the mathematical finance and economics literature [7, 39], product demand uncertainty is also described by the intercept θ. That essentially describes random shifts of the demand curve – with the same slope λ. This approach would require, in addition to historical demand data, pricing data that are unavailable. In the treatment of the present chapter we decided not to use hypothetical data, thus maintaining the credibility of actual results in the context of the assumptions made.

Collecting Eqs. (17.51), (17.53), (17.54), (17.56) and (17.59) for the PC segment we get the complete calculation of the monthly profit. A similar set of equatio ns holds for the SUV segment:

$$\pi_t^{\text{PC}*} = P^{\text{PC}}(q_t^{\text{PC}})^S - \xi(q_t^{\text{PC}})^S$$

$$(q_t^{\text{PC}})^S = \min\left\{\left(q_t^{\text{PC}}\right)^D, q^{\text{PC}*}\right\},$$

$$q_t^{\text{PC}*} = \max\left(\frac{\left(\frac{\theta^{\text{PC}}}{\lambda^{\text{PC}}} - \frac{\theta^{\text{SUV}}}{\lambda^{\text{SUV}}}\right) + 2\frac{1}{\lambda^{\text{SUV}}}K + \left(\frac{1}{\lambda^{\text{PC}}}\boldsymbol{\lambda}_\alpha^{\text{PC}^T}\boldsymbol{\alpha}^{\text{PC}} + \frac{1}{\lambda^{\text{SUV}}}\boldsymbol{\lambda}_\alpha^{\text{SUV}^T}\boldsymbol{\alpha}^{\text{SUV}}\right) - \left(\xi^{\text{PC}} - \xi^{\text{SUV}}\right)}{2\left(\frac{1}{\lambda^{\text{PC}}} + \frac{1}{\lambda^{\text{SUV}}}\right)}, \frac{Kc_{\text{CAFE}}^{\text{SUV}}\left(\mathbf{x}^{\text{SUV}}\right)}{c_{\text{CAFE}}^{\text{SUV}} - c_{\text{CAFE}}^{\text{PC}}}\right),$$

$$\xi^{\text{PC}} = c_0^{\text{PC}} + c_{\text{CAFE}}^{\text{PC}},$$

$$c_{\text{CAFE}}^\tau q^\tau = \begin{cases} c_{\text{CAFE}}^\tau q^\tau, & \sum_1^m c_{\text{CAFE}}^\tau q^\tau \geq 0 \\ c_{\text{CAFE}}^\tau q^\tau + \dfrac{u(c_{\text{CAFE}}^\tau)c_{\text{CAFE}}^\tau q^\tau}{\sum_1^m (u_{\text{CAFE}}^\tau)\left|c_{\text{CAFE}}^\tau\right| q^\tau} \sum_1^m c_{\text{CAFE}}^\tau q^\tau & \sum_1^m c_{\text{CAFE}}^\tau q^\tau < 0 \end{cases} \qquad \text{Eq. (17.64)}$$

From Eq. (17.51) after substituting q^τ for $(q_t^\tau)^S$ using Eq. (17.59) and C^τ using Eq. (17.53), we get the monthly profit $\pi_t^{\tau *}$ over the 84-month sales period for product τ.

$$\pi_t^{\tau *} = P^{\tau}(q_t^{\tau})^S - (\xi^{\tau})(q_t^{\tau})^S \quad \text{Eq. (17.65)}$$

During the first 24 months we have null profits.

Recall that we assume the decision to develop the new engines and transmissions has already been made, and so the decision facing the firm now is one of resource allocation. Upon determining the optimal production ratio, this decision will be implemented immediately. Hence, the decision contains no embedded real option, which simplifies the net present value calculation. The time period T for both products is estimated to be seven years or 84 months, and includes the product development, production start-up and sales periods. The present value, PV, of discounted future payoffs $\pi_t^{\tau *}$ is formally represented as an integral over the space of sample paths of the underlying stochastic processes X and Y

$$PV \approx \frac{\sum_1^n \left[\int_0^H \left(\sum_{\tau=1}^2 \pi_{t_n}^{\tau *} \right) e^{-(WACC)t} dt \right]}{n} \quad \text{Eq. (17.66)}$$

where $n = 100,000$. The exponent r, the weighted average cost of capital, is estimated as in Eq. (17.43)

$$WACC = r_d(1-t_c)\frac{D}{E+D} + r_e\frac{E}{E+D} \quad \text{Eq. (17.67)}$$

where r_d = firm's cost of debt; t_c = tax rate; D = market value of debt; E = equity; and r_e = cost of equity estimated by the CAPM ([8] and Section 17.4):

$$r_e = r_f + \beta(r_m - r_f) \quad \text{Eq. (17.68)}$$

where r_f = risk-free rate; $(r_m - r_f)$ = market risk premium; and β = firm stock's sensitivity to fluctuation of the market as a whole. Using the values in Table 17.5 we estimate the weighted average cost of capital of a publicly traded U.S. automotive manufacturer to be 9.4%. By using a single r for all 84 months we are making the assumptions that the firm's capital structure (the debt to equity ratio D/E) and the risk of the firm to its shareholders would remain the same for all 84 months.

To estimate the PV we generate 100,000 random walks, resulting in a 100,000×84 matrix. Discounting all future payoffs $\pi_{t_n}^{\tau *}$ across the probability space we get a 100,000×1 matrix. The present value is the average of those 100,000 numbers (see also Fig. 17.3). Note that Eq. (17.66) does not take into account working capital.

TABLE 17.5 PARAMETERS VALUES IN EQS. (17.67, 17.68)

Parameter	Value
t_c	36%
D	\$81.3B
r_f	6.7%
β	1.1
Long-term interest	\$2.8B
Long-term debt	\$31.3B
Number of shares	756M
Stock price	\$58
Market risk premium	8.4%

Subtracting the fixed capital investment needed (I = \$3B), we calculate the NPV:

$$\text{NPV} = \text{PV}-I$$
$$\approx -I + \frac{\sum_1^n \left[\int_0^H \left(\sum_{\tau=1}^2 \pi_{t_n}^{\tau *} \right) e^{-WACCt} dt \right]}{n} . \quad \text{Eq. (17.69)}$$

Other investment costs are ignored; for example, the cost of building the production facility plant is considered a sunk cost because we assume the plant has already been built.

The NPV expression in Eq. (17.69) is the stochastic calculation of the objective in the model of Eq. (17.57). This NPV criterion is based on the optimal economic conditions [7; 47, p. 291] computed in Eq. (17.56), the uncertainty of future cash flows Eqs. (17.60) and (17.61), and the engineering performance Eq. (17.33).

Bounds on vehicle performance attributes define the constraints for each product and its corresponding market segment. These "engineering" constraints are expressed in terms of the design variables using the ADVISOR program cited earlier.

The model in Eq. (17.70) involves four variables and 16 constraints (eight each for the PC and SUV segments of the engineering design model). The complete model of Eq. (17.70) is now assembled using the expressions derived in the preceding sections.

$$\text{maximize} - \$\text{B} + \frac{\sum_1^n \left[\int_0^I \left(\sum_{\tau=1}^2 \pi_{t_n}^{\tau *} \right) e^{-(WACC)t} dt \right]}{n}$$

with respect to $\{\mathbf{x}\}$
subject to
$(\text{final drive})_{PC} \geq 2.5$
$(\text{final drive})_{PC} \leq 4.5$
$(\text{final drive})_{SUV} \geq 2.5$
$(\text{final drive})_{SUV} \leq 4.5$
$(\text{engine size})_{PC} \geq 50$ (kW)
$(\text{engine size})_{PC} \leq 150$ (kW)
$(\text{engine size})_{SUV} \geq 150$ (kW)
$(\text{engine size})_{SUV} \leq 250$ (kW)
$(\text{fuel economy})_{PC} \geq 27.3$ (mpg)
$(\text{acceleration 0 to 60})_{PC} \leq 12.5$ (s)
$(\text{acceleration 0 to 80})_{PC} \leq 26.3$ (s)
$(\text{acceleration 40 to 60})_{PC} \leq 5.9$ (s)
$(\text{5-sec distance})_{PC} \geq 123.5$ (ft)
$(\text{max acceleration})_{PC} \geq 13$ (ft/s^2)
$(\text{max speed})_{PC} \geq 97.3$ (mph)
$(\text{max grade at 55 mph})_{PC} \geq 18.1$(%)
$(\text{fuel economy})_{SUV} \geq 12.8$ (mpg)
$(\text{acceleration 0 to 60})_{SUV} \leq 9.8$ (s)
$(\text{acceleration 0 to 80})_{SUV} \leq 22.8$ (s)
$(\text{acceleration 40 to 60})_{SUV} \leq 5.0$ (s)
$(\text{5-sec distance})_{SUV} \geq 154.5$ (ft)
$(\text{max acceleration})_{SUV} \geq 15.4$ (ft/s^2)
$(\text{max speed})_{SUV} \geq 100.4$ (mph)
$(\text{max grade at 55 mph})_{SUV} \geq 18.6$(%)

Eq. (17.70)

We now present the results obtained by solving the valuation problem posed in Eq. (17.70). The divided rectangles (DIRECT) optimization algorithm [25] was used. DIRECT can locate global minima efficiently without derivative information, when the number of variables is small, as in this case. It is often inefficient at refining local minima, and so a sequential quadratic programming algorithm was combined with DIRECT to find solutions in a more efficient manner.

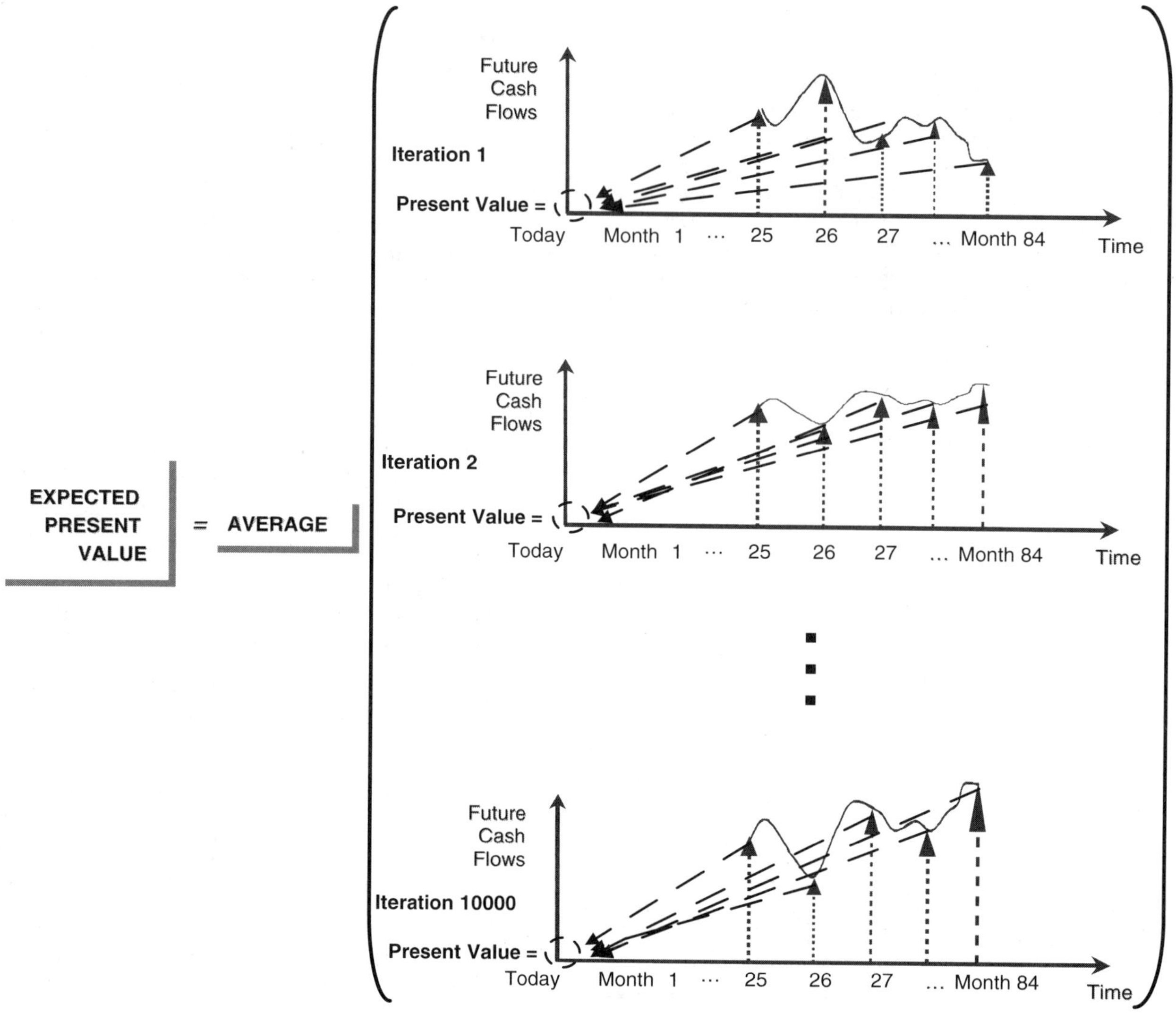

FIG. 17.3 EXPECTED PRESENT VALUE

Optimization problems were solved for two different production capacities: 50,000 and 20,000. The optimal solutions found are shown in Table 17.6.

Prior to interpreting the results, recall that Eq. (17.70) is a portfolio design problem with scarce resources. That is, the sum of the optimal quantities of each segment is greater or equal to the manufacturing resources of the firm. One can interpret the results by paying attention to production quantity q, the regulatory penalty (or credit) per vehicle and the vector of design variables (i.e., engine size, final drive). Specifically:

- At the small capacity level of 20,000, the small quantity of produced compact vehicles calls for high fuel efficiency of the compact car and average performance of the SUV.
- At the high capacity level of 50,000, the quantity of compact vehicles has enough scale to mitigate the regulatory penalty, while improving the performance of the sport utility vehicle by a 28% increase in engine size.

From a public policy point of view, the CAFE regulation can be interpreted as an active constraint to consumer preferences. As long as the consumer asks for more horsepower, and the producer can materialize this preference in terms of profit, the fuel efficiency regulatory constraint would be active. A question of interest is whether or not new technologies can make the CAFE constraint inactive and at what cost.

We proceed in the next section with the synthesis of the economic, investment and engineering decision models.

17.5 THE DESIGN DECISION MODEL OF THE PRODUCT DEVELOPMENT FIRM

For a single segment and a single product affecting demand, at time instance t the inverse demand curve is from Eq. (17.30)

$$P_t = \frac{\theta_t}{\lambda_P} - \frac{1}{\lambda_P} q_t + \frac{1}{\lambda_P} \boldsymbol{\lambda}_\alpha^T \boldsymbol{\alpha} \quad \text{Eq. (17.71)}$$

Revenue R_t is equal to price times quantity

$$R_t = \frac{\theta_t}{\lambda_P} q_t - \frac{1}{\lambda_P} q_t^2 + \frac{1}{\lambda_P} \boldsymbol{\lambda}_\alpha^T \boldsymbol{\alpha} q_t \quad \text{Eq. (17.72)}$$

and profit π_t is equal to revenue minus cost

$$\pi_t = \frac{\theta_t}{\lambda_P} q_t - \frac{1}{\lambda_P} q_t^2 + \frac{1}{\lambda_P} \boldsymbol{\lambda}_\alpha^T \boldsymbol{\alpha} q_t - C_t(q_t) q_t \quad \text{Eq. (17.73)}$$

TABLE 17.6 SOLUTIONS OF THE ENTERPRISE MODEL FOR DIFFERENT CAPACITIES

Variable	Solution for K = 20, 000	Solution for K = 50, 000
Quantity^{PC}	**5,286**	**28,675**
w^{PC}	26 (%)	58 (%)
$CAFE^{PC}$	−$508	−$510
Engine^{pc}	**72.85** (kW)	**72.50** (kW)
(Final drive)^{PC}	**3.49**	**3.53**
(Fuel economy)^{PC}	37.67 (mpg)	37.70 (mpg)
(Acceleration 0 to 60)^{PC}	12.43 (s)	12.4 (s)
(Acceleration 0 to 80)^{PC}	26.17 (s)	26.23 (s)
(Acceleration 40 to 60)^{PC}	5.7 (s)	5.62 (s)
(5-sec distance)^{PC}	130.46 (ft)	130.62 (ft)
(max acceleration)^{PC}	16.05 (ft/s^2)	16.05 (ft/s^2)
(max speed)^{PC}	110.7 (mph)	111.09 (mph)
(max grade at 55 mph) ^{PC}	18.37 (%)	18.52 (%)
Quantity^{SUV}	**14,714**	**21325**
w^{SUV}	74 (%)	42 (%)
$CAFE^{SUV}$	$183	$314
Engine^{SUV}	**194** (kW)	**250** (kW)
(Final drive) ^{SUV}	**3.97**	**2.58**
(Fuel economy) ^{SUV}	16.65 (mpg)	14 (mpg)
(Acceleration 0 to 60) ^{SUV}	8.24 (s)	7.54 (s)
(Acceleration 0 to 80) ^{SUV}	19.09 (s)	18.3 (s)
(Acceleration 40 to 60) ^{SUV}	4.06 (s)	3.55 (s)
(5-sec distance) ^{SUV}	193.29 (ft)	196.7 (ft)
(max acceleration)^{SUV}	19.24 (ft/s^2)	19.24 (ft/s^2)
(max speed) ^{SUV}	132.07 (mph)	119.18 (mph)
(max grade at 55mph)^{SUV}	26.57 (%)	36.93 (%)
Fleet CAFE	0	−$7.9M
NPV(7-year period)	**$7.3B**	**$9.36B**

where the cost is time-dependent due to learning curves and economies of scale and is also a function of the production quantity. Let us assume that the cost function is a quadratic function of the quantity produced [39; 40, p. 239; 47, p. 290; 7]:

$$C(q_t) = c_0 + c_1 q_t + c_2 q_t^2 \qquad \text{Eq. (17.74)}$$

Product observed attributes $\boldsymbol{\alpha}$ are either functions of design decisions $\boldsymbol{\alpha}(\boldsymbol{x}_d)$ or design decision themselves $\boldsymbol{\alpha}(\boldsymbol{x}_d) = \boldsymbol{x}_d$. Eq. (17.73) then becomes

$$\pi_t = \frac{\theta_t}{\lambda_P} q_t - \frac{1}{\lambda_P} q_t^2 + \frac{1}{\lambda_P} \boldsymbol{\lambda}_\alpha^T \boldsymbol{\alpha} q_t - c_0 q_t - c_1 q_t^2 - c_2 q_t^3 \qquad \text{Eq. (17.75)}$$

The profit maximization problem is as follows:

$$\max_{q_t} \pi_t = \frac{\theta_t}{\lambda_P} q_t - \frac{1}{\lambda_P} q_t^2 + \frac{1}{\lambda_P} \boldsymbol{\lambda}_\alpha^T \boldsymbol{\alpha} q_t - c_0 q_t - c_1 q_t^2 - c_2 q_t^3 \qquad \text{Eq. (17.76)}$$

Setting the first derivative with respect to quantity equal to zero, the optimum is found to be

$$q_t^* = \frac{2\left(c_1 + \frac{1}{\lambda_p}\right) + \sqrt{4\left(c_1 + \frac{1}{\lambda_p}\right)^2 + 12c_2\left(\frac{\theta_t}{\lambda_p} + \frac{1}{\lambda_p}\boldsymbol{\lambda}_\alpha^T\boldsymbol{\alpha} - c_0\right)}}{-6c_2} \qquad \text{Eq. (17.77)}$$

The second derivative of Eq. (17.76) is negative and Eq. (17.77) maximizes profits.

Let us analyze further Eq. (17.77). Based on the random values of θ_t the decision-maker will respond with q_t*. One of the assumptions here is that product design decisions are irreversible. That means product observable attributes $\boldsymbol{\alpha}$ cannot be substantially altered once design decisions **x** are made. However, as one can see from Eq. (17.77), **x** have partially determined the optimal production decisions q_t*. Under what condition will the design decisions not affect q_t*? One can easily infer that from Eq. (17.77). Whenever $(1/\lambda_P)\boldsymbol{\lambda}_\alpha^T\boldsymbol{\alpha}$ equals to c_0 then q_t* is not affected by **x**.

The latter can be explored further. Let us assume a scalar α. Also,

$$\frac{1}{\lambda_P}\lambda_\alpha = \frac{1}{\frac{\Delta Q}{\Delta P}}\frac{\Delta Q}{\Delta \alpha}, \text{ or}$$

$$\frac{1}{\lambda_P}\lambda_\alpha = \frac{\Delta P}{\Delta Q}\frac{\Delta Q}{\Delta \alpha} \qquad \text{Eq. (17.78)}$$

$(\Delta P/\Delta Q)(\Delta Q/\Delta\alpha)$ denotes how much a change in the product attribute will affect product demand that in turn affects price. Therefore, the product $(\Delta P/\Delta Q)(\Delta Q/\Delta\alpha)\alpha$ measures the demand side for product differentiation. The cost-coefficient c_0 is the only coefficient in Eq. (17.74) that is independent of the unit amount of production q. It simply denotes how much the first unit of production costs. It is reasonable to assume that this is affected from design decisions **x**. $c_0(\mathbf{x})$ measures the supply side of differentiation. Therefore, when the capacity to supply differentiation equals the demand for differentiation, engineering decisions do not affect business ones:

$$\frac{\Delta P}{\Delta Q}\frac{\Delta Q}{\Delta \alpha}\alpha(\mathbf{x}) = c_0(\mathbf{x}) \qquad \text{Eq. (17.79)}$$

Under Eq. (17.79) and in the case where the market conditions are such that the firm can pass all the cost of differentiation to a price increase, then business and engineering decision-making are independent of each other.

Let us assume that the time period as to when the design decisions are realized is known. The present value of the future cash flows stream π_t realized from the commercialization time $t_0 = 0$ until the end of the life cycle H is:

$$\text{PV} = \mathbb{E}\left(\sum_0^H \frac{\frac{\theta_t}{\lambda_P} q_t^* - \frac{1}{\lambda_P}(q_t^*)^2 + \left(\frac{1}{\lambda_P}\right)\lambda_\alpha^T\alpha(\mathbf{x})q_t^* - c_0(\mathbf{x})q_t^* - c_1 q_t^{*2} - c_2 q_t^{*3}}{(1+W\,\text{ACC})^t}\right) \qquad \text{Eq. (17.80)}$$

where $t \in \{0 \ldots H\}$ The product life-cycle period is $(H - 0)$. The net present value is the difference between the present value future cash flows and the required investment cost:

$$\text{NPV} = -I + \mathbb{E}\left(\sum_{0}^{H} \frac{\frac{\theta_t}{\lambda_P} q_t^* - \frac{1}{\lambda_P}(q_t^*)^2 + \left(\frac{1}{\lambda_P}\right)\boldsymbol{\lambda}_\alpha^T \boldsymbol{\alpha}(\mathbf{x}) q_t^* - c_0(\mathbf{x}) q_t^* - c_1 q_t^{*2} - c_2 q_t^{*3}}{(1 + W\text{ACC})^t}\right)$$

Eq. (17.81)

At the product commercialization time t optimal design decisions need to be made

$$\text{maximize } -I + \text{E}\left(\sum_{0}^{H} \frac{\frac{\theta_t}{\lambda_P} q_t^* - \frac{1}{\lambda_P}(q_t^*)^2 + \frac{1}{\lambda_P}\boldsymbol{\lambda}_\alpha^T \boldsymbol{\alpha}(\mathbf{x}) q_t^* - c_0(\mathbf{x}) q_t^* - c_1 q_t^{*2} - c_2 q_t^{*3}}{(1 + W\text{ACC})^t}\right)$$

with respect to $\mathbf{x}$ Eq. (17.82)

subject to $\boldsymbol{g}(\mathbf{x}) \leq 0$

where $\boldsymbol{g}(\mathbf{x})$ = functional engineering constraints, which are part of a standard engineering decision model.

For initial resources R the firm will form a portfolio of n products $\tau \in \{1, 2, \ldots, n\}$.We let the portfolio be defined by $\mathbf{w} = w^1, w^2, \ldots, w^n$, which denotes the resources allocated to each product, and corresponding designs $\boldsymbol{x} = \boldsymbol{x}^1, \boldsymbol{x}^2, \ldots, \boldsymbol{x}^n$. The portfolio problem of the firm is

$$\begin{aligned} &\text{maximize} && \mathbf{w}_n^T \mathbf{NPV}_n \\ &\text{with respect to} && \mathbf{w}, \mathbf{x} \\ &\text{subject to} && \sum w_n = 1 \\ &&& \mathbf{w}_n^T \mathbf{I}_n \leq R \\ &&& \mathbf{g}(\mathbf{x}) \leq 0 \end{aligned}$$

Eq. (17.83)

17.6 CONCLUSION

Product design involves different disciplines that are linked with each other. The decision-maker needs to acknowledge this linking for optimal design decision-making. Engineering decisions and investment decisions are linked with respect to time. Both are long-term irreversible decisions (with respect to economic decisions) that will affect the firm during the product life cycle. Engineering decisions and economic decisions are linked with respect to customer's purchasing choice. Price, technical characteristics, macroeconomic conditions and demographics will interplay in shaping consumer's behavior. Product technical characteristics are outcomes of engineering decisions and need to be treated simultaneously with all other factors affecting final choice. Although technical characteristics will remain unchanged during the life cycle, it is most likely that economic decisions and macroeconomic conditions will change. Therefore, engineering decisions will affect economic decisions across the life cycle.

The main assumption behind the presented synthesis is that economic, investment and engineering design decisions do not take place in a vacuum. They take place simultaneously and affect each other, either implicitly or explicitly. The main motivation behind this synthesis is that when a product decision is made the decision-maker wants to gather as much data as possible, as fast as possible, and utilize them in an effective way. For these three reasons all the models presented rely on historical data. The reader should bear in mind the following, quoted by Arnold Harberger, an economist from the Chicago School of Economics: "... I think we must take it for granted that our estimates of future costs and benefits (particularly the latter) are inevitably subject to a fairly wide margin of error, in the face of which it makes little sense to focus on subtleties aimed at discriminating accurately between investments that might have an expected yield of 10½ percent and those that would yield only 10 percent per annum. As the first order of business we want to be able to distinguish the 10 percent investments from those yielding 5 or 15 percent..." [20]. This type of thinking is consistent with that of engineers who build models of physical artifacts to predict their behavior under uncertainty.

We conclude by recalling the hypothesis stated in the seminal paper of Stabell and Fjeldstad [42]. According to their proposition, firms that are focusing on problem-solving activities can be modeled as "value shops": Their flow of activities is not linear, but iterative between activities and cyclical across the activity set. In this chapter we validated their hypothesis. Optimal decisions in one discipline depend on the optimal decisions of the other. Therefore product design is not a linear forward-looking process but rather an iterative process among disciplines.

ACKNOWLEDGMENTS

The work described here was partially supported by the Antilium Project of the H.H. Rackham Graduate School, the Automotive Research Center and the General Motors Collaborative Research Laboratory, all at the University of Michigan. This support is gratefully acknowledged. The opinions expressed are only those of the authors.

PROBLEMS

17.1 The standard quantity-price demand curve in microeconomics is assumed to be linear, as in Eq. (17.2). Its proposed extension to include design attributes, Eq. (17.29), is also linear. An alternative way to include the influence of design decisions on demand is to assume that the demand intercept and slope in Eq. (17.2) are functions of the design characteristics. Then the demand function will become nonlinear. Discuss advantages and disadvantages of the two approaches.

17.2 A nonlinear evaluation of demand is the use of the logit model Eq. (17.27). Discuss advantages and disadvantages of using this model versus the standard linear model. Derive an analytical expression for comparing the two models by linearizing the logit model using a Taylor Series expansion and comparing coefficients.

17.3 Demand elasticities can be calculated from historical sales data, as in Eq. (17.32). For relatively complex products with several attributes what are the challenges in extracting such elasticities from historical sales data? What assumptions must be made about the product over the time period under study?

17.4 Derive the optimal demand for maximum profit given in Eq. (17.35). Compute the sensitivity of optimum demand with respect to the CAFE penalty c_{CAFE}.

17.5 The presence of uncertainty makes economic models particularly difficult. List all sources of uncertainty involved in the financial models described in this chapter relevant to product design. Next consider and list all sources of uncertainty when predicting the engineering performance of a product based on common engineering analysis tools. Compare the two lists. Where do you think uncertainty plays a more important role—in engineering or in economic decisions? Which of these sources of uncertainty can the decision-maker control? When would you invest more resources in reducing uncertainty?

17.6 A common criticism of actual product development processes in (typically large) companies is that engineering and business decisions made about a product are segregated and often incompatible. Based on the discussion in Section 17.5, under what conditions would such segregation not lead to inferior product solutions? The primary business decision is setting product price. What is the key difference between price and other design attributes over the life cycle of the product? How does this difference justify focusing on price versus product attributes prior to product launch? How may this focus shift after the product launch?

REFERENCES

1. Allen, B., 2000. "A Toolkit for Decision-Based Design Theory," *J. of Engrg. Valuation and Cost Analysis*, 3(1), pp. 85–105.
2. Allen, B., 2001. "On the Aggregation of Preferences in Engineering Design," *ASME DETC, DETC2001/DAC-21015*, Pittsburgh, PA.
3. Berry, S., 1994. "Estimating Discrete Choice Models of Product Differentiation," *RAND J. of Eco.*, 23(2), pp. 242–262.
4. Berry, S., Levinsohn, J. and Pakes, A., 1995. "Automobile Prices in Market Equilibrium," *Econometrica*, 63(4), pp. 841–890.
5. Berry, S., Levinsohn, J. and Pakes, A., 1998. Differentiated Products Demand Systems From a Combination of Micro and Macro Data: The New Car Market,". *Tech. Rep. 6481*, National Bureau of Economic Research.
6. Besanko, D., Dranove, D. and Shanley, M., 2000. "*The Economics of Strategy, 2nd Ed.*, John Wiley and Sons, New York, NJ.
7. Bollen, N. P. B., 1999 "Real Options and Product Life Cycles," *Mgmt. Sci.*, 45(5), pp. 670–684.
8. Brealey, R. A. and Myers, S. C., 2000. *Principles of Corporate Finance*, McGraw-Hill, New York, NY.
9. Chen, W., Lewis, K. and Schmidt, L., 2000. "Decision-Based Design: An Emerging Design Perspective," *J. of Engrg. Valuation and Cost Analysis*, 3(1), pp. 57–66.
10. Clyde, P., 2001. *Business Economics 501*, University of Michigan Business School, Ann Arbor, MI.
11. Cook, H. E., 1997. *Product Management: Value, Quality, Cost, Price, Profits, and Organization*, Chapman & Hall, Hingham, MA.
12. Cooper, A.B., Georgiopoulos, P., Kim, H.M. and Papalambros, P. Y. 2003. "Analytical Target Setting: An Enterprise Context in Optimal Product Design," *ASME DETC, DETC2003/DAC-48734*, Chicago, IL.
13. Dixit, A., 1992. "Investment and Hysteresis," *J. of Eco. Perspectives*, 6(1), pp. 107–132.
14. Dixit, A. K., and Pindyck, R. S., 1994. Investment Under Uncertainty, Princeton University Press, Princeton, NJ.
15. Donndelinger, J. and Cook, H. E., 1997. "Methods for Analyzing the Value of Automobiles," *SAE Int. Cong., SAE Paper 970762*. Detroit, MI.
16. Georgiopoulos, P., 2003. "*Enterprise-wide Product Design: Linking Optimal Design Decisions to the Theory of the Firm.*" Ph.D. Thesis, University of Michigan, Ann Arbor, MI.
17. Georgiopoulos, P., Jonsson, M. and Papalambros, P. Y., 2005. "Linking Optimal Design Decision to the Theory of the Firm: The Case of Resource Allocation," *J. of Mech. Des.*
18. Grant, R. M., 1998. *Contemporary Strategy Analysis*, Blackwell.
19. Gu, X., Renaud, J. E., Ashe, L. M., Batill, S. M., Budhiraja, A. S. and Krajewski L.J., 2002. "Decision-Based Collaborative Optimization," *J. of Mech. Des.*, 124(1), pp. 1–13.
20. Harberger, A. C., 1972. *Project Evaluation*, MacMillan, London.
21. Hazelrigg, G. A., 1998. "A Framework for Decision-Based Engineering Design," *J. of Mech. Des.*, Vol. 120, pp. 653–658.
22. Hazelrigg, G. A., 2000. *J. of Engrg. Valuation and Cost Analysis*, Special Ed.
23. Hull, J., 2000. Options, Futures, and other Derivatives, Prentice-Hall, Upper Saddle River, NJ.
24. Irvine, F. O., 1983. "Demand Equations for Individual New Car Models Estimated Using Transaction Prices With Implications for Regulatory Issues," *S. Eco. J.*, 49(3), pp. 764 782.
25. Jones, D.R., 2001. *The DIRECT Global Optimization Algorithm*, Vol. 1, Kluwer Academic Publishes, Dordrecht, The Netherlands, pp. 431–440.
26. Kim, H.M., Kumar, D. K. D., Chen, W. and Papalambros, P. Y., 2004. "Target Feasibility Achievement in Enterprise-Driven Hierarchical Multidisciplinary Design," *10th AIAA/ISSMO Multidisc iplinary Analysis and Optimization Conf., AIAA-2004-4546*, Albany, NY.
27. Koujianou-Goldberg, P., 1998. "The Effects of the Corporate Average Fuel Economy Standards in the Automobile Industry," *J. of Industrial Eco.*, March, pp. 1–33.
28. Li, H. and Azarm, S., 2000. "Product Design Selection Under Uncertainty and With Competitive Advantage," *J. of Mech. Des.*, 122(4), pp. 411–418.
29. Li, H. and Azarm, S., 2002. "An Approach for Product Line Design Selection Under Uncertainty and Competition," *J. of Mech. Des.*, 124(3), pp. 385– 392.
30. Luenberger, D. G., 1998 *Investment Science*, Oxford University Press.
31. Markish, J. and Willcox, K., (2002a). "Multidisciplinary Techniques for Commercial Aircraft System Design," Vol. 2, *9th AIAA/ISSMO Symp. on Multidisciplinary Analysis and Optimization, AIAA- 2002-5612*, Atlanta, GA.
32. Markish, J. and Willcox, K., (2002b). A Value-Based Approach for Commercial Aircraft Conceptual Design," *23rd ICAS Cong.*, Toronto, Canada.
33. Marston, M., Allen, J. K. and Mistree, F., 2000. "The Decision Support Problem Technique: Integrating Descriptive and Norm ative Approaches in Decision-Based Design," *J. of Engrg. Valuation and Cost Analysis*, 3(1), pp. 107–129.
34. Marston, M. and Mistree, F., 1998. "An Implementation of Exp ected Utility Theory in Decision-Based Design," *ASME DETC, DETC1998/DTM-5670*, Atlanda, GA.
35. McConville, G. P. and Cook, H. E., 1996. "Estimating the Value Trade-Off Between Automobile Performance and Fuel Economy," *SAE Int. Cong., SAE Paper 960004*, Detroit, MI.
36. Michalek, J., Feinberg, F. and Papalambros, P. Y., 2004. "Linking Marketing and Engineering Product Design Decisions via Analy tical Target Cascading," *J. of Product Innovation Mgmt.: Special Issue on Des. and Marketing in New Product Dev.*
37. Nevo, A., 2000. "A Practitioner's Guide to Estimation of Random Coeffcients Logit Models of Demand," *J. of Eco. & Mgmt. Strategy*, 9(4), pp. 513–548.
38. Papalambros, P. Y. and Wilde, D. J., 2000. *Principles of Optimal Design*, 2nd Ed., Cambridge University Press, New York, NY.
39. Pindyck, R. S., 1988. "Irreversible Investment, Capacity Choice, and the Value of the firm," *The Amn. Eco. Rev.*, 78(5), pp. 969–985.
40. Pindyck, R.S. and Rubinfeld, D.L., 1997. *Microeconomics, 4th Ed.*, Prentice-Hall, New York, NY.
41. Scott, M.J. and Antonsson, E.K., 1999. "Arrow's Theorem and Engineering Design Decision Making," *Res. in Engirg. Des.*, Vol. 11, pp. 218–228.
42. Stabell, C.B. and Fjeldstad, Ø. D., 1998. "Configuring Value for Competitive Advantage: On Chains, Shops, and Networks," *Strategic Mgmt. J.*, Vol. 19, pp. 413–437.
43. Teece, D., 1986. "Profiting From Technological Innovation: Implications for Integration, Collaboration, Licensing and Public Policy," *Res. Policy*, Vol. 15, pp. 285–305.
44. Thurston, D.L., 1999. "Real and Perceived Limitations to Decision-Based Design," *ASME DETC, DETC1999/DAC-8750*, Las Vegas, NV.
45. Thurston, D.L., Carnahan, J. V. and Liu, T., 1994. "Optimization of Design Utility," *J. of Mech. Des.*, Vol. 116, pp. 801–808.

46. Thurston, D. L. and Locascio, A., 1994. "Design Theory for Design Economics," *Engrg. Econ.*, 40(1), pp. 41–72.
47. Trigeorgis, L., 1998. *Real Options, Managerial Flexibility and Strategy in Resource Allocation*, The MIT Press, Cambridge, MA.
48. Tseng, C. L. and Barz, G., 1999. "Short-Term Generation Asset Valuation," *32nd Hawaii Int. Conf.*
49. WardAuto, 1998. *Automotive Yearbook*, Ward's Communication, Southfield, MI.
50. Wassenaar, H. J. and Chen, W. 2003. "An Approach to Decision-Based Design With Discrete Choice Analysis for Demand Modeling," *J. of Mech. Des.*, Vol. 125, pp. 490–497.
51. Wassenaar, H. J., Chen, W., Cheng, J. and Sudjianto, A., 2003. "Enhancing Discrete Choice Demand Modeling for Decision-Based Design," *ASME DETC, DETC-DTM48683,* Chicago, IL.
52. Wassenaar, H. J., Chen, W., Cheng, J. and Sudjianto, A., 2004. "An Integrated Latent Variable Choice Modeling Approach to Enhancing Product Demand Modeling, *ASME DETC, DETC2004-57487,*v Salt Lake City, UT.
53. Whitcomb, C.A., Palli, N. and Azarm, S., 1999. "A Prescriptive Production Distribution Approach for Decision Making in New Product Design" *IEEE Trans. on Sys. Man and Cybernetics, Part C,* 29(3) pp. 336–348.
54. Wipke, K. and Cuddy, M., 1996. "Using an Advanced Vehicle Simulator (Advisor) to Guide Hybrid Vehicle Propulsion System Development." *Tech. Rep. S/EV96, NREL.*
55. WSJ, 2002. "Detroit Again Attempts to Dodge Pressures for a 'greener' Fleet, *Wall Street J.* January 1.

CHAPTER

18

Multilevel Optimization for Enterprise-Driven Decision-Based Product Design

Deepak K. D. Kumar, Wei Chen, and Harrison M. Kim

NOMENCLATURE

A = customer-product-selection attributes
A_{eng} = customer-product-selection attributes related to engineering performance
A_{ent} = customer-product-selection attributes related to enterprise product planning
AIO = all-in-one
ATC = analytical target cascading
ATS = analytical target setting
C = total product cost
DBD = decision-based design
DCA = discrete choice analysis
E = engineering design attributes
$E(U)$ = expected value of enterprise utility
E^* = utopia target of E
E^D = achievable product performance
MDO = multi-disciplinary design optimization
MNL = multinomial logit
P = product price
Q = product demand
RP = revealed preference
S = customer demographic attributes
SP = stated preference
T^U = targets of E set by the enterprise product planning problem
t = time interval for which demand/market share is to be predicted
U = enterprise utility
u_{in} = true utility of choosing alternative i by customer n
V = selection criterion used by the enterprise (e.g., profit, revenues, etc.)
W_{in} = deterministic part of the utility of choosing alternative i by customer n
X = design options
X_d = engineering design options
X_{ent} = enterprise planning options
Y = exogenous variables (represent sources of uncertainty in market)
ε_{in} = random unobservable part of the utility of choosing alternative i by customer n

18.1 INTRODUCTION

There is a growing recognition in the design community of the need for a rigorous design approach that considers the enterprise goal of making profits and the decision-maker's risk attitude, while also dealing adequately with engineering needs and various sources of uncertainty. Decision-based design (DBD) [1] is a collaborative design approach that recognizes the substantial role that decisions play in design and in other engineering activities. In recent years, we have seen many (DBD) related research developments in the field of engineering design (e.g., [2; 3; 4; 5; 6; 7; 8; 9; 10; 11; 12, 13; 14; 15; 16]). For profit-driven design under the DBD framework, ideally, all product design decisions, whether directly related to engineering or otherwise, are made simultaneously to optimize the enterprise-level design objective, i.e., to maximize the expected utility, expressed as a function of net revenue (profit), subject to various sources of uncertainty. The existing implementation of the profit-driven DBD approach [11; 12, 13] seeks to integrate *enterprise product planning and engineering product development* by using an all-in-one (AIO) approach and solving as a single optimization problem. The *enterprise* is defined here as the organization that designs and produces an artifact to maximize its utility (e.g., profit). Marketing, production planning and other enterprise-level activities are referred to as *enterprise-level product planning*; engineering-related design activities are referred to as *engineering product development.*

Designing a large-scale artifact typically involves multidisciplinary efforts in marketing, product design and production making. The AIO approach is often practically infeasible in such situations due to computational and organizational complexities. Optimization by decomposition, while alleviating the problem of having to deal with a large number of design variables and constraints at the same time, is made necessary by a number of factors. The decomposed approach helps enable simultaneous multidisciplinary optimization wherever possible and also addresses organizational needs to distribute the work over several groups of engineers/analysts. Historical evolution of engineering disciplines and the complexity of the multidisciplinary design optimization (MDO) problem suggest that disciplinary autonomy is a desirable goal in formulating and solving MDO problems. In MDO, several design architectures have been developed to support collaborative multidisciplinary design using distributed design optimization, e.g., concurrent subspace optimization (CSSO) [17], bilevel integrated system synthesis (BLISS) [18, 19], collaborative optimization (CO) [20; 15], and analytical target cascading (ATC) [21; 22, 23, 24, 25; 27]. A comprehensive review of the MDO architectures is provided by Kroo [28]. It should be noted that the choice of MDO formulations largely depends on whether the problem follows the hierarchical or non-hierarchical characteristics of decision flow. In most of the MDO approaches listed above, a complex engineering problem is non-hierarchically decomposed along disciplinary or other user-specified boundaries into a number of sub-problems. Then they are brought into multidisciplinary agreement by a system-level coordination process. In our opinion, the non-hierarchical MDO infrastructure is better suited to capture the interrelationships between multiple engineering disciplines in engineering-level product development; however, a hierarchical approach, such as ATC, is more appropriate in an enterprise-driven product design scenario where the enterprise decision-making is often done at a higher level to set up targets for engineering product development. To represent the organizational infrastructure in industry more accurately, the interrelationships between enterprise product planning and engineering product development, as well as the engineering product development itself at system, subsystem and component levels, should be treated as hierarchical. Such a hierarchical framework, as will be detailed later, is more representative of the hierarchical decision-making in the industry.

In this chapter, we present a DBD-based hierarchical approach to enterprise-driven design that treats enterprise-level product planning and engineering-level product development as two interrelated but separate optimization problems in a multilevel optimization framework. To fully integrate business and engineering decision-makings, we illustrate how a disaggregate probabilistic choice model can be used to establish the link between the decomposed enterprise product planning and engineering development models. Any hierarchical approach, like the one presented here, should ensure preference consistency. In other words, the optimization of the engineering objectives at the product development level needs to correspond to the maximization of the utility objective function at the enterprise product planning level. This is to guarantee that the solution from the multilevel optimization procedure will be close, if not identical, to the one that is obtained by solving the AIO-integrated enterprise and engineering problem. As will be discussed in Section 18.3, if the feasible domain imposed by the engineering product development is disconnected in the space of engineering performance attributes, achieving the design which corresponds to the maximum enterprise utility becomes more challenging. A search algorithm that can systematically explore attribute targets in the disconnected feasible domain to lead the engineering product design to feasible and optimal designs in the enterprise context is needed and such an algorithm is also presented here.

The organization of the chapter is as follows: In Section 18.2, a discussion on enterprise-driven design approaches that incorporate economic considerations and customer preferences is presented. The review in Section 18.2 outlines various approaches to establishing the link between engineering decisions and their business impact. Section 18.3 presents the transformation of an AIO DBD framework into a hierarchical optimization formulation, which combines enterprise-level planning and engineering product development efforts using a probabilistic choice modeling approach for demand modeling. The section also presents an optimization algorithm, which enables designers to handle problems when the feasible design is not continuous in the space of engineering performance attributes. An automotive suspension design case study is used to illustrate the approach in Section 18.4. Conclusions and future work are presented in Section 18.5.

18.1 DISCUSSION OF CONTEMPORARY ENTERPRISE-DRIVEN DESIGN APPROACHES

There have been a number of efforts in the design community to integrate economic considerations with the engineering product development efforts, e.g., [11; 12; 13; 15; 26; 27; 29, 30, 31; 3, 32, 33, 34]. The primary goal in all these approaches remains the same, i.e., to arrive at a design that is optimal with respect to enterprise-level objectives. While modeling the interaction between enterprise-level product planning and engineering-level product development, one widely used approach involves computing the *net revenue* of the enterprise based on the *demand* for the product as a function of product attributes; demand plays a critical role in assessing the profit as it contributes to the computation of both revenue and life-cycle cost. Among the various demand modeling approaches adopted by the design community, differences exist in the type of customer data used for the analysis, the modeling of uncertainty in customer preferences, as well as how the model uses customer preference data. Some models aggregate customer preferences while others examine this information at the individual level. Also, either stated preference (SP) data [35] or revealed preference (RP) [36] data may be used for demand modeling. RP data refers to actual choice, i.e., actual (purchase) behavior that is observed in real choice situations. SP surveys are used to learn about how people are likely to respond to new products or new product features through surveys. While a more detailed discussion of the multinomial logit (MNL) demand modeling approach is presented in Section 18.3.3, we provide below a brief review of demand modeling approaches that have been used for engineering design applications.

Azarm et al. [3, 32, 33, 34] employed a traditional conjoint analysis-based demand modeling approach within a multi-attribute instead of single-criterion DBD framework. They subsequently introduced a customer-centric utility metric that formed the basis for product design selection. Their selection procedure considered the preferences of both the customers' and designers' utility, while also factoring in market competition and uncertainties in product life, market size, cost, etc.

Cooper et al. [26] proposed a bilevel framework that links ATC [21; 22, 23, 24, 25; 27] and analytical target setting (ATS). In their approach, ATC is used for the hierarchical product development process, while the solution to the ATS problem optimizing enterprise-level objectives (e.g., profit) sets suitable targets for various engineering attributes. Under the ATS-ATC framework, initially, a simplistic linear model was used to capture demand [26]. A more sophisticated disaggregated choice modeling approach has been proposed in recent work [27], where a choice-based conjoint analysis approach is implemented within the MNL framework to analyze SP data. The "part worth" model [37, 38] was used to model changing customer perceptions for different ranges of product attributes. Such an approach helps capture the nonlinearity in customer preference over the entire range of the product attribute. One limitation of the existing

ATS-ATC framework is that the ATS formulation for enterprise planning has been a deterministic formulation that does not consider uncertainty or designer's risk attitude in decision-making.

Wassenaar and Chen [11, 12, 13] employed disaggregated probabilistic demand models based on discrete choice analysis (DCA) in a DBD framework. The DBD framework is a single-criterion and collaborative approach that can be used to obtain the optimal settings of product attributes at the enterprise level to maximize the net revenue of a firm with the consideration of uncertainty and risk attitude. They demonstrated the use of a MNL-based customer choice model [36; 39] to identify the optimal levels of relevant customer product selection attributes, i.e., product attributes that are of interest to customers. They showed how these attributes could be used to maximize the expected utility of a firm considering engineering needs, the socioeconomic and demographic background of customers. A major feature of the DCA-based disaggregate demand models is the use of data of individuals instead of group averages. This enables a more accurate representation of the heterogeneity among individuals and avoids paradoxes associated with aggregating preferences of a group. However, Wassenaar and Chen's implementation did not show how the optimal settings of product attributes should be further cascaded to the descriptions of engineering design options. Also, using the AIO approach to making simultaneous enterprise planning and engineering development decisions under the DBD framework is not practically feasible due to the reasons explained earlier. In this chapter, we illustrate how the DBD approach can be implemented using a multilevel optimization approach that treats enterprise planning and engineering development as two separate but closely related optimization problems. Our approach utilizes the disaggregated probabilistic demand modeling approach introduced in [11; 12, 13] for product design to establish the critical link between the enterprise planning and engineering product development models.

18.3 A MULTILEVEL OPTIMIZATION APPROACH TO DECISION-BASED DESIGN

Here, the DBD approach is implemented using an optimization approach that treats enterprise planning and engineering product development as hierarchical, but interactive, activities. In this section, the AIO DBD approach is introduced first and the proposed hierarchical optimization formulation is presented later. Some important features of the demand modeling approach employed in this work are also presented here. Finally, we present a search algorithm that can lead the engineering product design to feasible and optimal designs in the enterprise context.

18.3.1 AIO DBD framework

An AIO DBD framework, an extension of the DBD framework presented in [11, 12, 13], is shown in Fig. 18.1. Unlike the flowchart they proposed, we split the design options $\boldsymbol{X}$ into two groups here: the engineering design options, $\boldsymbol{X}_d$, and the enterprise planning options, $\boldsymbol{X}_{ent}$, to separate the decisions made in these two domains and to illustrate their impact separately. The engineering design options $\boldsymbol{X}_d$ represent engineering decisions made by product designers, while $\boldsymbol{X}_{ent}$ typically includes quantities like price, P, warranty options, annual percentage rate (APR) of auto loan, which are determined at the enterprise level. The arrows in the flowchart (Fig. 18.1) indicate the existence of relationships between the different entities (parameters) in DBD, instead of showing the sequence of implementation.

In our DBD framework, a distinction is made between customer-product-selection attributes $\boldsymbol{A}$ and engineering design attributes $\boldsymbol{E}$. The customer-product-selection attributes $\boldsymbol{A}$ are product features and financial attributes (such as service and warranty) that a customer typically considers when purchasing the product. Engineering design attributes $\boldsymbol{E}$ are any quantifiable product properties that are used in the engineering product development process. They are described as performance functions of engineering design options $\boldsymbol{X}_d$ through engineering analysis. To estimate the effect of design changes on a product's market share and consequently on the firm's revenues, DCA [36; 39] is used for demand modeling that establishes the relationship between the customer-product-selection attributes $\boldsymbol{A}$, the socioeconomic and demographic attributes $\boldsymbol{S}$ of the market population, price P, time t, and the demand Q. From market analysis point of view, the input $\boldsymbol{A}$ to the demand model could be attributes with physical units (e.g., fuel economy) or without (e.g., level of comfort). However, to assist engineering decision-making, customer-product-selection attributes $\boldsymbol{A}$, related to the engineering performance, need to be expressed in terms of quantifiable engineering design attributes $\boldsymbol{E}$ in demand modeling. Engineering design attributes $\boldsymbol{E}$, apart from including the quantifications of some of the attributes $\boldsymbol{A}$, also include attributes that are of interest only to design engineers, e.g., stress level of a structure. Likewise, some of the nonperformance-related attributes $\boldsymbol{A}$ are not influenced by the engineering design attributes $\boldsymbol{E}$, but by the enterprise planning options $\boldsymbol{X}_{ent}$. Therefore, $\boldsymbol{A}$ and $\boldsymbol{E}$ can be viewed as two sets that share a number of common elements.

The flowchart in Fig. 18.1 coincides with an optimization loop that identifies the optimal design options $\boldsymbol{X}$ (including both $\boldsymbol{X}_d$ and $\boldsymbol{X}_{ent}$) to maximize the expected utility $E(U)$. It should be noted that uncertainty is considered explicitly and the enterprise goal is expressed as the maximization of expected utility. The enterprise utility U is expressed as a function of the selection criterion V (e.g., profits, revenues, etc.) and is designed to reflect the enterprise's risk attitude. In enterprise-driven design, the selection criterion V could be the net profit Π for the enterprise and expressed as a function of product demand Q, price P, cost C, exogenous variables Y (the sources of uncertainty in the market) and time t. Based on these relationships, as shown in [Eq. (18.1)], demand Q, expressed as a function of customer-product-selection attributes $\boldsymbol{A}$, customer demographic attributes $\boldsymbol{S}$, price P and time t, can be further expanded as a function of engineering design attributes E, enterprise planning options $\boldsymbol{X}_{ent}$, as well as $\boldsymbol{S}$, P and t.

$$V = \Pi(Q, C, P, \mathbf{Y}, t) \qquad \text{Eq. (18.1a)}$$

$$V = Q[\mathbf{A}_{eng}(\mathbf{E}), \mathbf{A}_{ent}(\mathbf{X}_{ent}, \mathbf{Y}), \mathbf{S}, P, t] \times P - C(\mathbf{E}, \mathbf{X}_{ent}, \mathbf{Y}, Q, t) \qquad \text{Eq. (18.1b)}$$

$$V = V(\mathbf{E}, \mathbf{X}_{ent}, \mathbf{S}, P, \mathbf{Y}, t) \qquad \text{Eq. (18.1c)}$$

Here we assume that the attributes $\boldsymbol{A}$ can be split into two groups; those related to product performance $\boldsymbol{A}_{eng}$ determined by engineering design attributes $\boldsymbol{E}$; and those non-engineering

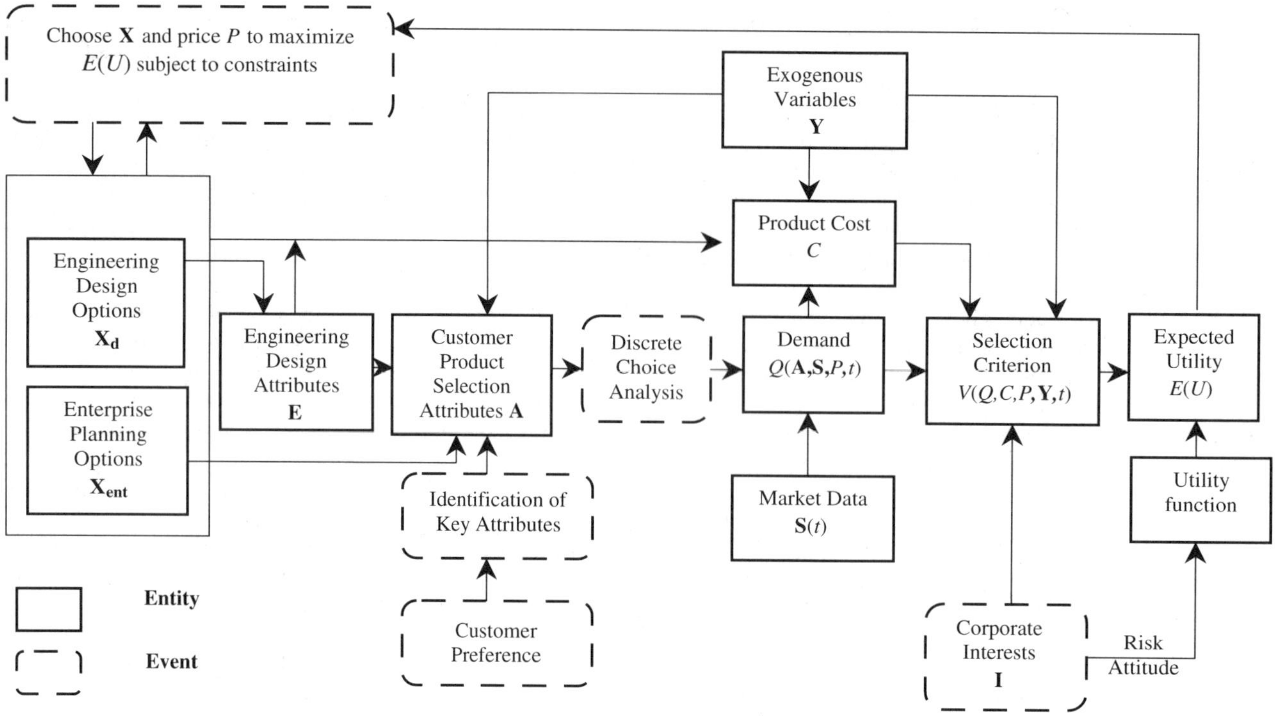

FIG. 18.1 ALL-IN-ONE (AIO) DBD FRAMEWORK

attributes A_{ent} determined by enterprise planning options X_{ent} and exogenous variables Y. If we assume that the total product cost C can be expressed as a function of engineering design attributes E, enterprise planning options X_{ent}, product demand Q, as well as variables Y and t, we can then transform the selection criterion V into a function of (E, X_{ent}, S, P, Y, t). This representation of the selection criterion V as a function of engineering design attributes **E**, instead of directly as a function of engineering design options X, greatly facilities the decomposition of the AIO DBD formulation into hierarchical enterprise planning and engineering product development as introduced next.

18.3.2 Multilevel Optimization Formulation to DBD

Figure 18.2 illustrates the difference between the AIO approach and the proposed multilevel optimization formulation to DBD. The AIO approach in Fig. 18.2(a) treats the problem of maximizing the expected value of enterprise-level utility $E(U)$ as a single optimization problem, where the decisions on product planning and product development are made simultaneously. Figure 18.2(b) illustrates the proposed decomposed hierarchical framework, representing our view of the interaction between enterprise-level product planning and engineering product development. Following the "target cascading" paradigm [21; 22, 23, 24, 25; 27] for hierarchical decision-making in industrial settings, we view the engineering product development as a process for meeting the targets set from the enterprise level.

Using a multilevel optimization formulation, at the upper level, the enterprise-level product planning problem maximizes the expected utility $E(U)$ with respect to the engineering design attributes E and the enterprise variables X_{ent} subject to enterprise-level design capability. Decisions made on the optimal levels of engineering design attributes E, represented as E^*, are then used as targets or T^U, passed to the lower-level engineering product development process. The objective of the lower-level engineering product development is to minimize the deviation between the performance target T^U and the achievable product performance response E while satisfying the engineering feasibility constraints g, with respect to engineering design options X_d. Restrictions on cost can be considered either as constraints or targets in engineering-level product development. The equation $E = r(X_d)$ stands for the engineering analysis models that capture the relationship between engineering design attributes and design options. The achievable product performance E^D is then transferred to the enterprise-level problem.

Under a multilevel design framework, an ideal product development scenario is when the targets corresponding to the optimal enterprise utility would lead to an engineering design matching the targets perfectly. Unfortunately, such a match is rare due to constraints introduced at the product development level. This enforces the adjustment of the targets set at the enterprise level. This adjustment may shift the enterprise utility value away from its original optimal value. In return, however, a consistent feasible design that satisfies the engineering constraints may be obtained. As will be detailed in Section 18.3.4, in the iterative procedure of solving optimization problems at both enterprise and engineering levels, an additional constraint $\|E - E^D\| \geq \Delta^D$ is added at the enterprise level to enforce setting an alternative target for the engineering problem. Based on the minimum deviation Δ^D from the utopia target E^*, the enterprise problem sets a geometric boundary constraint and a target outside the boundary. Then the engineering design problem is solved again based on a new target. An alternative target may be a guide to finding an alternative engineering design in a disconnected feasible domain of attribute targets that corresponds to a better enterprise-level utility. If the engineering product development comes up with the same design from the previous iteration, the radius Δ^D is expanded once again based on the slope information of the utility function and the engineering problem is solved once more. If the engineering problem continues to come up with the same feasible design, it means that there exists no disconnected feasible domain at least around the utopia

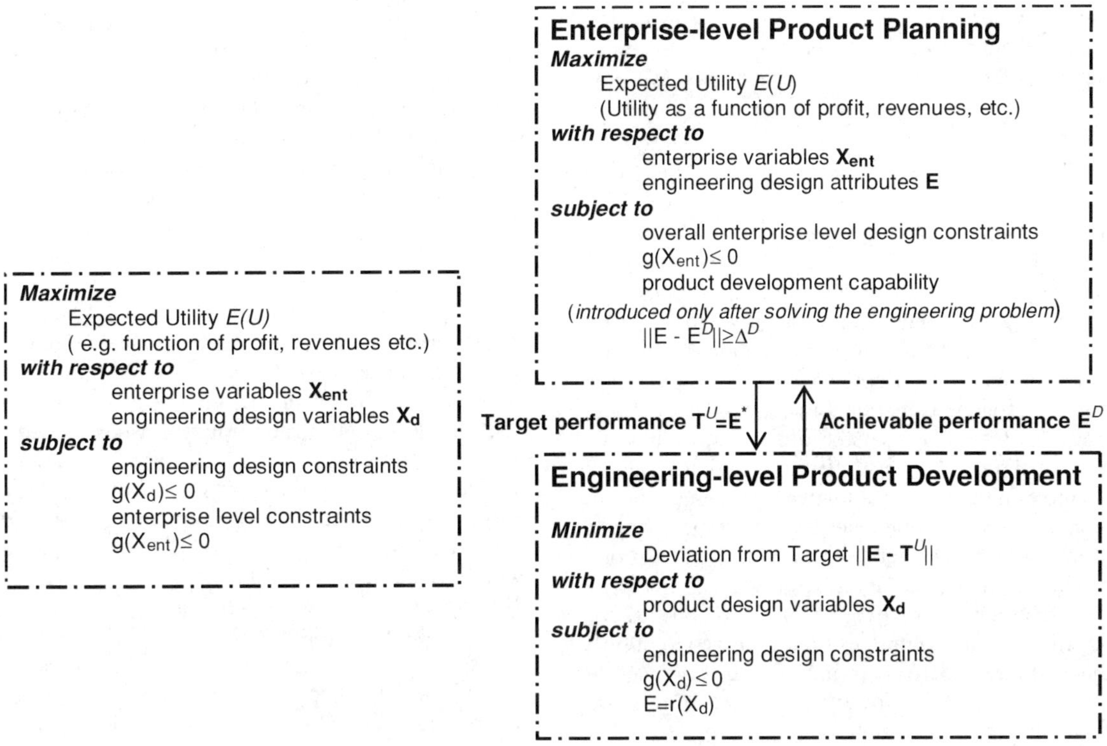

(a) All-In-One Approach (b) Hierarchical Approach

FIG. 18.2 COMPARISON BETWEEN AIO AND HIERARCHICAL APPROACH TO DBD

utility design. The formulation shows that the targets identified for $\boldsymbol{E}$ serve as the critical link between the optimization problems at two levels.

It should be noted that the engineering product development typically involves the design of multiple engineering systems. Therefore, the optimization problem at the engineering development level can be further decomposed and solved using multilevel optimization. Based on the nature of decomposition, either non-hierarchical or hierarchical, different multilevel optimization formulations can be used. Most of the work considered up to this point in MDO research, e.g., BLISS [19] and collaborative optimization (CO) [20], were concerned with decomposing a problem into a series of problems and solving them using bilevel optimization formulations. On the other hand, the ATC approach decomposes the original engineering problem hierarchically at multiple levels, and operates by formulating and solving a minimum deviation optimization problem (to meet targets) for each element in the hierarchy. Compared to the bilevel optimization formulations for engineering-level product development, we believe that the multilevel hierarchical modeling facilitated by the ATC approach better represents a multilayered organizational decision-making infrastructure. In this hierarchical model, subsystems and components can be supplied by different organizational units or outsourced to independent companies. In the following subsection, we discuss the demand modeling approach that helps establish the link between enterprise planning and engineering product development in a multilevel optimization formulation.

18.3.3 DCA for Demand Modeling

In order to separate enterprise-level product planning and engineering-level product development efforts, it is important to develop a demand model that can predict the economic impact of design changes as a function of engineering design attributes $\boldsymbol{E}$. As part of our hierarchical approach to the problem of maximizing the expected value of enterprise-level utility $E(U)$, demand modeling is used to capture the impact of quantitative engineering design attributes on customer choices, ultimately the product market share. Here, DCA is used to model the choice behavior of customers. While more details on DCA can be found in chapters 9 and 10, some key features of the method are presented here. Each customer has to choose one alternative from a finite set of mutually exclusive and collectively exhaustive competitive alternatives. A key concept in DCA is the use of random utility to address unobserved taste variations, unobserved attributes and model deficiencies. This entails the assumption that the individual's true utility u consists of a deterministic or observable part W and a random unobservable disturbance ε:

$$u_{in} = W_{in} + \varepsilon_{in} \qquad \text{Eq. (18.2)}$$

The deterministic part of the utility can be parameterized as a function of observable independent variables (customer-product-selection attributes A, customer socioeconomic and demographic attributes S and price P) and unknown coefficients β, which can be estimated by observing choices respondents make. The idea behind including the demographic attributes of customers is to capture the heterogeneous nature of customers. The utility function terms are represented with the double subscript in, representing the nth respondent and the i choice alternative:

$$W_{in} = f(\mathbf{A}_i, P_i, \mathbf{S}_n : \beta_n) \qquad \text{Eq. (18.3)}$$

To provide the link between enterprise-level product planning and engineering-level product development, all attributes $\boldsymbol{A}$ in

Eq. (18.3) need to be converted to quantifiable engineering design attributes $\boldsymbol{E}$, as well as nonengineering-related product attributes $\boldsymbol{A}_{ent}$ that are influenced by enterprise planning options. We now have:

$$W_{in} = f(\mathbf{E}_i, \mathbf{A}_{ent(i)}, P_i, \mathbf{S}_n : \beta_n) \qquad \text{Eq. (18.4)}$$

Various multinomial market analysis techniques [36; 39; 40] can be employed to obtain the model in Eq. (18.4) based on the collected market data. However, the MNL model [36] is used in this chapter. The primary difference among the DCAs (e.g., nested logit, mixed logit, etc.) is the degree of sophistication with which they model the unobserved error ε. More advanced techniques are also able to model heterogeneity better. In the MNL model, the coefficients of the utility function for the product attributes are identical across all customers. However, heterogeneity is modeled by considering demographic attributes (e.g., age, income, etc.) in the utility function. In an MNL model, any two customers with at least one demographic attribute different from each other (e.g., one of them is older than the other or has a different income) are going to have different degrees of preference (i.e., choice probabilities) for the same product. In contrast, techniques such as mixed logit [40] address heterogeneity more effectively by modeling utility function coefficients as random variables. However, such techniques are limited by their prohibitive computational expense.

MNL is a popular choice because it produces a closed-form probabilistic choice model, and hence is more computationally tractable. Also, the error distribution, which is assumed to be Gumbel or extreme value, closely approximates the normal distribution, a more reasonable assumption for the error. The form of the choice probability for the MNL model is shown below [Eq. (18.5)], where $\Pr_n(i)$ = probability that respondent n chooses alternative i; and J = choice set that is available to individual n.

$$\Pr_n(i) = \frac{e^{W_{in}}}{\sum_{j \in J} e^{W_{jn}}} \qquad \text{Eq. (18.5)}$$

It should be noted that MNL is characterized by the independence of irrelevant alternatives (IIA) property [36], which assumes that when one is choosing between two alternatives, all other alternatives are irrelevant, indicating that each alternative has the same unobserved error part, ε, in the utility. However, in many cases, choice alternatives from different market segments are likely to share common attributes. For example, to accurately model the entire automobile market, one would need to consider various market segments (e.g., midsize, sports, luxury, etc.) together. In such a case, it is more reasonable to assume that customers are likely to consider cars from a particular segment, for example, SUV, to be more similar to each other and different from cars from another segment, such as midsize sedans.

18.3.4 Engineering Design Target Exploration Algorithm

As shown in Fig. 18.2, our approach models the enterprise-level problem and the engineering-level problem as two separate problems in a multilevel optimization formulation. The enterprise product planning sets the targets for the engineering product development problem corresponding to the maximum utility. In most engineering design cases, it is uncommon to meet the utopia target perfectly due to the trade-off nature of multiple attribute target values or physical feasibility (i.e., no feasible design is available to meet the targets perfectly). If the engineering feasible domain is disconnected in the space of performance attributes (i.e., between words, multiple, discrete feasible designs are available), the task becomes more challenging. Disconnected feasible performance domains often occur in the design of complex systems where multiple engineering disciplines are involved and each discipline seeks distinctly different design alternatives in downstream engineering development. For example, the vehicle suspension design case study in Section 18.4 illustrates a case where a vehicle manufacturer attempts to maximize the enterprise utility, based on two disconnected feasible target performance domains imposed by suppliers of suspension components. In such cases where the feasible domain is disconnected, the engineering design with the minimum deviation from the attribute targets (i.e., the design which is a converged solution from the multilevel optimization) may not correspond to the maximum possible utility value, and a new set of targets for the engineering problem need to be assigned. To explore this disconnected feasible target space, a search algorithm that can systematically explore attribute targets to lead the engineering product design process to finding a feasible and optimal design in the enterprise context should be employed. Here, we present an algorithm, first proposed in [25]. The proposed algorithm guides the enterprise-level decision-maker to assign alternative targets so that the enterprise maximizes net revenue and the suppliers achieve targets as closely as possible. The adjustment of targets set at the enterprise level may shift the enterprise utility value away from its original utopia value. In return, however, a better (i.e., higher utility) feasible design can be obtained, satisfying the engineering constraints that may exist in other disconnected feasible domain. While the details on the algorithm are available in [25], some of the important features are presented here. The original enterprise-level utility optimization problem is

$$\begin{aligned} P^0_{ent}: \quad & \max_{\mathbf{X}_{ent}} E\left[U(\mathbf{T}, \mathbf{X}_{ent})\right], \\ \text{s.t.} \quad & g\left(\mathbf{X}_{ent}\right) \le 0 \end{aligned} \qquad \text{Eq. (18.6)}$$

where the objective is to maximize the expected utility $E(U)$, a function of engineering design attributes $\boldsymbol{E}$ and enterprise level variables $\boldsymbol{X}_{ent}$. The optimal value of engineering design attributes $\boldsymbol{E}^*$ obtained from the model [Eq. (18.6)] are assigned as the utopia targets $\boldsymbol{T}^*$ for engineering development. The engineering problem then finds an optimal response to the utopia target with the minimum deviations [see formulation in Fig. 18.3(b)]. Figure 18.3 illustrates 1-D and 2-D cases where the (feasible) minimum deviation from the utopia target does not match the best available utility. Points A and B are both engineering local optima with the minimum deviation from the target in each of the feasible spaces. The deviation for point A is smaller, but the corresponding utility is not higher than that of point B. Note that these plots represent the engineering target (attributes $\boldsymbol{E}$) domain instead of the design option space.

To enable the move from one feasible domain to another, a circular inequality constraint [Figure 18.3(b)) is imposed in the enterprise problem based on the achievable engineering response E^D, and the enterprise DBD problem is resolved, as shown in Eq. (18.7). The physical meaning of the additional constraint is that it imposes a minimum geometric distance from the utopia target so that the enterprise problem is forced to find an alternative target for the engineering problem. The idea of adding the constraint is to explore targets at a new domain that may potentially lead to feasible designs with a better value for the expected utility. The points inside the circular constraint are ruled out because they are infeasible (otherwise they should be identified as the solution in the previous iteration as their deviations from the utopia target is

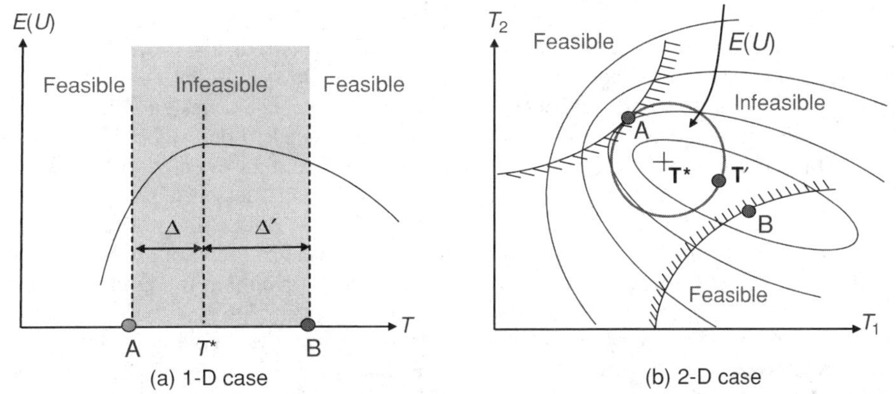

FIG. 18.3 UTILITIES WITH ENGINEERING FEASIBLE DOMAIN IMPOSED (POINTS A AND B ARE BOTH ENGINEERING LOCAL OPTIMA WITH THE MINIMUM DEVIATION FROM THE TARGET; DEVIATION FOR POINT A IS SMALLER, BUT THE CORRESPONDING UTILITY IS NOT HIGHER THAN THAT OF POINT B)

less). Note: For the discussion on the solution of the enterprise-level product planning problem, engineering design attributes $\boldsymbol{E}$ are referred to as targets $\boldsymbol{T}$. The modified enterprise problem P'_{ent} [Eq. 18.7)], based on the engineering design E^D generates a new target $\boldsymbol{T'}$ for the engineering problem. Based on the new target, the engineering problem finds point B as the optimum with the minimum deviation from the new target $\boldsymbol{T'}$. Point B is farther from the original utopia target; however, the corresponding utility is higher than that of point A. As a result, point B is selected as the optimal engineering design that has a better utility value.

$$P'_{\text{ent}} : \max_{\mathbf{T},\mathbf{X}_{\text{ent}}} E\left[U(\mathbf{T},\mathbf{X}_{\text{ent}})\right]$$

subject to Eq. (18.7)

$$C_{\text{aux}} : \left\|\mathbf{T}-\mathbf{E}^D\right\| \geq \left|\frac{\alpha}{\phi}\right| \Delta \text{ where } \Delta = \left\|\mathbf{T}^* - \mathbf{E}^D\right\|$$

To avoid returning to the previous solution, additional slope information is utilized to adjust the radius of the restricted feasible domain in the enterprise problem. Hence, $\left|\frac{\alpha}{\phi}\right|\Delta$ is used instead of Δ. Here, α and ϕ are the gradients of the utility function with respect to the current response $\boldsymbol{E}^D$ and new target $\boldsymbol{T'}$. The reader is referred to [25] for analytical case studies and a more detailed explanation of the algorithm. The proposed iterative procedure is terminated as soon as an engineering-level design is found with a better utility; the goal of the algorithm is to explore the target space of engineering design attributes $\boldsymbol{E}$ until a feasible engineering design with a better enterprise utility is identified. The proposed algorithm does not attempt to find the global optimum; instead, it explores the engineering feasible domain to find an alternative feasible design with a better utility if it exists in a disconnected feasible domain.

The hierarchical framework presented in Figure 18.2(b) shares a number of features with the ATC representation of the product planning and product development subproblems in [27]. However, there are some important differences. While the ATC formulation in [27] treats the enterprise problem as deterministic, we incorporate uncertainty and designer's risk attitude into the formulation by using expected utility $E(U)$ as the optimization criterion for the enterprise problem. In [27], the product planning problem not only considers the optimization of the enterprise-level objective (e.g., profit), but also attempts to minimize the deviation between achievable engineering design attributes and targets set by marketing. In our approach, enterprise and engineering objectives are treated separately in product planning and product development problems. Expected utility is the only optimization criterion at the enterprise level and the ATC approach is used only for the engineering product development problem. Such an approach not only preserves the essential distinction between product planning and product development functions, but it results in far fewer iterations between the enterprise and engineering-level problems, compared to Michalek et al. [27]. In fact, the engineering-level problem needs to be resolved only if the design space is disconnected and the enterprise-level targets are adjusted by the product planning problem.

18.4 SUSPENSION DESIGN CASE STUDY

An automotive suspension system design problem is used to demonstrate our approach. The assumptions made for the case study and the rationale behind, including the various engineering design attributes $\boldsymbol{E}$ to represent customer-product-selection attributes $\boldsymbol{A}$ in the demand model, are first explained. Also, details on the interpretation of the obtained demand model and examination of the suspension design obtained using the multilevel optimization framework are provided.

18.4.1 Hierarchical Approach to Suspension Design Problem

Here, we demonstrate the enterprise-driven DBD approach to the suspension system design of a midsize car. We demonstrate in this case study: (1) how the problem can be solved using a multilevel optimization formulation; (2) how the demand model is created to provide linking between enterprise planning and engineering development; and (3) how the new optimization algorithm is used to explore the targets of engineering design attributes that correspond to the disconnected feasible design space.

While treating the enterprise product planning of vehicle and engineering product development of the suspension system as hierarchical decision-making activities [Fig. 18.3(b)], we view the suspension design in engineering product development as a hierarchical design problem by itself and solve it using the ATC approach. The ATC formulation used here is based on the one in [23]. The schematic of the multilevel decision-based suspension design model is illustrated in Fig. 18.4. Here, targets for front and rear suspension stiffnesses ($\boldsymbol{T}^U = \boldsymbol{E}^*$, the top-level suspension system attributes), are set by solving the enterprise-level DBD product planning problem. These targets are then used to guide the subsystem engineering development of front and rear suspensions. By solving the problems at the subsystem level, the targets for front and rear coil spring stiffness are identified to guide the engineering development at the component level (i.e., front and rear coil spring designs).

18.4.2 Demand Model for Midsize Car Segment

To formulate the demand model that is useful for profit and cost estimations in the enterprise-level DBD problem, DCA is applied to study the impact of front and rear suspension characteristics and other important vehicle-level product attributes (e.g., engine performance, ride quality and comfort) on customers' choices of a midsize car. This case study is developed using market data provided by the Power Information Network group (PIN) at J.D. Power & Associates. Data on the vehicle attributes are from *Ward's Automotive Yearbook* [41]. The statistical software package STATA [43] is used to estimate the multinomial choice model coefficients based on the maximum likelihood criterion. For choice set selection in demand modeling, 12 vehicles (seven models and 12 trims) from the midsize segment are used to represent the entire market for midsize cars. The assumption is that customers only consider vehicles from the midsize car segment, specifically the 12 vehicle trims, when making their decisions.

Examples of customer-product-selection attributes $\boldsymbol{A}$ and engineering design attributes $\boldsymbol{E}$ considered in the suspension design case study are presented in Fig. 18.5. Customer-product-selection attributes $\boldsymbol{A}$ belong to system-level attributes. They are grouped into engineering-related customer attributes $\boldsymbol{A}_{eng}$ and nonengineering or enterprise-related customer attributes $\boldsymbol{A}_{ent}$. Examples of $\boldsymbol{A}_{eng}$ considered for demand modeling in the case study are performance, quality, comfort and handling. The relationship between $\boldsymbol{A}_{eng}$ and engineering design attributes $\boldsymbol{E}$ is illustrated in Fig. 18.5. Fuel economy and horsepower are examples of engineering design attributes that are related to performance, while vehicle length and suspension stiffnesses are related to handling. Front and rear suspension stiffnesses also influence customers' view of comfort. In this work, to facilitate engineering decision-making, the engineering design attributes E at the system level are modeled directly as the inputs of the demand model. Figure 18.5 also illustrates the cascading of engineering design attributes from vehicle system level to subsystem level, then to component level in suspension design. It should be noted that transfer relationships need to be established between the performance attributes at different levels of the hierarchy. For example, front/rear suspension stiffness is a performance attribute at the subsystem level and its target is identified as a design variable in the vehicle system-level optimization. Front/real coil spring stiffness is a performance attribute at the component level; its target is identified through optimization at the subsystem level. The demand model also considers various nonengineering-oriented design attributes $\boldsymbol{A}_{ent}$. APR of auto loan and resale value, which are grouped under $\boldsymbol{A}_{ent}$, coincide with the enterprise planning options $\boldsymbol{X}_{ent}$ for our case study. It should be noted that all design attributes unrelated to the suspension design, (e.g., horsepower and fuel economy) are assumed to be constant.

As a result of MNL analysis, demand Q (Table 18.1) is expressed as a function of demographic and product attributes such as income, age, retail price, resale value, vehicle dependability index (VDI; a quality measure expressed in terms of defects per 100 parts), APR of loan, fuel economy, vehicle length, front suspension stiffness and rear suspension stiffness. For the estimation of the demand model (customer), utility biases (Eq. 18.2) due to excluded variables are measured using alternative specific constants and alternate specific variables (see [36]). They measure the "average preference" of an individual relative to a "reference" alternative. One important consideration when evaluating MNL models is that demographic variables related to the customer's age and income can be included only as alternative specific variables. In our model, the income terms capture the interaction of the customer's income with different vehicle alternatives. Eleven income variables were estimated: each corresponded to a particular vehicle alternative, and income variable 1, corresponding to vehicle alternative 1, was used as reference. It is anticipated that the income variables corresponding to more expensive cars will have positive signs since people with higher incomes are likely to view expensive cars more favorably.

Table 18.1 includes results of the demand model estimation and observations on the model estimation results follow. Negative signs of retail price, VDI, APR and vehicle length mean that customers prefer lower values for these variables, i.e., customers prefer cheaper cars, lower interest rates, fewer defects and cars that facilitate easy parking. A positive sign for fuel economy means that customers

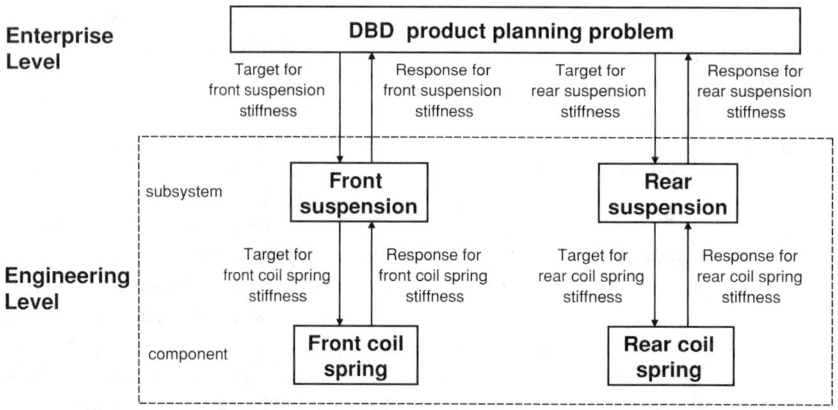

FIG. 18.4 SCHEMATIC OF MULTILEVEL DECISION-BASED SUSPENSION DESIGN MODEL

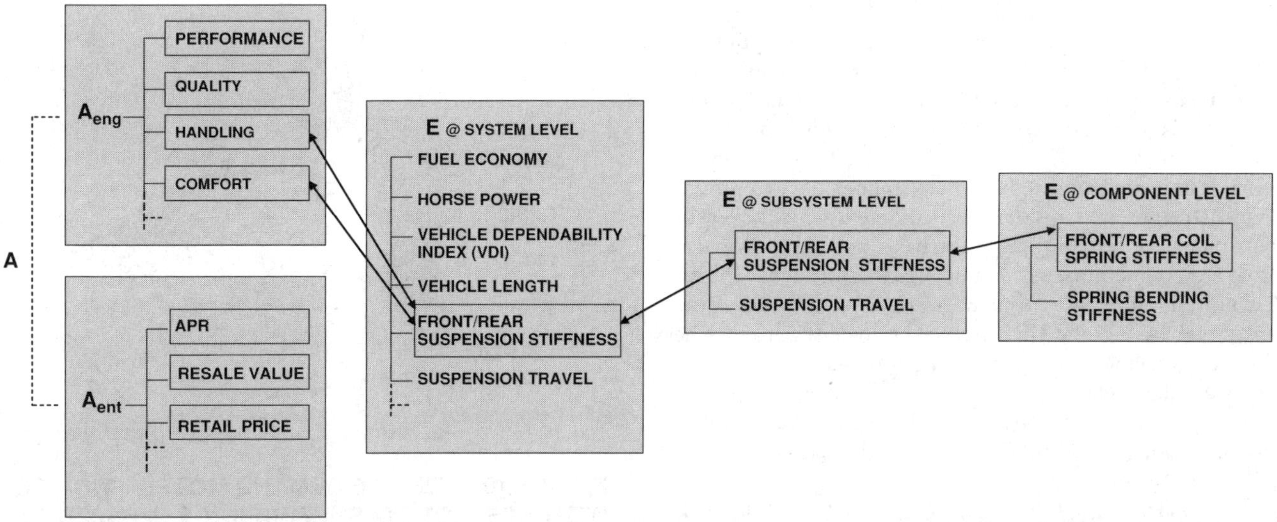

FIG. 18.5 EXAMPLES OF CUSTOMER-PRODUCT-SELECTION ATTRIBUTES *A* AND ENGINEERING DESIGN ATTRIBUTES *E* IN THE AUTOMOTIVE SUSPENSION DESIGN CASE STUDY. (THE ARROWS REPRESENT CORRELATIONS UTILIZED IN THE CURRENT CASE STUDY)

prefer higher gas mileage, and positive signs for the suspension stiffnesses mean that stiffer suspensions are preferred. Typically, a stiffer suspension generally translates to better handling and load-carrying abilities but also results in a harsher ride. Hence, the current choice model indicates that customers (in the present data set) "value handling characteristics more than ride quality." Also, since we are dealing with variables normalized with respect to their extreme values, the magnitude of the coefficients should reflect their relative importance. The statistical goodness-of-fit of the different MNL models developed for this purpose is evaluated using likelihood estimates. While choosing the final model, it is necessary to not only compare the statistical goodness of fit measures, but also pay attention to the signs and magnitudes of the different terms in the utility function.

18.4.3 Solving the Suspension Design Problem Using Multilevel Optimization

Even though our focus in this study is on suspension system design, it is necessary to compute the net profit of producing the whole vehicle to formulate the utility optimization model at the enterprise level. In this work, the net profit in [Eq. (18.8)] is used for this purpose. The price of a vehicle P is assumed to be constant

TABLE 18.1 RESULTS OF ESTIMATION OF THE MULTINOMIAL LOGIT MODEL

Attribute Type	Description	β Coefficient	t-value	95% Confidence Interval
Demographic attributes	Income × vehicle 2	0.13	6.41	(0.09, 0.18)
(Income variables capture the	Income × vehicle 3	0.01	0.48	(−0.03, 0.05)
interaction between customers' income	Income × vehicle 4	0.06	2.57	(0.01, 0.11)
and vehicle type. (*e.g., income ×*	Income × vehicle 5	−0.10	−4.20	(−0.15, −0.05)
vehicle 2 captures the interaction	Income × vehicle 6	−0.08	−3.37	(−0.13, −0.03)
between income and midsize vehicle 2)	Income × vehicle 7	0.07	2.99	(0.02, 0.11)
	Income × vehicle 8	0.08	3.22	(0.03, 0.12)
	Income × vehicle 9	0.08	3.25	(0.03, 0.14)
	Income × vehicle 10	0.19	9.38	(0.15, 0.23)
	Income × vehicle 11	0.05	2.29	(0.01, 0.10)
	Income × vehicle 12	0.04	1.18	(−0.02, 0.10)
Demographic attributes Variable captures interaction between age and country of origin of product.	Interaction between customers' age and country of origin of product	0.13	12.37	(0.11, 0.15)
Product attributes	Retail price	−1.57	−4.14	(−2.31, −0.82)
	Resale value	2.15	2.54	(0.49, 3.80)
	Vehicle dependability index	−1.69	−1.49	(−3.92, 0.53)
	Annual percentage rate (APR)	−1.05	−1.34	(−2.58, 0.49)
	Fuel economy	0.64	1.51	(−0.19, 1.46)
	Vehicle length	−0.60	−0.5	(−2.95, 1.74)
	Front suspension stiffness	1.75	3.11	(0.65, 2.85)
	Rear suspension stiffness	0.88	1.28	(−0.47, 2.24)

or unchanged from the current design, C_{susp} represents the unit cost for the suspension system and C_0 the unit cost for the rest of the vehicle system. In this work, C_0 is assumed to be constant, since changes will be made only in suspension parameters. The suspension system cost is assumed to be linearly proportional to the suspension stiffnesses [Eq. 18.9)]. In the context of design, among the engineering design attributes in the demand model, only front/rear suspension stiffnesses are considered as variables, while the rest of vehicle attributes are set constant at the baseline values. We also assume that the competitors will not change their designs. Uncertainty is captured in the cost variables a_f and a_r, considered as the exogenous variables $\boldsymbol{Y}$ in the DBD framework (see Fig. 18.1). The suspension cost variables a_f and a_r are both defined as normally distributed variables, with the mean $\mu_a = 0.05$ [\$/kNm^{-1}] and standard deviation $\sigma_a = 0.005$ [\$/kNm^{-1}]. For the current study, values for a midsize sedan, $P = \$20{,}000$, $C_0 = \$18{,}100$ are used.

$$\Pi = Q \times (P - C_{susp} - C_0) \qquad \text{Eq. (18.8)}$$

$$\Pi = Q \times (P - a_f k_{sf} - a_r k_{sr} - C_0) \qquad \text{Eq. (18.9)}$$

Also, a logarithmic function of profit Π, consistent with the principles of utility theory [42], is used as the enterprise utility $U(\Pi)$ to account for the risk-averse nature of the firm. Then the enterprise-level objective is to maximize the expected utility $E(U)$ where:

$$E(U) = \int Uf(a)da \qquad \text{Eq. (18.10)}$$

In this work, the enterprise planning problem is considered an unconstrained optimization problem. The expected utility function indicates that softer front suspension and stiffer rear suspension lead to the highest utility (Fig. 18.6). (In Fig. 18.6, the shaded regions in the target performance space indicate a disconnected feasible domain for the suspension design. The two shaded regions in the figure represent two suspension design options (i.e., softer vs. stiffer front/rear suspension designs) available to the designer.) For example, the suspension manufacturing supplier provides two alternatives for suspension design and the vehicle producer adjusts its product planning decision based on the availability of engineering designs. When applying the proposed multilevel optimization algorithm (Section 18.3.4), utopia targets $\boldsymbol{T}^*$ [$k_f = 30.2(k\text{N}/m), k_r = 19.5(k\text{N}/m)$, corresponding to an expected utility 14.27 and average profit \$80,460], are assigned for the suspension design problem after solving the original DBD at the enterprise level. Design A with the minimum deviation [$k_f = 25(k\text{N}/m), k_r = 19.5(k\text{N}/m)$ with expected utility 14.215 and average profit −\$8,935, i.e., loss] is found after solving the multilevel optimization using ATC at the engineering development level. Based on design A, additional geometric constraint is added at the enterprise level [Eq. 18.7)], which assigns an alternative target $\boldsymbol{T}'$ [$k_f = 30.2(k\text{N}/m), k_r = 24.83(k\text{N}/m)$ with expected utility 14.241 and average profit \$26,722] to the suspension design problem. Based on the new target $\boldsymbol{T}'$; the ATC finds design B [$k_f = 30.2(k\text{N}/m), k_r = 26(k\text{N}/m)$ with the improved expected utility 14.218 and average profit \$23,064] with the minimum deviation. The corresponding expected utility for design B is higher than that of design A. Design B is selected as the final design, with a better enterprise-level utility value.

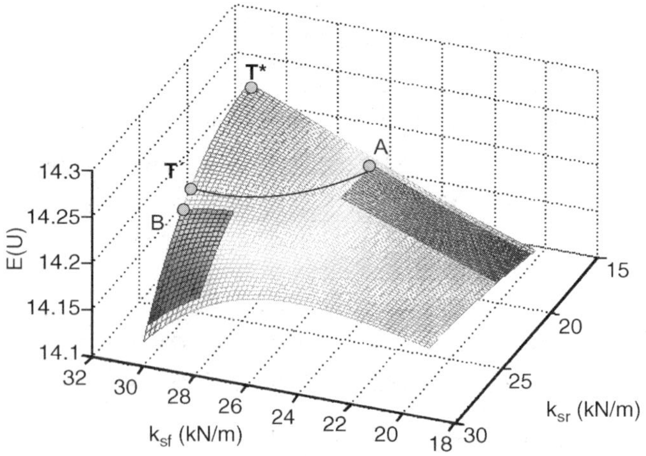

FIGURE 18.6 VEHICLE DEMAND MODEL: PROFIT CHANGE WITH RESPECT TO SUSPENSION STIFFNESS CHANGES SHADED AREAS REPRESENT DISCONNECTED FEASIBLE SUSPENSION DESIGN DOMAIN. AFTER ENGINEERING DESIGN A IS FOUND, A GEOMETRIC LIMITING CONSTRAINT (SOLID CIRCLE) IS ADDED IN THE UTILITY SPACE TO FIND AN ALTERNATIVE TARGET *T′* TO EXPLORE ANOTHER DISCONNECTED FEASIBLE SPACE.

Table 18.2 summarizes the iteration process using the proposed multilevel optimization algorithm. Detailed suspension designs at the subsystem and component level are summarized in Table 18.3 to Table 18.6. For example, the front suspension design model takes coil spring stiffness, spring-free length and torsional stiffness as inputs to the suspension model and returns suspension stiffness and suspension travel as outputs. Among the inputs, coil spring stiffness is cascaded to the front coil spring design model as a target. Based on the coil spring target from the suspension model, the coil spring design at the component level takes wire diameter, coil diameter and pitch as inputs and returns actual coil spring stiffness and spring bending stiffness as outputs.

TABLE 18.2 ITERATION HISTORY: MAXIMIZING EXPECTED UTILITY WITH VEHICLE SUSPENSION DESIGN CHANGE (SEE FIG. 18.6) (HERE THE VALUE OF PROFIT HAS BEEN COMPUTED FOR MEAN VALUES OF a_f AND a_{rh}).

Iteration	Target for Front Suspension Stiffness (N/mm) T_1^*	Target for Rear Suspension Stiffness (N/mm) T_2^*	Desired Profit (\$)	Desired $E[U]$	Response for Front Suspension Stiffness (N/mm)	Response for Front Suspension Stiffness (N/mm)	Profit Achieved (\$)	$E[U]$
1	30.2	19.5	80,460	14.272	25.0 (point A)	19.5 (point A)	−8,935	14.215
2	30.2	24.83	26,722	14.241	30.2 (point B)	26 (point B)	23,064	14.218

TABLE 18.3 FRONT SUSPENSION DESIGN

Front Suspension Subsystem Design	Type	Optimal Value	Lower Bound	Upper Bound
Coil spring stiffness (N/mm)	Input	117.53	30	160
Spring-free length (mm)	Input	372.9	300	650
Torsional stiffness (N-m/deg)	Input	30.0	20	85
Suspension stiffness (N/mm)	Output	30.2	19	30.2
Suspension travel (m)	Output	0.1	0.05	0.1

TABLE 18.5 FRONT COIL SPRING DESIGN

Front Coil Spring Design	Type	Optimal Value	Lower Bound	Upper Bound
Wire diameter(m)	Input	0.014	0.005	0.03
Coil diameter(m)	Input	0.074	0.05	0.20
Pitch	Input	0.04	0.04	0.10
Coil spring stiffness (N/mm)	Output	124.9	—	—
Spring bending stiffness (N-m/deg)	Output	16.1	—	—

18.5 CONCLUSIONS AND FUTURE WORK

In this chapter, a multilevel optimization approach that integrates enterprise-level product planning with engineering-level product design based on the DBD principles and ATC concept is presented. The hierarchical approach presented provides a systematic way to resolve the organizational and computational complexities involved in integrating design efforts across the enterprise. Based on the review of various existing demand modeling approaches in engineering design, a disaggregate probabilistic demand modeling approach is selected to model customer choices at the enterprise level. It is shown how customer preferences for product attributes can be translated into the market share of a product and used to guide the engineering product development process by setting up the targets for engineering design attributes. An effort is made to clearly distinguish between the various types of product attributes that are commonly used in an enterprise-driven design situation; some attributes are more relevant to customer desires, while others are used by engineers alone; some depend on engineering design options, while others are more related to enterprise planning options. Conventionally, the product development tries to match the utopia target with minimum deviation. In cases where the utopia target is unattainable, the minimum deviation design may not correspond to the highest possible enterprise utility if the design space is disconnected. The target exploration algorithm employed for the hierarchical product development scenario in this work sets alternative targets for the product development with higher corresponding enterprise-level utility. An automotive suspension design case study was presented to demonstrate the effectiveness of our approach. The case study uses the customer data for the midsize sedan market and obtains a demand model that studies the impact of the suspension variables on customer choices. Our approach explores the enterprise-level preference target space to assign the suspension design targets for higher utility, and the ATC successfully cascades the targets at the subsystem and component levels to achieve a design consistent with enterprise-level goals.

The proposed multilevel optimization model is potentially the most useful during the early stages of a design process when the iterative process can be used to revise, negotiate and validate design targets by engineering design and the enterprise-planning groups. In an actual implementation, some targets may be more loosely specified than others. Future formulations should accommodate the setting of a range of targets. It should be noted that while cost is not one of the targets set by the enterprise planning problem in our case study, it could be a target or modeled as a constraint in the engineering-level problem. Since complex systems design involves the design of a number of subsystems and components, and each of these designs may have different computational requirements for design evaluations due to the different nature of analysis models, future implementation should consider the optimal allocation of computational resources as a part of the multilevel optimization solution algorithm.

Future work will also involve introducing uncertainty in engineering product development and expanding the use of the MNL model to incorporate the use of both stated and revealed preference data. Such models can be more effectively incorporated to model multilevel decision-making scenarios as well as heterogeneous market segments (e.g., considering sedans with SUVs in the same choice set). In this work, the role of manufacturing on decisions made during the product planning and development processes is mainly reflected in the product cost modeling. Most industrial design problems involve manufacturing decision-making as a part of the engineering product development. To separate manufacturing and product design decision-making, the manufacturing aspects

TABLE 18.4 REAR SUSPENSION DESIGN

Rear Suspension Subsystem Design	Type	Optimal Value	Lower Bound	Upper Bound
Coil spring stiffness (N/mm)	Input	80.1	30	160
Spring-free length (mm)	Input	410.3	300	650
Torsional stiffness (N-m/deg)	Input	58.7	20	85
Suspension stiffness (N/mm)	Output	25.8	19	30.2
Suspension travel (m)	Output	0.1	0.05	0.1

TABLE 18.6 REAR COIL SPRING DESIGN

Rear Coil Spring Design	Type	Optimal Value	Lower Bound	Upper Bound
Wire diameter(m)	Input	0.02	0.005	0.03
Coil diameter(m)	Input	0.14	0.05	0.20
Pitch	Input	0.05	0.05	0.10
Coil spring stiffness (N/mm)	Output	84.5	—	—
Spring bending stiffness (N-m/deg)	Output	33.2	—	—

of the product development process need to be investigated further and their relationship with product design needs to be more rigorously modeled.

ACKNOWLEDGMENT

We are grateful to the specialists at the Power Information Network group (PIN) at J.D. Power & Associates for their thoughtful contributions and their efforts to gather data for the vehicle demand model. The support from the NSF grant DMI0335880 and Ford University Research Program are also acknowledged. The engineering development model of the suspension system design was created during Dr. Harrison Kim's Ph.D. study, supported by the Optimal Design Laboratory at the University of Michigan.

REFERENCES

1. Hazelrigg, G., 1996. *Systems Engineering: An Approach to Information-Based Design*, Prentice-Hall, Inc., Upper Saddle River, NJ.
2. Thurston, D. L., 2001. "Real and Misconceived Limitations to Decision Based Design with Utility Analysis," *Trans. ASME: J. of Mech. Des.*, 123 (June), pp. 176–182.
3. Li, H. and Azarm, S., 2000. "Product Design Selection under Uncertainty and With Competitive Advantage," *Trans. ASME: J. of Mech. Des.*, Vol. 122, pp. 411–418.
4. Callaghan, A. R. and Lewis, K. E., 2000. "A 2 -Phase Aspiration-Level and Utility Theory Approach to Large Scale Design," *Proc. DETC 2000 ASME Des.Engrg. Tech. Conf., DETC2000/DTM-14569*, Baltimore MD.
5. Scott, M. J. and Antonsson, E. K., 1999. "Arrow's Theorem and Engineering Design Decision Making," *Res. in Engrg. Des.*, **11**(4), pp. 218–228.
6. Roser, C. and Kazmer, D., 2000. "Flexible Design Methodology," Proc., *DETC 2000 ASME Des. Engrg. Tech. Conf., DETC2000/DFM-14016*, Baltimore, MD.
7. Marston, M., Allen, J. and Mistree, F., 2000. "The Decision Support Problem Technique: Integrating Descriptive and Normative Approaches in Decision Based Design," *Engrg. Valuation & Cost Analysis, Special Ed.* on *"Decision-Based Design: Status & Promise,"* **3**(2/3), pp. 107 –130.
8. Shah, J. J. and Wright, P. K., 2000. "Developing Theoretical Foundations of DFM," *Proc., DETC 2000 ASME Des.Engrg. Tech. Conf., DETC2000/DFM-14015*, Baltimore, MD.
9. Dong, H. and Wood, W. H., 2004. "Integrating Computational Synthesis and Decision-Based Conceptual Design," *Proc., ASME 2004 Des. Engrg. Tech.* Conf., Salt Lake City, UT.
10. Schmidt, L. C. and Herrmann, J. W., 2002. "Viewing Product Development as a Decision Production Problem," *Proc. ASME 2002 Des. Engrg. Tech. Conf.*, Montreal, Canada.
11. Wassenaar, H. J. and Chen, W., 2003. "An Approach to Decision-Based Design With Discrete Choice Analysis for Demand Modeling," *Trans., ASME: J. of Mech. Des.*, 125(3), pp. 490–497.
12. Wassenaar, H. J., Chen, W., Cheng, J. and Sudjianto, A., 2003. "Enhancing Discrete Choice Demand Modeling for Decision-Based Design," *Proc., DETC 2003 ASME Des. Engrg. Tech. Conf.*, (also, in *Trans., ASME: J. of Mech. Des.*), Chicago, IL.
13. Wassenaar, H. J., Chen, W., Cheng, J. and Sudjianto, A., 2004. "An Integrated Latent Variable Modeling Approach for Enhancing Product Demand Modeling," *Proc., DETC 2004 ASME Des. Engrg. Tech. Conf.*, Salt Lake City, UT.
14. Lewis, K. and See, T.-K., 2004. "Multiattribute Decision-Making Using Hypothetical Equivalents," *Trans., ASME: J. of Mech. Des.*, **26**(6), pp. 950–958.
15. Gu, X., Renaud, J., Ashe, L., Batill, S., Budhiraja, A. and Krajewski, L., 2002. "Decision-Based Collaborative Optimization," *Trans., ASME: J. of Mech. Des.*, **124**(1), pp. 1–13.
16. McAdams, D. A. and Dym, C. L., 2004. "Modeling and Information in the Design Process," *Proc., DETC 2004 ASME Des. Engrg. Tech. Conf., DETC2004-57101*, Salt Lake City, UT.
17. Stelmack, M. A. and Batill, S. M., 1997. "Concurrent Subspace Optimization of Mixed Continuous/Discrete Systems," *AIAA Paper 97-1229, AIAA/ASME/ASCE/AHS/ASC 38th Struc., Struc. Dyn. and Mat.* Conf., Kissimmee, FL.
18. Sobieszczanski-Sobieski, J., 1988. "Optimization by Decomposition: A Step from Hierarchic to Nonhierarchic Systems," *Tech. Rep. NASA CP-3031*, NASA Langley Res. Ctr., Hampton, VA.
19. Sobieszczanski-Sobieski, J., Agte, J. and Sandusky, R., 1998. "Bi-Level Integrated System Synthesis (BLISS)," *Tech. Memo., TM-1998-208715*, NASA Langley Res. Ctr., Hampton, VA.
20. Braun, R., 1996. *"Collaborative Optimization: Architecture for Large-Scale Distributed Design,"* Ph.D. thesis, Stanford University, Stanford, CA.
21. Michelena, N., Kim, H. M. and Papalambros, P.Y., 1999. "A System Partitioning and Optimization Approach to Target Cascading," *Proc., 12th Int. Conf. on Engrg. Des.*, Munich, Germany, Vol. 24, pp. 1109–1112.
22. Kim, H. M., 2001. *"Target Cascading in Optimal System Design,"* Ph.D. thesis, University of Michigan, Ann Arbor, MI.
23. Kim, H. M., Rideout, D. G., Papalambros, P. Y. and Stein, J. L., 2003. "Analytical Target Cascading in Automotive Vehicle Design," *Trans., ASME: J. of Mech. Des.*, **125**(3), pp. 481–489.
24. Kim, H. M., Michelena, N. F. and Papalambros, P. Y., 2003. "Target Cascading in Optimal System Design," *Trans. ASME: J. of Mech. Des.*, **125**(3), pp. 474–480.
25. Kim, H. M., Kumar, D., Chen, W. and Papalambros, P. Y., 2004. "Target Feasibility Achievement in Enterprise-Driven Hierarchical Multidisciplinary Design," *AIAA-2004–4546, Proc., 10th AIAA/ISSMO Multidisciplinary Analysis and Optimization Conf.*, Albany, NY.
26. Cooper, A. B., Georgiopoulos, P., Kim, H. M. and Papalambros, P. Y., 2003. "Analytical Target Setting: An Enterprise Context in Optimal Product Design," *Proc., 2003 ASME Des. Automation Conf., DAC-48734*, Chicago, IL.
27. Michalek, J., Feinberg, F. and Papalambros, P. Y., 2004. "Linking Marketing and Engineering Product Design Decisions via Analytical Target Cascading," *J. of Prod. Innovation Mgmt.: Special Issue on Des. and Marketing in New Prod. Dev.*
28. Kroo, I., 1997. "Multidisciplinary Optimization Applications in Preliminary Design—Status and Directions," *Proc., 38th AIAA/ASME/ASCE/HS/ASC, Struc., Struc. Dynamics and Mat. Conf.*, AIAA 97-1408, Kissimmee, FL.
29. Donndelinger, J. and Cook, H. E., 1997. *Methods for Analyzing the Value of Vehicles, Soc. of Automotive Engrg.*, Feb.
30. Cook, H. E., 1997. *Product Management: Value, Quality, Cost, Price, Profit and Organization*, Chapman and Hall.
31. Monroe, E. M. and Cook, H. E., 1997. "Determining the Value of Vehicle Attributes using a PC based tool," *SAE Tech. Paper Ser. 970763, Inte. Congr. and Exposition*, Detroit, MI.
32. Besharati, B., Azarm, S. and Farhang-Mehr, A., 2001. "On the Aggregation of Preferences in Engineering Design," *Proc., DETC 2001 ASME Des. Engrg. Tech. Conf.*, Pittsburgh, PA.
33. Besharati, B., Azarm, S. and Farhang-Mehr, A., 2002. "A Customer-Based Expected Utility Metric for Product Design Selection," *Proc., DETC 2002 ASME Des. Engrg. Tech. Conf.*, Montreal, Canada.
34. Besharati, B., Luo, L., Azarm, S. and Kannan, P. K., 2004. "An Integrated Robust Design and Marketing Approach for Product Design Selection Process," *Proc., DETC '04, ASME Des. Engrg. Tech. Conf. and Computers in Engrg. Conf.*, DETC2004-57405, Salt Lake City, UT.
35. Louviere, J. J., Hensher, D. A. and Swait, J. D., 2000. *Stated Choice Methods, Analysis and Application*, Cambridge University Press.
36. Ben-Akiva, M. and Lerman, S. R., 1985. *Discrete Choice Analysis*, The MIT Press, Cambridge, MA.
37. Green, P. and Wind, Y., 1975. "New Way to Measure Consumers' Judgments," *Harvard Bus. Rev.*, Vol. 53, pp. 107–117.

38. Green, P. and Srinivasan, V., 1990. "Conjoint Analysis Marketing: New Developments With Implications Research and Practice," *J. of Marketing*.
39. Koppelman, F. and Sethi, V., 2000. "Closed Form Discrete Choice Models," *Handbook of Transport Modeling*, D.A. Hensher and K.J. Button, eds., Pergamon Press, Oxford, pp. 211–228.
40. Train, K., 2003. *Discrete Choice Methods with Simulation*, Cambridge University Press.
41. *Ward's Automotive Yearbook*, 2000. Ward's Communications, Detroit, MI.
42. Keeney, R.L. and Raia, H., 1976. *Decisions With Multiple Objectives: Preferences and Value Tradeoffs*, Wiley, New York, NY.
43. STATA Corporation, 2003. *STATA Programming Reference Manual*, ISBN 1-881228-73-8.

PROBLEMS

18.1 Provide a product design and development example in which the enterprise-level objective conflicts with the engineering-level design objective.

18.2 Discuss the advantages of the decomposition-based hierarchical approach to DBD compared to the AIO approach.

18.3 List the assumptions made when converting the AIO DBD formulation to the decomposed multilevel optimization formulation.

18.4 Discuss how the manufacturing decisions should be separated from the product design decisions in the multilevel DBD framework.

18.5 Utopia target from the enterprise level is rarely achieved perfectly when it is assigned to the engineering design team. For the 1-D and 2-D engineering target space, show schematically how the proposed approach can achieve a better enterprise-level objective in case there exist two design alternatives at the engineering level. (None of the two design alternatives match the target perfectly, i.e., they are imposed in a disconnected feasible engineering design space.)

development processes as well as information linkages between product development team members [3, 4] and linkages between components in products [5]. In this research, the P-VDP has been represented using a task-based DSM model. This is an attractive representation because it is very compact and, as will be shown throughout this chapter, it lends unique insight into the relationships between the production of information, iteration and decision-making in the P-VDP.

A DSM is a symmetrical matrix in which the rows and columns represent the elements of a system. Each entry in the DSM indicates an interface between two elements in the system. For a task-based DSM, entries in a row represent the inputs that a given task receives from other tasks in the system and entries in a column represent the outputs that a task sends to other tasks in the process. In a task-based DSM, there are three archetypical patterns of information flow between tasks:

- Tasks are dependent, or conducted in series, if the information flow between them imposes a sequential order for completion of the tasks.
- Tasks are independent, or conducted in parallel, if there is no information flow between them.
- Tasks are interdependent, or coupled, if each task requires output from the other in order to be completed. Coupled tasks must either be executed simultaneously with continuous exchanges of information or be carried out in an iterative fashion.

The P-VDP DSM model is shown in an unprocessed form in Fig. 19.1 and with optimized task sequencing in Fig. 19.2. Cividanes [6] provides an in-depth discussion of the development and analysis of this model. In this chapter, the discussion of the P-VDP DSM model is focused on two critical points. First, work in the P-VDP is conducted primarily in a serial, feed-forward fashion—as shown in Fig. 19.1 by the relatively small number of marks above the diagonal and in Fig. 19.2 by the relatively small number and size of iteration blocks in the first two phases of the process. Second, the iterative activity in the P-VDP is confined to a small number of blocks that vary significantly in size, as shown in Fig. 19.2. When applying DSM to improve process efficiency, it is customary to focus on activity within the iteration blocks within the DSM model. In this chapter, it will be shown that examination of the iteration blocks in a DSM process model is also critical for understanding decision-making within a process. This will require both an examination of the activities occurring within the iteration blocks and a formal statement of the definition of decision-making.

19.2.3 Types of Iteration in the P-VDP

The iteration blocks observed in the P-VDP may be classified into two types: The first, shown in Fig. 19.3, contains small blocks of coupled tasks that are performed by individuals or subteams within the vehicle development team. Examples of these blocks include a marketing analyst determining the vehicle's feature content and providing a revenue estimate, design engineers developing computer-aided design (CAD) models and completing technical specification documents, and manufacturing engineers developing preliminary designs for stamping dies and providing estimates of investment cost. The second type, shown in Fig. 19.4, contains large blocks of coupled tasks that are distributed across the vehicle-development team. Two of these blocks were observed in the P-VDP: The first of these, detailed in Fig. 19.5, occurs early in the P-VDP and leads to the selection of a vehicle platform. The second, detailed in Fig. 19.6, occurs toward the end of the P-VDP and leads to the approval of a balanced preliminary vehicle design.

The iteration blocks shown in Figures 19.5 and 19.6 share two noteworthy similarities: First, each block contains a network of complex interactions across the entire vehicle development team, most notably the engineering, manufacturing, marketing and finance activities. Second, each block includes a gate review

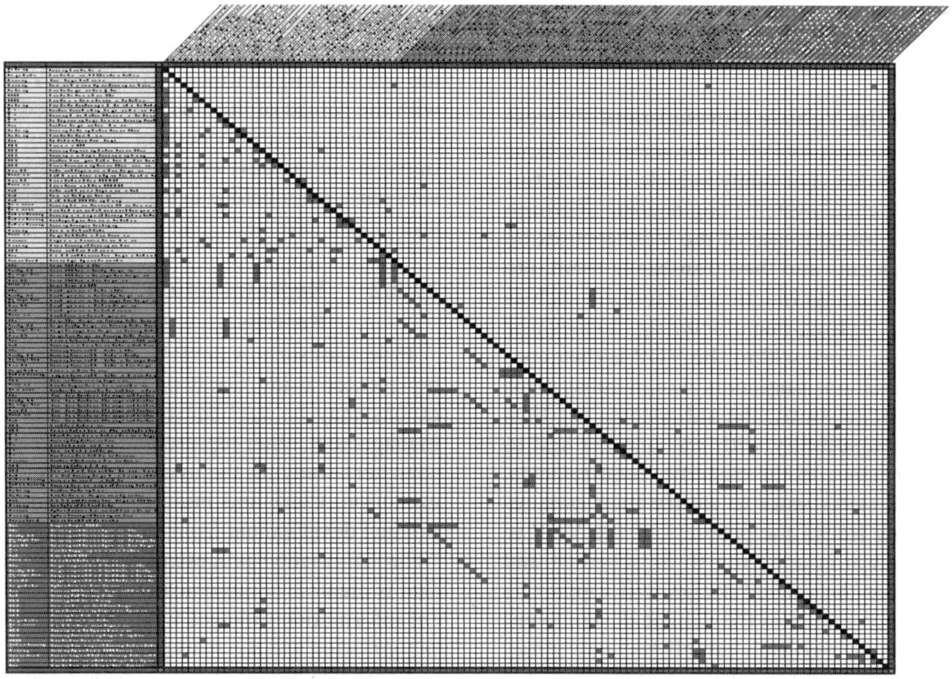

FIG. 19.1 AN UNPROCESSED DSM REPRESENTATION OF THE P-VDP

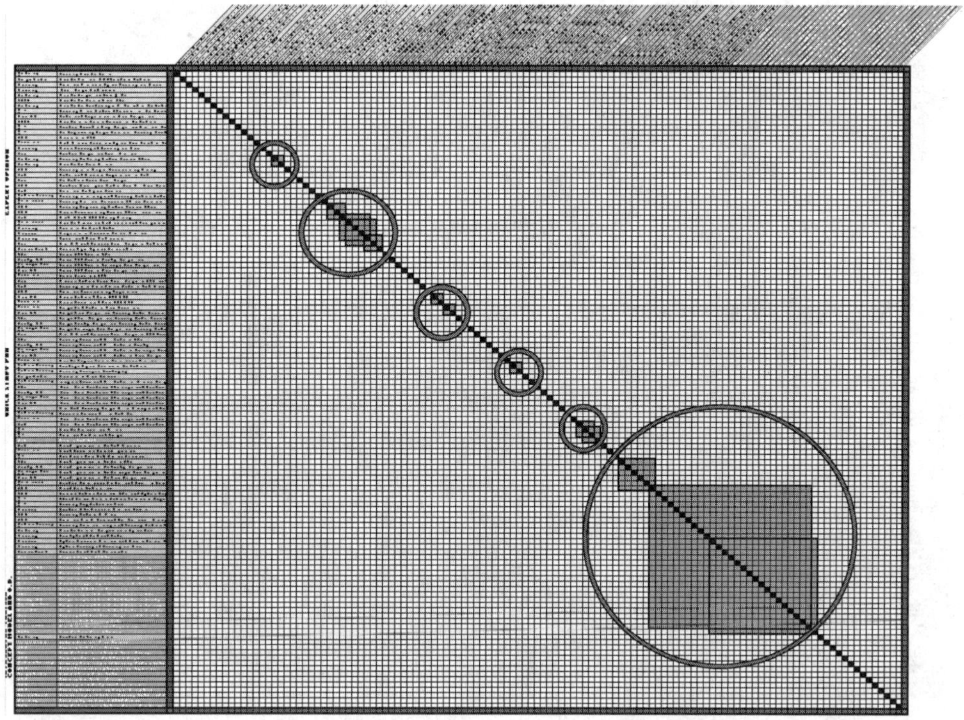

FIG. 19.2 ITERATION BLOCKS IN THE P-VDP

in which the vehicle development team presents its status and direction to a review board of senior managers and formally requests permission to proceed further. This observation leads to the interesting hypothesis that these two iteration blocks are the manifestation of the decision-making process in the P-VDP. However, before that hypothesis may be developed and explored, it is necessary to define exactly what a decision is and how it is made.

19.3 DECISION-MAKING

19.3.1 The Definition of a Decision

Although there has been a considerable amount of research into decision-making in engineering design, the engineering design community has not reached a consensus as to the definition of a decision. Conceptually, Herrmann and Schmidt [7] view decisions as value-added operations performed on information flowing

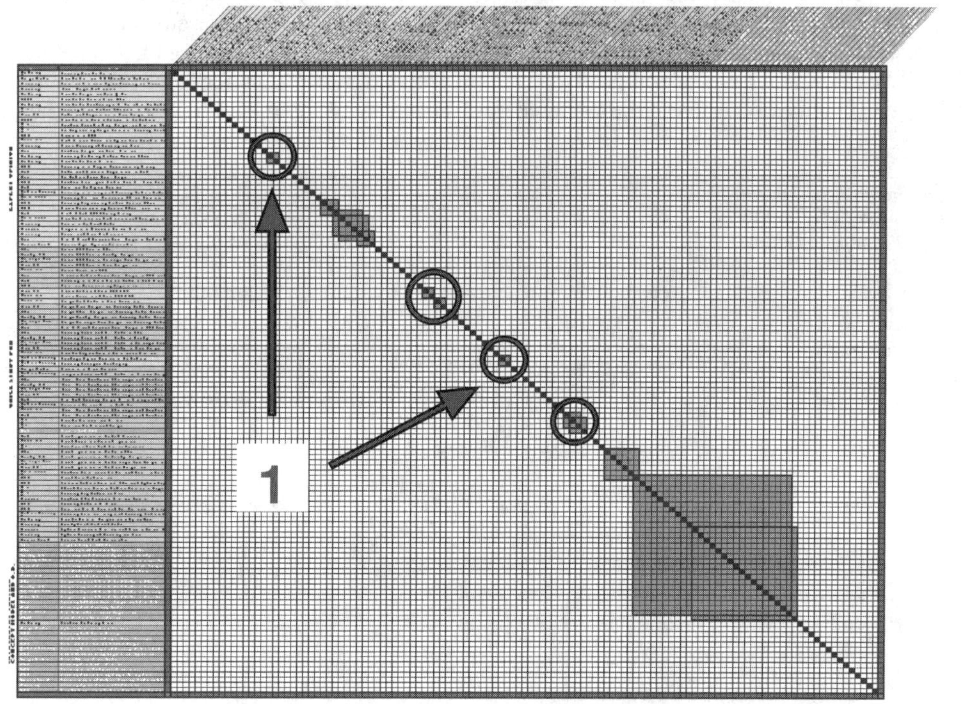

FIG. 19.3 LOCALIZED, INTERDEPENDENT TASKS IN THE P-VDP

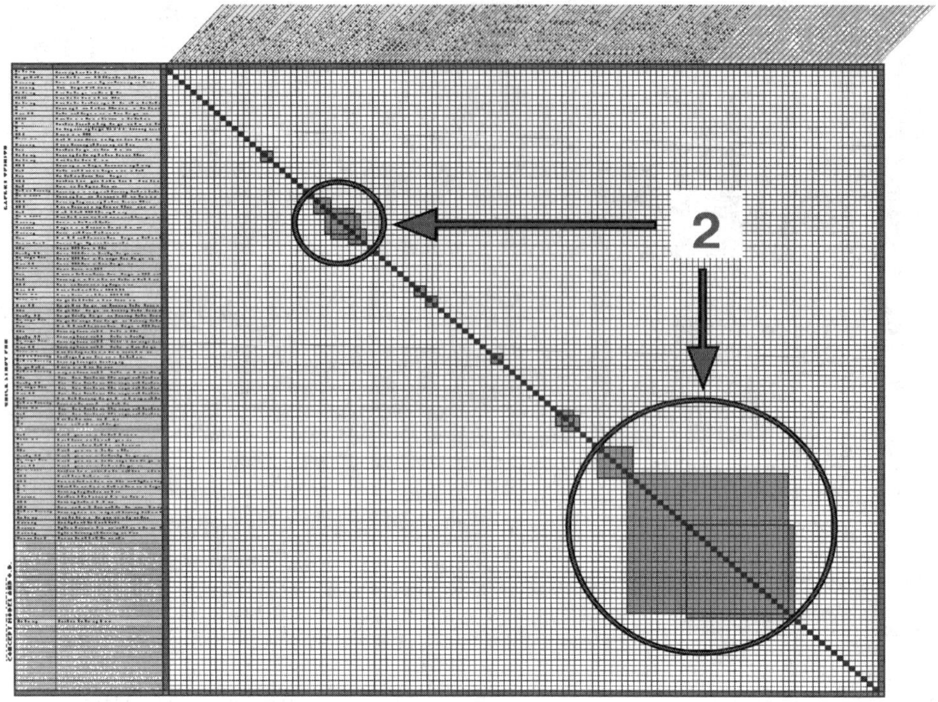

FIG. 19.4 TEAM INTERACTION DRIVEN BY THE P-VDP GATE REVIEW PROCESS

through a product development team, while Carey et al. [8] view decisions as strategic considerations that should be addressed by specific functional activities at specific points in the product development process to maximize the market success of the product being developed. There is definitely great potential for improving product development processes such as the P-VDP by applying the methods derived from these definitions. However, for reasons to be discussed later, the definition used in this research is the explicit and succinct definition from [9]: "A decision is an irrevocable allocation of resources, in the sense that it would take additional resources, perhaps prohibitive in amount, to change the allocation."

Commonly, we understand that a decision is a selection of one from among a set of alternatives after some consideration. This is illustrated well by Hazelrigg [10] in his discussion of the dialogue between Alice and the Cheshire cat in *Alice in Wonderland*. He notes that in every decision, there are alternatives. Corresponding to these alternatives are possible outcomes. The decision-maker weighs the possible outcomes and selects the alternative with the outcomes that he or she most prefers. Although apparently simple, this discussion contains several subtle but powerful distinctions: First, there is a clearly designated decision-maker who will select an alternative and will experience the consequences of that selection. Second, the decision is made according to the preferences of the decision-maker—not those of the decision-maker's stakeholders, or customers, or team members, or for that matter anyone's preferences but the decision-maker's. Third, the decision-maker's preferences are applied not to the alternatives, but to the outcomes. Finally, and perhaps most importantly, the

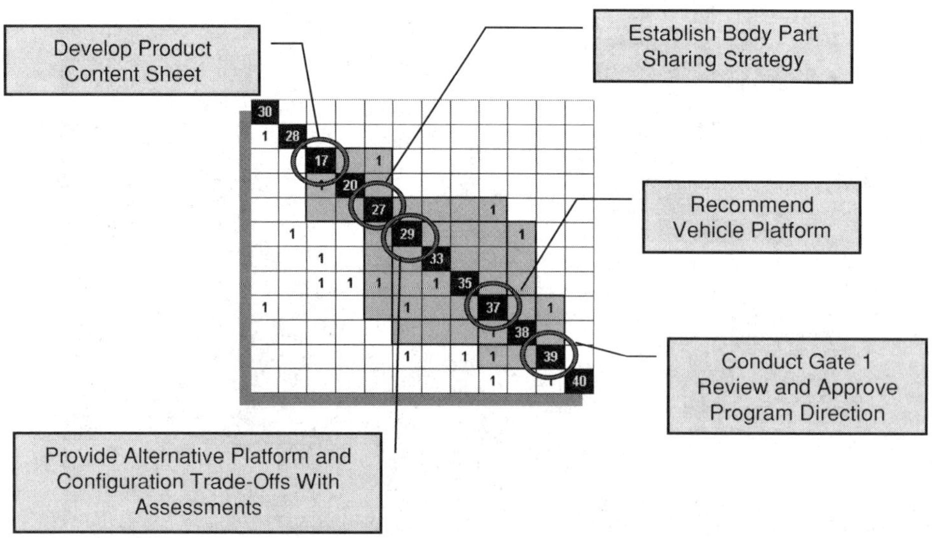

FIG. 19.5 INTERDEPENDENT TASKS IN VEHICLE PLATFORM SELECTION

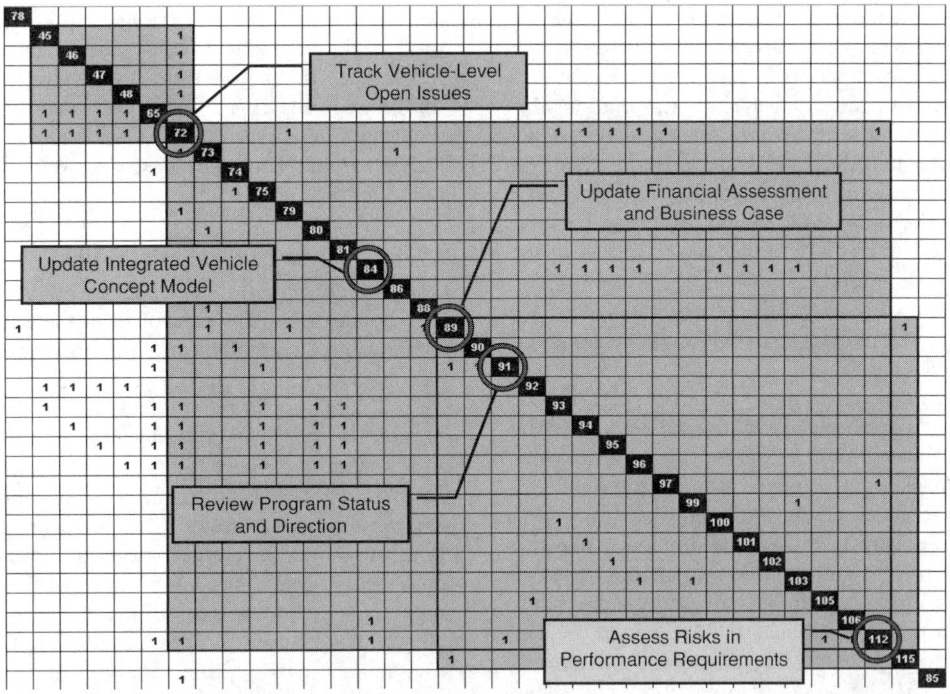

FIG. 19.6 INDERDEPENDENT TASKS IN PRELIMINARY VEHICLE DESIGN

decision is an action taken in the present to achieve a desired outcome in the future.

Future states cannot be known or predicted with absolute certainty. Therefore, it is imperative that uncertainty is both acknowledged and incorporated in decision-making processes (in this case, decision-making in the P-VDP). Nikolaidis [11] discusses the different types of uncertainty observed in vehicle design decisions. Although some types of uncertainty cannot be reduced, many types can. One of the foremost tasks in the P-VDP (as well as in any other product development process) is the reduction of uncertainty prior to decision-making. This occurs naturally through the Decision Analysis Cycle.

19.3.2 The Decision Analysis Cycle

Decision-makers are, by definition, people who have the authority to allocate an organization's resources. In a vehicle development program, these people are typically executives or senior managers. They make decisions (knowingly or not) using the Decision Analysis Cycle shown in Fig. 19.7 and described completely in [9].

The discussion of the phases in the Decision Analysis Cycle contains several precisely defined terms from the language of formal decision analysis [9]. The term "value" is used to describe a measure of the desirability of each outcome. For a business, the value is typically expressed as some measure of profit. The term "preferences" refers to the decision-maker's attitude toward postponement or uncertainty in the outcomes of his decision.

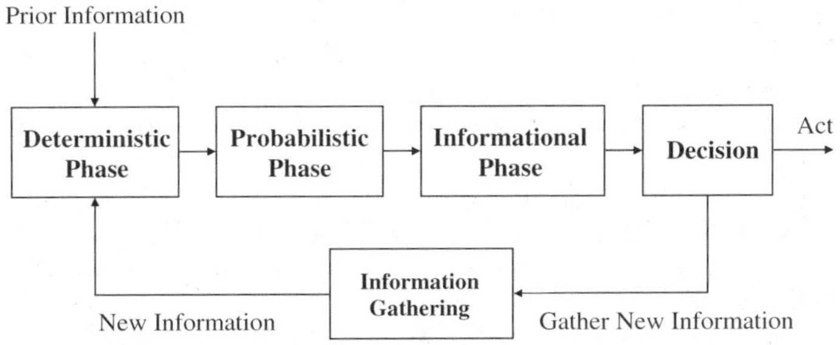

FIG. 19.7 THE DECISION ANALYSIS CYCLE

The three phases of the Decision Analysis Cycle that precede the decision are:

- Deterministic Phase : The variables affecting the decision are defined and related, values are assigned and the importance of the variables is measured without consideration of uncertainty.
- Probabilistic Phase : Probabilities are assigned for the important variables. Associated probabilities are derived for the values. This phase also introduces the assignment of risk preference, which provides the solution in the face of uncertainty.
- Informational Phase : The results of the previous two phases are reviewed to determine the economic value of eliminating uncertainty in each of the important variables of the problem. A comparison of the value of information with its cost determines whether additional information should be collected.

Application of the Decision Analysis Cycle in the P-VDP can only be discussed in the context of a decision. Cafeo et al [12] present a very detailed diagram of decision-making in the VDP. For the purpose of this discussion, the simplified decision diagram shown in Fig. 19.8 will be used.

The rectangular block in this diagram represents the decisions to be made. In the P-VDP, these are high-level vehicle design alternatives, such as the selection of the vehicle platform and the configuration of the various vehicle subsystems. The ovals in this diagram represent uncertain quantities. In the P-VDP, we represent the vehicle's functional attributes and its performance in the marketplace (in terms of sales volume and transaction price) as

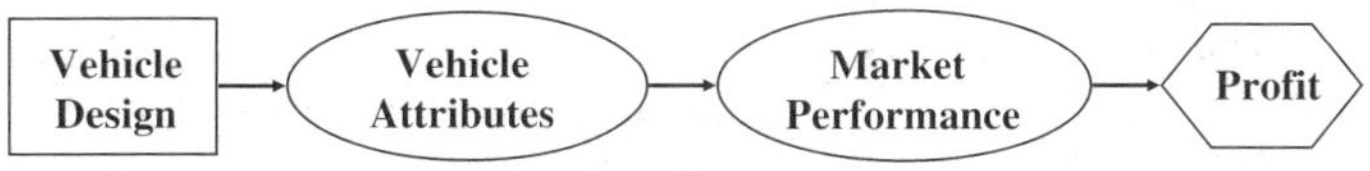

FIG. 19.8 A SIMPLIFIED P-VDP DECISION DIAGRAM

uncertainties. Finally, the hexagon in this diagram represents the decision-maker's value. As would generally be true in business, the value in the P-VDP is a measure of profit. The arrows in the diagram represent the decision-maker's beliefs about the relationships between the decisions, the relevant uncertainties and the value. Thus, the interpretation of this diagram is that profit depends on the vehicle's market performance, which depends on its functional attributes, which in turn depends on the design of the vehicle.

The decision presented in Fig. 19.8 may be made using the Decision Analysis Cycle shown in Fig. 19.7. In the Deterministic Phase, the specific set of design alternatives to be considered are identified and a corresponding set of analytical and experimental methods and tools are selected for relating the design alternatives to their effects on the vehicle's attributes, its market performance and, ultimately, to the profit generated by selling the vehicle. In the Probabilistic Phase, the uncertainties in the estimates of the vehicle's attributes and its market performance are characterized and propagated to examine their effects on profit. In the Information Phase, the decision-maker examines whether or not it is prudent to allocate additional resources to gather more information (usually through further studies with analytical tools) to increase the likelihood of realizing greater profit. Finally, in the Decision Phase, the decision-maker selects a design alternative and commits to executing the remainder of the VDP based on this selection.

19.4 DECISION-MAKING IN THE P-VDP

19.4.1 Manifestation of the Decision Analysis Cycle

Examination of the P-VDP DSM model suggests that, whether knowingly or unknowingly, the weightiest decisions in the P-VDP are made using the Decision Analysis Cycle. Each phase of the Decision Analysis Cycle may be identified within each phase of the P-VDP. The Deterministic Phase is represented by a long, predominantly serial chain of work tasks preceding a large iteration block. During this phase, engineers generate design alternatives and corresponding performance assessments to present to decision-makers. Other functional staffs, such as finance, marketing and manufacturing, also provide their assessments of outcomes corresponding to the engineers' design alternatives, such as potential changes in cost and revenue and required changes to manufacturing facilities. The Probabilistic Phase is represented by the gate review inside the large iteration block. Although few managers would be able to articulate it, observation of vehicle development teams reveals that managers update their beliefs concerning the possible outcomes resulting from selection of the various design alternatives and their likelihood of occurrence during these gate reviews. It should be noted that these decision-makers already hold some beliefs at the outset of the VDP based on their prior knowledge and experience. This prior information is then supplemented both by the information generated throughout the P-VDP collected and the outcomes of decisions made throughout the course of the P-VDP. The Informational Phase is represented by the large iteration block. In this phase, the decision-makers determine whether to gather additional information or to act based on the information available to them and on their tolerance for risk. This cycle is repeated until the decision-makers are prepared to commit to a course of action; the necessary decisions are then made and execution of the P-VDP proceeds to the next phase.

While it is definitely encouraging to note that many of the behaviors and methods prescribed by the disciplines of formal decision analysis and DBD may be observed in leading-edge product development processes such as the P-VDP, there are definitely opportunities to achieve a more ideal execution of the product development process. The current state of deployment of tools and methods from the discipline of DBD and the opportunities for future research are reviewed for each phase of the Decision Analysis Cycle. For an illustration of the integrated application of these tools and methods, please refer to the door seal design example presented in [12].

19.4.2 The Deterministic Phase

The Deterministic Phase is focused primarily on two activities. The first is the generation of design alternatives. Some excellent methods for accomplishing this task are presented in [13] and [14]. The second is quantification of the complex relationships between the decisions to be made (selection of the design alternatives to be used in the product) and the value (profit). A very simplified representation of these relationships is shown in Fig. 19.8. In reality, these relationships are much more complex; indeed, managing this complexity is one of the foremost challenges in VDP execution. Fundamentally, there are two approaches to managing this complexity. One approach is to decompose the product development system into smaller units; the other is to retain the interconnectivity within the product development system with a fully integrated cross-functional design environment. A considerable amount of research has been conducted on the development and application of each approach. In practice, both approaches have been successfully applied to reduce the burden of complexity in product development.

The product development system may be decomposed into smaller units by using product architecture, organizational structure or tasks in the product development process as the basis for the decomposition. Each of these types of decomposition may be performed using the Design Structure Matrix as demonstrated in [2–5]. Alternatively, a product may be decomposed according to its function structure using the methods developed by Stone et al. [15]. These methods have proven to be quite powerful for decomposing a system into lower-level units while maintaining relationships between them. For example, when a vehicle is decomposed into subsystems, these methods are very effective for maintaining the necessary interfaces between the subsystems. Application of decomposition methods in a decision-based environment, however, poses the additional challenge of decomposing the enterprise value into a set of specific objectives for each unit, such that all of the units may work in parallel to achieve a common enterprise-level goal. This decomposition can simplify decision-making by decentralizing it; however, it can also lead to suboptimal results if there is insufficient coordination of actions between the units. There is no guarantee that the objectives given to each unit will lead to the greatest benefit for the enterprise. Therefore, if the enterprise value is to be decomposed into lower-level objectives, mechanisms must be put in place for rebalancing the objectives allocated to each of the units. Kim et al. [16, 17] have made substantial progress toward developing these rebalancing methods, but this remains a very challenging problem.

The alternative approach is to retain the interconnectivity within the product development system using a fully integrated cross-functional design environment. The benefit of this approach is that decisions are consistently made in the best interest of the enterprise. Its drawback is that it requires a set of analysis tools from all of the product development disciplines, including engineering, manufacturing, marketing and finance, which are compatible with one another both in terms of their level of detail and their flow of information. While it can be challenging to conduct these analyses just by assembling a team of cross-functional experts to provide their judgments, recent research suggests that it is now possible to conduct these analyses in a model-based environment using a combination of well-established and emerging technologies.

A number of methods have been developed for linking the engineering and business domains by estimating the impacts of engineering designs on cost and revenue. The procedures published by Dello Russo et al. [18] are broadly applicable for estimating costs of new designs and the technical cost modeling process presented by Kirchain [19] is particularly well suited to assessing the cost impacts of implementing new materials and manufacturing systems. Cook [20] has developed a suite of compact, transparent, low-overhead tools that may be exercised by product development teams with very little overhead to assess the impacts of design decisions on a product's marketplace performance. Chen [21] has applied discrete choice modeling, a more computationally expensive but very powerful technique, for estimating the effects of design decisions on a product's market demand. Application of these methods has been extended beyond the realm of research and into private-sector application, as discussed in [22] and [23]. Although in practice the linkages between the engineering and business disciplines are established manually, the development and application of multidisciplinary design frameworks, such as those in [24] and [25] is becoming more prevalent among practicing product development teams.

19.4.3 The Probabilistic Phase

In the Probabilistic Phase, decision-makers estimate and encode the uncertainties that have been identified as relevant to the design alternative decisions [9]. A key distinction in the Probabilistic Phase is that between variation and uncertainty. Variation is an inherent state of nature; the resulting uncertainty may not be controlled or reduced. Conceptually, variation is easy to incorporate in mathematical models using Monte Carlo simulation. For any number of random variables, the input distributions are sampled and used in the model to calculate an output. The aggregate of the outputs are used to form a statistical description. However, there are challenges in implementation. Monte Carlo simulation requires many evaluations of the mathematical model. If the mathematical models are very complex (for example, finite-element models with hundreds of thousands of elements used for analyzing vehicle structures), it is often prohibitively expensive or impossible to perform a Monte Carlo simulation within the time allotted for the analysis within the P-VDP.

In contrast, uncertainty is a potential deficiency due to a lack of knowledge. In general, this is very difficult to handle. Acquiring and processing additional information, perhaps by conducting experiments or by eliciting information from experts, may reduce uncertainty; however, it is often difficult to determine a priori which critical pieces of information need to be obtained. Thus it is necessary to undertake an iterative process of discovery to reduce uncertainty. It may also be necessary to allocate resources (e.g. people, time, money) to reduce uncertainty, and these allocation decisions are generally difficult due to the resource constraints faced by product development teams.

A growing body of formal methodologies is available to aid the decision-makers in identifying key uncertainties and in mitigating the risks associated with them. The Design For Six Sigma (DFSS) methodology [26] is frequently applied to this end. DFSS may be applied to identify, prioritize and monitor both the product attributes critical to satisfying the customers' expectations as well as the uncertainties that pose the greatest challenges in consistently delivering them. Robust Design [27] principles may then be applied to develop subsystem and component designs that are less sensitive to these uncertainties, increasing the likelihood that the designs will satisfy customers' expectations.

The results generated using these methodologies (or, for that matter, the results generated using any mathematical model, whether deterministic or nondeterministic) are used to augment the decision-maker's state of information. However, they are not a direct substitute for the belief-based probability assessments that are used by the decision-maker as the basis for the decision. A decision-maker's beliefs about these uncertainties are ultimately based on his general state of information, which is augmented by new information presented in every gate review. For an interesting discussion of how the results of an analytical model affect a decision-maker's beliefs, see [12].

Arguably, the greatest remaining challenge in the Probabilistic Phase lies in training the product development community to explicitly manage uncertainty in the product development process. From observation of teams in preliminary vehicle development, it is clear that the decision-makers understand that their decisions are made under uncertainty and risk and that work in the P-VDP should be focused on reducing uncertainty. However, these uncertainties and risks are typically expressed implicitly or in vague semantic terms such as "high confidence." Training all product development professionals to express their beliefs explicitly and precisely using a formal mathematical language of uncertainty will result in higher quality of execution of the probabilistic phase by removing ambiguity and focusing the team on exploring only those uncertainties that are material to the decisions at hand. Perhaps more fundamentally, product development professionals must accept uncertainty as an inherent part of the product development process, whether or not it may be conveniently represented and modeled and whether or not it may be reduced. Generally speaking, both engineers and decision-makers are more comfortable with uncertainties that are easily quantified (such as variation in manufacturing) or easily reduced (such as performance attributes that may be estimated through analysis or testing); however, these are not necessarily the uncertainties that dominate the decision. Engineers and decision-makers alike must recognize all of the uncertainties relevant to their decisions in the decision-making process (including those that are difficult to represent or difficult to reduce). They must also recognize that uncertainty cannot be completely eliminated and they must grow comfortable making decisions in the face of this uncertainty.

19.4.4 The Informational Phase

In the Informational Phase, the decision-maker must decide either to continue collecting information or to commit to a course of action. It is typical for decision-makers to make several requests for additional information before arriving at a decision. In the P-VDP DSM model, this may be observed as the large iteration blocks surrounding the gate reviews in the process. The challenges for improving execution of the Information Phase are similar to those faced in the Probabilistic Phase. In both, the primary challenge is leading the product development team to understand and accept the concept of epistemic uncertainty, the type of uncertainty that cannot be reduced and that will always be present in decision-making. Teams that do not accept this reality are vulnerable to being trapped in the informational phase, performing ever more analyses and tests in a futile attempt to reduce uncertainties that can be reduced no further.

An even greater challenge in the Information Phase is leading the product development team to understand and accept the concept of the Value of Information. As discussed in [9], for any exercise conducted to reduce uncertainty there is a quantifiable cost and a corresponding quantifiable economic value of reducing the uncertainty. This economic value is measured in terms of the decision-maker's "value" (in this case, a measure of profit) and is determined by propagating the effects of selecting the various alternatives through the network of uncertainties in the decision diagram to the decision-maker's value. Applying the concept of "Value of Information," it is prudent for a decision-maker to request additional information to reduce uncertainty when this reduction in uncertainty may lead to the selection of a different course of action or when it leads to a change in the value that exceeds the cost of collecting the additional information.

19.4.5 The Decision

The Decision Analysis Cycle concludes once the decision-makers have gathered sufficient information and are willing to commit to a course of action. In the P-VDP, examples of decisions would be selection of the design alternatives to be developed and authorization of necessary changes to the manufacturing system. In keeping with the previously stated definition of a decision, the decision-makers must irrevocably allocate resources as the result of their actions. Thus, the selection of a design alternative would occur in a binding fashion by signing a letter of intent or a contract with a supplier. Likewise, authorization of changes to the manufacturing system would include signing documents authorizing the exchange of funds from the vehicle program's budget to the manufacturing facility. Until these actions have occurred, no decisions have been made.

While many decisions in product development are made predominantly based on the decision-makers' intuition, both awareness and application of structured decision-making methods are increasing. Application of formal Decision Analysis methods [9] are generally limited to strategic, corporate-level decisions, whereas lower-level decisions over product content and manufacturing process changes are frequently made using simpler, more intuitive methods such as Pugh's selection method [28] or Saaty's Analytic Hierarchy Process [29]. While these methods can bring clarity to otherwise confusing decisions, practitioners must take care to avoid known pitfalls when applying them. Lewis [30] presents an excellent discussion of the limitations of these methods and proposes a more mathematically robust alternative. Saari [31] has pursued a similar course of action for social choice methods and Scott and Antonsson [32] have done the same for mathematical methods of preference aggregation. This research is of great practical significance because its results can be applied immediately to improve decision-making in product development teams.

In addition to application of structured methods, decision-making in product development may also be improved through a cultural shift from focusing on good outcomes to focusing on good decisions: "A good decision is based on the information, values and preferences of a decision-maker. A good outcome is one that is favorably regarded by a decision-maker. It is possible to have good decisions produce either good or bad outcomes. Most persons follow logical decision procedures because they believe that these procedures, speaking loosely, produce the best chance of obtaining good outcomes ." [9]

This distinction is critical to the idea of a true learning organization. Recrimination based on undesirable outcomes is a natural behavior, but it must be avoided because it hampers the long-term effectiveness of the product development organization by reinforcing a culture of fear and excessive risk aversion. The types of questions that should be asked when evaluating the effectiveness of a decision are "Was all of the available information used?" or "Was sound logic used when the decision was made?" or "Were the preferences of the decision-maker properly aligned with the organization's overall business goals?" These questions promote a healthy attitude toward decision-making under uncertainty and risk. The decisions in the product development process will be made under uncertainty and risk regardless of how the uncertainty is characterized and managed; thus the attitude toward decision-making must be shifted from avoiding risk and focusing on outcomes to confronting risk directly with the best available methods and concentrating on the quality of execution of the decision-making process.

19.5 CONCLUSIONS

The notion that the product development process is a series of tasks has been firmly established; however, there is a growing body of evidence supporting the notion that the product development process is also a series of decisions. This research shows that these two views of the product development process are consistent with one another and that they complement one another. Certainly some of the iteration in the product development processes is attributable to the complexities of the interactions between the product's components, the product development team members and the tasks in the product development process. However, much of the iteration observed in the product development process, particularly the iteration surrounding high-impact decisions, is a natural manifestation of the Decision Analysis Cycle—an iterative process of generating alternatives and forming beliefs about the possible consequences of selecting them, culminating in selection of one of the alternatives under uncertainty and risk.

The distinction between these two types of iteration is crucial because the means of addressing these two different types of iteration are fundamentally different. Iteration due to complexity may be reduced by applying logistical tools or formal decomposition methods. However, reduction of iteration in the Decision Analysis Cycle requires a new suite of tools from the DBD community. Research into the development of nondeterministic simulation and optimization methods, the integration of interdisciplinary tools into enterprise-wide design integration frameworks, and the exploration of the relative merits and limitations of available decision support methods promotes healthy execution of the Decision Analysis Cycle in product development. In addition, a healthy attitude toward uncertainty and risk also improves the quality of execution of the Decision Analysis Cycle. This extends beyond learning to communicate objectively about uncertainty using a formal mathematical language to accepting the simple truth that uncertainty and risk are inherent parts of every decision made in the product development process. Adoption of these methods and attitudes will not eliminate iteration in the product development process, but it will ensure that the Decision Analysis Cycle, although iterative by nature, is a value-added process that produced value-focused results. Indeed, the developers of DBD may claim success when product development projects are no longer measured in terms of the magnitude of the resources they consume, but in terms of the amount of value that they have created.

ACKNOWLEDGMENTS

The author gratefully acknowledges the host of researchers and product development professionals whose collaboration has made this chapter possible. The P-VDP DSM model was developed collaboratively with Prof. Steven Eppinger, Dr. Daniel Whitney,

Dr. Ali Yassine, Dr. William Finch and Alberto Cividanes of the MIT Center for Innovation in Product Development and with Dr. Devadatta Kulkarni and Dr. Alexandra Elnick of the GM Research & Development Center. Applications of Decision Analytic principles to the P-VDP were investigated with Prof. Ronald Howard, Prof. Ross Shachter, Ali Abbas, Apiruk Detwarasiti, Christopher Han and John Feland of the GM—Stanford Collaborative Research Laboratory in Work Systems and with Dr. John Cafeo and Dr. Robert Lust of the GM Research & Development Center. Finally, special thanks to Prof. Kemper Lewis, Prof. Wei Chen and Prof. Linda Schmidt for sparking the interest in studying the design process as a sequence of decisions, as well as to Prof. Jeffrey Herrmann for providing the impetus to formally integrate the task-based and decision-based views of the P-VDP.

REFERENCES

1. Clark, K. and Fujimoto, T., 1991. *Product Development Performance—Strategy, Organization, and Management in the World Auto Industry*, Harvard Business School Press, Boston, MA.
2. Eppinger, S. et al., 1994. "A Model-Based Method for Organizing Tasks in Product Development," *Res. in Engrg. Des.*, 6 (1), pp. 1–13.
3. Sosa, M., Eppinger, S. and Rowles, C., 2003. "Identifying Modular and Integrative Systems and Their Impact on Design Team Interactions," *ASME J. of Mech. Des.* Vol. 125, pp. 240–252.
4. Eppinger, S., 2001. "Innovation at the Speed of Information," *Harvard Bus. Rev.,* 79 (1), pp. 149-158.
5. Pimmler, T. and Eppinger, S., 1994. "Integration Analysis of Product Decompositions," *Proc., ASME Des. Engrg. Tech. Conf.*, Minneapolis, MN.
6. Cividanes, A., 2002. "*Case Study: A Phase-Gate Product Development Process at General Motors*," S.M. thesis, Massachusetts Institute of Technology, Cambridge, MA.
7. Herrmann, J. and Schmidt, L., 2005. "Viewing Product Development as a Decision Production System," *Decision Making in Engineering Design*, ASME Press, New York, NY, Chapter 20.
8. Carey, H., et al., 2002. "Corporate Decision-Making and Part Differentiation: A Model of Customer-Driven Strategic Planning," *Proc., ASME Des. Engrg. Tech., Conf.*, Montreal, Canada.
9. Matheson, J. E. and Howard, R. A., 1989. "An Introduction to Decision Analysis," *Readings on the Principles and Applications of Decision Analysis,* Strategic Decisions Group, Palo Alto, CA.
10. Hazelrigg, G. A., 2001. "The Cheshire Cat on Engineering Design," *ASME J. of Mech. Des.*
11. Nikolaidis, E., 2005. "Types of Uncertainty in Design Decision Making," *Engineering Design Reliability Handbook,* CRC Press, Boca Raton, FL, Chapter 8.
12. Cafeo, J. et al., 2005. "The Need for Nondeterministic Approaches in Automotive Design: A Business Perspective," *Engineering Design Reliability Handbook*, CRC Press, Boca Raton, FL, Chapter 5.
13. Keeney, R., 2005. "Stimulating Creative Design Alternatives Using Customer Values," *Decision Making in Engineering Design*", ASME Press, New York, NY, Chapter 7.
14. Otto, K. and Wood, K., 2000. "Generating Concepts," *Product Design*, Prentice-Hall, Upper Saddle Hill, NJ, Chapter 10.
15. Stone, R. et al., 2000. "Product Architecture," *Product Design*, Prentice-Hall, Upper Saddle Hill, NJ, Chapter 9.
16. Kim, H., et al., 2003. "Analytical Target Cascading in Automotive Design," *ASME J. of Mech. Des.*, Vol. 125, pp. 474–480.
17. Kim, H. et al., 2003. "Target Cascading in Optimal System Design," *ASME J. of Mech. Des.*, Vol. 125, pp. 481–489.
18. Dello Russo, F., Garvey, P. and Hulkower, N., 1998. "Cost analysis," *Tech. Paper No. MP 98B0000081,* MITRE Corporation, Bedford, MA.
19. Kirchain, R., 2001. "Cost Modeling of Materials and Manufacturing Processes," *Encyclopedia of Materials: Science and Technology,* Elsevier, New York, NY, pp. 1718–1727.
20. Cook, H., 2005. "The Role of Demand Modeling in Product Planning," *Decision-Making in Engineering Design*, ASME Press, New York, NY, Chapter 11.
21. Chen, W., 2005. "Discrete Choice Demand Modeling for Decision-Based Design," *Decision-Making in Engineering Design,* ASME Press, New York, NY, Chapter 9.
22. Donndelinger, J. and Cook, H., 1997. "Methods for Analyzing the Value of Automobiles," *Proc., SAE Int. Cong. and Exposition*, Detroit, MI.
23. Wassenaar, H. et al., 2004, "Demand Analysis for Decision-Based Design of Automotive Engine," *Proc., SAE Int. Cong. and Exposition*, Detroit, MI.
24. Fenyes, P., Donndelinger, J. and Bourassa, J.-F., 2002. "A New System for Multidisciplinary Analysis and Optimization of Vehicle Architectures," *Proc., 9th AIAA/ISSMO Symp. on Multidisciplinary Analysis and Optimization*, Atlanta, GA.
25. Wassenaar, H. and Chen, W., 2003. "An Approach to Decision Based Design with Discrete Choice Analysis for Demand Modeling," *ASME J. of Mech. Des.*, Vol. 125, pp. 490–497.
26. Chowdhury, S., 2002. *Design for Six Sigma: The Revolutionary Process for Achieving Extraordinary Profits*, Dearborn Trade Publishing, Chicago, IL.
27. Phadke, M., 1989. *Quality Engineering Using Robust Design*, Prentice Hall, Englewood Cliffs, NJ.
28. Pugh, S., 1991. *Total Design: Integrated Methods for Successful Product Engineering*, Addison-Wesley, Reading, MA.
29. Saaty, T., 1980. *The Analytic Hierarchy Process*, McGraw-Hill, New York, NY.
30. Lewis, K., 2005. "Multiattribute Decision-Making Using Hypothetical Equivalents and Inequivalents," *Decision-Making in Engineering Design*, ASME Press, New York, NY, Chapter 14.
31. Saari, D., 2005. "Fundamentals and Implications of Decision-Making," *Decision Making in Engineering Design*, ASME Press, New York, NY, Chapter 4.
32. Scott, M. and Antonsson, E., 1999. "Arrow's Theorem and Engineering Design Decision Making," *Res. in Engrg. Des.*, 11(4), pp. 218–228.

PROBLEMS

19.1 Under what circumstances would engineering design be an iterative process? Under what circumstances would it be a serial process?

19.2 What are some of the high-impact decisions in vehicle development that would be brought to an executive decision panel? What kinds of decisions would be made locally by engineers?

19.3 Consider the decision diagram in Fig. 19.8:

a. What other decisions could be added to this diagram?
b. What other uncertainties might be relevant to these decisions?
c. Would decision-makers consider any additional values besides profit in vehicle development decisions?

19.4 What types of uncertainties in the vehicle development process are reducible? What types are not?

19.5 Using the decision diagram in Fig. 19.8, illustrate how two different decision panels could make two different selections given the same set of design alternatives.

CHAPTER

20

PRODUCT DEVELOPMENT AND DECISION PRODUCTION SYSTEMS

Jeffrey W. Herrmann and Linda C. Schmidt

20.1 INTRODUCTION

A product development organization faces many challenges as it strives to become a market leader. Goals of attaining superior quality, rapid time-to-market and low cost are often thwarted by poor decisions based on incomplete or missing information. A company's existing product portfolio will become less profitable over time due to changes in manufacturing technology, consumer preferences and government regulations (to name a few influences). The sales volume graph of a typical product starts slowly, accelerates, becomes flat and then steadily declines. Although there may be a few products that remain profitable for many years, vibrant firms must continually develop new products that will generate more profits.

Product development is a complex and lengthy process of identifying a need, designing, manufacturing and delivering a solution—often in the form of a physical object—to the end-user. Product development is a difficult task made more difficult by the challenges inherent in complex, open-ended and ill-defined tasks. A successful product development process incorporates information inputs from seemingly unrelated and remote areas of an organization into the decision-making process [27].

A *product development organization* includes the accountants, purchasing agents, marketing specialists, customer liaisons, data managers, engineers, managers and other personnel who make process and product engineering decisions and perform these activities. (Note: In this discussion, the term *new product* covers the redesign of an existing product as well.)

Economists have proposed a model for the decision-making operations of businesses, like product development organizations, intent on making a profit in a market economy. It is called the theory of "the firm," and it suggests that businesses make decisions on production (both product mix and production levels) based on the goal of maximizing net revenue. This optimal production position is a point of equilibrium between the firm and the markets in which it operates. The firm predicts revenue according to the principles of a perfect market economy, known conditions of production (past, present and future), as well as predictable reaction of consumers and competitors (the market) to changes in the firm's production mix.

The theory of the firm depicts the firm as an organization that is acting with a single motivation under conditions of perfect knowledge. The idealized theory of the firm has been modified by modern economists, organizational theorists, social psychologists, social behavior scientists, political theorists and administrative theorists, to name some of the fields of specialists that study the operation of a business enterprise. The firm is an entity whose operation will continue to be studied as long as it exists. One thing we can say with certainty is that the idealized firm is actually a group of people making real-time decisions based on available information. That makes this one of the messiest problems in the world.

20.1.1 Overview of Product Development

Product development is a necessary and important part of the activities performed by an organization that seeks profit by delivering an artifact or service to some part of the buying public. Product development determines what the firm will manufacture and sell. That is, it attempts to design products that customers will buy and to design manufacturing processes that meet customer demand profitably. Fundamentally, then, a product development organization transforms information about the world (e.g., technology, preferences and regulations) into information about products and processes that will generate profits for the firm. It performs this transformation through the decision-making of individuals working with specialists. Poor decisions during product development lead to products that no one wants to buy and products that are expensive to manufacture in sufficient quantity.

A *product development process* is the set of activities needed to bring a new product to market. For a variety of reasons (which will be discussed in Section 20.3), product development teams decompose the problem of finding a profitable design into a product development process, which provides the mechanisms for linking a series of design decisions that do not explicitly consider profit. A complete product development process includes product design, engineering design, process design (both manufacturing and assembly) and production planning.

A *product development project* is the set of actual activities that are performed during the development of a specific new product. Typically, the product development project will follow the product development process, but the project will deviate from the process as circumstances warrant, just as the actual operation of an airline deviates from the published schedule as bad weather, equipment failures and other unexpected events occur.

The following nine steps are the primary activities that many product development teams accomplish [24]:

Step 1. Identify the customer needs.
Step 2. Establish the product specification.
Step 3. Define alternative concepts for a design that meets the specification.
Step 4. Select the most suitable concept.
Step 5. Design the subsystems and integrate them.
Step 6. Build and test a prototype; modify the design as required.
Step 7. Design and build the tooling for production.
Step 8. Produce and distribute the product.
Step 9. Track the product during its life cycle to determine its strengths and weaknesses.

This list (or any other description that uses a different number of steps) is an extremely abstract depiction that not only conveys the scope of the process, but highlights the inherent (but unquestioned) decomposition. Product design (Steps 1 and 2) and engineering design (Steps 3 to 6) are the dominant activities in the early stages of a product development process.

Product design is the first step in the product development process. Product design starts with looking at the social, economic and technological factors in the world today, and using those to identify an opportunity. This effort requires (or, at least, is aided by) interdisciplinary work to fully understand the marketing, engineering and design aspects that go into a product. Once an opportunity is identified, understanding the target market and end-user, as well as the stakeholders involved, is needed. Throughout the product development process, supervisors are continuously looking back to the stakeholders for feedback on the emerging design.

Engineering design begins as the needs and wants of the user are translated into specific opportunities for new or improved products. The engineering of a product is the articulation of the performance criteria for the product and the description of a (usually) physical artifact that will provide the performance. These criteria are translated into concepts and further refined into a complete detailed design.

Product development requires a wide variety of decisions (as discussed in Section 2). Because making good decisions requires expertise and an organization of people can be experts in only a few things, a manufacturing firm specializes in a certain class of products. It focuses its attention on the market for that class of products, the technologies available to produce that class, and the regulations relevant to that class.

Important product development information can come from a wide variety of individuals separated by their duties in the organization. For example, a warranty unit may have data about an unanticipated battery defect, while the vice president for environmental affairs has a report on pending legislation that will require battery recycling in some European markets. Many other factors effect the passage of information to and within an organization [1]. For example, the status of an individual within a unit determines, to some extent, the scope of influence of new data introduced by that person. A successful product development process is one that can unite the efforts of these individuals to ensure that the organization develops profitable products.

20.1.2 A Decision-Making View of Organizations

Before looking at product development organizations more closely, it may be useful to discuss the importance of decision-making in organizations of any kind.

Organizations are decision-making systems. Simon [27], who describes decision-making systems in the administration of manufacturing firms and government agencies, defines an organization as "the pattern of communications and relations among a group of human beings, including the process for making and implementing decisions." Here Simon acknowledges that the single entity "firm" view of a business enterprise is not as accurate a model for understanding complex decisions. As producing goods becomes a smaller part of the economy and organizations become more aware of the indirect consequences of their activity, the decision-making process in an organization becomes more and more important. Of course, each organization is unique. Different organizations have different structures and decision-making and information flow patterns due to the differences in their history, environment and objectives.

Decision-making systems often go through three stages of evolution [17]. In the first stage, when an organization is small, skilled managers make decisions as situations arise. In the second stage, the complexity of the operations increases and the firm installs a system of decision-making. For routine decisions, heuristics or simple rules guide decision-making. (General heuristics will be

Exhibit A

Simon [27] provides an excellent example of a composite decision: the design and production of a new battleship for the British Navy. We shall briefly summarize the process here.

1. The First Sea Lord and the Assistant Chief of Naval Staff determine the features that the ship should have (speed, radius of action, offensive qualities, armor protection).
2. The Director of Naval Construction and the Controller develop ideas for the ship and estimate the size and cost of these.
3. The Sea Lords select one of these alternatives.
4. The Director of Naval Construction determines the ship's approximate dimensions and shape.
5. The Engineer-in-Chief arranges the equipment needed to move the ship while the Director of Naval Ordnance determines the positions of the weapons systems.
6. The Director of Torpedos designs the torpedo armament, and the Director of Electrical Engineering designs the electrical machinery, lighting and other systems.
7. The Director of Naval Equipment decides on the boats that the ship will carry and the anchors and cables; the Director of the Signal Department designs the communications equipment; the Director of Navigation designs the means for navigating the ship; and other groups design other parts of the ship (altogether, 14 departments are involved).
8. When conflicts appear between different systems, the directors meet, discuss the problems and agree on compromises.
9. The Board approves the completed design.

presented in Section 20.4.2). In the third stage, the firm seeks to improve decision-making by implementing decision support tools (discussed in Section 20.2.5).

A key feature of a decision-making system is the recognition that many decisions are "composite decisions" that require many other decisions to be made (see Exhibit A). Decisions are based on information provided by others in the organization. A decision, once made, influences future decisions to some degree. The decisions are interconnected in a complex web. Top-down communication is one form of the information flow, as a supervisor gives policies and directions to subordinates. Other flows are also important for getting the relevant facts to a decision-maker, as in gathering intelligence from where operations are occurring (on the factory floor or on the battlefield). One decision leads to information that is used to make another and so forth, until each decision is an assembly of many decisions. The decisions made by specialized units will reflect the more general goals of the organization through schedules, systems of rewards and penalties, and limits. The decisions of the unit are loosely coupled with the decisions of the organization through these mechanisms. For instance, profit maximization is one goal of a manufacturing firm, but that is not considered explicitly in many decisions.

20.1.3 A Decision Production System View of Product Development

Product development organizations, like other organizations, are decision-making systems.

The simplest decision-making task is to select the best option from a set of alternatives that is given to decision-makers when each alternative is completely defined and precise predictions of the results from choosing any alternative are also known. This scenario rarely, if ever, exists in product development. Yet, we know that products are designed and corporations who market them often survive and sometimes thrive. Developing a method for studying decision-making practices in an organization will identify good practices and create more opportunities for success.

Herrmann and Schmidt [16] introduced the term *decision production system* to describe an information flow governed by decision-makers who operate under time and budget constraints to produce new information. The term is relevant because a product development organization creates new product designs and other information that are the accumulated results of complex sequences of decisions.

Because decision-making requires information, generates information and determines who gets what information, employees on different hierarchical levels will be exchanging information at different points in the product development process. Yet all are involved in the processing of information and knowledge at the same level. In this way they resemble operators on the same shop floor. Section 20.7 will elaborate on this similarity.

As will be discussed in Section 20.6, because decisions are often composite decisions, some participants in the decision-making process make decisions and some do not. A decision-maker gets some information, makes a decision, and consequently generates new information. Part of the "makes a decision" step may involve exchanging information with others. For example, as part of a test process, a designer in New Product Development sends a solid model of a component to the Life-Cycle Testing unit, where a finite-element analysis expert determines how the part will behave and returns a report to the designer.

The decision production system (DPS) perspective looks at the organization in which the product development process exists and considers the decision-makers and their information processing tools (like databases) as units of a manufacturing system that can be viewed separately from the organization structure. As a result the hierarchical view and decision production system view of a product development organization are quite different. This was observed by Simon [27], who notes that an organization's "anatomy" for information processing and decision-making is naturally different from the departmentalization displayed in an organization's chart. The greater the interdependence between decision-makers, the less the DPS will resemble an organization chart.

The decision production system perspective (first described in [16]) is a unique way to view a product development organization. This perspective views product development organizations as a network of decision-makers, information processors, knowledge repositories (e.g, handbooks, websites, industry catalogs) and interactive databases (e.g., product data management systems) through which information flows. By viewing organizations in this manner, one can understand how information flows and who is making the key decisions.

The DPS perspective is an overarching framework to map product development activities (with an emphasis on decisions) within an organization in such a way as to illustrate current decision-making practices. The DPS representation of a product development organization provides a meta-level view of the actual decision-making processes taking place in an organization, which are not necessarily the processes that management may have prescribed. The DPS perspective enables problem identification in decision-making practices that will lead to a more effective deployment of resources, including decision support tools.

The DPS perspective enables a deeper understanding of the organization than typical hierarchical organization charts of a firm or Gantt charts of product development projects. Understanding the real process (as opposed to the corporate guide for the design process) is a key step in improving product development. Furthermore, recognizing a designer as a "knowledge agent" and the designing activity as a crucial organizational knowledge process can improve an organization's ability to innovate within their competitive environment [2]. The need for research on new work practices [3] and the need for developing new representation schemes for product development [20] are additional motivations for considering the DPS perspective.

20.1.4 How this Chapter is Organized

The remainder of this chapter attempts to answer the following questions:

- How do product development organizations make decisions? (Section 20.2)
- Why do they do it that way? (Section 20.3)
- Is it rational? (Section 20.4)
- What problems exist? (Section 20.5)
- How can we improve these decision-making systems? (Section 20.6)

Section 20.7 concludes the chapter.

20.2 DECISION-MAKING IN PRODUCT DEVELOPMENT ORGANIZATIONS

This section starts by considering the organization itself and then discusses the role of teams, the types of decisions that are made, the objective of profitability and the types of decision support tools used.

Exhibit B: Design and Management Decisions

Kidder [19] describes the development of a mini computer by Data General. As part of the narration, the book describes many of the decisions that the development team made during the computer's development. The lists below highlight some of those decisions. (Many of the design decisions that were made during development are not listed because either the book did not describe them or describing them would require too much room.) Each item in the lists describes the decision made and who made it. References are to the pages in the book where the decision is described. Both types of decisions occur at different levels in the organization structure. Higher management decisions effect more people and more of the process, while higher design decisions effect more of the computer.

Selected Management Decisions

1. The Vice President of engineering approved the project (page 47).
2. West decided to hire inexperienced engineers who had just graduated (page 59).
3. West decided to have two teams: one for designing the hardware, one for designing the microcode (pages 59, 105).
4. West decided that Wallach should be the architect (page 68).
5. Wallach decided to begin designing the architecture by organizing the memory (page 76).
6. West reviewed the designs (page 119).
7. Rasala created the debugging schedule (pages 130, 145).
8. West approved using microdiagnostic programs (page 134).
9. West approved building a simulator for testing microcode (page 161).
10. Alsing picked Dave Peck and Meal Firth to write simulators (page 163).
11. West decided who would work on which new projects (page 232).
12. Rasala decided to work in the lab to increase morale (page 256).

Selected Design Decisions

1. West decided that the new computer should be a 32-bit computer that can run older programs written for another computer (page 42).
2. Wallach decided to worry about preventing accidental damage, not malicious theft (page 78).
3. Wallach decided that the memory protection scheme should use the segment number as the security level (page 80).
4. Wallach defined the instruction set (page 83).
5. Engineers negotiated the design details (page 116, 159).
6. West decided that the computer would use PAL integrated circuits (pages 118, 121, 268).
7. The engineers wrote the microcode and the schematics (page 121).
8. Holland organized the microcode (page 158).
9. West and Rasala decided to keep the ALU on one board by limiting its functionality (pages 213, 255).
10. West decided which cables and connectors the computer should use (page 230).
11. West decided how the machine should be started (page 230).

costs and sales (changes that result from design decisions) effect profitability. For instance, delaying the time-to-market may cost the firm $5,700 per day. In addition to helping a product development team make good decisions, the rules remind the team members that their actions effect the firm's profitability.

Some firms have developed models for understanding the impact that high-level decisions have on profitability. Stonebraker [31] describes a decision analysis model that Bayer uses to decide whether or not to start development of a drug. This model evaluates the net present value of a drug development project and describes the data collection and influence diagram used in the evaluation. Although this was not linked directly to the design decisions, it does give a good example of evaluating profitability in the presence of uncertainty. The amount of effort required to create a model to evaluate a small number of decisions shows that larger models that evaluate the impact of many decisions will require much more effort.

20.2.5 Decision Support Tools

Decision support tools are common in product development organizations. These tools range from spreadsheet algorithms predicting product performance, to extensive prototype building and testing under the guidelines of a Taguchi-style designed experiment, to programs for in-field testing of specially hardwired products to understand the use cycle. Computer-aided design systems are also a type of decision support tool because they allow search over a portfolio of product information. Product data management systems are an important repository for information that results from design decisions and is used to make future decisions.

The goal of some decision support tools is to help decision-makers treat problems in a more integrated fashion. In particular, some researchers (following [14]), have begun studying design optimization problems that integrate customer preferences, revenue and profits. See, for instance, [21, 34] as well as other chapters in this book.

20.3 DECOMPOSITION

Product development organizations seek to make good decisions. In practice, they decompose a design problem into a series of subproblems, and design engineers and other members of the team must try to satisfy a variety of constraints and make tradeoffs between multiple competing objectives. The primary question that must be raised is why is decomposition popular?

Product development, if viewed as an optimization problem ("find the most profitable product"), is an extremely large and complex puzzle. Decomposition is a natural strategy for attacking large problems. Cognitive limitations force human decision-makers to decompose problems into subproblems. Many writers have documented these cognitive limitations (see [26], for example).

Decomposition occurs in a wide variety of problem domains. In manufacturing facilities, the manufacturing planning and control systems are decomposed into modules that solve a variety of problems that range from aggregate production planning to master production scheduling and material requirements planning as well as down-to-detailed-shop-floor scheduling. Closely related to product development processes are the typical systems engineering processes, including top-down and bottom-up approaches, all of which involve some kind of decomposition.

Related to this issue is the question of constraints. Constraints exclude solutions that are infeasible with respect to one or more of the many different conditions that a successful design must satisfy. It must be understood, however, that their role is to reduce the search effort. If the objective is to maximize profit, one can formulate a design problem with no constraints. In this approach, the evaluation of profit must penalize any unreasonable solution. For instance, if the power tool is too heavy, few customers will buy the tool and sales and profit will be low. While theoretically possible, this approach clearly results in a huge search space and a complex objective function. Thus, the computational effort will be extremely large.

By contrast, including constraints (such as an upper bound on weight) limits the search space and simplifies the objective function, which makes solving the problem much easier. Hazelrigg [13] gives additional examples of constraints that simplify design problems (see Exhibit C).

Exhibit C: Examples of Constraints

Hazelrigg [13] argues that many constraints are design decisions made at higher levels. These constraints, which make the resulting design problem easier, are actually proxies for system-level objectives. Hazelrigg provides the following examples:

1. **Computing the optimal trajectory of a spacecraft that is visiting another planet:** Constraint: *Avoid crashing into the surface of the planet.* This constraint helps maximize the value of the scientific data that the mission produces.
2. **Designing an aircraft autopilot for automated landings:** Constraint: *The maximum acceleration should be within certain bounds.* This constraint helps the autopilot achieve a safe and comfortable landing.
3. **Designing a signal processing algorithm for TV image compression:** Constraint: *The probability of information loss in the reconstructed image should be essentially zero.* This constraint helps the algorithm produce a high-quality image.

20.4. RATIONALITY

A product development organization is a decision production system in which decision-makers must make decisions with limited information, with limited resources and under time constraints. This fact clearly indicates that design engineers will make decisions using heuristics. Idealized optimization and decision analysis techniques are usually infeasible in the real world. Does this mean that product development organizations are acting irrationally? Answering this question first requires considering the nature of rationality and then looking at the ways in which decisions are made in practice.

20.4.1 Perspectives on Rationality

There are different ways to view rationality. The dictionary defines rational as "consistent with or based on reason." Stirling [30] gives the following definition: "*A rational decision* is one that conforms either to a set of general principles that govern preferences or to a set of rules that govern behavior." There are different types of rationality, and decision-makers choose and use different types of rationality in different situations. The following discussion is based on [30].

Substantive Rationality: "Nothing but the best will do." The first type of rationality deals with principles about preferences. First, for all of the possible options, the decision-maker has a total ordering over them. Second, the decision-maker should choose the option that is most preferred. That is, the decision-maker optimizes. This total ordering can be quantified as a utility function U so that the decision-maker prefers A to B if and only if $U(\mathrm{A}) > U(\mathrm{B})$.

The selected option is a function of choosing this approach and the utility function. Most of the work involves collecting information and performing calculations to evaluate the utility function. (Of course, determining the utility function is not easy either.) Substantive rationality is the paradigm that guides formal decision analysis.

Unfortunately, this approach requires a complete understanding of the situation and extensive computational power to calculate the utility and find the optimal solution. Moreover, most decision-makers do not use this approach in practice.

Procedural Rationality: "It has always worked before." This type of rationality deals with decision-making processes. Procedural rationality implies that a decision-maker uses specific rules or procedures to make a choice. The procedures may be certain choice strategies (like those discussed below). A firm's product development process may describe a certain type of financial analysis for making a go/no-go decision.

Rules-of-thumb are a form of procedural rationality. For instance, when constructing a manufacturing system, one must decide how many persons or machine tools are needed for each step in the manufacturing process. Suppose that the system should produce 2000 parts per day, and each worker can produce 300 parts per day. The lower bound is 7 people. But this will lead to 95% utilization (2,000/2,100), which can cause large delays and congestion. A rule-of-thumb is to have enough capacity so that utilization equals approximately 85% (in this case, the capacity should be 2,000/0.85 = 2,352 parts per day, which requires 8 people). One could form an optimization procedure to trade off the cost of persons versus the cost of congestion (work-in-process inventory). But procedural rationality implies that the rule-of-thumb should be used.

Procedural rationality is context specific. The rules or procedures that make sense in one domain may be poor choices in another. Gigerenzer and Todd [10] expand on this in detail, showing that a fast and frugal heuristic like recognition is good in a certain domain. In general, choice strategies that don't use optimization may be more powerful in complex and messy problems. However, the rules may yield solutions with poor quality because there is no guarantee of optimality.

Bounded Rationality: "This is the best we could do in the time available." Bounded rationality starts with the observation that information and computational power (be it computers or people) are limited in the real world, and this prevents complete optimization. In this paradigm satisficing is seen as an appropriate strategy. Many of the choice strategies reflect this paradigm.

There are, however, two types of bounded rationality. In the first, the choice to stop searching (for information or alternatives) is viewed as part of a more comprehensive optimization problem.

This returns the decision-maker to substantive rationality, where the time or computational limits are part of the decision and the decision-maker needs to optimize the whole thing.

In the second, simple rules are used for stopping the search. For example, the "lexicographic" choice strategy that looks at one attribute and then another until only one alternative is left. The "satisficing" choice strategy stops as soon as it finds an acceptable alternative.

Intrinsic Rationality: "You get what you pay for." Intrinsic rationality looks at each alternative by itself and considers whether the expected benefits of the alternative exceed the expected losses. Ben Franklin's prudential algebra (described in Exhibit D) compares the pros and cons of a single alternative, which follows this approach. If the alternative is a net gain, then keep it, else discard it. Intrinsic rationality allows a decision-maker to create a set of (intrinsically) rational solutions (instead of just one optimal or satisfactory solution). The overall quality of the solutions may vary, since some will have small benefits (with small costs) and some will have large gains (with large costs).

(3) **Additive Difference:** Consider two alternatives at a time, compare attribute by attribute, estimating the difference between the two alternatives, and sum up the differences across the attributes to provide a single overall difference score. Carry the winner to the next viable alternative and make the same comparison. At the end of this process, the best alternative is the one that has "won" all the pairwise comparisons.

(4) **Satisficing:** Set "acceptability" cutoff points on all important attributes; then look for the first alternative that is at least as good as the cutoff values on all important attributes, or use the strategy to select a set of good-enough alternatives for further consideration.

(5) **Disjunctive:** Set "acceptability" cutoff points on all important attributes; then look for the first alternative that is at least as good as the cutoff values on any important attribute, or use the strategy to select a set of alternatives that are each very good on at least one dimension for further consideration.

Exhibit D: Franklin's Prudential Algebra

One of the classic examples of a decision-making process is found in the following letter from Benjamin Franklin to Joseph Prestly (this text found online at http://member.nifty.ne.jp/highway/dm/franklin.htm; see also [30]).

London, Sept. 19, 1772

Dear Sir,

In the affair of so much importance to you, wherein you ask my advice, I cannot, for want of sufficient premises, advise you what to determine, but if you please I will tell you how. When those difficult cases occur, they are difficult, chiefly because while we have them under consideration, all the reasons pro and con are not present to the mind at the same time; but sometimes one set present themselves, and at other times another, the first being out of sight. Hence the various purposes or inclinations that alternatively prevail, and the uncertainty that perplexes us. To get over this, my way is to divide half a sheet of paper by a line into two columns; writing over the one Pro, and over the other Con. Then, during three or four days consideration, I put down under the different heads short hints of the different motives, that at different times occur to me, for or against the measure. When I have thus got them all together in one view, I endeavor to estimate their respective weights; and where I find two, one on each side, that seem equal, I strike them both out. If I find a reason pro equal to some two reasons con, I strike out the three. If I judge some two reasons con, equal to three reasons pro, I strike out the five; and thus proceeding I find at length where the balance lies; and if, after a day or two of further consideration, nothing new that is of importance occurs on either side, I come to a determination accordingly. And, though the weight of the reasons cannot be taken with the precision of algebraic quantities, yet when each is thus considered, separately and comparatively, and the whole lies before me, I think I can judge better, and am less liable to make a rash step, and in fact I have found great advantage from this kind of equation, and what might be called moral or prudential algebra.

Wishing sincerely that you may determine for the best, I am ever, my dear friend, yours most affectionately.

B. Franklin

This type of reasoning is useful for selecting a small set of alternatives for an optimization approach or a set of alternatives from which a decision-maker can pick and choose (e.g., which new products should be developed?).

20.4.2 Decision-Making Strategies

There are numerous strategies used to make a decision, especially when there are many alternatives and multiple criteria on which to compare them. The following list and the accompanying descriptions are from [12]:

(1) **Dominance:** Search for an alternative that is at least as good as every other alternative on all important attributes and choose it, or find an alternative that is worse than any other alternative on all attributes and throw it out.

(2) **Additive Linear:** Weigh all the attributes by their importance. Then consider each alternative one at a time and calculate a global utility by valuing each attribute, weighing it by its importance and adding up the weighted values.

(6) **Lexicographic:** First, review the attributes and pick the most important one; then choose the best alternative on that attribute. If there are several "winners" on the first attribute, go on to the next most important attribute and pick the best remaining alternative(s) on that attribute. Repeat until only one alternative is left.

(7) **Elimination by Aspects:** Pick the first attribute that is salient and set a cutoff "acceptability" point on it. Throw out all alternatives that are below the cutoff on that one attribute. Then pick the next most attention-getting attribute, set an "acceptability" point on it and again throw out all alternatives that are below the cutoff. Repeat until only one alternative is left.

(8) **Recognition Heuristic:** Choose the first alternative that is recognized.

In general, these strategies provide a trade-off between the comprehensiveness of the search and the mental effort involved. All can be considered rational in one way or another, which accounts for their popularity. As mentioned before, strategies that make sense in one domain may be poor choices in another. In the area of product development, it is unlikely that successful decisions for one product will carry over into another market, which may have different end-users and competitors.

An alternative strategy for product development decision-making is to create the best set of processes for getting information and a robust decision-making infrastructure that will be effective for a wide range of decision-making. This strategy again promotes the meta-level view of an organization and its operation that are uncovered by creating a decision-production system model.

It is also worth mentioning that Benjamin Franklin, when asked for advice, did not pick an alternative but instead described a decision-making process (see Exhibit D).

20.5 PROBLEMS IN DECISION-MAKING

Although decomposition may be rational, that does not imply that product development organizations are behaving optimally. There is always room for improvement. This section will discuss problems with poor decision-making performance, the distance from corporate objectives, decision-making confusion, the absence of models and the cost of maintaining models.

20.5.1 Poor Decision-Making Performance

There are certain limits that are obstacles to efficient decision-making performance [27]:

- Decision-makers have a limited ability to perform their job because they don't know proper decision-making methods.
- Decision-makers have a limited ability to make correct decisions, due to conflicts between loyalties to the individual, the unit and the organization.
- Decision-makers have limited knowledge about the facts and considerations that are relevant.

Busby [4] identified common failures that occur during decision-making in product development:

- Not involving others in decisions (which limits the information used to make the decisions).
- Not telling others the assumptions that they can make, the normal requirements and the exceptional circumstances that can occur.
- Not considering other's goals or requirements.
- Not knowing the effect of one's action on another, not knowing the effect of a change on another.
- Not defining the scope of tasks allocated to others, and not determining the scope of tasks assigned to oneself.

Many of the errors listed above stem from not understanding the information flow and decision-making in the product development organization and not seeing one's role in the decision production system. That is, they are failures to maintain information responsibility [8].

20.5.2 Distance from Corporate Objectives

Generally, the mechanisms linking decision-makers to the overall corporate goals are constraints and incentives such as schedules, rewards and penalties. That is, the decision-making system is "loosely coupled" [27]. Therefore, profit is an indirect influence on most decision-makers in the firm. Instead, the more influential factors are the work assignments, the practices that the firm establishes, the authority of supervisors, the formal and informal communication channels, and the skills and knowledge of each person.

20.5.3 Presence Decision-Making Confusion

The multiple factors that influence decision-making can lead to confusion and different interpretations of what designers should be doing. A study of Volvo engineers responsible for the final development of new engines revealed that some engineers believed their job was to make the engine meet performance specifications, others thought that they needed to resolve trade-offs between performance categories, and a third set wanted to make the engine provide the customer with a good driving experience [23].

20.5.4 Missing Models

The scarcity of useful models is another set of problems. It is difficult to understand how detailed design decisions affect profitability. Profitability is determined by a huge number of variables, many of which are beyond the firm's control. Managers and researchers are still trying to understand how high-level design decisions effect expected profitability. Product profit models can estimate the total profit that a new product will yield [29]. Such models include estimates and projections that are based on the firm's experience with similar products. This type of model clearly shows how unit cost, sales price, sales and development costs effect expected profitability. Although it is certain that the product's design (along with its price) will effect these measures (unit cost, sales price, sales and development costs), knowledge of these relationships is incomplete. One can start with an educated guess about sales volume and refine this as more information becomes available.

Since a manufacturing firm pays for the labor, material and components and has extensive knowledge in this domain, the relationship between product design and unit cost is the one area where the most work has been done, and there exists a large amount of research on technical cost modeling and manufacturability analysis. Manufacturing system performance also effects profitability (see, for instance, [7]). In other cases, experience is needed to estimate how a design change will alter, say, total sales.

Practical models that relate lower-level design decisions to profitability do not yet exist, and researchers have studied only some simple cases (see, for instance, [11]).

The early phase of new product development is often called the "fuzzy front end." This phase typically involves qualitative methods for understanding customer needs, wants and desires, including scenario development, new product ethnography, ergonomics and lifestyle reference. (For more about this topic, see [5].) It is very difficult (if not impossible) to formulate an optimization problem that includes qualitative methods, since these have extremely complex interactions among a variety of qualities.

20.5.5 The Cost of Model Maintenance

Product development organizations have to spend time to construct, validate and maintain the models that they do use in decision support tools. As the models get more complex, the maintenance becomes more expensive. It is easy for models to become outdated since updating the models with the latest information may be viewed as a low priority.

To illustrate the effort involved, consider the decision analysis model that Bayer used to decide whether or not to start preclinical development of a specific drug [31]. The modeling and analysis project required a six-month effort by a team of 10 persons. Moreover, this project looked primarily at a go/no-go decision and did not evaluate design alternatives.

20.6 IMPROVING DECISION-MAKING IN PRODUCT DEVELOPMENT

Like other systems, a decision production system should be effective and efficient. In today's post-industrial society, designing a decision production system requires organizing how decisions are made and a key issue is how information is processed. Decision-making is shared between different persons and between persons and machines, and sharing decision-making requires sharing information.

Holt *et al.* [17] describe an ideal decision-making system: "First, management wants good decisions—the goal is to select those that are less costly and have the more desirable outcomes. Second, since making decisions takes time, talent and money, we do not seek the very best decision without some regard to the cost of research. Rather, management wants decision-making methods that are easy and inexpensive to operate. Third, it would be desirable, if the techniques were available, to handle large and complex problems more nearly as wholes, in order to avoid the difficulties that occur when problems are treated piecemeal. Fourth, it is certainly advantageous to use fully the knowledge and experience available within the firm. Intimate knowledge of the decision problem is indispensable to improvement in decision-making methods."

The above quote illustrates the tension between integration and decomposition. Decomposition transforms a large problem into a series of small decisions that are less difficult to make. But integration treats the problem more nearly as a whole, in order to avoid the problems that decomposition introduces. It is possible that improvements in information technology will make further integration feasible and desirable, if organizations can collect the necessary data and define the relevant relationships.

Product development processes are not explicitly designed to optimize profitability. Still, the never-ending quest to improve processes leads managers to change them, first hoping to improve this metric, then hoping to improve another, always seeking changes that improve all metrics simultaneously. Because different firms find themselves in different positions, they seek different things from their processes. More precisely, there exists a large set of objectives, and each organization prioritizes these objectives differently. Thus, each firm finds a different process most desirable for itself, in the same way that different families looking at the houses for sale in the same city choose different houses based on their own priorities on location, price, number of bedrooms and so forth.

20.6.1 A DPS-Based Improvement Methodology

Simon [27] argues that systematic analysis of the decision-making in a product development process would be useful for implementing changes to the product development organization in a timely and profitable manner, and he proposes the following technique for designing an organization:

- Examine the decisions that are actually made, including the goals, knowledge, skills and information needed to make them.
- Create an organization pattern for the tasks that provide information for these decisions.
- Establish (or change) the pattern of who talks to whom, how often and about what.

Of course, this must be repeated for the more specific decisions that form the more general decisions.

Viewing a product development organization as a decision-making system leads to a systems-level approach to improving product development. In particular, this perspective is not concerned primarily with formulating and solving a design optimization problem. Moreover, the problem is not viewed only as helping a single design engineer make better decisions (though this remains important). Instead, the problem is one of organizing the entire system of decision-making and information flow.

As with other efforts to improve manufacturing operations or business processes, improving product development benefits from a systematic improvement methodology. The methodology presented here includes the following steps in a cycle of continuous improvement (as shown in Figure 20.4), which is based in part on ideas from [6]:

(1) Study the product development decision-making system.
(2) Build, validate and analyze one or more models of this decision-making system.
(3) Identify feasible, desirable changes.
(4) Implement the changes, evaluate them and return to Step 1.

The important features of the decision-making system are the persons who participate in it, the decisions that are actually made, including the goals, knowledge, skills and information needed to make those decisions. Also relevant are the processes used to gather and disseminate information. It will also be useful to study other processes that interact with product development, including marketing, regulatory compliance, manufacturing planning and customer service.

An especially important part of studying product development is determining the sources that provide information to those making decisions. If they are not documented, changes to the system may eliminate access to these sources, which leads to worse decision-making. In addition, like any group of tools accumulated over time, it is critical to review how and when each decision support tool is applied to the product development process. This requires a meta-level understanding of decision-making during all phases of product development.

Many process improvement approaches begin with creating a map or a flowchart that shows the process to be improved. For instance, in organizations adopting lean manufacturing principles, it is common for a team that plans to improve a production

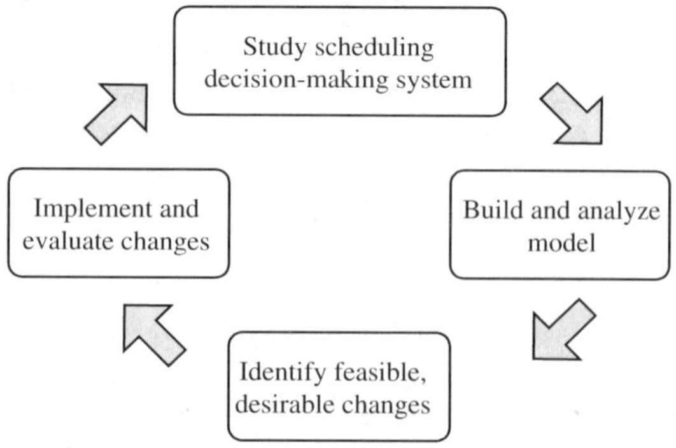

FIG. 20.4 THE METHODOLOGY FOR IMPROVING DECISION-MAKING SYSTEMS

line to begin the improvement event with a value stream mapping exercise.

Creating a model of the as-is product development organization has many benefits. Though it may be based on preexisting descriptions of the formal product development process, it is not limited to describing the "should be" activities. The process of creating the model begins a conversation among those responsible for improving the organization. Each person involved has an incomplete view of the system, uses a different terminology and brings different assumptions to the table. Through the modeling process, these persons develop a common language and a complete picture. Validation activities give other stakeholders an opportunity to give input and also to begin learning more about the system. Even those who are directly involved in product development benefit from the "You are here" information that a model provides.

No particular modeling technique is optimal. There are many types of models available, and each one represents a different aspect of the decision-making system. It may be necessary to create multiple models to capture the scope of the system and its essential details. Swimlane diagrams [25] can be useful, as discussed below. As in other modeling efforts, wasting time on unneeded details or scope is a hazard. The purpose of the model should guide its construction and the selection of the appropriate level of detail.

In general, representing decision-making systems is a difficult task. A decision-making system may involve a complex social network. The information that decision-makers collect, use and exchange comes in many forms and is not always tangible. Some decisions are routine, while others are unique. The documentation of decision-making systems usually does not exist. If it does, it is typically superficial. (Notable exceptions are the decisions made by government bureaucracies, as when a state highway administration designs a new highway. In such cases, the decision-making process is well documented.)

Analyzing such models quantitatively is usually not possible. However, a careful review of the model will reveal unnecessary steps or show how one group's activities are forcing others to behave unproductively. The model can show the impact of implementing proposals to change decision-making. For example, if the firm wants to add an environmental impact review to the product development process, will the results of that review provide timely and relevant information that can be used to redesign the product?

Evaluating the feasibility and desirability of potential changes and selecting the ones to implement requires time and effort to build consensus among the stakeholders. A "to-be" model of the product development process can show how the system will operate after the changes are implemented.

Changes that are implemented should be evaluated to determine if the product development process has improved. That is, is the decision quality increasing? Do engineers have the information needed to make the most profitable decisions? Is less time spent to develop new products? Are the products more successful? The questions asked depend upon the problems that motivated the improvement effort.

Ideally, product development organizations should undergo a continuous cycle of improvement. The organization and its environment are always changing. People come and go. Proposals that were infeasible become possible, and changes that were ignored become desirable. Money becomes available for software, or the existing software vendor goes out of business. Better information is appearing, and decisions that were easy become hard.

20.6.2 Swimlanes

Representing decision-making systems is a difficult task. The most typical representation is an organization chart, which lists the employees of a firm, their positions and the reporting relationships. However, this chart does not explicitly describe the decisions that these persons are making or the information that they are sharing. Another representation is a flowchart that describes the life cycle of an entity by diagramming how some information (such as a customer order, for example) is transformed via a sequence of activities into some other information or entity (such as a shipment of finished goods).

Swimlanes [25] are a special type of flowchart that adds more detail about who does which activities, a key component of a decision-making system. A swimlane diagram yields a structured model that describes the decision-making and information flow most efficiently and clearly shows the actions and decisions that each participant performs. A swimlane diagram highlights the who, what and when in a straightforward, easy-to-understand format. Unlike other forms, they identify the actors in the system. There are other names used to describe this type of diagram, including process map, line of visibility chart, process responsibility diagram, and activity diagram. Figure 20.5 shows a swimlane diagram of a technical service process for a power tool manufacturer.

A swimlane diagram includes the following components:

- *Roles* that identify the persons who participate in the process.
- *Responsibilities* that identify the individual tasks each person performs.
- *Routes* that connect the tasks through information flow.

Sharp and McDermott [25] present techniques for modeling branching, optional steps, the role played by information systems, steps that iterate, steps that are triggered by the clock and other details. The following summarizes some key points.

A single diagram is the path of a single item (e.g., form or schedule) as it goes through a process. Each person gets a row from left to right. An organization, a team, an information system or a machine can have a row. In the row go boxes—one box for each task that the person performs. Arrows show the flow of work from one task to another and also indicate precedence constraints (what has to be done before another task can start).

Tasks can involve multiple actors, so the task should span the different actors' rows. While there are multiple flowchart symbols available, Sharp and McDermott recommend a simple box with occasional icons to represent an inbox or a clock. Boxes should be labeled with verb-noun pairs (e.g., "create schedule" rather than "new schedule"). Transportation steps and other delays should be included.

Flow should go generally from left to right, with backward arrows for iteration. A conditional flow should have one line that leaves an activity and then splits into two lines. Flow from an activity to two parallel steps should have two lines.

Managing detail requires multiple diagrams. The highest level shows one task per person per handoff. This clarifies the relationships and flow of information between persons. Another diagram can show the tasks that are key milestones that change the status of something—decisions, communication activities (passing and receiving information) and iteration. An even more detailed diagram can describe the specific ways in which the tasks are done (via fax or e-mail, using specific tools or other special resources). As in other modeling efforts, keep in mind the purpose of the model and the need to satisfy this (so the

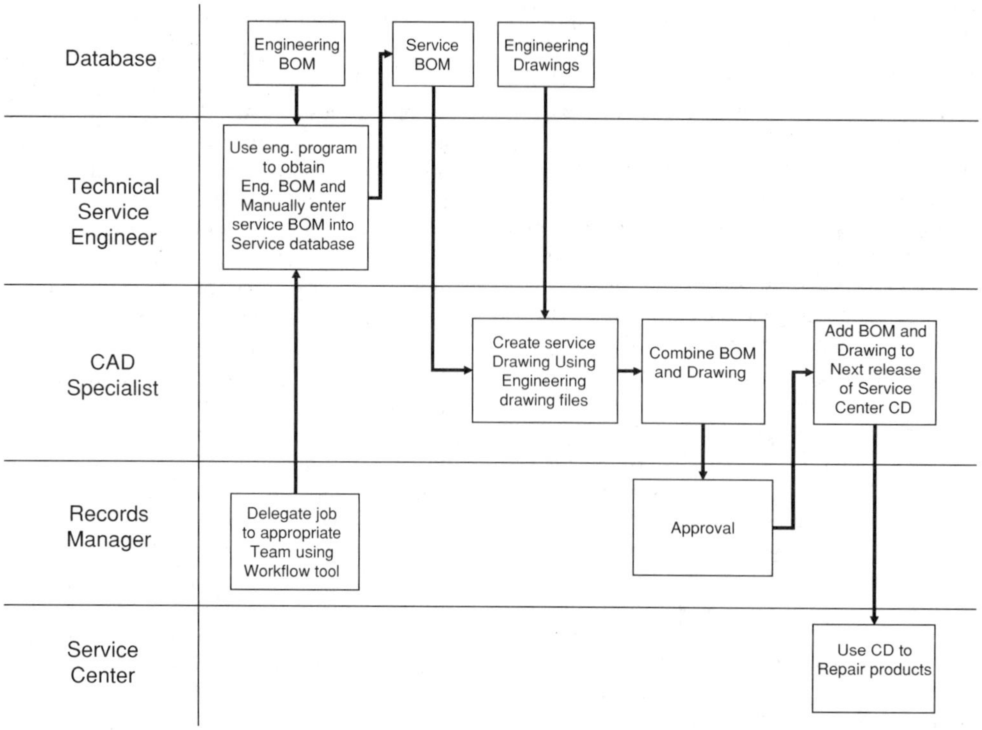

FIG. 20.5 A SWIMLANE DIAGRAM FOR CREATING TECHNICAL SERVICE REPORTS (CREATED BY DANIEL FITZGERALD)

model is good enough) without wasting time on unneeded details or scope.

20.6.3 DPS Network Model and Related Heuristics

Decision production systems can be represented in many different ways. Each type of representation allows the user to focus on a particular aspect of the system. This section illustrates the use of a graph with directed edges as the model of a decision production system network. The graphlike representation emphasizes the connections between processors in the network. The graph view of a DPS also enables equivalent transformations, such as into an adjacency matrix. Key processors can be identified by the number and type of connections they maintain with others in the network.

The DPS perspective is useful in identifying an information and decision flow path currently existing within an enterprise. Consider how an electronics board fabrication enterprise (hereinafter called FABCO) would implement a system for creating material disclosure statements (MDS) that customers request. The implementation of environmental regulations that control the content of potentially hazardous materials and the designation of restricted materials (e.g., those that can no longer be included in products sold in certain countries) is affecting product manufacturers in the United States. Supplying an MDS upon request is likely to be a requirement for selling products in other countries, and the scope of regulation is growing both with respect to materials of concern and effected market locations. Compliance with regulations will require the establishment of a decision-making system within an existing enterprise.

The initial step in mapping the DPS that creates an MDS at FABCO is to identify the people currently involved in relevant activities (or representatives from each appropriate department). These people form a network of information processors and decision-makers. The network extends outside the enterprise to customers, suppliers, regulatory agencies and contacts within peer industry groups and professional organizations. The next step in modeling FABCO's MDS DPS is to graph the network of relationships between these people.

FABCO's DPS initial information processor network is shown in Fig. 20.6. The network was created by investigating the relevant information flow. This modeling step can be accomplished by starting anywhere in the network, tracking information flows and their directions and recording it all in the form of a graph. A quick review of the network in Fig. 20.6 leads to the following observations:

(1) The key information processor in this DPS is FABCO's Environmental Engineer. This processor is represented by a highly connected node depicted inside the right-hand boundary of FABCO's internal network graph.
(2) The process is initiated by a request from an external source (the customer).
(3) The process relies on information inputs from other external sources interacting with both the Environmental Engineer and the Health & Safety group.
(4) There is an independent flow of information that bypasses the key processor. In this flow, the Purchasing group obtains material safety data sheets (MSDS) from the company's suppliers.

Observations 1 and 4 reveal important characteristics about the MDS process. It is possible that the key figure is not using all the information already flowing into FABCO. There is likely some duplication of information-gathering effort.

Like many enterprises, FABCO has deployed a number of decision-support tools throughout its organization. These tools (like business groups and human specialists) are repositories of information and must be added to the DPS network. The resulting network is a DPS, which is shown in Fig. 20.7.

Reviewing FABCO's MDS DPS reveals inefficiencies. For example:

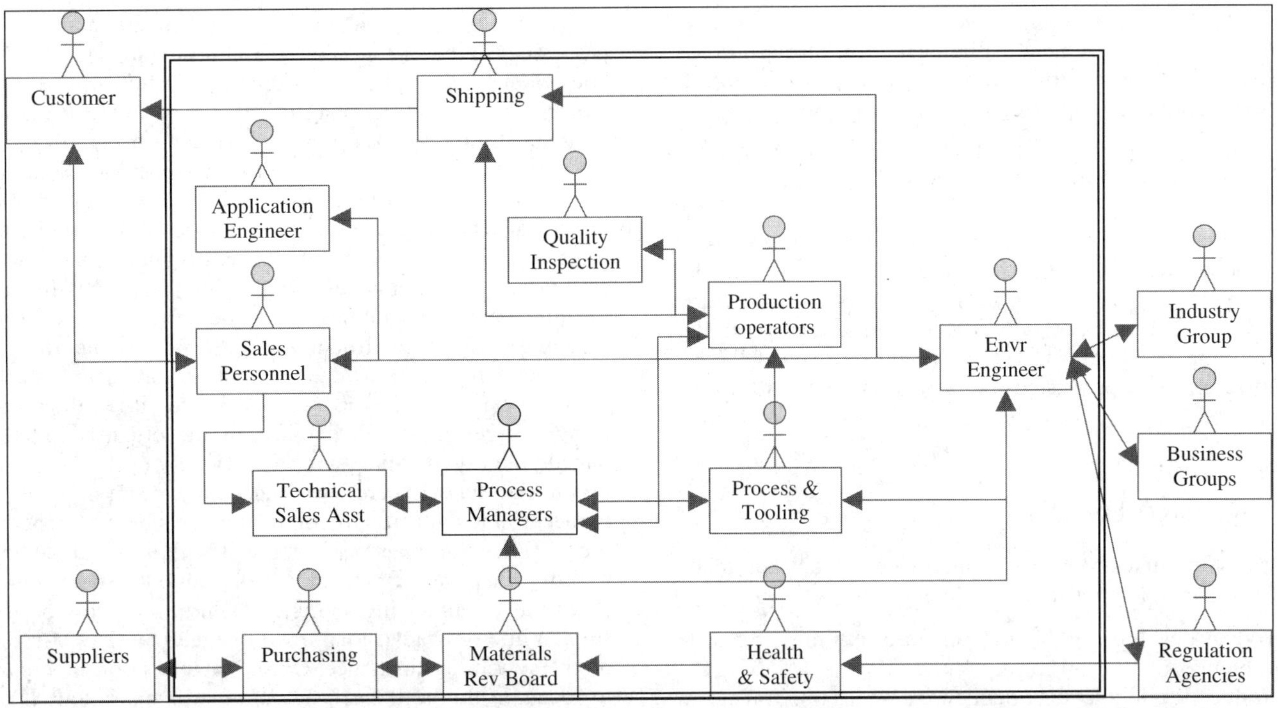

FIG. 20.6 NETWORK OF FABCO'S MATERIAL DISCLOSURE STATEMENT (MDS) CREATION PROCESSORS

(1) The Environmental Engineer has created and is maintaining two separate databases for writing an MDS for a particular customer. If this person were to leave the organization, it is likely the information will be stripped from the network.

(2) The link between Purchasing and Process Managers is the decision-support tool MRP2, a material requirements planning system. This connection in the network is a tool, not a person.

(3) The Process Managers maintain a separate material requirements planning system that is used for preparing quotes for Sales and recording process changes initiated by Production Operators.

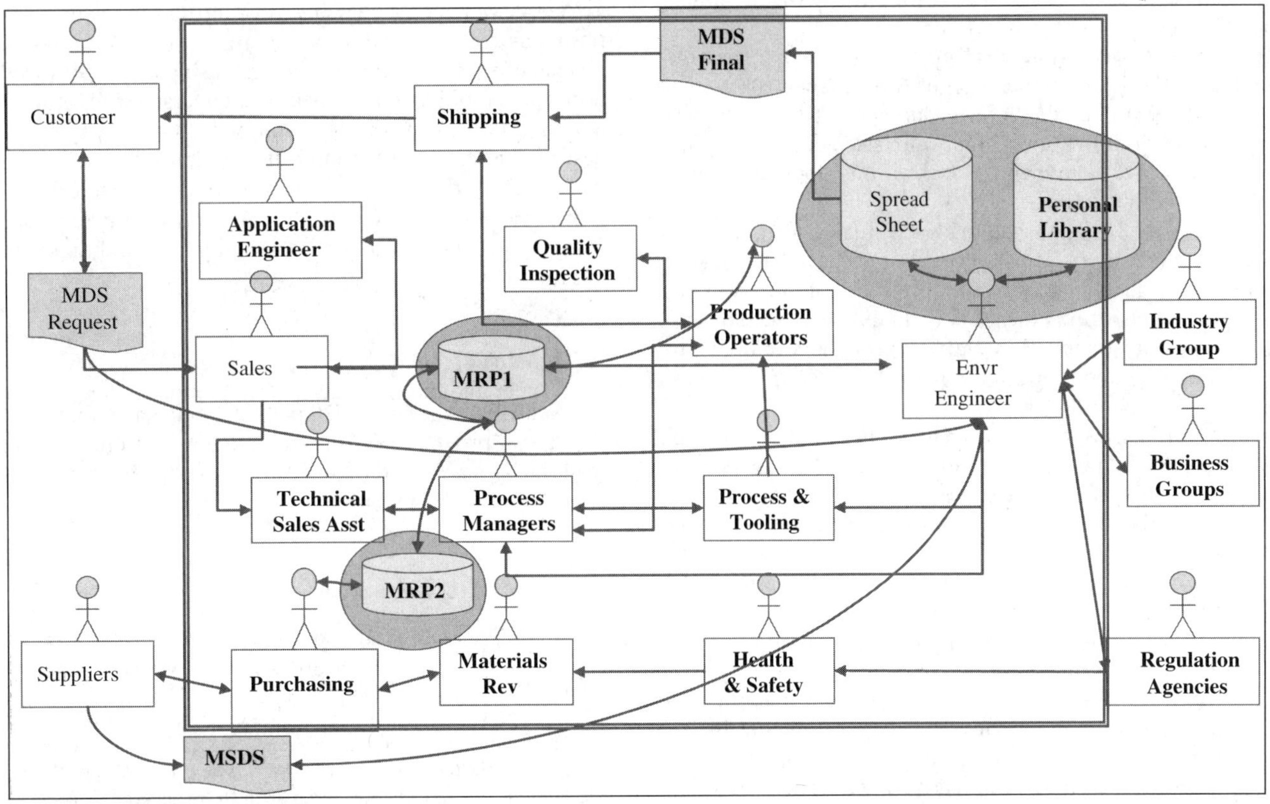

FIG. 20.7 DPS OF FABCO'S MATERIAL DISCLOSURE STATEMENT (MDS) CREATION PROCESSORS

(4) The MSDS information that is obtained from suppliers is sent to the Environmental Engineer to be stored in case it is needed for a future MDS, bypassing the Materials Review Board and the Health & Safety group.

The analysis of FABCO's MDS DPS led to recommendations to create a materials database that would be maintained by both the Materials Review Board and the Health & Safety group. New MSDS information would be reviewed by Materials Review Board and then added to the materials database. The Environmental Engineer would fold the existing personal library of information into this database, retain access to it and share the responsibility for maintaining it. Other suggestions to improve the DPS for the organization were suggested by this analysis but will not be discussed here.

20.7 CONCLUSIONS

This chapter's discussion of product development has made two observations that appear contradictory:

(1) Product development should solve a profit-maximization problem.
(2) Product development is a sequence of steps that transform customer requirements into a satisfactory product design.

This chapter reconciled this contradiction and synthesized these ideas by highlighting the role of decomposition in product development. Because the profit-maximization problem is extremely complex, it cannot be solved directly. Instead, product development organizations decompose the problem into a set of subproblems that form the product development process. The tendency toward decomposition (which reduces search effort) is checked by the desire for integration (which improves the quality of the solution).

The performance of the system is limited by the people, money and time available. A key implication of this perspective is that engineers, in the presence of these resource constraints, do not have the luxury of optimization. Thus, heuristics play an important role in product development. However, this does not imply that the decisions are irrational.

Understanding the DPS can lead to identifying opportunities to improve the organization's decision-making process. Of course, good individual decision-making is required, but it is important to create a process that links these individuals together. Individual decisions are not isolated. Especially in product development, decisions are based upon information that results from other decisions.

There are similarities between a DPS and a factory. Hopp and Spearman [18] define "manufacturing system" as "an objective-oriented network of processes through which entities flow." We believe that this accurately describes product development organizations as well as factories. Both are organizations concerned with generating profits. The process of manufacturing an item is decomposed into many manufacturing, assembly and testing steps. In the same way, the process of creating a new product design must be decomposed into many activities and decisions. Within a factory, parts flow from one machine to another. This view is analogous to product development organizations since information flows from one decision-maker to another (or between information processors who transform data for decision-makers).

In a factory, at a given point in time, there are many items being processed. Some of the items are almost-completed final assemblies. Some are complete subassemblies, waiting for final assembly. Others are components that are being used to make these subassemblies. Similarly, in a product development organization, there are many product development projects underway. Some projects are almost done, and products will be available to be sold soon. Other projects are in the middle of the detail design phase. Other projects are just beginning, and the product development teams are considering customer needs and searching for successful concepts.

Unfortunately, quantitatively modeling a decision production system is much more difficult due to the iterative nature of a product development process, the preemption that occurs as engineers interrupt one task to work on another, the difficulties in identifying sources of knowledge within the organization [27] and outside the organization, and the difficulty of defining the scope of a design task. (There have been some initial attempts to gain managerial insight from theoretical models based on the decision production system perspective; see [15, 9], for example.)

From this perspective, improving individual decision-making is similar to enhancing a single machining process or assembly step. While desirable, that small change may have little impact on the system performance. On the factory floor, it is reasonable to focus on the bottleneck process or the operations that introduce many defects. But the firm must also consider the layout of the factory, planning and scheduling activities and other system-level issues, all of which impact the flow of parts through the factory and the ability to satisfy customer orders on time and in a cost-effective manner. Similarly, a product development organization must consider its decision-making processes, which impact the flow of information and the ability of the organization to get products to market quickly and use engineering resources productively.

One advantage of viewing product development as a DPS is the focus on information processing and decision-making flows instead of personnel-reporting relationships. The DPS view can be used to help organization members understand the flows of information and decisions in the same way that an organization chart describes administrative authority relationships and a process plan (routing) describes the flow of material through a factory.

It is the authors' hope that the DPS perspective will help those studying product development organizations and those struggling to improve them.

ACKNOWLEDGMENTS

Joseph Donndelinger, Daniel Fitzgerald and many other collaborators provided assistance and many useful insights. This material is based upon work supported by the National Science Foundation under grant number 0225863.

REFERENCES

1. Argote, L., McEvily, B. and Regans, R., 2003. "Managing Knowledge in Organizations: An Integrative Framework and Review of Emerging Themes," Mgmt. Sci., 49 (4), pp. 571–582.
2. Bertola, P. and Teixeira, J. C., 2003. "Design As a Knowledge Agent: How Design As a Knowledge Process Is Embedded Into Organizations to Foster Innovation," *Des. Studies*, Vol. 24, pp. 181–194.
3. Brown, J. S., 1998. "Research That Reinvents the Corporation," *Harvard Business Review on Knowledge Management*, Harvard Business School Press, Boston, MA.

4. Busby, J.S., 2001. "Error and Distributed Cognition in Design," *Des. Studies*, Vol. 22, pp. 233–254.
5. Cagan, J. and Vogel, C. M., 2002. *Creating Breakthrough Products: Innovation from Product Planning to Program Approval*, Prentice Hall PTR, Upper Saddle River, NJ.
6. Checkland, P., 1999. *Systems Thinking, Systems Practice*, John Wiley & Sons, Ltd., West Sussex.
7. Chincholkar, M. M. and Herrmann, J. W. "Incorporating Manufacturing Cycle Time Cost in New Product Development," Proc., *DETC '01, ASME 2001 Des. Engrg. Tech. Conf. and Computers and Information in Engrg. Conf., DETC2001/DFM-21169*, Pittsburgh, PA.
8. Drucker, P. F.,1998. "The Coming of the New Organization," *Harvard Business Review on Knowledge Management*, Harvard Business School Press, Boston, MA.
9. Fitzgerald, D. P., Herrmann, J. W., Sandborn, P. A., Schmidt, L. C., and Gogoll, T., 2005. "Beyond Tools: a Design for Environment Process," *Int. J. of Performability Engrg.*, 1 (2), pp. 105–120.
10. Gigerenzer, G., Todd, P. M. and the ABC Res. Group, 1999. *Simple Heuristics That Make Us Smart*, Oxford University Press. 1999; a precis available online at http://www-abc.mpib-berlin.mpg.de/users/ptodd/SimpleHeuristics.BBS/.
11. Gupta, S. K. and Samuel, A. K., 2001. "Integrating Market Research With the Product Development Process: A Step towards Design for Profit," *Proc. DETC. 01, ASME 2001 Des. Engrg Tech. Conf. and Computers and Information in Engrg Conf., DETC2001/DFM-21202*, Pittsburgh, PA. September 9–12, 2001.
12. Hastie, R. and Dawes, R.M., 2001. *Rational Choice in an Uncertain World*, Sage Publications, Thousand Oaks, CA.
13. Hazelrigg, G. A., 1996. *Systems Engineering: An Approach to Information-Based Design*, Prentice Hall, Upper Saddle River, NJ.
14. Hazelrigg, G. A., 1998. "A Framework for Decision-Based Engineering Design," *J. of Mech. Des.*, Vol. 120, pp. 653–658.
15. Herrmann, J. W., 2004. "Information Flow and Decision-Making in Production Scheduling," *2004 Ind. Engrg. Res. Conf.*, Houston, TX.
16. Herrmann, J. W. and Schmidt, L. C., 2002. "Viewing Product Development as a Decision Production System," *Proc. DETC '02 ASME 2002 Des. Engrg. Tech. Conf. and Computers and Information in Engrg. Conf., DETC2002/DTM-34030*, Montreal, Canada. Sep 29-Oct. 2, 2002
17. Holt, C. C., Modigliani, F., Muth, J. F. and Simon, H. A., 1960. *Planning Production, Inventories, and Work Force*, Prentice-Hall, Inc., Englewood Cliffs, NJ.
18. Hopp, W. J. and Spearman, M. L., 2001. *Factory Physics, 2nd Ed.*, Irwin McGraw-Hill, Boston, MA.
19. Kidder, T., 1981. *The Soul of a New Machine*, Little, Brown, Boston, MA.
20. Krishnan, V. and Ulrich, K. T., 2001. "Product Development Decisions: A Review of the Literature," *Mgmt. Sci.*, 47 (1), pp. 1–21.
21. Li, H. and Azarm, S., 2001. "Product Line Design Selection Under Uncertainty and With Competitive Advantage," *Proc. DETC '01, ASME 2001 Des. Engrg. Tech. Conf. and Computers and Information in Engrg. Conf., DETC2001/DAC-21022*, Pittsburgh, PA.
22. Otto, K. and Wood, K. 2001. *Product Design*, Prentice Hall, Upper Saddle River, NJ.
23. Sandberg, J., 2001. "Understanding Competence At Work," *Harvard Bus. Rev.*, 79 (3), pp. 24–28.
24. Schmidt, L.C., Zhang, G., Herrmann, J. W., Dieter, G. and Cunniff, P.E., 2002. *Product Engineering and Manufacturing,* 2nd Ed., College House Enterprises, Knoxville, TN.
25. Sharp, A. and McDermott, P., 2001. *Workflow Modeling*, Artech House, Boston, MA.
26. Simon, H. A., 1981. *The Sciences of the Artificial*, 2nd Ed., MIT Press, Cambridge, MA.
27. Simon, H. A., 1997. *Administrative Behavior*, 4th Ed., The Free Press, New York, NY.
28. Sleazak, S. L. and Khanna, N., 2000. "The Effect of Organizational Form on Information Flow and Decision Quality," *J. of Eco. & Mgmt. Strategy*, 9 (1), pp. 115–156.
29. Smith, P. G. and Reinersten, D. G., 1991. *Developing Products in Half the Time*, Van Norstrand Reinhold, New York, NY.
30. Stirling, W. C., 2003. *Satisficing Games and Decision-Making*, Cambridge University Press, Cambridge.
31. Stonebraker, J. S., 2002. "How Bayer Makes Decisions to Develop New Drugs," *Interfaces*, 32 (6), pp. 77–90.
32. Ulrich, K. T. and Eppinger, S. D., 2004. *Product Design and Development*, McGraw Hill, New York, NY.
33. Walton, M., 1997. *Car*, W.W. Norton & Company, New York, NY.
34. Wassenaar, H.J., Chen, W., Cheng, J. and Sudjianto, A., 2005. "Enhancing Discrete Choice Demand Modeling for Decision-Based Design," *J. of Mech. Des.*, 127 (4), pp. 514–523.

PROBLEMS

20.1. Search for an article or book that discusses the role of search in decision-making. An acceptable article should have appeared in a peer-reviewed scholarly journal or a research conference.

a. As you search, record your search process: What actions did you take? What decisions did you make as you conducted your search?
b. Read the article or book about search in decision-making and answer the following question: How well do the concepts or models presented in your article or book apply to the search that you conducted.

20.2. Describe a complex engineering decision that was decomposed into subproblems.

a. What were the subproblems that were solved? In what order where the subproblems solved? How was the solution to each subproblem used to solve another subproblem? For each subproblem, what was the goal or objective of that subproblem? What constraints had to be satisfied during the solution of that subproblem?
b. As much as possible, express the overall decision as a single problem. What are the objectives and constraints of this integrated problem? What are the benefits and disadvantages of the current decomposition? Why was this decomposition chosen? What are its strengths? What are the weaknesses?
c. Propose a different decomposition of the decision (using different subproblems, a different sequence or other changes). How is it better than the existing decomposition? How is it worse than the existing decomposition?

20.3. Select an organization with which you are quite familiar. The organization should perform a transformation process that chiefly involves decision-making.

a. Clearly define the activity of the organization by providing a statement of what the organization does, how it does this and why it does this. You may want to include information about the organization's customers, actors, owners and environment as well as the key performance measures related to effectiveness and efficiency.
b. Construct and describe a swimlane diagram that identifies the key persons in the organization, the key decisions and activities and the information that flows between them. (Note that this is a model of the organization's current processes.)
c. Identify the activities that the organization must do to achieve the definition from (a) and draw a conceptual model. (Note that these are not necessarily how the firm operates.) The model does not describe any actual organization, but it must be reasonable. There

should be approximately five to nine verbs, connected in a logical fashion. Include monitoring or supervisory activities as appropriate.

20.4. Repeat Problem 3 using a network graph representation of the organization as described in Section 20.6.3. Use the network DPS model to make observations as to the flow of information through the organization (e.g., identify key information processors and point out the types of information required from outside the organization).

SECTION

7

Decision Making in Decentralized Design Environments

INTRODUCTION

The trend in many product design scenarios is toward decentralization of the decisions and tasks involved in the design process. Issues like the globalization of product design, supplier-driven design and service outsourcing have all created an economic culture of decentralizing the expertise necessary to develop and bring a product to market. This decentralization creates a network of distributed, yet interactive decision-makers whose collective task is to effectively design products or systems. While it may increase the efficiency of a process by tackling a set of smaller, distributed decisions, this efficiency may be realized only at the significant expense of product quality, feasibility, optimality and stability because of the complex dynamics involved in decentralized decision processes. In this section of the text, some of the prominent issues in the complex dynamics of decentralized decision processes are studied from different perspectives and fundamental principles behind decentralized design processes are presented.

A note of terminology clarification is warranted. It is assumed that the terms *decentralized* and *distributed* are, for discussion purposes, interchangeable. *Distributed design* implies a design process that is spread out, scattered or divided up. *Decentralized design* implies a design process that is withdrawn from a center of concentration. It is the partitioning of a product design process to various suppliers, divisions, departments, and/or teams that creates a decentralized network of distributed decisions made by a diverse collection of decision-makers. The term *collaborative design* is quite common in systems design as well, and implies a decentralized design process where decision makers can collaborate and make joint decisions, if necessary, effectively bridging the gap between distributed decision-makers. All of these scenarios are discussed in this section.

One of the fundamental tools that has been used to model and study decentralized design is *game theory*. It is the study of the theory of rational behavior for interactive decision problems. In engineering design, it is certainly paramount that the decision-makers behave rationally and many decisions in design are not made in isolation, but are made based on other decisions and when made, affect other decisions. This is precisely why game theory has brought some valuable insight into decentralized design problems. In Chapter 21, the fundamentals of game theory in the context of decision-making are presented. The fundamentals are presented in such a way to give the reader a basic understanding of the primary constructs and mathematics in game theory. In the chapters following, these fundamentals, along with others, are built upon in more advanced studies and applications in decentralized design. In Chapter 22, some of the basic *architecture issues* in distributed design are presented and the use of *modern information technologies* to coordinate a distributed design process is discussed and demonstrated. In Chapter 23, the basic issue of whether or not a distributed design process will *converge* or simply *diverge* without any rational resolution is studied and presented. A number of different classes of problems are studied and *convergence criteria* are developed to aid decision-makers in making effective decisions regarding problem structure and outcome. In Chapter 24, a value aggregation perspective is taken to expand *distributed design* into *collaborative design* where joint decisions are possible among decision-makers. The focus of the chapter is on design objective structuring and aggregation for distributed decision-making.

CHAPTER

21

GAME THEORY IN DECISION-MAKING[1]

Charalambos. D. Aliprantis and Subir K. Chakrabarti

21.1 INTRODUCTION TO GAMES

Classical decision theory and optimization theory have often exclusively focused on situations in which a single decision-maker needs to find an optimal decision that maximizes or minimizes an objective function that depends on some given parameters and the decision variable of the decision-maker. In many situations, however, the well-being of an individual depends not only on what he or she does but on the outcome of the choices that other individuals make. In some instances, this element of *mutual interdependence* is so great that it must be explicitly taken into account in describing the situation.

For example, in discussing the phenomenon of *Global Warming* it would be ludicrous to suggest that any one country could, by changing its policies, affect this in a significant way. Global warming is precisely that: a global phenomenon. Therefore, in any analysis of global warming we have to allow for this. But then this raises questions about what is the right strategy[2] to use in tackling the problem. How should any one country respond? What will be the reaction of the other countries and so on. Clearly, actions taken must be strategically analyzed and it is not as clear as to what is an optimal strategy.

Let us take a look at a situation in which strategic play is important. The following excerpt taken from the *New York Times*[3] reported on a settlement made by airlines on a price fixing lawsuit.

- "Major airlines agreed to pay $40 million in discounts to state and local governments to settle a price fixing lawsuit. The price fixing claims centered on an airline practice of announcing price changes in advance through the reservations systems. If competitors did not go along with the price change, it could be rescinded before it was to take effect."

It seemed that the airlines were trying to coordinate price increases by using a signaling scheme. If the other airlines did not go along with the change, the price increase would not take effect. Why would an airline be interested in knowing how the other airlines would respond? Why were the airlines so wary about changing prices unilaterally? The reasons are not immediately obvious. Some of the incentives for doing what the airlines were doing can be surmised from the following description of the situation.

21.1.1 Example

Suppose US Air and American Airlines (AA) are thinking about pricing a round-trip airfare from Chicago to New York. If both airlines charge a price of $500, the profit of US Air would be $50 million and the profit of AA would be $100 million. If US Air charges $500 and AA charges $200 then the profit of AA is $200 million and US Air makes a loss of $100 million. If, however, US air sets a price of $200 and AA charges $500 then US Air makes a profit of $150 million while AA loses $200 million. If both charge a price of $200 then both airlines end up with losses of $10 million each. This information can be depicted in the form of Table 21.1, shown below:

TABLE 21.1 THE FARE SETTING GAME

	American Airlines		
US Air	Fare	$500	$200
	$500	(50, 100)	(−100, 200)
	$200	(150, −200)	(−10, −10)

The example is illustrative of what was happening in the airline industry. It is worth noting that it would be best for both airlines to coordinate price changes, because without such coordination the airlines would end up making fairly serious losses. In situations of this kind the following three elements always seem to be present:

(1) There are two or more participants.
(2) Each participant has a set of alternative choices.
(3) For each outcome there is a payoff that each participant gets.

These are the essential ingredients that constitute what is called a game in strategic form. In more formal language, a **strategic form game** consists of a *set of players*; for each player there is a *strategy set*; and for each *outcome* (or strategy combination) of the game there is a payoff for each player.

It would be nice if we could find certain central principles that would allow us to analyze the solution to games, in the same way that we were able to find general principles for solving optimization problems, as we did in the last lecture. One might start by asking what is most likely to happen in a game once the players are completely informed about the game they are playing. In other words, given a situation that can be modeled as a game, what guiding principles should we use in deciding the most plausible outcome of the game? We shall discuss some of these principles in this lecture.

[1]The research of Charalambos D. Aliprantis is supported in part by NSF grants EIA-0075506, SES-0128039, DMS-0437210 and ACI-0325846.

[2]The word "strategy" is the Greek word "στρατηγική," which means a plan or an approach.

[3]Source: "Suit Settled by Airlines," 1994. *New York Times,* October 12. p. D8.

21.2 TWO-PERSON MATRIX GAMES

The most elementary depiction of a game is the one featured in the fare setting game. In that example, we gave a description of the payoffs or the profits that the airlines would make for every possible outcome of the game using a table. We can use such a *matrix* format for many interesting games. We start our discussion with one of the most well-known matrix games called the Prisoner's Dilemma (see Table 21.2). The game illustrates a social phenomenon, which is best understood using game theoretic ideas. It describes a situation in which the players would do better by cooperating but nevertheless seem to have an incentive not to do so!

21.2.1 Example

This game—which perhaps has been the most widely analyzed game—is given by the following matrix.

TABLE 21.2 THE PRISONER'S DILEMMA

		Player 2	
Player 1	Strategy	Mum	Fink
	Mum	(−1, −1)	(−10, 0)
	Fink	(0, −10)	(−5, −5)

The matrix game shown here is best described as a situation where two individuals who have committed a crime have a choice of either confessing the crime or keeping silent. In case one of them confesses and the other keeps silent, then the one who has confessed does not go to jail whereas the one who has not confessed gets a sentence of 10 years. In case both confess, then each gets a sentence of five years. If both do not confess, then both get off fairly lightly, with sentences of one year each.

The matrix game shows clearly that there are two players (the two prisoners) and the strategy set of each player is {Mum, Fink}. The payoffs are given by the pairs (a, b) for each outcome, with a being player 1's payoff and b player 2's payoff; here, of course, $-a$ and $-b$ represent years in jail. The matrix completely describes a game in strategic form. In examining the game, one notices the following features:

(1) Both players have a stake in keeping mum as they both get a sentence of one year each.
(2) Given that a player is going to keep mum, the other player has an incentive to fink.

These are precisely the sort of paradoxes that are so inherent in playing games. The central issue is not only about the choice that a player makes but also about the choices of the other players.

A close examination of the game shows that if player 1 uses the "confess" (Fink) strategy, then he gets a better payoff for each choice that player 2 makes. To see this, let $u_1(\cdot, \cdot)$ denote the utility function of player 1 and note that if player 2 plays "Mum," then: $u_1(\text{Fink, Mum}) = 0 > -1 = u_1(\text{Mum, Mum})$, while if player 2 plays "Fink," then: $u_1(\text{Fink, Fink}) = -5 > -10 = u_1(\text{Mum, Fink})$.

That is, *no matter what the choice of player 2, it is best for player 1 to play the strategy* **Fink**. We say that the strategy *Fink* is a **strictly dominant** strategy of player 1. A similar examination of player 2's strategies reveals that the strategy *Fink* is a strictly dominant strategy for player 2.

In the absence of any communication or any coordination scheme, rational players are expected to play their strictly dominant strategies since a strictly dominant strategy gives a player an unequivocally higher payoff. A solution to the "Prisoner's Dilemma," could, therefore, end up being (Fink, Fink). This is the solution using *strictly dominant strategies.*

We note that the solution using strictly dominant strategies will give each player a sentence of five years, which, of course, is a worse outcome than if each prisoner could trust the other to keep mum. This conflict between playing non-cooperatively, in which case the strictly dominant strategy solution seems so persuasive, and playing so as to coordinate to get the better payoff is what makes predicting the outcome of a game difficult.

Going back to the fare-setting game, we notice that setting the fare of $200 is a strictly dominant strategy for both airlines. Hence, the strictly dominant strategy solution causes both airlines to make a loss of $10 million. This then provides airlines with an incentive to try and reach some form of a price-fixing agreement.

The two games that we have discussed so far are examples of matrix games. They are formally defined as follows.

21.2.2 Definition

A matrix game *is a two-player game such that*:
(1) Player 1 has a finite strategy set S_1 with m elements.
(2) Player 2 has a finite strategy set S_2 with n elements.
(3) The payoffs of the players are functions $u_1(s_1, s_2)$ and $u_2(s_1, s_2)$ of the outcomes.

$$(s_1,s_2) \in S_1 \times S_2 \qquad \text{Eq. (21.1)}$$

The matrix game here is played as follows: at a certain time player 1 chooses a strategy $s_1 \in S_1$ and simultaneously (and independently) player 2 chooses a strategy $s_2 \in S_2$ and once this is done each player i receives the payoff $u_i(s_1, s_2)$. If $S_1 = \{s_1^1, s_2^1, \ldots, s_m^1\}$, $S_2 = \{s_1^2, s_2^2, \ldots, s_n^2\}$ and we put

$$a_{ij} = u_1(s_i^1, s_j^2) \text{ and } b_{ij} = u_2(s_i^1 s_j^2) \qquad \text{Eq. (21.2)}$$

then the payoffs can be arranged in the form of the $m \times n$ matrix shown in Table 21.3.

TABLE 21.3 THE TWO-PERSON MATRIX GAME

		Player 2			
Player 1	Strategy	s^2_1	s^2_2		s^2_n
	s^1_1	(a_{11},b_{11})	(a_{12},b_{12})		(a_{1n},b_{1n})
	s^1_2	(a_{21},b_{21})	(a_{22},b_{22})		(a_{2n},b_{2n})
	$\vdots$	$\vdots$	$\vdots$	$\ddots$	$\vdots$
	s^1_m	(a_{m1},b_{m1})	(a_{m2},b_{m2})		(a_{mn},b_{mn})

The idea of a "solution" of a game is usually identified by the concept of the *Nash equilibrium*, which is defined below.

21.2.3 Definition

A pair of strategies $(s_1^*, s_2^*) \in s_1 \times s_2$ is a **Nash Equilibrium**[4] *of a matrix game if*:

(1) $u_1(s_1^*, s_2^*) \geq u_1(s, s_2^*)$ for each $s \in S_1$

(2) $u_2(s_1^*, s_2^*) \geq u_2(s_1^*, s)$ for each $s \in S_2$

In other words, a Nash equilibrium is an outcome (i.e., a pair of strategies) of the game from which none of the players have an incentive to deviate as, given what the other player is doing, it is optimal for a player to play the Nash equilibrium strategy. In this sense, a Nash equilibrium has the property that it is *self-enforcing*. That is, if both players knew that everyone has agreed to play a Nash equilibrium, then everyone would indeed want to play his Nash equilibrium strategy for the simple reason that it is optimal to do so.

The Nash equilibrium has been widely used in applications of game theory. Perhaps a reason for this popularity of the Nash equilibrium is that when one looks at an outcome which is not a Nash equilibrium, then there is at least one player who is better off playing some other strategy if that outcome is proposed. An outcome that is *not* a Nash equilibrium is, therefore, not going to be self-enforcing.

21.3 STRATEGIC FORM GAMES

We saw that a game between two players can be written as a matrix game. In many applications the games are often played between more than two players. Also, the strategy sets of the players may be such that the games do not have a nice matrix representation. Fortunately, however, most of the ideas about how to solve matrix games can be easily extended to a more general class of games—the class of strategic form games. We start by defining strategic form games in a more formal way.

21.3.1 Definition

A strategic form game or a game in normal form *is simply a set of n persons labelled 1, 2, . . . , n (and referred to as the* **players** *of the game) such that each player i has*:

(1) A choice set S_i (also known as the strategy set of player i and its elements are called the strategies of player i).

(2) A **payoff function** u_i: $S_1 \times S_2 \times \cdots \times S_n \rightarrow \Re$.

The game is played as follows: each player k chooses simultaneously (and independently of each other) a strategy $s_k \in S_k$ and once this is done each player i receives the payoff $u_i(s_1, s_2, \ldots, s_n)$. A strategic form game with n players, strategy sets $S_1, \ldots, S_n$ and payoff functions $u_1, \ldots, u_n$ will be denoted by:

$$G = \{S_1, \ldots, S_n, u_1, \ldots, u_n\} \qquad \text{Eq. (21.3)}$$

So in order to describe a strategic form game G, we need the strategy sets and the payoff functions of the players.

One should notice immediately that each payoff function u_i is a real-function of the n variables $s_1, s_2, \ldots, s_n$; where each variable s_k runs over the strategy set of player k. The value $u_i(s_1, s_2, \ldots, s_n)$ is interpreted as the **payoff** of player i if each player k plays the strategy s_k.

The Cartesian product $S_1 \times S_2 \times \cdots \times S_n$ of the strategy sets is known as the **strategy profile set** or the **set of outcomes** of the game and its elements $(s_1, s_2, \ldots, s_n)$ are called **strategy profiles** or **strategy combinations**. Of course, the payoff $u_i(s_1, s_2, \ldots, s_n)$ for a player i might represent a monetary gain or loss or any other type of "satisfaction," which is of importance to the player.

21.3.2 Example

[*A Strategic Form Game*] This is a strategic form game with three players 1, 2, 3. The strategy sets of the players are:

$$S_1 = S_2 = S_3 = [0, 1] \qquad \text{Eq. (21.4)}$$

Their payoff functions are given by

$$u_1(x, y, z) = x + y - z,\ u_2(x, y, z) = x - yz \text{ and } u_3(x, y, z) = xy - z \qquad \text{Eq. (21.5)}$$

where for simplicity we let $s_1 = x$, $s_2 = y$ and $s_3 = z$.

If the players announce the strategies $x = (1/2)$, $y = 0$ and $z = (1/4)$, then their payoffs will be

$$u_1\left(\tfrac{1}{2}, 0, \tfrac{1}{4}\right) = \tfrac{1}{4},\ u_2\left(\tfrac{1}{2}, 0, \tfrac{1}{4}\right) = \tfrac{1}{2} \text{ and } u_3\left(\tfrac{1}{2}, 0, \tfrac{1}{4}\right) = -\tfrac{1}{4} \qquad \text{Eq. (21.6)}$$

Clearly, the strategy profile (1, 1, 0) gives a better payoff to each player.

When a strategic form game is played, a player's objective is to maximize her payoff. However, since the payoff of a player depends not just on what she chooses, but also on the choices of the other players, the issue of optimizing one's payoff is a lot more subtle here than in the case of the simpler decision problem when there is just one decision-maker. An individual player may, if she or he knows the choices of the other players, choose to maximize her payoff given the others' choices. But then, all the other players would want to do the same. Indeed, it seems quite natural to look for an outcome that results from the simultaneous maximization of individual payoffs. Such a strategy profile is usually called—as in the case of matrix games—a Nash equilibrium and is defined as follows.

21.3.3 Definition

A Nash equilibrium *of a strategic form game*

$$G = \{S_1, S_2, \ldots, S_n, u_1, u_2, \ldots, u_n\} \qquad \text{Eq. (21.7)}$$

is a strategy profile $(s_1^*, s_2^*, \ldots, s_n^*)$ such that for each player i we have:

$$u_i(s_1^*, \ldots, s_{i-1}^* s_i^*, s_{i+1}^*, \ldots, s_n^*) \geq u_i(s_1^*, \ldots, s_{i-1}^*, s, s_{i+1}^*, \ldots, s_n^*) \qquad \text{Eq. (21.8)}$$

for all $s \in S_i$.

The appeal of Nash equilibrium stems from the fact that if a Nash equilibrium is common knowledge, then every player would indeed play the Nash equilibrium strategy, thereby resulting in the Nash equilibrium being played. In other words, a Nash equilibrium strategy profile is *self-enforcing*. Hence, if the players are searching for outcomes or solutions from which no player will have an incentive to deviate, then the only strategy profiles that satisfy such a requirement are the Nash equilibrium points.

[4] This equilibrium concept was introduced by John Nash in 1951. For this and related work, John Nash was awarded the Nobel Prize in Economics in 1994.

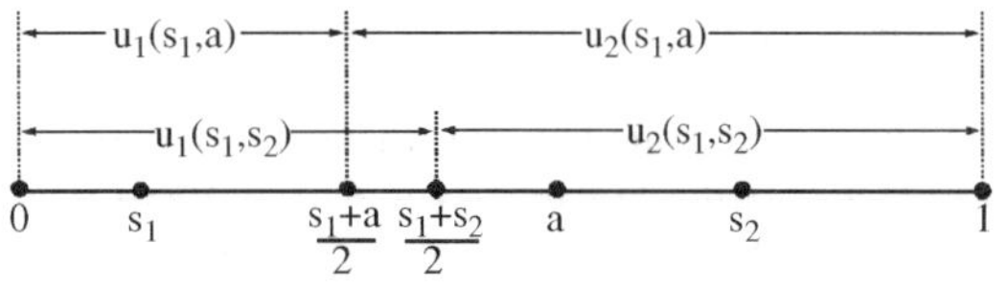

FIG. 21.2 PAYOFFS FOR POSITIONS ON THE IDEOLOGICAL SPECTRUM

see Fig. 21.2. This clearly contradicts the fact that (s_1, s_2) is a Nash equilibrium, and so $s_1 = s_2 = ½$ must be true.

The preceding two steps show that the strategy profile $(s_1, s_2) = [(1/2), (1/2)]$ is the only possible candidate for a Nash equilibrium of the game. To complete the argument, we shall show that $[(1/2), (1/2)]$ is indeed a Nash equilibrium.

Step III: The strategy profile $[(1/2), (1/2)]$ is a Nash equilibrium.

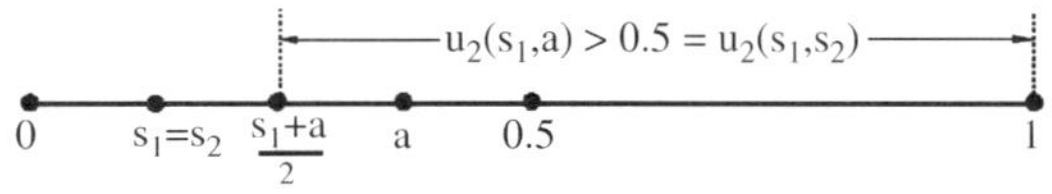

FIG.21.3 PAYOFFS AND POSITIONS

From Figure 21.3 it should be clear that if candidate 2 keeps the strategy ½, then candidate 1 cannot improve his utility $u_1[½, ½] = 0.5$ by choosing any strategy $a \neq 0.5$ (see Fig. 21.4).

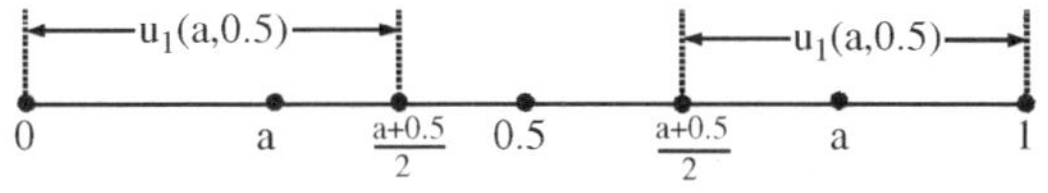

FIG. 21.4 PAYOFFS AND IDEOLOGIAL POSITIONS

This model's prediction is, therefore, that each candidate will seek to appeal to the *median voter*, the voter who is exactly in the middle of the distribution of the ideological spectrum.

The next example is in some ways perhaps one of the more interesting applications of game theory. It shows how perverse incentives can sometimes work against what is in the common interest. While the example focuses on the exploitation of a commonly owned resource, like the world's fishing grounds, a little reexamination of the example shows that it has implications for *Global Warming* and the exploitation of the world's rain forests; to mention just a few of the situations that would fit into this general mold. It brings to surface an element that is present in many games, including the prisoner's dilemma: the Nash equilibrium, which describes what happens when the players play *noncooperatively*, may lead to an outcome in which each player gets less than what they could get by adhering to a cooperative agreement, like e.g., treaties among countries on fishing rights.

21.4.3 Example

[*Use of Common Property Resources*] Suppose that there are n countries that have access to fishing grounds in open seas. It is widely accepted that the fishing grounds of the world, which may be viewed as common property resources, have been overfished, i.e., the amount of fishing has been so intensive that there is a sense that in the near future the fish population will reach levels so low that some species may be in danger of extinction.

One of the major achievements of game theory—from a practical standpoint—has been to show why such common property resources will always be exploited beyond the point that is the most desirable from the collective viewpoint. The argument, which we make in some detail here, is that the Nash equilibrium of the game that is played between the consumers of the resource will always lead to an outcome which is worse than the socially most desirable.

We do this by using a simple model of a strategic form game. Let there be n players with player i using r_i amount of the resource. The total resource used is then $R = \sum_{i=1}^{n} r_i$. The following now describe the chief features of the game.

(1) The cost to player i of getting r_i units of the resource depends not only on the amount r_i used by the player but also on the amount $R - r_i = \sum_{j \neq 1} r_j$ used by the other players. This cost is denoted by $C(r_i, R - r_i)$. We shall assume that the cost function $C: (0,\infty) \times (0,\infty) \to (0,\infty)$ satisfies the following properties:

a. $$\frac{\partial C(r,R)}{\partial r} > 0, \quad \frac{\partial C(r,R)}{\partial R} > 0, \quad \frac{\partial^2 C(r,R)}{\partial r^2} > 0$$

and $$\frac{\partial^2 C(r,R)}{\partial R^2} > 0 \ \textit{for all } r > 0 \textit{ and } R > 0.$$

That is, the marginal cost of using a resource increases with the total amount of the resource used.[6] Hence, as the countries catch more, the marginal cost of catching additional fish goes up.

b. The marginal cost function satisfies

$$\lim_{R\to} \frac{\partial C(r,R)}{\partial r} = \infty \quad \text{and} \quad \lim_{R\to} \frac{\partial C(r,R)}{\partial R} = \infty$$

Indeed, it is not unreasonable to assume that the marginal cost starting from some small number greater than zero increases monotonically without bound. These properties of the cost function are consistent with the intuition that as more fish is caught the harder it becomes to catch additional amounts.

c. To simplify matters, the cost function C will be taken to be a *separable function* of the form $C(r,R) = k(r) + K(R)$. In this case, the properties in part (a) can be written as:

$$\kappa'(r) > 0, \kappa''(r) > 0, K'(R) > 0, \text{ and } K''(R)$$

for all $r > 0$ and $R > 0$. An example of a separable cost function of the above type is given by $C(r, R) = r^2 + R^2$.

(2) The utility that a player receives from r_i units of the resource is $u(r_i)$. We suppose that the function $u: (0,\infty) \to (0,\infty)$ satisfies $u'(r) > 0$ and $u''(r) < 0$ for each $r > 0$. This simply means that, as the amount of r consumed increases, the value of an additional unit of r falls. (In mathematical terms, u is a strictly increasing and strictly concave function.) We also assume that the marginal utility at zero is greater than the marginal cost at zero, i.e.:

$$\lim_{r \to o^+} u'(r) > \lim_{r \to o^+} \kappa'(r).$$

[6]Recall that the marginal cost of a cost function $C(x)$ is the derivative $C'(x)$. As usual, $C'(x)$ is interpreted as the cost of producing one additional unit of the product when x units have already been produced.

The situation we have just described can be written as an n-person game in strategic form as follows:

- There are n players.
- The strategy set of player i is $(0,\infty)$, the open interval of all positive real numbers. (In fact, $S_i = (0, R_{max})$, where R_{max} is a certain maximum amount of the resource.)
- The payoff of player i is

$$\pi_i(r_1,r_2,\ldots,r_n)=u_i(r_i)-C(r_i,R-r_i)$$
$$=u(r_i)-[\kappa(r_i)+K(R-r_i)].$$

By the Nash Equilibrium Test, the Nash equilibria of the game are the solutions $(r_1^*,\ldots,r_n^*)$ of the system $\partial\pi_i(r_1,r_2,\ldots,r_n)/\partial r_i=0, i=1,2,\ldots,$ subject to $\partial^2\pi_i(r_1^*,\ldots,r_n^*)/\partial r_i^2<0$ for each i = 1,..., n. Taking into account that $R=\sum_{j=1}^{n} r_j$ and, $R-r_i=\sum_{j\neq 1} r_j$, a direct computation of the partial derivatives gives:

$$\frac{\partial\pi_i(r_1,r_2,\ldots,r_n)}{\partial r_i}=u'(r_i)-\kappa'(r_i)=0, 1, 2\ldots, n \qquad \text{Eq. (21.23)}$$

and $\partial^2\pi_i(r_1^*,\ldots,r_n^*)/\partial r_i^2=u''(r_i)-\kappa''(r_i)<0$ for each i = 1,...,n. (For this conclusion, we use the fact that $u''(r)<0$ *and* $k''(r)>0$ *for each* $r>0$.)

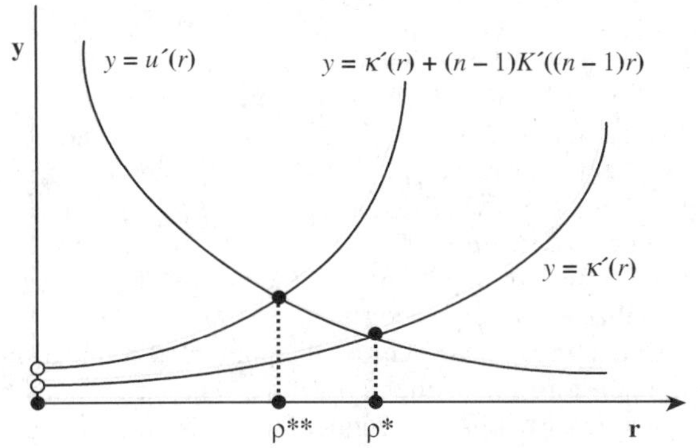

FIG. 21.5 EQUILIBRIUM OF THE RESOURCE EXTRACTION GAME

The geometry of the situation guarantees $r_1=r_2=\cdots=r_n=\rho^*$.[7] (See Fig. 21.5.) That is, at a Nash equilibrium $(r_1^*,\ldots,r_n^*)$ each player consumes exactly the same amount of the resource

$$r_1^*=r_2^*=\ldots=r_n^*=\rho^*=\frac{R^*}{n} \qquad \text{Eq. (21.24)}$$

where $R^*=r_1^*+r_2^*+\ldots+r_n^*=\eta\rho^*$. *So*, $[(R^*/n),\ldots,(R^*/n)]$ is the only Nash equilibrium of the game. Hence, the amount R^* of the resource consumed at the Nash equilibrium is the unique solution of the equation

$$u'\left(\frac{R^*}{n}\right)=\kappa'\left(\frac{R^*}{n}\right). \qquad \text{Eq. (21.25)}$$

[7]Since $u''(r)<0$ for each $r>0$, we know that u' is a strictly decreasing function. Since $k''(r)>0$ for each each $r>0$, the function k' is strictly increasing. So, $u'(r) = k'(r)$ has a unique solution ρ^*; see Fig. 21.5.

[8]The social optimum is the amount that leads to the maximum *joint payoff*. Hence, if society is made up of the players in the game, then the social optimum gives us the amount that would lead to the most desirable outcome from the social viewpoint.

In contrast to the condition for a Nash equilibrium given above, the social optimum[8] R^{**} solves

$$\max_{R>0}\left(n\left[\left(\tfrac{R}{n}\right)-[\kappa(\tfrac{R}{n})+K(R-\tfrac{R}{n})]\right]\right) \qquad \text{Eq. (21.26)}$$

That is, the social optimum R^{**} is chosen to maximize the total payoff to all the members of society. The First-Order Test for this gives

$$n\left\{\tfrac{1}{n}u'(\tfrac{R^{**}}{n})-[\tfrac{1}{n}\kappa'(\tfrac{R^{**}}{n})+(1-\tfrac{1}{n})K'(R^{**}-\tfrac{R^{**}}{n})]\right\}$$

Eq. (21.27)

which, after some algebraic simplifications, yields

$$u'\left(\frac{R^{**}}{n}\right)=\kappa'\left(\frac{R^{**}}{n}\right)+(n-1)K'\left(\frac{n-1}{n}R^{**}\right) \qquad \text{Eq. (21.28)}$$

Again, we leave it as an exercise for the reader to verify that Eq.(21.28) has a unique solution $R^{**}=\eta\rho^{**}$. From examining Eqs. (21.25), and (21.28) and Figure 21.5, we see that $R^*>R^{**}$. Clearly, the amount of resource R^* that is used in a Nash equilibrium is *strictly greater* than the amount R^{**} of consumption of the resource that is best for the common good. One wonders at this point about the intuition behind this rather remarkable result. A moment's thought shows that if the game is played independently by the players, then the private incentives are to use the resource as much as is justified by the cost of consuming the resource to the individual player. In a Nash equilibrium, a player is concerned about the impact of his consumption of the resource only on his cost, and ignores the cost imposed on the others. The cost to the individual, however, is a lot less than the cost imposed on society collectively. For the socially optimum amount of consumption of the resource however, the cost imposed on everyone is taken into consideration, and as a result the amount of consumption justified by the overall cost to society is less.

The next example is based on a model of a "Second Price Auction." The issue here is the amount that an individual at the auction should bid in order to maximize her surplus from the auction. Obviously, an immediate complication is that the surplus that a bidder receives depends on whether she has the winning bid. Since, whether an individual wins depends on the bids that the others make, we see that the payoff of an individual depends on the entire array of bids. Auctions, therefore, can be written as n-person strategic form games. We see in this example that thinking of auctions in the form of a game can lead us to very interesting and sharp insights.

21.4.4 Example

[*Second Price Auction*] A seller has an expensive painting to sell at an auction that is valued at some amount by n potential buyers. Each buyer k has his own valuation $v_k>0$ of the painting. The buyers must simultaneously bid an amount; we denote the bid of buyer i by $b_i\in(0,\infty)$. In a second price auction the highest bidder gets the painting and pays the second highest bid. If there is more than one buyer with the highest bid, the winner is decided by a drawing among the highest bidders and she pays the highest bid. The rest receive a payoff of zero.

We can formulate this auction as a strategic form game in which there are:

(1) n players (the n buyers; the auctioneer is not considered a player)
(2) The strategy set of each player is $(0,\infty)$

(3) The payoff of a player k is the following expected utility function:

$$\pi_k(b_1,...,b_n) = \begin{cases} v_k - s & \text{if} \quad b_k > s \\ 0 & \text{if} \quad b_k < s \\ \frac{1}{r}(v_k - s) & \text{if } k \text{ is among } r \text{ buyers with higest bid} \end{cases}$$

Eq. (21.29)

where s = second highest bid.[9]

We claim that the strategy profile $(v_1, v_2, \ldots, v_n)$ is a Nash equilibrium for this game. We shall establish this in two steps:

A player i never gains by bidding $b_i > v_i$.

To see this, assume $b_i > v_i$ and let $b_i = max_{j\neq i} b_j$. We distinguish five cases:

Case 1: $b_{-i} > b_i$

In this case, some other bidder has the highest bid and so player i gets zero, which he could get by bidding v_i.

Case 2: $v_i < b_{-i} < b_i$

In this case, bidder i wins and gets $v_i - b_{-i} < 0$. However, if he would have bid v_i, then his payoff would have been zero—a higher payoff than that received by bidding b_i.

Case 3: $b_{-i} = b_i$

Here bidder i is one among r buyers with the highest bid and he receives $(v_i - b_{-i} / r) < 0$. But by bidding v_i he can get 0, a higher payoff.

Case 4: $b_{-i} < v_i$

In this case bidder i gets $v_i - b_{-i}$, which he could get by bidding v_i.

Case 5: $b_{-i} = v_i$

Here again bidder i is one among r buyers with the highest bid and he receives $\frac{v_i - b_{-i}}{r} = 0$. But by bidding v_i, he can also get 0.

A player i never gains by bidding $b_i < v_i$.

If $b_{-I} > v_i$, then bidder i would have a zero payoff, which is the same as the payoff she would get if she bid v_i. On the other hand, we leave it as an exercise for the reader to verify that if $b_{-i} < v_i$, then player i would do at least as well if she bid v_i.

We have thus shown the following:

The strategy profile $(v_1, v_2, \ldots, v_n)$ is a Nash equilibrium.

Therefore, it is reasonable to expect that every bidder will bid his or her true valuation of the painting and the bidder with the highest valuation wins. Note that this is true even if the bidder's do not know the valuation of the other bidders.

21.5 SEQUENTIAL DECISIONS

In all that we have seen so far, decisions had to be made once, and the decision-makers then received the rewards. In many contexts, however, decisions have to be taken sequentially and the rewards are received only after an entire sequence of decisions has been taken. For instance, in manufacturing, the product usually has to go through a sequence of steps before it is finished and at each step the manufacturer has to decide which of several alternative processes to use. Before becoming established in one's career or profession an individual has to take a sequence of decisions, which leads to a final outcome. Similarly, financial planning over a lifetime is done via a sequence of decisions taken at various points of an individual's life span.

By now we have a fairly good grasp of how optimal decisions are made when a decision has to be made once. Sequential decision-making is different because the decision-making process is more involved. A choice made initially has an impact on what choices can be made later. For instance, in choosing a career, if an individual decided not to go to school then the choice of a career is limited to those who require only a high school education. Similarly, if one chooses not to save very much in the early years of one's life, then the choice of how much to accumulate for retirement in the later years is much more constrained. This fact that choices made in the initial stages affect the alternatives available in the later stages, is an element of decision-making that is central to sequential decision-making.

In every situation that we encountered so far, the payoff to the individual depended on the sequence of decisions made by the individual. In many other contexts, however, the payoff to the individual may depend not just on what the individual does, but also on the sequence of decisions made by other individuals.

Thus, we may have a game that is being played by a number of individuals, but instead of taking decisions simultaneously, the players may have to play the game sequentially. For instance, if an investor makes a takeover bid, then the bid has to be made before the management of the firm can respond to the bid. Such a situation is best analyzed as a game in which the investor makes his move in the first stage and the management then responds in the second stage. Obviously, the players in this game are not moving simultaneously, but rather in two stages. Games that are played in stages are variously called *multistage games, games in extensive form* or *sequential games.* In our case, we will use the term **sequential game** for any game in which moves by more than one player are made in a sequence.

In this chapter we shall outline the analytical foundation of sequential decisions and sequential games. The basic mathematical notions needed to illustrate sequential decisions and sequential games is those of a graph and of a tree. To simplify the discussion, we shall consider only trees.

21.6 GRAPHS AND TREES

In this section we will lay down the basic framework for the discussion of sequential decisions and sequential games. We start by introducing the concept of a graph and a tree.

21.6.1 Definition

A **directed graph** *is a pair G = (V, E), where V is a finite set of points (called the* **nodes** *or the* **vertices** *of the graph) and E is a set of pairs of V (called the* **edges** *of the graph).*

A directed graph is easily illustrated by its diagram. The **diagram of a directed graph** consists of its vertices (drawn as points of a plane) together with several oriented line segments corresponding to the pairs of the edges. For instance, if (u, v) is an edge, then in the diagram of the directed graph we draw the line segment $\vec{uv}$ with an arrowhead at the point v. The diagram shown in Fig. 21.6(a) is the diagram of the directed graph with vertices

$$V = \{u, v, w, x, y\} \qquad \text{Eq. (21.30)}$$

and edges

[9]Note that if player k is the only buyer with the highest bid, then $s = max_{i\neq k} b_i$.

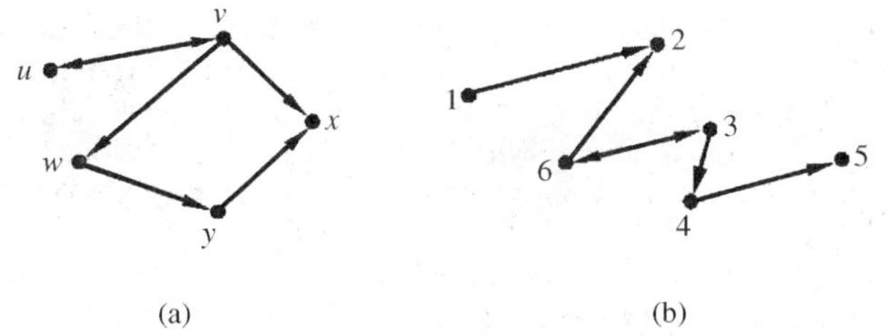

FIG. 21.6 AN EXAMPLE OF A GRAPH

$$E = [(u, v), (v, u), (v, w), (v, x), (w, y), (y, x)] \quad \text{Eq. (21.31)}$$

21.6.2 Definition

A directed graph T is said to be a **tree** if

(1) There exists a distinguished node R (called the **root** of the tree) which has no edges going into it.

(2) For every other node v of the graph there exists **exactly one path** from the root R to u. An example of a tree is shown in Fig. 21.7.

There is a certain terminology regarding trees that is very convenient and easy to adopt.

- If (u, v) is an edge of a tree, then u is called the **parent** of the node v and node v is referred to as a **child** of u.
- If there is a path from node u to node v, then u is called an **ancestor** of v and node v is known as a descendant of u.

With the above terminology in place, the root R is an ancestor of every node and every node is a descendant of the root R.

Here are some other basic properties of trees; we leave the verification of these properties as an exercise for the reader.

21.6.3 Theorem

In any tree,

(1) There is at most one path from a node u to another node v.

(2) If there is a path from u to v, then there is no path from v to u.

(3) Every node other than the root has a unique parent.

(4) Every nonterminal node has at least one terminal descendant node.

The unique path joining a node u to another node v in a tree will be denoted by $P(u, v)$. For instance, in the tree of Fig. 21.8, we have $P(u, 4) = u \rightarrow 1 \rightarrow 3 \rightarrow 4$. Notice that the path $P(u, c)$ is itself a tree having root u and terminal node c.

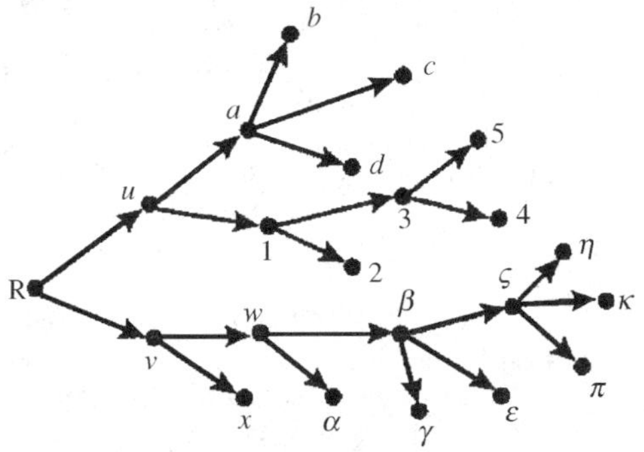

FIG. 21.7 AN EXAMPLE OF A TREE

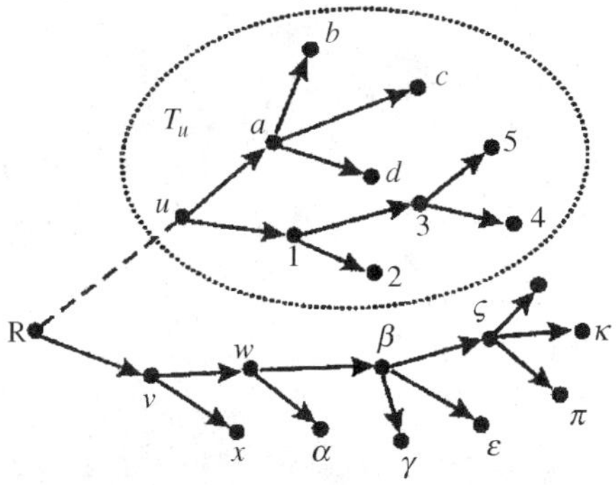

FIG. 21.8 A TREE WITH A BRANCH

A **branch** of a tree T is a directed graph having nodes starting at a node u and containing all of its descendants together with their original edges. We shall denote by T_u the branch starting at u. It should not be difficult to see that T_u is itself a tree whose root is u. The branch T_u using the directed graph of Fig. 21.7 is shown in Fig. 21.8.

21.7 UNCERTAINTY AND SINGLE-PERSON DECISIONS

Uncertainty is introduced in sequential decision problems by adding nodes at which nature chooses. The following examples indicate how uncertainty can be handled in sequential decision problems.

21.7.1 Example

A pharmaceutical firm X faces a decision concerning the introduction of a new drug. Of course, this means that there is an initial decision about how much to spend on research and development, the possibility that the drug may fail to be developed on schedule, and the fact that the drug may not be quite successful in the market. At each stage of this decision-making process, we notice the presence of uncertainty. A decision tree of this problem is shown in Fig. 21.9.

At the initial stage firm X has to decide whether to spend a large amount "Hi" or a small amount "Lo" on research and development. The result of this investment could either lead to success S or failure F with the probability p of success being higher in the case of Hi expenditure on research and development. Even when the drug is successfully produced, the firm may decide not to market it. The uncertainty about whether the drug can be produced or not is handled here by introducing the node "Nature," at which nature

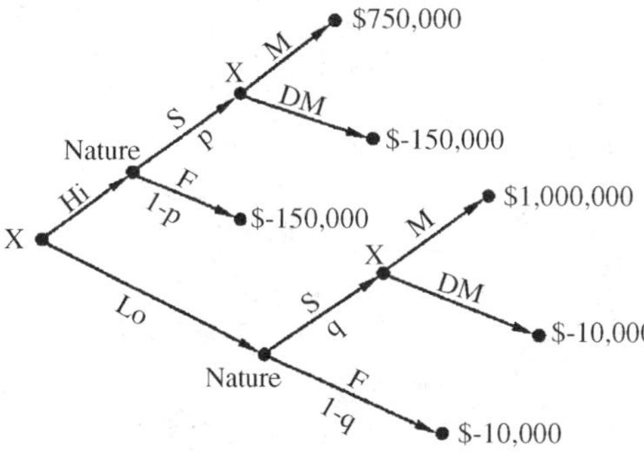

FIG. 21.9 A DECISION TREE

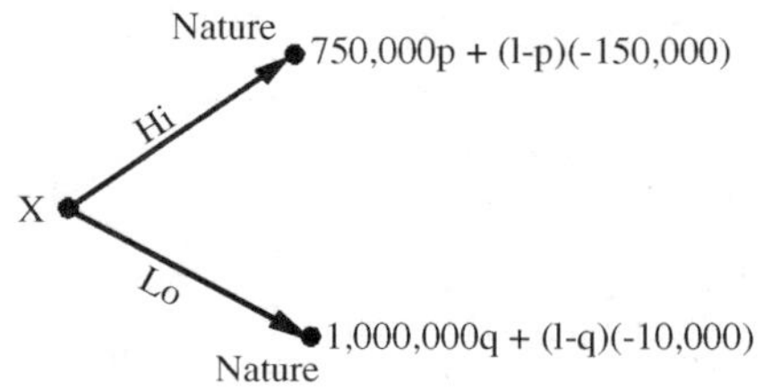

FIG. 21.10 A DECISION NODE

chooses. The edges M and DM stand for "Market" and "Do not Market" the drug.

We can solve this decision problem by using the method of "backward induction." In the present case, with the payoffs as shown in Fig. 21.9, the firm has to decide at the nodes of the First Stage of the backward induction whether to market (M) or not to market (DM) the drug; the firm always chooses to market the drug. But then this leads to the truncated version of the decision tree shown in Fig. 21.10, in which case the payoffs are expressed in the form of expected payoffs.

The firm now has to compare two lotteries involving a Hi expenditure choice and a Lo expenditure choice. If the firm is risk neutral the choice, of course, is the lottery with the highest expected value, otherwise, the choice would depend on the von Neumann–Morgenstern utility function of the firm. If the firm is risk neutral and the expected profits are negative, then the firm will not proceed with the marketing of the product.

The firm can, however, face a slightly more complex problem if it is unsure about how successful the drug will be once it is marketed. Firms will often want to resolve such uncertainty by trying to gather some information about the marketability of their products, and on the basis of this information would revise their estimates of how well their products will do in the market. The processing of such information into the decision-making process is of great importance to any firm. To illustrate this we go back to our previous example.

21.7.2 Example

We consider the same pharmaceutical firm X as in Section 21.7.1. However, we now expand the original decision tree so as to include the event that the drug once marketed may not do very well. This decision tree is now shown in Fig. 21.11.

The two added edges G (good) and B (bad) at the nodes where "Nature" interferes allow for the possibility (with probability s) for the produced drug to be a real money maker and also for the possibility (with probability $1 - s$) to be a complete failure.

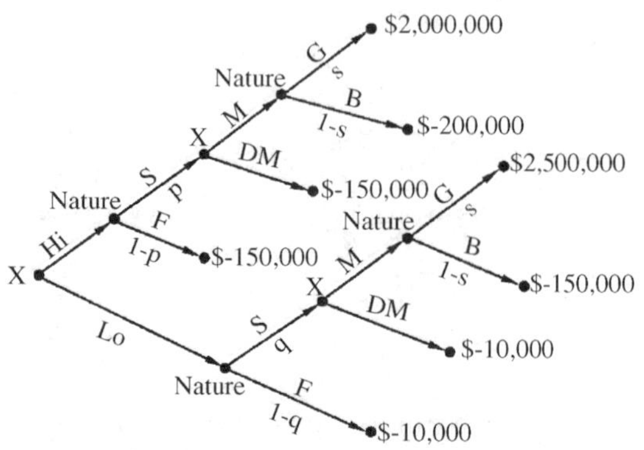

FIG. 21.11 A DECISION TREE WITH UNCERTAINTY

The *prior probability* that the drug will do well in the market is given by s. It is interesting to observe that after the firm gathers information about the market, this prior probability is revised to a *posterior probability*. This is usually done by using Bayes' formula from probability theory, which we describe below.

Bayes' formula—one of the most famous and useful formulas in probability theory and statistics—provides an answer to the following important question: *If an event B is known to have occurred what is the probability that another event A will happen?*

21.7.3 Theorem

(**Bayes' Formula**) *If A and B are two events in a probability space (S, P), then*

$$P(A/B)=\frac{P(B/A)P(A)}{P(B/A)P(A)+P(B/A^{c})P(A^{c})}.^{10} \qquad \text{Eqs. (21.32)}$$

As usual, the event A^c is the complementary event of A, i.e.,

$$A^{c}=X\setminus A=\{\chi\in S:\chi\notin A\} \qquad \text{Eq. (21.33)}$$

and so $P(A^c) = 1 - P(A)$. The non-negative numbers $P(U/V)$ appearing in Bayes' formula are known as conditional probabilities. We say that $P(U/V)$ is the **conditional probability** of the event U given the event V and define it by:

$$P(U/V)=\frac{P(U\cap V)}{P(V)} \qquad \text{Eq. (21.34)}$$

provided that $P(V) > 0$. Therefore, a useful way to interpret Bayes' formula is to think of it as the conditional probability of event A given that event B is observed.

Bayes' formula is useful whenever agents need to revise or update their probabilistic beliefs about events. The following example provides an illustration of Bayes' formula and indicates its usefulness and wide applicability.

21.7.4 Example

It is known that a certain disease is fatal 40% of the time. At present a special radiation treatment is the only method for curing the disease. Statistical records show that 45% of the people cured took the radiation treatment and that 20% of the people who did not survive took the treatment. What is the chance that a person suffering from the disease is cured after undergoing the radiation treatment?

We set up the problem as follows. First, in the sample space of all persons suffering from the disease, we consider the two events:

A = The person is cured from the disease
B = The person is taking the radiation treatment

Our problem is confined to finding $P(A/B)$.

Notice that A^c = The person did not survive. To apply Bayes' formula, we need to compute a few probabilities. From the given information, we have:

---

[10] This theorem is essentially due to Thomas Bayes (1702–1761), an English theologian and Mathematician. This famous formula which immortalized Bayes, was included in his article "*Essays Towards Solving a problem in the Doctrine of Chances*." It was published posthumously in 1763 in the *Philosophical Transactions of the Royal Society of London*, vol. **53**, 370–418.

$P(A) = 0.6$
$P(B/A) = 0.45$
$P(A^c) = 0.4$
$P(B/A^c) = 0.20$

Consequently, according to Bayes' formula, the desired probability is

$$P(A/B)=\frac{P(B/A)P(A)}{P(B/A)P(A)+P(B/A^c)P(A^c)} = \frac{0.45\times 0.6}{0.45\times 0.6+0.2\times 0.4}=0.771 \quad \text{Eq. (21.35)}$$

In other words, a person having the disease has a 77.14% chance of being cured after undergoing the radiation treatment.

21.7.5 Example

[*Revising the Prior Probability*] Going back to the decision problem of the pharmaceutical firm *X* (21.7.2), the **prior probability** that the drug will do well in the market (i.e., the good outcome *G* occurs) is given by $P(G) = s$. The firm, in order to find out more about how the market will receive the drug, may perform a test *I*; for instance, study what a sample of potential buyers think of the drug.

Based on this study the firm may want to revise its probability $P(G)$. If the test is successful, then the firm infers that the market condition is better than originally thought and would want to revise $P(G)$ accordingly. However, if it is not successful, then the inference should go the other way. Bayes' formula provides the tool for revising this prior probability $P(G)$ conditioned on the new information *I* obtained from the test. The **posterior probability**, as the revised probability is called, is given by Bayes' formula:

$$P(G/I)=\frac{P(I/G)P(G)}{P(I/G)P(G)+P(I/B)P(B)} \quad \text{Eq. (21.36)}$$

where $P(I/G)$ = probability that the test indicates success if indeed the market situation is *G*; and $P(I/B)$ = probability that the test indicates success when the market situation is *B*.

It is of interest to note that if the new information is good and reliable, then the posterior (or revised) probability should predict the state of the market with a high degree of accuracy, which usually means that the revised probability would be close to zero or one depending on the state of the market. Bayes' formula is, therefore, a nice way of using relevant information to "update beliefs about events."

Now suppose that $P(I/G) = 0.9$ and $P(I/B) = 0.2$. If $s = 0.6$, then after a test of the market which gave a positive result, the revised posterior probability $P(G/I)=(0.9\times 0.6)/(0.9\times 0.6+0.2\times 0.4)$ 0.87. This is a lot higher than the prior probability of 0.6. The firm, therefore, revises its belief about the state of the market being good after observing a positive result from the test. The information from the test is used to revise the probability upward. In the decision tree this will have consequences as the expected payoff from marketing the drug changes drastically.

21.8 SEQUENTIAL GAMES

In this section we build on the material in the previous section but now analyze situations in which multiple decision-makers are now involved in making sequential decisions. We thus have both the elements of game theory and sequential decisions present in the analysis. We start with the definition of a sequential game.

21.8.1 Definition

A tree *T* is said to be an *n*-player (or *n*-person) **game tree** (or a **game tree** for *n*-players $P_1, \ldots, P_n$), if

(1) Each nonterminal node of the tree is "owned" by exactly one of the players
(2) At each terminal node *v* of the tree an *n*-dimensional "payoff" vector

$$p(v) = [p_1(v), p_2(v), \ldots, p_n(v)] \quad \text{Eq. (21.37)}$$

is assigned

We emphasize the following two things regarding game trees:

(1) No terminal node is owned by any player.
(2) There is no guarantee that each player "owns" at least one nonterminal node of the tree. That is, in an *n*-person game there might be players who do not own any nonterminal nodes (These are known as nonactive players.)

A node *N* owned by a player *P* is also expressed by saying that the node *N* **belongs to** player *P*. The nonterminal nodes of a game tree are called **decision nodes**.

A strategy in a "sequential" game thus seems to be a fairly subtle concept. Briefly, a strategy s_i for a player *i* in a sequential game consists of the choices that the player is going to make at the nodes he owns. Therefore, a strategy for a player in a sequential game is a *complete plan* of how to play the game and prescribes the choices at every node owned by the player. In other words, a player's strategy will indicate the choices that the player has planned to make *a priori*, i.e., before the game starts. A **strategy profile** for an *n*-person sequential game is then simply an *n*-tuple $(s_1, s_2, \ldots, s_n)$, where each s_i is a strategy for player *i*. It is useful to note here that once a strategy profile $(s_1, \ldots s_n)$ is given in a sequential game, a terminal node of the game tree will be reached automatically. In other words, as mentioned before, a sequential game is understood to be played as follows: The player (say P_i) who owns the root *R* chooses a node according to his selected strategy s_j; here he chooses the node $s_j(R)$. Then the player who owns the node $s_j(R)$ chooses according to his strategy and the game continues in this fashion until a terminal node *v* is reached and the game ends. Subsequently, each player *i* gets the payoff $p_i(v)$. Notice that the strategy profile $(s_1, s_2, \ldots s_n)$ uniquely determines the terminal node *v* that is reached. Hence, the payoff (or utility) of each player is a function u_i of the strategy profile $(s_1, s_2, \ldots, s_n)$. That is, we usually write:

$$u_i(s_1, s_2, \ldots, s_n) = p_i(v) \quad \text{Eq. (21.38)}$$

Thus, in sum, a sequential game is represented by a game tree with players moving sequentially. At each information set, the player who needs to choose has determined a priori a choice (i.e., an edge) at each of the nodes in the information set, which is exactly the same for each node in the same information set. After the players have chosen their actions at their information sets a terminal node is reached and the outcome of the game is realized.

A solution of a sequential game is understood to be a Nash equilibrium, and is defined as follows.

21.8.2 Definition

In an *n*-player sequential game (with perfect or imperfect information) a strategy profile $(s^*, s^*, \ldots, s^*)$ is said to be a **Nash equilibrium** (or simply an **equilibrium**) if for each player *i*, we have

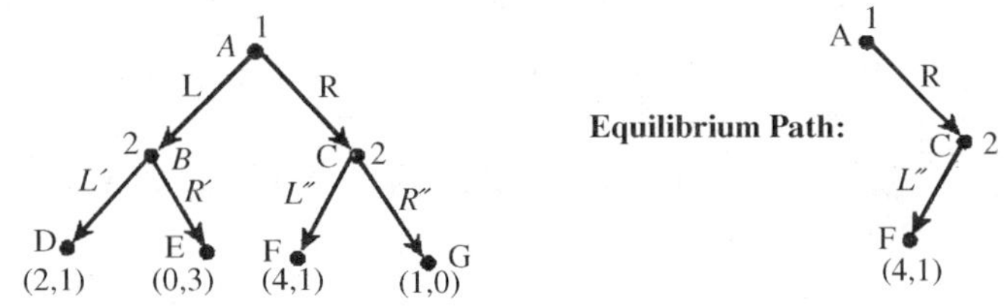

FIG. 21.12 THE EQUILIBRIUM PATH

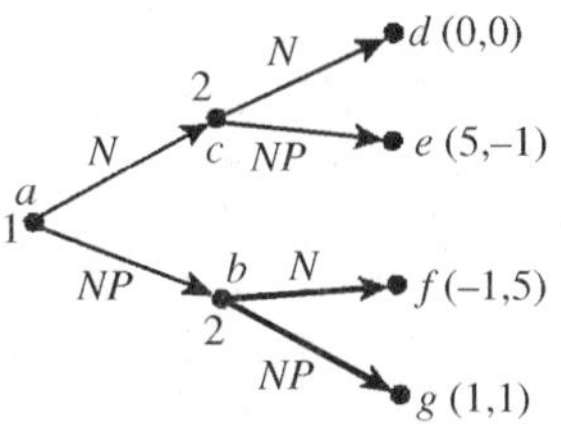

FIG. 21.13 A NUCLEAR DETERRENCE GAME

$$u_i(s_1^*,\ldots,s_{i-1}^*,s_i^*,s_{i+1}^*,\ldots,s_n^*)=\max_{s\in S_i} u_i(s_1^*,\ldots,s_{i-1}^*,s,s_{i+1}^*,\ldots,s_n^*)$$

Eq. (21.39)

In other words, as in the previous cases, a Nash equilibrium is a strategy profile $(s_1^*,s_2^*,\ldots,s_n^*)$ such that no player can improve his payoff by changing his strategy if the other players do not change theirs.

Let us illustrate the preceding discussion with an example.

21.8.3 Example

Consider the following simple two-person sequential game with perfect information whose game tree is shown in Fig. 21.12.

If player 1 plays L then node B is reached and player 2 will play R', in which case player 1 gets zero. If player 1 plays R then node C is reached and player 2 plays L'' (and player 1 gets 4). The solution path is, therefore, $A \to C \to F$, which leads to the terminal node F at which the payoff vector is (4, 1).

Now if we think of the strategies that the players use, we find that player 1 has choices at one node (the node A) at which he can choose either R or L. Player 2, however, has to choose at the two different nodes B and C. Player 2's strategy is, therefore, a function from $\{B,C\}$ to $\{L', R', L'', R''\}$ with the feasibility restriction that from node B one can only choose R' or L' and a similar restriction on choices from node C. What strategies are then equilibrium strategies?

The reader should verify that the strategy profiles $(\{R\}, \{R', L''\})$ and $(\{R\}, \{L', L''\})$ are the only two Nash equilibria of the game. They both support the equilibrium path $A \to C \to F$.

Do *sequential games have equilibrium points*? The answer is "Yes" if the sequential game is of "perfect information." This important result was proved by H. W. Kuhn[11].

21.8.4 Theorem

(Kuhn) Every sequential game with perfect information has a Nash equilibrium.

To prove Theorem (21.8.4) one must employ the so-called **Backward Induction Method**. We now present two examples of sequential games.

21.8.5 Example

[*Nuclear Deterrence*] Two nuclear powers are engaged in an arms race in which each power stockpiles nuclear weapons. At issue is the rationality of such a strategy on the part of both powers.

Let us examine the question by looking at a stylized version of the game that the two powers are engaged in. Country 1 moves in the first stage and may choose between nuclear weapons N or non-proliferation (NP). Country 2 in stage 2 of the game observes the choice that country 1 has made and chooses between N and NP. A representative game tree of the situation is shown in Fig. 21.13.

According to the payoffs shown in Fig. 21.13, country 2 likes the option N whether country 1 chooses NP or N. If country 1 chooses NP, then country 2 by choosing N guarantees for itself a very powerful position vis-a-vis country 1. If country 1 chooses N, then country 2 would like to choose N as this allows it a credible deterrence against a possible nuclear attack by country 1.

Knowing country 2's thinking on this issue, country 1 knows that it is optimal for it to choose N. It is easy to see that the backward induction solution of this game is the following.

- *Country 2 chooses N irrespective of whether country 1 chooses N or NP.*
- *Country 1 chooses N.*

In other words, the path $a \to c \to d$ is the only Nash equilibrium path of the game.

While the example is quite clearly highly stylized, it brings to the fore the incentives that countries have in engaging in arms races. In the game, it is clearly rational for the two countries to build up their nuclear arsenal. And left to themselves the countries would do exactly what the model predicts.

It is also clear that both countries would be better of without having to spend on an arms race, but the equilibrium solution predicts differently. This is precisely why arms races are so prevalent and why it is so difficult to dissuade countries from pursuing other strategies.

The next example is very well known in economics. We revisit the scenario of the duopoly game of Section 21.4.1, but instead of having the firms move simultaneously, we now have one firm making its move before the other firm. That is, one of the firm sets its quantity before the other firm. This, of course, changes the entire game. The game has now become a sequential game with perfect information as the quantity choice of the firm that sets its quantity first is known to the second firm when the second firm decides what quantity to produce. This duopoly model was first analyzed by von Stackelberg.

21.8.6 Example

[*The Stackelberg Duopoly Model*] The Stackelberg duopoly game is played as follows. There are two firms producing identical products; firm 1 and firm 2. Firm 1 chooses a quantity $q_1 \geq 0$ firm 2 observes q_1 and then chooses q_2. The resulting payoff or profit of firm i is:

$$\pi_i(q_1,q_2)=q_i[p(q)-c_i] \qquad \text{Eq. (21.40)}$$

where $q = q_1 + q_2$, $p(q) = A-q$ = market clearing price when the total output in the market is q; and c_i = marginal cost of production of the product by firm i. That is, the profit of each firm i is

[11]Harold W. Kuhn is Professor Emeritus of Mathematical Economics at Princeton University. He made many contributions to Game Theory.

$$\pi_i(q_1,q_2) = q_i(A-q_1-q_2-c_i) \qquad \text{Eq. (21.41)}$$

Note that the game is a two-person sequential game with two stages and with perfect information. If we use the Backward Induction Method to solve the game, we must first find the reaction of firm 2 to every output choice of firm 1. Hence, we must find the output q_2^* of firm 2 that maximizes firm 2's profit given the output q_1 of firm 1. That is, $q_2^* = q_2^*(q_1)$ solves:

$$\pi_2(q_1,q_2^*) = \max_{q_2 \geq 0} \pi_2(q_1,q_2) \qquad \text{Eq.(21.42a)}$$

$$= \max_{q_2 \geq 0} q_2(A-q_1-q_2-c_2) \qquad \text{Eq.(21.42b)}$$

Since $\pi_2(q_1,q_2) = -(q_2)^2 + (A-q_1-c_2)q_2$, taking the first and second derivatives with respect to q_2, we get

$$\frac{\partial \pi_2}{\partial \pi_2} = -2q_2 + A - q_1 - c_2 \text{ and } \frac{\partial^2 \pi_2}{\partial q_2^2} = -2 < 0 \qquad \text{Eq. (21.43)}$$

So, according to the First- and Second-Order Tests, the maximizer q_2^* is the solution of the equation $(\partial \pi_2/\partial q_2) = -2q_2 + A -$ $q_1 - c_2 = 0$. Solving for q_2, we get

$$q_2^* = q_2^*(q_1) = \frac{A-q_1-c_2}{2} \qquad \text{Eq. (21.44)}$$

provided $q_1 < A - c_2$.

Firm 1 should now anticipate that firm 2 will choose q_2^* if firm 1 chooses q_1. Therefore, firm 1 will want to choose q_1 to maximize the function

$$\pi_1(q_1,q_2^*) = q_1(A-q_1-q_2^*-c_1) \qquad \text{Eq.(21.45a)}$$

$$= q_1\left(A - q_1 - \frac{A-q_1-c_2}{2} - c_1\right) \qquad \text{Eq.(21.45b)}$$

$$= \frac{1}{2}[-(q_1)^2 + (A+c_2-2c_1)/q_1] \qquad \text{Eq.(21.45c)}$$

subject to $q_1 \geq 0$. Using again the First- and Second-Order Tests, we get

$$\frac{\partial \pi(q_1,q_2^*)}{\partial q_1} = -q_1 + \frac{A+c_2-2c_1}{2}$$

$$\text{and } \frac{\partial^2 \pi(q_1,q_2^*)}{\partial q_1^2} = -1 < 0 \qquad \text{Eq. (21.46)}$$

Therefore, the maximizer of $\pi(q_1,q_2^*)$ is $q_1^* = (A + 2c_1 - 3c_2)/4$ Substituting this value in Eq. (21.44), we get

$$q_2^* = \frac{(A+2c_1-3c_2)}{4} \qquad \text{Eq. (21.47)}$$

This is the Backward Induction solution of the Stackelberg game. The equilibrium strategy of firm 1 is $q_1^* = (A+c_2-2c_1)/2$ while the equilibrium strategy of firm 2 is $q_2^* = (A + 2c_1 - 3c_2)/4$.

21.9 AUCTIONS AND BARGAINING

Auctions have been used to sell and buy goods since prehistory and even today auctions are used quite frequently. Sotheby's of London, with branches in most of the wealthy metropolitan centers of the world, is in the business of auctioning rare art and antiques to wealthy buyers. Local municipalities use some form of auctioning to hire contractors for specific projects. Offshore oil leases are regularly auctioned to the major oil companies as well as independent wildcatters. One of the largest auctions, with billions of dollars changing hands, took place in July 1994. The United States government "auctioned licenses to use the electromagnetic spectrum for personal communications services: mobile telephones, two-way paging, portable fax machines and wireless computer networks."[12] As we see, auctions are used in many different contexts.

Auctions, as we briefly discussed earlier, can be written as games. There, we analyzed an auction in which bids were made simultaneously for a single good by bidders who knew the worth of the good. But this is only one kind of an auction. There are auctions in which only a single unit of an indivisible good is sold, and there are auctions in which a single seller sells n different goods. The airwaves auction was of the latter kind. Auctions can also be classified according to whether the winner pays the winning bid or the second highest bid. In some auctions the bidders know what the value of the item is worth to the individual, whereas in other auctions there is a great deal of uncertainty about it. In auctions of offshore drilling rights, the bidders have only some estimate of the true value of the lease.

An auction can take many different forms. The bidders can make their bids *simultaneously* and put them in sealed envelopes, or they may *bid sequentially* with the auctioneer calling out the successive bids. An auction may also be a *first price auction* in which the winner pays the highest bid, or it might very well be a second price auction where the winner pays only the second highest bid.

We now indicate how auctions can be viewed as games. An auction is a gathering of n persons (called bidders or players numbered from 1 to n) for the sole purpose of buying an object (or good). The winner of the object is decided according to certain rules that have been declared in advance. We assume that the good for sale in an auction is worth v_i to the ith bidder. The value v_i is called the valuation of the object by bidder i. As expected, the valuations play a crucial role in the analysis of auctions as games. Without any loss of generality, we can suppose that $v_1 \geq, v_2 \geq \ldots \geq v_n$.

Auctions are classified according to their rules as well as according to the information the bidders have about the **valuation vector** $(v_1, v_2, \ldots, v_n)$. An auction in which every bidder knows the vector $v = (v_1, v_2, \ldots, v_n)$ is called an auction with **complete information**, otherwise it is an auction with **incomplete information**. As usual, most auctions are those with incomplete information.

21.10 INDIVIDUAL PRIVATE VALUE AUCTIONS

An individual private value auction is an auction in which the bidders only know their own valuation of the item, though they may have some idea about the valuation of the other bidders. For instance, when you are at an auction of a rare piece of art, you know how much you are willing to pay for it, but you have only a vague idea about how much the others value it.

In this section, we study in detail an auction with two bidders who use "linear rules" as their bidding strategies. In this first-price sealed-bid individual private value auction, each bidder i has her own valuation v_i of the object, and the bidder with the highest bid

[12]McAfee, R. P. and McMillan, J., 1996. "Analyzing the Airwaves Auction," *J. Eco. Perspectives*, Vol. **10**, pp. 159–175.

wins. In case both bid the same amount, the winner is decided by a draw. So, as before, the payoff functions of the players are given by

$$u_1(b_1,b_2)=\begin{cases} v_1-b_1 & \text{if } b_1>b_2 \\ \frac{v_1-b_1}{2} & \text{if } b_1=b_2 \\ 0 & \text{if } b_1<b_2 \end{cases} \quad \text{Eq. (21.48)}$$

and

$$u_2(b_1,b_2)=\begin{cases} v_2-b_2 & \text{if } b_2>b_1 \\ \frac{v_2-b_2}{2} & \text{if } b_2=b_1 \\ 0 & \text{if } b_2<b_1 \end{cases} \quad \text{Eq. (21.49)}$$

Here, as mentioned above, we are assuming that if the bidders make the same bid, then the winner is decided by the toss of a coin so that the probability of winning is 1/2. Thus, the utility in this case is the expected payoff from winning the auction.

Here the bidders do not know the true valuation of the object by the other bidder. Though each player is uncertain (due to lack of information) about the true valuation of the other player, each player has a belief (or an estimate) of the true valuation of the others. Since player i does not know player j's true valuation v_j of the object, she must treat the value v_j as a random variable. This means that the belief of player i about the true value of v_j is expressed by means of a distribution function F_i. That is, player i considers v_j to be a random variable with a distribution function F_i. Thus, player i believes that the event $v_j < v$ will happen with probability

$$P_i(v_j \le v)=F_i(v) \quad \text{Eq. (21.50)}$$

We note from the outset that since this is a game, each bidder can arrive at an optimal bid only after guessing the bidding behavior of the other players. Naturally, the bids b_1 and b_2 of the players must be functions of the two valuations v_1 and v_2. In other words, $b_1 = b_1(v_1)$ and $b_2 = b_2(v)$.

Given the lack of information on the part of the players, the best that any player can do is to choose a bid that maximizes her expected payoff. Notice that the expected payoff of the players are given by

$$\begin{aligned} E_1(b_1,b_2) &= P_1(b_1>b_2)u_1(b_1,b_2)+P_1(b_1=b_2)u_1(b_1,b_2) \\ &\quad + P_1(b_1<b_2)u_1(b_1,b_2) \quad \text{Eq.(21.51a)} \\ &= (v_1-b_1)P_1(b_1>b_2)+\tfrac{1}{2}(v_1-b_1)P_1(b_1=b_2) \quad \text{Eq.(21.51b)} \end{aligned}$$

and

$$E_2(b_1,b_2)=(v_2-b_2)P_2(b_2>b_1)+\tfrac{1}{2}(v_2-b_2)P_2(b_2=b_1) \quad \text{Eq. (21.52)}$$

Observe that the first term in the formula $E_1(b_1,b_2)$ describes the possibility that bidder 1 wins and receives the payoff $v_1 - b_1$, and the second term gives the payoff when there is a tie, in which case bidder 1's expected payoff is $1/2(v_1-b_1)$.

So, in this auction, the strategy of a bidder, say of bidder 1, is simply her bidding function $b_1(v_1)$ and her objective is to maximize her expected payoff given the bidding function $b_2 = b_2(v_2)$ of the second bidder. Thus, the expected payoff functions can be written as:

$$Eu_1(b_1 \mid b_2)=(v_1-b_1)P_1(b_1>b_2)+\tfrac{1}{2}(v_1-b_1)P_1(b_1=b_2) \quad \text{Eq. (21.53)}$$

and

$$Eu_2(b_2 \mid b_1)=(v_2-b_2)P_2(b_2>b_1)+\tfrac{1}{2}(v_2-b_2)P_2(b_2=b_1) \quad \text{Eq. (21.54)}$$

This now naturally leads to our old concept of a Nash equilibrium. A pair of bidding functions $[b_1^*(v_1),b_2^*(v_2)]$ is said to be a **Nash equilibrium** for the individual private value auction if for every bidding function $b_1(v_1)$ of player 1 we have:

$$Eu_1(b_1 \mid b_2^*)\le Eu_1(b_1^* b_2^*) \quad \text{Eq. (21.55)}$$

and for each bidding function $b_2(v_2)$ of player 2 we have:

$$Eu_2(b_2 \mid b_2^*)\le Eu_2(b_2^* b_1^*) \quad \text{Eq. (21.56)}$$

We now work out the details in a specific case. Assume that both players know that the valuation of the object lies between a lower value $\underline{v}\ge 0$ and an upper value $\overline{v} > \underline{v}$. Assume further that each bidder knows that the valuation of the other bidder is uniformly distributed on the interval $[\underline{v},\overline{v}]$. That is, bidder i knows only that the true valuation v_j of bidder j is a random variable whose density function $f_i(v)$ is given by:

$$f_i(v)=\begin{cases} \frac{1}{\overline{v}-\underline{v}} & \text{if } \underline{v}<v<\overline{v} \\ 0 & \text{otherwise.} \end{cases} \quad \text{Eq. (21.57)}$$

In other words, player i believes that the likelihood of v_j having at most the value v is given by:

$$P_i(v_j\le v)=\int_{-\infty}^{v} f_i(t)\,dt=\begin{cases} 0 & \text{if } v<\underline{v} \\ \frac{v-\underline{v}}{\overline{v}-\underline{v}} & \text{if } \underline{v}\le v\le\overline{v} \\ 1 & \text{if } v<\overline{v}. \end{cases} \quad \text{Eq. (21.58)}$$

It should be clear that the following two "rationality" conditions must be satisfied:

$$\underline{v}\le b_2(\underline{v}) \text{ and } \underline{v}\le b_1(\underline{v})\le\overline{v} \quad \text{Eq. (21.59)}$$

As both bidders have symmetric information about each other's valuation, each should use essentially the same reasoning to choose an optimal strategy. We have the following result.

Bidding Rules in the Two-Bidder Case

Assume that in a Two-bidder individual private value auction the valuations of the bidders are independent random variables uniformly distributed over an interval $[\underline{v},\overline{v}]$. *Then the* ***linear bidding rules***

$$b_1(v_1)=\frac{1}{2}\underline{v}+\frac{1}{2}v_1 \text{ and } b_2(v_2)=\frac{1}{2}\underline{v}+\frac{1}{2}v_2 \quad \text{Eq. (21.60)}$$

form a symmetric Nash equilibrium.

The graph of a linear rule is shown in Fig. 21.14.

Let us now see how these linear bidding rules work out in a simple example.

21.10.1 Example

Suppose two bidders are bidding for a painting that each knows is worth between \$100,000 and \$500,000, and that each bidder's

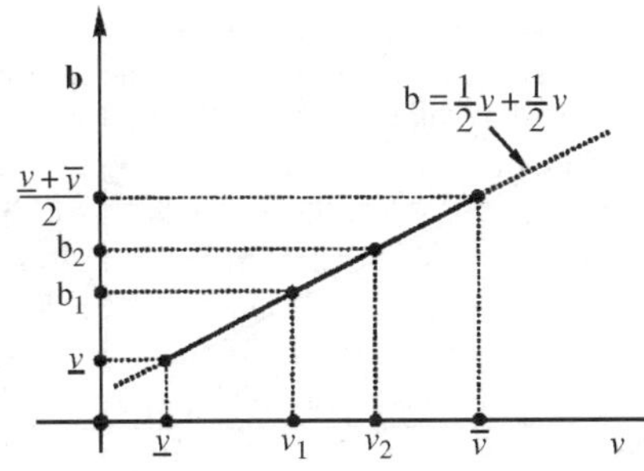

FIG. 21.14 A LINEAR BIDDING RULE

valuation of the painting is uniformly distributed over the interval [100,000, 500,000]. Thus, in this case v = 100,000 and $\bar{v}$ = 500,000. The equilibrium bidding rules in this case are:

$$b_i(v_i)=\frac{1}{2}v_i+50{,}000,\quad i=1,2 \qquad \text{Eq. (21.61)}$$

If bidder 1's true valuation is \$200,000 then she bids b_I = \$150,000, and if bidder 2's true valuation is \$250,000, then bidder 2 bids b_2 = \$175,000. The auctioneer in this case collects \$175,000 and bidder 2 gets the painting for \$175,000. The bidding rules apply to any number of bidders. Specifically, we have the following general result.

Bidding Rules in the *n*-bidder Case

Assume that in an n-bidder individual private value auction the valuations of the bidders are independent random variables uniformly distributed over an interval $[\underline{v},\bar{v}]$. *Then the* **linear bidding rules**

$$b_i(v_i)=\frac{1}{n}\underline{v}+\frac{n-1}{n}v_i,\quad i=1,2,3 \qquad \text{Eq. (21.62)}$$

form a symmetric Nash equilibrium.

21.11 ENGLISH AUCTIONS

One of the most popular type of auctions is the one in which the auctioneer uses a sequential bidding procedure. There are quite a few variants of sequential bid auctions. The most widely used is a variant of the **English auction** in which the auctioneer calls successively higher bids and a bidder then indicates whether she is willing to make that bid. The bidder who makes the last bid in this sequence of bids then wins the auction and pays that bid. In Japan a slightly different form of the English auction is used. The price is posted using an electronic display and the price is raised continuously. A bidder who wishes to be active at the current price depresses a button. When she releases the button she has withdrawn from the auction. The Dutch often use a sequential bidding procedure to auction tulips and tulip bulbs. The auction, however, starts with a high price and the price is continuously lowered until a bidder agrees to pay the bid. These auctions are called **Dutch auctions** and are obviously quite different from the English one. Here we analyze the standard version of the English auction. Such an auction is again a gathering of n persons for the sole purpose of buying an object under the following rules.

(1) The auctioneer (the person in charge of the auction) starts the bidding by announcing a price b_0 for the object. This is round (or stage) zero of the auction. The quoted price b_0 is the floor price of the object at round zero. We assume that $b_0 > 0$.

(2) Once the price b_0 is announced by the auctioneer, the players start bidding in a sequential fashion, i.e., in succession one after the other. Successive bids must be higher than the prevailing floor price. Thus, the first person who announces a price $b_1 > b_0$ brings the auction to round 1 and the price b_1 is now the floor price of round 1. The next player who bids a price $b_1 > b_2$ brings the auction to round 2 and to the floor price b_2, and so on. At each stage of the auction every player has the right to bid again, even if she had bid in earlier rounds.[13] Consequently, the floor price b_k at stage k is the result of the successive bids:

$$0<b_0<b_1<b_2<\cdots<b_k \qquad \text{Eq. (21.63)}$$

(3) If at some round k no one bids higher, then the player with the last bid b_k is declared to be the winner and the auction ends. The player with the last bid then pays the amount b_k to the auctioneer and gets the object.

Since the process of bidding in an English auction is drastically different from a sealed-bid auction, it is, of course, quite natural to wonder whether the final bid would be different from the sealed-bid auction. Again, we start discussing English auctions by assuming that each bidder i has a true valuation v_i of the item. As before, without loss of generality, we may assume that the valuations are ranked in the order:

$$v_1\geq v_2\geq \ldots \geq v_n \qquad \text{Eq. (21.64)}$$

Since an English auction is still an individual private value auction, the bidders do not know the valuations of the others and know only their own valuations of the item. We shall see that one of the advantages of an English auction is that the bidders do not really need to know the distribution of possible valuations to bid optimally. We go on to examine the nature of an optimal strategy in an English bid auction.

Claim 1: *No bidder will bid more than her valuation.*
In order to justify this claim, we must interpret it in the framework of expected payoffs. Assume that a bidder i bids $b_k > v_i$ at the kth round of bidding. Then her belief about her chances of winning the auction is expressed by a number $0 \leq p \leq 1$, where p = the probability that b_k is the highest bid, and, of course, $1 - p$ is the probability that some other bidder will bid a higher price at the $(k + 1)$ round. So, player i, by bidding $b_k > v_{ix}$ the kth round expects a payoff of

$$p(v_i-b_k)+(1-p)\cdot 0=p(v_i-b_k)\leq 0 \qquad \text{Eq. (21.65)}$$

This is negative if p is not zero. However, notice that she can have an expected payoff which is at least as high, by bidding no more than her evaluation v_i.
Claim 2: Bidder i will bid as long as the last bid is below v_i.
To establish this claim, there are two cases to consider. First, if bidder i made the last bid b_k, then bidder i will not bid as long as there are no further bids, in which case bidder i wins and receives the payoff $v_i - b_k$. However, if $b_k > v_i$, then this payoff is negative and she would have been better of at the kth round,

[13]It is understood here that rational bidders will not make two consecutive bids since by doing so they simply lower their expected payoffs of winning the auction. Two successive bids by the same player is tantamount to bidding against oneself.

either by not bidding at all, or, in case $b_{k-1} < v_i$, by bidding b_k such that $b_{k-1} < b_k \leq v_i$.

If the floor price after k rounds of bidding is $b_k < v_i$, and bidder i did not make the last bid, then bidder i will bid an amount b_{k+1} on the $(k + 1)$th round such that $b_k < b_{k+1} \leq v_i$ as the expected payoff from bidding at the $(k + 1)$th round is:

$$p(v_i - b_{k+1}) + (1 - p) \cdot 0 = p(v_i - b_{k+1}) \geq 0 \qquad \text{Eq. (21.66)}$$

This expected payoff is positive if bidder i thinks that there is a positive probability p that b_{k+1} is the highest bid. In this case, the expected payoff from not bidding is zero, irrespective of the beliefs of player i.

We can now use the preceding two claims to determine the winning bid in an English auction. Clearly, the bidding stops as soon as the floor price b_t at the tth round of bidding exceeds or is equal to v_2, the second highest bid. Since the second highest bidder has no incentive to bid $b_t > v_2$, the bid must have been made by bidder 1, and hence, $b_t \leq v_1$. Therefore, in an English auction, the winning bid b^* must always satisfy $v_2 \leq b^* \leq v_1$. We emphasize here that the winning bid is independent of the information or beliefs that players have about each others valuations. The final bid is simply a consequence of the true valuations of the bidders.

One needs to compare the preceding conclusion with the outcome in a sealed-bid auction. Recall that in a sealed-bid auction the bid made by the bidders is not independent of their beliefs about the valuations of the others. One thus faces the following intriguing question: *Given a choice of the two forms of auctions, which one of the two would an auctioneer choose?*

The answer, as we shall see below, depends on the valuations of the players as well as on their beliefs about each others true valuations.

21.11.1 Example

Let us go back to Example 21.10.1 in which there are two bidders with valuations v_1 = \$250,000 and v_2 = \$200,000. If the auction is an English auction the bidding would stop as soon as the bid went over \$200,000. Thus, the auctioneer will net a little over \$200,000 for the item. In the case of the sealed-bid auction, where the beliefs of the bidders about the valuations of the others are uniformly distributed between \$100,000 and \$500,000, the winning bid is only \$175,000. Thus, in this case, the English auction generates significantly more revenue for the auctioneer than the sealed-bid auction.

In contrast, if we now change the parameters to v = \$200,000, v_1 = \$300,000 and v_2 = \$200,000, then the sealed-bid auction would get a winning bid of \$250,000 and the English auction could get a winning bid of only \$200,000. Thus, in this case the sealed-bid auction generates substantially more revenue than the English auction.

21.12 COMMON-VALUE AUCTIONS

A common-value auction is a first-price sealed-bid auction in which

- The underlying true value of the object is the same for all bidders (hence the name *common-value auction*).
- The bidders receive information about the true value of the object by means of "signals."

In a common-value auction, the bidders have the least amount of information. In addition to not knowing the valuations of the others, they are also uncertain about their own valuations. In such auctions, each bidder receives a "noisy" signal about the true value of the object and on the basis of this signal she forms an estimate of its value. Consequently, in a common-value auction, the valuation of bidder i is viewed as a random variable not only by the other bidders but also by bidder i herself.

Typical examples of common-value auctions are auctions of offshore oil leases. In these auctions, the bidders, who are typically the big oil producing firms and some independent wildcatters, do not have a precise idea of the value of the leases. They form an estimate of the value of the lease on the basis of some signal they observe. The U.S. government, which auctions these tracts of ocean, provides a legal description of the location of the area being leased. The bidders are responsible for gathering whatever information they can about the tract. In this case, the information provided by geologists and seismologists is usually the noisy signal observed by the bidders.

In order to make our model as simple as possible, we assume that:

- The bidders observe signals that are independent realizations of random variables that are uniformly distributed on the interval [0, 1].
- The object that is being auctioned is known to take only two possible values; a high value v_h and a low value v_ℓ.
- The joint density function of the random value v of the object and the signal ω is given by:

$$f(v\ \omega) = \begin{cases} \omega & \text{if } v = v_h \\ 1 - \omega & \text{if } v = v_\ell \end{cases} . \qquad \text{Eq. (21.67)}$$

Under these assumptions, we can establish the following bidding rules.

Bidding Rules in an n-Bidder Common-Value Auction
Assume that in an n-bidder common-value auction the bidders observe signals that are independent random variables that are uniformly distributed over the interval [0, 1]. Then the linear bidding rules

$$b_i = v_\ell + \frac{n-1}{n}(v_h - v_l)\omega_i,\ i = 1, \ldots, n \qquad \text{Eq. (21.68)}$$

is a symmetric Nash equilibrium for the common-value auction.

21.13 BARGAINING

In the previous discussion we used game-theoretic arguments to understand auctions of various kinds. We saw that auctions are special types of markets in which buyers bid for an object. However, there are many other forms of markets in which, instead of buyers simply bidding for the good, buyers and sellers actually make offers and counteroffers. To analyze and understand such markets, we need a different approach from the one used in the previous sections. Here we shall discuss a model of bargaining and trade. The housing market as well as the market for automobiles are good examples of markets in which the good is traded only after the buyer and the seller have reached an agreement on the price. In these cases, the agreement is reached only after a certain amount of bargaining.

When one enters the housing market, say as a buyer, the individual looks at houses that are for sale at some listed price. The buyer then makes a decision about which of these houses is the most desirable and within the individual's budget. Once the decision is made, the buyer makes an offer to the seller, usually at a price lower than the listed price. The seller then either accepts the offer or makes a counteroffer that is somewhere between the original list price and the offer of the buyer. The buyer either accepts the counteroffer or makes another counteroffer or possibly terminates the bargaining process. Another example of a market that uses such a bargaining process is the automobile market, which also starts the bargaining process with a list price quoted by the seller. Clearly, such markets are quite different from auctions and, as we shall see, can be sequential games.

The bargaining problem is associated with the following classical problem:

- How should a number of individuals divide a pie?

Stated in this way, the problem seems to be fairly narrowly defined. However, understanding how to solve it provides valuable insights into how to solve more complex bargaining problems. When a buyer and a seller negotiate the price of a house they are faced with a bargaining problem. Similarly, two trading countries bargaining over the terms of trade, a basketball player discussing his contract with the owners of a team, or two corporations arguing over the details of a joint venture, are all examples of such two-person bargaining. In all these bargaining situations, there is usually a set **S** of alternative outcomes and the two sides have to agree on some element of this set. Once an agreement has been reached, the bargaining is over, and the two sides then receive their respective payoffs. In case they cannot agree, the result is usually the status quo, and we say there is **disagreement**. It is quite clear that the two sides will not engage in bargaining unless there are outcomes in S, which give both sides a higher payoff than the payoffs they receive from the status quo. Thus, if (d_1,d_2) are the payoffs from the disagreement point, then the interesting part of S consists of those outcomes that give both sides higher payoffs than the disagreement payoffs. We can thus define a bargaining problem as follows.

21.13.1 Definition

A two-person bargaining problem (*or* **game**) *consists of two persons (or players) 1 and 2, a set S of* **feasible alternatives** (*or* **bargaining outcomes** *or simply* **outcomes**), *and a utility function* u_i *on S for each player i, such that:*

(1) $u_1(s) \geq d_1$ *and* $u_2(s) \geq d_2$ *for every* $s \in S$
(2) *At least for one* $s \in S$ *we have* $u_1(s) > d_1$ *and* $u_2(s) > d_2$

Notice that condition 2 guarantees that there is a feasible alternative, which makes both players strictly better off relative to the disagreement point. This condition makes the bargaining problem nontrivial. Formally, we can write a bargaining problem as a triplet:

$$\beta = [S,(u_1,d_1),(u_2,d_2)] \qquad \text{Eq. (21.69)}$$

where S, u_1 and u_2 satisfy properties 1 and 2 of Definition 21.13.1.

Now notice that to every alternative $s \in S$ there corresponds a pair of utilities $[u_1(s), u_2(s)]$. Such a pair will be called a **utility allocation**. Thus, with every bargaining game, we can associate its set of utility allocations

$$U = \{[u_1(s),u_2(s)] : s \in S\} \qquad \text{Eq.(21.70)}$$

In case we need to the designate the game to which U belongs, we shall write U_S instead of U. Clearly, U is a subset of the u_1u_2-plane.

As with any game, here too, we are interested in finding a satisfactory "solution" to the bargaining game. For our brief exposition to the subject our solution will be confined to the Pareto optimal "solutions."

21.13.2 Definition

(Pareto Optimality or Pareto Efficiency) *An outcome* $s^* \in S$ *is said to be Pareto optimal or Pareto efficient) if there is no other outcome* $s \in S$ *satisfying:*

(1) $u_1(s) \geq u_1(s^*)$ and $u_2(s) \geq u_2(s^*)$
(2) $u_i(s) > u_i(s^*)$ for at least one player i.

We now proceed to describe a solution rule, which is Pareto optimal and satisfies some other additional important properties that we shall mention in the basic theorem without defining them here. We start by associating to each bargaining game $\beta = [S,(u_1,d_1),(u_2,d_2)]$ the function $g_\beta : S \rightarrow \Re$ defined by

$$g_\beta(s) = [u_1(s) - d_1][u_2(s) - d_2] \qquad \text{Eq. (21.71)}$$

and let $\sigma(\beta)$ be the set of all maximizers of the function g_β, i.e.,

$$\sigma(\beta) = \{s \in S : g_\beta(s) = \max_{t \in S} g_\beta(t)\} \qquad \text{Eq. (21.72)}$$

Note here that when g_β does not have any maximizer over the set S, then $\sigma(\beta) = \emptyset$, the empty set. We shall call the members of $\sigma(\beta)$ the Nash solutions of the bargaining game.

Here is the basic result regarding the Nash solutions.

21.13.3 Theorem

(Nash) *If a bargaining game has a compact set of utility allocations (i.e., U is closed and bounded), then Nash solutions exist and every one of them is Pareto optimal, independent of irrelevant alternatives and independent of linear transformations.*

The notions of convexity and symmetry are usually associated with the Nash solution and they are defined as follows.

21.13.4 Definition

The set of utility allocations U of a bargaining game is said to be:

(1) **Convex**, if it contains every point on the line segment joining any two of its points
(2) **Symmetric**, if $(u_1,u_2) \in U$ implies $(u_2,u_1) \in U$.

Geometrically, symmetry means that the set U is symmetric with respect to the bisector line $u_1 = u_2$. These properties are illustrated in the sets shown in Fig. 21.15.

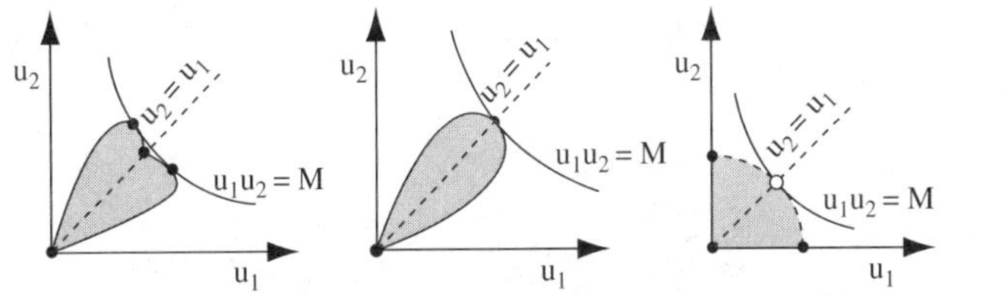

(a) A symmetric, compact and nonconvex set (b) A symmetric, compact and convex set (c) A symmetric, nonclosed, bounded and convex set

FIG. 21.15 EXAMPLES OF BARGAINING SETS

Next, we exhibit three examples to show how one applies the Nash solution to bargaining games.

21.13.5 Example

Suppose two individuals are bargaining over a sum of money; say \$100. If they cannot agree on how to divide the money, none of them gets any money. The bargaining set S in this case consists of all pairs (m_1, m_2) of non-negative real numbers such that $m_1+m_2 \leq 100$, where m_i denotes the amount of money that player i receives. That is:

$$S = [(m_1, m_2): m_1 \geq 0,\ m_2 \geq 0, \text{ and } m_1 + m_2 \leq 100] \quad \text{Eq. (21.73)}$$

The utility that any individual gets is measured by the amount of money she receives. Therefore, the utility functions of the players are

$$U_1(m_1,m_2) = m_1 \quad \text{and} \quad u_2(m_1,m_2) = m_2 \quad \text{Eq. (21.74)}$$

Notice that if there is disagreement, the players get $d_1 = d_2 = 0$. It is clear from this that the bargaining game is convex and symmetric. Also, notice that for this bargaining game, we have

$$g(m_1,m_2) = u_1(m_1,m_2)u_2(m_1,m_2) = m_1m_2 \quad \text{Eq. (21.75)}$$

Here there exists a unique maximizer of g, which is the only Nash solution of the bargaining game. This solution is Pareto optimal, independent of irrelevant alternatives, independent of linear transformations and symmetric. This unique maximizer of g is $m_1^* = m_2^* = 50$. Consequently, the Nash solution is to give \$50 to each; see Fig. 21.16.

21.13.6 Example

An individual has listed her house at \$120,000. Her reservation price for the house is \$100,000. She knows that at any price less than \$100,000 she is better off not selling the house. A potential buyer looks at the house and is willing to buy it at the price of \$120,000, which also happens to coincide with his reservation price. However the buyer would, of course, be better off by getting the house at less than \$120,000.

We clearly have a bargaining problem. In this case there are two individuals who can make a potential net gain of \$20,000 and so the question is how should the two divide this among themselves. If the payoffs of the individuals are simply the money they receive, then (according to Nash's solution) the two individuals would agree to divide the amount \$20,000 equally and complete the transaction at a price of \$110,000.

Thus, Nash's bargaining solution provides an intuitively satisfactory and sharp answer to a pricing problem in the housing market.

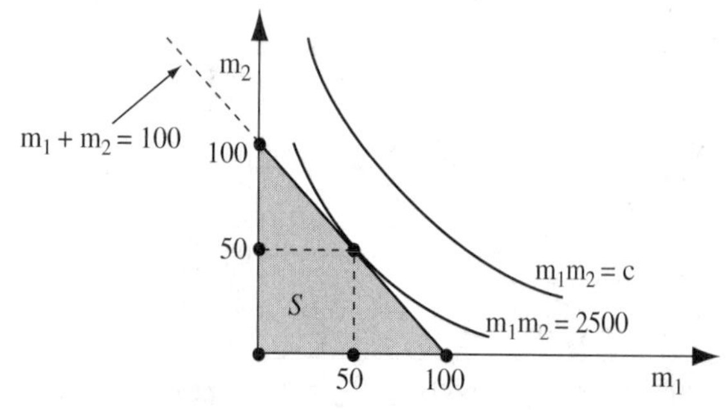

FIG. 21.16 BARGAINING OVER A SUM OF MONEY

In the example that we just saw, there is a unique Nash solution to the bargaining problem that satisfies the conditions of Pareto efficiency, independence of irrelevant alternatives, independence from linear transformations and symmetry. Indeed, in the above example, the Nash solution gives us exactly what we think the solution ought to be. In many cases, however, this approach to the bargaining problem fails to provide a satisfactory solution. The following example shows why this may happen. The example also highlights the importance of convexity.

21.13.7 Example

Suppose a couple is trying to decide whether they should go to a football game or to a Broadway show. The set of outcomes is thus given by

$$S = \{\text{Go to football, Go to Broadway, Disagreement}\}$$

In case they go to the Broadway show, the utility of individual A is $u_A = 4$, and the utility of individual B is $u_B = 1$. If they go to the football game, their utilities are reversed and $u_A = 1$ and $u_B = 4$. In case they disagree the payoffs are $u_A = u_B = 0$.

Clearly, when we use the approach of the theorem in Section 21.13.3 to find the solution to the bargaining problem, we end up with two answers:

(1) Either both go to the Broadway show
(2) Both go to the football game

This is all we can say if we use the theorem in Section 21.13.3. But, in this case one can argue that the two individuals should really toss a coin to determine where they should go. But then, should the coin be a fair coin? That is, should they decide to go to one or the other place with a probability of one-half?

If the coin chooses the alternative of going to the Broadway show with a probability p, and the alternative of going to the football game with a probability of $1 - p$, then the expected payoffs of the two individual are given by

$$\begin{aligned} Eu_A &= 4p + (1-p) = 3p+1 \\ \text{and } Eu_B &= p + 4(1-p) = 4 - 3p \end{aligned} \quad \text{Eq. (21.76)}$$

Now if we choose p to maximize $(Eu_A - 0)(Eu_B - 0)$, then p maximizes the function

$$g(p) = (3p + 1)(4 - 3p) = -9p^2 + 9p + 4 \quad \text{Eq. (21.77)}$$

The maximum is obtained when p satisfies the first-order condition

$$g'(p) = -18p + 9 = 0 \quad \text{Eq. (21.78)}$$

which gives $p = 1/2$. Thus individuals should indeed choose a fair coin. This seems to be a reasonable way of solving the bargaining problem, and we find that allowing individuals to extend the set of alternatives to include *joint randomization* or *correlation* leads to a more satisfactory solution to the bargaining problem.

PROBLEMS

21.1 Find the Nash equilibria of the Fare Setting Game.

21.2 Find the Nash equilibrium of the Prisoner's Dilemma. Also, find a strategy profile that gives a higher payoff than the payoff the players get in the Nash equilibrium.

21.3 Show that if a matrix game can be solved by using iterated elimination of dominated strategies, then the solution is a Nash equilibrium.

21.4 Consider a two-person strategic form game in which $S_1 = S_2 = \Re$. The utility functions of the two players are $u_1(x,y) = xy^2 - x^2$ and $u_2(x,y) = 8y - xy^2$.
Find the Nash equilibrium of the game. [Answer:(2, 2)]

21.5 Find the Nash equilibrium of a two-person strategic form game with strategy sets $S_1 = S_2 = \Re$ and utility functions

$$u_1(x,y) = y^2 - xy - x^2 - 2x + y$$

and $u_2(x,y) = 2x^2 - xy - 3y^2 - 3x + 7y$

[Answer: $(-\frac{19}{16}, \frac{16}{11})$]

21.6 Two firms (call them 1 and 2) produce exactly identical products. Firm one produces q_1 units of the product and firm 2 produces q_2 units so that the total number of units of the product in the market is $q = q_1 + q_2$.
We assume that:

(1) The market price of the product is $p(q) = 100 - 2\sqrt{q}$
(2) The production cost of producing q_1 units by firm 1 is $C_1(q_1) = q_1 + 10$
(3) The production cost of producing q_2 units by firm 2 is $C_2(q_2) = 2q_2 + 5$

Set up a strategic form game with two players whose payoff functions are the profit functions of the firms.
Determine the following.

(1) The profit functions $\pi_1(q_1,q_2)$ and $\pi_2(q_1,q_2)$ of the firms.
(2) The Nash equilibrium of the game.
(3) The market price of the product at the Nash equilibrium.
(4) The profits of the firms at the Nash equilibrium.

[Hints: (a) $\pi_1(q_1,q_2) = (99 - 2\sqrt{q_1 + q_2})q_1 - 10$ and

$$\pi_2(q_1,q_2) = (98 - 2\sqrt{q_1 + q_2})q_2 - 5$$

(b) The Nash equilibrium can be found by solving the system

$$\frac{\partial \pi_1(q_1,q_2)}{\partial q_1} = 0 \quad \text{and} \quad \frac{\partial \pi_2(q_1,q_2)}{\partial q_2} = 0$$

or (after computing derivatives and simplifying)

$$3q_1 + 2q_2 = 99\sqrt{q_1 + q_2} \qquad (21.1)$$

$$2q_1 + 3q_2 = 98\sqrt{q_1 + q_2} \qquad (21.2)$$

Dividing Eq. (21.1) and Eq. (21.2) and simplifying yields $q_2 = 96/101\, q_1$. Substituting this value in Eq. (21.1) and working the algebra, we get $q_1 = 795.88$. This implies $q_2 = 756.48$. So, the Nash equilibrium is $(q_1^*, q_2^*) = (795.88, 756.48)$.
(c) The market price is $p = 21.2$.
(d) $\pi_1(795.88, 756.48) = 16,066.78$;
$\pi_2(795.88, 756.48) = 14,519.42$

21.7 Consider the "Common Property Resources" problem in the example in Section 21.4.3 with functions $u(r) = \sqrt{r}, \kappa(r) = 2r^2$ and $K(R) = R^2$. Show that these functions satisfy the required properties and compute the Nash equilibrium R^* and the social optimum R^{**}.

[Answers: $R^* = \frac{n}{4}$ and $R^{**} = \frac{n}{\sqrt[3]{4[4 + 2(n-1)^2]^2}}$]

21.8 Verify the properties listed in Theorem 21.6.3.

21.9 Show that any path $P(u, v)$ of a tree is itself a tree with root u and terminal node v. In particular, show that every path from the root of a tree to a terminal node is a subtree.

21.10 Verify that every branch T_u of a tree is itself a tree having root u.

21.11 Verify that the remaining part of a tree T after removing all the descendants of a node is a subtree of T.

21.12 Consider the decision problem in Section 21.8.3 assuming that the firm is risk neutral.

(1) Solve the firm's decision problem.
(2) Express the solution in terms of p, q and s.
(3) What happens if $p = 0.9$ and $q = 0.4$?
(4) At what value for s will the firm decide to market the drug? Does this depend on p and q?

21.13 Consider the example in Section 21.8.6 and the values given there for the conditional probabilities. Assume also that the firm is risk neutral.

(1) Solve the firm's decision problem in terms of p and q.
(2) What happens if $p = 0.9$ and $q = 0.5$?
(3) If the test costs \$50, 000, will the firm want to pay for it?
(4) What is the maximum amount the firm will pay for the test?

21.14 Estimates show that 0.3% of the U.S. population is carrying the sexually transmitted HIV virus, known to cause the deadly disease AIDS. In order to study the spread of the HIV virus in the population, it was suggested that the U.S. Congress pass a law requiring that couples applying for a marriage licence should take the blood test for the HIV virus. The HIV blood test is considered very effective, since:

(1) A person with the HIV virus has a 95% chance to test positive.
(2) An HIV virus-free person has a 4% chance to test positive.

After several lengthy discussions, it was decided that the HIV blood test was ineffective for determining the spread of the AIDS disease and its implementation was abandoned. Can you figure out what argument persuaded the legislators of the ineffectiveness of the HIV virus test for determining the spread of the AIDS disease? [Hint: Consider the events "A = a person taking the HIV virus test has the disease" and "B =the test is positive." Using Bayes' formula determine that P (A/B) ≈ 6.67%!]

21.15 Verify that the path $a \to c \to d$ is the only Nash equilibrium path of the game in the example in section 21.8.5. Also, find all the Nash equilibrium strategy profiles that support the path $a \to c \to d$.

21.16 Consider the two-person sequential game shown in Fig. 21.17.

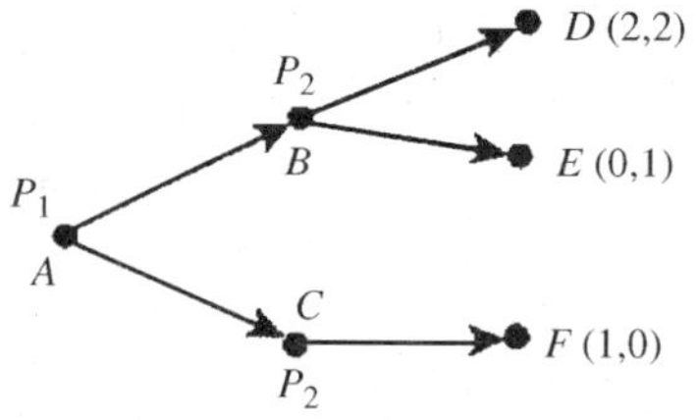

FIG. 21.17 A SEQUENTIAL GAME

(1) Show that the only equilibrium path given by the Backward Induction Method is the path $A \longrightarrow B \longrightarrow D$.

(2) Show that the path $A \longrightarrow C \longrightarrow F$ is an equilibrium path supported by the Nash equilibrium $(\{AC\}, \{CF, BE\})$.

21.17 Suppose in a market there are two firms, namely firm 1 and firm 2, that produce an identical product. The firms face an inverse demand curve $p(q) = 10 - q$, where p is the price and q is the total output produced by the firms. The marginal cost of each firm is a constant \$2.

(1) Find the equilibrium output of the two firms when they decide on their output simultaneously as in the example in Section 21.4.1.

(2) Find the equilibrium output of the firms if firm 1 produces its output first and firm 2 follows as in the example in Section 21.8.6.

(3) Compare the profits of firm 1 in 1 and 2. Comment on the result.

21.18 Consider a second-price sealed-bid auction and view it as a strategic form game with n players, where the strategy set S_i of player i is $[0, \infty)$; the set of all possible bids b_i. Assume that $v_1 > v_2 \geq v_3 \geq v_4 \geq \cdots \geq v_n$. Show that any vector of bids $(v_2, b_2, b_3, \ldots, b_n)$, where $v_2 > b_2 \geq v_3$ and $v_i \geq$ and $v_i \geq b_i$ for $3 \leq i \leq n$ is a Nash equilibrium.

21.19 A local municipality is floating a tender for the construction of a park. There are five local contractors who want to bid for the contract. The bidder who makes the lowest bid gets the contract. Write down the strategic form game for this and explain how the contractors will bid if they know each other's cost for constructing the park.

21.20 A rare fossil has been discovered in West Africa. It has been decided that the fossil will be auctioned. It is known to the auctioneer that two museums attach the same value of \$5 million to this fossil while the next possible buyer values it at \$4 million. Should the auctioneer use a first-price sealed-bid auction or a second-price auction? What does the auctioneer expect to get?

21.21 Consider an individual private value auction with two bidders. Each player knows that the valuation of the other player is a uniformly distributed random variable on an interval $[\underline{v}, v]$. Bidder 1 also knows that bidder 2's bidding function is given by: $b_2(v_2) = (v_2 - \underline{v})^2 + \underline{v}$, but bidder 2 does not know the bidding function of player 1. Find the best response bidding function $b_1(v_1)$ of player 1. [Answer: $b_1 = 2/3\underline{v} + 1/3 v_1$]

21.22 Consider an auction with two bidders in which player 1 knows that player 2's valuation is a random variable, which is uniformly distributed on the interval $[\underline{v}, \overline{v}]$, and player 2 knows that player 1's valuation is also a random variable, which is uniformly distributed on the interval $[v_*, v^*]$. Assume that $v_* < v < v^* < v$. Find the equilibrium linear bidding rules of the players in this auction. [Answers:

$$b_1(v_1) = 1/6v_* + 1/3\underline{v} + 1/2v_1 \text{ and } b$$
$$+1/3v_* + 1/2v_2]$$

21.23 We have seen that the final bid in an English auction is $b^* \geq v_2$, where v_2 is the valuation of the second highest bidder. How does this compare to the winning bid in a second-price sealed-bid auction?

21.24 Assume that in an n-bidder common-value auction the bidders observe signals that are independent random variables uniformly distributed over the interval [0, 1]. Show that the symmetric linear bidding rules
$\mathrm{b}_i = v_\ell + (n-1)/n\,(v_h - v_\ell)\omega_i,\ i = 1, \ldots, n,$
form a Nash equilibrium for the common-value auction.

21.25 Show that every Pareto optimal bargaining outcome is independent of linear transformations.

21.26 Consider the function $g(m_1, m_2) = m_1 m_2$ of the example in Section 21.13.5. Show that on the set of feasible alternatives $S =$
$[(m_1, m_2)\text{: } m_1 \geq 0,\ m_2 \geq 0, \textit{ and } m_1 + m_2 \leq 100]$
the function g attains its maximum value only at the pair $(m_1^*, m_2^*) = (50, 50)$.

21.27 If you examine the bargaining game of the example in Section 21.13.5 carefully you will notice that the players have utility functions that are linear in money. Suppose instead that player 1 has the utility function $u_1(m_1, m_2) = \sqrt{m_1}$. Find the Nash solution for this bargaining game. Discuss the implications for the Nash solution rule.

21.28 Suppose you are bargaining over the price of a Toyota Camry. The list price of the version that you want is \$20,000. The invoice price is \$18,000.

(1) Assuming that the utility functions are proportional to the money received, set up the bargaining game and find its Nash solution.

(2) Now suppose that a dealer located 60 miles away has agreed to sell the car at \$18,500. Reformulate the bargaining game with this outside option and find its Nash solution.

CHAPTER

22

ANALYSIS OF NEGOTIATION PROTOCOLS FOR DISTRIBUTED DESIGN

Timothy Middelkoop, David L. Pepyne, and Abhijit Deshmukh

Design is generally a collaborative activity involving multiple individuals and multiple design tools. Different domain experts and domain specific tools are needed to bring specialized knowledge and analysis to bear on different aspects of the design problem. Moreover, for large-scale design problems, decomposing the overall design problem and assigning different parts of it to separate individuals or design teams is the only way to manage the problem's complexity. Since design decisions made by the different individuals are not independent, the team must coordinate and integrate their decisions in order to reach a final design agreeable to all participants. Even when all the participants can be brought together in one place, reaching the best compromise to a constrained multicriteria decision problem can be challenging. When the participants work for different organizations located at different geographic locations, the problem takes on a new dimension of difficulty. Using modern information technologies to solve these problems is the subject of this chapter.

22.1 DISTRIBUTED DESIGN

One way to view a design problem is as an optimization problem involving a search over a high-dimensional *design space* for the design that gives performance that is "best" in some sense. What the different dimensions of the design space represent depends on the specific design problem. For the design of an airplane they might, for example, represent such factors as the number and size of the engines, the passenger capacity and physical dimensions such as length, width and weight. The choices made along each of these dimensions contribute in different ways to the resulting performance of the final design, which for a commercial aircraft would probably center around the expected income of the resulting design. Since performance is generally not a separable function of the different design choices, they cannot be made independently, but rather require coordinated selection and compromise. For example, the number of passengers impacts physical dimensions, which impacts weight, which impacts wingspan and engine type, which limits which airports the plane can fly to and determines fuel efficiency. All of this impacts manufacturing and other life-cycle costs, which impacts selling price, which impacts customer demand, which impacts expected income of the final design.

Being a bit loose mathematically, let us define a design problem as an optimization problem of the form,

$$\min_{x \in X}[y = f(\boldsymbol{x})] \qquad \text{Eq. (22.1)}$$

subject to

$$g(x) \leq 0 \qquad \text{Eq. (22.2)}$$

where $\boldsymbol{x}$ = a vector of numerical design parameters; X = a space of feasible values for the design parameters; and y = a scalar performance measure. The challenge of the design problem is that while the space of possible designs might be huge, the space of designs meeting all of our performance requirements might be quite small, particularly when all of the constraints are taken into account. Moreover, it is often the case in complex design problems that the function f is not known explicitly. In this case, designers must build prototypes or use numerical tools such as finite-element analysis or simulation in order to evaluate different design alternatives. Similarly, the constraint function g may also not be known in explicit form, requiring complicated computations to check design feasibility. These considerations generally preclude a centralized solution to a design problem, and we must resort to *distributed decision-making* (DDM) methods.

22.2 METHODS FOR DISTRIBUTED OPTIMIZATION

Many methods have been developed for distributed optimization. Each of these has its own advantages and disadvantages. We next give a brief literature review to explain the methods in the context of the design problem in Eq. (22.1).

22.2.1 Parallel Methods

Parallel methods involve decomposing a problem and/or the algorithm for solving the problem and using a distributed collection of processors to solve it. Extensive coverage of parallel, distributed methods for continuous variable problems is given in [1]. For example, for linear equations the Jacobi algorithm, Gauss-Seidel algorithm and Richardson's method are all fixed-point solution techniques that have parallel versions. When the decision variables are discrete, parallel methods for combinatorial optimization is also well developed, for example, [2]. Many of the parallel techniques are based on the basic sequential methods: depth first, branch and bound, iterative deepening A* (IDA*) and so on.

Hard problems, such as nonlinear continuous variable problems with many local optima and combinatorial scheduling and routing problems, have resulted in the development of many so-called *meta-heuristics*. These methods are often based on ideas found in nature. Simulated annealing, ant colony optimization are meta-heuristics that admit parallel implementations. Simulated annealing models the process by which metals when they are cooled reach a minimum energy configuration. Simulated annealing is essentially a local gradient search method with a random restart based on a timing/temperature schedule. The book edited by Azencott [3] discusses many methods for parallelizing the simulated annealing algorithm, including periodically interacting searches, multiple trials and partitioning of configurations. Simulated annealing has also been applied to combinatorial problems. Ant colony optimization is a meta-heuristic modeled after the way colonies of ants forage for food. One of its first applications was to the traveling salesman problem [4], but it has been shown experimentally to perform well on many other practical problems [5]. The ant algorithm is a probabilistic search method with global information encoded as *pheromone*. The level of pheromone indicates the quality of the search path and is used to control the trade-off between exploration of the search space and exploitation of the best solutions found so far. The pheromone is then updated based on the quality of the ant's tour. Techniques for parallelizing ant colony optimization range from simple, high-level parallel implementations that take the best result, to complex, parallel combinations of ants and parallel evaluation of solutions that assign multiple processors to single ants. Results from [6] show encouraging experimental speedup results (>3×) for large traveling salesperson problems (>200 cities). However, as the authors state, theoretical results show that parallelization may not scale well with problem size.

The main disadvantage with most parallel methods is that they tend to assume that there are no communication limitations between the distributed processors, i.e., error-free, infinite bandwidth communication channels are assumed. This is clearly a limiting assumption when some of the computational nodes may be humans, as in a design problem, and when communication is over an unreliable, bandwidth-limited, communication infrastructure, such as the Internet. Another limitation of parallel methods can be the difficulty in parallelizing the problem. In particular, a parallel method must not only consider the algorithm, but the underlying computational architecture on which the parallel algorithm will be executed. For example, parallelization requires partitioning the search space and distributing it onto the underlying hardware resources [7]. Moreover, often this partitioning must be done dynamically as the search proceeds. Partitioning and scheduling is a very hard problem. De Bruin, Kindervarter and Trienekens [8] discuss load balancing and review various loading balancing approaches.

22.2.2 Decision Theoretic Techniques

The parallel and distributed optimization methods discussed above were methods attempting to solve the overall optimization problem in Eq. (22.1) explicitly by decomposing it in a way that allows it to be worked on in parallel. In contrast, distributed decision-making involves decomposing the overall problem in such a way that each decision has its own utility associated with it. This converts the problem in Eq. (22.1) into a multicriteria and multi-objective decision problem. Multicriteria and multi-objective methods provide a way to combine conflicting criteria and objectives into a single integrated decision that satisfies (but may not optimize) our original design problem.

The extensive literature on decision-making techniques covers a wide variety of applications. There are several main methods discussed for solving multicriteria, multi-objective optimization problems, including preference aggregation, the compromise solution method, bargaining, fair division and heuristics. In the design context, [9] studies the properties of several alternative preference aggregation procedures. The work in [10, 11] regards a multicriteria technique to find a solution as close as possible to an ideal point: the point where group utility is maximized and the maximum of individual is minimized. The technique seeks to find the compromise solution (*satisficing* solution) that minimizes the distance to the ideal point, where distance is measured in terms of ℓ_p norms and penalty functions. The work in [12] characterizes the Euclidean compromise solution.

Bargaining provides a way to reach consensus in multicriteria, multi-objective problems. Conley, McLean and Wilkie [13] relate the compromise solution approaches to the bargaining approaches and show that many compromise solution techniques have a dual bargaining approach. More specifically, they show that the Nash bargaining solution has a dual, which is the Euclidean Yu compromise solution. Related to the compromise solution method is fair division. The goal in fair division is to split issues of contention (or resources) in the most efficient way. In many cases the structure of fair division is such that there is no mediator, making the method ideal for decentralized implementation. The book [14] by Brams and Taylor covers many fair division methods with known solution properties such as envy-free and equitability. One such method from this book is the *adjusted winner* procedure (also extensively covered in [15]), which splits a pool of indivisible objects between parties by a procedure after simultaneous revelation. Under the assumption that players do not strategically misrepresent their preferences, the results are efficient (in the sense that no agent can make a unilateral decision that will make it better off without making another agent worse off), equitable and can be implemented without a mediator. These methods are valuable when participation is mandatory. However, it is more difficult to model larger systems where participation is optional. Moreover, most of the problems solved are small-scale or even limited to two parties.

Heuristic methods for multicriteria, multi-objective decision problems include multi-objective evolutionary algorithms and genetic algorithms. Multi-objective evolutionary algorithms [16] are a class of solution methods that have been parallelized. Parallelization can occur by task decomposition (the algorithm), objective decomposition (the utility function) and data decomposition (database storage of domain data) with measurable speedup. Work in [17] uses various distribution techniques such as master slave, island and diffusion to distribute the population to different servers; however, according to the authors, the effectiveness of the parallelization and its interaction with the underlying domain problem is not well researched. Genetic algorithms can be used to compute large multi-objective problems [18]. There are many techniques for parallelizing genetic algorithms; however, fine-grained parallelism (also known as diffusion or cellular genetic algorithms) is best suited to multi-objective problems [19]. Fine-grain genetic algorithms are similar to their traditional counterparts with the exception that individuals are distributed over a mesh and only interact with individuals within neighbor hoods, with neighborhoods overlapping to allow the diffusion of good solutions throughout the population. This type of parallelism is well suited to systems with a large number of nodes.

22.2.3 Economic and Game Theoretic Techniques

General equilibrium theory describes a body of techniques for modeling economic systems with large numbers of economic agents and predicting the collective outcome of their economic interactions. In particular, general equilibrium models the balance between supplier-production and consumer-demand in large markets. Through appropriate mapping of a design problem into a problem involving suppliers and consumers, these techniques are well suited to designing systems requiring the coordination of large numbers of individuals. Classic texts such as [20, 21] provide answers to the question of "What is the outcome of a system that contains a large number of interacting rational entities, where every player always maximizes his utility, thus being able to perfectly calculate the probabilistic result of every action?" Although this work provides general characterization of the outcome, it largely ignores the specifics of the implementation. Work done in [22, 23] attempts to use general equilibrium models to build market-based multi-agent systems. This work extends market models by relaxing an auction algorithm to allow asynchrony in the bidding process. Again, however, details about how to distribute algorithm are not addressed.

General equilibrium provides a means to look at problems on a large scale but fail to model individual interactions. Bargaining theory attempts to address this shortcoming by using a number of models to represent inter-agent interaction. Although many economists believe that strategic bargaining by its very definition cannot prescribe an outcome, these techniques attempt to limit, or even eliminate, the range of indeterminacy [24]. Hence bargaining occupies an important place in economic theory where the pure bargaining problem is important. A common technique used in bargaining is game theory, which in its purest form models the decision-making processes and interactions between ideal individuals [25]. In this form it is ill-equipped to handle large-scale interactions. It can be used, with the proper modeling assumptions, to model the bargaining process and, in a limited way, to model larger bargaining systems. A game theoretical model for strategic bargaining was first developed by Rubinstein [26]. Rubinstein's original work took a simple model of two agents bargaining over a single continuous unary item and showed elegantly that when agents are impatient, the cost of delay is sufficient to yield an immediate result. Later, Rubinstein along with others derived the Nash bargaining solution from this strategic approach [27]. Strategic bargaining has been extended in various directions over the years. A set of extensions based on the core model of the two-party single-issue model can be found in [28]. These extensions include multiple parties, multiple issues, options and others. Some of the first extensions relaxed the assumption that agents have **perfect information**. Chatterjee and Samuelson [29] first addressed the problem of uncertainty in the bargaining game using the Nash bargaining solution, and later it was extended to the **strategic form** by Fudenberg and Tirole [30]. As models begin to address more issues (multi-issue and multiparty), the **bargaining agenda** becomes an important factor in the overall analysis. The role of the agenda is addressed in [31] and well as in [32], which showed that the agenda has a direct influence on the existence and number of equilibria in a system with multiple issues. Busch and Horstmann [33] show that the agenda itself can be used as a signaling device. Work by John and Raith [34] develops an *n*-stage bargaining model in which the agenda is optimized based on the risk of breakdown at the end of each bargaining session. Coles and Muthoo [35] extend the general model by addressing evolving utility functions over time and their influence on the equilibrium. More recently, work has been done on extending the original models to include multiple extensions. Multidimensional issues and **asymmetric information** are addressed in [36]. It is important to explicitly model the information-sharing behavior because inter-bargaining information can produce multiple noncompetitive subgame **perfect equilibria** as shown in [37].

Bargaining in markets extends the standard bargaining model to include a market process. In this case, the bargaining is connected by a market mechanism, which replaces portions of the **extensive form game** with a **matching technology**. The market decomposes a potentially large **game tree** into one of many independent bargaining games connected via an expectation. This makes the analysis tractable and facilitates modeling the dynamics of the system. Various forms of market equilibrium can be obtained from a market setting by altering the matching technology and bargaining structure. Osborn and Rubinstein provide a rigorous treatment of bargaining markets in their book [38]. This material covers a wide range of bargaining markets and characterizes their equilibrium behavior. In this work, agents are treated as sequential decision makers, which is beneficial when using extensive form games in the analysis by reducing the number of parallel moves. Recent work by Trefler [39] on bargaining markets builds on this work by developing a model that addresses many issues such as nonstationary markets with asymmetric information. Implementation and search theory has also been used by Jackson and Palfrey [40] to study the conditions for voluntary implementation and attainability in bargaining markets and under which conditions efficient trading rules can and cannot exist.

22.2.4 Multi-Agent Systems

Multi-agent systems are a relatively recent approach to solving optimization problems by building computational solutions. A multi–agent system is a collection of software agents that can sense and act on their environment. These systems have been used to solve a large number of distributed problems combining many disciplines that use a broad range of techniques from formal verification to socioeconomics [41]. The quality of these systems range from ad-hoc solutions to systems based on formal methods. Rigorous methods include formal specification and other logic-based approaches. These methods provide results that combine theory with an underlying software implementation. Belief desire and intention (BDI) architectures, such as the work by Rao and Georgeff [42], is one example. These systems provide formal, structured and implementable methods for modeling multi-agent systems. The dMARS system [43] is an implementation of a BDI architecture that is a formal specification based on the Z specification language [44], which can be easily implemented. Extending this further leads to directly executable specifications and executable logic systems such as [45], which can be used as both an analysis technology and in the final implementation. Other approaches [46] use mechanism programming languages to model and verify subgame perfect mechanisms. All these techniques provide tools for building rigorous multi-agent problem-solving systems.

Part of the approach for solving distributed problems is task decomposition and joint planning. Durfee [47] covers the use of these techniques in multiagent systems, making the important distinction between the three approaches to solving distributed problems (in this case in terms of planning). Centralized planning for distributed plans, distributed planning for centralized plans, and finally the most difficult, distributed planning for distributed plans. Distributed hierarchical planning is an example of this type of problem-solution technique. Corkill [48] first investigated the advantages of distributed hierarchical planning on the *Tower of Hanoi problem*.

Distributed constraint satisfaction is another technique to solve distributed problems and is used in multi-agent systems. This method is similar in ways to other techniques as it can be seen as a hierarchical search (search is used since this problem is known to be NP-complete) with advanced heuristics to direct it. The paper by Yokoo and Hirayama [49] is a good review of this material and in it they state that the difference between distributed constraint satisfaction and plain constraint satisfaction is that the distributed solution should not require global information. In their formulation the community of agents search for consistency among local actions. They distribute control of variables to agents for which the agents must find a solution that solves both local and inter-agent constraints. Many of the distributed constraint satisfaction algorithms have been shown to be *complete* (the algorithm is guaranteed to find a solution if one exists or stop if one does not), which is important in most design settings. Multi-agent systems use techniques from various fields to solve real problems. The general areas covered here highlight the use of methods similar to those in the previous section on optimization techniques.

There is existing research that addresses the need for strong results for implementable protocols. However, large-scale systems for the most part are being ignored. The multi-agent community is beginning to use formal game theoretic models of negotiation and bargaining [50]. The work by Kraus [51] has also taken a more theoretical approach to modeling negotiation in multi-agent systems using techniques based on bargaining and game theory. This work formalizes allocation and negotiation techniques that can be used in applications and simulations; however it does not address large-scale systems or nonseparable utility. Fatima et al. in [52] also develop formal bargaining models for multi-issue multi-agent negotiation, but again use separable utility.

22.2.5 Summary

The above literature provides many of the tools required in the construction of a distributed design architecture. The remainder of the chapter discusses some of the many aspects of developing a design system. We start this discussion with an overview of the product realization process.

22.3 ISSUES IN DISTRIBUTED DESIGN

In order to achieve design objectives, intelligent synthesis environments, which incorporate design tools, high-fidelity analysis, manufacturing and cost information, have been proposed for reducing product realization time. The term *product realization*, in the context of this discussion, is defined as the description, design, manufacture, testing and life-cycle support of a product.

Typically, product realization takes place in distinct phases, such as requirements definition, design, manufacture and product support. Each phase involved activities by distinct engineering disciplines, such as fluid dynamics, material science, structural analysis, machine design, electrical design and process design. Each phase and engineering discipline was characterized by different time scales, ontologies, data formats, analysis techniques and communication media. The differences among project phases and engineering disciplines can result in inefficiencies in the product realization process. The fundamental cause of the resulting inefficiency is due to insufficient communication and difficulty of coordination among the various product realization phases and engineering disciplines.

The disconnect between the project phases, e.g., between design and manufacturing, can be a major impediment to efficient product development. This inefficiency is reflected in a cycling through the product redesign iterations. During the early conceptual phase, for example, sensitivity to new technology and design changes is the greatest. With better communication and coordination between design team members, design errors can be caught early, greatly reducing downstream reengineering costs. Better communication can also result in better technology transfer, resulting in cost-saving innovations and infusion of new technology in the designs.

22.3.1 Team Coordination

The communication and coordination among project phases and engineering disciplines is a difficult and active research challenge. A fundamental research goal is to improve the design of complex systems by improving communication and coordination among the various project phases during early stages of the product life cycle. Realizing such a system involves research into several elements:

(1) Immersive virtual engineering environments, wherein the entire product team has access to all the product and process information needed for each team member to complete his or her task.
(2) Common user interfaces, such as a web browser interface, to coordinate access to project information gathered from diverse sources.
(3) Common data formats, which permit the integration and translation of different information from different disciplines.
(4) Integrated analytical and process management tools for collaborative engineering.

The primary emphasis of an integrated synthesis environment is on the distributed collaborative nature of large-scale designs. This requirement forces the research to address the issues of geographical separation, heterogeneous computing platforms and diverse design goals.

22.3.2 Tool Coordination

One of the most difficult issues is the issue of scalability. As the number of tools in an integrated design environment increase in both numbers and sophistication, the ability to use these tools effectively becomes increasingly difficult [53]. Thus, members of a design team must not only be able to communicate with each other, they must also be able to communicate with their analysis tools, which themselves may be geographically dispersed over heterogeneous computing platforms. This requires the ability to locate the right tools dynamically during a design iteration, since different tools may be appropriate at different stages of design, or a tool may be unavailable due to machine failure or excessive load.

The ability to effectively communicate between diverse tools is strewn with technical difficulties and ontological issues that impede the integration process. A major difficulty is the number of proprietary software systems and data formats. With the advent of the world wide web and related technologies the barriers to data transfer have been significantly reduced. Data interoperability, however, still remains a major roadblock in any large-scale design activity. The ability to transfer information, not just raw data, from one analysis domain to another is complicated by differences in semantics and time scales. From the design perspective, we need to ensure semantic interoperability, which includes design rationale, ontological issues and other meta data. This data is either not captured by present systems or often lost in data translation.

As the product design process becomes more decentralized and there is proliferation of new software and hardware systems, the need for a *flexible* solution, capable of handling the variety and changes, for integration is acute. Furthermore, in case of large-scale systems it is essential to be able to decompose the problem into smaller subproblems that are tractable. Agent-based design, which we describe next, provides an information framework for intelligent synthesis environment by enlisting the use of mobile software agents.

22.4 AGENT-BASED DESIGN

The strength of a multi-agent system approach is largely due to the fact that it explicitly deals with implementation issues such as communication and synchronization while at the same time using formal methods. Because design will typically involve human agents as well as software agents, a multi-agent approach seems a very natural one because it follows the way that design problems are actually decomposed. Given the size of the search space and the specialized knowledge and tools required to make decisions along the various dimensions, design problems are typically decomposed into subproblems, with each assigned to different individuals, teams of individuals or different tools. If we assign utilities to each decision-maker, then the design problem becomes a multi-objective multicriteria optimization problem to combine the individual utilities into the best compromise solution.

Conceptually, such a decomposition results in a number of agents searching over different dimensions of the high-dimensional search space for the best design. The design decisions of the different agents must be coordinated and integrated in order to reach a final design agreeable to all participants. Instead of performing a centralized search over a design space, the multi-agent methodology is desirable to deal with the asynchronous and distributed nature of the underlying design process. The search model presented in this chapter is based on a collection of individual decision-making entities. These entities, or agents, make choices based on limited information and preferences. In particular, this chapter presents a distributed design methodology where designs emerge as a result of negotiations between the different stakeholders in the design process. In the proposed methodology each decision-maker is represented as an autonomous agent. An agent in the context of this chapter is an autonomous computational entity that is capable of migrating across computing environments asynchronously. An agent may be a human or a software program. These agents have intelligence in the form of individual goals, beliefs and learning mechanisms, and interact cooperatively to accomplish overall product design objectives. Specially, each agent influences a specific subset of the design parameters in an effort to maximize its individual utility. Since design parameters are coupled, agents generally need to negotiate with other agents in order to settle on a agreeable compromise that achieves an accept able utility value. We now discuss ways in which to model and analyze the negotiation required to reach a consensus.

22.5 ANALYZING NEGOTIATION

When using a multi-agent approach for design, agents representing different subproblems negotiate in order to reach agreement on design parameters of common interest. Negotiation provides a conceptually appealing way to coordinate decision-making in large-scale design problems. In this section we develop a modeling framework for a qualitative study of several important issues related to distributed negotiations, including: What collective agreement will the different decision-makers converge to? Will they converge at all? How good are the resulting decisions and consequently the resulting design?

The modeling framework we develop can be viewed as a linearized approximation of a complicated dynamic multi-agent negotiation process involving the agent's individual negotiation abilities and inter-agent relationships and interactions. The framework is motivated by the systems sciences where it is standard practice to linearize a dynamical system about an operating point and do the analysis with the linearized model. The power of a linear model is that it allows the use of the many powerful results from linear systems theory, and hence allows us to study in a mathematically rigorous way the multi-agent negotiation process. Our main reference for this section is [1].

22.5.1 Formulation

The setup for our modeling framework for multi-agent negotiations is the following. Assume a population of n agents negotiating in an attempt to reach agreement over the value of some real valued variable x. In a design problem, x is one of the design parameters.

To approximate the dynamics of the negotiation process, assume that the agents negotiate in rounds $t = 1, 2, \ldots$. To begin the negotiation, each agent i submits an initial offer $x_i(0)$ for the value of x. Then on subsequent rounds, each agent adjusts its offer by taking a convex combination of its offer and the offers made by the other agents who are participating in the negotiation. Thus, mathematically we approximate the negotiation process with the dynamical equation:

$$x(t+1) = (1-g)x(t) + gAx(t) \qquad \text{Eq. (22.3)}$$

where $x(t)$ n-dimensional vector of the agent's offers at step t; $g \in (0, 1)$ = a relaxation parameter; and $\mathbf{A}$ = an n-by-n matrix whose entries are nonnegative scalars $a_{i,j} \geq 0$. For reasons that will become clear later, we assume the $a_{i,j}$ satisfy:

$$\sum_{j=1}^{n} a_{i,j} = 1 \qquad \text{Eq. (22.4)}$$

(i.e., $\mathbf{A}$ = a matrix whose rows all sum to 1). Here we note that taken together, the non–negativity assumption $a_{i,j} \geq 0$ and Eq. (22.4) imply that $0 \leq a_{i,j} \leq 1$ for all i, j.

In the context of negotiation the $a_{i,j}$'s model in qualitative terms an agent's influence on the final negotiated value. The relative values of the diagonal entries, $a_{i,i}$, reflect agent i's individual preferences and negotiation abilities. The off-diagonal entries, $a_{i,j}$, reflect the inter-agent relationships in terms of the relative influence of one agent over another. These entries capture, for example, situations where some agents are leader agents and others are follower agents, and situations where agents collaborate to increase their collective bargaining power. To reflect its role in capturing the nature of agent interactions during negotiation, we call the $\mathbf{A}$ matrix the *interaction matrix*.

To illustrate how $a_{i,j}$ can capture different negotiation dynamics, we present the following examples. First, let $g = 1$ in Eq. (22.3) and consider the two agent example:

$$x(t) = \begin{bmatrix} 1 & 0 \\ 0 & 1 \end{bmatrix} x(0)$$

Clearly, these two agents will never reach agreement, since neither will budge from their initial offers, i.e., $x(t) = x(0)$ for all $t \geq 0$. On the other hand, suppose

$$x(t) = \begin{bmatrix} 0.9 & 0.1 \\ 0.8 & 0.2 \end{bmatrix} x(0)$$

In this case it can be shown that (assuming synchronous updating, to be discussed later):

$$x(\infty) = \begin{bmatrix} 0.89 & 0.11 \\ 0.89 & 0.11 \end{bmatrix} x(0)$$

in which case the agents reach agreement, i.e., $x_1(\infty) = x_2(\infty) = 0.89x_1(0) + 0.11x_2(0)$, but the final value will be much closer to agent one's initial offer than it will be to agent two's initial offer. This example, therefore, represents a situation where agent one is a more influential negotiator than agent two, as reflected by $a_{1,1} >> a_{2,2}$. We also note that this example satisfies Eq. (22.4), a required condition for convergence to a consensus solution where $x_1(\infty) = x_2(\infty)$. As a final example consider three agents negotiating according to:

$$x(t) = \begin{bmatrix} 0.8 & 0.1 & 0.1 \\ 0.25 & 0.5 & 0.25 \\ 0.4 & 0.4 & 0.2 \end{bmatrix} x(0)$$

in which case

$$x(\infty) = \begin{bmatrix} 0.61 & 0.24 & 0.15 \\ 0.61 & 0.24 & 0.15 \\ 0.61 & 0.24 & 0.15 \end{bmatrix} x(0)$$

that means that agent one is able to dominate the final outcome of the negotiation, i.e., $x_1(\infty) = x_2(\infty) = x_3(\infty) = 0.61x_1(0) + 0.24x_2(0) + 0.15x_3(0)$. However, suppose agent three teams up with agent two to collaborate against agent one. This increased strength of interaction between agents three and two can be captured by increasing the value of the off-diagonal entry $a_{3,1}$. For example, setting

$$x(t) = \begin{bmatrix} 0.8 & 0.1 & 0.1 \\ 0.25 & 0.5 & 0.25 \\ 0.2 & 0.6 & 0.2 \end{bmatrix} x(0)$$

results in:

$$x(\infty) = \begin{bmatrix} 0.54 & 0.30 & 0.16 \\ 0.54 & 0.30 & 0.16 \\ 0.54 & 0.30 & 0.16 \end{bmatrix} x(0)$$

or $x_1(\infty) = x_2(\infty) = x_3(\infty) = 0.54x_1(0) + 0.30x_2(0) + 0.16x_3(0)$. Thus by collaborating (as reflected by a large relative value of $a_{3,2}$), agents two and three are able to increase their collective bargaining power. With larger numbers of agents, much more complicated agent interrelationships can be modeled.

22.5.2 Basic Properties

Negotiation is a success when the agents are able to reach consensus on the value x. In the context of Eq. (22.3), this is a solution vector x whose components are all equal in value. We call such a solution a *consensus solution*. Next we establish the existence of such consensus solutions. Specifically, suppose there exists a vector $\overline{x}$ such that:

$$\overline{x} = (1-g)\overline{x} + gA\overline{x} \qquad \text{Eq. (22.5)}$$

such a vector is called a (*fixed point*) of Eq. (22.3). Under Eq. (22.4), it is not hard to show that any vector whose components are all equal is a fixed point of Eq. (22.3). To do so, let us rewrite Eq. (22.5) in terms of the components of $\overline{x} = (\overline{x}_1, \ldots, \overline{x}_n)$, i.e.:

$$\overline{x}_i = (1-g)\overline{x}_i + gA\overline{x}_i$$

Now suppose that $\overline{x}_i = \overline{x}_j$ for all i, j, and denote this common value by z. Then

$$z = (1-g)z + g\sum_{j=1}^{n} a_{i,j}z = (1-g)z + gz\sum_{j=1}^{n} a_{i,j} = (1-g)z + gz = z$$

where the third equality follows from Eq. (22.4). Thus, we have the following result.

Lemma 1 (Existence of consensus solutions): For the negotiation model defined by Eqs. (22.3) and (22.4), *every* vector $\overline{\boldsymbol{x}}$ whose components are all equal is a fixed point.

Lemma 1 assures us that the negotiation model always has at least one consensus solution. In order to ensure that consensus solutions are the *only* fixed points of Eq. (22.3) we need an additional restriction on the interaction matrix. In particular, we need to require that the matrix $\boldsymbol{A}$ is *irreducible.*[1] Technically irreducible means the following. Given an n-by-n non-negative matrix $\boldsymbol{A}$ ($n \geq 2$) form a directed graph $G = (V, E)$ with vertices $V = \{1, \ldots, n\}$ and edges $E = [(i, j)| i \neq j,\ a_{i,j} \neq 0]$, where $a_{i,j}$ is the i, jth entry of A. Then the matrix $\boldsymbol{A}$ is irreducible if for every pair of vertices i, j, there exists a positive path through the graph G leading from i to j. What this means is that every agent in a negotiation must have some nonzero influence, either directly or indirectly, over every other agent. Irreducibility precludes, for example, the $\boldsymbol{A}$ matrix being the identity matrix, which satisfies Eq. (22.4) (and hence **Lemma 1**), but which represents agents who simply refuse to budge from their initial offers. It also precludes block matrices in which independent clusters of agents negotiate among themselves but never communicate their negotiated settlement with other clusters. We present the following without proof.

Lemma 2: If $\boldsymbol{A}$ is an irreducible matrix, then the *only* fixed points of the negotiation model defined by Eqs. (22.3) and (22.4) are consensus solutions, i.e., vectors whose components are all equal in value.

According to **Lemma 2**, if negotiation converges, it converges to a consensus solution. However, there are uncountably many such solutions. While we cannot say a priori which one of them the negotiation will converge to, we can say the following.

[1] An n-by-n ($n \geq 2$) non-negative matrix M is irreducible ***if and only if*** the matrix $(I + M)^{n-1}$ has only positive entries.

Lemma 3 (Compromise consensus): If $\boldsymbol{A}$ is irreducible and the negotiation model defined by Eqs. (22.3) and (22.4) converges, then it converges to a consensus solution $\bar{x}$ whose value falls between the smallest and largest of the agents initial offers, i.e., $\min_i [x_i(0)] \leq \bar{x}\ \max_i [x_i(0)]$.

The proof of **Lemma 3** follows from the fact that all intermediate negotiated values are convex combinations. Such a compromise consensus is a reasonable requirement of any negotiation protocol; each agent being asked to compromise on its initial starting offer, the agent with the highest initial offer having to compromise downward, the agent with the lowest initial offer having to compromise upward.

Remark: When the only fixed points of the negotiation model are vectors whose components are all equal, we will sometimes use $\boldsymbol{x}$ to represent the *vector* of the agent's offers and other times we will use $\boldsymbol{x}$ to represent the *scalar* common agreed on outcome of the negotiation.

22.5.3 Social Welfare Measure

We assume that each agent i participating in the negotiation enters with some target value c_i for the design parameter in question. In design, the value of c_i comes, for example, from the agent's preferences (e.g., the preference of a customer agent), an agent's expert opinion, or an agent's engineering analysis. We evaluate the outcome $\bar{x}$ of the negotiation by comparing it against the value x^* that minimizes the sum of squared deviations from each individual agent's target value. That is, we evaluate the quality of the outcome $\bar{x}$ of a negotiation in terms of the *social welfare function:*

$$S(\bar{x}) = \frac{1}{2}\sum_{i=2}^{n} v_i (c_i - \bar{x})^2 \qquad \text{Eq. (22.6)}$$

where $\bar{x}$=(scalar) outcome of the negotiation; c_i= agent i's target for the value for $\bar{x}$; and v_i=a weighting factor to reflect situations where it is more important to satisfy the preference of some agents than it is to satisfy the preferences of other agents (e.g., one of the agents may be the customer who is commissioning the design). Our definition of the social welfare function is related to *deviation variables* in *goal programming*. That is, similar to deviation variables, our social welfare function gives a measure of the discrepancy between the feasible design space and the aspiration space.

Using simple calculus we can compute the (scalar) value x^* that minimizes Eq. (22.6). Specifically, x^* minimizes Eq. (22.6) if and only if:

$$\frac{dS(x^*)}{dx^*} = 0 = \frac{d}{dx^*}\left[\frac{1}{2}\sum_{i=1}^{n} v_i (c_i - x)^2\right] = -\sum_{i=1}^{n} v_i c_i + x^* \sum_{i=1}^{n} v_i \qquad \text{Eq. (22.7)}$$

where sufficiency comes from the strict convexity of $S(x)$. Solving Eq. (22.7) yields

$$x^* = \frac{\sum_{i=1}^{n} v_i c_i}{\sum_{i=1}^{n} v_i}$$

which, if we assume the v_i's have been normalized such that:

$$\sum_{i=1}^{n} v_i = 1 \qquad \text{Eq. (22.8)}$$

reduces to

$$x^* = \sum_{i=1}^{n} v_i c_i \qquad \text{Eq. (22.9)}$$

Thus, we have just proved the following.

Lemma 4: Under assumption Eq. (22.8):

$$x^* = \arg\min_{x \in \mathcal{R}+} S(x) = \sum_{i=1}^{n} v_i c_i$$

Remarks: Social welfare is a *multicriteria* optimization problem. Central to multicriteria optimization is the concept of *Pareto efficient* solutions. A change that can make at least one agent better off without making any other agent worse off is called a Pareto improvement. A solution is called Pareto efficient when no further Pareto improvements can be made. Thus, if a solution is Pareto efficient, then it is the case that it is not possible to make any agent better off without making some other agent worse off. In this sense, Pareto efficient solutions can be said to achieve social welfare; they achieve the best compromise insofar as the population as a whole is concerned. It is easy to show that the value x* is Pareto efficient, since at this value no agent can improve its situation without making some other agent worse off. This follows directly from the uniqueness of x^*. Under assumption Eq. (22.8), it is also not hard to show that the optimal value x^* always falls between the smallest and the largest of the agents target values, i.e., $\min_i[c_i] \leq x^* \leq \max_i[c_i]$. This follows because, under Eq. (22.8), Eq. (22.9) is a convex combination.

22.5.4 Convergence Analysis

As discussed, agent i's observable negotiation behavior is captured in our model by the values of the $a_{i,j}$'s in row i of the interaction matrix $\boldsymbol{A}$. To understand how different negotiation behaviors will impact the outcome, we next study the relationships between the interaction variables $a_{i,j}$ and the outcome that the negotiation converges to. The difficulty as we will see is that in general the outcome of a negotiation can depend not only on the agents negotiation behaviors (the $a_{i,j}$'s), but also on the agent's initial offers [the $x_i(0)$], and on the frequency with which the agents update their offers.

During negotiation, agent i is responsible for updating the offer x_i. There are a number of ways this updating can be done. The first updating scheme would have the agents update their offers simultaneously according to:

$$x_i(t+1) = (1-g)x_i(t) + g\sum_{j=i}^{n} a_{i,j} x_j(t) \qquad \text{Eq. (22.10)}$$

This scheme is sometimes referred to as *Jacobi updating.* A second updating scheme would have the agents update their offers round-robin in some fixed order according to:

$$x_i(t+1) = (1-g)x_i(t) + g\sum_{j=1}^{i-1} a_{i,j} x_j(t+1) + g\sum_{j=1}^{n} a_{i,j} x_j(t) \qquad \text{Eq. (22.11)}$$

In this case, the agents update their offers in a sequential, round-robin manner. This form of updating is called *Gauss-Seidel updating.*

We notice that both updating schemes Eqs. (22.10) and (22.11) assume that the agents are *synchronized*, i.e., in both cases the agents must wait for every other agent to update their offers before proceeding to the next update. In contrast to these is *asynchronous updating* in which the agents update their offers at random arbitrary times. Following [1], let T_i be the set of times at which agent i updates its offer x_i and let $0 \le \tau_j^i(t) \le t$ be the last time that a value for x_j was received by agent i (agent i may not have the most current value of x_j). Then

$$x_i(t+1) = \begin{cases} (1-g)x_i(t) + g\sum_{j=1}^{i-1} a_{i,j} x_j[\tau_j^i(t)] & t \in T^i \\ x_i(t) & t \notin T^i \end{cases} \quad \text{Eq. (22.12)}$$

The asynchronous case is the most realistic for a large multi-agent system since it describes situations in which agent's computational capabilities are not equal and in which updating messages are delayed, lost or even arrive out of order. In the case where some of the agents are people, for example, they may not have attended all of the design meetings and hence their preferences were not expressed at every updated stage of the design negotiations.

We now establish the conditions for the negotiation model defined by Eq. (22.3) under Eq. (22.4) to converge to a fixed point. The conditions we establish will be sufficient to ensure convergence for all the updating schemes Eqs. (22.10) to (22.12). We begin our analysis by defining

$$f[x(t)] = (1-g)x(t) + gAx(t) \quad \text{Eq. (22.13)}$$

in which case we can write Eq. (22.3) as $x(t+1) = f[x(t)]$. Let $\bar{X}$ be the set of fixed points of $f(x)$, i.e., the set of points such that $\bar{x} = f(\bar{x})$. We know from **Lemma 1** that this set is not empty. It should also be clear that the mapping f is continuous.

A necessary condition for the negotiation to converge is that the mapping Eq. (22.13) is *nonexpansive* with respect to the *maximum norm* [1]. For a vector $\mathbf{y}$ the *maximum norm* is defined as, $\|\mathbf{y}\|_\infty = max_i|\mathbf{y}_i|$ For a square n-by-n matrix $\boldsymbol{M}$, the maximum norm induces the *matrix norm:*

$$\|\boldsymbol{M}\|_\infty = \max_{y\neq 0}\frac{\|My\|_\infty}{\|y\|_\infty}$$

A mapping $x := f(x)$ is said to be *nonexpansive* with respect to the maximum norm if it satisfies

$$\|f(x) - x^*\|_\infty \le \|x - x^*\|_\infty \text{ for all } x \in \Re^n \text{ and all } x^* \in X^*$$

In words, if a mapping f is nonexpansive with respect to the maximum norm, then every component x_i of x progressively moves toward a fixed point $\bar{x}$ with every update, or at least moves no further away.

Following [1] let us define $h(x) = x - (I - A)x$. Then $f(x) = (1 - g)x + gh(x)$. It is not hard to show that if $h(x)$ is nonexpansive, then $f(x)$ is. Thus, let us determine the conditions for $h(x)$ to be nonexpansive. Specifically, suppose $\bar{x}$ is a fixed point of f. Then we have $(I - \boldsymbol{A})\bar{x} = 0$ in which case we can write:

$$\begin{aligned} \|h(x) - \bar{x}\|_\infty &= \|x - (I - \boldsymbol{A})x - \bar{x}\|_\infty \\ &= \|x - (I - \boldsymbol{A})x - \bar{x} + (I - \boldsymbol{A})\bar{x}\|_\infty \\ &= \|\boldsymbol{A}(x - \bar{x})\|_\infty \\ &\le \|\boldsymbol{A}\|_\infty \|x - \bar{x}\|_\infty \end{aligned}$$

which implies that if $\|\boldsymbol{A}\|_\infty \le 1$ then $\|h(x) - \bar{x}\|_\infty \le \|x - \bar{x}\|_\infty$. In other words, if $\|\boldsymbol{A}\|_\infty \le 1$ then $h(x)$ [equivalently $f(x)$] is nonexpansive. It turns out that there is a simple relationship between the entries of a matrix and the induced matrix infinity norm. In particular, for an n-by-n matrix $\boldsymbol{M}$ an equivalent definition of the matrix maximum norm is

$$\|\boldsymbol{M}\|_\infty = \max_i \sum_{j=1}^{n} |m_{i,j}|$$

With this definition we can prove the following.

Theorem 1: Consider the negotiation model defined by Eq. (22.3) with interaction matrix $\boldsymbol{A}$ satisfying Eq. (22.4). A necessary condition on the interaction parameters a_{ij} for the negotiation to converge is

$$\sum_{j=1}^{n} |a_{i,j}| \le 1 \; \forall i \quad \text{Eq. (22.14)}$$

Another way to view a negotiation is as discrete-time dynamical system. A basic result from linear systems theory is that the discrete-time linear system $x := \boldsymbol{A}x$ is stable if the spectral radius of $\boldsymbol{A}$ (i.e., the largest eigenvalue of $\boldsymbol{A}$) satisfies $\rho(\boldsymbol{A}) \le 1$; asymptotically stable if $\rho(\boldsymbol{A}) < 1$. A standard result from linear algebra is that the spectral radius of a matrix satisfies $\rho(\boldsymbol{M}) \le \|\boldsymbol{M}\|$ for any induced norm on $\boldsymbol{M}$. Thus, satisfaction of Eq. (22.14) implies that $\rho(\boldsymbol{A}) \le 1$. In fact,

Lemma 5: Let $\boldsymbol{A}$ be an n-by-n matrix satisfying Eq. (22.4). Then $\rho(\boldsymbol{A}) = 1$.

Proof: Let $\boldsymbol{e}$ be a column vector with all components equal to 1. Then from Eq. (22.4) we must have $\boldsymbol{A}e = e$. Thus, $\lambda = 1$ is an eigenvalue of $\boldsymbol{A}$ and $\rho(\boldsymbol{A}) \ge 1$. On the other hand, $\rho(\boldsymbol{A}) \le \|\boldsymbol{A}\|_\infty = 1$, which establishes the result.

22.5.5 Social Welfare Outcome

When a negotiation converges we would like to be able to make some statements about how the resulting outcome compares to the optimal social welfare outcome. In a *goal programming* approach to design, for example, we would like to be able to guide the negotiations to a design solution that meets the design goals with minimum deviation from the design targets. In other words, we would like to understand how one might guide a negotiation to the social welfare solution in Eq. (22.9).

For the discussion to follow, let us suppose that $g = 1$, in which case we can write the update equation as, $x(t+1) = \boldsymbol{A}x(t)$. Then assuming synchronous (i.e., Jacobi) updating we have $x(t) = \boldsymbol{A}^t x(0)$. Now clearly the only way $x(t)$ will converge is if $\boldsymbol{A}^t$ converges to a constant matrix $\bar{\boldsymbol{A}}$, in which case we get $\bar{x} = \bar{\boldsymbol{A}}x(0)$.

Let us compute $\bar{\boldsymbol{A}}$ for the $n = 2$ case. Suppose we write:

$$\boldsymbol{A}^t = a\boldsymbol{A} + bI \quad \text{Eq. (22.15)}$$

Then, because the eigenvalues of $\boldsymbol{A}$ are distinct, $\boldsymbol{A}$ is diagonalizable and we can write, $\boldsymbol{A}^t = \boldsymbol{N}\Gamma^t\boldsymbol{N}^{-1}$, where $\boldsymbol{N}$ is the matrix that diagonalizes $\boldsymbol{A}$ and Γ is a diagonal matrix of the eigenvalues of $\boldsymbol{A}$. Plugging this into Eq. (22.15) we get, $\boldsymbol{N}\Gamma^t\boldsymbol{N}^{-1} = a\boldsymbol{N}\Gamma\boldsymbol{N}^{-1} + \mathbf{b}\boldsymbol{N}I\boldsymbol{N}^{-1}$, which (premultiplying by $\boldsymbol{N} - 1$ nd postmultiplying by $\boldsymbol{N}$) is equivalent to $\Gamma^t = a\Gamma + bI$, which in the 2-by-2 case is

$$\begin{bmatrix} \lambda_1^t & 0 \\ 0 & \lambda_2^t \end{bmatrix} = a\begin{bmatrix} \lambda_1 & 0 \\ 0 & \lambda_2 \end{bmatrix} + b\begin{bmatrix} 1 & 0 \\ 0 & 1 \end{bmatrix}$$

Solving the above for a and b we get:

$$a(t) = \frac{\lambda_1^t - \lambda_2^t}{\lambda_1 - \lambda_2}, \quad b(t) = \lambda_2^t - a(t)\lambda_2 \qquad \text{Eq. (22.16)}$$

We already know that, under Eq. (22.4), the largest eigenvalue of A, $\lambda_1 = 1$. Thus we must have that the second largest eigenvalue of A, $\lambda_2 < 1$. Moreover, since the sum of the eigenvalues of A equals the trace of A, we also have $\lambda_1 + \lambda_2 = 1 + \lambda_2 = a_{1,1} + a_{2,2}$ from which we get $\lambda_2 = a_{1,1} + a_{2,2} - 1$. Assuming there exists an $a_{i,i} > 0$ we have $-1 < \lambda_2$, in which case, letting $t \to \infty$ in Eq. (22.16) we get

$$a(\infty) = \frac{1}{1-\lambda_2}, \quad b(\infty) = \frac{-\lambda_2}{1-\lambda_2}$$

Putting it all together we have the following.

Lemma 6: Let A be a 2-by-2 non-negative matrix satisfying Eq. (22.4). Moreover assume that at least one of the diagonal entries $a_{i,i} > 0$. Then

$$\lim_{t\to\infty} A^t = \bar{A} = \frac{1}{2-(a_{1,1}+a_{2,2})}\begin{bmatrix} 1-a_{2,2} & 1-a_{1,1} \\ 1-a_{2,2} & 1-a_{1,1} \end{bmatrix} \qquad \text{Eq. (22.17)}$$

Noting that $\bar{A}$ depends only on the eigenvalues of $\bar{A}$, we have the following corollary.

Corollary 1: Let A be a 2-by-2 matrix satisfying Eq. (22.4). Moreover assume that at least one of the diagonal entries $a_{i,i} > 0$. Then, $\bar{x} = \lim_{t\to\infty} A^t x(0) = \bar{A}x(0) =$ is uniquely determined by the vector of initial offers $x(0)$.

Remarks: Note that the rows of A^t will sum to one for any t. In particular, the rows of Eq. (22.17) sum to 1. This reinforces the fact that the outcome of the negotiation is a consensus solution. We also note that finding the solution by raising the matrix A to powers of t is equivalent to synchronous updating. While in the case of synchronous updating the outcome is uniquely determined by the starting offers, this is not true in the asynchronous updating case. In fact, with asynchronous updating the negotiation outcome will generally also depend on the frequency with which the offers are updated.

In the case of synchronous updating, we can derive conditions on the entries of the interaction matrix A that will ensure the negotiation converges to the social optimal value x^* given by Eq. (22.9). In particular, suppose that each agent i submits its target value as its initial offer, i.e., suppose $x_i(0) = c_i$ for each i. Then the social welfare solution is obtained if

$$\bar{A} = \begin{bmatrix} v_1 & v \\ v_1 & v_2 \end{bmatrix}$$

since then

$$x(\infty) = \bar{A}c = \begin{bmatrix} v_1 & v_2 \\ v_1 & v_2 \end{bmatrix}\begin{bmatrix} c_1 \\ c_2 \end{bmatrix} = \begin{bmatrix} v_1c_1 + v_2c_2 \\ v_1c_1 + v_2c_2 \end{bmatrix} = \begin{bmatrix} x^* \\ x^* \end{bmatrix}$$

Thus, from **Lemma 6** we have the following.

Lemma 7: Let A be an irreducible 2-by-2 non-negative matrix with entries satisfying Eq. (22.4). Assume synchronous updating and that each agent i submits its target c_i as its initial offer. Then, for a given $a_{1,1}$ the social welfare solution is achieved when $a_{2,2}$ satisfies:

$$a_{2,2} = \frac{v_1(2-a_{1,1})-1}{v_1 - 1} \qquad \text{Eq. (22.18)}$$

Proof of **Lemma 7** is left for Problem 22.8.

22.5.6 Numerical Examples

Here we give examples to illustrate how the linearized model of negotiation can be used to understand different agent negotiation behaviors. The idea is that we have some number of agents, members of a design team, who are negotiating over the value of some design parameter x. Based on its own analysis, each agent has some preference c_i for the value of the parameter. In general there is conflict between the agents' preferences so that $c_i \neq c_j$ for different agents i and j.

Example 1: Consider a case with two agents. Suppose agent one has preference $c_1 = 1$ and proposes $x_1(0) = c_1$. Suppose agent two has preference $c_2 = 0$ and proposes $x_2(0) = c_2$ such that $c_1 \neq c_2$. Clearly these two agents are in conflict over the value of the design variable. So typically they will negotiate over different design options to reach agreement. Consider the 2-by-2 interaction matrix:

$$A = \begin{bmatrix} 0.8 & 0.2 \\ 0.2 & 0.8 \end{bmatrix} \qquad \text{Eq. (22.19)}$$

Here the entries in the first row are intended to capture the negotiation behavior of agent one and the entries in the second row the negotiation behavior of agent two. Note that the rows sum to 1 as required by Eq. (22.4). Notice also that the matrix is irreducible and also satisfies **Theorem 1**. Moreover, since $-1 < \lambda_2$ for this example ($\lambda_2 = a_{1,1} + a_{2,2} - 1 = 0.6$), convergence holds with $g = 1$ and we can write the updated equation in Eq. (22.3) as

$$x(t+1) = Ax(t) = \begin{bmatrix} 0.8 & 0.2 \\ 0.2 & 0.8 \end{bmatrix} x(t) \qquad \text{Eq. (22.20)}$$

Thus, with $x(0)$ denoting the vector of the agent's initial offers, the value of the design parameter at step t of the negotiation (in the synchronous updating case) is given by $x(t) = A^t - x(0)$ in which case the equilibrium final negotiated value for the design parameter can be determined from $\lim_{t\to\infty} A^t x(0) = \bar{A}x(0)$. Since this is a 2-by-2 example, we use **Lemma 6** to find

$$\bar{A} = \frac{1}{2-(a_{1,1}+a_{2,2})}\begin{bmatrix} 1-a_{2,2} & 1-a_{1,1} \\ 1-a_{2,2} & 1-a_{1,1} \end{bmatrix} = \begin{bmatrix} 0.5 & 0.5 \\ 0.5 & 0.5 \end{bmatrix} \qquad \text{Eq. (22.21)}$$

Hence, in the synchronous updating case, the negotiation converges to the value $\bar{x} = 0.5$. In other words, matrix A represents a case where both agents are "equally accommodating" in terms of the value of the parameter they are willing to agree on, and converge to the value that splits the difference between their initial offers.

Example 2: Consider again an example with two agents. As before, suppose agent one makes an initial offer of two and agent one an initial offer of 0. However, suppose for this case that agent one has a very strong preference for the value of the parameter it will agree on. Suppose the second agent has a weaker preference for the value. These preferences can be reflected in the ***A*** matrix by setting $a_{1,1} > a_{2,2}$. In particular, if

$$\begin{bmatrix} 0.9 & 0.1 \\ 0.3 & 0.7 \end{bmatrix} \qquad \text{Eq. (22.22)}$$

then the negotiation model with $g = 1$ converges to an equilibrium parameter value of $\bar{x} = 0.75$. Thus, in this case agent one is able to drive the negotiation.

As we see from the above examples, those agents i with relatively larger values for $a_{i,i}$ are those that are able to keep the final negotiated value near to their initial offer—with the extreme case $a_{i,i} = 1$ reflecting an agent who simply refuses to negotiate. Thus, the $a_{i,i}$ can be interpreted as reflecting such things as an agent's preferences, negotiation skills, market dominance and so on.

The next example will give interpretation to the off-diagonal entries $a_{i,j}$ in the interaction matrix.

Example 3: For this example we assume three agents. Suppose that initially agent one offers a value of 1 and the other two agents both offer values of 0. Here we are going to show how the $a_{i,j}$ can be used to model agents who collaborate in order to drive down the final outcome. Consider first the matrix

$$\begin{bmatrix} 0.8 & 0.1 & 0.1 \\ 0.25 & 0.5 & 0.25 \\ 0.4 & 0.4 & 0.2 \end{bmatrix} \qquad \text{Eq. (22.23)}$$

Here (according to the relative values of the $a_{i,i}$) agent one dominates the negotiation, and agent two is more influential than agent three. For $g = 1$ this is reflected by the convergence of the parameter value to $\bar{x} = 0.61$. Now suppose that the weaker agent, agent three, teams up with agent two. The off diagonal entries $a_{i,j}$ can be used to reflect this collaboration. In particular, increasing $a_{3,2}$ strengthens the degree of cooperation between agents two and three. For example, if we set

$$\begin{bmatrix} 0.8 & 0.1 & 0.1 \\ 0.25 & 0.5 & 0.25 \\ 0.1 & 0.8 & 0.1 \end{bmatrix} \qquad \text{Eq. (22.24)}$$

then agents two and three work together to drive the equilibrium value x^* from 0.61 down to 0.51, and they are able to do this without any change in agent one's negotiation behavior.

Our final examples will illustrate some of the theoretical properties we developed.

Example 4: This example shows how the frequency of updating may change the resulting outcome. Consider again the interaction matrix ***A*** from Example 2. As we already saw, in the synchronous case with $g = 1$ this converges to an equilibrium value of $\bar{x} = 0.75$. But suppose agent one makes two updates for every one update of agent two. Then the outcome is $\bar{x} = 0.56$. That is, by simply changing the frequency with which some agents update their positions, the final outcome of a negotiation can be changed. In the case of multiple fixed points, a property of asynchronous updating is that the timing of the offer updates can determine which fixed point the algorithm converges to. In general, the more often an agent negotiates (e.g., compromises on its offer), the less influence it ends up having on the final outcome. This agrees with empirical evidence that suggests that if you are in a strong negotiating position, you give away nothing, i.e., you adjust your offer as few times as possible.

Example 5: Finally, we want to show how to achieve the social welfare solution. Suppose $v_1 = 0.25$ and $v_2 = 0.75$, that is, agent two's preference is more important than agent one's (e.g., agent two might be the customer). Let $c_1 = 1.0$ and $c_2 = 0.0$ (agent one wants a high value, agent two a low one). Applying Eq. (22.9) we get the social optimal value of 0.25. Choosing $a_{1,1} = 0.8$ and using **Lemma 7** gives $a_{2,2} = 0.9333$, in which case:

$$\begin{bmatrix} 0.8 & 0.2 \\ 0.0667 & 0.9333 \end{bmatrix} \qquad \text{Eq. (22.25)}$$

which for $x(0) = [c_1 c_2]'$ converges to $\bar{x} = 0.25 = x^*$ as desired.

These examples illustrated via small examples how a linear approximation would otherwise involve an analysis of complicated agent strategic behavior and complicated agent interaction can qualitatively capture many important issues in multi-agent negotiations. These included issues related to the dependence of the outcome on the agent's negotiation abilities and interaction interrelationships, the agent's initial starting offers as well as on the frequency with which the agents update their offers. The examples also discussed the conditions required to obtain the "socially optimal" outcome.

22.6 AGENT-BASED DESIGN ARCHITECTURES

Integrated or *agent-based architectures* are increasingly being used to coordinate and control complex systems. Many applications of agent architectures have been reported in the areas of robotics, manufacturing and distributed computations (e.g. [54, 55, 56]). The task of coordinating a large design team and its tools is certainly a problem in the coordination and control of a complex system. In our context, agents are defined as autonomous entities, either human, hardware or software, which perform tasks in order to achieve an overall system objective (see also [57, 58]).

In the design context, agents have many advantages. They provide a means for developing a flexible, intelligent architecture for integrating design, analysis and manufacturing tools. Agents provide customizable interfaces to large and diverse data sources. They also provide ways to standardize data access by placing a software layer between the user and the data. The need for proprietary interfaces to information sources can thus be avoided by abstracting the user from the data. By using network–based agents, the information residing on a remote host or a server can be accessed by multiple users. We now present an implemented design architecture used in the aerospace industry.

22.6.1 Multi-Agent Design Architecture

Multi-agent design architecture (MADA) is a multi-agent system for distributed decision-based design (DBD). The main idea behind MADA is to provide an automated and distributed approach to solving a wide range of *parametric design problems.*

MADA decomposes a design problem into distributed tasks, and then searches the design space in a distributed manner in a way that takes advantage of the underlying structure of the decomposed problem. By focusing on distributing both the decomposition and the search parts of the design problem, the resulting system is more robust and scalable to large design projects.

Design using MADA can be divided into three broad phases: The first involves the decomposition of a design into a *parametric form*, whereby a design instance is completely defined by a finite set of parameters. The second phase focuses on the integration of the design, analysis and manufacturing tools to provide an automated system for repeated analysis of design alternatives. Finally, the third phase focuses on the development and analysis of distributed search methodologies to evaluate a large number of design alternatives.

Parameter Maps: An overall design problem is represented in MADA by a set of parameters. These parameters represent such things as constituent dimensions, tolerances, materials and other information required to completely define the design. Information related to the performance of the design is also expressed in parametric form. The relationships between design parameters are expressed in terms of *parameter maps*, which define the sequencing of design and analysis tasks for a given domain. These maps also encode the conversion of one parameter to another by analysis tools. As such these parameter maps define how information must flow during the design process.

Traversing the Parameter Map: Searching the design space with MADA is done by traversing the parameter map. This requires first integrating all the tools needed to completely traverse the parameter map. This involves three tasks: incorporating design and analysis tools in an information architecture; coordinating the design tasks specified in the parameter map; and transferring information between tools. MADA uses three broad categories of agents to accomplish these tasks, as shown in Fig. 22.1. *Tool agents* interface with analysis tools at specific locations, such as a tool agent interacting with a CAD tool. *Facilitator agents* coordinate the design process of a specific object. They do this by sequencing the different tasks in the parameter map. Finally, *worker agents* are responsible for transferring, transforming and storing information for different tool agents and facilitator agents.

Searching the Feasible Design Space: Design is a process of searching through the space of feasible designs to identify the best (or at least a *satisficing* alternative). MADA uses a *constraint satisfaction and search* (CSS) tool to identify feasible design alternatives. Multiple design instances can be evaluated in parallel by creating independent facilitator agents that manage parameter maps for each alternative.

Key Architectural Features: Key features of MADA are the ability to decompose the overall design problem into an ordered sequence of tasks; the capability of adding new tools in the system without affecting the entire system; the flexibility of using new resources introduced in the agent environment; asynchronous operation of different tools; and the distributed problem-solving capability of the design search process.

Key Computational Features: MADA integrates design team members and their tools using a community-of-agents paradigm. The agent-based paradigm offers many advantages in terms of flexibility and fault tolerance over the usual "point-to-point" integration methods. Another important advantage of MADA is in its underlying approach to accessing the product information reservoir. Traditionally, one would use a single shared database to hold all the common information for a product design (e.g., blueprints,

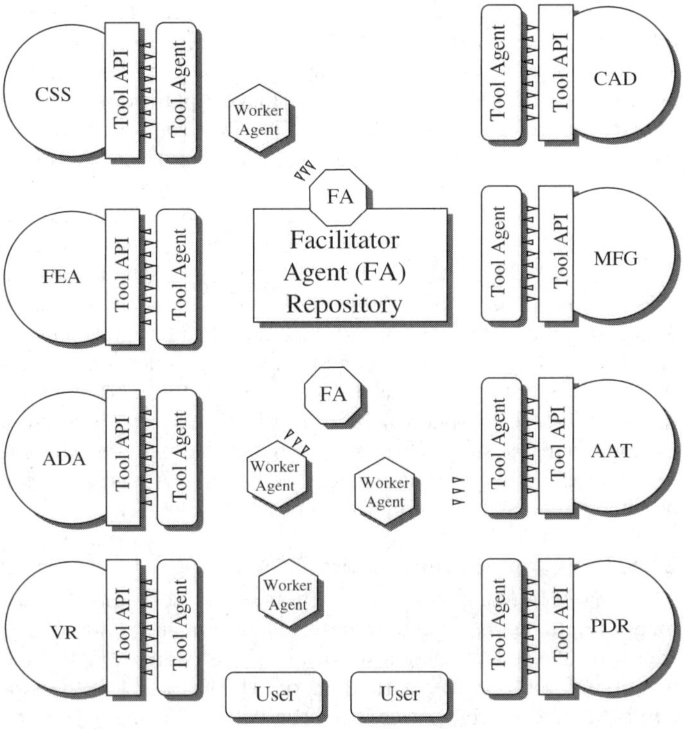

CSS: Constraint Satisfaction and Search
FEA: Finite-Element Analysis Tools
ADA: Aerodynamic Design Analysis Tools
AAT: Acoustic Analysis Tools
CAD: Computer-Aided-Design Tools
MFG: Manufacturing Analysis Tools
VR: Virtual Reality Environment
PDR: Product Data Repository
FA: Facilitator Agent
WA: Worker Agent
TA: Tool Agent
API: Application Program Interface

FIG. 22.1 THE MADA MODEL

etc.). In MADA, in contrast, the information remains distributed in the analysis tools and the environment in which it is generated. The exchange of information is not conducted through a central database, but is brokered among distributed tools by worker agents. Thus MADA provides an open and scalable architecture in which the details of application programming interfaces and application-specific data formats can be made transparent to the end user.

Through an agent approach, MADA provides seamless integration, coordination and cooperation among diverse analytical tools, which can be changed with minimal impact on the system operation. MADA also presents a single user-friendly interface for accessing the entire range of design processes and data.

22.6.2 System Architecture

The overall functioning of MADA is shown in Fig. 22.1. MADA is a community of agents that exist in an environment or a multi-agent facility [59], which represents the collection of tools, knowledge and procedures required for collaborative design and analysis tasks. The MADA framework contains three distinct software entities: (1) facilitator, worker and tool agents; (2) middleware and the associated application program interfaces (APIs); and (3) the application or analysis software. The facilitator agents and the worker agents provide means of converting overall design goals into manageable and coordinated tasks. Whereas the tool agents and back-end APIs provide an abstraction layer, which allows the application programs to consume and produce information in a common data and command/communication infrastructure.

An application software may conduct analysis that requires input information from multiple sources. For instance, a manufacturing analysis tool may need both material as well as geometry information. The tool agents provide the necessary intelligence to ensure complete and valid data before the analysis is initiated. The tool agents present information to the rest of the MADA agents though a common agent communication language, or middleware. The tool agents reside at a particular node or application location, and manage the input/output requirements of application software in order to accomplish tasks required by the design process. The tool agent resolves any missing data by utilizing the services provided by the worker agents and the product data repository. Once a data set is complete for a specific input-output combination, it is cached and is available for retrieval or future use. If any other agents have requested the information, they are notified when the information set is complete. Each tool agent is concerned only with a specific instance of information and its internal dependencies. The tool agent does not contain any design knowledge or rationale. The tool agents interact with analysis tools via standard interfaces, thereby isolating their proprietary requirements from the multi-agent system. Thus, tools with similar analysis capabilities, such as different finite-element analysis (FEA) software, can be interchanged with minimal system redesign.

22.6.3 Computational Framework

The MADA computational environment is composed of domain-specific tools, agents and the environment in which the agents reside. In this section we detail the software design of the underlying multi-agent environment on which the rest of the system is built. The agent system has been designed using object oriented and layered programming techniques. Close attention was paid to conforming to standards and isolating proprietary system requirements where possible. This approach gives the system the flexibility to change tools and applications, and even underlying implementation technologies, without disrupting other parts of the system. The results of this approach on the MADA environment can be seen in Fig. 22.2.

MADA also needs to operate in heterogeneous computing networks, and therefore must utilize technologies that facilitate interaction between diverse operating platforms. The MADA environment uses Java as its core programming language and inherits

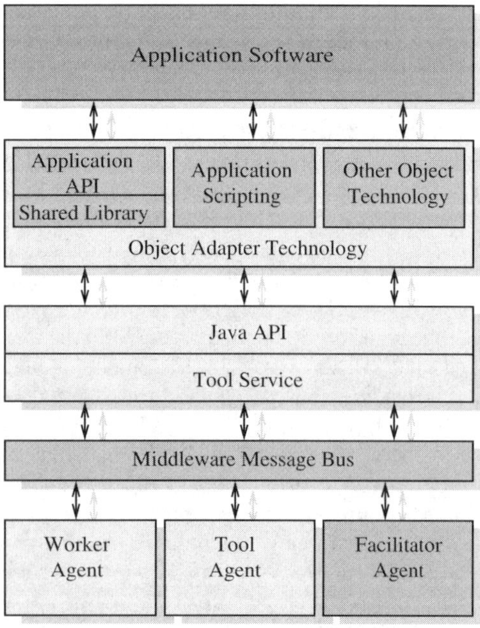

FIG. 22.2 MADA LAYERS AND AGENTS

many key features from it, including the ability to execute the environment any operating system that has a Java virtual machine (JVM) [60]. There are other middleware systems available with similar properties; the choice of middleware system is a function of the information technology environment associated with the design community. The middleware system or a multi-agent facility is what provides a common set of resources and services for mobile agents, such as mobility tracking, data persistence and serialization, message passing, naming services and life-cycle support. MADA uses a combination of standard technologies in a custom environment. Careful attention to the interface design has allowed the environment to successfully evolve with the state-of-the-art. Unfortunately, the decision regarding which systems to use is complex and is beyond the scope of this chapter.

22.6.4 Design Example

We now discuss the use of the MADA framework for a specific product. The design of an exhaust nozzle operating at supersonic Mach numbers, reminiscent of a high speed civil transport (HSCT) nozzle, is chosen as an example. The individual tools used in the integrated design system are kept simple enough to run efficiently on desktop computers with low cycle times. However, the MADA framework is capable of accommodating more complex tools, such as computational fluid dynamics, NASTRAN advanced FEA product, etc. Even with the simple product chosen, the communication and coordination between different tools remains significantly complex. The issues of scalability of the MADA architecture to complex products and assemblies, such as the space station or the next generation space shuttle, are currently being examined.

The different tools used in the integrated system for the HSCT nozzle are as follows: (1) aerodynamics tool; (2) geometric design tool; (3) FEA tool; and (4) manufacturing tool. The purpose of the aerodynamic tool is to determine the changes of flow properties inside the nozzle. The ambient conditions, nozzle pressure ratio (NPR: stagnation pressure/ambient pressure) and the temperature ratio (TR: stagnation temperature/ambient temperature) are specified as the initial nozzle parameters and are the first tools used in the design. The parametric definition of the nozzle is then used to create a geometric representation of the nozzle by the design tool Pro-Engineer. The part geometry file provides the base model for other analysis tools. Material selection is done mostly by experience and the information is encoded by a database, which also provides all the required thermo-mechanical data. Material properties and the nozzle geometry are then combined and used by the analysis software MARC and its graphical front-end MENTAT. MARC analyzes a given design configuration, and evaluates the maximum stress, strain, displacement and the specific location. The actual stresses in the nozzle compared to the allowable stress supplied by the MATDAT tool then determine the next design iteration based on the goals specified in the constraints module. The manufacturing planning tool in Pro-Engineer provides a time estimate, which in addition to the material selection module can be used to estimate the manufacturing cost for the design instance.

22.6.5 Computational Results

We now present computational results for a supersonic nozzle design. Table 22.1 shows the inputs and outputs of the system for the nozzle design. In order to control the search of the design space, MADA allows the specification of various input parameters and constraints. The parameters that determine the

Table 22.1 Design Instances Generated by MADA for a Supersonic Nozzle

Input	Case 1	Case 2
Nozzle pressure ratio	3.4	3.4
Ambient pressure (Pa)	1.00E+05	1.00E+05
Ambient temperature (K)	294	294
Temperature ratio	3.0	3.45
Mach number at exit	1.46	1.66
Starting area (m^2)	0.125	0.125
Thickness upper-bound (m)	0.01	0.01
Output		
Geometry	Axisymmetric	Axisymmetric
Thickness (m)	0.005	0.0075
Length (m)	0.1652	0.3206
Material	Aluminum	Aluminum
Cost coefficient	7.2684	7.2702
Number of iterations	9	75
Avg. run time for failed runs (msec)	4.78	4.42
Avg. run time for successful runs (msec)	91,444.25	15,198.4
Geometry	Rectangular	Rectangular
Thickness (m)	0.005	0.005
Length (m)	0.1652	0.3206
Material	Aluminum	Aluminum
Cost coefficient	13.4896	24.3036
Number of iterations	150	75
Avg. run time for failed runs (msec)	4.91	4.68
Avg. run time for successful runs (msec)	110,353.7	115,578.3

performance characteristics of the nozzle are listed in the input section of Table 22.1. The thickness parameter has the additional capability of specifying a range. Two nozzle input specifications were selected from [61] and are listed as Case 1 and Case 2 in Table 22.1. These two cases produce a total of four resultant configurations, axisymmetric and rectangular geometry for each case. The design search for these configurations was constrained by the material selection and thickness parameters. The results of the design search by MADA are summarized in the output section of the table.

The metrics listed in the output section of Table 22.1 are a subset of the design details and performance metrics generated during the design cycle. Thickness, length and material are the output parameters for each configuration. The cost coefficient is the metric used to drive the search routine. Note that the cost coefficient is a dimensionless metric. Actual cost values depend on the manufacturing equipment used and the raw material costs. The final three outputs in the table are performance metrics of MADA as a design system. The number of iterations is the number of individual design attempts made by the search routine. This number does not reflect the number of invalid designs immediately rejected by the facilitator agent. The processing time of these invalid designs is listed as failed runs. The time for successful runs represents the instances when the design cycle was completed and resulted in a feasible design. The final configurations for the axisymmetric geometry for Case 1 is shown in Fig. 22.3, and that for the rectangular geometry for Case 2 is shown in Fig. 22.4.

22.7 SUMMARY

The analysis of negotiation protocols for distributed design happens at many levels, from the underlying computational architecture to the core decision protocols that determine design parameters. This chapter covered distributed design and optimization, multi-agent design, the analysis negotiation protocols and an implementation, the multi-agent design architecture (MADA). All of these topics represent important aspects of the overall analysis of a design system.

It is important to note that the key intellectual question in realizing multiagent decision systems revolve around distributed

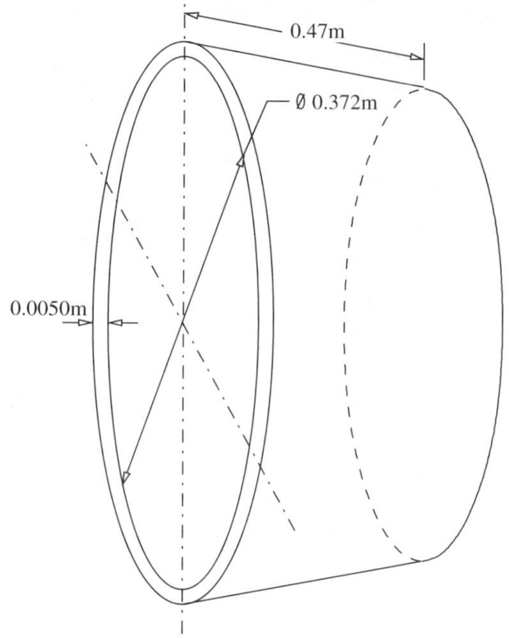

FIG. 22.3 AXISYMMETRIC NOZZLE DESIGN INSTANCE

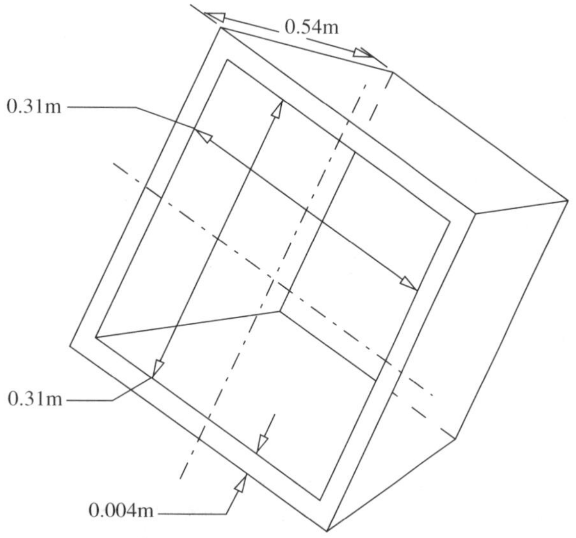

FIG. 22.4 RECTANGULAR NOZZLE DESIGN INSTANCE

decision-making. As of today, the design of a society of agents that is ideally suited to solving a complex design problem remains an open problem. It is hoped that this chapter provides some insight into the issues related to developing rigorous and robust protocols for negotiated design.

FURTHER READING

Systems that achieve the above are sometimes referred to as *integrated synthesis environments* (ISE). Several prototype environments have been developed, for example, in the aerospace and automotive industries. For details, the interested reader is referred to [62, 63, 64, 65, 66, 67].

ACKNOWLEDGMENTS

The research presented in this chapter was supported in part by NSF Grant # DMI-9978923, NASA Ames Research Center Grants # NAG 2-1114, NCC 2-1180, NCC 2-1265, NCC 2-1348, and DARPA/AFRL Contract #F3060299-2-0525.

REFERENCES

1. Bertsekas, D. P. and Tsitsiklis, J. N., 1997. *Parallel and Distributed Computation: Numerical Methods*, Athena Scientific.
2. Ferreira, A. and Pardelos, P., eds., 1996. *Solving Combinatorial Optimization Problems in Parallel, Methods and Techniques, Lecture Notes in Computer Sci*, Vol. 1054, Verlag, New York.
3. Azencott, R., ed., 1992. *Simulated Annealing:Parallelization Techniques*, Wiley, New York, NY.
4. Dorigo, M. and Gambardella, L. M., 1997. "Ant Colony System: A Cooperative Learning Approach to the Traveling Salesman Problem," *IEEE Trans. on Evolutionary Computation*, **1**(1), pp. 53–66.
5. Leibkuchler, K. H., 2000. "Ant Colony Heuristics for the Capacitated Alternate Path Vehicle Routing Problem," Master's thesis, Univ. of Massachusetts, Amherst, MA.
6. Randall, M. and Luis, A., 2002. "A Parallel Implementation of Ant Colony Optimization," *J. of Parallel and Distributed Computing*, **62**(9), pp. 1421–1432.
7. Lueling, R., Monien, B., Reinefeld, A. and Tschoeke, S., 1996. "Mapping Tree-Structured Combinatorial Optimization Problems Onto Parallel Computers," *Lecture Notes in Computer Sci.*, Vol. 1054, pp. 115–144.
8. de Bruin, A., Kindervater, G. A. P. and Trienekens, H. W. J. M. 1996. "Towards an Abstract Parallel Branch and Bound Machine," *Lecture Notes in Computer Sci.,* Vol. 1054, pp. 145–170.
9. Scott, M. J., 1999. "Formalizing Negotiation in Engineering Design," PhD thesis, California Institute of Technology, Pasadena, CA.
10. Yu, P.-L., 1973. "A Class of Solutions for Group Decision Problems," *Magmt. Sci.*, Vol. 19, pp. 936–946.
11. Yu, P.-L., 1985. *Multiple-Criteria Decision-Making: Concepts, Techniques, and Extensions*, Perseus Publishing.
12. Voorneveld, M. and van den Nouweland, A., 2000. "An Axiomatization of the Euclidean Compromise Solution," Preprint 1145.
13. Conley, J. P., McLean, R. and Wilkie, S., in press. "Axiomatic Foundations for Compromise Theory: The Duality of Bargaining Theory and Multi-Objective Programming," *Games and Economic Behavior.*
14. Brams, S. J. and Taylor, A. D., 1996. *Fair Division: From Cake-Cutting to Dispute Resolution*, Cambridge University Press.
15. Raith, M. G., 1999. "The Structure of Fair-Division Problems and the Design of Fair-Negotiation Procedures," *Game Practice: Contributions From Applied Game Theory, Vol. 23*, Kluwer Academic Publishers, Chapter 14, p. 288.
16. Veldhuizen, D. A. V. and Lamont, G. B., 2000. "Multiobjective Evolutionary Algorithms: Analyzing the State-of-the-Art," *Evolutionary Computation*, **8**(2), pp. 125–147.
17. Veldhuizen, D. A. V., Zydallis, J. B. and Lamont, G. B., 2002. "Issues in Parallelizing Multiobjective Evolutionary Algorithms for Real World Applications," *Proc., 2002 ACM Symp. on App. Computing*, ACM Press, pp. 595–602.
18. Collette, Y. and Siarry, P., 2003. *Multiobjective Optimization: Principles and Case Studies, Springer*, Chapter 5.
19. Cantáu-Paz, E., 2000. *Efficient and Accurate Parallel Genetic Algorithms*, Kluwer Academic Publishers.

20. Quirk, J. and Saposnik, R., 1968. *Introduction to General Equilibrium Theory and Welfare Economics*, McGraw-Hill, New York, NY.
21. Rader, T., 1972. *Theory of General Economic Equilibrium*, Academic Press.
22. Wellman, M. P. and Wurman, P. R., 1997. "Market-Aware Agents for a Multi-Agent World," *Robotics and Autonomous Sys.*, Vol. 24, pp. 115–125.
23. Wellman, M. P., Walsh, W. E., Wurman, P. R. and MacKie-Mason, J. K., 1998. "Auction Protocols for Decentralized Scheduling," *Submission to Games and Economic Behavior*, July.
24. Roth, A. E., ed., 1985. *Game Theoretic Models of Bargaining*, Cambridge University Press.
25. Osborne, M. J. and Rubinstein, A., 1994. *A Course in Game Theory*, MIT Press.
26. Rubinstein, A., 1982. "Perfect Equilibrium in a Bargaining Model," *Econometrica*, **50**(1), pp. 97–110.
27. Binmore, K., Rubinstein, A. and Wolinsky, A., 1986. "The Nash Bargaining Solution in Economic Modelling," *The RAND J. of Eco.*, **17**(2), pp. 176–188.
28. Muthoo, A., 1999. *Bargaining Theory With Applications*, Cambridge University Press.
29. Chatterjee, K. and Samuelson, W., 1983. "Bargaining Under Incomplete Information," *Operations Res.*, **31**(5), pp. 835–851.
30. Fudenberg, D. and Tirole, J., 1983. "Sequential Bargaining with Incomplete Information," *The Rev. of Eco. Studies*, **50**(2), pp. 221–247.
31. Inderst, R., 2000. "Multi-Issue Bargaining With Endogenous Agenda," *Games and Eco. Behavior*, **30**(1), pp. 64–82.
32. In, Y. and Serrano, R., 2003. "Agenda Restrictions in Multi-Issue Bargaining," *J. of Eco. Behavior and Org.*
33. Busch, L.-A. and Horstmann, I. J., 1999. "Signaling via an Agenda in Multi-Issue Bargaining With Incomplete Information," *Eco. Theory*, **13**(3), pp. 561–575.
34. John, R. and Raith, M. G., 2001. "Optimizing Multi-Stage Negotiations," J. *of Eco. Behavior and Org.*, **45**(2), pp. 155–173.
35. Coles, M. G. and Muthoo, A., 2003. "Bargaining in a Non-Stationary Environment," *J. of Eco. Theory*, **109**(1), pp. 70–89.
36. Sen, A., 2000. "Multidimensional Bargaining Under Asymmetric Information," *Int. Eco. Rev.*, **41**(2), pp. 425–450.
37. Rubinstein, A. and Wolinsky, A., 1990. "Decentralized Trading, Strategic Behaviour and the Walrasian Outcome," *Rev. of Eco. Studies*, **57**(1), pp. 63–78.
38. Osborne, M. J. and Rubinstein, A., 1990. *Bargaining and Markets*, Elsevier Science & Technology.
39. Trefler, D., 1999. "Bargaining With Asymmetric Information in Non-Stationary Markets," *Eco. Theory*, **13**(3), pp. 577–601.
40. Jackson, M. O. and Palfrey, T. R., 1998. "Effciency and Voluntary Implementation in Markets With Repeated Pairwise Bargaining," *Econometrica*, **66**(6), pp. 1353–1388.
41. Weiss, G., ed., 1999. *Multiagent Systems: A Modern Approach to Distributed Artificial Intelligence*, MIT Press.
42. Rao, A. S. and George, M. P., 1991. "Modeling Rational Agents Within a BDI Architecture," *Proc., 2nd Int. Conf. on Principles of Knowledge Representation and Reasoning.*
43. d'Inverno, M., Kinny, D., Luck, M. and Wooldridge, M., 1998. "A Formal Specification of dMARS," *Intelligent Agents IV: Proc., 4th Int. Workshop on Agent Theories, Architectures and Languages*, Singh, R. and Wooldridge, eds., Vol. 1365, Lecture Notes in AI, Springer-Verlag, pp. 155–176.
44. Spivey, J. M., 1992. *The Z Notation: A Reference Manual, 2nd Ed*, Prentice Hall International Series in Computer Science.
45. Müller, M., Müller, T. and van Roy, P., 1995. "Multi-Paradigm Programming in Oz," *Visions for the Future of Logic Programming: Laying the Foundations for a Modern successor of Prolog*, D. Smith, O. Ridoux and P. V., Roy, eds., Workshop in Association with ILPS '95.
46. Pauly, M., 2002. "Programming and Verifying Subgame Perfect Mechanisms," *Tech. Rep. ULCS-02-018*, The Univ. of Liverpool.
47. Durfee, E. H., 1999. *Multiagent Systems: A Modern Approach to Disttributed Artificial Intelligence*, MIT Press, Chapter 3.
48. Corkill, D. D., 1982. "A Framework for Organizational Self-Design in Distributed Problem Solving Networks," PhD thesis, Univ. of Massachusetts.
49. Yokoo, M. and Hirayama, K., 2000. "Algorithms for Distributed Constraint Satisfaction: A Review," *Autonomous Agents and Multi–Agent Sys.*, **3**(2), pp. 185–207.
50. Gerding, E. H., van Bragt, D. D. B. and Poutre, J. A. L., 2000. Scientific Approaches and Techniques for Negotiation. A Game Theoretic and Artificial Intelligence Perspective, *Tech. Rep. SEN-R0005*, CWI.
51. Kraus, S., 2001. *Strategic Negotiation in Multiagent Environments*, MIT Press.
52. Fatima, S. S., Wooldridge, M. and Jennings, N. R., 2004. "An Agenda Based Framework for Multi-Issue Negotiation," *Artificial Intelligence*, **152**(1), pp. 1–45.
53. Lander, S. E., 1997. "Issues in Multiagent Design Systems," *IEEE Expert*, **128**(3), pp. 18–26.
54. Parker, L., 1994. "ALLIANCE: An Architecture for Fault Tolerant, Cooperative Control of Heterogeneous Mobile Robots," *Proc., Int. Conf. on Intelligent Robots and Sys.*, pp. 776–783.
55. Smith, R., 1980. "The Contract Net Protocol: High-Level Communication and Control in a Distributed Problem Solver," *IEEE Trans. on Computers*, **23**(12), pp. 1104–1113.
56. Lewis, W., 1981. "Data Flow Architectures for Distributed Control of Computer Operated Manufacturing Systems: Structure and Simulated Applications," PhD thesis, Dept. of Industrial Engrg., Purdue Univ., West Lafayette, IN.
57. Garvey, A. and Lesser, V., 1995. "Representing and Scheduling Satisficing Tasks," *Imprecise and Approximate Computation*, Kluwer Academic Publishers, pp. 23–34.
58. Jennings, N. R., Sycara, K. P. and Wooldridge, M., 1998. "A Roadmap of Agent Research and Development," *J.* of *Autonomous Agents and Multi-Agent Sys.*, **1**(1), pp. 7–36.
59. Crystaliz Inc., General Magic Inc, GMD FOKUS and IBM Corporation., 1997. "Mobile Agent Facility Specification," *Tech. Rep. OMG TC cf/97-06-04*, The Object Management Group, June.
60. Lindholm, T. and Yellin, F., 1999. *The Java™ Virtual Machine Specification*, 2nd Ed., Addison-Wesley Pub. Co.
61. Krothapalli, A., Soderman, P., Allen, C., Hayes, J. A. and Jaeger, S. M., 1997. "Flight Effects on the Far-Field Noise of a Heated Supersonic Jet," *AIAA* J., **35**(6), pp. 952–957.
62. Mecham, M., 1997. "Lockheed Martin Develops Virtual Reality Tools for JSF," *Aviation Week & Space Technology*, McGraw-Hill, Oct., pp. 51–53.
63. Valenti, M., 1998. "Re-engineering Aerospace Design," *Mech.* Engrg., 120 (Jan.), pp. 70–72.
64. Bloor, S. and Owen, J., 1995. *Product Data Exchange*, UCL Press, Ltd.
65. Regli, W. C. and Gaines, D. M., 1997. "National Repository for Design and Process Planning," *Computer-Aided Des.*, **29**(12), pp. 895–905.
66. Sriram, R., Gorti, S., Gupta, A., Kim, G. and Wong, A., 1998. "An Object-Oriented Representation for Product and Design Processes," *J. of CAD*, **30**(7), pp. 489–501.
67. Goldin, D. S., Venneri, S. L. and Noor, A. K., 1999. "Ready for the Future?", *Mech. Engrg.*, **121**(11), pp. 61–64.

PROBLEMS

22.1 Think of a large-scale design problem, such as the design of a new cruise ship, a new commercial passenger airliner, or a large civil engineering project like a skyscraper, a mall or a dam. Write down all of the individuals you would include on the project design team. What decisions would these different individuals be responsible for? How would the decisions made by the different individuals impact one another?

22.2 Show that under Eq. (22.4) $\|A\|_\infty = 1$, which implies that Eq. (22.3) is always nonexpansive.

22.3 Give an example of an interaction matrix that satisfies Eqs. (22.4) and (22.14) but is not irreducible. This suggests that irreducibility of the interaction matrix, while necessary to ensure convergence to a consensus outcome, is not necessary to ensure convergence of the negotiation model Eq. (22.3).

22.4 Consider the interaction matrix

$$\boldsymbol{A} = \begin{bmatrix} 0 & 1 \\ 1 & 0 \end{bmatrix}$$

Show that this matrix is irreducible and that $\rho(A) = 1$. Also show, using simulation, that Eq. (22.3) does not converge with $g = 1$, but will converge with $g < 1$.

22.5 Show Eq. (22.16).

22.6 Prove Eq. (22.17).

22.7 Write a simulator to illustrate the remark of **Corollary 1**.

22.8 Prove **Lemma 7**.

CHAPTER

23

The Dynamics of Decentralized Design Processes: The Issue of Convergence and its Impact on Decision-Making

Vincent Chanron and Kemper E. Lewis

23.1 INTRODUCTION

Most complex systems, including engineering systems such as cars, airplanes and satellites, are the results of the interactions of many distinct entities working on different parts of the design. Decentralized systems constitute a special class of design under distributed environments. They are characterized as large and complex systems divided into several smaller entities that have autonomy in local optimization and decision-making. The issue of *decentralized design processes* is to have the designers involved in the process converge to a single design solution that is optimal and meets the design requirements, while being acceptable to all the participants. This is made difficult by the strong interdependencies between the designers, which are usually characteristic of such systems.

Other chapters have focused on the modeling of design-related issues, the generation of design alternatives and the decision-making in different environments. This chapter introduces the issue of convergence of decentralized design processes. Why do some decentralized design problems converge to a final design while other design teams involved in another process cannot seem to find an agreement? Is it possible to predict the convergence of such processes beforehand or does one have to wait for the final stages of the design processes to realize the failure of the design process to find a final unique design? Is the final solution found by these processes optimal? Are there ways to improve those processes to speed up convergence and ensure optimality? Those questions are all related to current research issues [1, 2] and the state-of-the-art of these topics is presented in this chapter.

23.2 DECENTRALIZED DESIGN: THE PROBLEM

The focus of this chapter is a theoretical study of the design of complex engineering systems, or those systems that necessitate the decomposition of the system into smaller subsystems in order to reduce the complexity of the design problems. Most of these systems are very large and multidisciplinary in nature, and therefore have a great number of subsystems and components. This creates issues in understanding the interactions between all these subsystems, in order to create more efficient design processes. In this chapter, we focus on the dynamics of distributed design processes and attempt to understand the fundamental mechanics behind these processes in order to facilitate the decision process between networks of decision-makers.

The complexity of engineering products has been growing steadily, and this trend is bound to continue. A few figures can demonstrate easily the sharp increase in the complexity in engineering products. In the automotive industry, for example, *The Economist* reports that "it took 700 parts to make the model Ford T, while modern cars pack many more in their radio alone" [3]. In the aerospace industry, products are even more complex; for example, "there are 3 million parts in a Boeing 777 provided by more than 900 suppliers" [4]. Similarly the software industry is facing increasing complexity in the millions of lines of codes it comprises [5]. Companies face similar complexity in their production planning, administration, resources planning, among others. That led to the development of companies such as SAP (Germany) or PeopleSoft (U.S.) that develop software and platforms to help companies manage this complexity. Similarly, in engineering design, optimization software such as i-SIGHT propose a solution for managing the complexity of problems with several technical objectives, and involving several evaluation platforms [computer-aided design (CAD), fluid mechanics, finite elements, etc.]. But true complex problems (think of a plane, for example) are beyond the capabilities of such software, and companies have to find other solutions to manage complexity. Outsourcing, one form of decentralization, is one of them, and its use is increasing. In this section we explain why companies make this choice of distributing the design of their products, and we also present the issues that are created.

The complexity of design problems is one reason for the decentralization of decisions; another related one is the multidisciplinary nature of these systems, which makes it impossible for one designer, or even a single design team, to consider the entire system as a single design problem. Typically, in complex systems, breaking it up into smaller units or subsystems will make the system more manageable [6, 7].

The decentralization of decisions is unavoidable in a large organization where having only one centralized decision-maker is usually not applicable [8]. A more effective way is to delegate decision responsibilities to the appropriate person, team or supplier. In fact, decentralization is recommended as a way to speed up product development processes and decrease the computational time and the complexity of the problem [9].

In a business prospective, decentralization also has several benefits. The decentralization of manufacturing tasks has long been popular among American companies due to the shrinking cost of transportation (rail, road, airplane). As an example, the cost of air freight (as measured by average revenue per ton-kilometer) fell by 78% between 1955 and 1996 [10]. But even bigger advantages can be achieved by decentralizing the design, not just the manufacturing. Indeed, *"For example, the V6 car engines that Toyota sends from Nagoya (Japan) to Chicago take anywhere between 25 and 37 days to arrive, forcing the car company to hold costly stocks. The movement of white-collar work, on the other hand, is subject to no physical constraints. Communications are instant and their cost is declining rapidly towards becoming free"* [3]. One of the advantages could be to reduce the development time of a product by having design teams spread around the world in order to achieve "24-hour design": one design team in America, one in Europe and one in Asia, working on the same product and communicating the information to the next team at the end of its eight-hour shift, for example. But a bigger, and more important, advantage is the notion of "risk-share partners," in order to share the development costs of a product with other companies in order to minimize the risks of this investment. Indeed, when a company outsources the design of a part of its products to a supplier, it does not only buy this part from the company. Oftentimes, the suppliers are asked to become "risk-share partners," where they will share the loss if the final product is not successful, but also share the benefits in case it is successful. This happens very often in the aerospace industry, where the costs of designing and developing a new product are so large that a single company can usually not invest all the money without risk-share partners. In the civil aircraft segment, for example, Boeing—though the world largest aerospace firm—outsources the design of a great number of components to its risk-share partners. It is said, for example, that the Japanese companies will design and build more than 35% of the structure of the new Boeing 7E7—including the wings and fuselage parts; foreign content might even run as high as 70% [11]. The development costs of those parts is of course paid by the Japanese companies (which will, in return, get benefits if the 7E7 is a commercial success), thus cutting the development costs for Boeing.

While the decomposition of complex problems certainly creates a series of smaller, less complex problems, it also creates several challenging issues associated with the coordination of these less complex problems. The origin of these problems is the fact that the less complex subproblems are usually coupled. Systems are said to be coupled if their solution is dependent upon information from other subproblems. The ideal case would be a system that could be broken up into subsystems without interdependence. Unfortunately, design variables and parameters usually have an influence on several subproblems. Design variables and parameters that are controlled within a subsystem are called local, while nonlocal information is controlled by another subsystem [12].

In order to solve these problems, previous work has been done on the decomposition of the system into smaller ones; using design structure matrices [13], a hierarchical approach [14], or by effectively propagating the desirable top-level design specifications to appropriate subsystems [15, 16], their efficiency has been compared [17]. For more complex problems, however, the decomposition is natural, as it follows the areas of competencies and/or the physical characteristics of the product to be designed. A good example is the European civil aircraft manufacturer Airbus, which designs and builds its airplanes all around Europe. The first decomposition is made following the main parts of the airplane and assigned depending on the area of expertise of its subdivisions: The design and manufacturing of the wings, for example, is traditionally assigned to Airbus UK. However, even a subsystem such as a wing needs to be further decomposed into smaller subsystems as it is multidisciplinary. The decomposition is then made along "centers of excellence" and "centers of competence", reflecting the multidisciplinarity of the system to be designed. Decomposition techniques can then be used to determine the allocation of design variables and of resources to these centers, which are further responsible for the interaction with the external suppliers [18].

The decentralization of decisions seems to be the common trend in several industries in order to deal with the complexity and financing of the products. However, having several distributed design teams creates coordination issues. One reason for this is the fact that the individual teams have a limited vision of the overall product and process because of poor management and communication (in the case of the subsidiary of a company) or because of communication obstacles such as technology privacy (in the case of external suppliers). As a result, the individual design teams tend to privilege the optimality of their own subsystem, rather than the optimality of the overall product.

This has been noted in the design of engineering products [19] but this phenomenon is not inherent to engineering design. This is the classic "tragedy of the commons" problem from human economics [20]. Although prudent cooperation among design teams would increase overall optimality of the product, maximization of individual subsystems is standard. Cooperation increases collective success but usually at the cost of individual success. Penalization of individual optimality, as seen in some biological systems [21], can prevent defection from cooperative duties but such coercive measures must not infringe on the creative process.

Architecture is also a field where different entities have to work together to design a complex product: a building. The inefficiencies of building processes have been studied by Phillip Bernstein, an architect and professor at Yale University, and these inefficiencies are in part because the construction industry is so fragmented, he says. "Designers, architects, engineers, developers and builders each make decision that serve their own interests, but create huge inefficiencies overall" [22]. Therefore, the coordination of distributed design teams seems to be important to improve the design process of several industries, and possibly to avoid potential design failures.

There are numerous other examples of design failures for complex systems. For example, America's Internal Revenue Service (IRS) had to spend $4 billion on a multi-year effort to overhaul its computer system that failed completely in 1997 [5]. The major problem for these failures is mostly that they cost a lot of money. Indeed, even if design projects do not fail completely, they are still delayed, canceled, or over budget. In the software industry, for example, a study by the Standish Group, a technology consultancy, estimated that 30% of all software projects are canceled, nearly half come in over budget, 60% are considered failures by the organization that initiated them and nine out of ten come in late [5].

This section showed that something needs to be done about managing and understanding the complex interactions involved in the design of complex systems. The next section presents a background to understand better the challenges faced in engineering design before solutions to tackle them are introduced.

23.3 DECENTRALIZED DESIGN: THE BACKGROUND

In order to improve design processes and avoid design failures, this chapter tries to formally describe the dynamics and interactions involved in such design scenarios. We believe that, in order to be able to design better, those dynamics have to be well understood. As Tufte [23] puts it: *"An essential analytic task in making decisions based on evidence is to understand how things work—mechanism, trade-offs, process and dynamics, cause and effect. That is, intervention thinking and policy-thinking demand causality-thinking."*

Therefore, explaining and understanding the dynamics involved will help us make better decisions in design, and it is the goal of this chapter. This section presents the background for this work, in terms of problem formulation for decentralized decision processes, and numerically shows some of the design failures that can occur during the design process.

A common approach to solving those design problems with interacting subsystems is to use Game Theory. As introduced in the Chapter "Fundamentals of Game Theory in Decision-Making" in Section 2, Game Theory provides a mathematical framework that models the interaction between decision-makers, also called *players* [24]. It was mainly used in the fields of economics and social sciences before applications were found in other areas of interest, from the stock exchange to engineering design. The main goal of using Game Theory in engineering design is to try to improve the quality of the final solution in a multiobjective, distributed design optimization problem [25]. Previous work in Game Theory includes work to model the interactions between the designers if several design variables are shared among designers [26]. In [27], Game Theory is formally presented as a method to help designers make strategic decisions in a scientific way. In [28], distributed collaborative design is viewed as a noncooperative game, and maintenance considerations are introduced into a design problem using concepts from Game Theory. In [29], the manufacturability of multi-agent process planning systems is studied using Game Theory concepts. In [30], noncooperative protocols are studied and the application of Stackelberg leader/follower solutions is shown. Also in [31], a Game Theory approach is used to address and describe a multifunctional team approach to concurrent parametric design. This set of previous work has established a solid foundation for the application of game theory in design, but has not directly studied the mechanisms of convergence in a generic decentralized design problem, which is what this chapter proposes to do.

Game Theory is a mathematical representation of real situations. In engineering design, those mathematical models are called design scenarios. We present here the main scenarios that are used in the study of decentralized design [32]. Table 23.1 shows the general problem formulation for an optimization design problem involving several design teams (in this case two designers, also called players). This formulation is used later to explain the design scenarios.

In Table 23.1, $\mathbf{x}_1$ represents the vector of design variables controlled by designer 1, while designer 2 controls design variable vector $\mathbf{x}_2$. We denote $\mathbf{x}_{1c}$ and $\mathbf{x}_{2c}$ the nonlocal design variables, variables that appear in a model but are controlled by the other player. In some decomposed problems, one variable may be local to many subsystems. This kind of problem is not investigated in this chapter, but is part of the current work of the research community. We now present the three main types of design scenarios, or protocols, used for modeling decentralized decision-making problems [33].

TABLE 23.1 MULTIPLAYER OPTIMIZATION PROBLEM FORMULATION

Player 1's Model	Player 2's Model
Minimize	*Minimize*
$\mathbf{F}_1(\mathbf{x}_1,\mathbf{x}_{2c})=\{F_1^1,F_1^2 \ldots F_1^p\}$	$\mathbf{F}_2(\mathbf{x}_{1C},\mathbf{x}_2)=\{F_2^1,F_2^2 \ldots F_2^q\}$
subject to	*subject to*
$g_j^1(\mathbf{x}_1,\mathbf{x}_{2c})\le 0 \quad j=1\ldots m_1$	$g_j^2(\mathbf{x}_{1C},\mathbf{x}_2)\le 0 \quad j=1\ldots m_2$
$g_k^1(\mathbf{x}_1,\mathbf{x}_{2c})=0 \quad k=1\ldots l_1$	$h_k^2(\mathbf{x}_{1C},\mathbf{x}_2)=0 \quad k=1\ldots l_2$
$\mathbf{x}_{1L}\le\mathbf{x}_1\le\mathbf{x}_{1U}$	$\mathbf{x}_{2L}\le\mathbf{x}_2\le\mathbf{x}_{2U}$

23.3.1 Cooperative Protocol

In this protocol, both players have knowledge of the other player's information and they work together to find a Pareto solution. A pair ($\mathbf{x}_{1P}$, $\mathbf{x}_{2P}$) is Pareto optimal [34] if no other pair ($\mathbf{x}_1$, $\mathbf{x}_2$) exists such that:

$$F_i(\mathbf{x}_1,\mathbf{x}_2)\le F_i(\mathbf{x}_{1P},\mathbf{x}_{2P}) \quad i=1,2 \qquad \text{Eq. (23.1a)}$$

and $$F_j(\mathrm{x}_1,\mathrm{x}_2)\le F_j(\mathrm{x}_{1P},\mathrm{x}_{2P}) \quad \text{for at least one } j=1,2 \qquad \text{Eq. (23.1b)}$$

Systems thinking is the key to full cooperation in modern organizations where a shared vision is common and subscribed to by all members of an organization [35]. However, shared vision does not suggest that the designers will necessarily fully cooperate. Mathematical and model cooperation are required to assume full cooperation and that the final design will be Pareto optimal. Unfortunately, this is rarely the case in distributed environments, as there are several obstacles to this full cooperation.

23.3.2 Noncooperative Protocol

This protocol occurs when full coalition among players is not possible due to organizational, information or process barriers. Players must make decisions by assuming the choices of the other decision-makers. In an iterative approach, the final solution would be a Nash equilibrium. A strategy pair ($\mathbf{x}_{1N}$, $\mathbf{x}_{2N}$) is a Nash solution if:

$$F_1(\mathbf{x}_{1N},\mathbf{x}_{2N})=\min_{\mathbf{x}_1} F_1(\mathbf{x}_1,\mathbf{x}_{2N}) \qquad \text{Eq. (23.2a)}$$

and $$F_2(\mathbf{x}_{1N},\mathbf{x}_{2N})=\min_{\mathbf{x}_2} F_2(\mathbf{x}_{1N},\mathbf{x}_2) \qquad \text{Eq. (23.2b)}$$

In other words, *"A point is said to be a Nash Equilibrium or a Nash Solution if no designer can improve unilaterally his/her objective function"* [36]. This solution has the property of being individually stable, but is not necessarily collectively optimal, meaning that, at this point, each designer will perceive the design point to be optimal [37], whereas the solution is not necessarily Pareto optimal. This is because any unilateral decision to change a design variable value by either designer cannot, by definition, result in a better objective function value for the designer who makes the change. The Nash Equilibrium also has the property of being the fixed point of two subsets of the feasible space:

$$(\mathbf{x}_{1N}, \mathbf{x}_{2N}) \in \mathbf{X}_{1N}(\mathbf{x}_{2N}) \times \mathbf{X}_{2N}(\mathbf{x}_{1N}) \qquad \text{Eq. (23.3)}$$

where $$\mathbf{X}_{1N}(\mathbf{x}_2) = \{\mathbf{x}_{1N} / \mathbf{F}_1(\mathbf{x}_{1N}, \mathbf{x}_2) = \min_{\mathbf{x}_1} \mathbf{F}_1(\mathbf{x}_1, \mathbf{x}_2)\}$$

$$\mathbf{X}_{2N}(\mathbf{x}_1) = \{\mathbf{x}_{2N} / \mathbf{F}_2(\mathbf{x}_1, \mathbf{x}_{2N}) = \min_{\mathbf{x}_2} \mathbf{F}_2(\mathbf{x}_1, \mathbf{x}_2)\}$$

are called the *rational reaction sets* of the two players. The **rational reaction set (RRS)** of a player is a function that embodies his reactions to decisions made by other players.

23.3.3 Leader/Follower protocol

When one player makes his decision first, he has a leader/follower relationship [38]. This is a common occurrence in a design process when one discipline plays a large role early in the design, or in a design process that involves a sequential execution of interrelated disciplinary processes. Player 1 is said to be the leader if he/she declares his/her strategy first, by assuming that Player 2 behaves rationally. Thus the model of Player 1 as a leader is the following:

$$\text{Minimize} \quad \mathbf{F}_1(\mathbf{x}_1, \mathbf{x}_2) \qquad \text{Eq. (23.4a)}$$

$$\text{subject to} \quad \mathbf{x}_2 \in \mathbf{X}_{2N}(\mathbf{x}_1) \qquad \text{Eq. (23.4b)}$$

where $\mathbf{X}_{2N}(\mathbf{x}_1)$ = RRS of Player 2.

For the reasons explained in the previous section, assuming that designers who are distributed interact with others in full cooperation is a utopian state. That would require the designers to share every single detail of their model with the other designers involved in the design of the same product. This is not achievable and a recent study even shows that it is not even necessarily desirable [39]. Therefore, the cooperative scenario cannot be used to model the interactions of designers in a decentralized environment; it is only used as a test bed to compare the final solutions, since it leads to Pareto optimal solutions. A more realistic approach is to consider that designers are in a situation of *limited cooperation*, meaning that they are eager to cooperate, but only to a certain extent (the limit is defined by the existing communication barriers or the willingness to cooperate). Therefore, the noncooperative scenario is used to model the relationships between designers in order to reflect the imperfect information and cooperation that exist, even within the same corporation. In other words, we focus on decentralized design scenarios where full and efficient exchange of all information among subsystems is not possible.

Even though most companies are trying to break down the walls between the different disciplines, many decisions are taken in a sequential manner. Companies should of course strive for cooperation, but noncooperation is an involuntary result of organizational or informational barriers among decision-makers. In particular, competitive suppliers designing parts for the same overall product are usually not willing to share their analysis models, thus also resulting in noncooperation.

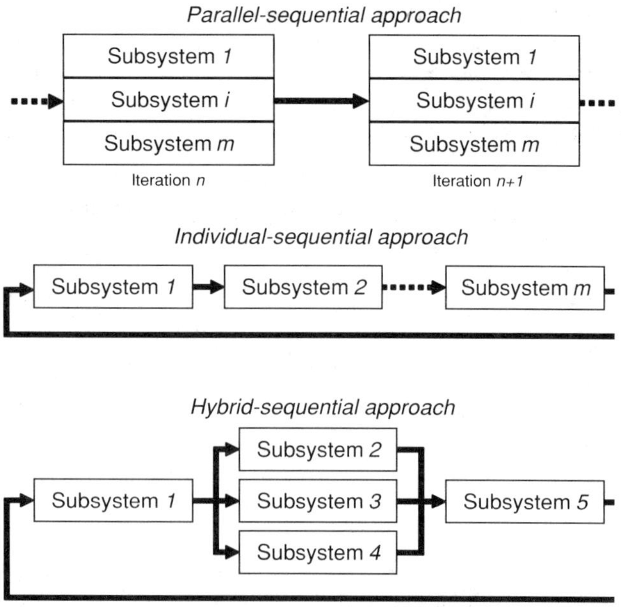

FIG. 23.1 SEQUENTIAL APPROACHES TO DESIGN

The presence of nonlocal variables in the model of subsystems requires a certain level of communication between the design teams. In a sequential approach, for example, this information flow goes back and forth between the design teams until they reach an agreement on a particular design point. This iterative approach is not necessarily the ideal process to design a product, but is, in fact, widely used. It is even recommended in the development of certain products, such as the development of software products [5]. The sequential process can take several forms. It can be parallel-sequential (at each time step, every subsystem solves its own model using the design variables' values obtained at the previous time step), or individual-sequential (one discipline goes after another, in a specified order) or even hybrid (a combination of the two). This is depicted in Fig. 23.1.

The final design point of the sequential process is known as a Nash equilibrium, whose properties are shown in Eqs. (23.2) and (23.3). The fact that the designers agree on a final design is known as *convergence* of the design process [19]. The issue of divergence in an engineering design process was noted as early as in [25], and remains an issue to be solved [27]. What happens in those cases is that the sequential approach taken by the designers is endless [40]. Exchanging design variable values back and forth, the design teams cannot agree on a final design because, at each iteration, at least one designer will not be satisfied by the point chosen.

This phenomenon is best explained by a simple example. The following example is derived from [25] and presents two simple decentralized decision problems involving two designers. Figure 23.2 illustrates a situation in which the two designers have the following objective functions, to be minimized:

$$F_1 = x^2 - 3x + xy \qquad \text{Eq. (23.5a)}$$

$$F_2 = \frac{y^2}{2} - xy \qquad \text{Eq. (23.5b)}$$

where designer 1 controls x and minimizes F_1, and designer 2 controls y and minimizes F_2, with $x \geq 0$ and $y \geq 0$. Figure 23.2

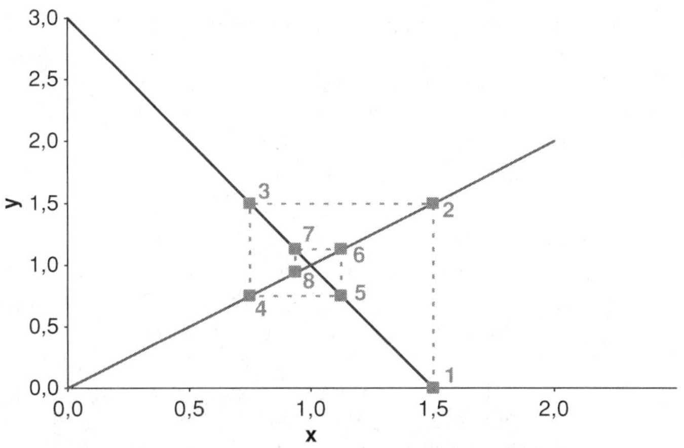

FIG. 23.2 A CONVERGENT DECENTRALIZED DESIGN EXAMPLE

illustrates the RRS for each designer as well as the first iterations of the iterative process.

According to Fig. 23.2, the designers move back and forth between their RRS. This is the typical behavior in an iterative process: The designers will always have a tendency to come back to their RRS as it is the point that minimizes their objective function. By continuing the iterative process, the designers finally reach the Nash solution, intersection of their RRSs:

$$(x_N, y_N) = (1,1) \qquad (F_{1N}, F_{2N}) = (-1, -0.5) \qquad \text{Eq. (23.6)}$$

However, this Nash solution is nonoptimal since there are some points in the design space where both designers could improve their objectives. For example, at $x = 2$ and $y = 1/3$, the values of the two objective functions are:

$$F_1 = -1.33 \qquad F_2 = -0.61 \qquad \text{Eq. (23.7)}$$

which are better for both designers. Therefore, the Nash solution is dominated by this point since both objectives have been improved.

Therefore, when using this iterative decision process in a decentralized environment, the final solution is not necessarily optimal since both objectives could be improved. But what is perhaps even more surprising is that this process does not necessarily have to converge to the Nash solution. Consider the case where two designers have the following objective functions:

$$F_1 = \frac{x^2}{4} - 1.5x + xy \qquad \text{Eq. (23.8a)}$$

$$F_2 = \frac{y^2}{2} - xy \qquad \text{Eq. (23.8b)}$$

With these objective functions, whichever designer goes first, and whatever the starting point is, the decentralized decision system will always diverge. As an example, designer 2 starts with the tentative solution ($x = 0$, $y = 0.8$), and passes this information to designer 1 who adjusts x. This information is then passed back to designer 2 to adjust y, and so on in an iterative approach. Carrying out this process results in a divergent process, as illustrated in Fig. 23.3.

This issue of a divergent decentralized decision process is challenging. Indeed, in the case of convergence, designers agree on a

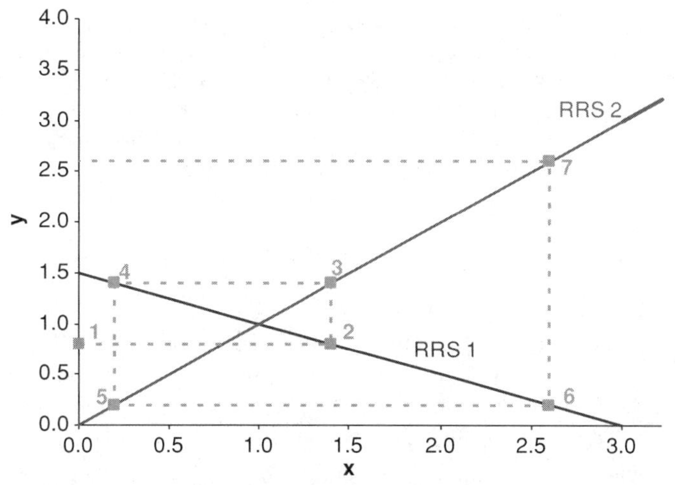

FIG. 23.3 A DIVERGENT DECENTRALIZED DESIGN EXAMPLE

final combination of the design variables even though the solution is not necessarily optimal. However, in the case of divergence, designers will never agree on a final design since one of them will always be able to change the value of his/her design variables and improve his/her objective function. How the two designers might go about choosing the final design is then difficult to predict. But in the absence of any additional information or intervention by a third party, it seems obvious that the solution will not be optimal.

Since this problem of convergence is crucial to the design process, a way to determine whether there is convergence or not would be beneficial to studying the dynamics of decentralized decision processes. The next section presents the basic mathematical formulation and develops the convergence criteria for a large range of decentralized decision processes.

23.4 HOW TO DETERMINE THE CONVERGENCE OF A DECENTRALIZED DECISION PROCESS?

The previous sections explained the advantages but also the issues of decentralizing the decision-making in engineering design. It has also been shown that it would be very useful to be able to determine whether a decentralized decision process is convergent or divergent. This section introduces several methods in order to determine the convergence of decentralized decision processes; the methods all have the same scope, but each is applied to a specific kind of problem.

First, some assumptions have to be made as to what kind of problems we are trying to solve. We assume that the design problem has already been subdivided into smaller subsystems, either naturally because several different companies interact on the design of the same product, or because the system has been subdivided into smaller subsystems using one of the techniques described in the previous section. Other assumptions are listed below:

- There is *any number* of subsystems, with a minimum of two.
- The allocation of the design variables is *mutually exclusive*: every design variable is controlled by one and only one design team.
- Every design team has control over *any number* of design variables, with a minimum of one.

- The model of every subsystem is an optimization problem with an objective function to minimize, but no constraints. If the initial problems have constraints, they have to be included in the objective function using penalty functions [41]: The new "pseudo objective function" is the sum of the initial objective function and a penalty term, which is typically the sum of the squares of each constraint:

$$\Phi(x) = F(x) + P(x) \quad \text{Eq. (23.9)}$$

This formulation is also useful for subsystems that are not written as optimization statements, i.e., without, any objective function and constraints. Those subsystems, designated as "black boxes" can be put into the desired form by approximating the latent objective function of the subsystem [42].

Those are the main assumptions that have to be met by the problem formulation in order to be able to study its convergence. We now present the convergence results for a subclass of these problems, namely the problems where designers have only quadratic objective functions. Problems with more complex formulations have been studied [2] but the results are not presented in this chapter.

The objective function (or pseudo-objective function) of subsystem i is denoted $\mathbf{F}_i$. Equation (23.10) shows the most general form of a quadratic equation, which is used as the mathematical representation of the objective function of every designer.

$$\mathbf{F}_i = \mathbf{x}_i^T \mathbf{A}_i \mathbf{x}_i + \mathbf{x}_{-i}^T \mathbf{B}_i \mathbf{x}_{-i} + \mathbf{x}_{-i}^T \mathbf{C}_i \mathbf{x}_i + \mathbf{D}_i \mathbf{x}_i + \mathbf{E}_i \mathbf{x}_{-i} + \mathbf{F}_i \quad \text{Eq. (23.10)}$$

where $\mathbf{x}_i$ = vector of design variables controlled by designer i; and $\mathbf{x}_{-i}$ = vector of design variables *not* controlled by designer i:

$$\mathbf{x}_{-i} = \mathbf{x} \backslash \mathbf{x}_i = \{x_j \in \mathbf{x},\, x_j \notin \mathbf{x}_i\} \quad \text{Eq. (23.11)}$$

The matrix $\mathbf{C}_i$ embodies the coupling between subsystem i and all the other subsystems. In order to make more visible the coupling of the subsystem i with every particular other subsystem, the coupling term of Eq. (23.10) is rewritten as shown in Eq.(23.12). The matrix $\mathbf{C}_i$ is essentially subdivided in a series of smaller submatrices $\mathbf{C}_{ij}$, each of them expressing the coupling between subsystem i and j.

$$\mathbf{x}_{-i}^T \mathbf{C}_i \mathbf{x}_i = \sum_{\substack{j=1 \\ j \neq i}}^{m} \mathbf{x}_j^T \mathbf{C}_{ij} \mathbf{x}_i \quad \text{Eq. (23.12)}$$

where $\mathbf{C}_{ij}$ = a set of smaller matrices embodying the coupled terms of the design variables of designer j into designer i's model.

With this mathematical representation for the objective function of all the designers, the convergence can be studied. It can be decomposed into several steps, which are described next.

(1) ***Find the Rational Reaction Sets***
The first step for analyzing the stability properties of a design process is to find the equilibrium points of the design space. As mentioned earlier in Eq. (23.3), they lie at the intersection of m subsets of the design space, the RRSs, where m is the number of designers or design teams involved in the design process. We denote n_i the number of design variables controlled by designer i. We also denote $\mathbf{x}$ the state vector or vector of all the design variables, grouping all the design variables of every designer, i.e, $\mathbf{x} = \{\mathbf{x}_i, i = 1 \ldots m\}$. Therefore, the length of x is defined as

$$N = \sum_{i=1}^{m} n_i \quad \text{Eq. (23.13)}$$

Finding the RRSs in those conditions is done by setting to zero the first partial derivative of the objective function (or pseudo-objective function). The optimization problem being unconstrained, the global minimum of an objective function in terms of the other designers' design variables is found by setting the first partial derivatives to zero. Practically, this is done by holding constant the design variables controlled by all the other designers and taking the partial derivative with respect to the design variables he or she is controlling (to study the influence of changing their values). Therefore, the equation of the RRS of designer i (RRS_i) is shown in Eq. (23.14):

$$\text{RRS}_i: \quad \frac{\partial \mathbf{F}_i}{\partial \mathbf{x}_i} = 0 \quad \text{Eq. (23.14)}$$

Finding the RRS for every designer will therefore provide us with m sets of equations representing the rational behavior of every designer. Each set is a vector of n_i scalar equations. They give the values of the design variables of a designer at an iteration, as a function of the values of the design variables of the other designers at the previous iteration. They can also be rewritten as N scalar equations, one for each design variable.

Carrying out the partial derivative of Eq. (23.14) using the mathematical representation of the objective function shown in Eq. (23.10), we can find a unique equation for the RRSs of every designer. The equation of the RRS of designer i is shown in Eq. (23.15) It is valid only if $\mathbf{A}_i$ is invertible; in some situations, it might not be invertible, but those cases are not interesting for this study, and are discussed in [33].

$$\mathbf{x}_i = -\frac{1}{2} \mathbf{A}_i^{-1} \sum_{\substack{j=1 \\ j \neq i}}^{m} \mathbf{C}_{ij}^T \mathbf{x}_j - \frac{1}{2} \mathbf{A}_i^{-1} \mathbf{D}_i^T \quad \text{Eq. (23.15)}$$

A set of m different equations can be written similar to Eq. (23.15), representing the RRS of every designer. This set of equations representing the RRSs can then be used to find the equilibrium points of the design space, which is the next logical step.

(2) ***Find the equilibrium points***
The equilibrium points lie at the intersection of the RRSs of every designer. This can be calculated using the set of N equations defined by Eq. (23.15). Since we are considering quadratic objective functions, these N equations are linear, because they are obtained by taking the first derivative of the quadratic pseudo-objective function. Therefore, to find the equilibrium points of the design space, we need to solve a system of N linear equations with N unknowns (the design variables). This system has either no solution (meaning that there is no Nash equilibrium), an infinite number of solutions (a line of Nash equilibriums, for example) or a

unique solution. An infinite number of Nash solutions is unlikely, because it would require every designer to have the same RRS in some region of the design space. Therefore, a quadratic distributed optimization problem will primarily have either one Nash equilibrium or none. This (potential) Nash equilibrium point is the only final design attainable by distributed designers using a sequential approach, but it does not necessarily mean that the designers will converge to it. It depends on the stability of this equilibrium, which is the point of study of this chapter, and which is the next logical step in our approach.

(3) ***Study the stability of the equilibrium***
Similar to the notion of equilibriums in physics, equilibrium points in the design space of an engineering design problem can be either stable or unstable. Consider, for example, a pendulum. There are two obvious equilibrium positions: $\theta = 0$ and $\theta = \pi$. One of them is known to be a stable equilibrium ($\theta = 0$) since the pendulum will always have a tendency to come back (i.e., *converge*) to this equilibrium position. On the other hand, the other one ($\theta = \pi$) is an unstable equilibrium, because, if it is perturbed from this equilibrium position, the pendulum will never come back to it (i.e., *diverge*). The same is true in decentralized decision-making problems, and the equilibrium can be either stable or unstable, which is what we study here.

A quadratic distributed decision-making problem is defined as a stable system if, independent of the values of the initial conditions, it goes to a steady state in a finite time [43]. In our quadratic environment, the steady-state point would naturally be the Nash equilibrium found at the previous step.

In order to study the stability of those systems, we use concepts from *Linear System Theory*. It analyses the mathematical description of physical systems. Similarly, in this chapter, we describe mathematically the interactions of designers acting in a distributed environment, our physical system. Linear System Theory concentrates on quantitative analysis (where the responses of systems excited by certain inputs are studied) and on qualitative analysis (which investigates the general properties of systems, such as stability). Qualitative analysis is very important, because design techniques may often evolve from this study [44]. We propose next a qualitative analysis of distributed problems, and further analogies with Linear System Theory are made later on.

In order to be able to use tools from Linear System Theory, we have to make the analogy between the set of equations of the RRSs shown in Eq. (23.15) and the main form of discrete update equation in Linear System Theory, called the state-space equation, and shown in Eq. (23.16).

$$\mathbf{x}(k+1) = \mathbf{\Phi}\mathbf{x}(k) + \mathbf{\Gamma}\mathbf{u}(k) \qquad \text{Eq. (23.16)}$$

where $\mathbf{x}$ = state vector (vector of variables that we are studying); and $\mathbf{u}$ = input vector. In this study, since we are not influencing the design process in any way and are just studying its dynamics, we set the input vector equal to the unity vector (corresponding to no special outside influence). The matrix $\mathbf{\Phi}$, the state matrix, represents the dynamics of the system, how it updates from one iteration to the next. The matrix $\mathbf{\Gamma}$, the input matrix, embodies the influence of outside intervention, or, in our case, of initial conditions.

First, we need to write Eq. (23.15) as a discrete-time update equation; this represents the sequential approach to the design process and is shown in Eq. (23.17).

$$\mathbf{x}_i(k+1) = -\frac{1}{2}\mathbf{A}_i^{-1}\sum_{\substack{j=1\\ j\neq i}}^{m}\mathbf{C}_{ij}^T\,\mathbf{x}_j(k) - \frac{1}{2}\mathbf{A}_i^{-1}\mathbf{D}_i^T \qquad \text{Eq. (23.17)}$$

We can now identify the set of m equations similar to Eq. (23.17) with Eq. (23.16). To do so, the coefficient of $\mathbf{x}_j(k)$ with the summation is identified with the matrix $\mathbf{\Phi}$, while the constant term is identified with $\mathbf{\Gamma}$. Equations (23.18) and (23.19) show the expressions for the matrices $\mathbf{\Phi}$ and $\mathbf{\Gamma}$; $\mathbf{\Phi}$ can be written as the multiplication of two block matrices, block i being of size n_i.

$$\mathbf{\Phi} = -\frac{1}{2}\begin{bmatrix} \mathbf{A}_1^{-1} & & & \\ & \mathbf{A}_2^{-1} & & \mathbf{0} \\ \mathbf{0} & & \ddots & \\ & & & \mathbf{A}_m^{-1} \end{bmatrix} \cdot \begin{bmatrix} \mathbf{0} & \mathbf{C}_{12}^T & \cdots & \mathbf{C}_{1m}^T \\ \mathbf{C}_{21}^T & \mathbf{0} & & \vdots \\ \vdots & & \ddots & \vdots \\ \mathbf{C}_{m1}^T & \cdots & \cdots & \mathbf{0} \end{bmatrix} \qquad \text{Eq. (23.18)}$$

$$= -\frac{1}{2}\operatorname{diag}\left(\mathbf{A}_i^{-1}\right) * \mathbf{\Lambda} \quad \text{with} \quad \mathbf{\Lambda} = \begin{cases} \boldsymbol{\lambda}_{ij} = \mathbf{C}_{ji}^T \\ \boldsymbol{\lambda}_{ii} = \mathbf{0} \end{cases}$$

$$\mathbf{\Gamma} = -\frac{1}{2}\begin{bmatrix} \mathbf{A}_1^{-1}\mathbf{D}_1^T \\ \vdots \\ \mathbf{A}_m^{-1}\mathbf{D}_m^T \end{bmatrix} \qquad \text{Eq. (23.19)}$$

The formulations of these matrices look fairly complicated, but, in fact, they are straightforward, as they are only functions of the matrices involved in the objective functions of the designers. From Eqs. (23.18) and (23.19), $\mathbf{\Phi}$ is a square matrix comprising of blocks and its size is the sum of the size of every block that is of size n_i, which is N from Eq. (23.13). Thus, $\mathbf{\Phi}$ is of size $N \times N$; similarly, $\mathbf{\Gamma}$ is of size $N \times 1$.

Therefore, we now have the formulation for the two matrices $\mathbf{\Phi}$ and $\mathbf{\Gamma}$ and a new update equation for the state vector $\mathbf{x}$ shown in Eq. (23.20).

$$\mathbf{x}(k+1) = \mathbf{\Phi}\mathbf{x}(k) + \mathbf{\Gamma} \qquad \text{Eq. (23.20)}$$

Once Eq. (23.20) has been derived, it is possible to find the steady state and the stability of the problem. The steady-state solution corresponds to the equilibrium point of the physical system studied, the design space in our case. If it exists, Linear System Theory ensures its uniqueness, given by Eq. (23.21).

$$\mathbf{x}^* = [\mathbf{I}_N - \mathbf{\Phi}]^{-1} \cdot \mathbf{\Gamma} \qquad \text{Eq. (23.21)}$$

where $\mathbf{I}_N$ = identity matrix of size N.

However, it is important to study the stability of this equilibrium. According to Linear System Theory, the definition of asymptotic stability is used [44]:

Theorem: *The equation* $\mathbf{x}(k+1) = \mathbf{A}\,\mathbf{x}(k)$ *is asymptotically stable if and only if all eigenvalues of* $\mathbf{A}$ *have magnitudes less than 1.*

Therefore, the stability of the equilibrium point of the design space can be expressed as a function of the spectral radius of the state matrix $\mathbf{\Phi}$. The spectral radius of a matrix $\mathbf{\Phi}$, denoted $r_\sigma(\mathbf{\Phi})$, is the largest absolute value of the eigenvalues of the matrix $\mathbf{\Phi}$. This is expressed in Eq. (23.22).

$$r_\sigma(\mathbf{\Phi}) = \max\{|\lambda| \setminus \lambda = \text{eigenvalue of } \mathbf{\Phi}\} \qquad \text{Eq. (23.22)}$$

The convergence analysis of the design process can therefore be captured as follows. The design process converges to the equilibrium point found thanks to Eq. (23.21) if and only if:

$$r_\sigma(\mathbf{\Phi}) < 1 \qquad \text{Eq. (23.23)}$$

This result is very important, as it means that building the state matrix $\mathbf{\Phi}$ and calculating its spectral radius gives insightful information on the convergence of the design process to the equilibrium point, which can also be calculated using Eq. (23.21). The matrix $\mathbf{\Phi}$ is also called the **characteristic matrix**.

Equations (23.21) and (23.23) thus give us the equilibrium point and the convergence (or divergence) of the design process, in the case of decentralized systems with quadratic objective functions. What happens when the objective functions are not quadratic is more complex, and *Nonlinear Control Theory* has to be used [2]. The next section discusses how those results can influence how designers go about decision-making in decentralized environments, and how we can help them make better decisions.

23.5 APPLICATION TO A DESIGN PROBLEM

In the previous section, we introduced methods to be able to determine whether a decentralized design problem is convergent or divergent. We show here, step by step, how these methods can be applied to a design problem.

Step 1. Write the objective functions of the designers as a matrix equation.

If the objective function of a designer is known and is quadratic, then it can be easily written in a matrix format. Take the example of the following quadratic objective function to be minimized, where designer 1 controls $\mathbf{x} = [x_1\ x_2]^T$, and $\mathbf{y} = [y_1\ y_2\ y_3]^T$ is controlled by other designers.

$$F_1 = 2x_1^2 + x_2^2 + 2x_1x_2 + 3x_1y_1 - 2x_1y_2 + 2x_1y_3 + x_2y_2 - x_2y_3 + 2x_1 + 4x_2 \qquad \text{Eq. (23.24)}$$

From Eq. (23.10), we know we have to write this objective function as:

$$F_1 = \mathbf{x}^T\mathbf{A}_1\ \mathbf{x} + \mathbf{y}^T\mathbf{B}_1\ \mathbf{y} + \mathbf{y}^T\mathbf{C}_1\ \mathbf{x} + \mathbf{D}_1\ \mathbf{x} + \mathbf{E}_1\ \mathbf{y} + \mathbf{F}_1 \qquad \text{Eq. (23.25)}$$

We can therefore rewrite Eq. (23.24) as shown in Eq. (23.26) and easily identify the matrices $\mathbf{A}_1$, $\mathbf{C}_1$ and $\mathbf{D}_1$:

$$F_1 = \mathbf{x}^T\begin{pmatrix} 2 & 1 \\ 1 & 1 \end{pmatrix}\mathbf{x} + \mathbf{y}^T\begin{pmatrix} 3 & 0 \\ -2 & 1 \\ 2 & -1 \end{pmatrix}\mathbf{x} + (2\quad 4)\mathbf{x} \qquad \text{Eq. (23.26)}$$
$$= \mathbf{x}^T\mathbf{A}_1\mathbf{x} + \mathbf{y}^T\mathbf{C}_1\mathbf{x} + \mathbf{D}_1\mathbf{x}$$

Step 2. Calculate the characteristic matrix $\mathbf{\Phi}$.
Equation (23.18) gives a straightforward formula to calculate the matrix $\mathbf{\Phi}$ after having identified all the matrices from all the objective functions.
Step 3. Calculate $r_\sigma(\mathbf{\Phi})$, the spectral radius of matrix $\mathbf{\Phi}$.
Equation (23.23) gives the convergence criterion for any decentralized decision problem. If this is a spectral radius, we can conclude that the design process will be convergent; otherwise, it will be divergent.
Step 4. Calculate the final values of the design variables.

If necessary (i.e., if the process is convergent), we can calculate the final values of the design variables of the designers. Equation (23.21) gives the formula to be used, which gives the values of all the design variables.

After developing the methods to determine the convergence of a decentralized decision problem, we showed here how it could be applied to a real design problem. We refer the reader to the homework section for practical exercises on complete design problems.

The next section discusses how those results can influence how designers go about decision-making in decentralized environments, and how we can help them make better decisions.

23.6 CONCLUSION: HOW DOES THIS INFLUENCE DECISION-MAKING IN DESIGN?

In this chapter, we studied engineering systems that are multidisciplinary in nature and therefore require knowledge from several design teams. This, along with other constraints, usually forces the decentralization of decisions. While centralization of decisions is sometimes a preferred approach from a technical perspective, in this chapter, we study the scenario where centralization is not possible because of geography, cost, time, resources, organizational structure, etc. This happens in many product design processes as complete centralization and communication among design groups, engineering teams, suppliers, manufacturers and other relevant parties are not feasible. Therefore, the decision-makers involved in a design process cannot fully cooperate, and are in a state of limited cooperation. In order to find a final optimal design, they need to understand the fundamental mechanisms of the process.

This chapter presents a formal mathematical formulation, based on concepts from Game Theory, in order to understand the dynamics of the design process. The convergence of such systems, as well as the final values of the design variables are found and understood using this method.

Even though in industrial design problems the decentralized designers may not have complete knowledge of the other designers' design objectives, the underlying dynamics of the process will help either upper-level managers or the decision-makers themselves make better decisions when decomposing, modeling and solving complex design problems. In addition, by understanding the fundamental dynamics, coordination of decision-support tools and infrastructures can be more effectively applied.

PROBLEMS

23.1 Two designers, one design variable. Develop the convergence criterion for decentralized decision processes involving only two designers, each controlling only one design variable, and with each quadratic objective function. ***Hint***: Write the most general form of a quadratic equation for the objective function of both designers. They should be function of both design variables (x and y). Do not forget the coupled terms! You can then follow the same steps as presented in this chapter, in order to find the new convergence criterion.

23.2 Two designers, several design variables. Develop the convergence criterion for decentralized decision processes involving only two designers, each controlling several design variables, but with only quadratic objective function.

Hint: Write the most general form of a quadratic equation for the objective function of both designers. They should be function of both vectors of design variables ($\boldsymbol{x}$ and $\boldsymbol{y}$). Do not forget the coupled terms! It should be a matrix equation similar to Eq. (23.10), but involving only two designers. You can then follow the same steps as presented in this paper, in order to find the new convergence criterion.

23.3 Decentralized design example. In this problem, we give two simple decentralized design examples. For each problem, you should determine whether an iterative approach would yield a convergent or a divergent process. In case of convergence, you should also determine the final values of the design variables of every designer.

Hint: For each problem, we give the objective function of each designer. You should write those objective functions in the matrix form shown in this chapter. You can then build the characteristic matrix and answer the questions of convergence and final values of design variables. You can use a software such as Matlab to calculate the spectral radius of the matrix $\boldsymbol{\Phi}$.

a. First design example:
Designer 1 controls the vector of design variable $\mathbf{x} = [x_1\ x_2]^T$ and has the following objective function:

$$F_1 = 2x_1^2 + x_2^2 + 2x_1x_2 + 3x_1y_1 - 2x_1y_2 + 2x_1y_3 + x_2y_2 - x_2y_3 + 2x_1 + 4x_2$$

Designer 2 controls the vector of design variable $\mathbf{y} = [y_1\ y_2\ y_3]^T$ and has the following objective function:

$$F_2 = 2y_1^2 + y_2^2 + \frac{1}{3}y_3^2 + y_1x_1 - 3y_1x_2 + 4y_2x_1 - y_2x_2 - 2y_3x_1 - x_2y_3 + y_1 + 2y_2 + 3y_3$$

b. Second design example:
Designer 1 controls the vector of design variable $\mathbf{x} = [x_1\ x_2]^T$ and has the following objective function:

$$F_1 = 2x_1^2 + x_2^2 + 2x_1x_2 + x_1y_1 + 2x_1y_2 + x_2y_2 - x_2y_3 - 2x_1 + 3x_2$$

Designer 2 controls the vector of design variable $\mathbf{y} = [y_1\ y_2\ y_3]^T$ and has the following objective function:

$$F_2 = 2y_1^2 + y_2^2 + \frac{1}{3}y_3^2 + y_1x_1 - y_2x_2 - 2y_3x_1 + 3y_1 + 2y_2 + 2y_3$$

REFERENCES

1. Chanron, V. and Lewis, K. 2004. "Convergence and Stability in Distributed Design of Large Systems," *Proc., ASME Des. Engrg. Tech. Conf., DETC2004-57344*, ASME, NewYork, NY.
2. Chanron, V., Singh, T. and Lewis, K., 2004. "Equilibrium Stability in Decentralized Design Systems," *Int. J. of Sys. Science*.
3. "Outsourcing," 2004. *The Economist*, Nov. 13–19.
4. Boeing, 2004. "Boeing 777 Facts," http://www.boeing.com/commercial/777family/pf/pf_facts.html.
5. "Managing complexity," 2004. *The Economist*, Special Report on Software Development, Nov. 27–Dec. 30.
6. Krishnamachari, R. and Papalambros, P., 1997. "Hierarchical Decomposition Synthesis in Optimal Systems Design," *J. of Mech. Des.*, **119**(4), pp. 448–457.
7. Kusiak, A. and Wang, J., 1993. "Decomposition of the Design Process," *J. of Mech. Des.*, **115**(4), pp. 687–695.
8. Lee, H. and Whang, S., 1999. "Decentralized Multi-Echelon Supply Chains: Incentives and Information," *Mgmt. Sci.*, **45**(5), pp. 633–640.
9. Prewitt, E., 1998. "Fast-Cycle Decision Making," *Harvard Mgmt. Update*, **3**(2), pp. 8–9.
10. Krueger, A., 2002. "Economic Growth in a Shrinking World: The IMF and Globalization," Speech at the *Pacific Council on Int. Policy*, San Diego, CA.
11. Pritchard, P. and MacPherson, A., 2004. "Outsourcing US Commercial Aircraft Technology and Innovation: Implications for the Industry's Long Term Design and Build Capability," Occasional Paper No. 29, Canada–United States Trade Center, Univ. at Buffalo, NY.
12. Balling, R. J. and Sobiezczanski-Sobieski, J., 1994. "Optimization of Coupled Systems: A Critical Overview of Approaches," *Proc., 5th AIAA/USAF/NASA/ISSMO Symp. on Multidisciplinary Analysis and Optimization*, AIAA-94-4330-CP.
13. McCulley, C. and Bloebaum, C. L., 1996. "A Genetic Tool for Optimal Design Sequencing in Complex Engineering Systems," *Stru. Optimization*, **12**(23), pp. 186–201.
14. Sobieszczanski-Sobieski, J., James, B. J. and Dovi, A. R., 1985. "Structural Optimization by Multilevel Decomposition," *AIAA J.*, **23**(11), pp. 1775–1782.
15. Michalek, J. and Papalambros, P., 2004. "An Efficient Weighting Update Method to Achieve Acceptable Consistency Deviation in Analytical Target Cascading," *Proc., ASME Des. Engrg Tech. Conf., DETC2004-57134*.
16. Michelena, N., Park, H. and Papalambros, P., 2003. "Convergence Properties of Analytical Target Cascading," *AIAA J.*, **41**(5), pp. 897–905.
17. Braun, R., Gage, P., Kroo, I. and Sobieski, I., 1996. "Implementation and Performance Issues in Collaborative Optimization," *Proc., 6th AIAA/USAF/NASA/ISSMO Symp. on Multidisciplinary Analysis and Optimization, AIAA 96-4017.*
18. European Association of Aerospace Industries, 2001. "Aerospace Within the European Research Area–Centres of Excellence," *AECMA Pub. PP193*.
19. Chanron, V. and Lewis, K., 2003. "A Study of Convergence in Decentralized Design," *Proc., ASME Des. Engrg. Tech. Conf., DETC03/DAC-48782*, ASME, NewYork, NY.
20. Hardin, G. 1968. "The Tragedy of the Commons," Sci., Vol. 162, pp. 1243–1248.
21. Kiers, E. T., Rousseau, R. A., West, S. A. and Denison, R. F., 2003. "Host Sanctions and the Legume-Rhizobium Mutualism," *Nature*, Vol. 425, pp. 78–81.
22. "The Rise of the Green Building," 2004. *The Economist*, Dec. 4–10, Technology Quarterly.
23. Tufte, E. R., 1997. *Visual Explanations: Images and Quantities, Evidence and Narrative*, Graphics Press, Cheshire, CT.
24. von Neumann, J. and Morgenstern, O., 1944. *Theory of Games and Economic Behavior*, Princeton University Press, Princeton, NJ.
25. Vincent, T. L. 1983. "Game Theory as a Design Tool," *J. of Mech., Transmissions, and Automation in Des.*, Vol. 105, pp. 165–170.

26. Lewis, K. and Mistree, F., 1997. "Modeling the Interactions in Multidisciplinary Design: A Game Theoretic Approach," *AIAA J. of Aircraft*, **35**(8), pp. 1387–1392.
27. Marston, M. and Mistree, F., 2000. "Game-Based Design: A Game Theoretic Extension to Decision-Based Design," *Proc., 12th Int. Conf. on Des. Theory and Methodology, DETC2000/DTM-14578.*
28. Hernandez, G., Seepersad, C. C. and Mistree, F., 2002. "Designing for Maintenance: A Game Theoretic Approach," *Engrg. Optimization*, **34**(6), pp. 561–577.
29. Allen, B., 2001. "Game Theoretic Models of Search in Multi-Agent Process Planning Systems," *Proc., Special Interest Group on Manufacturing (SIGMAN) Workshop, Int. Joint Conf. on Artificial Intelligence.*
30. Jagannatha Rao, J. R., Badhrinath, K., Pakala, R. and Mistree, F., 1997. "A Study of Optimal Design Under Conflict Using Models of Multi-Player Games," *Engrg. Optimization*, **28**(1–2), pp. 63–94.
31. Chen, L. and Li, S., 2001. "Concurrent Parametric Design Using a Multifunctional Team Approach," *Proc., ASME Des. Engrg. Tech. Conf., DETC 2001/DAC-21038,* ASME, NewYork, NY.
32. Lewis, K. and Mistree, F., 1998. "Collaborative, Sequential and Isolated Decisions in Design," *ASME J. of Mech. Des.*, **120**(4), pp. 643–652.
33. Chanron, V., 2002. "A Study of Convergence in Decentralized Design," Master's thesis, State Univ. of New York at Buffalo, NY.
34. Pareto, V., 1906. *Manuale di Economica Polittica*, Societa Editrice Libraia, Milan, Italy; translated into English by A.S. Shwier, *Manual of Political Economy*, 1971, Macmillan, New York, NY.
35. Senge, P. M., 1990. *The Fifth Discipline*, Currency Doubleday.
36. Thompson, G. L., 1953. "Signaling Strategies in *n*-Person Games," *Contributions to the Theory of Games, Vol. II,* H.W. Kuhn, and A.W. Tucker, eds. Princeton University Press, Princeton, NJ, pp. 267–277.
37. Friedman, J. W., 1986. *Game Theory With Applications to Economics*, Oxford University Press, New York, NY.
38. von Stackelberg, H., 1952. *The Theory of the Market Economy*, Oxford University Press, Oxford.
39. Wetmore, W. and Summers, J., 2004. "Influence of Group Cohesion and Information Sharing on Effectiveness of Design Review," *Proc., ASME Des. Engg. Tech. Conf., DETC2004-57509*, ASME, New York, NY.
40. Loch, C., Mihm, J. and Huchzermeier, A., 2003. "Concurrent Engineering and Design Oscillations in Complex Engineering Projects," *Concurrent Engrg.*, **11**(3), pp. 187–199.
41. Vanderplaats, G. N., 1999. *Numerical Optimization Techniques for Engineering Design*, 3rd Ed., VR&D.
42. Wang, G. G., 2003. "Adaptive Response Surface Method Using Inherited Latin Hypercube Design Points," *J. of Mech. Des.*, Vol. 125, pp. 210–220.
43. Lee, T. S. and Ghosh, S., 2000. "The Concept of 'Stability' in Asynchronous Distributed Decision-Making Systems," *IEEE Trans. on Sys., Man, and Cybernetics-Part B: Cybernetics*, **30**(4), pp. 549–561.
44. Chen, C.-T., 1999. *Linear System Theory and Design*, Oxford University Press.

CHAPTER

24

VALUE AGGREGATION FOR COLLABORATIVE DESIGN DECISION-MAKING

Yan Jin and Mohammad Reza Danesh

24.1 INTRODUCTION

In recent years, engineering collaboration and teamwork has been a hot topic within the research community of engineering design. With the increasing complexity of design problems and shorter design lead time, companies are pressed to conduct product development projects with a team of experts from various technical domains who are oftentimes from different geographical locations.

Collaboration in engineering design can be viewed as activities of planning design tasks, exchanging design information with other team members, detecting and resolving design conflicts, generating designs for subtasks and integrating them into an overall design. In a typical collaborative design scenario, the overall design task is divided into multiple subtasks. Each member of the team takes the responsibility of one or more subtasks and tries to develop solutions for them. Ideally, once solutions for all subtasks are found, team members can put them together and form a total solution for the overall design task. In practice, however, design solutions for subtasks often don't match one another because of either physical inconsistencies and/or functional conflicts. In many cases, the subsolutions are developed in isolation and are at most suboptimal with respect to the overall design objectives. As a result, excessive backtracking and design revisions are needed to complete the overall design. Hence, although the concept of collaboration and distribution is intriguing, in many situations it is not efficient and may even prolong the design cycle.

A simple solution to this problem is to arrange frequent meetings and telephone conversations, or use other communication methods from the early stages of product development so that local design activities of the team members can be more coordinated and the potential downstream conflicts can be avoided. The disadvantage of using such methods is that they add another layer of complexity to the process that may incur enormous cost and lead to unsatisfactory results. Designers, especially at the conceptual design stage, often are not clear about with whom they should coordinate and what kind of information they should acquire or provide. Even with the most advanced communication tools, sharing design information alone is not a solution for effective collaboration. Designers must understand the reason or rationale behind the design information received from others. Without this understanding, the meaning of the information may be misinterpreted and its implication in a bigger design context may not be realized.

In our research, we take a decision-based approach to collaborative design. We view collaborative design as a collaborative and distributed decision-making process in which designers work together to frame design decisions, clarify design values, share design information and make their design decisions with the consideration of the decisions made by others. There are two issues that must be addressed to realize this approach: First, methods and tools are needed to allow designers to make design decisions based on explicitly stated and consistent design values. Second, collaboration should go beyond coordinating design activities and exchanging design information. It should encourage designers to share their design values and make sure their decisions contribute to the overall values of the design rather than merely local ones.

Researchers in the area of engineering design have done extensive work on modeling design processes and providing guidelines for design decision-making. The axiomatic design model [1] proposed a zigzag design process and identified two axioms to support decision-making. The independence axiom suggests maintaining independence between functional requirements when choosing function requirements and design parameters, and the information axiom suggests minimizing information content, or maximizing success probability, for choosing from among the alternatives that already satisfy the independence axiom. Systematic design [2] advocates that engineering design must be carefully planned and systematically executed. Besides a detailed process model including four phases and their corresponding steps, a number of general rules, basic principles and specific guidelines are suggested for design decision-making [2]. Decision theoretic models use the axiomatic framework of classical decision theory as the basis for engineering decision-making [3, 4] and emphasize the development of consistent utility functions by designers [5, 6, 7]. Game theory [8] based approach has been proposed to support collaborative design [9]. Design problems have also been defined in various algebraic forms and solved as optimization problems [10, 11, 12].

In the field of decision theory, the research on multi-objective decision-making provides a framework to generate rankings and ratings for alternates based on decision-makers' preferences. Researchers [13, 14, 15, 16, 17, 18, 19] have developed mathematical methods to systematically define a decision problem, develop objectives, assess desirability for alternatives, and deal with uncertainties and risks. Furthermore, value-focused thinking proposed by Keeney [20] emphasizes the importance of the front end of decision-making and provides general methods for eliciting decision-makers' values and generating otherwise unthinkable alternatives. Although these models usually cannot be directly applied to model the collaborative design process, they provide a solid theoretical foundation for collaborative design decision-making.

Our research on value aggregation for collaborative design attempts to develop a model and mechanism that allow designers to take into consideration the values and decisions of other designers while making their own decisions [21]. For successful collaborative design, it is imperative that engineers work closely with each other from the early stages of the design and strive toward increasing the overall value of design, rather than just local ones. Coordinating with others about design values or objectives not only leads to the design results of higher overall values, but improves collaboration efficiency by avoiding potential downstream conflicts caused by inconsistency in design values or objectives.

The basic idea behind the value aggregation approach to collaborative design is the following: the designers' lack of knowledge of other designers' design values prevents them from choosing globally best design alternatives. If one can provide a value-based framework that captures design values of designers and arrange them in a way that they can be effectively considered, managed and negotiated, then the overall design quality would be increased, the conflicts among design outcomes of different designers minimized and, consequently, the collaboration process would be more effective and efficient. The focus of our research, hence, is on developing a value-based design process model that explicitly captures design values and allows designers to aggregate their design values for effective design decision-making. In this chapter, we first introduce the concept of design value and a value-based design process in Section 24.2. After that, we discuss the issues and methods for value aggregation in Section 24.4. A simple case example is presented in Section 24.5 to illustrate the benefit of the proposed approach. Finally, Section 24.6 presents concluding remarks.

24.2 DESIGN VALUE AND VALUE-BASED DESIGN

The decision-based approach to engineering design emphasizes the decision-making aspect of the design process. At every step of design, a designer faces a decision problem that must be resolved effectively and efficiently. As for many decision problems in other areas, there are two fundamental issues involved in design decision-making. One is concerned with specifying what a designer wants to achieve—i.e., what is the designer's design value?—and the other is related to predicting how well a chosen alternative's outcome might contribute to the specified design value -i.e., how to deal with uncertainties and risks associated with the decision. In an ideal "decision-based design" (DBD) situation, a designer, when solving a design problem, should first clarify what he/she wants by developing an explicit value function that can be applied to evaluate all potential alternatives at all stages of design. During the design process, the designer should make subjective yet best judgment about the probabilities of the outcomes of specific alternatives. The possible outcomes together with their occurring probabilities are applied to the value function to determine whether or not a given alternative is the most valuable one worth choosing. When multiple designers from various functional domains are involved in the design process, a set of value functions should be developed in a "centralized" way so that the consistency can be maintained among them.

The premise of the above ideal decision-based design situation is that: (1) it is manageable to develop a holistic value function that is applicable throughout the design process—i.e., every decision situation in the process can be explicitly mapped into the value function for effective evaluation; and (2) all potential alternatives are readily available and their attributes well understood, since otherwise it would not be possible to predefine the value function that can be applied to every possible design decision. Although in many economic decision situations, such as consumers' choosing desirable products from among alternatives, and many policy decision situations, such as whether or not investing in a specific public project, the premise is true, decision-making in engineering design is more complicated. Except for certain routine parametric design problems, the design alternatives are to be explored and discovered, rather than being available. In addition, although high-level performance attributes are introduced at the beginning of design, the lack of knowledge of details of the product to be designed makes it difficulty to develop a complete value function for the whole design process. The multidisciplinary nature of modern design problems further complicates the issue. Moreover, engineering design problems are often multi-attribute, meaning that the desirability of a product is determined by multiple attributes, some are technical and others more economical. There has been research advocating using a single attribute, such as profit, as the measure of the desirability of all choices made during a design process [3, 22]. While these models are effective for highly routine design problems and for product planning decision-making [2], for less-routine and non-routine design problems, it is very difficult if not impossible to develop a practically meaningful profit-centered value function that can be applied to guide decision-making in *every* decision situation during design. More graspable concepts such as functionality, quality and cost are often used to guide decisions in design.

The difficulty of pursuing the ideal DBD process does not mean one should give up on following the decision theoretic principles. Rather, it demands development of practically effective approaches to engineering design that can guide designers to maximize design values when they make either individual or collaborative design decisions. Our approach of value-based design and value aggregation for collaboration is an attempt to develop such a practical method for collaborative design.

24.2.1 Design Objectives

A designer's design value in general specifies what is important to him/her. It is used to evaluate the potential consequences of specific design alternatives. Generally speaking, design value can be expressed as general, or context independent, principles such as *minimize use of materials* and *choose shortest path of force*. For a given design problem, however, design values can be made explicit by identifying design objectives within that design context. A design objective can be a statement of some aspect associated with the design product that the designer desires to achieve. Designers often have to consider multiple, and sometimes competing, design objectives. For example, in designing bicycle frames, *maximize strength* and *minimize weight* can be two important objectives for a designer. Using the concept of design objective, a designer's design value can be manifested by a set of design objectives.

Following [20], we define a design objective to have a specific *purpose* to be strived for, a *direction* of preference, and a design *context* in which the objective is defined. A design objective can also be elaborated to include a number of *attributes* that serve as measures to indicate how much the objective is achieved and an evaluation *function* that can be used to transform the *attributes* into a single scalar value to be used for alternative evaluation. Following are the five properties that define a *design objective*.

- ***Purpose***: Indicates the area of concern in a given design situation. For example, *maximum speed* and *0–60 mph acceleration time* are two areas of concern in a vehicle design problem.
- ***Direction***: Indicates designer's preferred direction for the purpose. Direction can be either *maximize* or *minimize*, e.g., "*maximize* maximum speed," "*minimize* 0–60 mph acceleration time."
- ***Context***: Indicates a subspace of the design space to which the design objective corresponds. For the above example, "vehicle design" is a broad context. Other subcontexts may include "engine design," "body design," etc.
- ***Attributes***: An attribute serves as a measurement used to evaluate the degree to which an alternative achieves the design objective. For a given design objective, there can be one or more attributes for achievement evaluation. Each attribute must be associated with a measurable unit. For example, "maximize comfort" can be measured by "noise level (db)," "vibration level (Hz)" and "leg space (inch)."
- ***Function***: Contains the value function or value model that maps the measured attributes' levels into a single scalar number indicating the relative desirability of the achievement of the objective and can be used to derive preferences for design alternatives.

Using these five properties, a design objective can be expressed as:

<Design Objective> ::=
<Direction><Purpose>in<Context>measured-by<Attributes> evaluated-by<Function>.

For example, the design objective of "maximize passenger comfort" can be expressed as:

<maximize><passenger comfort> in <car design> measured-by <noise level (db), vibration level (Hz), leg room (inch)> evaluated-by<u (noise level, vibration level, leg room)>

In practice, it is often the case that the definition of a specific design objective is initially incomplete because identifying attributes and assigning a value function both require designers to make judgments as part of design decisions. In these cases, a design objective can take short forms, e.g.,

<DesignObjective> ::= <Direction><Purpose>in<Context>
or
<DesignObjective> ::= <Purpose>in<Context>
when *<direction>* is obvious, or simply
<DesignObjective> ::= <Purpose>
when both *<direction>* and *<context>* are obvious.

Design objectives are different from other two related concepts, namely *design requirements* and *design constraints*. Design requirements are often ill-defined. They may reflect what are demanded and hoped by customers but do not fully capture what a designer may desire. Design constraints are often set to restrict the design space so that the search for solutions can be made more efficient. While most hard constraints, such as those defined based on the laws of physics, must be obeyed, many soft constraints are often circumstantial and their validity is relative to how the decision situation is defined and what design objectives are set. For example, "keep manufacturing cost under $1,000" is a soft constraint. If a designer can create a design that cost more than $1,000 to manufacture but can generate much more profit, then the constraint should certainly be relaxed or ignored. It is worth mentioning that our value-based design approach is very much against any constraint-based design methods. Although imposing constraints can increase the search efficiency by reducing the search space, it also screens out potentially good alternatives.

In ideal design decision situations described above, all the design objectives are explicitly developed and the associated attributes and value functions specified. In these cases, multiple attribute design-making methods [7, 23, 25] can be applied to assist the DBD process. In practice, however, identifying design objectives and developing a design objective structure that clearly captures the relations among the design objectives, both within individual designers and among them, can be a major part of the design process itself. To assist design objective identification and structuring, design objectives can be further *categorized* as follows based on the meaning of "Purpose."

- ***Economical objectives***: These define a designer's expected profit or cost of the design consequence. For example, <minimize><cost>in<vehicle manufacturing> is an economical design objective.
- ***Functional objectives***: These define expected performance features of a design. For example, <maximize><rigidity> in<vehicle structure design> or <maximize><accuracy of movement>in<robot arm design> are functional objectives.
- **Physical objectives:** They represent expectations on the physical and embodiment features of a design outcome. For example, <minimize><curb weight>in<vehicle design> is a physical objective.
- ***Ergonomic and environmental objectives***: These objectives define the expected behavior of the designed product with respect to its interactions with the users and the environment. For example <maximize><passenger comfort> and <minimize emission> are ergonomic and environmental objectives, respectively.
- ***Managerial objectives***: They are related to the management aspects of the product design process. Examples of manage-rial objectives include <minimize><lead time>in<vehicle design> and <maximize><learning opportunities for young designers>in<car design>.

The five design objective categories identified above are intended to represent a mutually exclusive and collectively exhaustive set of possible design objectives. While it is not necessary to have design objectives from each of the above categories in a design process, the categories do provide directions for designers to look for their design objectives. There are many design examples where designers do not explicitly consider the managerial or ergonomics aspects of their developed concepts, but it is always desirable to make all relevant design objectives explicit throughout the design process. Designers should consider identification and elicitation of design objectives as a major part of their design work.

In his value-focused thinking framework, Keeney classified objectives into three levels: strategic objectives, fundamental objectives and means objectives [20]. Strategic objectives usually reflect the strategic directions of a company or a specific product development project. They are often economical and sometimes ergo and environmental objectives. For example, <maximize><profit> and <maximize><safety> can be two strategic design objectives for new car development. Functional objectives are usually fundamental objectives for engineering design because functions are the essence for which the product is being developed. Sometimes managerial objectives may be fundamental if they reflect the realization of the company or project strategies. Physical objectives are means objectives that are related to the achievement of functional or fundamental objectives.

24.2.2 Design Objective Structuring

In our proposed value-based design approach, design value is at the center of the design process. In this approach, the major steps of design are:

(1) Identify the design problem.
(2) Define design objectives and develop structures of design objectives so that they are concrete enough to be linked to design contexts for creating alternatives.
(3) Generate design alternatives.
(4) Analyze and evaluate alternatives based on the structured design objectives.
(5) Select the alternative with the highest level of desirability.

The mathematical representation of the above process applied to ideal design situations is described in Appendix I. Here we discuss the value-based feature of the process. In this design process, step 2 signifies the value-based thinking for design decision-making. It not only clarifies designers' design value, but provides a framework for the designers to create and identify potential design alternatives. We call this step "design objective structuring" in which design objectives are defined and their relationships identified. It takes a list of design requirements as input and generates a design objective hierarchy or network as output.

In engineering design, the objective structuring process can help designers clarify what he/she is looking for, guide designers to avoid random search for design alternatives, and provide means for designers to expand the design context by drawing relevance to the new areas in the design space. Design needs are often given to designers in the form of design requirements. It is the designer's job to convert these requirements to design objectives. A designer should first define the overall design problem by abstracting from the list of requirements as introduced in [2]. After that a number of initial design objectives should be identified based on the objective categories. Figure 24.1 illustrates the initial design objectives developed for a beverage can design problem.

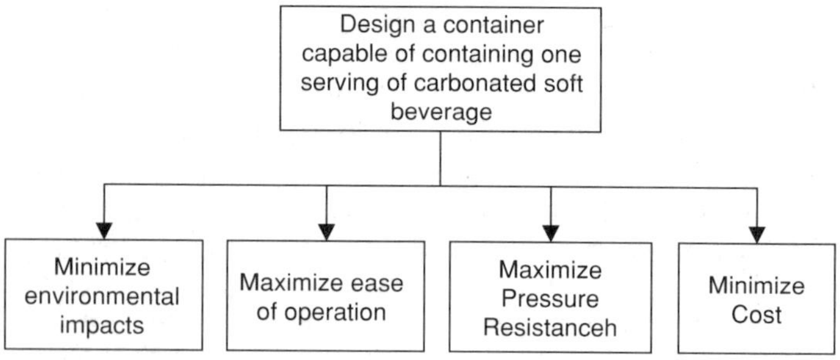

FIG. 24.1 INITIAL DESIGN OBJECTIVES FOR BEVERAGE CAN DESIGN PROBLEM

The next step in design objective structuring is to elaborate the initial design objectives, e.g., those in Fig. 24.1, into more concrete and applicable ones for which both measurement *attributes* and evaluation *functions* can be found. We identified two basic ways for design objective elaboration. They are decomposition—i.e., the union of sub-objectives forms the super-objective—and refinement, i.e., the sub-objective is the focus of the super-objective. These two elaboration methods can be applied to either *purpose* or *context* dimensions of a design objective. In our research, following design objective structuring methods are introduced (see Table 24.1). Our ongoing research attempts to develop a library of design objectives and rules to facilitate design objective structuring.

- ***Purpose decomposition*:** In this type of elaboration, sub-objectives can be found by decomposing the high-level *purpose* into more specific low-level ones. For example, in design objective "<minimize><vehicle noise> in <passenger car design>," the *purpose* is <vehicle noise>. It can be decomposed into two subpurposes: <engine noise> and <body noise>. So the decomposed objectives include <minimize><engine noise> and <minimize><body noise>.

TABLE 24.1 DESIGN OBJECTIVE STRUCTURING METHODS AND EXAMPLES

Definition	Examples
Purpose decomposition: $<P_1><C_1> =>$ $<P_{1-1}><C_1>\cap<P_{1-2}><C_1>\cap \ldots \cap<P_{1-n}><C_1>$ *where* $<P_1> = <P_{1-1}>\cap<P_{1-2}>\cap \ldots \cap<P_{1-n}>$	<minimize><vehicle noise> in < vehicle development > => <minimize><engine noise> in < vehicle development > <minimize><body noise> in < vehicle development >
Purpose refinement: $<P_1><C_1> => <P_{1-1}><C_1>$ *where* $< P_{1-1}>$ *causes* $<P_1>$	<minimize><body noise> in < vehicle development > => <minimize><body vibration> in < vehicle development >
Context decomposition: $<P_1><C_1> =>$ $<P_1><C_{1-1}>\cap <P_1><C_{1-2}>\cap \ldots \cap<P_1><C_{1-n}>$ *where* $<C_1> = <C_{1-1}>\cap<C_{1-2}>\cap \ldots \cap<C_{1-n}>$	<minimize><cost> in <vehicle development> => <minimize><cost> in <vehicle design> <minimize><cost> in <vehicle manufacturing> <minimize><cost> in <vehicle materials>
Context refinement: $<P_1><C_1> => <P_1><C_{1-1}>$ *where* $<C_{1-1}>\subset<C_1>$	<minimize><cost> in <vehicle development> => <minimize><cost> in <vehicle materials> Note: If the designer thinks both design and manufacturing are fixed costs, he/she may focus on only the material cost when making design decisions.

- ***Purpose refinement*:** Sometimes, the *purpose* of an objective is too general to be clear. In these cases, the sub-objective can be identified by pinpointing to a more specific area of concern. For example, when elaborating "<minimize><body noise>," the purpose <body noise> may be refined into <body vibration>. In this case, <body vibration> is considered as the sole source of <body noise>.
- ***Context decomposition*:** This method is applied to the cases where the purpose of the design objective remains constant but the context can be made more specific by going into more details. For example, in design objective "<minimize> <cost> in <vehicle development>," the *context* is <vehicle development>. It can be decomposed into the following three sub-contexts: <vehicle design>, <vehicle manufacturing> and <vehicle materials>. So if one wants to minimize vehicle cost, he/she must make sure that the costs of design, manufacturing and materials are all minimized.
- ***Context refinement*:** This method elaborates a higher-level design objective into a lower-level one by focusing the context on a more specific one. For example, for ship design problems, both design and manufacturing costs are fixed costs. A designer may pay attention to variable costs by focusing specifically on materials. So he/she may elaborate the design objective "<minimize><cost> in <ship development>" into "<minimize> <cost> in <materials>." In this case, the context <materials> is much more focused than <ship development>.

24.3 AGGREGATION OF DESIGN VALUES IN COLLABORATIVE DESIGN

In collaborative design, when a designer generates alternatives for a local design task, it is most likely that the decision of choosing from among these alternatives will impact one or more other tasks. If the impacted or dependent design tasks are within the scope of this designer's responsibility, then the designer must deal with the dependencies based on local information and knowledge. On the other hand, if the dependent tasks belong to another designer, then the designer must estimate or acquire needed information on the dependent tasks and maintain the dependencies in the local design process. From a value-based design perspective, considering and maintaining such dependencies between design tasks at both the value level, i.e., related to design objectives, and task level, i.e., related to design parameters and (hard) constraints, are the key to successful collaboration.

24.3.1 Design Value Distribution and Aggregation

In an ideal collaborative design situation, designers of a design team should work together from the very beginning and all along on defining design problems, identifying design objectives, creating design solutions and finally developing an overall design. When a design task is decomposed and the subtasks are assigned to individual designers, it is essential that the design objectives are managed in a centralized fashion and distributed to corresponding designers when needed. Centralized management is the key to maintain the consistency among distributed design objectives. We call this approach of managing design objectives *value distribution*. In this approach, design objective structuring is centralized and value distribution is carried out to allow multiple designers to simultaneously generate, evaluate and select design alternatives based on the given objectives.

In practice, however, due to the time limitation and the availability of expert designers, it is rarely the case that designers can develop all design objectives *together*. Often, after several high-level design objectives are developed, each designer will work separately to develop his/her local design objectives and solutions for local design tasks. During the process of design, designers much coordinate with each other to make sure that their local designs will eventually work together. We argue that in addition to information exchange, the coordination should include sharing and critiquing each other's design objectives. By taking into account the design objectives of the tasks that are related to the local task, a designer can make decisions that add values not only to local design, but to those of others. In addition to increasing the overall value of design, this approach can also reduce potential conflicts between downstream design activities. We call this the *value aggregation* approach to collaborative design since the design objectives are developed by each individual designer and aggregated when being applied to collaborative decision-making.

24.3.2 Levels of Value Aggregation and Coordination Cost

In the value aggregation based collaborative design, in order for designers to make globally effective decisions, two or more designers need to *aggregate* their separate design objectives, if their tasks are dependent on each other. Depending on to what degree designers integrate other dependent designers' design objectives into their local decision-making, there can be three different levels of aggregation: *zero aggregation*—i.e., totally distributed tasks without sharing objectives; *partial aggregation*—i.e., only part of dependent design decisions are made based on shared design objectives; and *complete aggregation*—i.e., all design decisions involving dependencies with other designers' tasks are made by sharing design objectives.

To illustrate the different levels of value aggregation, consider three designers A, B and C in a team collaborating on a design problem, as shown in Fig. 24.2. Each designer is responsible for several design tasks. We define the union of these tasks as the responsibility boundary of that designer. In Fig. 24.2, each dark dot represents

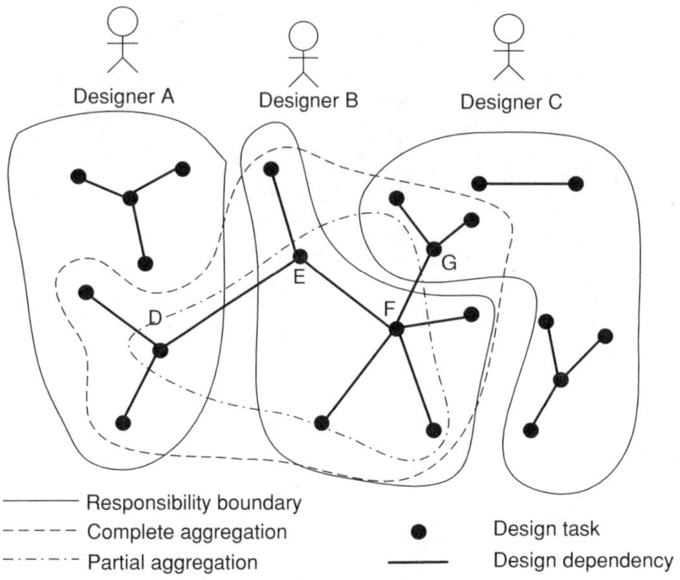

FIG. 24.2 LEVELS OF VALUE AGGREGATION IN COLLABORATIVE DESIGN

a design task. Solid lines linking the design tasks represent the dependencies between them. We refer to the complete graph of all design tasks with the links as design *task dependency network* that can be used to illustrate how tasks are dependent on each other and to find what other tasks need to be considered when solving a specific design task. In real design practices, obtaining dependency network graph at the early stage of design process can be difficult. Usually, the network evolves over time as the design process progresses and when designers generate new design subtasks.

In Fig. 24.2, *zero aggregation* refers to the situations where multiple designers are working on their own design tasks without sharing design objectives. With zero aggregation of design objectives, designers work independently at the design value level, but they are still passing information to each other at the task level when requested. They consider the impact of their decisions on other members of the team only when design conflicts are identified. It is usually the case that conflicts are recognized only after decisions are made and when design progresses to the point where the integration of solutions of subtasks is carried out. When such conflicts happen, designers have to backtrack to their previous decisions and reevaluate other design alternatives in order to resolve the conflicts. With no value aggregation, the coordination between designers is only at task level.

Partial aggregation, which is represented by dotted lines in Figure 24.2, refers to the situations where design objectives are shared between several but not all dependent tasks. Partial aggregation may be due to the fact that design dependencies between design tasks are not completely established, either because the design process is not complete or the designers cannot recognize the dependency. It can also be because engaging in value aggregation is too time consuming.

Complete aggregation, represented by broken lines in Figure 24.2, is a case where design dependencies among design tasks are completely identified and design objectives are shared by designers for making their responsible design decisions. In both partial and complete aggregation, designers move from *task-level* coordination to *value-level* coordination. Value aggregation is desirable because it helps increase the overall design value and can reduce the potential conflicts that may occur later in the design process.

From the above discussion, it is apparent that to ensure good collaboration, designers should attain complete aggregation of design objectives. In practice, however, coordination at the value level entails cost. Balancing the amount of *task-level* coordination and that of *value-level coordination* may lead to the most efficient design process.

Task-level coordination includes exchanging task information, e.g., parameters, and resolving design conflicts between designers. *Value-level coordination* on the other hand refers to working together, either face-to-face or virtually, to develop and share design objectives. Through value-level coordination, designers align their design objectives and take into consideration other designers' objectives when making local design decisions. Our experience has shown that the alignment of design objectives may reduce the likelihood of task-level conflicts among dependent design tasks [26]. As a result, more value aggregation may reduce total cost of task-level coordination, as shown in Figure 24.3. Achieving higher-level value aggregation, however, will incur value-level coordination cost. More value aggregation requires more exchange and negotiation about design objectives, as shown in Fig. 24.3. From a coordination efficiency point of view, it can be speculated that a balance between task and value-level coordination is more desirable. However, the benefit of value aggregation is not just the reduction of total coordination cost. Rather, it is for increasing the quality of design, as shown in the next example. Therefore, the added value-level coordination cost should be compensated by increased design quality. Figure 24.3 illustrates the desirable range of value aggregation. Our ongoing research attempts to develop an agent-supported, argumentation-based multilevel negotiation protocol to facilitate value-level coordination [27, 28].

24.4 A CASE EXAMPLE

To illustrate the value aggregation approach to collaborative design decision-making, consider a poppet-relief valve design problem. Two designers (mechanical engineer and manufacturing engineer) have to collaborate with each other in order to satisfy the design requirements. The general schematic of the valve is given in Fig. 24.4. The poppet-relief valve allows flow of fluid from the

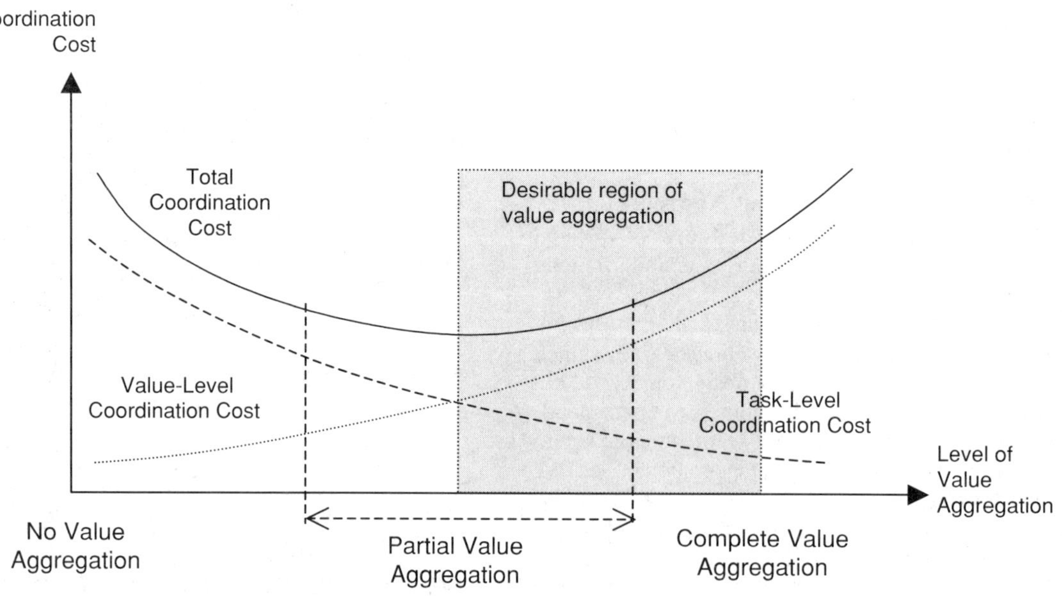

FIG. 24.3 COORDINATION COST AND LEVEL OF VALUE AGGREGATION

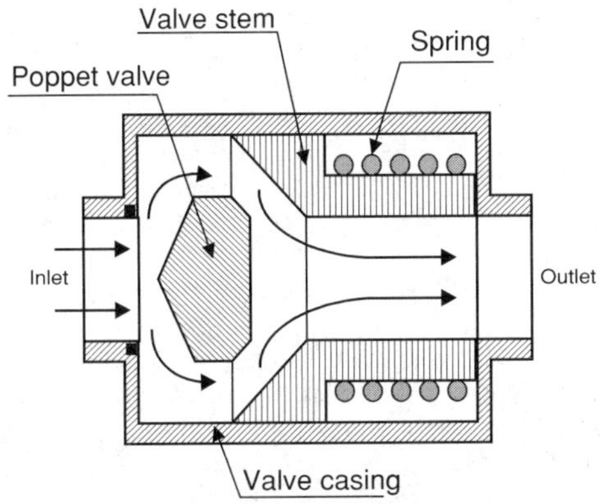

FIG. 24.4 SCHEMATIC OF POPPET-RELIEF VALVE

inlet to the outlet when the pressure of the fluid exceeds a certain threshold pressure called the "cracking" pressure. If the pressure is greater than the cracking pressure, the fluid opens the poppet valve and holds it in equilibrium against a helical compression spring. At pressures below cracking pressure, the poppet valve is held against a seal due to the helical spring, and thereby cuts off fluid flow from the inlet to the outlet. The design includes a poppet valve, poppet valve stem, a helical compression string enclosed in the valve casing and the valve casing.

Let us assume that at some point during the design process, the mechanical designer is working on "design valve casing" task. The functional or fundamental objective of the design task for the mechanical designer is <maximize><function of material>in<valve casing design>. Applying the "purpose decomposition" objective structuring method, the mechanical designer introduces the following three sub-objectives as his desired performance of casing materials:

- *<minimize><casing weight>in< casing design>measured by<weight(kg)>*
- *<maximize><casing strength>in<casing design>measured by<load(N)>*
- *<minimize><corrosion effects>in< casing design>measured by<corrosion index(index number)>*

For the manufacturing engineer, the functional or fundamental objective for manufacturing the valve casing is <minimize><cos t>in<casing manufacturing>. Following the "context decomposition" method for objective structuring, the manufacturing engineer identified the following three more detailed sub design objectives.

- *<minimize><material cost>in<casing design>measured by <unit-cost($)>*
- *<maximize><manufacturability>in<casing design>measured by<unit-manufacturing-cost($)>*
- *<maximize><ease of assembly>in<casing design>measured by<level(high/medium/low)>*

To illustrate the difference between the value aggregation and no value aggregation, let us assume the mechanical designer is initially considering three materials as design alternatives, namely: steel, aluminum and brass. If the designer doesn't consider the objectives of the manufacturing engineer, the summary of his desirability toward each design alternative is illustrated as in Table 24.2. In this table, each number represents the desirability of a design alternative with respect to the corresponding design objective normalized to a range from 0 to 100; 0 represents the least desirable alternative and 100 represents the most desirable one.

TABLE 24.2 DESIRABILITY OF CASING MATERIAL ALTERNATIVES (MECHANICAL DESIGNER)

	Design Alternatives		
Design Objectives	**Steel**	**Aluminum**	**Brass**
<Weight>	12	56	45
<Strength>	92	45	55
<Corrosion>	75	70	65
Overall desirability	100	60	0

TABLE 24.3 DESIRABILITY OF CASING MATERIAL ALTERNATIVES (MANUFACTURING ENGINEER)

	Design Alternatives		
Manufacturing Objectives	**Steel**	**Aluminum**	**Brass**
<Material Cost>	10	40	55
<Manufacturability>	30	65	55
<Ease of assembly>	45	65	65
Overall desirability	0	94.4	100

In the next step, the designer calculates the overall desirability of each alternative with respect to the corresponding designer.

Similarly for the manufacturing engineer, the three proposed design alternatives have the following desirability (Table 24.3).

From the above two tables, it is apparent that the mechanical designer prefers to use steel as the valve casing material and the manufacturing engineer prefers to use brass. By establishing the dependency between the mechanical designer's task and the manufacturing engineer's task (i.e., designed artifact should be suitable for the manufacturing process), the two members of the team realize that they have to collaborate on selecting an alternative that best suits their "aggregated value." If we aggregate the relevant design objectives of both members into one complete set of design objectives and evaluate the available alternatives with respect to them, the results will be as given in Table 24.4.

Table 24.4 reveals an interesting conclusion. By considering a complete set of design objectives from all dependent designers, a new alternative is ranked as the best alternative that was not considered in the previous cases as a candidate. In the above simple example, in both cases where designers were making their decision in isolation, they didn't consider aluminum as the best candidate. Once the design objectives were aggregated, a new option revealed itself as the most appropriate solution to satisfy all requirements of both the mechanical designer and manufacturing engineer.

Several observations can be made from the above example. First, aggregation of design objectives is a fundamental part of collaboration. In engineering practice, designers must consider the impact

TABLE 24.4 AGGREGATION OF DESIGN OBJECTIVES

		Design Alternatives		
Combined Objectives		**Steel**	**Aluminum**	**Brass**
Complete set of design objective	<Weight>	12	56	45
	<Strength>	92	45	55
	<Corrosion>	75	70	65
	<Material cost>	10	40	55
	<Manufacturability>	30	65	55
	<Ease of assembly>	45	70	65
Aggregated desirability		0	100	92.7

of their decisions on other team members' decisions. By doing so, they can increase the value of the overall design and reduce the potential conflicts that can arise because of poor early decisions.

Second, by clearly addressing the shared design objectives, designers get more insights into how their design tasks that are dependent on each other. Hence, if needed, they can revise their design objectives. Further, the relation of value aggregation between two design tasks can be kept during the design process and used later to facilitate future collaborations. For example, if a previously chosen alternative has to be changed for some reason, then from the value aggregation relations the designer would know which other design tasks must be notified of the design change.

Finally, design value and design objectives are designer-dependent. Both identification and structuring of design objectives are based on designers' subjective judgments. Without a set of consistent and shared methods and tools, it is likely that different designers may develop completely different design objectives that can hardly be shared. Our ongoing research attempts to address this issue by introducing a library of meta design objectives, a set of rules and methods for objective structuring, and a negotiation protocol for both task-level and value-level coordination.

24.5 CONCLUDING REMARKS

In this chapter, an approach to value-based design and value aggregation is proposed to support collaborative design decision-making. The definition of design objectives, the methods of design objective structuring and the dependency-based design objective aggregation are the core concepts of this approach. The case example presented has shown how value aggregation can increase the overall value of design decisions and avoid downstream design conflicts.

Research in engineering design, including collaborative design, in the past decade has mostly focused on providing analysis methods and evaluation guidelines for design decision-making. Early efforts such as systematic design and axiomatic design provide decision support based on engineering principles gained either from physics or experience. Other methods including Quality Function Deployment (QFD) attempt to provide tools to help designers organize and manage their design information more effectively so that design decisions can be made by considering all relevant information. Although not so rigorous, the practical basis of these methods has made them relatively easy to apply in real design situations. The decision theoretic approaches to engineering design, on the other hand, are based on rigorous mathematical theories. The vast amount of modeling and calculation effort needed to apply these approaches to real problems, however, has limited their attractiveness. While efforts should be made to extend the envelop of the rigorous decision theory in the design field, another approach is to adapt the principles of design theory to create practically applicable design methods. The research presented in this chapter is our first step in this endeavor that aims at a decision-theory-based and practically applicable framework for design and collaborative design.

This research was supported in part by National Science Foundation under a CAREER grand DMI-9734006. Additional support was provided by industry sponsors. The authors are grateful to NSF and the industry sponsors for their support.

REFERENCES

1. Suh, N. P., 1990. *The Principles of Design*, Oxford University Press, Oxford, MA.
2. Pahl, G. and Beitz, W., 1996. *Engineering Design, A Systematic Approach*, Springer.
3. Hazelrigg, G. A., 1998. "A Framework for Decision-Based Engineering Design," *J. of Mech. Des.*, Vol. 120, pp. 653–658.
4. Hazzelrigg, G., A., 1996. *Systems Engineering*, Prentice Hall.
5. Thurston, D. L., 1999. "Real and Perceived Limitations to Decision Based Design," *Proc., of ASME Des. Engrg. Tech., Conf.*, ASME, New York, NY.
6. Thurston, D. L., 1994. "Optimization of Design Utility," *J. of Mech. Des.*, Vol. 116, pp. 801–808.
7. Thurston, D. L., 1991. "A Formal Method for Subjective Design Evaluation With Multiple Attributes," *Res. in Engrg. Des.*, Vol. 3, pp. 105–122.
8. Luce, R. D. and Raiffa, H., 1957. *Games and Decisions*, Dover Publications, Inc., New York, NY.
9. Lewis, K. and Mistree, F., 2001. "Modeling Subsystem Interactions: A Game Theoretic Approach," *J. of Des. and Manufacturing Automation*, 1 (1) pp. 17–36.
10. Bras, B. and Mistree, F., 1991. "Designing Design Processes in Decision-Based Concurrent Engineering," *SAE J. of Mat. and Manufacturing*, Vol. 100, pp. 451–458.
11. Mistree, F. and Allen, J. K., 1997. "Optimization in Decision-Based Design," *Position Paper: Open Workshop on Decision-based Des.*, Orlando, FL.
12. Mistree, F., Smith, W. F. and Bras, B., 1993. *A Decision-Based Approach to Concurrent Engineering*, Chapman & Hall, New York, NY.
13. von Neumann, J. and Morgenstern, O., 1953. *Theory of Games and Economic Behavior*, 3rd Ed., Princeton University Press, Princeton, NJ.
14. Arrow, K. J., 1986. *Social Choice and Multi-Criterion Decision-Making*, The MIT Press, Cambridge, MA.
15. Fishburn, P. C., 1965. "Independence in Utility Theory With Whole Product Sets," Operations Res., Vol. 13, pp. 28–45.
16. Fishburn, P. C., 1970. *Utility Theory for Decision-Making*, John Wiley and Sons, Inc., New York, NY.
17. Keeney, R. L., 1975. "Group Decision-Making Using Cardinal Social Welfare Functions," *Mgmt. Sci.*, 22 (3), pp. 430–437.
18. Hammond, J, S., Keeney, R. L. and. Raiffa, H., 1999. *Smart Choices: A Practical Guide to Making Better Decisions*, Harvard Business School Press, Cambridge, MA
19. Keeney, R. L. and Raiffa, H., 1976. *Decisions With Multiple Objectives: Preferences and Value Tradeoffs*, John Wiley and Sons, Inc., New York, NY.
20. Keeney, R., 1992. *Value-Focused Thinking–A Path to Creative Decision-Making*, Harvard University Press, Cambridge, MA.
21. Danesh, M. R. and Jin, Y., 1999. "ADN: An Agent-Based Decision Network for Concurrent Design and Manufacturing," *Proc., ASME Des. Engrg. Tech., Conf.*, ASME, New York, NY.
22. Wassenaar, H. J. and Chen, W., 2003. "An Approach to Decision Based Design," *ASME J. of Mech. Des.*, 125 (Sept.).
23. Sen, P. and Yang, J., 1998. *Multiple Criteria Decision Support in Engineering Design*, Springer.
24. Keeney, R. L., 1976. "A Group Preference Axiomatization With Cardinal Utility," *Mgmt. Sci.*, 23, (2), pp. 140–145.
25. Vlacic, L., Ishikawa, A., Williams, T. J. and Tomizawa, G., 1997. "Applying Multi-attribute-Based Group Decision Making Techniques in Complex Equipment Selection Tasks," *Group Decisions and Negotiations*, Vol. 6, pp. 529–556.
26. Danesh, M. R. 2001. "A Value Based Design Framework for Conceptual Design," Ph.D. thesis, Univ. of Southern California, CA.
27. Jin, Y. and Levitt, R., 1993. "I-Agents: Modeling Organizations Problem Solving in Multi-agent Teams," *Intelligent Sys. in Accounting, Finance and Mgmt.*, Vol. 2, pp. 247–270.
28. Jin, Y. and Lu, S. 2004. "Agent-Based Negotiation for Collaborative Design Decision-Making," *CIRP Annals*.

APPENDIX I: DECISION-BASED DESIGN PROCESS

Following is a general DBD process used in our value aggregation approach to collaborative design decision-making.

Step 1. Identify the design problem:

A design task T is initially given as a high-level objective q_0 associated with a list of requirements (r_1, r_2, ... r_n):

$T_0 = \{q_0; r_1, r_2, \ldots r_n\}$.

Step 2. Design objective structuring:

From the initial task T_0, the overall objective q_0 is decomposed and structured into an design objective hierarchy DoH:

$q_0 \Rightarrow \text{DoH} = \{ q_0; q_{1\text{-}1}, q_{1\text{-}2}, \ldots, q_{i\text{-}j}, \ldots, q_{m\text{-}n}; R_{\text{qij-qkl}}\}$.

Step 3. Generate design alternatives:

For all the leaf objectives, q_{f1}, q_{f2} ... q_{fk}, identify or generate one or more design elements $ei = \{ei1, ei2, .., eig\}$, $1<=I<=k$, where each of the design elements can contribute to achieving the corresponding objective. All the design elements for all the leaf design objectives forms a morphology chart for creating design alternatives.

Select feasible alternatives $t_g = \{e1i, e2j \ldots ekl\}$ and form a feasible alternative set

$A = \{tf1, \ldots tfh\}$, where h = numbe of feasible alternatives.

Step 4. Analyze and evaluate alternatives:

In order to make decisions on selecting one alternative from A, the designer should use desirability of each alternative as his/her criteria. If the design objectives are not dependent of each other, and the designer has complete control over design objectives, and enough information about design consequences, then for given design alternative $A=\{t_1, t_2, \ldots, t_n\}$ $n \geq 2$ a multilinear utility function can be written as:

$$U(t_1,t_2,\ldots,t_n) = \sum_{i=1}^{n} k_i u_i(t_i) + \sum_{i=1}^{n}\sum_{j \succ i}^{n} k_{ij} u_i(t_i) u_j(t_j) + \sum_{i=1}^{n}\sum_{j \succ i}^{n}\sum_{h \succ j}^{n} k_{ijh} u_i(t_i) u_j(t_j) u_h(t_h) + \ldots$$

where k=a normalization factor.

If design objectives are "additively independent" of each other, the above equation can be simplified to:

$$U(t_1,t_2,\ldots,t_n) = \sum_{i=1}^{n} k_i u_i(t_i)$$

After assigning desirability to each design alternative, the designer should select the design alternative, that has the maximum monetary value.

In collaborative decision-making, M designers (stakeholders) make decisions on one instance S_m, $m = 1,2,\ldots, M$ with utility functions U_m, $m = 1,2,\ldots, M$. In this case, we can define a "status quo" level for the utility of the group to be $U^0 = (U_1^0, U_2^0, \ldots, U_m^0)$. Each designer prefers to maintain his/her own utility. The goal here is to find a design alternative "t" such that $U_m(A) \succ U_m^0$ for all m. The total desirability of a design alternative is equal to the summation of desirability of all design objectives of local and dependent designers. In this case, design values of all dependent designers are aggregated in one single value, which will be used to rate the alternatives.

Step 5. Selection and decomposition:

The last step in the design process is to select the design alternative, that has the maximum utility. This can be easily done by sorting the design alternatives by rating and selecting an alternative with highest monetary value. The above-mentioned process will be executed at every level of the design process. The design process will terminate once all design solutions are found.

PROBLEMS

24.1 Based on the design objective definition and categories described in this chapter, develop a design objective structure of a design problem of your choice. Do you think the five categories can cover everything you want to achieve? If not, what categories need to be added?

24.2 Discuss, with examples, why it is ideal to manage design values in a centralized way and why this is not always the case in practice.

24.3 Discuss what are similarities and differences between the "value aggregation" approach and the "game theory" based approach.

SECTION

8

VALIDATION OF DESIGN METHODS

INTRODUCTION

In the preceding sections of this book several different perspectives and approaches for decision-making in engineering design were covered. The views cover a variety of topics such as preference modeling, concept generation, demand modeling, and centralized and distributed decision-making. It has been noted that some of these views are not entirely consistent with each other. Assessing the validity of different decision-based design (DBD) approaches requires a scientific validation framework that should be generic yet useful for validating all design methods, including decision-based ones. In this section, the challenge of validation is discussed and recommended approaches are presented. Due to the inherently diverse nature of design activities, currently there is no single, accepted method to perform validation of design methods. In this section, different perspectives on design method validation are articulated.

In Chapter 25 the authors explore the historical roots of modern epistemology (the theory of knowledge) and present a new validation procedure, namely, the Validation Square, which includes *structural* validity and *performance* validity as primary requirements for a sound method. In the chapter the authors also offer examples and advice for practical application of the validation square in engineering design research. In Chapter 26 past efforts in creating validation frameworks are reviewed and a model-based technique to design method validation is proposed, where design scenarios are simulated repeatedly over a whole range of uncertain parameters to assess the expected utility of a design method. The approach is demonstrated by evaluating and comparing some robust parameter design methods. In Chapter 27 a set of validation criteria for design methods, specifically as they are used to promote design decisions, are introduced. The criteria are then applied to a well-known design method tool, the house of quality (HoQ). Through application of the validation criteria to the HoQ, limitations of the HoQ in supporting design decisions are uncovered through rigorous statistical analysis and the implications to any design method are presented and discussed.

In essence, the three chapters in this section address validation from different philosophical perspectives as implied by the different definitions adopted for "validation." In Chapter 25 the authors define scientific knowledge within the field of engineering design as "socially justifiable belief, " according to the Relativistic School of Epistemology. They draw the conclusion from the open nature of design method synthesis for which subjective judgments of usefulness are important for critically evaluating design methods. A design method must not only be theoretically sound but also useful in practice to be valid; either one alone is not sufficient. In Chapter 26 the validation of design methods is presented as a process of examining objective evidence that a design method fulfills stated requirements for a specific intended use in the design of an engineering system. In Chapter 27 a valid design method is defined as one that provides results that are logical, uses meaningful information, assesses the probability of success of a decision, does not bias the designer, and provides a sense of robustness in the results.

There are important consequences for the proposed methods for validation depending on which definition is adopted. For example, defining knowledge as "socially justifiable belief" (Chapter 25) challenges the notion that theoretical or objective validation is sufficient and proposes that it must be coupled with subjective evaluations of its usefulness, whereas the definition in Chapter 26 supports a wholly objective perspective, but excludes an investigation as to whether the requirements of the method are stated correctly in the first place. In Chapter 27, while a set of criteria is discussed, useful design methods may exist that satisfy only a subset of the criteria, while not violating the remaining criteria.

We encourage the readers to explore the breadth of perspectives spanning the three chapters. An effective way to explore differences is to identify a current point of controversy in design theory (e.g., multicriteria approach vs. single-criterion approach), and apply different validation philosophies to each position. Finally, we note that the chapters included in this section focus on the validation of design methods, but not the validation of predictive engineering models. The latter is a subject being studied for model-based design decision-making.

CHAPTER

25

The Validation Square: How Does One Verify and Validate a Design Method?

Carolyn C. Seepersad, Kjartan Pedersen, Jan Emblemsvåg, Reid Bailey, Janet K. Allen, and Farrokh Mistree

Validation[1] of engineering research has traditionally been anchored in the context of scientific inquiry that demands formal, rigorous and quantitative validation. Logical induction and deduction play key roles in this formalism, making it particularly useful for validating internal consistency within the framework of the scientific method. Since much engineering research is based on mathematical modeling, this kind of validation has worked–and still works–very well. There are, however, other areas of engineering research that rely on subjective statements as well as mathematical modeling, making formal, rigorous and quantitative validation problematic. One such area is that of design methods within the field of engineering design. In this paper, we explore the question: "*How does one validate design research in general, and design methods in particular, given that many proposed designs will never be realized and that it is often infeasible to follow the realized designs through their complete life cycles?*"

Anchored in the tradition of scientific inquiry, research validation is strongly tied to a fundamental problem addressed in epistemology: "*What is scientific knowledge, and how is new knowledge confirmed?*" Thus, we first look to epistemology for: (1) reasons why the traditional approach of formal, rigorous and quantifiable validation is problematic for engineering design research; and (2) an alternative approach to research validation. We present a new validation procedure, namely, the Validation Square, and offer advice for applying it in an engineering design research context.

We recognize that no one has *the* answer to the questions we pose. To help us converge on an answer to these questions, we think aloud and invite you to join us. It is our hope that the ensuing discussion will enrich all of us as members of the design community.

[1]As a philosophical term, *validation* refers to internal consistency (i.e., a logical problem), whereas *verification* deals with justification of knowledge claims. In modeling literature, on the other hand, these terms are swapped, and in this paper we use the terms as used in the modeling literature; i.e., *verification* refers to the internal consistency, while *validation* refers to justification of knowledge claims [1].

25.1 THE SEARCH FOR A RIGOROUS APPROACH TO VALIDATE DESIGN METHODS

How should a design method be validated? As scholars, what approach should we follow to validate our work rigorously? As educators, how should we teach the next generation of researchers to establish the validity of their own contributions to the state of knowledge in engineering design? As advisors, what are the characteristics of a high-quality M.S. thesis/Ph.D. dissertation?[2] These questions are important to both faculty and students who perform research and document their findings in the scholarly literature, and they provided the impetus for writing this chapter.[3]

Validating the internal consistency of a design method in the tradition of scientific inquiry does not guarantee that it is externally relevant, i.e., that it is *useful for its intended purpose*. Because subjective judgments of usefulness are important for critically evaluating design methods, we need to *augment* traditional validation methods to ensure the external relevance of proposed solutions. To do so, we go to the roots of epistemology for alternative ways of conceptualizing design knowledge.

25.1.1 The Historical Roots of Modern Epistemology

Epistemology (the theory of knowledge) began in ancient Greece with Phyrro, Plato and Aristotle who held a foundationalist view of knowledge. According to this view, knowledge of the world rests on a foundation of indubitable beliefs. From this foundation, further propositions can be inferred to create a superstructure of known truths that are absolute and innate [3].

Modern epistemology emerged from this foundationalist basis in the 17th century with the introduction of rationalism by Descartes [4] and empiricism by Locke [5]. The foundationalist views were advanced in the 20th century with the introduction of positivism by Wittgenstein [6]. Positivism was centered on the verification principle that statements are meaningless unless they

[2]An answer to this question is presented in the appendix.

[3]The work presented here has evolved from a paper by Pederson, Emlemsvåg and co–authors [2].

can be formalized for analytical and/or empirical investigation. It shares the fundamental assumption that rational knowledge is the only valid knowledge.

Although positivism became less popular in the second half of the 20th century, many of the basic ideas of foundationalism persisted in the school of thought known as reductionism. Methodological reductionism has been the most influential reductive approach in modern science. Methodological reductionists postulate that the properties of the whole are the sum of the properties of its parts. Hence, analysis of the parts is sufficient to gain knowledge about the whole system. Reductionism incorporates the assumption that knowledge is absolute and logically verifiable. It is based on objective quantification and the fundamental assumption that *objectivity exists*.

From this historical perspective, we see that formal, rigorous and quantitative validation is anchored in the foundationalist/formalist/reductionist school of epistemology. Accordingly, this school of thought is based on the fundamental assumptions that: (1) *truths (knowledge) are innate and absolute*; (2) *only rational knowledge is valid*; and (3) *objectivity exists*.

These three fundamental assumptions make "formal, rigorous and quantitative" validation problematic for engineering design research because it is based on *subjective* statements. Challenges to these three assumptions have given rise to a relativist and fundamentally different school of epistemology.

25.1.2 The Relativistic/Holistic/Social School of Epistemology

The notion of *innate and absolute truths* was first challenged by Kant [7], followed by Hegel [8], Kuhn [9] and Sellars [10]. From their perspective, knowledge is socially, culturally and historically dependent. Hence, there are no neutral foundations of knowledge, and entirely objective verification of knowledge claims is not possible.

Relativist philosophers challenge the assumption that *only rational knowledge* is valid knowledge. Kuhn [9] and Quine [11] observed that science progresses when prevailing theories cannot provide adequate explanations to scientific problems under investigation, making way for new theories. The new theories are then accommodated to experiments not because they satisfy some absolute scientific principles, but because they are convenient, causing minimal disturbance in the existing theory. Hence, our ability to be rational depends on a basic ability to exercise intelligent judgment that cannot be completely captured in systems of rules and is not entirely accessible to investigation through the senses or formal calculations.

Furthermore, the very *existence of objectivity* was challenged by Wittgenstein [6], Einstein [12] and Gödel [13], among others. Einstein stated that: "One may compare these rules [related to the scientific method] with the rules of a game in which, while the rules are arbitrary, it is their rigidity alone which makes the game possible. However, the fixation will never be final. It will have validity only for a special field of application."

Wittgenstein addressed the issue of objectivity in mathematics, and claimed that, "mathematics is merely a tool consistent only within itself and hence content free." This view was supported by Kurt Gödel who claimed that, "every formal number theory contains an indecisive formula, i.e., neither the formula nor its negation is provable in the theory."

As charted in Fig. 25.1, a new school of epistemology, namely, the relativist/holistic/social school, grew out of the refutation of the fundamental assumptions underlying the foundationalist/reductionist/formalist school of epistemology. The different views of knowledge have a significant impact on research validation.

25.1.3 The Impact of Different Views of Knowledge on Research Validation

Foundationalist/formalist/reductionist validation is a strictly formal, algorithmic and confrontational process in which new knowledge is either true or false. The validation becomes a matter of formal accuracy rather than practical application. This approach is appropriate for closed problems that have unique, right or wrong

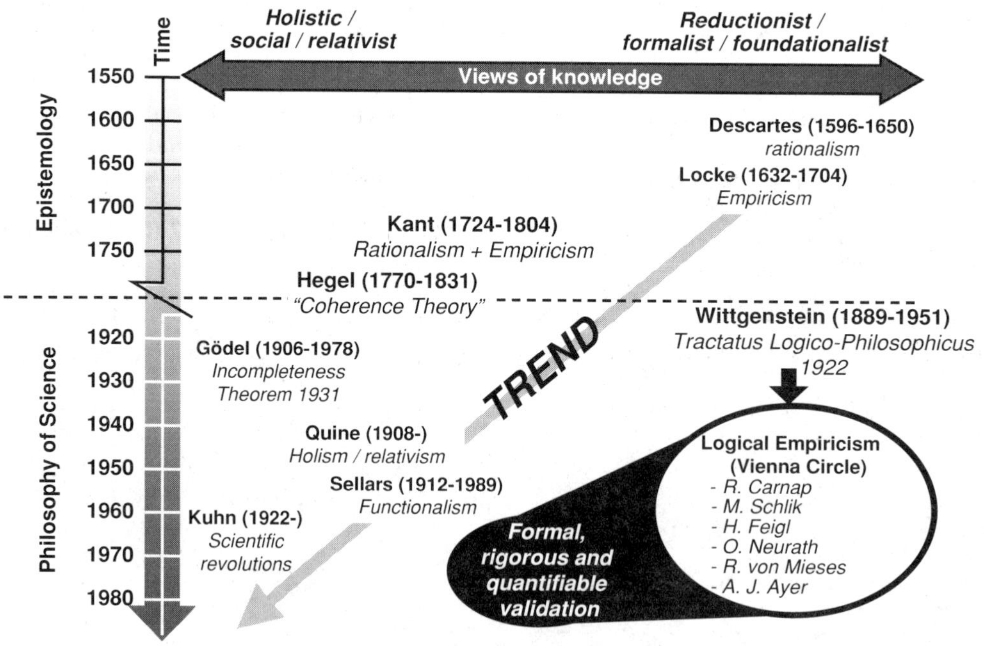

FIG. 25.1 THE EVOLUTION OF THOUGHT

answers associated with them, such as mathematical expressions or algorithms.

Relativistic/holistic/social validation, on the other hand, is a semiformal and conversational process in which validation is viewed as a gradual process of building confidence in the usefulness of the new knowledge with respect to a purpose. This approach makes a relativist validation procedure appropriate for design methods that are focused on open problems. Open problems, for which there may be many acceptable solutions, necessitate design methods that are not only logical and rigorous, but also demonstrably useful for achieving their intended purposes. Accordingly, design methods often incorporate human judgment and nonprecise representations as well as mathematical modeling.

To ensure that validated design research is not only mathematically and logically sound but also relevant and useful for its intended purpose, a relativist validation procedure is asserted, based on the following definition: "We define scientific knowledge within the field of engineering design as socially justifiable belief according to the Relativistic School of Epistemology. We are motivated by the open nature of design method synthesis, for which valid new knowledge must be not only logically and/or mathematically rigorous but also capable of yielding useful, intended results. Thus, validation becomes a process of building confidence in the usefulness of design research with respect to a purpose."

25.2 THE "VALIDATION SQUARE"– A FRAMEWORK FOR VALIDATING ENGINEERING DESIGN RESEARCH

The purpose of this chapter is to introduce a rigorous framework for validating engineering design methods. Based on the preceding epistemological discussion, we propose two primary tasks for such a framework:

(1) Establish the ***structural*** validity of a design method. The method should be logically rigorous, internally consistent and mathematically correct. The context for which the method is valid should be clearly articulated as a set of underlying assumptions and examples should be chosen carefully to reflect the intended domain of application. The process of structural validation is predominantly *qualitative*.

(2) Establish the ***performance*** validity of a design method. The method should provide useful results with respect to its intended purpose. The purposeful effectiveness of the method should be demonstrated with carefully chosen examples that reflect the intended domain of application. Whenever possible, quantitative measures should be established for evaluating the method's performance. Grounded in the results of the example applications, arguments should be supplied for the anticipated extent of the method's usefulness *beyond* the examples provided. Performance validation is primarily *quantitative*.

These two primary aspects of validation are incorporated in the Validation Square, a rigorous validation framework for engineering design research, illustrated in Fig. 25.2. As shown, the Validation Square is divided into four quadrants. The left and right halves are associated with structural and performance validation, respectively. Each half of the square is divided further into a domain-specific and a domain-independent quadrant, associated with the validity of the method for the domain-specific examples investigated in the research and for broader domains of application, respectively.

25.2.1 Structural Validation–A Qualitative Process

As illustrated in Fig. 25.2, *structural validity* has three complementary facets: (1) the internal consistency of each of the individual constructs constituting the method; (2) the internal consistency of the method itself, as an integration of parent constructs; and (3) the appropriateness of the example problems used to verify the performance of the method.[4]

1. *Internal consistency of each parent construct:* Design methods often build upon existing constructs. If so, the internal consistency of each of the parent constructs must be established. We suggest critically evaluating the literature for evidence that the constructs offer structural validity (i.e., internal consistency, logical rigor and mathematical correctness) and performance validity (i.e., delivery of results that are useful with respect to the construct's intended purpose). Small-scale empirical studies and formal or informal proofs or arguments are appropriate for supplementing the evidence available in the literature. Convincing evidence must be provided that the constructs are broadly acceptable and appropriate for the intended purpose and domain of application.

2. *Internal consistency of the method:* To establish the internal consistency of the method as an assemblage of constructs, it is appropriate to establish the internal consistency of the integrated method. Formal or informal proofs, logical arguments and flowchart representations of information flow are appropriate techniques for establishing the structural validity of the method. With flowcharts, it is easy to outline the steps (constructs) of a method and the requisite input and output information, as well as to determine whether adequate input information is available for each step. At this step, it is important to articulate clearly and explicitly the intended purpose of the method and to argue convincingly that the method is *theoretically* capable of achieving its intended goals.

3. *Appropriateness of the example problems:* Example problems should be identified for verifying the performance of the method. To build confidence in the appropriateness of the example problems, we suggest adopting several different viewpoints. First, document that the example problems are similar to the problems for which the parent constructs are generally accepted. Then, document that the example problems are representative of actual problems for which the method is intended. Finally, document that the data associated with the example problems is adequate to support a conclusion.

Establishing the validity of the individual parent constructs (1) and the integrated method (2) addresses the structural "soundness" of the method in a general sense; therefore, it is denoted domain-independent structural validity. Establishing the appropriateness of the example problems for testing the performance of the method (3) deals with the structural soundness for some particular instances; therefore, it is denoted domain-specific structural validity. Both of these types of structural validity are evaluated with predominantly qualitative techniques.

[4]The numbered items correspond to the numbers in brackets in each quadrant of the Validation Square.

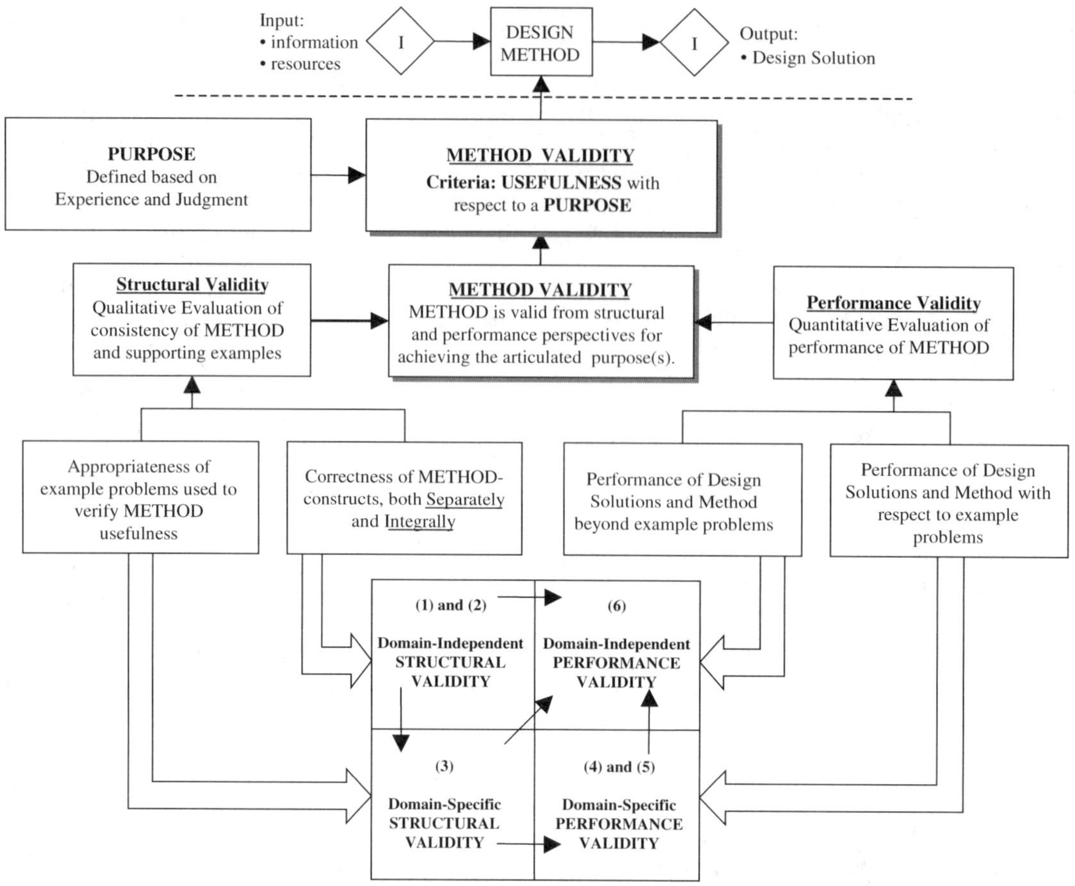

FIG.25.2 DESIGN METHOD VALIDATION: A PROCESS OF BUILDING CONFIDENCE IN USEFULNESS WITH RESPECT TO A PURPOSE

25.2.2 Performance Validation – A Quantitative Process

As illustrated in Fig. 25.2, performance validation has three complementary facets: (4) establishing that the outcome of the method is useful with respect to its intended purpose for the chosen example problem(s); (5) establishing that the demonstrated usefulness is linked to applying the method itself; and (6) reasoning that the method is useful for domains that are broader than the chosen examples.

4. *Establishing the usefulness of the method for the chosen example problems:* To establish the usefulness of the method, it should be applied to representative example problems. The example problems should be deemed appropriate for this purpose according to the domain-specific structural validation quadrant of the Validation Square. Metrics for usefulness should be identified to measure the extent to which the method achieves its articulated purpose(s). For example, from an industrial perspective the purpose of a design method is linked typically to reducing cost and/or time and/or improving quality. From a scholarly perspective, the purpose may also include increasing the stock of scientific and engineering knowledge.

5. *Establishing that the demonstrated usefulness is linked to applying the method:* To establish that the demonstrated usefulness is linked to applying the method itself, we suggest evaluating the individual contributions of each step (construct) individually, as well as together. It is also helpful to compare solutions obtained with and without the method, allowing a quantitative comparison with benchmark results.

6. *Reasoning that the method is useful for domains that are broader than the example problems:* To build confidence in the generality of the method, we suggest induction based on the following line of argument. In (1) we demonstrate that the individual constructs are generally accepted for some limited applications. In (2) we demonstrate the internal consistency of the way the constructs are integrated into the overall method. In (3) we demonstrate that the constructs are applied within their accepted domains. In (4) we demonstrate the usefulness of the method for some chosen example problems, which in (3) have been demonstrated to be appropriate for testing the method. And finally, in (5) we demonstrate that the demonstrated usefulness is related directly to applying the method. Based on this line of reasoning, we claim generality, i.e., that the method is useful beyond the example problems that were tested. Specifically, we claim that the method is appropriate for problems that are similar to the chosen example problems, i.e., within the accepted domain of application of the proposed design method. Every validation rests ultimately on justified belief, as described in Section 25.1. Hence, the purpose of applying the Validation Square is to present evidence that justifies belief in the general usefulness of the method with respect to an articulated purpose.

If the method is deemed useful for some limited instances (4) and (5), we denote this as *domain-specific* performance validity. Similarly, if the method is deemed useful in a more general sense (6), we denote this as *domain-independent* performance validity.

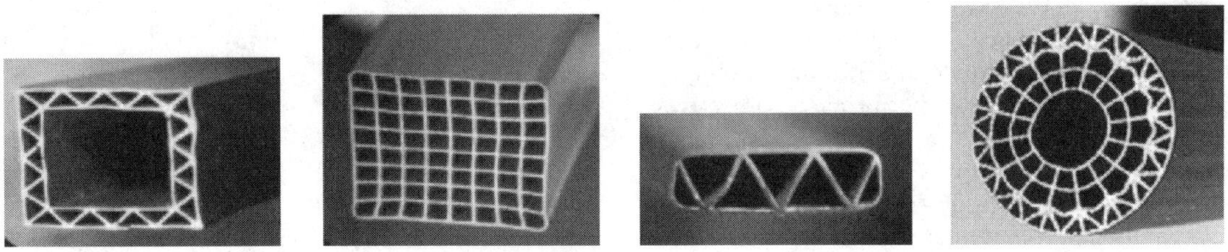

FIG. 25.3 ORDERED, PRISMATIC CELLULAR MATERIALS

We have proposed the Validation Square as a framework for validating design methods. In the next section, we provide some practical guidelines for application of the Validation Square.

25.3 APPLYING THE VALIDATION SQUARE FOR VALIDATING ENGINEERING DESIGN RESEARCH

In this section, we offer examples and advice for practical application of the Validation Square to engineering design research. The examples are drawn from research to establish methods for designing robust, multifunctional cellular materials (c.f., [14–16]). As illustrated in Fig. 25.3, the materials are ordered, metallic cellular materials with extended prismatic cells. The materials can be produced with nearly arbitrary 2-D topologies and dimensions, metallic base materials and wall thicknesses as small as 50 microns via a thermo-chemical extrusion fabrication process developed at Georgia Tech by the Lightweight Structures Group [17]. The design challenge is to tailor the topology and dimensions of the materials to multifunctional applications, such as the gas turbine engine combustor liner illustrated in Fig. 25.4. These applications require adequate performance in multiple, functional domains—such as heat transfer and structural load bearing—that place conflicting demands on the material structures. A robust topological preliminary design exploration method (RTPDEM) has been established for exploring and generating robust, multifunctional cellular topology and other preliminary design specifications [16]. Aspects of the validation process for the RTPDEM are highlighted in this section as examples of applying each facet of the Validation Square strategy. As illustrated in the Validation Square diagram in Fig. 25.2, validation is a four-phase process in which we establish that the design method provides solutions correctly (structural validity) and provides correct solutions (performance validity). This must be shown for the example problems of interest (domain-specific) and for broader classes of problems or applications (domain-independent). Each phase is discussed sequentially in the following sections.

25.3.1 Domain-Independent Structural Validation

The primary consideration for domain-independent structural validity is the logical consistency of the proposed design method. Often, a design method is at least partially a synthesis or assembly of parent methods or constructs. In this case, internal consistency must be established not only for the overall method but also for the *individual* parent constructs that comprise it.

The first step is to determine the requirements for the design method. At least two categories of requirements should be enumerated:

(1) Requirements for the *outcomes* of the method, such as the functional, behavioral and structural characteristics or

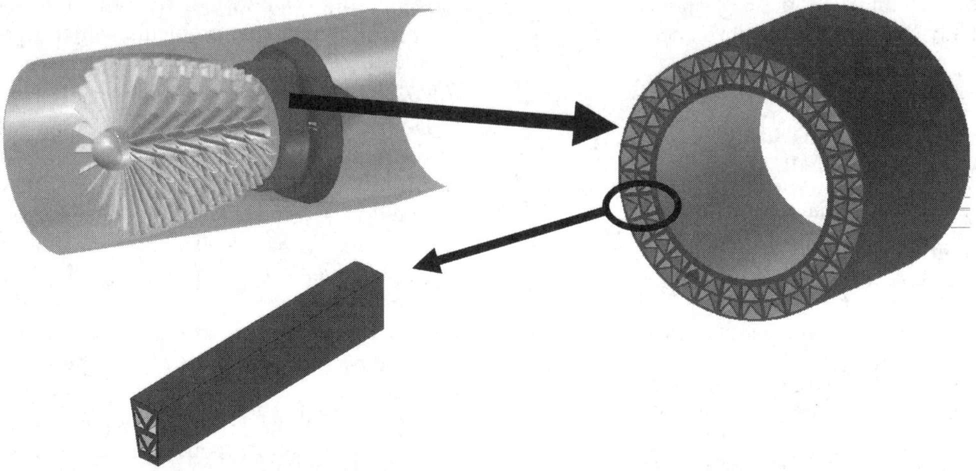

FIG. 25.4 CELLULAR MATERIALS FOR A MULTIFUNCTIONAL APPLICATION WITHIN THE COMBUSTOR LINER OF A NEXT-GENERATION GAS TURBINE ENGINE [16]

quality of the resulting products. For example, the RTP-DEM is intended to facilitate the realization of families of designs that are manufacturable and exhibit a range of trade-offs between multifunctional performance objectives and robustness to dimensional and topological variation.

(2) Requirements for the *process* by which the method generates the outcomes. Examples include the efficiency of the method, computational requirements such as distributed versus local computing or supercomputers versus desktop PCs, the knowledge and experience levels required of the intended user, the ability to accommodate multiple designers, as well as intended designer interactions with one another and with the computing framework.

These requirements provide the foundation for metrics that are used to evaluate the usefulness of the method throughout the validation process. Sample high-level requirements for the RTPDEM are listed in Table 25.1 [16]; these high-level requirements are decomposed into a hierarchical set of more specific requirements. Often, the requirements are identified most easily by considering the intended context for application of the method (e.g., multifunctional cellular materials). Characteristics of the intended domain of application should be enumerated and may include details of the intended physical domains (e.g., structural mechanics, thermodynamics, electromagnetics), types of performance parameters, classes of variables (i.e., continuous, discrete, binary) and product architectural characteristics (e.g., degree of modularity, size, user interfaces).

After the requirements for the design method have been established, the next step is to search the technical literature related to each parent construct and critically evaluate it with respect to its established advantages, limitations and accepted domains of application. For example, because the RTPDEM is founded on topology design techniques, robust design methods and multi-objective decision support, relevant literature is critically reviewed in each field in the context of the design method requirements documented in a previous step. The limitations of currently available constructs and methods can be used to confirm the need for the proposed design methods.

At this stage, it may be necessary to apply the parent constructs or methods to small example problems to build familiarity with the methods and more concretely establish their capabilities and limitations. This is especially important for two sets of conditions: (1) if the parent methods/constructs have not been used for applications similar to the intended domain of application of the proposed design method; or (2) if the performance of the parent methods/constructs have not been adequately documented in the literature. For example, topology design methods for the RTPDEM are applied to a standard example problem and coupled with various search/optimization algorithms to ascertain the relative performance of each alternative algorithm. A sample diagram of processor time versus problem size or resolution for three different algorithms is illustrated in Fig. 25.5.

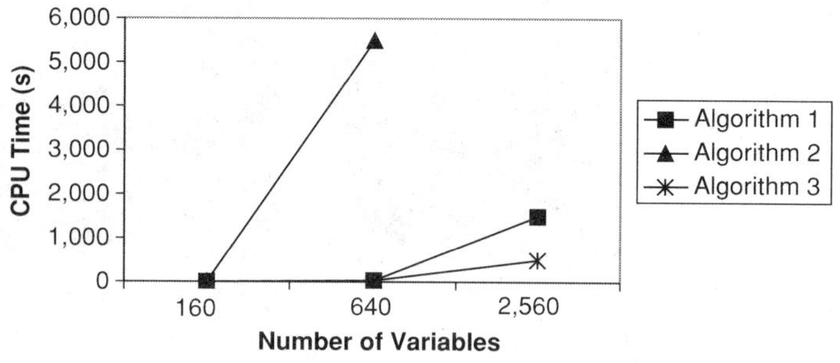

FIG. 25.5 INVESTIGATING PERFORMANCE CAPABILITIES TO SUPPORT THEORETICAL STRUCTURAL VALIDATION OF ALGORITHMS THAT ARE FOUNDATIONAL CONSTRUCTS WITHIN A DESIGN METHOD [18]

Next, it is important to establish the internal consistency of the proposed design method in its entirety. This can be accomplished both logically and empirically. Techniques include logical arguments, formal or informal mathematical proofs and flowcharts. Flowcharts are especially useful for verifying that there is adequate input for each step in a design process and that adequate output is provided for the next step. Empirical techniques include small example problems designed to test a specific capability of the method. These experiments are especially useful when empirical results can be compared with well-established or theoretical data.

Finally, it is important to compare the capabilities and limitations of the proposed method and its parent constructs with the design method requirements established previously. Based on this exercise, the structural validity of the design method is confirmed independently of *specific* example problems or domains of applications. However, the intended domain of application serves an important role of providing context for domain-independent structural validation—a role that is particularly prominent in prompting the requirements by which the design method is evaluated.

TABLE 25.1 HIGH-LEVEL REQUIREMENTS FOR A SAMPLE DESIGN METHOD

Number	Requirement
1	The method should facilitate exploration and generation of *families* of multifunctional or multi-objective compromise solutions.
2	The method should facilitate systematic modification of topology.
3	The method should facilitate consideration and maximization of robustness and flexibility with respect to many sources of variation, including tolerances and topological imperfections.
4	The method should be systematic, efficient and effective relative to ad hoc methodologies.

25.3.2 Domain-Specific Structural Validation

Domain-specific structural validation involves building confidence in the appropriateness of the example problems selected for illustrating and verifying the performance of the design method.

The first step in establishing domain-specific structural validity is to document that the example problems are similar to the problems for which the design method and its constituent constructs are generally accepted or intended. An important aspect of the previous validation phase—domain-independent structural validation—is establishing the accepted domain of application of the design method. The domain of application is typically described as a list of characteristics of the design problems for which the design method is intended. In the domain-specific structural validation phase, the characteristics of the example problems are documented and compared with those for which the design methods are intended.

TABLE 25.2 DESIGN CAPABILITIES DEMONSTRATED IN EACH EXAMPLE [16]

	Example 1 Structural Heat Exchanger	Example 2 Robust Structural Materials	Example 3 Combustor Liners
(a) Multifunctional Design Exploration			
Single Domain	✔	✔	
Multiple Domains	✔		✔
Distributed Multifunctional Synthesis	✔		✔
(b) Robust Design Exploration			
Variation in control factors		✔	✔
Variation in topology		✔	✔
Variation in material properties		✔	
Variation in operating conditions			✔
Robust design methods to support distributed, multifunctional design			✔
(c) Topology Design			
Structural		✔	
Multifunctional (Structural and Thermal)			✔
Coupled with robust design methods		✔	✔
Distributed			✔

The investigator should check that the example problems together exhibit *all* of the critical characteristics of the design problems for which the method is intended. It is appropriate to have multiple example problems—several of which possess subsets of the characteristics and one or more of which possess a broader or unified set of characteristics. Such an approach enables detailed, independent investigation of specific aspects of the design method with targeted, small-scale examples as well as broader investigation of important interactions and other system-level issues with holistic, all-encompassing examples. Table 25.2 includes a sample list of design problem characteristics for which the RTPDEM is intended and indicates the subset of characteristics shared by each of three example problems used to validate the RTPDEM. It is extremely important for each example to fulfill a role in the validation process that is not fulfilled by other example problems. Otherwise, efforts are simply duplicated, and the example should be discarded. Conversely, if one or more of the critical characteristics of the design problems (for which the method is intended) are missing from the example problems, it is important to identify additional examples or to expand one or more of the planned examples to include the missing characteristics.

The next step involves documenting the fact that the data from the examples can be used to support conclusions with respect to the performance of the design methods. One aspect of this task is to determine whether the example problems represent *actual* problems for which the design method is intended. Simplifying assumptions are made in any design example with respect to the quantity of data, the number and type of variables, the extent to which broader aspects of the system are considered and many other characteristics. The investigator should document the simplifying assumptions embedded in the example problems and confirm that the assumptions will not affect his/her ability to draw conclusions from the examples. For example, when making assumptions, an investigator must not simplify away a critical characteristic for which the design method is intended. A second aspect of this task is documenting that each example will yield qualitative and/or quantitative data that can be compared, contrasted and otherwise processed to evaluate the performance of the proposed design method.

25.3.3 Domain-Specific Performance Validation

Domain-specific performance validation involves building confidence in the usefulness of a method using example problems and case studies. Representative example problems are used to evaluate the results of applying the design method in terms of the outcome- and process-related design method requirements documented in the domain-independent structural validation phase. For example, one of the outcome-related design method requirements for the RTPDEM is the exploration and generation of families of designs that exhibit ranges of multifunctional performance trade-offs. The RTPDEM is used to generate the family of cellular material designs presented in Fig. 25.6. The designs are intended to serve as structural heat exchangers and balance thermal and structural performance, as illustrated on the horizontal and vertical chart axes, respectively.

It is also important to establish that the resulting usefulness is, in fact, a result of applying the method. For example, solutions obtained with and without the construct/method can be compared and/or the contribution of each element of the method can be evaluated in turn. When validating the RTPDEM, the multifunctional performance of the designs illustrated in Fig. 25.6 is compared with those generated with conventional, single-objective optimization techniques as well as with ad hoc designs that are generated using engineering intuition without the benefit of systematic design methods or search techniques. The objective of the comparisons is to determine whether utilizing the RTPDEM actually improves the robustness and/or multifunctional performance or provides an improved balance between multifunctional objectives compared with single-objective or ad hoc designs. Also, important performance measures of parametrically tailored materials are compared with the same

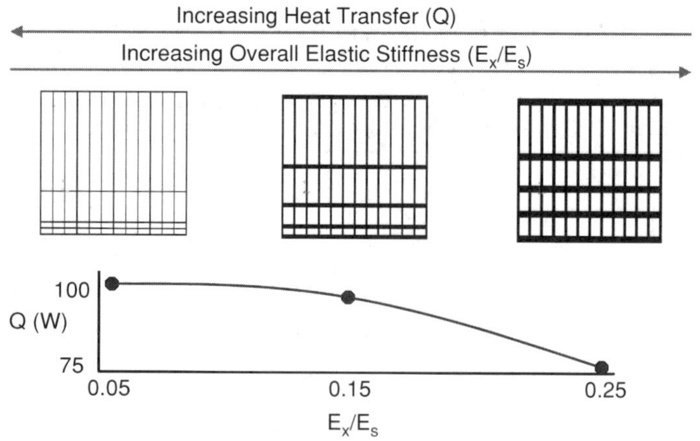

FIG. 25.6 SAMPLE MULTIFUNCTIONAL DESIGNS GENERATED WITH THE RTPDEM [16]

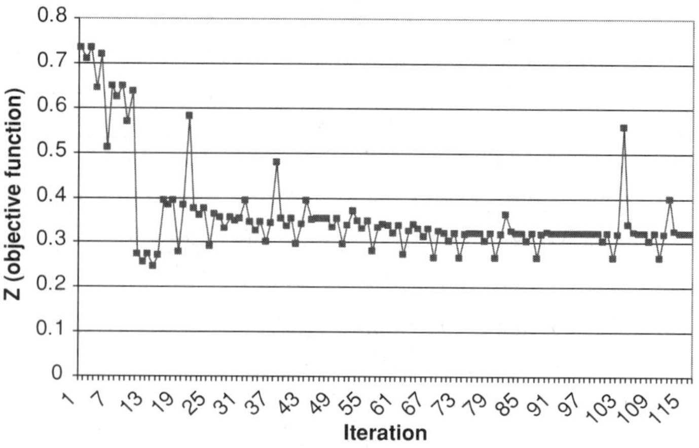

FIG. 25.7 SAMPLE CONVERGENCE PLOT FOR A TRIAL RUN OF THE RTPDEM [16]

measures for parametrically *and topologically* tailored materials, as well as for materials designed with robustness considerations to gauge the impact of each aspect of the RTPDEM.

An important part of domain-specific performance validation is careful review of the data used to support any conclusions. This involves establishing its accuracy, internal consistency and quality. For example, in optimization exercises, multiple starting points, active constraints and goals, and convergence can be documented to verify that the solution is stationary and robust. In Fig. 25.7, a sample convergence plot is illustrated for the RTPDEM for one of the designs in Fig. 25.6.

If computational models are required for the examples, it is important to verify that data obtained from the models accurately represents aspects of the problem relevant to the design method being tested. Data or results from the models should be compared with experimental data, well-established theoretical results or more comprehensive computational models. A sample comparison between a fast, approximate thermal finite-element (FE) model (utilized in a thermal application of the RTPDEM) and detailed FLUENT results for three different heat source temperatures is illustrated in Fig. 25.8. Based on a comparison of the data, the model should be observed to react to inputs in an expected manner, similar to the reaction of an actual system. A similar step is performed for the design method and its constituent constructs in the domain-independent structural validation phase; here, it is performed for the computational models and data needed for the specific example problem(s).

25.3.4 Domain-Independent Performance Validation

Domain-independent performance validation involves building confidence in the generality of the method and accepting that the method is useful beyond the example problems. The first step is to revisit the intended domain of application established in other validation phases. The characteristics of the application domain are enumerated as part of domain-independent structural validation and used to evaluate the appropriateness of the example problems as part of domain-specific structural validation. The investigator can argue logically that the design method is useful for examples with the *precise* characteristics of the example problems used to validate the design method. An intuitive argument must be made that the design method is useful for a *more general* class of problems. The investigator should use his/her judgment and experience with the examples to clearly indicate the characteristics that a design problem *should* have and those it *should not* have in order to be eligible for utilization of the design method. At this stage, it is appropriate to list design problem characteristics and conditions for which the design method *may* be applicable but have not been explicitly tested; extending the design method and establishing its

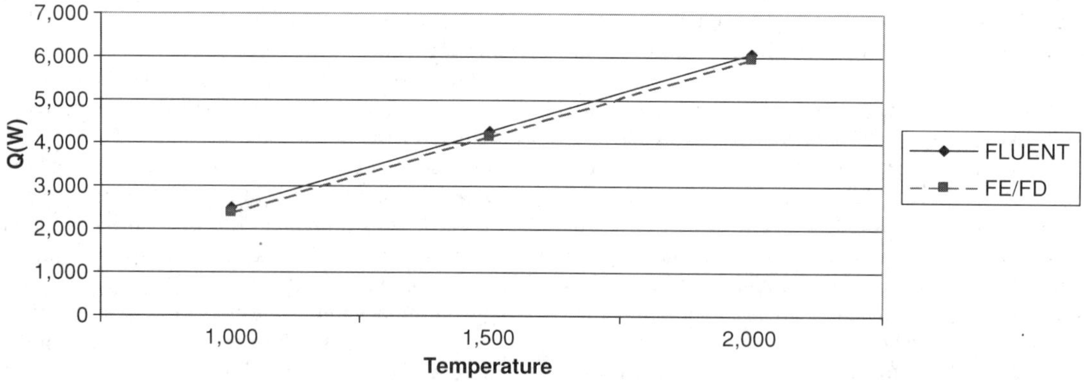

FIG. 25.8 A COMPARISON BETWEEN FAST FINITE ELEMENT (FE/FD) AND DETAILED FLUENT HEAT TRANSFER (Q) PREDICTIONS FOR A RANGE OF HEAT SOURCE TEMPERATURES [16]

TABLE 25.3 A NEW VIEW OF KNOWLEDGE VALIDATION

Old View on Knowledge Validation	Fundamental Assumptions	Refutation Based on	New Emerging Assumptions	New View on Knowledge Validation
Foundationalist	Knowledge is absolute/ innate	Kant, Hegel, Sellars, Quine, Kuhn	Knowledge is socially justifiable belief	Relativist
Reductionist	Rationality only valid basis for knowledge	Honderich, Einstein	Intuition valid basis for defining purpose for application of knowledge	Holistic
Formalist	Objectivity exists.	Hegel, Kuhn, Wittgenstein, Gødel, Einstein	Research validation linked to usefulness	Social and conversational

capabilities for these applications represent opportunities for future work. For example, using design problem examples, the usefulness of the RTPDEM is demonstrated for thermal and structural functional domains and for problems with observed variations in dimensions and topology. Further work is required to extend it to other functional domains and other sources of variation such as boundary conditions and material properties.

25.4 CLOSURE

In this paper we have questioned the adequacy of "formal, rigorous and quantitative" validation for engineering design research, and we have articulated a set of assumptions that leads us to a new view of knowledge validation, namely, a relativist/holistic/social view (see Table 25.3).

Based on the changed view, we assert that validating a design method is a process of demonstrating usefulness with respect to a purpose. Based on this assertion we present a framework for guiding this process, namely, the Validation Square (see Fig. 25.9). This framework builds on research in systems dynamics as well as a tradition of using posits in engineering design. However, the Validation Square as presented in this paper extends all these efforts by offering a prescriptive approach that is more comprehensive and systematic.

We assert that the Validation Square is appropriate for validating research results in general, as long as the proposed research can be subjected to qualitative and quantitative evaluation as outlined in Section 25.2.

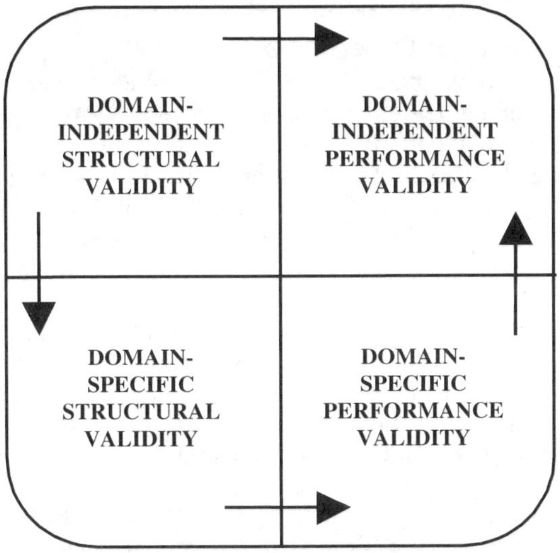

FIG. 25.9 THE VALIDATION SQUARE

In summary, we recognize that no one has *the* answer to the challenging question of how to validate research in engineering design. We trust that you enjoyed thinking aloud with us. We now invite you to comment upon what we have presented so that together we can create something of value for the engineering design community and for the next generation of researchers!

ACKNOWLEDGMENTS

How should a proposed design method be validated? In our laboratory, this question was first addressed by Jon Shupe in his dissertation (Ph.D. 1998, [19]). Warren Smith (Ph.D. 1992, [20]), Reid Bailey (M.S. 1997, [21]), Jesse Peplinksi (Ph.D. 1997, [22]), Jan Emblemsvåg (Ph.D. 1999, [23]), Kjartan Pedersen (Ph.D. 1999, [24]), Reid Bailey (Ph.D. 2000, [25]) and Carolyn Conner Seepersad (Ph.D. 2004, [16]) have contributed to the answer presented in this chapter.

We acknowledge George Hazelrigg for posing the question *vis-à-vis* validation of a proposed design method to the design research community. This provided the impetus for us to provide an initial response at the ASME Design Theory and Methodology Conference in 2000.

REFERENCES

1. Barlas, Y. and Carpenter, S., 1990. "Philosophical Roots of Model Validation: Two Paradigms," *Sys. Dyn. Rev.*, Vol. 6, pp. 148–166.
2. Pedersen, K., Emblemsvåg, J., Bailey, R., Allen, J.K. and Mistree, F., 2000. "The Validation Square: Validating Design Methods," *Proc., ASME Des. Theory and Methodology Conf., DETC2000/DTM-14579*, ASME, New York, NY.
3. Honderich, T., ed., 1995. *The Oxford Companion to Philosophy*, Oxford University Press, New York, NY.
4. Descartes, R., 1641, 1931. "Meditations on First Philosophy," *The Philosophical Works of Rene Descartes*, Cambridge University Press, Cambridge.
5. Locke, J., 1690, 1894. *An Essay Concerning Human Understanding*, Clarendon Press, Oxford.
6. Wittgenstein, L., 1921, 1961. *Tractatus Logico-Philosophicus*, Routledge and Kegan Paul, London, U.K.
7. Kant, I., 1781, 1933. *Critique of Pure Reason*, St. Martins Press, London, U. K.
8. Hegel, G. W. F., 1817, 1959. *Encyclopedia of Philosophy*, Philosophical Library, New York, NY.

9. Kuhn, T., 1962, 1970. *The Structure of Scientific Revolutions*, University of Chicago Press, Chicago, IL.
10. Sellars, W., ed., 1963. *Empiricism and the Philosophy of Mind: Science, Perception and Reality*, Humanities Press, New York, NY.
11. Quine, W. v. O., 1953. *Two Dogmas of Empiricism: From a Logical Point of View*, Harvard University Press, Cambridge, MA.
12. Einstein, A., 1950. *The Theory of Relativity & Other Essays*, MJF Books, New York, NY.
13. Gödel, K., 1931. "Über Formal Unentscheidbare Sätze der *Principia Mathematica* und Verwandter Systeme," *Monatshefte für Math. u. Physik* Vol. 38, PP. 173–198.
14. Seepersad, C. C., Dempsey, B. M., Allen, J. K., Mistree, F. and McDowell, D. L., 2004, "Design of Multifunctional Honeycomb Materials," *AIAA J.*, Vol. 43, PP. 1025–1033. Vol. 11, No. 2–3, pp. 163–181.
15. Seepersad, C. C., Kumar, R. S., Allen, J. K., Mistree, F. and McDowell, D.L., 2004. "Multifunctional Design of Prismatic Cellular Materials," *J. of Computer-Aided Mat. Des.* Vol. 11, No. 2–3, pp. 163–181.
16. Seepersad, C. C., 2004, *"A Robust Topological Preliminary Design Exploration Method With Materials Design Applications,"* Ph.D. dissertation, The G.W. Woodruff School of Mech. Engrg., Georgia Ins. of Tech., Atlanta, GA.
17. Cochran, J. K., Lee, K. J., McDowell, D. L., and Sanders, T. H., 2000. "Low Density Monolithic Honeycombs by Thermal Chemical Processing," *Proc., 4th Conf. on Aerospace Mat., Processes, and Env. Tech.*, Huntsville, AL.
18. Fernåndez, M. G., 2003. *"A Comparative Study of Optimization Algorithms for the Topological Design of Structures," ME6103 Final Proj. Rep.*
19. Shupe, J. A., 1998. *"Decision-Based Design: Taxonomy and Implementation,"* Ph.D. dissertation, Dep. of Mech. Engrg., Univ. of Houston, Houston, TX.
20. Smith, W., 1992. "Modeling and Exploration of Ship Systems in the Early Stages of Decision-Based Design," Ph.D. dissertation, Dep. of Mech. Engrg., Univ. of Houston, Houston, TX.
21. Bailey, R.R., 1997. *"Designing Robust Industrial Ecosystems,"* M.S. thesis, The George W. Woodruff School of Mech. Engrg., Georgia Inst. of Tech., Atlanta, GA.
22. Peplinski, J., 1997. *"Enterprise Design: Extending Product Design to Include Manufacturing Process Design and Organization,"* Ph.D. dissertation, The George W. Woodruff School of Mech. Engrg., Georgia Inst. of Tech., Atlanta, GA.
23. Emblemsvåg, J., 1999. *"Activity-Based Life-Cycle Assessments in Design and Management,"* Ph.D. dissertation, The George W. Woodruff School of Mech. Engrg., Georgia Inst. of Tech., Atlanta, GA.
24. Pedersen, K., 1999. *"Designing Platform Families: An Evolutionary Approach to Developing Engineering Systems,"* Ph.D. dissertation, The George W. Woodruff School of Mech. Engrg., Georgia Inst. of Tech., Atlanta, GA.
25. Bailey, R.R., 2000. *"Input-Output Modeling of Material Flows in Industry,"* Ph.D. dissertation, The George W. Woodruff School of Mech. Engrg., Georgia Inst. of Tech., Atlanta, GA.

25.5 APPENDIX: CHARACTERISTICS OF WELL-WRITTEN M.S. THESES AND Ph.D. DISSERTATIONS IN DESIGN FARROKH MISTREE

25.5.1 Rationale For This Appendix

At first glance, this appendix may seem disconnected from the body of the material presented in this chapter. We debated whether to include this material in this chapter–and after some reflection decided to do so as a singular contribution.

In Section 25.1 we posed the following questions: "How should a design method be validated? As scholars, what approach should we follow to validate our work rigorously? As educators, how should we teach the next generation of researchers to establish the validity of their own contributions to the state of knowledge in engineering design? As advisors, what are the characteristics of a high-quality M.S. thesis/Ph.D. dissertation?"

All but the last question has been touched upon in the body of this chapter. An answer to the last question is particularly important since we educators are routinely called on "to validate" the work presented in M.S. and doctoral dissertations. Hence, for completeness, in this appendix a singular viewpoint with respect to the last question is presented for consideration by our colleagues in the academic community.

25.5.2 Preamble

Each academic unit has a different vision of itself and the standards it sets for itself. Within an academic unit there is a diversity of opinions *vis-à-vis* expectations and standards. Over the years, I have observed that there is a vast difference in expectations *vis-à-vis* what constitutes an M.S. thesis. At some institutions, the M.S. thesis involves undertaking a project and the outcome is a tad more than a term paper. At other institutions, an M.S. thesis is substantial. I belong to the latter category.

25.5.3 What is the Difference Between a Doctoral Dissertation and an M.S. Thesis?

I expect both to be well-written and have value. The value may differ. I expect something new to emerge from a doctoral dissertation or a new interpretation given to existing data. For an M.S. thesis I am comfortable with a problem being solved and well-documented. Although a student may solve an industrial problem as part of the M.S. work, the problem must be set in a scholarly context–preferably with an explanation of the intellectual context of the problem, critical review of the literature and thorough verification/validation of the work performed (see body of text in this chapter).

I recognize that there is a difference between research and development. I expect this distinction to be respected both in an M.S. thesis and a Ph.D. dissertation. Finally, I expect students to have learned how to identify, formulate and resolve problems associated with research/development. I have seen some doctoral dissertations, particularly in design, where the dissertation is essentially a lot more of the same–a person is being given a doctorate for five years (instead of two) of master's-level work and there is no contribution to advancing knowledge.

25.5.4 Characteristics of a High-Quality M.S. Thesis/ Doctoral Dissertation

Content/value: Foundational material for one or more conference papers and at least one paper in a quality journal should be embodied in an M.S. thesis.

Framing the thesis: The question to be investigated is substantiated by the review and the question is framed appropriately within the context of the state-of-art or state-of-practice. There is a scholarly review of the literature. Commentary on the literature reviewed is insightful and is anchored in cited papers. A commentary that exemplifies the need to pursue a particular line of investigation is expected. The research question/or the question that is foundational to development is clearly articulated and is anchored in the critical review of the literature.

Development of theme: Presentation of the "story" in a manner that is connected, logical, and consistent.

Explanations: There needs to be cross-referencing between chapters and also between sections in a chapter. My comments pertaining to "stand-alone" chapters read "put in context of preceding" or "put in context of research question" or "dies." Text needs to bring figures alive; figures with limited related text result in comments from me such as "talk to figure." Text without figures result in comments from me such as "cannot visualize." Poor Tables of Content that do not "tell the story" of the thesis when read elicit comments from me such as "particularize title," "jumps," "disconnected," "dies."

Body: A systematic, logical, connected progression toward answering the research question/development question.

Verification and validation: This is a weak spot in both M.S. theses and Ph.D. dissertations in design. Typically, a method is proposed, an example problem is solved and conclusions are drawn. Solving one or two examples does not, in my opinion, allow one to claim that the method has been verified. At a minimum I expect key assumptions, limitations and ramifications to be spelled out. I expect a clear statement of why each of the examples has been chosen and what is being verified through the exercise of each of these examples. I expect a clear statement as to why the results are right and argumentation to support claims/conclusions. And isn't this what the Validation Square is all about?

Closure: Claims at the end of the thesis must be consistent with what have been stated in the abstract and introduction and what is supported by the material presented in the text. Commentary associated with closing the loop between that which was promised and delivered must be included. I expect comments on next logical steps to be taken with this project; these need to be warranted by what is presented in the text.

References: The citations must be complete and accurate. The format must be self-consistent within each category (e.g., journal articles, chapters in books, books, articles in conference proceedings, etc.). I expect the names of all authors to be included and the style to be consistent—not first names for some and only initials for others.

Appendices: These must be self-contained and add value at the point where referenced in the text. An appendix is not a place to dump a solitary figure or table.

Nomenclature/glossary: Included on an as-needed basis.

Acknowledgment: Credit must be given to all those who assisted and financial sources. If students work together to produce more substantial results than either one alone could produce, then an acknowledgement of collaboration and a clear discussion of the relationship between the projects is expected in the thesis.

Grammar and punctuation: I expect my students to be educated not just trained. I look for correct spellings and the correct use of words. Some common mistakes: principle/principal, method/methodology, example/case study, etc. Correct use of mathematical symbols: For example, there is a significant difference between ≤ and <. I expect students not to use active verbs when inanimate objects are the subject, e.g., figures do not show anything. I expect to see a key word/key phrase being used consistently throughout the thesis; minimize the use of different words to refer to the same thing. For example, if a distinction has been made between a "design problem" and an "optimization problem," I expect the right phrase to be used at the right place. If there is no distinction between the two phrases, then I expect use of one or the other.

Personality: The technical/academic personality of the author is discernable through his/her writing. In a high-quality thesis I become aware of a student's curiosity, willingness to pursue leads, willingness to take risks, attention to detail, ability to frame questions and draw conclusions, spark, broader contexts, etc.

PROBLEMS

25.1. Identify a widely accepted design method (e.g., house of quality). Imagine that you were the original architect of the method. Describe how you would have applied the Validation Square to validate the method. What is the outcome?

25.2. Provide an example of a hypothetical design method that is mathematically correct and internally consistent but lacks usefulness with respect to a stated purpose. What are the lessons of value?

25.3. Engineering design researchers are usually limited practically to a small number of examples for validating their design methods. In other fields (e.g., econometrics, experimental physics or biology), many more data points may be available. How does this impact our ability to validate design methods?

25.4. Often, students are told to design products based on the "voice of the customer." What are the core assumptions on which these methods are based? What information would be needed to validate the voice of the customer? If, for whatever reason, this data is not available and cannot easily be obtained, what should the careful designer do then?

25.5. Often successive experimentation can statistically improve data accuracy. On the other hand, experimentation can be costly. In the context of validation, please discuss the trade-off between obtaining accurate data and the cost of obtaining this data. What are the lessons learned?

25.6. Must all methods used in a design process be equally valid? Is a design process just as valid as its weakest method? As with error propagation, is the validation of a multistage process cumulative?

25.7. If you are working as a member of a design team and, for example, you performed the structural computations required for the RTPDEM, what should you tell the thermal designer about the validity of your methods?

25.8. If you were establishing a database full of design methods, how would you represent the validity of a particular method? How would you associate validity information with a method and retrieve this information if you wished to apply this method to a different problem? How would you describe the context in which this method is valid? What are the key lessons to be learned?

CHAPTER

26

MODEL-BASED VALIDATION OF DESIGN METHODS

Dan Frey and Xiang Li

26.1 INTRODUCTION

This chapter will discuss validation of design methods and introduce a model-based approach to validation that is intended to make the process more objective. First, we describe the challenge of validation as it applies to design methods. Then we describe some of the past efforts to create frameworks and concepts for validation in design. Subsequently, we explore new ideas about validation of design methods and their relationship to decision theory. Finally, we propose a model-based technique and use it to evaluate and compare some robust parameter design methods.

26.2 THE CHALLENGE OF VALIDATION

Frequently, designers are compelled to choose from among various known approaches to some aspect of their work. To illustrate some of the methodological choices designers face, consider the following sequence of alternatives:

Scenario #1: A market opportunity was identified. A designer seeks to develop concepts for this opportunity and is considering either:

a. Leading a team in a brainstorming session as described in [1].
b. Having the team apply the Theory of Inventive Problem Solving as described in [2].
c. Giving the team members time to develop concepts individually and allowing each person to approach the task in any way he chooses.

Scenario #2: After some concepts have been developed, a designer faces a choice between:

a. Having the team engage in Pugh's method of controlled convergence (aka Pugh concept selection) [3].
b. Having the team work together to estimate the expected values of utility for each design alternative and subsequently choosing the design with the highest estimate of expected utility.
c. Choosing the concept himself, based on his personal judgment and experience.

Scenario #3: After a concept has been selected, the designer seeks to set targets for attributes of the design and seeks to choose between:

a. Having the team use the house of quality (HoQ) to set targets and priorities.
b. Having the team develop a market model that maps from cost and performance to demand and profit.
c. Choosing the market leader's performance as the target.

Scenario #4: After a concept is selected and a prototype is developed, the designer may face a choice between:

a. Improving the robustness of the design using a crossed arrangement of orthogonal arrays as pioneered by Taguchi and described by Phadke [4].
b. Improving robustness of the design using single array designs and response modeling as described by Wu and Hamada [5].
c. Not doing robust parameter design and thereby saving the investment in labor, experimental apparatus and other resources necessary.

These scenarios illustrate that designers make a variety of methodological choices. When faced with these scenarios, designers often seek some evidence that will inform their choice. How can an engineer know whether a design method is likely to accomplish a particular objective? How can she compare two methods with regard to their effectiveness in a specific context? How can such evidence be collected within realistic constraints of cost, time and other resources? These are central issues in validation of design methods. This chapter cannot deal with all these scenarios in detail, but it suggests a framework for dealing with validation in general and gives a specific procedure for dealing with the last scenario.

26.3 LITERATURE REVIEW

A review of major contributions to validation must include discussion of philosophy, especially epistemology—the branch of philosophy concerned with the nature of knowledge, the justification of knowledge and the nature of rationality. There are three prominent contemporary views of the justification of knowledge claims [8]:

- *Foundationalism* holds that some instances of knowledge are basic and that the remaining instances are justified by relating them to basic beliefs (e.g., by deduction from axioms).
- *Relativism* argues that knowledge cannot be validated in an objective way and that individual, subjective preferences and

rules of fraternal behavior among scientists must be considered a part of validation processes.

- *Naturalistic* epistemology promotes empirical study of how subjects change behavior based on sensory data, in other words, how they learn from experience.

Some of the foundations of design method validation were first laid out by Schön and Argyris in the 1970s and 80s. In the tradition of naturalistic epistemology, Schön conducted field studies of engineers and other professionals such as architects, and managers [9]. One of his key results was that skilled practitioners frequently rely on *tacit knowledge*, that is, knowledge the designer demonstrates through action but cannot articulate with complete fidelity. This creates a serious challenge for foundationalist approaches to validation in engineering design. One cannot rely exclusively on deduction from basic knowledge instances if some of the key elements of basic knowledge cannot be expressed in natural language or mathematics. Based on these observations, Schön suggested there is a need for an *epistemology of practice*.

Argyris introduced an important epistemological concept applicable to professional practice—a distinction between an *espoused theory*, which is reported by the practitioner as guiding his work, and a *theory-in-action*, which is consistent with what a practitioner actually does [10]. Because much knowledge used by professionals is tacit, espoused theory and theory in action are often different for a particular subject engaging in a particular task. Building upon these concepts, Schön and Argyris [11] proposed a framework for evaluating theories of professional practice. Their framework included checks for: (1) internal consistency; (2) congruence with the espoused theory; (3) testability of the theory; and (4) effectiveness of the theory.

Recently, a framework quite similar to that of Schön and Argyris was proposed by Pedersen et al. [12], who described a Validation Square including four quadrants; (1) theoretical structural validity involving checks on internal consistency; (2) empirical structural validity; (3) empirical performance validity involving tests of effectiveness; and (4) theoretical performance validity. Details of the Validation Square can be found in Chapter 25.

A key difference between the Validation Square and the Schön and Argyris framework is that the former is explicitly founded on relativist epistemology. Pedersen et al. [12] define knowledge as "socially justifiable belief." To provide a contrast to the previous chapter, we invite readers to ask whether a relativist philosophy is an adequate basis for validation. If experts agree by a social process that something is valid, does that necessarily make it valid? In other fields, such as in medicine, when important choices are made concerning validity, a process is put in place with safeguards that promote greater objectivity. Can we adapt practices from other fields that tend to promote objectivity rather than defaulting to relativism?

An important step toward objective validation is suggested by researchers in cognitive psychology. Todd and Gigerenzer [13] proposed a means of validating decision methods comprising the following steps: (1) proposing computational models of candidate methods that are realistically based on human competences, and testing whether they work via simulation; (2) mathematically analyzing when and how the methods work with particular environmental structures; and (3) experimentally testing when people use these methods. Todd and Gigerenzer used this method to generate extensive evidence that simple heuristic methods are valid in the sense that they are more effective in real world scenarios than procedures that are theoretically superior. Todd and Gigerenzer's work is strongly aligned with naturalistic epistemology since it was derived from observation of many different kinds of subjects in various decision environments.

In developing our approach to design method validation, we rely primarily on the Todd and Gigerenzer framework and a naturalistic epistemology. A key step in describing this framework is a definition of key terms.

26.4 A DEFINITION OF DESIGN METHOD VALIDATION

We define validation of design methods as "confirmation by examination and provision of objective evidence that the design method fulfills stated requirements for a specific intended use in the design of an engineering system." This definition closely matches the one proposed by the Institute of Electrical and Electronics Engineers (IEEE) for validation of software: "confirmation by examination and provision of objective evidence that the particular requirements for a specific intended use are fulfilled" [14]. The definition is also akin to language in the 1962 amendment of the Food, Drug, and Cosmetics Act, which requires drug manufacturers to provide "evidence consisting of adequate and well-controlled investigations . . . that the drug will have the effect it purports or is represented to have under the conditions of use prescribed, recommended or suggested in the labeling or proposed labeling thereof" [15].

How should we go about "examination and provision of objective evidence" related to design methods? How can we say that a design method fulfills "requirements for a specific intended use" given that every time we design we are making a foray into uncharted territory? As the next section will show, an approach consistent with decision theory can be developed.

26.5 DECISION THEORY AND VALIDATION

When a designer chooses a method from among alternative methods, she makes a decision. It is therefore interesting to consider how decision theory may be relevant. In this chapter, we propose that the designer should choose from among the known set of design methods the one that has the largest expected value of utility. This approach is different from other applications of decision theory to design. Many researchers propose that the designer should choose the *design* that has the largest expected value of utility from among alternative designs. We are proposing that the designer might instead choose a *design method* that has the largest expected value of utility from among alternative methods. We propose that applying decision theory directly in design may not provide the highest expected value of utility as compared to other methods that provide a better fit with human cognitive abilities and the structure of specific design scenarios.

To illustrate the difference between these two ways of applying decision theory, consider scenario #2, discussed previously, in which the designer chooses from among different ways of selecting design concepts. The usual approach discussed in decision-based design (DBD) is to estimate the expected value of utility for all the design concepts and then pick the one with the largest value. But the approach we suggest involves first evaluating the various methods of selecting design concepts to determine their effectiveness. Having engineers apply decision theory might lead to the highest expected value of utility because it helps engineers align their choices with their preferences and the available information. On the other hand, Pugh's method leads to a particular type of social interaction among engineers because it prompts them to compare two concepts head-to-head and perhaps this yields information that would otherwise not come to light. On the other hand, perhaps it is best to let one person (e.g., the team leader) make the decision because the individual human brain has evolved to make good decisions under uncertainty. Because of these considerations, deciding which method is best will probably require knowledge beyond decision theory such as social psychology and cognitive science. But how can we use multiple areas of scientific knowledge to estimate the expected value of utility of a design method? To explore this question, it will be necessary to define and discuss several terms such as "design method," "expected value" and so on.

Design Method

A schema which determines choices a designer will make in every relevant set of circumstances for every set of information he may possess at the moment each choice is made.

The definition above is a paraphrase of the definition of a "strategy" in game theory—". . . a plan which specifies which choices he (a player) will make in every possible situation, for every possible actual information he may possess at that moment in conformity with the pattern of information which the rules of the game provide for him for that case" [16].

The above definition of design method includes the term "information." We are using the term "information" in much the same way it is used in game theory. The "information set" is what the player knows about the state of the game and the opponent's strategy at a given node of a game and is what the player may use in deciding on the current move. Similarly, a designer has some input data with which to inform his choices at various points within a design scenario.

It is important to note that information does not *uniquely* determine a choice among alternatives. Such choices are also determined by the decision-maker's preferences and decision procedures. However, our definition of "design method" states that the choices must be specified for every possible state of information (the same is true of "strategies" in game theory). Thus, a "design method" as defined here completely codifies the designer's preferences and decision procedures insofar as they are reflected in choices made within the method (again, the same is true of "strategies" in game theory). This is a necessity for our purposes since many existing design methods do restrict the preferences of the decision-maker and we wish to be able to include them among the alternative methods being evaluated. To include methods that **do not** restrict the decision-maker's preferences, it is necessary to create variants of those methods that specify the preference structure (for example, by assuming that the designer prefers to maximize expected profit).

As discussed above, our definition of "design method" constrains the designer's preferences during application of the method. However, a decision-maker somewhere in the organization is free to choose the design method. The choice of method is in some cases made by the designer himself, in other cases the choice is made by a project team leader and in other cases the choice is made for an entire corporation by an executive who prescribes a standardized work process. That decision-maker will express his preferences through the choice among all known alternative methods. These preferences are then reflected in the actions of the designers who implement the design method. This may be one of the reasons that organizations seek codified design methods—they afford a mechanism for management to ensure that particular values are reflected in the actions of its designers.

According to our definition of validation, a design method can be validated only in light of a "specific intended use." To be more rigorous in the definition of the "specific intended use," we define a "design scenario" to which a design method is applied.

Design Scenario

A specification of a finite set of all possible actions a designer may take, an objective function for evaluating the outcomes and so on.

This definition is a paraphrase of the definition of a "game" in game theory [16]: "A complete system of rules determined by specification of a finite set of all possible plays, a function for evaluating the plays, etc." Just as a strategy concerns how a game is played, a design method describes how a design scenario is played out. A design scenario may be a real-world design scenario, but this definition also covers computational models in which the actions of the "designer" are implemented by software (just as a game such as the prisoner's dilemma may be implemented in software).

This chapter will address validation of robust parameter design methods. In this set of methods, experiments play an important role as a source of information to designers. Therefore, experiments must be specified as part of the design scenarios. We use the term "experiment" in the usual sense used by statisticians—an operation upon a system and the associated observations. This definition includes a wide variety of design activity, including measurements made with prototype hardware and runs of computer simulations. In all these cases, the resulting observations are subject to some uncertainty, which we defined as:

Uncertainty

A lack of precise knowledge of some quantity. Uncertainty about a continuous numerical quantity can be quantified mathematically by modeling the quantity as a random variable x with a probability density function $p(x)$ defined over a support set S [17].

Having defined uncertainty, we may now define the term "expected value":

Expected Value

A function $E(x)$ of a random variable x with a probability density function $p(x)$ defined over a support set S and defined by the integral $E(x) = \int_S xp(x)dx$ [17].

We have proposed that the designer should choose a design method that maximizes the expected value of utility. By "utility" we mean a real scalar, which provides a preference ordering for all alternatives. How one can compute an expected value of utility for a design method? At a minimum, it will be necessary to define some random variables affecting the utility. Let us imagine a design method is adopted by an organization for a particular purpose. During the choice of the design method, the outcomes (such as profit made) are yet unknown. The outcome of any particular method will be affected by many uncertain or variable parameters including:

- **Uncertain parameters affecting the design method:** An example is pure experimental error in Taguchi methods. The designer runs experiments and each observation may vary from trial to trial even under apparently identical conditions.
- **Uncertain parameters defining the design method:** An example is assignment of control factors to identically constructed columns of orthogonal arrays in robust design. Consider those in Table 26.1. Each factor's main effect is orthogonal to every main effect but is aliased with three two-factor interactions. Since each column is essentially the same, the assignment of seven different physical parameters

TABLE 26.1 A FRACTIONAL FACTORIAL DESIGN (2^{7-4})

Trial	A	B	C	D	E	F	G
1	−1	−1	−1	−1	−1	−1	−1
2	−1	−1	−1	+1	+1	+1	+1
3	−1	+1	+1	−1	−1	+1	+1
4	−1	+1	+1	+1	+1	−1	−1
5	+1	−1	+1	−1	+1	−1	+1
6	+1	−1	+1	+1	−1	+1	−1
7	+1	+1	−1	−1	+1	+1	−1
8	+1	+1	−1	+1	−1	−1	+1

to the columns marked A through G is an arbitrary choice. Thus, the assignment may be modeled as a random process.

- **Other uncertain parameters defining the design scenario:** For example, designers may have uncertain knowledge of the competitors' strategies. The quality of this information will affect the design outcomes.

All the types of uncertain parameters listed above may be modeled as random variables (or sets of random variables). The outcome of the design process (such as the profit earned) is a function of these random variables and it is therefore a random variable itself. The distribution of the design outcome will affect the expected value of utility based on the preferences of the designer (for example the designer's attitude regarding risk). Therefore, there exists an expected value of utility for any design method applied to any design scenario. The next section proposes a way to make an estimate of the expected value of utility of a design method as applied to a design scenario.

26.6 INTRODUCING A MODEL-BASED APPROACH TO VALIDATION

How can we determine what outcomes a design method is likely to produce when it is applied to a design scenario? One approach is to try out different methods under a single instance of application and observe the results. For example, Kunert et al. [18] evaluated two different methods for robust parameter design. Specifically, they used crossed array designs and combined arrays to improve the consistency of a sheet metal spinning process. In effect, this was a paired comparison experiment where the experimental "treatment" was the design method employed. The result of the experiment was that the crossed array method led to process settings with a more consistent profile of the sheet metal parts as compared with the single array method.

The experiment described above provides some objective evidence validating the crossed array method. However, this evidence is not conclusive. First, the experiment included only one replication of the paired comparison. If the exact experiment were repeated, the single array might have beaten the crossed array simply due to random variations. Further, if the same two methods were applied to some other engineering system, say a lithography process rather than a sheet metal spinning process, the single array may have prevailed instead of the crossed array. A paired comparison in a single instance of application does not provide as much information as we would like to have regarding the validity of a design method.

Given the variability of the outcomes of design methods, how can we hope to validate a method based on "evidence that the design method fulfills stated requirements for a specific intended use?" In addressing this question, it is useful to draw an analogy with medicine. Patients are typically faced with a range of different treatments. For example, a person with occasional, mild aches may choose from among aspirin, acetaminophen or ibuprofen. Each of these has been demonstrated to be effective by a formal process defined explicitly by the U.S. Food and Drug Administration. This does not mean that every person who takes ibuprofen (for example) will experience relief of pain. It does mean that ibuprofen has been shown to have a beneficial effect across a population of subjects. In addition, the validation process may identify risks and side effects (e.g., stomach upset). In some cases, the validation process supports specific claims of superiority of one treatment over another (e.g., ibuprofen is superior to aspirin for relieving arthritic pain) or for having less of a side effect (e.g., less stomach upset). In other words, the medical profession has developed validation processes providing significant value to patients and doctors by providing objective evidence.

Is it reasonable to hope that design methods might be validated in much the same sense that medicines are currently validated? Can we know that a particular method is effective for a particular purpose? Can we identify risks and side effects of methods and include those in our "advertisements" of design methods? Can we compare two methods and determine objectively which one is superior for a class of scenarios? We propose that the answer is "yes" to all these questions (although it requires great effort). In addition, we propose that some of the same validation techniques from medicine may be applied to design methods. Here we focus on just one example – the use of "models."

In medical research and development it is common to use animals to test medical treatments that are being developed for humans. In fact, some of the animals used have been genetically modified specifically to enhance their usefulness as test subjects. Interestingly, these animals are often referred to as "models of human disease." A significant advantage of these models is that replication is made simpler by the fact that multiple instances of the model can be made by breeding them. Another advantage is that close control of experimental conditions and close observation of variables is simpler with lab mice than it is with humans. Such advantages led us to define an analogous entity for design method validation:

Design Method Validation Model

A design scenario simulation useful for design method validation because it has specific characteristics that resemble real-world design scenarios. People can create such models by deliberately building regularities into them.

This definition is a paraphrase of the National Human Genome Research Institute's definition of an animal model [19]: "A laboratory animal useful for medical research because it has specific characteristics that resemble a human disease or disorder. Scientists can create animal models, usually laboratory mice, by transferring new genes into them."

We propose that design scenario simulations can be used much in the same way that animal models are used in medical research. First, a large number of replicates are made representative of a class of engineering design scenarios. Different design methods are applied to this population of design scenarios. The outcomes of the design scenarios are recorded and analyzed. These data can then be used as objective evidence as part of a decision among alternative design methods.

In medical research, an animal model does not provide conclusive evidence since there are differences between the animal model and the reality it is intended to represent (a disease in a human). However, animal models are considered to be an indispensable tool for medical research and development. They are used to accumulate evidence as a precursor to clinical trials where the effectiveness of the treatment will be demonstrated even more convincingly. A similar process might be used in validating design methods. To provide an illustration of this approach, the following sections describe and then apply a model-based approach for validating robust design methods.

26.7 A MODEL USEFUL FOR VALIDATION OF ROBUST PARAMETER DESIGN METHODS

One of the key things the designer does in robust parameter design is to perform a series of experiments on an engineering system while changing control factor and noise factor settings. For our design method validation model we need a model of an engineering system that responds to changes in control and noise factors. The conclusions we wish to draw should be applicable not only to one specific engineering system, but to broad classes of engineering systems. What parameters of engineering systems should we include in our model?

As previously discussed, Kunert et. al. [18] carried out a paired comparison of cross array and single array designs by applying them to real experiments on sheet metal spinning hardware. That study suggested that crossed array methods are more effective than single array methods for making a response more consistent in the presence of noise. Kunert et al. made a conjecture that the reason single arrays worked relative poorly is that they relied too greatly on the assumption of "effect sparsity." This is probably a factor we should represent in the model. It turns out that there are two other factors frequently discussed in Design of Experiments: "hierarchy" and "inheritance." These three regularities in the response of engineering systems are defined below:

- *Sparsity of effects*: among several experimental effects examined in any experiment, a small fraction will usually prove to be significant.
- *Hierarchy*: main effects are generally larger than two-factor interactions, two-factor interactions are generally larger than three-factor interactions and so on.
- *Inheritance*: an interaction is more likely to be significant when its "parent" factors are significant.

These properties of engineering systems have been expressed in mathematical form in a hierarchical prior probability model developed in [20]. An adaptation of this model is expressed in Eqs. (26.1) to (26.10)

$$y(x_1, x_2, \quad , x_n) = \sum_{i=1}^{n} \beta_i x_i + \sum_{i=1}^{n} \sum_{\substack{j=1 \\ j>i}}^{n} \beta_{ij} x_i x_j + \sum_{i=1}^{n} \sum_{\substack{j=1 \\ j>i}}^{n} \sum_{\substack{k=1 \\ k>j}}^{n} \beta_{ijk} x_i x_j x_k + \varepsilon \qquad \text{Eq. (26.1)}$$

$$x_i \sim NID(0, w^2) \quad i \in 1 \quad m \qquad \text{Eq. (26.2)}$$

$$x_i \in \{+1, -1\} \quad i \in m+1 \quad n \qquad \text{Eq. (26.3)}$$

$$\varepsilon \sim NID(0, w^2) \qquad \text{Eq. (26.4)}$$

$$f(\beta_i | \delta_i) = \begin{cases} N(0,1) & \text{if } \delta_i = 0 \\ N(0,c^2) & \text{if } \delta_i = 1 \end{cases} \qquad \text{Eq. (26.5)}$$

$$f(\beta_{ij} | \delta_{ij}) = \begin{cases} N(0,1) & \text{if } \delta_{ij} = 0 \\ N(0,c^2) & \text{if } \delta_{ij} = 1 \end{cases} \qquad \text{Eq. (26.6)}$$

$$f(\beta_{ijk} | \delta_{ijk}) = \begin{cases} N(0,1) & \text{if } \delta_{ijk} = 0 \\ N(0,c^2) & \text{if } \delta_{ijk} = 1 \end{cases} \qquad \text{Eq. (26.7)}$$

$$\Pr(\delta_i = 1) = p \qquad \text{Eq. (26.8)}$$

$$\Pr(\delta_{ij} = 1 | \delta_i, \delta_j) = \begin{cases} p_{00} & \text{if } \delta_i + \delta_j = 0 \\ p_{01} & \text{if } \delta_i + \delta_j = 1 \\ p_{11} & \text{if } \delta_i + \delta_j = 2 \end{cases} \qquad \text{Eq. (26.9)}$$

$$\Pr(\delta_{ijk} = 1 | \delta_i, \delta_j, \delta_k) = \begin{cases} p_{000} & \text{if } \delta_i + \delta_j + \delta_k = 0 \\ p_{001} & \text{if } \delta_i + \delta_j + \delta_k = 1 \\ p_{011} & \text{if } \delta_i + \delta_j + \delta_k = 2 \\ p_{111} & \text{if } \delta_i + \delta_j + \delta_k = 3 \end{cases} \qquad \text{Eq. (26.10)}$$

This hierarchical probability model allows any desired number of response surfaces to be created such that the population of response surfaces has the desired properties of sparsity of effects, hierarchy and inheritance. Equation (26.1) represents a response y whose standard deviation might be reduced via robust design methods. The independent variables x_i are either noise factors or control factors depending on the index. Equation (26.2) shows that the first set of independent variables ($x_1, x_2, \ldots x_m$) represent noise factors and are assumed to be normally distributed. Equation (26.3) shows that the other independent variables ($x_m+1, x_m+2, \ldots x_n$) represent control factors and are assumed to be two-level factors. The variable ε represents the pure experimental error in the observation of the response, which was assumed to be normally distributed. Since control factors are usually explored over a wide range compared to the noise factors, the parameter w is included to set the ratio of the control factor range to the standard deviation of the noise factors. The parameter w is also used to set the standard deviation of the pure experimental error.

The response surface is assumed to be a third-order polynomial in the independent variables x_i. The coefficients β_i are the main effects. The coefficients β_{ij} model two-way interactions, including control by noise and noise-by-noise interactions. Similarly, the coefficients β_{ijk} model three-way interactions, including control-by-control-by-noise and control-by-noise-by-noise interactions. The model originally proposed in [20] did not include three-way interaction effects, but their addition is essential for validating robust design methods.

The values of the polynomial coefficients (β) are determined by a random process that models the properties of effect sparsity, hierarchy and inheritance. Equation (26.5) determines the probability density function for the first-order coefficients. Factors can be either "active" or "inactive," depending on the value (0 or 1, respectively) of their corresponding parameters δ_i. The parameter strength of active effects is assumed to be c times that of inactive effects. Similarly, Eqs. (26.6) and (26.7) determine the probability density function for the second-order and third-order coefficients, respectively.

Equation (26.8) enforces sparsity of effects. There is a probability p of any main effect being active. Equations (26.9) and (26.10) enforce inheritance. The likelihood of any second-order effect being active is low if no participating factor has an active main effect and is highest if all participating factors have active main effects. Thus generally one sets $p_{11} > p_{01} > p_{00}$ and so on.

Note that the model of Eqs. (26.1) to (26.10) uses normal distributions throughout. Clearly, other distributions can be assumed in these equations and the model can still be used to evaluate methods under the newly defined scenario. However, an advantage of modeling system responses as polynomials and the noise factors 1 through k as independent normal variables, is that the transmitted variation of the response due to the noise factors can be determined in closed form

$$\sigma^2(x_{m+1}, x_{m+2}, \quad , x_n) = \sum_{i=1}^{m}\left[\beta_i + \sum_{\substack{j=m+1\\j>i}}^{n} \beta_{ij}\cdot x_j + \sum_{\substack{j=m+1\\j>i}}^{n}\sum_{\substack{k=m+1\\k>j}}^{n} \beta_{ijk}\cdot x_j \cdot x_k\right]^2 \cdot w^2 + \sum_{i=1}^{m}\sum_{\substack{j=1\\j>i}}^{m}\left[\beta_{ij} + \sum_{\substack{k=m+1\\k>j}}^{n} \beta_{ijk}\cdot x_k\right]^2 \cdot w^4 + \sum_{i=1}^{m}\sum_{\substack{j=1\\j>i}}^{m}\sum_{\substack{k=1\\k>j}}^{m} \beta_{ijk}{}^2 \cdot w^6 \quad \text{Eq. (26.11)}$$

The model presented here is, in effect, our "lab mouse." It is certainly not the only model one might use for validating robust-parameter-design methods. Someone may propose a better model in the future (just as medical researchers constantly refine animal models). For now, this model provides the easiest way we know of to "breed" lots of engineering systems and test robust parameter design methods on them. How can we use this capability to study robust design methods?

26.8 USING THE MODEL TO SIMULATE ROBUST PARAMETER DESIGN

In using the model proposed above, one of the first questions to resolve is what set of alternative methods we seek to compare. Below is a list of methods to be evaluated. Each method is either popular in industry, well regarded in the literature or provides an interesting alternative to the others:

$2^{7-1}_{VII} \times 2^{3-1}_{III}$ with response modeling—a resolution VII fractional factorial 2^{7-1} inner array of control factors was crossed with a resolution III fractional 2^{3-1} factorial outer array of noise. The data from the design were used to calculate all noise main effects and control-by-noise interactions as well as control-by-control-by-noise interactions. Based on these parameters, the standard deviation was estimated based on Eq. 26.11 and the control factors were set to those discrete levels with the lowest estimated standard deviation.

$2^{7-4}_{III} \times 2^{3-1}_{III}$ with response modeling—a resolution III fractional factorial 2^{7-4} inner array of control factors was crossed with a resolution III fractional factorial 2^{3-1} outer array of noise. The data from the design were used to calculate all noise main effects and control by noise interactions. Based on these parameters, the standard deviation was estimated based on Eq. 11 and the control factors were set to those discrete levels with the lowest estimated standard deviation.

$2^{7-4}_{III} \times 2^{3-1}_{III}$ with type II S/N ratio—a resolution III fractional factorial 2^{7-4} inner array of control factors was crossed with a resolution III fractional factorial 2^{3-1} outer array of noise. For each row of the inner array, the type II signal-to-noise ratio $-\log(\sigma^2)$ was computed using the four observations in the outer array. The main effects of the control factors on the signal-to-noise ratio were computed and used to select from among the two discrete levels of the control factors.

2^{10-5}—a 32 run single array approach was employed as described in [5] with design generators A = 1, B = 2, C = 3, D=4, E = 234, F = 134, G = 123, a = 5, b = 124, c = 1,245. The single array was executed and the resulting data were used to calculate the main effect of the noise factors and control by noise interactions. Based on these parameters, the standard deviation was estimated based on Eq. (26.11) and the control factors were set to those discrete levels with the lowest estimated standard deviation.

$OFAT \times 2^{3-1}_{III}$—an adaptive one-factor-at-a-time (OFAT) plan as described in [21] was used to modify the control factors. At a randomly selected baseline configuration of the control factors, a resolution III fractional factorial 2^{3-1} outer array of noise was executed and the variance of the sample was calculated. Following this baseline assessment, a randomly selected control factor was varied and the outer array of noise was executed again. The variance at the current control factor setting was compared to the baseline variance. If the current settings resulted in the lowest observed variance, then the change was adopted and the next control factor was toggled. This adaptive process repeated until all seven control factors were changed. In each case, the change in control factor settings was adopted only if it resulted in the lowest variance observed so far in the experiment.

Before evaluating the methods above, let's define some variants of the model so that we can evaluate the effect of the model on the inferences drawn. The model described in the last section has lots of parameters, including several describing the probability of interactions. How can we decide what probabilities are reasonable? We propose it is reasonable to use values based on similar experiments conducted in the past. We collected data from engineering experiments at Ford Motor Company. We have a total of 90 response variables from 30 full factorial experiments at Ford. An engineer may choose to assume that future experiments will be statistically similar to a population of experiments his own company has conducted previously. Therefore we fitted our model to the data set from 90 responses. The resulting parameter values are listed in the first row of Table 26.2. In the second row of Table

TABLE 26.2 VALUES OF PARAMETERS IN THE MODEL VARIANTS

Parameter	p	p_{11}	p_{01}	p_{00}	p_{111}	p_{011}	p_{001}	p_{000}	c	w
Model #1	26%	18%	6.3%	1.4%	1.7%	11%	0.51%	0.8%	3.1	0.1
Model #2	26%	18%	6.3%	1.4%	0%	0%	0%	0%	3.1	0.1
Model #3	26%	18%	6.3%	1.4%		Not applicable—$\beta_{ijk}=0$			3.1	0.1
Model #4	26%		Not applicable—$\beta_{ij}=0$			Not applicable—$\beta_{ijk}=0$			3.1	0.1

26.2 is a set of parameters implying there are no "active" three-way interactions. In the third row is a set of parameters implying all three-way interactions equal zero. In the fourth row is a set of parameters implying all two-way and three-way interactions equal zero.

Table 26.3 presents the results of simulating the five alternative methods on the four variants of the model; 100 systems were generated from the models and the different methods were applied to each. The average reduction in standard deviation is tabulated for each method/model pair. The inter-quartile range of reduction in standard deviation is also tabulated. Note that the inter-quartile range of "reduction" in standard deviation is sometimes negative. This implies that, in many cases, a robust parameter design method can result in a confirmed performance of the system *less* robust than if no robust design method had been used at all. When confirmation experiments reveal this, the designer will generally just choose a good setting from among the ones tested and thereby get some benefit. This practice is not frequently discussed in the literature, but is often applied in the field. An advantage of the model-based validation approach is that it gives one a realistic sense of the uncertainty in the outcomes (which is higher than one might guess).

Now, let us focus on just the rows concerning the single array and crossed resolution III arrays. As Kunert et al. suggested [18], a crossed array is generally preferred over the single array method. Regardless of which model is assumed, the crossed array methods provide a greater mean reduction in standard deviation and the inter-quartile range is generally better as well. This means that if greater consistency of the response is the sole concern of the designer, the cross array will be preferred regardless of his risk attitudes. On the other hand, the single array may provide other advantages such as better estimation of factor effects (especially control factor main effects). A further observation regarding crossed arrays is that using a signal-to-noise ratio is more effective than using response modeling when three-factor interactions are present (even if they are generally small). When three-factor interactions are completely eliminated, the situation is reversed and response modeling provides an advantage over signal-to-noise ratios.

Finally, let us consider the adaptive OFAT method, $OFAT \times 2_{III}^{3-1}$. This approach, according to this model, represents an excellent option providing almost as much improvement as the $2_{VII}^{7-1} \times 2_{III}^{3-1}$ method but at 1/8 of the experimental costs. The OFAT-based method also provides more improvement than any of the other 32 run alternatives unless one assumes that all three-way interactions are zero, in which case a crossed array method with response modeling is preferred. The $OFAT \times 2_{III}^{3-1}$ method is a prime illustration of the value of model-based evaluation because it was developed using model-based evaluation as feedback during an iterative process of method design.

26.9 SUMMARY

This chapter has defined validation as examination and provision of objective evidence that a design method fulfills stated requirements for a specific intended use in the design of an engineering system. In other words, to the extent that a design method has been validated, a designer can have confidence that the method provides a specific set of benefits when used appropriately and can justify and communicate that confidence to others by pointing to concrete data. Ideally, the validation process also indicates that a particular method is superior to other alternatives being considered. This is a challenging standard for the validation process to meet, but other fields of human endeavor meet this standard. Every time you buy medicine from a drugstore, you have a similar objective—assurance of its effectiveness over a population of uses (although you generally do not have a guarantee of effectiveness in any specific case).

It is not a simple matter to validate the effectiveness of a design method over a population of different uses. One approach is to seek data from applications in the field. Field studies are an important source of data, but they have some drawbacks. A model-based approach is an important complement to field data. In a model-based approach to design method validation, design scenarios are simulated repeatedly over a whole range of uncertain parameters. A model-based approach provides estimates of both the expected value of design outcomes and measures of variability of the outcomes. By making such data available to the designer, model-based approaches allow designers to reflect their risk preferences in the choice of design methods.

In the example application to robust parameter design, there was a high degree of variability in outcomes across the population of uses of the methods. This was an important insight provided by the model-based approach. We suspect that many design methods exhibit a similar degree of variability in their outcomes once realistic factors regarding their implementation are taken

TABLE 26.3 PERCENT REDUCTION IN STANDARD DEVIATION FOR VARIOUS METHODS APPLIED TO VARIOUS MODELS

Method	Runs	Model Type			
		Model #1 (Fit to Data)	Model #2 (No Active Three-Factor Interactions)	Model #3 (Absolutely No Three-Factor Interactions)	Model #4 (Absolutely No Two-Factor or Three-Factor Interactions)
$2_{VII}^{7-1} \times 2_{III}^{3-1}$ Response modeling	256	Mean = 79% IQR = 74% to 85%	Mean = 81% IQR = 78% to 86%	Mean = 79% IQR = 74% to 84%	Mean = 0% IQR = 0% to 0%
$2_{III}^{7-4} \times 2_{III}^{3-1}$ Response modeling	32	Mean = 7% IQR = – 21% to 33%	Mean = 12% IQR = – 9% to 38%	Mean = 74% IQR = 66% to 81%	Mean = 0% IQR = 0% to 0%
$2_{III}^{7-4} \times 2_{III}^{3-1}$ S/N ratio (type II)	32	Mean = 18% IQR = – 14% to 58%	Mean = 15% IQR = – 18% to 50%	Mean = 33% IQR = 17% to 59%	Mean = 0% IQR = 0% to 0%
2^{10-5} Response modeling	32	Mean = 2% IQR = – 26% to 34%	Mean = 8% IQR = – 9% to 35%	Mean = 32% IQR = 15% to 54%	Mean = 0% IQR = 0% to 0%
$OFAT \times 2_{III}^{3-1}$	32	Mean = 58% IQR = 46% to 71%	Mean = 57% IQR = 45% to 68%	Mean = 53% IQR = 34% to 66%	Mean = 0% IQR = 0% to 0%

into account. The information about variability of outcomes may be one of the most useful facets of the model-based approach to design method evaluation. For a variety of reasons, information about variable outcomes of design methods is difficult to collect. Such information might help set more realistic expectations and prevent abandonment of methods that are generally good but occasionally ineffective. It might also enable the design researcher to develop methods that are more consistent in their outcomes.

The ideas presented here about design method validation are distinct from the perspectives most often voiced in the literature. Others have emphasized that design methods should be logical or self-consistent. By contrast, our approach does not directly evaluate the inner workings of design methods. Instead, we propose that "the proof of the pudding is in the eating."

One possible counterpoint to the approach presented here is—"outcomes cannot be used to evaluate decisions." It is true that *individual* outcomes cannot reliably evaluate *individual* decisions. However, it should be possible to use scientific and statistical methods to show that better decision-making methods produce better outcomes across a *population* of decisions.

PROBLEMS

26.1 Imagine you work for a designer faced with scenario #2 as described in section 26.2. Imagine that you, as part of his design team, have been asked to review existing data on the effectiveness of Pugh's method of controlled convergence (aka Pugh concept selection). Can you find any evidence regarding effectiveness of the method when used in authentic engineering practice? Can you find any evidence from a model-based approach to validation? Can you find any theoretical critique of the method or theoretical support for the method? What does the totality of evidence suggest?

26.2 Imagine you work for a designer faced with scenario #3 as described in section 26.2. Imagine that you, as part of his design team, have been asked to review existing data on the effectiveness of robust design methods. Can you find any evidence regarding the effectiveness of Taguchi's methods when used in authentic engineering practice? Can you find any theoretical critique of Taguchi's method? What does the totality of evidence suggest?

26.3 Below is a list of different circumstances under which people need to evaluate alternative methods. For each scenario, discuss the approaches typically used to evaluate the alternatives. Comment on the applicability of these approaches to evaluating engineering design methods.

a. The senior leaders in the Navy need to consider adopting new tactics for defending an aircraft carrier against submarines, including different maneuvers and different mixes of platforms, weapons and sensors.
b. An artist seeks to evaluate different ways to create the appearance of ocean waves by applying different types of paint, brushes and strokes.
c. A body of lawmakers (e.g., Parliament or Congress) considers various energy bills that are aimed at ensuring a supply of energy that is cost-effective, environmentally responsible and large enough to adequately meet projected levels of demand.

26.4 The coach of the New England Patriots football team, Bill Belichick, makes a lot of decisions. He decides, under time pressure, whether to challenge a referee's call on the field. He decides, under less time pressure, how to prepare for next week's game. He decides, under even less time pressure, when to trade a player for another player during the off-season. The following questions relate to Bill Belichick's decision-making:

a. Do you think Bill Belichick is a skilled decision-maker? Is it relevant that his team, the New England Patriots, has experienced great success (at least around the time this chapter was written)?
b. To what extent do you think Bill Belichick conforms to the norms of decision theory?
c. Is it possible to learn anything about effective decision-making by studying Bill Belichick?
d. Write a paragraph that defends or refutes the following proposition—"Bill Belichick is a designer."

26.5 Below are descriptions of three different philosophical positions:

a. Knowledge is socially justified belief. Knowledge is whatever set of propositions is generally agreed to be useful by a group of people at a particular time and place. Therefore, knowledge is relative.
b. Reality exists externally, independent of the human mind. When you and I interact with an object, our senses provide us information about the object. If we disagree concerning a statement about the object, we can resolve that disagreement through a process of interaction with the object (and with each other). Therefore, objective knowledge is possible.
c. Man is the measure of all things. When you and I interact with an object, that object is to each person as it appears to that person. If we disagree concerning a statement about the object, that disagreement is due to our different perspectives. Therefore, knowledge is subjective.

Imagine you get to decide which philosophical position is adopted to deal with the following situations. Which one would you apply to each?

- You arrive in a country that is new to you. You have to decide which side of the road to drive on.
- Two engineers are evaluating the design of an aircraft's wing. One argues that the wing will deflect excessively under the required load. The other argues that the design is stiff enough as it is.
- Two people test drive a car together. One says that the car is excellent and represents a terrific value at its current price. The other says the car is not very good and substantially overpriced.
- Two engineers from a company in North America are discussing a new design method currently being used in another continent. One argues that the method has been very successful and should be adopted by their company. The other argues that the method is flawed and will not work.

26.6 Read [11] and [12], paying particular attention to the validation frameworks they describe? Provide a comparative analysis of the two frameworks. What are the novel elements that Pedersen et al. contribute beyond the framework previously laid out by Schön and Argyris? Which framework do you find more useful? Why?

26.7 Consider the description of a model-based approach to design method validation in section 26.6. In particular, consider the suggestion that ". . . Design scenario simulations can be used much in the same way that animal models are used in medical research. First, a large number of replicates are made representative of a class of engineering design scenarios. Different design methods are applied to this population of design scenarios. The outcomes of the design scenarios are recorded and analyzed . . ." Does this description match the general procedure used by Olewnik and Lewis in chapter 27?

26.8 Table 26.1 lists the fractional factorial design 2^{7-4}. A key property of the design is that every column of the array is orthogonal to every other column. Two columns are orthogonal if the inner product of the two columns is zero. Select a pair of columns and verify that this property holds.

26.9 Imagine an engineering system has inputs A, B, C, D, E, F and G and that the response of the system is $y = A + B + F + G + 0.5AF$. Simulate the completion of a fractional factorial design 2^{7-4} by computing the resulting response for each row of the array in Table 26.1 (substitute the value $+1$ or -1 from each column into the associated variable in the equation for y). Then compute the estimates of the main effects based on the data. This step can be done using software such as Minitab or by averaging the response across all the rows for which each variable takes a value of +1 and subtracting the average of the response across all the rows for which each variable takes a value of –1. Now, relabel the columns in Table 26.1 in reverse order and repeat the process. Explain the results you observed.

26.10 A system has n significant main effects and zero insignificant main effects. How many significant three-way interactions is the system likely to have under the assumptions listed below? Please round the number of interactions to the nearest integer.

a. $n = 7$ and $p_{111} = 5\%$
b. $n = 7$ and $p_{111} = 15\%$
c. $n = 20$ and $p_{111} = 5\%$

26.11 Imagine that warrantee costs associated with a product are proportional to the standard deviation of the product's response y caused by variation of noise factors. Imagine that the estimated warrantee costs for a proposed design are \$5M (give or take \$1M). Roughly how much would you be willing to invest in a robust design method that cuts the standard deviation by 50% on average, but has an inter-quartile range of realized reduction in standard deviation of 40% to 60%?

REFERENCES

1. Osborne, A., 1979. *Applied Imagination*, Scribners, New York, NY.
2. Altshuller, G., 1984. *Creativity as an Exact Science*, Gordon and Breach, New York, NY.
3. Pugh, S., 1991. *Total Design: Integrated Methods for Successful Product Engineering*, Addison-Wesley, Reading, MA.
4. Phadke, M., 1989. *Quality Engineering Using Robust Design*, Prentice Hall, Englewood Cliffs, NJ.
5. Wu, C. F. J. and Hamada, M., 2000. *Experiments, Planning, Analysis, and Parameter Design Optimization*, John Wiley and Sons, New York, NY.
6. IEEE, 1998. *IEEE Standard for Software Verification and Validation, Std. 1012-1998*, IEEE Inc, New York, NY.
7. *U. S. Federal Food, Drug, and Cosmetic Act*, Chapter 9.V, Sec. 355(d), http://www.access.gpo.gov/uscode/title21/chapter9_.html.
8. Audi, R., ed., 1995. *The Cambridge Dictionary of Philosophy*, Cambridge University Press, Cambridge, U.K.
9. Schön, D. A., 1983. *The Reflective Practitioner: How Professionals Think in Action*, Basic Books, New York, NY.
10. Argyris, C., 1991. "Teaching Smart People How to Learn," *Harvard Bus. Rev.*, Reprint No. 91301.
11. Schön, D. A. and Argyris, C., 1975. *Theory in Practice: Increasing Professional Effectiveness*, Jossey-Bass, San Fransisco, CA.
12. Pedersen, K., Emblemsvag, J., Bailey, R., Allen, J. K. and Mistree, F., 2000. "Validating Design Methods & Research: The Validation Square," *Proc., ASME Des. Engrg. Tech. Conf.*, Baltimore, MD.
13. Todd, P. M. and Gigerenzer, G., 2003. "Bounding Rationality to the World," *J. of Eco. Psych.*, Vol. 24, pp. 143–165.
14. Institute of Electrical and Electronics Engineers, 1998. *IEEE Standard for Software Verification and Validation*, IEEE Std 1012-1998.
15. U. S. Federal Food, Drug, and Cosmetic Act, Chapter 9.V, Sec. 355(d), http://www.access.gpo.gov/uscode/title21/chapter9_.html.
16. von Neumann, J. and Morgenstern, O., 1953. *Theory of Games and Economic Behavior, 3rd Ed.*, Princeton University Press, Princeton, NJ.
17. Drake, A.W., 1988. *Fundamentals of Applied Probability Theory*, McGraw-Hill, Inc., New York, NY, pp. 1–277.
18. Kunert, J., Corinna, A., Erdbrügge, M. and Göbel, R., 2006. "A Comparison of Taguchi's Product Array and the Combined Array in Robust-Parameter-Design," *Proc., 11th Annual Spring Res. Conf. (SRC) on Statistics in Industry and Tech.*, Gaithersburg, MD, (Accepted to *Journal of Quality Technology*).
19. National Human Genome Research Institute, National Institutes of Health, http://www.genome.gov/glossary.cfm?key=animal%20model.
20. Chipman, H. M., Hamada, M. and Wu, C. F. J., 1997. "Bayesian Variable-Selection Approach for Analyzing Designed Experiments With Complex Aliasing," *Technometrics*, 39(4), pp. 372–381.
21. Frey, D. D., Engelhardt, F. and Greitzer, E. M., 2003. "A Role for One-Factor-at-a-Time Experimentation in Parameter Design," *Res. in Engrg. Des.*, 14(2), pp. 65–74.

CHAPTER

27

DEVELOPMENT AND USE OF DESIGN METHOD VALIDATION CRITERIA

Andrew Olewnik and Kemper E. Lewis

27.1 INTRODUCTION

Many researchers have embraced the decision-based design (DBD) approach and the need for concurrent design methods within industry has helped the growth of the DBD research community. In the wake of this growth, a number of new design decision support (DDS) frameworks/methodologies have been introduced in recent years [1–9], many of which are used specifically for concurrent decision-based engineering applications. This list is by no means exhaustive, and include: quality function deployment (QFD) [10], specifically, the house of quality (HoQ) [11], Pugh's concept selection matrix [12], scoring and weighting methods, analytical hierarchy process (AHP) [13], multi-attribute utility theory [14, 15], physical programming [16], Taguchi loss function [17] and Suh's axiomatic design [18]. The ultimate goal in all these methods is to lead the engineering designer to a final, "best" design. The difference is that each method has a unique way of defining "best".

The seemingly endless list of design-decision methodologies above goes a long way in proving that "a lack of agreement still exists on the exact implementation of DBD in engineering design" [9]. However, it could be suggested that there may never be one all-encompassing decision support methodology considering that companies that make use of such tools have different objectives and philosophies. Nevertheless, there should still be some criteria by which to judge these proposed decision support tools to ensure that their use will consistently yield the correct decision, i.e., that these methods are valid.

In this chapter, validation criteria for design methods, specifically as they are used to promote design decisions, are introduced. Upon introduction of those criteria, they are applied to a well-known and increasingly utilized design method, the HoQ. Through application of the validation criteria to the HoQ, it is possible to uncover limitations of the HoQ in supporting design decisions and promote discussion on the impact such limitations can have on the design process.

27.2 THE ROLE OF VALIDATION

Shupe et al. [19] are often cited with formalizing the paradigm of DBD, in which the fundamental premise is that "the principal role of the designer is to make decisions" [20]. However, while the DBD philosophy has provided a formal foundation for this approach to the design process, portraying engineering design as a decision-making process goes back at least as far as Tribus in the 1960s [21]. In the preface of his text, Tribus states that "the purpose of this book is to provide a formal means for making engineering design decisions." So applying fundamentals of decision theory to the engineering design process is a notion that has been around for quite some time. The formalization of this notion in the DBD philosophy and work under this perspective in the last 15 years has seen the number of researchers who prescribe to the DBD line of thinking grow dramatically.

Given the development of the DBD paradigm and the subsequent growth in DDS models born of this philosophy, now more than ever, there is a need to develop criteria to validate those models. Many researchers and the DBD community at large are aware of this need, as the topic of such validation has come in vogue in recent years. Many methodologies and models utilized in a DDS role have come under scrutiny for flaws in their fundamental mechanics or assumptions [22–25]. Specifically, Barzali [22] and Saari [25] showed the problems associated with pairwise comparisons and the conflicting decision results generated with methodologies that use such comparisons, like the AHP. Hazelrigg [23] and Olewnik [24] review the validity of other popular design decision tools and discuss criteria for the validation of such tools in general. Understanding and classifying design models [26], specifically the topic of validation of those models with respect to engineering design, is growing in importance both pragmatically [23] and philosophically [27]. The need for validation extends from the physical models utilized by designers [28] to validation of design practices like robust experimental design [29]. The focus in this research is related to the validation of design methods intended to support decision-making in the design process.

Validation in the context of DDS models and methods is vital because of the intended end-users: designers in industry. It is unlikely that designers in industry will have a background in decision theory. However, the development and utilization of methods/tools/models that are built upon such theory is likely to grow. Take, as an example, the implementation of a DBD style process at Praxair, Inc. [30], which relies on tools such as the HoQ and Pugh's concept selection matrix, to help designers make decisions in the development of new products and processes. Further examples of design-decision methods include work at Ford [31], J.D. Power [9] and Caterpillar [32]. The expectation that designers utilize design-decision methods that they neither created nor studied extensively provides obvious motivation for validation criteria. In the next section, the validation criteria for such methods are introduced.

27.3 THE VALIDATION CRITERIA

The goal in the development of validation criteria for DDS models is to balance recognized components of normative decision theory [33] and the cognitive concerns encountered by behavioral researchers who study decision-making processes [34]. In achieving this balance, the validation criteria draw heavily on the elements of decision theory laid out by von Neumann and Morgenstern [33] in order to provide a means of understanding the limitations of DDS models given their prescriptive nature and the cognitive decision concerns that designers bring to the design decision process. The elements that are critical to decision theory and choice problems are: (1) options; (2) the potential outcomes (or possibilities) associated with the options; (3) probabilities (or realizations) of each potential outcome; and (4) a measure of value for making choices based upon preferences of the decision-maker(s). With reference to the decision theory elements the validation criteria are introduced from the aspect that design-decision models should:

(1) **Be logical:** This simply means that the results that come from the model make sense with intuition. Testing for this can be accomplished by using test cases for which the results are intuitive and checking if the model results agree with intuition [35]. This is easier said than done and may not be immediately apparent when one considers some of the current models utilized. However, decision support methodologies should be constructed under the assumption that changes may need to be made in the future in order that they agree with logic. For example, work by Arrow [36] showed that combining transitive preferences of individual decision-makers can result in intransitive preferences if those decision-makers attempt to group their individual preferences. Looking at a method like "group HEIM," however, one finds that designers and design researchers have found a way to overcome such logical inconsistencies in order that designers may indeed group their preferences to arrive at a compromise decision [37].

This criterion is an obvious one, much as the presence of "options" is an obvious necessary component of a choice problem. Another important aspect of this criterion, however, is that the use of a particular design decision method makes sense with a company's design philosophy or infrastructure. With the multitude of design decision methods that have been developed and will continue to be developed, it is necessary for designers to utilize tools that are appropriate for their design scenarios. If adopting a particular design model has a high "cost" (education cost, implementation cost, etc.) associated with it, it defies logic to implement the model simply because it has worked for other companies.

(2) **Use meaningful, reliable information:** Any model utilizes information. In engineering design, "information enables us to make effective choices." [35]. The information that is incorporated into any design decision model should be meaningful in the sense that it provides insight into interdependencies among system variables or input-output relationships.

To be reliable, the information should come from appropriate sources [38]. An appropriate source refers to "people in the know." For instance, if information regarding a potential product market is needed, then someone with marketing expertise should be consulted. Another important consideration regarding the reliability of information is the level of uncertainty associated with it. Understanding the uncertainty in information leads to a better understanding of the possible errors in the achieved results and gives a feeling for the level of confidence one can have in the results. So much detailed analysis goes into the science of engineering design (experiments, analysis, etc.); thus, all the information used in the models that facilitate design decision-making should have the same rigorous foundation and not an arbitrary origin. It is difficult to imagine overcoming the conflict associated with design decision-making if the information used to process decisions is neither meaningful nor reliable. An example of potential flaws resulting from violation of this criterion is presented in the HoQ analysis later in this chapter.

(3) **Assess probability of success:** There should be some attempt to quantify the probability that a particular option will be realized, or the probability of achieving expected performance in the specifications of a concept. In every decision there is some analysis of being successful with a particular choice, though it may not be quantified. An example is crossing the street. A person does not actually assign a probability to being successful in crossing the street, yet when the choice is made to cross the street, the person has analyzed his/her chance of crossing safely and assessed it to be relatively high.

In engineering design, there should be a quantification of probability of success built into the model. The question the designers should be asking of each option is: "Can this option be realized and how confident are we of this realization?" This means applying a probability estimate to the realization of concepts in order to make design decisions. Of course, early in the design process (e.g., during concept selection), such probability assessments are likely to have a relatively high level of uncertainty associated with them. However, such information should still be a consideration before design decisions are processed. An example of a probability of success assessment, which some companies do perform, comes in the form of a risk assessment like "failure mode and effects analysis" (FMEA) as is done at Praxair [30]. Such an assessment is done multiple times throughout the design process as the uncertainty of information necessary for processing the FMEA improves. Using such a tool, even early in the process, allows designers to rule out particular options due to high risk of failure in necessary life cycle. Adding a probability assessment to some often-used design decision models, even at a high level, would be a useful addition that would better align the models with accepted decision theory.

(4) **Not bias the designer:** No matter the methodology, the preferences of the designer utilizing the methodology should not be set by the method itself. Forcing a preference structure on the designer parallels the notion that the process used in decision-making can influence the outcome, as shown by Saari [25]. Rather than imposing preferences, the decision method should allow the designer(s) to use their own set of preferences, which may change over time. Changing preferences are seen all the time in industry, as companies constantly change their goals and philosophies to remain competitive in ever-changing markets.

(5) **Provide a sense of robustness in results:** In the end, it is likely designers will have a rank order of their options. For Hazelrigg, this is a criterion of validation in itself [23]. Here,

however, it is desired that the design decision model provide a means of understanding how sensitive the rank order is to change. The need for such a criterion is important, especially early in the design process where uncertainty in information used in any model is relatively high. Such uncertainty should limit the confidence designers have in the results generated through the method until the uncertainty in information is decreased. The effect of such uncertainty could change the rank order of options; thus, as designers utilize any design decision model they should be aware of such a possibility.

The remainder of this chapter is spent investigating the HoQ under the validation criteria introduced above. However, it is worth noting at this point that the development of these five criteria for validation of design decision methods is not an arbitrary one. Their development represents an evolution of understanding in terms of decision theory and cognitive concerns of human beings that complicate the decision-making process. Discussion of this evolution and the perspective from which these criteria are developed can be found in [39]. Those interested in discussion on validation criteria for decision methods in general and design decision methods specifically should also seek work from [40] and [23], respectively.

27.4 APPLICATION OF THE CRITERIA: VALIDATION AND THE HOUSE OF QUALITY

By now, most people working in engineering design are aware of the management philosophy known as QFD and the primary tool of the philosophy, the HoQ [11]. At its root, HoQ is a conceptual tool for mapping attributes from one phase of the design process to the next. Referring to Fig. 27.1 as one representation of the design process, an example might be to utilize the HoQ in order to convert a set of "process design" specifications to "manufacturing" specifications in order to produce a particular product. The conceptual mapping provided by the HoQ within the design process is the transfer of information (arrows in Fig. 27.1) from one node of the design process to the next. This conceptual mapping allows a clear flow of information on a node-by-node basis in the design process from the identification of "perceived need" node all the way through the "manufacturing" node. This is a valuable tool in helping understand the role of different entities (management, engineering, marketing, etc.) and the general flow and type of information within the design process of Fig. 27.1. However, there is a serious deficiency in the HoQ with potential to affect decisions so early in the design process that later failures in the design or success of the product are unlikely to be traced to this issue. This deficiency results from the attempt to specify quantitative relationships in the mapping of customer attributes to technical attributes, i.e., mapping from the "perceived need" node to the "specification" node in Fig. 27.1. This deficiency is related to two of the validation criteria discussed in the previous section and the focus of this section is to discuss this deficiency and to explore its effect on design decision-making.

QFD began as a management tool in Japan in the early 1970s [11] and in short time became popular within industry in North America at companies like General Motors [11], Ford [11], Xerox [41] and many smaller firms [42]. QFD's main component, the HoQ, is utilized as both a stand-alone tool, as exemplified in [43], and as a tool integrated in larger design processes, as in [30], to support product and process design. With such far reaching use and application, the HoQ might be assumed a fundamentally valid design tool.

While a valid decision process does not guarantee desirable outcomes, a flawed decision process confounds information used in the decision process and the process itself, leaving no means of identifying what is at the root of the bad outcome, the information or the process. The validity of the HoQ and QFD in general has been challenged in [24] and [23], respectively. The main focus here is to explore the HoQ under the validation criteria described previously and the confounding of information and the process that occurs in the HoQ. Specifically, HoQ is found to violate two of the validation criteria, as it does not *use meaningful, reliable information* and it does not *provide a sense of robustness in results.* In subsequent sections, these claims are supported to some degree through empirical means. Further, an experimental exploration of the HoQ mechanics is performed to add rigorous support to these claims and to provide a classification of the HoQ as a *qualitative* tool that represents itself in a *quantitative* manner, which is potentially a dangerous representation for designers.

27.4.1 Background on the House of Quality

To support this exploration and subsequent discussion it is necessary to provide a brief background on the mechanics of the HoQ. Besides a conceptual mapping, the HoQ also functions as a model for understanding how attributes in one design node affect attributes in the subsequent design node. Consider Fig. 27.2, which shows a standard HoQ as taken from [44] and provides obvious explanation for its reference as a "house." The Customer Attributes (CAs) represent what the customer wants in the product. CAs are posed in customer language. The Technical Importance section represents the weight the customer assigns to each CA. The Customer Ratings section represents the customer's perception of how well a current

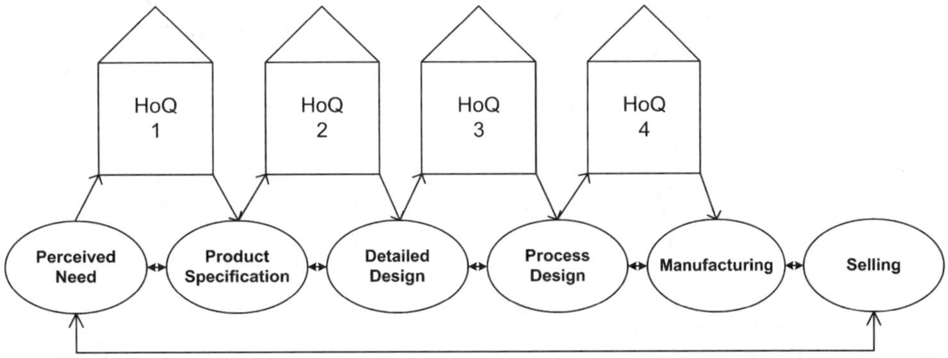

FIG. 27.1 THE DESIGN PROCESS AND THE HoQ

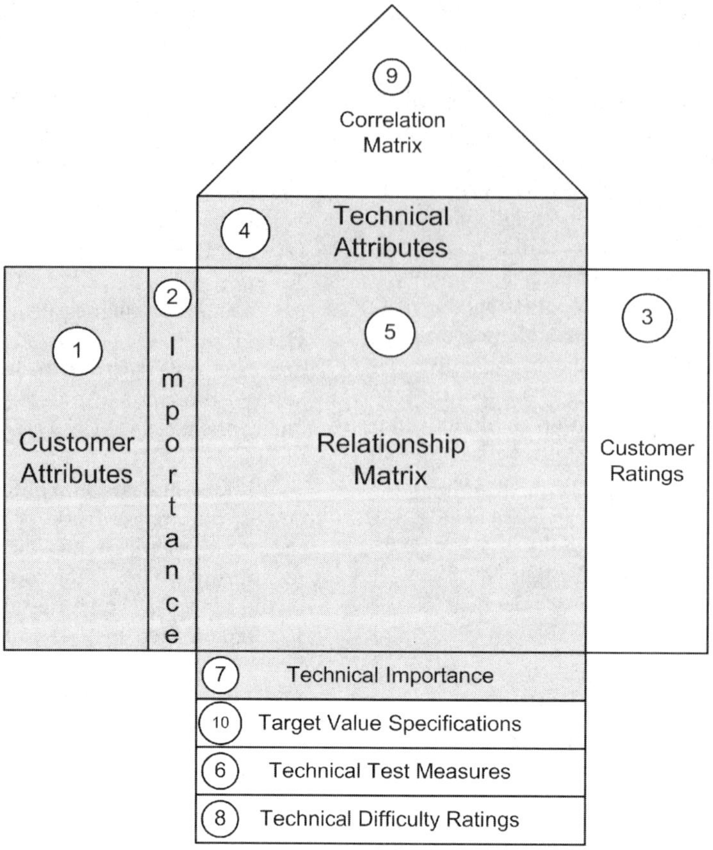

FIG. 27.2 STANDARD HoQ

product performs on each CA. The ratings may also compare competitor products. Technical Attributes (TAs) represent the product characteristics necessary to meet the CAs. The TAs, however, are in engineering design language. The Relationship Matrix is where relationships between CAs and TAs are identified and given a "weak," "medium" or "strong" relationship value. The Technical Test Measures and Technical Difficulty Ratings sections represent designer evaluations among the TAs. Target Value Specifications represent the target level the designers want each TA to reach. The Technical Importance section contains the calculated importance of each TA, which is a function of the Importance values and the values in the Relationship Matrix. Finally, the Correlation Matrix represents a matrix of the interrelationship among TAs.

Taking our starting point as the beginning of the design process in Fig. 27.1, the goal is to translate the "fuzzy voice of the customer" into measurements in the company language [45]. The steps to follow to complete this "translation" are provided by Breyfogle [44]. These steps are labeled in the HoQ of Fig. 27.2 and are as follows:

(1) Make a list of customer attributes. This list is usually identified through customer interviews and/or surveys.
(2) Identify the importance of each customer attribute. This information is also determined from customer surveys.
(3) Obtain customer ratings on existing design and competitor design.
(4) Designers compile a list of technical attributes to meet the customer attributes. These attributes should be scientifically measurable terms that can be assigned target values [44] and designers should avoid concept specific terms [45].
(5) Relationships should be identified in the relationship matrix and assigned qualitative value (weak, medium, strong). These qualitative relationships are later replaced by a quantitative three-number scale.
(6) Technical tests should be performed on existing design and competitor designs to gauge objective measures of difference.
(7) Importance of each technical attribute should be calculated in either absolute values or relative weights. This is done using Eqs. (27.1) or (27.2), respectively, where there are m CAs and n TAs and w_i represents the customer importance for the ith CA.

$$\text{raw score}\Big|_{j=1}^{n} = \sum_{i=1}^{m} \text{score}_{i,j} \times w_i \quad \text{Eq. (27.1)}$$

$$\begin{matrix}\text{relative}\\ \text{weight}\end{matrix}\Bigg|_{j=1}^{n} \frac{\text{raw score}_j}{\sum_{j=1}^{n} \text{raw score}_j} \quad \text{Eq. (27.2)}$$

(8) Difficulty of engineering each TA should be assessed.
(9) The correlation matrix should be filled out.
(10) Target values for each technical attribute should be set. This may be based on customer ratings from step 3.
(11) Select TAs to focus on based upon technical importance calculations of step 7 and technical difficulty assessment of step 8.

These primary steps can be carried out in subsequent HoQs used between other stages of the design process. These are the steps for a standard HoQ. Of course simplified and more complex

HoQs can be constructed depending on the designers and company utilizing the tool. With this essential background in mind, discussion can turn to the violations of the validation criteria and the resulting limitations previously described.

27.4.2 The House of Quality and the Designer

The HoQ is most commonly applied between the "perceived need" and "product specification" nodes of the design process, i.e., the phase described specifically in the steps described above. The role of the HoQ here is critical, as it is meant to model the relationship between the customer attributes of a product and the technical attributes of the product. This "language translation" and subsequent characterizations made about the importance of technical attributes based upon that translation is vital to the potential success of the product. That is, the HoQ model is meant to identify the most important technical attributes, i.e., help designers *decide* which technical attributes are most critical. As long as those technical attributes are the center of the product design decisions, the customer attributes will be satisfied to a level that makes the product desirable and ultimately successful. On a conceptual level, the fundamental mechanics behind the HoQ are well-suited to this goal; however, there are two complexities that arise in the implementation of the methodology that raise suspicion about the ultimate value of the results. To investigate these difficulties, it is beneficial to first discuss the implicit assumptions behind the HoQ model as it is implemented between the first two nodes of the design process in Fig. 27.1.

To aid this discussion of model assumptions, a reduced HoQ is shown in Fig. 27.3 (including only the gray components of Figure 27.2). This section of the house represents the components necessary to support discussion and empirical study in this paper. Note that the representation shown in Fig. 27.3 is representative of the form of all examples used in the paper. In order to fill out this HoQ, only a subset of the 11 steps described are necessary. Those steps are (1, 2, 4, 5, 7, 11). The assumptions behind these steps are critical to the results of Technical Importance.

- The first assumption is that the CAs and their individual importance (steps 1 and 2) are truly representative of the potential customer base. The validity of this assumption is a matter of marketing study and not contended herein. However, confidence in these two components is paramount.
- The second assumption is that the TAs (step 4) are the appropriate, measurable product characteristics to meet the CAs. This assumption might be considered the very crux of design. That is, the ability to understand perceived need and convert that understanding into a product or system seems to be the most basic function of a designer. Consider that one definition of design is "to create or contrive for a particular purpose or effect" [46]. Thus, to take issue with this assumption would be to take issue with the fundamental notion of design. That brings us to the third and fourth assumptions (step 5).
- The third assumption is that designers can indeed identify when a particular TA relates to a particular CA and the qualitative strength of that relationship, i.e., "weak," "medium" or "strong." It is likely that for the most part designers will be able to identify the existence of relationships, especially since they generated the TAs. However, it is possible that some "weak" relationships could be missed due to their subtle nature. The importance of this assumption will become evident in the experimental study. It is also reasonable to believe that designers can distinguish the qualitative level of the relationships.
- The fourth assumption is that the designer can later assign quantitative values to represent the qualitative level of identified relations and, further, the quantitative values are always the same. As suggested by Breyfogle [44], an example quantitative scale to utilize might be 1 for "weak," 3 for "medium" and 9 for "strong." However, Breyfogle indicates that this is an example possibility, not necessarily *the* scale to use. It is also not dictated that one quantitative three-number scale be used. Instead the choice of quantitative scale(s) is left up to the designers. It is worth noting, however, that throughout the literature on QFD and the HoQ the most common scale seen is (1-3-9) and that only one three-number scale is typically used in any given example.

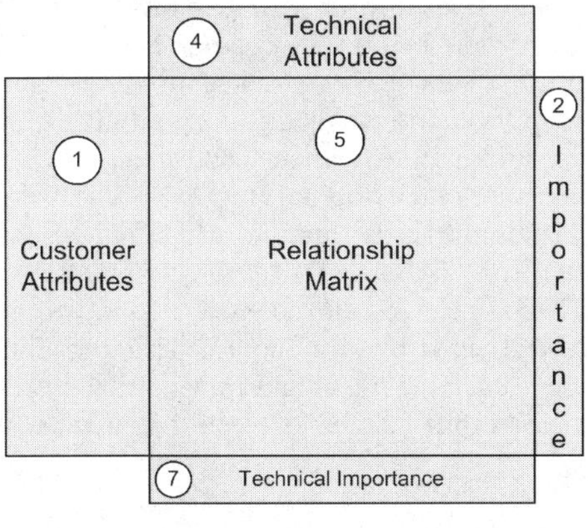

FIG. 27.3 REDUCED HOQ FOR CHAPTER DISCUSSION

All these assumptions of course provide sources of potential flaws that may result from the HoQ model. However, the potential for flaws as a result of assumptions is a difficulty associated with any model used in design; therefore, most of these assumptions are acceptable if not necessary altogether. The fourth assumption, however, is viewed as a critically flawed assumption that can result in disastrous design decisions depending on the level of credence the designers lend to the quantitative results of the HoQ model. This assumption provides a starting point for discussion with regard to the violated validation criteria and as motivation for empirical investigation of the HoQ.

The use of one three-number scale is an understandable simplification in the HoQ model. That is, it would be confusing and difficult for designers to try to apply multiple three-number scales throughout. For example, using (1-3-9) across one row of CAs and (2-5-8) across another. However, this simplification points to the larger issue, i.e., that designers have no reason to choose a particular quantitative relationship represented by a three-number scale. The fact that the scale consistently appears as (1-3-9) in the literature is further suggestive of this. The assumption that designers can choose an appropriate scale means that they know ahead of time both the range on which the relationship scale lies and the relative difference between weak, medium and strong. Put another way, this assumes the designers can put a quantitative value to reflect how a given TA will affect the perception of customers. It is difficult to accept that designers could indeed make this kind of assessment, yet this is exactly what they must do to generate the final "Technical Importance" as per step 7 from Breyfogle [44]. This assumption is the primary deficiency that leads to the violations of the validation criteria for design decision methods previously discussed. These violations are explored and proven through experimentation and empirical study of an HoQ example beginning in the next section.

Customer Attributes	Technical Attributes									Customer Importance
	Air Flow	Air Temperature	Balance (Torque)	Weight	Volume	Number of Parts	Physical Lifetime	Energy Consumption	Noise, Vibration, Electromagnetic Wave	
Dries quickly	9	9						9	9	9
Quiet	9							9	9	3
Operates easily			3	1					1	1
Operates safely	1	9	**3**				9		9	3
Comfortable to hold		1	9	9	9		3		1	9
Reliable	1	1				3	9	1	1	3
Portable				3	9					1
Energy efficient	9	9						9		9
	195	201	93	85	90	9	81	192	148	Raw score
	17.8%	**18.4%**	**8.5%**	**7.8%**	**8.2%**	**0.8%**	**7.4%**	**17.6%**	**13.5%**	Relative weight
	2	1	5	7	6	9	8	3	4	Rank

FIG. 27.4 EXAMPLE HoQ FOR DESIGN OF A HAIR-DRYER

27.4.3 Empirical Investigation of the House of Quality

Recall that the goal in conducting an empirical investigation of the HoQ is to explore the deficiency of assumption 4 discussed above and the validation criteria violations that are attributed to this assumption. To conduct this empirical investigation an example HoQ from the literature is utilized; it is shown in Fig. 27.4. The example in Fig. 27.4 is an HoQ for the design of a hair-dryer adapted from an example in [47]. The example represents an instance of potential design decision-making using the HoQ. The goal is to show how erroneous conclusions and decisions could be made regarding this product example due to the assumption of a quantitative relationship scale.

Consider how conclusions might be drawn from a given HoQ. Step 11 of the procedure for utilizing the HoQ suggests using the results from step 7, i.e., to look at the raw score (rank) or the relative weight calculated for each TA. The raw score, rank and relative weight are given for each TA in the example of Fig. 27.4. Designers must now draw conclusions based either on the ranked priority or the relative weight, as per step 11. The choice between using rank to prioritize and relative weights to make decisions provides several possible courses of action for designers. It is likely that every company that utilizes the HoQ has different approaches for handling this information, which may even change for each new design. To support the empirical investigation here, two possible approaches are discussed per each violation of the validation criteria.

27.4.3.1 A Lack of Meaningful, Reliable Information in the HoQ The first option is that the designers could utilize the relative weights to determine how resources should be allocated in the course of the design. Perhaps the designers could allow the relative weights to roughly influence the percentage of resources to spend in designing around each TA. The difficulty here, however, is that the designers do not truly know if the range and relative difference in the relationship scale is representative of the actual relationship between CAs and TAs. In this way, the quantitative scale information does not represent *meaningful, reliable information*. Thus, the relative weights could be potentially no better than those generated by some random process.[1] In order to investigate this idea, random processes that work within the framework of the HoQ were designed and used to "simulate" results. Three different random processes were compared to the results in the example of Fig. 27.4. The empirical results were generated as follows:

(1) **Insertion of discrete uniform random number:** In this recreation method, random numbers from a discrete uniform distribution (range 1 to 9) are inserted wherever a relationship exists in the original HoQ relationship matrix. The relative weight of each TA is calculated for each of the 1,000 recreated HoQs, and the average relative weight for each TA over all recreations is calculated. One thousand recreations are used to ensure that the true average of the possible permutations is represented. *The goal of this simulation is to observe whether using random numbers where relationships are known to exist yields results similar to the original HoQ example.* By recreating the HoQ in this way, the qualitative assessment (step 5 for filling out the HoQ) is lost.

(2) **Arbitrary insertion of a three-number scoring scale:** In this recreation method, a three-number scale consistent with the example is used. However, a score (zero, or another number consistent with the example scale) is arbitrarily inserted in the relationship matrix, without knowledge of where the actual relationships of the original HoQ exist. The controlling factor is a "column density" metric, which is calculated from the original HoQ for each TA (each column of the HoQ), using Eq. (27.3). This "column density" measures the percentage of CAs affected by any one TA. Using a uniform random number generator [0, 1] and moving down each column of the relationship matrix for each TA, a random number is

[1]This notion was put forth in open discussion at the Decision-Based Design Open Workshop held at the 2002 DETC.

generated and if it is greater than the column density for that TA, a zero is inserted; otherwise, a relationship is assumed to exist and another random number is generated. If the number is less than one-third, a low score is inserted (1 for the hair-dryer example); if the number is greater than or equal to one-third but less than two-thirds, a medium score is inserted (3 for the hair-dryer example); and if the number is greater than two-thirds, a high score is inserted (9 for the hair-dryer example). Once this procedure is completed for every position in the relationship matrix, the relative weight for each TA is calculated for each HoQ recreation and the average relative weight for each TA over the total number of recreations is calculated. *The goal of this simulation is to observe whether reducing the certainty of where relationships exist and the quantitative level of that relationship yields results similar to those of the original HoQ.*

$$\left.\begin{array}{c}\text{column}\\\text{density}\end{array}\right|_{j=1}^{n} = \frac{\text{no. of CAs affected}}{m} \qquad \text{Eq. (27.3)}$$

(3) **Arbitrary insertion of a discrete uniform random number:** In a manner similar to the previous recreation method, this random process also uses the "column density" to control the number of relationship scores inserted for each TA. However, when a relationship is assumed to exist in this case [i.e., if a uniform random number (0, 1) is less than or equal to the column density], a discrete random number from a uniform distribution is inserted (range 1 to 9). Again, the relative weight of each TA for each HoQ recreation is calculated and the average relative weight for each TA over all recreations calculated. *The goal of this simulation is similar to the previous approach but the certainty of the quantitative level of the relationship has been reduced further since the quantitative scale is represented by a discrete uniform distribution rather than a single three-number scale.*

The distribution of relative weights that result from each of these three random procedures for the hair-dryer example of Figure 27.4 are shown in Figs. 27.5, 27.6 and 27.7, where the circle represents the actual relative weight from the original HoQ and the triangle represents the average of the distribution. Each of the nine distributions is representative of the nine TAs from the original hair-dryer HoQ. Note that the TAs listed left to right in the hair-dryer HoQ appear left to right and row by row in the figures. The averages of each random process are shown for the hair-dryer example in Table 27.1. Numerically, the average relative weights generated using the random procedures appear similar to the relative weights from the original hair-dryer HoQ. The Wilcoxon signed rank test, which can be used to test whether the median of a distribution is equal to a scalar value [48], gives no verification that the distribution means were the same as the actual relative weights (scalar values) in the original HoQs. So, while a random number generator does not behave exactly as the HoQ method, the numerical proximity of results is hard to ignore.

It is not the case that every resulting relative weight from the random processes is numerically similar to the original HoQ. For example, in the case of the "noise, vibration, electromagnetic wave" TA for the hair-dryer, the results of the random processes is quite different from the original HoQ relative to the other TA results. This appears to result from the high "column density" associated with this particular TA, making it difficult to replicate the original HoQ results reasonably with a purely random process. In other examples not explored herein, similar deviations can also be explained by a high dominance in "qualitative tendency," i.e., the dominance of "weak, "medium" or "strong" relationships for one TA (column) in a given HoQ.

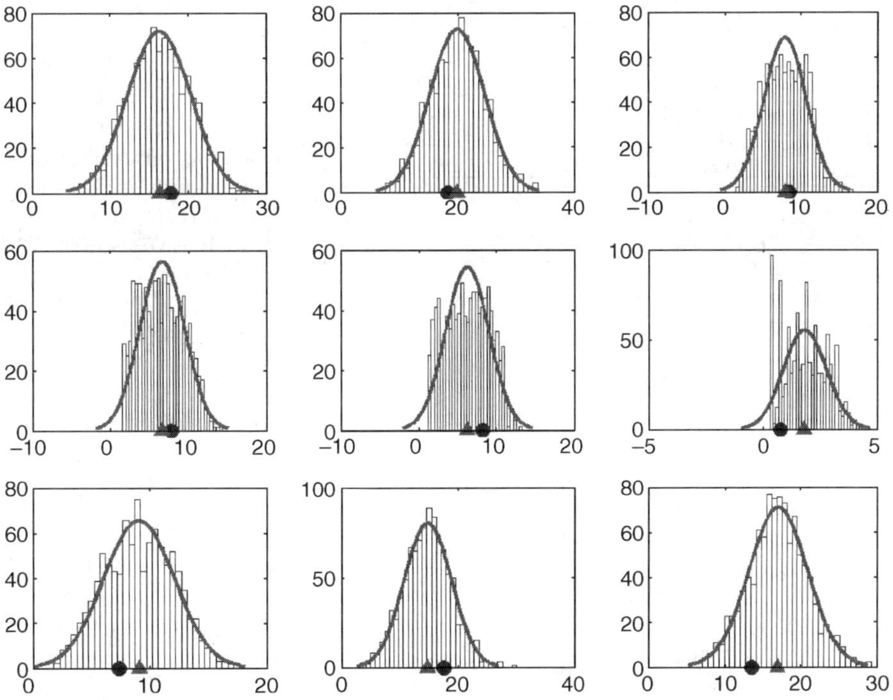

FIG. 27.5 RESULTS FOR HoQ RANDOM PROCESS 1

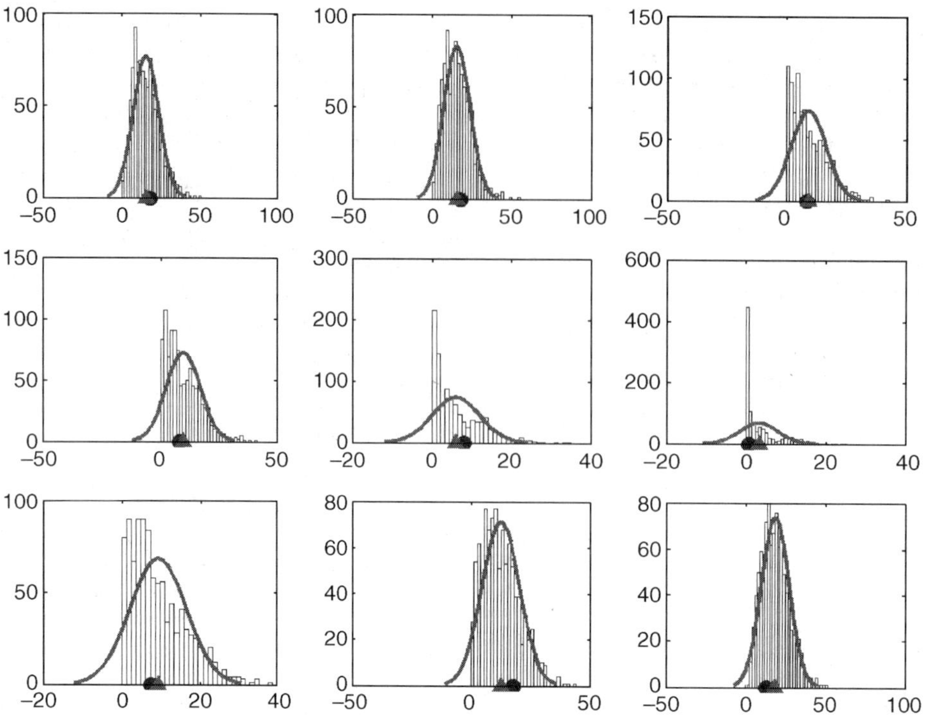

FIG. 27.6 RESULTS FOR HoQ RANDOM PROCESS 2

The results generated through treatment of the HoQ as a random process are not limited to the hair-dryer example alone. Similar results were found for other HoQ examples tested, which can be found in [39]. From the random process results there is a general consistency between the results of the original HoQ and random process approach 1. Any inconsistencies can typically be explained by extremes in "column density" and in other examples' "qualitative tendency," which provide interesting "factors" for exploring the mechanics of the HoQ experimentally. It is also interesting to note that in each of the examples explored above, random approaches 1 and 3 provide similar results in every case (except the refrigerator example) that also tend to be farther from the original HoQ results as compared to approach 1. This result implies the importance of knowing the location of relationships. Recall that relationships are not maintained in approaches 2 and 3, rather "column density" is used to control the number of relationships. This speaks to the

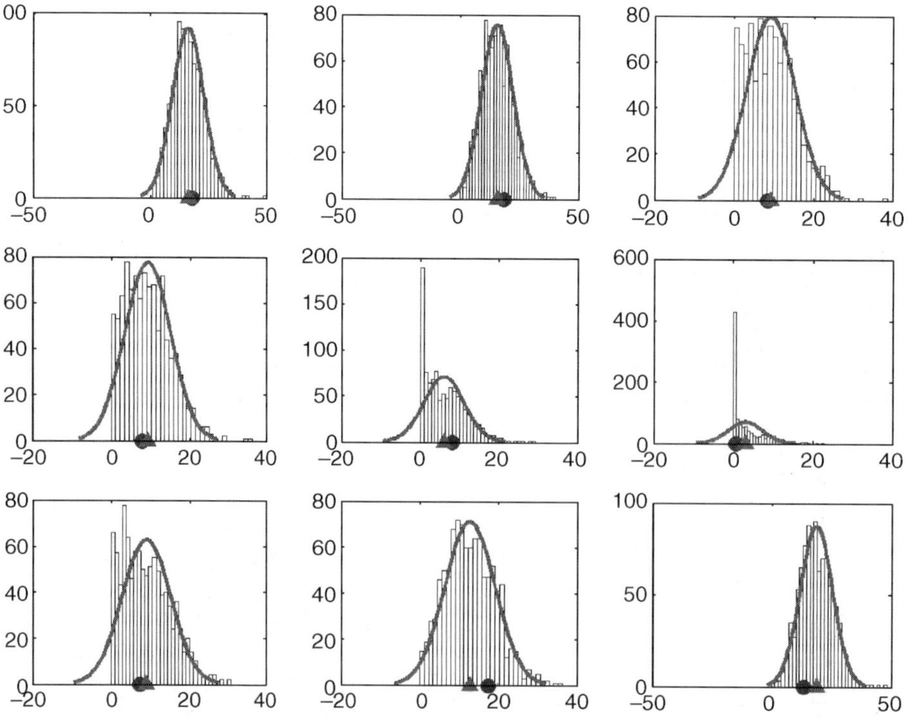

FIG. 27.7 RESULTS FOR HoQ RANDOM PROCESS 3

TABLE 27.1 AVERAGES FROM RANDOM PROCESS HoQs

Average Relative Weight %	Air Flow	Air Temperature	Balance (torque)	Weight	Volume	Number of Parts	physical lifetime	Energy Consumption	Noise, Vibration, Electromagnetic Wave
Original HoQ	17.8	18.4	8.5	7.8	8.2	0.8	7.4	17.6	13.5
Approach 1	16.4	20.0	7.9	6.7	6.2	1.9	9.1	14.8	17.0
Approach 2	15.5	15.8	9.3	9.4	6.1	3.3	9.2	12.6	18.8
Approach 3	15.9	15.7	9.3	9.3	6.1	3.2	9.1	12.9	18.6

obvious importance of knowing where relationships exist and also highlights the importance of having confidence in the importance specified by the customers for each CA.

The implication resulting from this lack of meaningful information in terms of the quantitative scale is the limited confidence designers should have in the relative weight for each TA that results from the HoQ model. This lack of confidence makes subsequent phases of the design process and the processing of design decisions based on the relative weights difficult and potentially wrong as a result of the process. For instance, after determining the relative importance of each TA, the designers must generate concepts and select one for detailed design. The concepts the designers generate and the one they ultimately choose is dependent upon the information (relative weights) that results from the HoQ model. The fact that these subsequent steps of the design process are based on meaningless information is a circumstance most designers (it is assumed) would prefer to avoid. Further, the primary motivation behind the HoQ methodology (designing based on the "voice of the customer") may be lost in subsequent steps of the process due to the meaningless information those steps are based upon.

The behavior of the HoQ being similar to random processes extends to examples of all size [39]. Though the evidence provided is anecdotal rather than statistical in nature, it should still raise concern among users of the HoQ method since the final results are not necessarily meaningful as presented in terms of relative importance. Further, the anecdotal evidence seen through this empirical investigation provides motivation for more statistically significant study of the HoQ, which is provided in subsequent sections. Now, however, the empirical investigation continues with reference to the other violated criteria.

27.4.3.2 A Lack of Robustness in HoQ Results The empirical investigation of the previous section assumed that the designers were looking at relative weights for the TAs that result from the HoQ model in order to compare and identify the most important TAs. In this section, it is assumed the designers utilize the rank order of the TAs to determine a subset of most important TAs to consider in the remainder of the design process. Referring again to the hair-dryer HoQ example of Fig. 27.4, note that the rank order of TAs from most to least important is provided at the bottom of the HoQ. In the case of the hair-dryer, "air temperature" is the most important TA and "number of parts" is the least important. If the designers are selecting a subset of TAs to focus on for the concept generation and selection steps to follow, it would be useful to know how robust this rank order is. However, the fact that the quantitative scale (e.g., 1-3-9) is used to differentiate perception of "weak," "medium" and "strong" rather than provide true information on relationships, implies that it is difficult to quantify or *provide a sense of robustness in results*; violating the fifth validation criteria.

To get a sense of the potential impact that a choice in quantitative scale has on the final rank order, consider the use of different scales. While it appears common practice in the literature to use (1-3-9) as the quantitative scale in the HoQ, there is no mandate that says this scale is *the* scale to use. In the steps described by [44], no quantitative scale is specified. This suggests that the quantitative scale is up to the designers using the methodology. If this is indeed the case, with no reason to suspect one scale is necessarily better than another, it would be useful to see how quantitative scale choice and uncertainty about the true quantitative relationship could affect rank order. Specific interest lies in potential drastic rank changes due to different scales or uncertainty in the scales.

Focusing first on the choice of quantitative scale, it is possible to see how a different three-number scale choice changes the TA rank order in any given example. Looking at the hair-dryer HoQ of Fig. 27.4, note how changing the scale choice affects the rank order of TAs shown in Table 27.2. The scales chosen are meant to represent other possible three-number scales in the range of 1 to 9 that could reflect designer perceptions of quantitative difference in qualitative relationships. Though it is clear that the rank of some TAs is robust to scale choice (e.g., "air flow"), it is also clear that some TAs can vary widely in the rank order depending on the scale choice. For the hair-dryer example: If the designers were looking for the four most important TAs, then the scale choice does not prove to be a problem as the top four ranking TAs remain so, independent of the scale choice. However, if the designers were looking for the five most important TAs, clearly the choice of scale would affect their top five. For example, using (1-3-5) rather than (1-3-9) makes "physical lifetime" the fifth most important TA rather than "balance," representing a rank change of three positions (from eighth place).

TABLE 27.2 EFFECT OF QUANTITATIVE SCALE CHOICE ON TA RANK ORDER

Quantitative Scale Choice	Air Flow	Air Temp.	Balance (Torque)	Weight	Volume	Number of Parts	Physical Lifetime	Energy Consumption	Noise, Vibration, Electromagnetic Wave
1,3,9	2	1	5	7	6	9	8	4	3
1,4,9	2	1	5	8	6	9	6	4	3
1,3,5	2	1	5	8	7	9	5	4	3
2,5,8	2	1	6	8	7	9	5	4	3

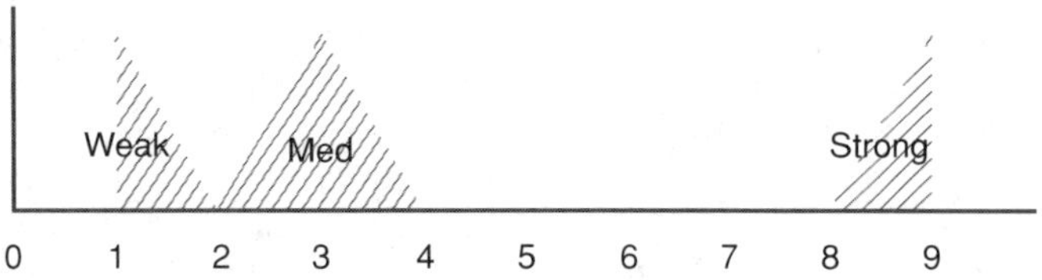

FIG. 27.8 1-3-9 SCALE WITH UNCERTAINTY

Even if a particular scale choice is assumed to be the best representation in the HoQ model, uncertainty in the "true" relationship value can lead to an inability to quantify the robustness of final TA rank order. Consider a possible example of uncertainty in relationship levels represented by the triangular distribution of Fig. 27.8 for a (1-3-9) scale choice. The uncertainty of the "true" relationship is represented by the triangular distribution around each value, where the vertex of each triangle represents the most likely value in a probabilistic sense.

Through the distributions, it is possible to see the potential final rank order of TAs through recreations of the HoQ examples. In each "recreation," the values in the original HoQ are replaced with values from the triangular distributions. For example, a relationship in the original HoQ identified as "weak" is replaced from a number drawn from the triangular distribution whose vertex is "1" in Figure 27.8. The resulting shift in rank order of TAs for the hair-dryer are shown in Fig. 27.9 for 100 recreations (simulations). Each line in the figures represents one of the TAs from the hair-dryer HoQ. Looking at the rank shifts shows how dependent some TA ranks are on uncertainty in the value of the relationship strength in the scale choice of the designers. For example, looking at Fig. 27.9, the hair-dryer TAs that rank between fifth and eighth place in the original HoQ are shown to change position often in the simulated recreations. This displays the lack of robustness of the rank order in these particular TAs due to the uncertainty in the quantitative scale. The implication for designers is that if they wanted to keep the five most important TAs, they would not truly know which TAs are the five most important.

The results for rank shift of TAs due to scale uncertainty are similar to the lack of knowledge in the scale itself (e.g., 1-3-9 or 2-5-8?). In each case, the true rank order is unknown, thus there is *no sense of robustness in results*. The implication in representing

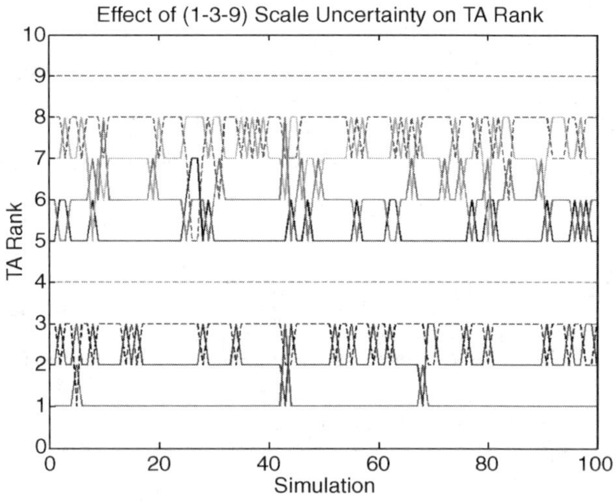

FIG. 27.9 TA RANK SHIFTS DUE TO SCALE UNCERTAINTY

hypothetical uncertainty in an assumed scale choice is that there is no way for designers to quantify the true uncertainty in the rank order of results. This difficulty arises because the quantitative scale choice is not a result of rigorous understanding of the true relationship between CAs and TAs, but rather an arbitrary scale selected to represent designer perceptions.

The empirical investigations presented here demonstrate the potential flaws of the HoQ as per the validation criteria. These flaws limit the HoQ model's ability to provide a true *quantitative* assessment of the relationship between CAs and TAs. This, in turn, limits the conclusions designers can draw about the importance of TAs, either relative to one another through relative weights or absolutely through rank. The empirical investigations suggest that, at best, the HoQ model is a qualitative assessment tool rather than a quantitative one. However, the fact that the HoQ relies on quantitative information (scale choice) as an input from designers and provides quantitative information as an output could lead designers to put too much confidence in the results. To avoid this situation, it is necessary to prove that the HoQ is a qualitative assessment tool rather than a quantitative one. This proof is provided through an experimental investigation of the mechanism of the HoQ from which the results are derived.

27.4.4 Experimenting With the House of Quality

The empirical results from the previous section suggest that the HoQ method should be limited to qualitative assessment. This limitation arises from the arbitrary nature of the quantitative scale, which inherently leads to results that are potentially meaningless and lack robustness. To support this conclusion, taking the empirical evidence and supporting it statistically is necessary.

In order to establish statistical evidence, an experiment based upon assumptions 3 and 4 necessary to fill out the HoQ model is conducted. Recall, assumption 3 is that the designers can identify where relationships exist in the "Relationship Matrix" and the qualitative nature of that relationship, and assumption 4 is that the designers can appropriately identify a three-number scale that captures the relationships quantitatively. Thus, an experiment would need to make factors that represented what the designers control from these two assumptions. The factors are identified as *column density, qualitative tendency* and *quantitative scale.*

To aid in describing the experiment setup, consider the HoQ of Fig. 27.10. The *column density* represents the number of CAs that a given TA effects and is calculated using Eq. (27.3). For example, the second TA in the HoQ of Fig. 27.10 has a *column density* of one-fifth, represented by the asterisk at the intersection of CA3 and TA2 in the relationship matrix.

Customer Attributes	Technical Attributes					Customer Importance
	TA1	TA2	TA3	TA4	TA5	
CA1			*		*	*
CA2			*	*	*	*
CA3		*		*		*
CA4					*	*
CA5				*	*	*
						Raw score
						Relative weight
						Rank

FIG. 27.10 EXPERIMENTAL HoQ

The *qualitative tendency* represents the most common qualitative relationship for a given TA. Thus, if a TA has a "weak" tendency it will have one more than half of the active cells in the column designated as "weak." For the example HoQ in Fig. 27.10, a TA with a column density of four-fifths and a "weak" qualitative tendency has at least three cells with a weak score inserted for calculation of technical importance. In the case of a tie in *qualitative tendency* (e.g., a TA column with one "weak," "medium" and "strong" relationship), there is some initial evidence that suggests using the lowest qualitative score to dictate the qualitative tendency for the TA column in question. However, this issue is still under study and is not a focus in the experiment discussed herein, as it is ensured that ties do not occur.

Finally, the *quantitative scale* is the three-number scale that is utilized to replace the qualitative relationships identified for calculation of the technical importance. It is important at this point to discuss the fundamental difference between the *qualitative tendency* and the *qualitative scale* factors. Looking at each of these factors individually through the experiment represents the differentiation of assumptions 3 and 4 that are integral to filling in the HoQ. Further, it is this differentiation between the qualitative and quantitative assessment of engineers that proves that the HoQ can support qualitative assessment, but **not** quantitative assessment. Such proof is shown in the experimental results.

The experiment was performed on a five-by-five HoQ similar to that in Fig. 27.10. To perform the experiment and study the effect of each factor, only one TA was varied on all three factors. In this case, TA1 of Fig. 27.10 was varied on all three factors. The levels and their corresponding values for each factor are shown in Table 27.3. A full factorial experiment was performed (using MATLAB) yielding 48 experiments (four levels × three levels × four levels), each representing a different HoQ configuration. The remaining TA columns were held constant in column density and are denoted by asterisks in Fig. 27.10. For example, TA3 has a constant column density of 2/5 in each design, thus two asterisks in the column. At each experiment, 500 simulations were performed allowing the relationship locations, score from current *quantitative scale* and customer weights (from a 1 to 5 rating scale) to be randomly selected (except for TA1 where the qualitative tendency controlled some of the score selections). Essentially, this treated these other components as noise in the experiment.

There is an obvious expectation for the effects of *column density* and *qualitative tendency*. Namely, any TA that affects multiple CAs, i.e., has a high *column density*, will naturally have a high relative weight and favorable rank position, since it will have more relationships with CAs than other TAs. Similarly, the more often a TA has a high *qualitative tendency*, the more likely it is to have high relative weight and improved rank position, since its quantitative scores will be higher than average. In the results of this experiment, as each of these factors increases for TA1, the relative weight should increase and rank should improve (first place is best). The primary goal then in this experiment is to study the effect of scale choice on these two importance metrics.

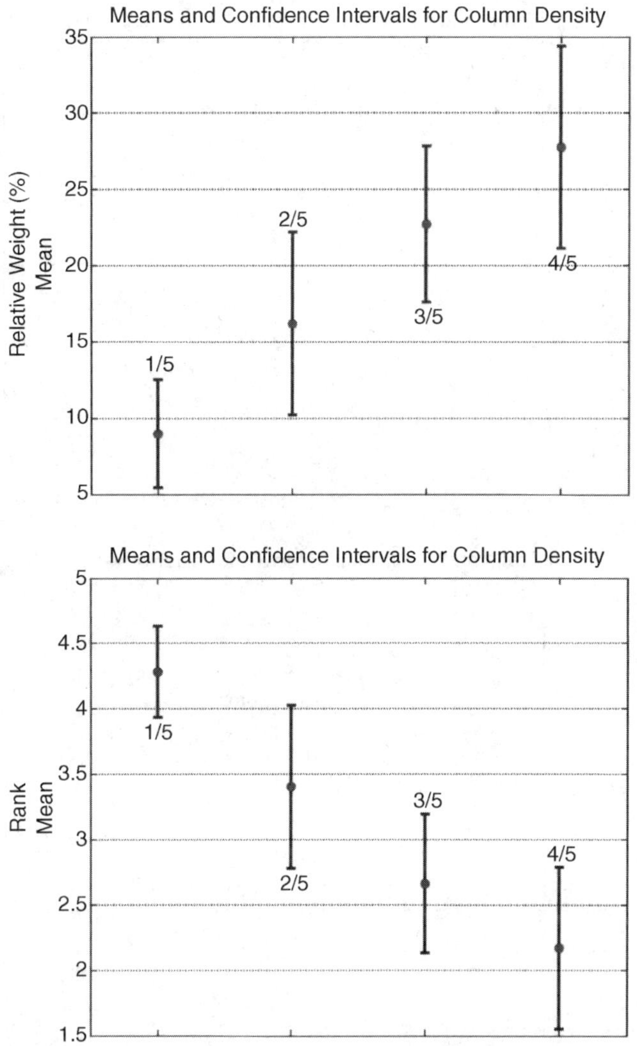

FIG. 27.11 EFFECT OF COLUMN DENSITY FACTOR IN 5×5 HoQ EXPERIMENT

Resulting main effects plots with mean and 95% confidence intervals for the rank and relative weight of TA1 are shown in Figs. 27.11, 27.12 and 27.13. The figures show the effect of a particular level for a given factor while the other factors vary on all levels. For example, in Fig. 27.11, when the level is "1/5" the *column density* for TA1 is held at this level while the other two factors (*qualitative tendency* and *quantitative scale*) vary over all levels. Thus, for this case when the *column density* is 1/5, the average relative weight (represented by the circle) is 9% as the other factors go through

TABLE 27.3 FACTORS AND LEVEL SETTINGS FOR HoQ EXPERIMENT

Factors	Column Density				Qualitative Tendency			Quantitative Scale			
Levels	1	2	3	4	1	2	3	1	2	3	4
Settings	1/5	2/5	3/5	4/5	Weak	Med	Strong	1 3 9	2 5 8	1 2 3	1 50 100

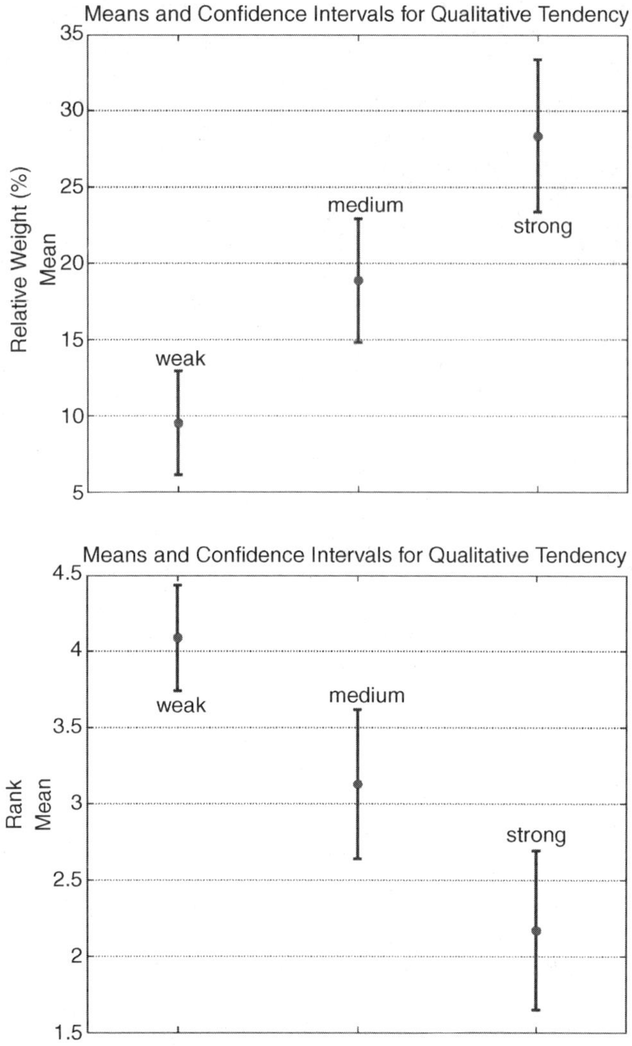

FIG. 27.12 EFFECT OF QUALITATIVE TENDENCY FACTOR IN 5×5 HoQ EXPERIMENT

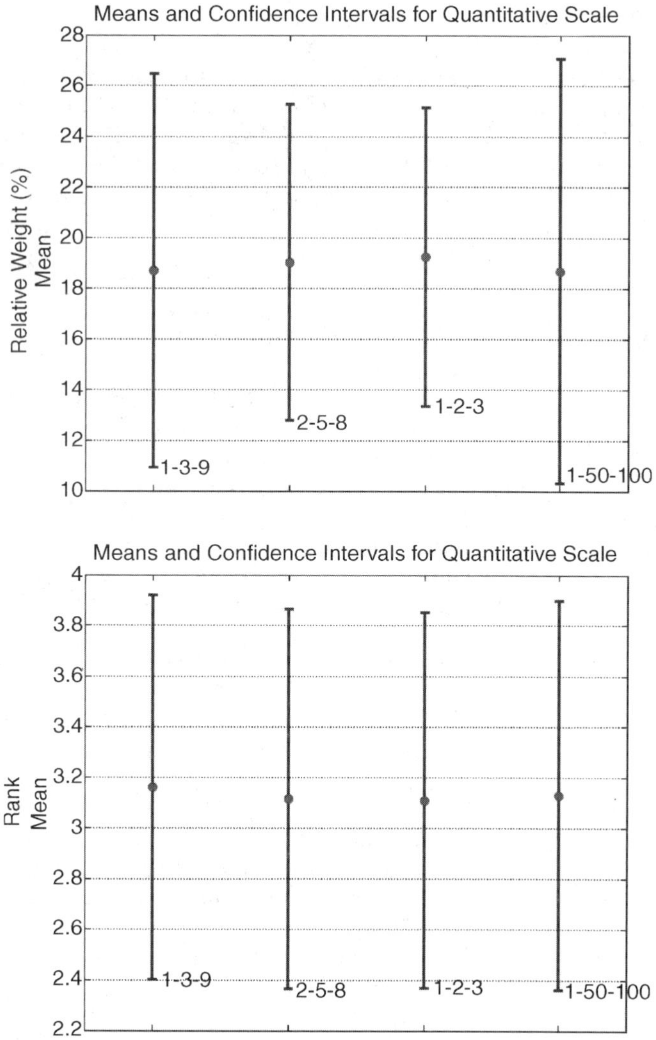

FIG. 27.13 EFFECT OF QUANTITATIVE SCALE FACTOR IN 5×5 HoQ EXPERIMENT

all 12 possible combinations (3 levels for *qualitative tendency* × 4 levels for *quantitative scale*) 500 times for each combination (to factor out noise as described above). The range represents the 95% confidence interval, essentially marking where the relative weight is expected to fall 95% of the time when the *column density* is 1/5. Similar explanation results for each factor as well for the rank order representation of TA1 results.

The value of looking at main effects plots like those of Figs. 27.11, 27.12 and 27.13 is that one can quickly assess which factors are important and which are not in an experiment. Note: Figs. 27.11 and 27.12 show the results expected for *column density* and *qualitative tendency*. As the *column density* increases in Fig. 27.11, the mean relative weight for TA1 increases from 9% to 27% and the mean rank decreases from 4.3 to 2.2. Similarly, as the *qualitative tendency* increases in Fig. 27.12, the mean relative weight increases from 10% to 28% and the mean rank decreases from 4.2 to 2.2. However, based on the results in Fig. 27.13, there is evidence that the choice of a three-number quantitative scale has no effect on the final relative weight and rank of a given TA. Effectively, the use of a three-number scale pushes the importance calculations to the expected average, in this case a relative weight of 20% and rank of three, for a given five-by-five HoQ.

The purpose of using unreasonable quantitative scales such as (1-2-3) and (1-50-100) was to show that even if the range is changed there is no effect on the mean and little effect on the confidence interval. These scales are thought of as "unreasonable" because they do not represent perceptual distinctions that a choice like (1-3-9) is intended to represent. For example, the relative difference between (1, 2 and 3) is so slight, it does little to differentiate "weak," "medium" and "strong" qualitative relationships that designers perceive. Similarly, (1-50-100) is too extreme in its relative difference. While the scales typically utilized are meant to reflect expert knowledge, they are nothing more than the designers' best guess to the quantitative level of the relationship. Further, the limitations applied through simplification of the process, i.e., use of one three-number scale that is assumed to exist on a range from one to nine and is typically dictated by practice, severely limits the extent to which conclusions may be drawn from the process as suggested by Fig. 27.13.

In order to add statistical significance to the experimental evidence presented thus far, comparison of the resulting relative weight and rank distributions is performed. The comparison is facilitated through a t-test performed on the resulting distributions of relative weight and rank order for each factor at each level. For example, looking at the *column density* factor there are four distributions of

12 data points (48 total experiments, 12 for each of the four levels). Similarly, for *qualitative tendency* there are three distributions with 16 data points and for *quantitative scale*, there are four distributions with 12 data points. The t-test allows a comparison of distributions per each factor to assess differences in response (relative weight or rank order) due to the factor level. In using the t-test, it is assumed that the distributions are normally distributed and have equal variance [49]. The null hypothesis for the t-tests performed is that the distributions have equal means, or *Ho*: $\mu_1 = \mu_2$, where μ_1 and μ_2 represent the means of the two distributions in question. In other words, the null hypothesis is that there is no effect due to changing the factor levels. The t-tests were performed at a significance level of $\alpha = 0.05$. From the t-test, a *P* statistic is calculated. If the value of *P* is greater than α, the null hypothesis is rejected (i.e., there is a difference in the distributions due to the changing levels). If the value of *P* is less than α, we fail to reject *Ho* (there is no difference in the distributions due to the changing levels).

The results of performing the t-tests are shown in Table 27.4. Note that a test of equality for the variances of the distributions for each factor level was performed and it was found that the assumption of equal variance is valid. For both column density and qualitative tendency the value of *P* is almost always less than α, indicating a rejection of the null hypothesis, *Ho*. This provides evidence that changing the levels for these two factors indeed affects the final relative weight and rank order of TAs in the HoQ. There are only two cases in which the value of *P* is larger than α for column density level comparisons. The first case occurs when comparing 2/5 and 3/5. However, since the *P*-value is only slightly greater than α, it suggests that there is some difference in the distributions, i.e., there is an effect due to changing from 2/5 to 3/5. Only in the case of changing *column density* from 3/5 to 4/5 is there clear statistical proof that there is little or no effect on the final relative weight and rank of TA1. This is understandable since a change from 3/5 to 4/5 only represents a 33% increase in *column density*, the number of CAs affected for the TA already exceeds half, thus reducing its relative effect.

For the *quantitative scale* factor, however, the value of *P* is much greater than α in every case, giving statistical creditability for accepting *Ho*. Namely, it can be concluded that there is no affect on the final quantitative results in the HoQ due to *quantitative scale* choice.

Given the conceptual limitations discussed under the validation criteria and the evidence provided here through experiment, both qualitative and quantitative, two conclusions can be drawn: The first is that the results generated from the HoQ are generally robust to scale choice. This explains why changing the scale choice does not generally affect the rank order of all TAs in the hair-dryer example of Table 27.2. However, this robustness means that the scale choice does not represent "expert knowledge" on the part of designers as is sometimes implied. Thus, designers should not utilize the relative weights as a reflection of true relative importance of one TA over another. At best, the results from the experiment suggest that it may be possible to get a *sense* of the rank importance of one TA over another, since it is evident that the *column density* and *qualitative tendency* of a TA seem to have dominating effect. That is, qualitative assessment is possible since the *qualitative tendency* resulting from the designer is important to final results, but the quantitative assessment in the form of TA relative importance is **not** possible since the *quantitative scale* choice has no effect on the results.

There is still a danger in the qualitative assessment as is shown in the empirical studies of the effect of uncertainty in the scale on TA rank order, which makes it difficult for designers to identify the subset of most important TAs in a robust sense.

Based on the statistical and empirical evidence in this chapter, it seems that the HoQ model is limited in its ability to support design decision-making. The statistical evidence also supports the notion that "extremes" in *column density* and *qualitative tendency* differentiate the random process results presented in the empirical

TABLE 27.4 STATISTICAL SIGNIFICANCE OF FACTORS IN HoQ EXPERIMENT

Results of t-Test for 5x5 HoQ Experiment			
Factor	**Levels Compared**	**Relative Weight *P* Value**	**Rank *P* Value**
Column Density	1/5 vs. 2/5	0.038	0.014
	1/5 vs. 3/5	0.000	0.000
	1/5 vs. 4/5	0.000	0.000
	2/5 vs. 3/5	0.078	0.061
	2/5 vs. 4/5	0.010	0.006
	3/5 vs. 4/5	0.220	0.200
Qualitative Tendency	Weak vs. medium	0.000	0.002
	Weak vs. strong	0.000	0.000
	Medium vs. strong	0.003	0.008
Quantitative Scale	1-3-9 vs. 2-5-8	0.930	0.927
	1-3-9 vs. 1-2-3	0.909	0.917
	1-3-9 vs. 1-50-100	0.984	0.950
	2-5-8 vs. 1-2-3	0.979	0.991
	2-5-8 vs. 1-50-100	0.915	0.977
	1-2-3 vs. 1-50-100	0.895	0.968

investigation and the original HoQ for some TAs. Further, the statistical results suggest that it may be possible to predict to some degree the output from a given HoQ based on several factors like those studied in the experiment. In the next section, such a predictive model for the HoQ is discussed, generated and compared to the example HoQs from this chapter.

In all, the results of this section show that designers should limit the importance placed on results from the HoQ method, especially those regarding quantitative value. Of course, it is realized that some of the limitations laid out in this chapter are likely known to varying degree by designers who utilize the HoQ regularly. However, it is important that these limitations are studied and reported in a rigorous fashion to ensure that they represent global rather than local knowledge. Specifically, this chapter showed that the HoQ is limited in its ability to support design decision-making because *it does not utilize meaningful, reliable information* (violating validation criterion 2) and *it does not provide a sense of robustness of results* (violating validation criterion 5).

The experimental study provides a clear picture of the limitation of the HoQ to support quantitative assessment. While the methodology can provide a qualitative assessment, as designers try to provide quantitative information (through scales) for quantitative assessment, the methodology loses its value. That is, to a degree the designers can use the HoQ model to process a subset of most important TAs (qualitative assessment), but beyond that point the quantitative assessment becomes no better than a random process that was also seen in this section.

27.5 CONCLUSION: THE ROLE OF VALIDATION CRITERIA ON DESIGN METHODS

The validation criteria introduced and applied in this chapter are not intended to invalidate design decision methods in a closed-form mathematical sense. Rather, the criteria provide a basis for evaluation of design methods from a decision-making perspective. Such evaluation allows for an understanding of the limitations of design methods as was performed on the HoQ. Such understanding is necessary in order that designers avoid flawed decision-making due to design methods utilized throughout the course of the design process.

It is not necessarily expected that the design methods used would address all of the decision issues that inspired the validation criteria. However, it would be beneficial for designers to understand which of these issues are not addressed, i.e., understand the limitations of design methods in terms of decision theory. Thus, these five criteria are intended to point out the elements of decision-making that may be missing from their particular methods of choice. Further, when the criteria are applied to some of these design methods, the potential flaws that could impact the design decisions are understood. The role of the validation criteria is to provide a perspective, namely, a decision theory perspective, from which designers can explore and understand the design methods they utilize to support design decision-making on a day-to-day basis.

REFERENCES

1. Danesh, M.R. and Jin, Y., 2001. "An Agent-Based Decision Network for Concurrent Engineering Design," *Concurrent Engrg.* **9**(1), pp. 37–47.
2. Gu, X., Renaud, J.E., Ashe, L.M. and Batill, S.M., 2000. "Decision-Based Collaborative Optimization Under Uncertainty," *Proc. ASME Des. Engrg. Tech. Conf., DETC2000/DAC-14297,* ASME, New York, NY.
3. Hazelrigg, G.A., 1998. "A Framework for Decision-Based Engineering Design," *J. of Mech. Des.,* Vol. **120**, pp. 653–658.
4. Marston, M. and Mistree, F., 1998. "An Implementation of Expected Utility Theory in Decision Based Design," *Proc., ASME Des. Engrg. and Tech. Conf., DETC98/DTM-5670,* ASME, New York, NY.
5. Olewnik, A., Brauen, T., Ferguson, S. and Lewis, K., 2004. "A Framework for Flexible Systems and its Implementation in Multi-Attribute Decision-Making," *ASME J. of Mech. Des.*, **126**(3), pp. 412–419.
6. Olewnik, A. and Lewis, K., 2005. "A Decision Support Framework for Flexible System Design," *J. of Engrg. Des.*
7. Roser, C. and Kazmer, D., 2000. "Flexible Design Methodology," *Proc., Des. for Manufacturing Conf.: Des. Engrg. Tech. Conf., DETC00/DFM-14016.*
8. Wassenaar, H., Chen, W., Cheng, J. and Sudjianto, A., 2004. "An Integrated Latent Variable Choice Modeling Approach for Enhancing Product Demand Modeling," *Proc., ASME Des. Engrg. Tech. Conf., DETC 2004-57487,* ASME, New York, NY.
9. Wassenaar, H.J. and Chen, W., 2003. "An Approach to Decision-Based Design With Discrete Choice Analysis for Demand Modeling," *ASME J. of Mech. Des.*, **125**(3), pp. 490–497.
10. Terniko, J., 1996. *Step by Step QFD: Customer Driven Product Design,* Responsible Management Inc. Nottingham, NH.
11. Hauser, J. and Clausing, D., 1988. "The House of Quality," *Harvard Busi. Rev.*, 66(3), pp. 63–74.
12. Pugh, S., 1996. *Creating Innovative Products Using Total Design*, Addison-Wesley Publishing Company Reading, MA.
13. Saaty, T., 1980. *The Analytical Hierarchy Process*, McGraw Hill, New York, NY.
14. Keeney, R. and Raiffa, H., 1993. *Decisions With Multiple Objectives: Preferences and Value Tradeoffs*, Cambridge University Press, U.K.
15. See, T.-K., Gurnani, A. and Lewis, K., 2004. "Multi-Attribute Deci sion Making Using Hypothetical Equvalents and Inequivalents," *J. Mech. Des.*, **126**(6), pp. 950–958.
16. Messac, A., 1996. "Physical Programming: Effective Optimization for Computational Design," *AIAA J.,* Vol. **1**, pp. 149–158.
17. Taguchi, G., 1986. *Introduction to Quality Engineering*, Asian Productivity Organization (distributed by American Supplier Institute, Inc.) Dearborn, MI.
18. Suh, N., 1990. *The Principles of Design*, Oxford University Press, New York, NY.
19. Shupe, J.A., Muster, D., Allen, J.K. and Mistree, F., 1988. "Decision-Based Design: Some Concepts and Research Issues," *Expert Systems, Strategies and Solutions in Manufacturing Design and Planning*, A. Kusiak, Editor Soc. of Manufacturing Engrs., Dearborn, MI, pp. 3–37.
20. Mistree, F., Smith, W.F., Bras, B.A., Allen, J.K. and Muster, D., 1990. "Decision-Based Design: A Contemporary Paradigm for Ship Design," *Proc., Annual Meeting of the Soc. of Naval Architects and Marine Engrs.*, San Francisco, CA.
21. Tribus, M., 1969. *Rational Descriptions, Decisions and Designs*, Pergamon Press, Inc. Elmsford, NY.
22. Barzilai, J., 1997. "A New Methodology for Dealing with Conflicting Engineering Design Criteria," *Proc., 18th Annual Meeting of the Am. Soc. for Engrg. Mgmt.*
23. Hazelrigg, G., 2003. "Validation of Engineering Design Alternative Selection Methods," *J. of Engrg. Optimization*, **35**(2), pp. 103–120.
24. Olewnik, A. and Lewis, K., 2005. "On Validating Engineering Design Decision Support Tools," *J. of Concurrent Engrg. Des. Res. and Appl.*, 13(2), pp. 111–122.
25. Saari, D., 2000. "Mathematical Structure of Voting Paradoxes. I: Pairwise Vote. II: Positional Voting," *Economic Theory*, Vol. **15**, pp. 1–103.
26. Dym, C. and McAdams, D., 2004. "Modeling and Information in the Design Process," *Proc., ASME Des. Engrg. Tech. Conf., DETC2004-57101*, ASME, New York, NY.
27. Pedersen, K., Emblemsvag, J., Bailey, R., Allen, J.K. and Mistree, F., 2000. "Validating Design Methods and Research: the Validation

Square," *Proc., ASME Des. Engrg. Tech. Conf., DETC2000/DTM-14579*, ASME, New York, NY.
28. Malak, R.J. and Paredis, C.J.J., 2004. "On Characterizing and Assessing the Validity of Behavioral Models and Their Predictions," *Proc., ASME Des. Engrg. Tech. Conf., DETC2004-57452*, ASME, New York, NY.
29. Frey, D. and Li, X., 2004. "Validating Robust Parameter Design Methods," *Proc., ASME Des. Engrg. Tech Conf., DETC2004-57518*, ASME, New York, NY.
30. Olewnik, A., Hammill, M. and Lewis, K., 2004. "Education and Implementation of an Approach for New Product Design: An Industry-University Collaboration," *Proc., ASME Des. Engrg. Tech. Conf., DETC2004-57320*, ASME, New York, NY.
31. Du, X., Sudjianto, A. and Chen, W., 2004. "An Integrated Framework for Probabilistic Optimization Using Inverse Reliability Strategy," *J. of Mech.* Des., **126**(4), pp. 562–570.
32. Strawbridge, Z., McAdams, D.A. and Stone, R.B., 2002. "A Computational Approach to Conceptual Design," *Proc., ASME Des. Engrg. and Tech. Conf., DETC2002/DTM-34001*, ASME, New York, NY.
33. von Neumann, J. and Morgenstern, O., 1947. *Theory of Games and Economic Behavior, 2nd Ed.*, Princeton University Press, Princeton, NJ.
34. Yates, J.F. and Estin, P.A., 1998. "Decision Making," *A Companion to Cognitive Science*, W. Bechtel and G. Graham, eds., Blackwell Publishers Ltd., Malden, MA, pp. 186–196.
35. Hazelrigg, G., 1996. *Systems Engineeri°ng: An Approach to Information-Based Design,* Prentice Hall Upper Saddle River, NJ.
36. Arrow, K., 1951. *Social Choice and Individual Values*, John Wiley & Sons, New York, NY.
37. See, T.-K. and Lewis, K., 2004. "A Formal Approach to Handling Conflicts in Multiattribute Group Decision Making," *Proc., ASME Des. Engrg. Tech. Conf., UT, DETC2004-57342,* ASME, New York, NY.
38. Matheson, D. and Matheson, J., 1998. *The Smart Organization: Creating Value Through Strategic R&D*, Harvard Business School Press, Boston, MA.
39. Olewnik, A., 2005. "Validating Design-Decision Support Models," Ph.D. dissertation, Univ. at Buffalo-SUNY, NY.
40. Howard, R.A., 1992. "In Praise of the Old Time Religion," *Utility Theories: Measurements and Applications*, W. Edwards, ed., Kluwer Academic Publishers, Boston, MA, pp. 27–55.
41. Hauser, J., 1993. "How Puritan-Bennet Used the House of Quality," *Sloan Mgmt. Rev.*, **Spring**.
42. QFDI, 2005. *Abstracts from Symposia on QFD*, access http://www.qfdi.org/books/, February.
43. Kaldate, A., Thurston, D., Emamipour, H. and Rood, M., 2003, "Decision Matrix Reduction in Preliminary Design," *Proc., ASME Des. Engrg. Tech. Conf., DETC2003/DTM-48665*, ASME, New York, NY.
44. Breyfogle, F.W., 1999. *Implementing Six Sigma: Smarter Solutions Using Statistical Methods*, John Wiley & Sons, Inc., New York, NY.
45. Hofmeister, K., 1995. "QFD in the Service Environment," *Quality Up, Costs Down: A Manager's Guide to Taguchi Methods and QFD*, W. Eureka and N. Ryan, eds., ASI Press, New York, NY, pp. 57–78.
46. *The American Heritage Dictionary of the English Language, 4th Ed.,* 2000. Houghton Mifflin Company.
47. Masui, K., Sakao, T., Aizawa, S. and Inaba, A., 2002. "Quality Function Deployment for Environment (QFDE) to Support Design for Environment (DFE)," *Proc., ASME Des. Engrg. Tech. Conf., DETC2002/DFM-34199,* ASME, New York, NY.
48. Gibbons, J.D., 1985. *Nonparametric Methods for Quantitative Analysis, 2nd Ed.,* American Sciences Press, Inc., Columbus, OH.
49. Montgomery, D., 2001. *Design and Analysis of Experiments,* 5th Ed., John Wiley & Sons, Inc., New York, NY.

PROBLEMS

27.1 **The validity of Suh's axiomatic design:** Nam Suh advocates the use of two axioms for achieving the best possible design when moving from the *functional requirements* (FRs) domain to the *design parameters* (DPs) domain.[2] This mapping of FRs to DPs can be thought of as moving from design specification to the detailed design phases of the design process in Fig. 27.1. Those axioms are the *independence* and *information* axioms. According to Suh, the design process should be carried out by first maintaining the independence of FRs (independence axiom) and then by minimizing information content (information axiom). The axioms must be applied in this order if designers are to achieve the best possible design, according to Suh. Under the context of the validation criteria for design decision methods described in this chapter, discuss Suh's theory for design. Are there any obvious violations? If so, what problems might result from these violations?

27.2 **Empirical investigation of the validity of Suh's axiomatic design:** Suh has developed metrics for both the independence and information axioms. To understand the implication of the independence axiom, it is easier to think of the mapping in terms of a linear transformation of the form: $\{FR\} = [DM]\{DP\}$, where [DM] represents the design matrix that maps the functional requirements to the design parameters. The ideal design (completely independent) is one in which the number of FRs equals the number of DPs (square design matrix) and the design matrix is identity. For any given design matrix, it is possible to find a quantitative measure of the independence. Suh gives two measures of independence: reangularity, R, and semangularity, S. Reangularity can be thought of as a measure of the interdependence among DPs and semangularity can be thought of as a measure of the correlation between one FR and any pair of DPs. Each has a maximum value of unity, which corresponds to a completely independent design. As the level of coupling increases, the reangularity and semangularity decrease. Information, I, is related to the designers' specifications (captured by DPs) and the capability of the manufacturing system. The more capable the manufacturing system is of meeting the designers' specifications (in terms of tolerances), the lower the information measure. If a manufacturing system is completely capable of meeting the designers' specifications, the measure of I is zero.

Now, consider a scenario in which designers must select from among four designs. The designs to select among have the metrics for independence and information indicated in the table below. Which design would Suh choose? What must the designer preferences be in order to agree with Suh? Does this analysis agree with your discussion from problem 1?

27.3 **Role of axiomatic design in concurrent engineering practice:** Given the previous questions and accompanying discussion, do you feel that axiomatic design is still a useful

Design	**Independence**		**Information**
	R	S	I
1	0.612	0.500	0.117
2	0.612	0.500	0.126
3	0.707	0.354	0.087
4	0.707	0.500	1.331

[2]Suh actually advocates the use of these axioms to map all four of his design domains to one another.

method for design? Is there any value to the axioms as outlined by Suh?

27.4 **Improving the house of quality:** Given the violations of the HoQ described in this chapter, should this design method still be applied by designers? How might the tool be changed to overcome the limitations described?

27.5 **Open question about the validation of design decision methods:** What other methods support design decision-making? What limitations are apparent when these methods are explored under the validation criteria? Are there other validation criteria that should be part of the five described in this chapter?

INDEX